The Weather and Climate: Emergent Laws and Multifractal Cascades

Turbulent and turbulent-like systems are ubiquitous in the atmosphere, but there is a gap between classical models and reality. Advances in nonlinear dynamics, especially modern multifractal cascade models, allow us to close the gap, to investigate the weather and climate at unprecedented levels of accuracy, comparing theories, models and experiments over huge ranges of space-time scales.

Using new stochastic modelling and data analysis techniques, this book provides an overview of the nonclassical, multifractal statistics. The authors demonstrate that by generalizing the classical turbulence laws it is possible to obtain emergent laws of atmospheric dynamics. These higher-level laws are empirically validated from weather and macroweather scales to climate scales, and over length scales of millimetres to the size of the planet. By generalizing the notion of scale, atmospheric complexity is reduced to a manageable scale-invariant hierarchy of processes, thus providing a new perspective for modelling and understanding the atmosphere. This new synthesis of state-of-the-art data and nonlinear dynamics is systematically compared with other analyses and global circulation model outputs. Applications of the theory are graphically demonstrated with many original multifractal simulations.

This thorough presentation of the application of nonlinear dynamics to the atmosphere is an important resource for atmospheric science researchers new to multifractal theory. It will also be of use to graduate students in atmospheric dynamics and physics, meteorology, oceanography and climatology.

Shaun Lovejoy is a professor of physics at McGill University, Montréal, and has been a pioneer in developing and applying new ideas in nonlinear dynamics to the geosciences since the late 1970s. This includes multifractals, generalized (anisotropic) scale invariance, universal multifractals and space-time multifractal modelling of geofields (especially clouds, precipitation and topography). He has published over 200 papers applying these ideas to the earth and environmental sciences. The unifying theme of this work is that when the notion of scaling is generalized to include anisotropy and multifractality, many key geofields display scaling behaviour over enormous ranges of scale – and that this nonclassical extreme variability is a new paradigm for the geosciences. In addition to these scientific contributions, Professor Lovejoy has actively promoted nonlinear processes in geophysics by cofounding the Nonlinear Process section at the European Geosciences Union (EGU) and the journal *Nonlinear Processes in Geophysics*. He has been vice-chair and subsequently chair of the Nonlinear Geophysics focus group at the American Geophysical Union (AGU) since 2006 and is currently the President of the Nonlinear Process Section of the European Geoscience Union.

Daniel Schertzer is a professor at École des Ponts ParisTech, Université Paris-Est, and scientific director of the chair in Hydrology for Resilient Cities, sponsored by Veolia Water. His research introduced multifractals and related techniques in hydrology, after having contributed to their theoretical developments in turbulence, in particular with the definition of a codimension formalism, the concepts of generalized scale invariance and universal multifractals. His work has covered many domains of geophysics and the environment, with a particular emphasis on atmospheric dynamics, precipitation extremes and remote sensing. His publications include three books and 135 ISI-indexed publications, which have received more than 4000 citations, and he is executive editor of the journal *Nonlinear Processes in Geophysics*, which he cofounded, as well as the Nonlinear Geophysics divisions of AGU and EGU. Professor Schertzer has been a union officer of both the AGU and EGU, a bureau member of the International Association for Hydrological Sciences and a board member of the European Academy of Wind Energy. He is also vice-president of the French National Committee of Geodesy and Geophysics and a member of the Higher Council of Meteorology (France).

The Weather and Climate: Emergent Laws and Multifractal Cascades

Shaun Lovejoy
McGill University, Montréal
Daniel Schertzer
Université Paris-Est, Ecole des Ponts Paris Tech

CAMBRIDGE
UNIVERSITY PRESS

CAMBRIDGE
UNIVERSITY PRESS

University Printing House, Cambridge CB2 8BS, United Kingdom

One Liberty Plaza, 20th Floor, New York, NY 10006, USA

477 Williamstown Road, Port Melbourne, VIC 3207, Australia

4843/24, 2nd Floor, Ansari Road, Daryaganj, Delhi - 110002, India

79 Anson Road, #06-04/06, Singapore 079906

Cambridge University Press is part of the University of Cambridge.

It furthers the University's mission by disseminating knowledge in the pursuit of education, learning and research at the highest international levels of excellence.

www.cambridge.org
Information on this title: www.cambridge.org/9781108446013

First published 2013
First paperback edition 2017

A catalogue record for this publication is available from the British Library

ISBN 978-1-107-01898-3 Hardback
ISBN 978-1-108-44601-3 Paperback

Contents

Preface

Few would argue that quantum and statistical mechanics do not apply to ordinary fluid flows, yet the latter involve such huge numbers of particles that these fundamental theories are rarely useful for solving practical – or even theoretical – fluid problems. Instead, one typically exploits the fact that when the number of particles is large enough, new continuum properties and notions such as fluid particle, fluid temperature and fluid velocity emerge which are governed by the higher-level equations of continuum mechanics and thermodynamics. While the latter can perhaps be obtained from the former, the derivations are not mathematically rigorous and the foundations of continuum mechanics are still actively researched. The continuum laws that emerge are indeed qualitatively new.

In a similar way, one expects new higher-level laws to emerge from the chaos of sufficiently strong hydrodynamic turbulence. While the latter presumably continue to obey the laws of continuum mechanics, their direct application is impractical and one searches for the emergence of new, even higher-level laws. Indeed "developed turbulence as a new macroscopic state of matter" (Manneville, 2010), appearing at high Reynolds number (*Re*), has been considered as a form of matter with properties that cannot simply be reduced to – nor simply deduced from – the governing Navier–Stokes equations. Consequently, it is not surprising that over the years new types of models and new symmetry principles have been developed in order to directly study, model and understand this hypothetical emergent state. Of particular relevance to this book are cascade models and (anisotropic) scale invariance symmetries.

The study of fully developed turbulence remains largely academic and has only had a rather peripheral impact on atmospheric science. This is ironic, since the atmosphere provides an unrivalled strongly nonlinear natural laboratory with the ratio of nonlinear to linear terms – given by the Reynolds number – that is typically $\approx 10^{12}$. Although the atmosphere certainly differs from incompressible hydrodynamics in several important ways, we may nevertheless expect higher-level laws to emerge. Furthermore, it is reasonable to expect that they will share at least some of the features of fully developed turbulence. This was indeed the belief of many of the pioneers of classical turbulence: L. F. Richardson, A. N. Kolmogorov, A. Obukhov, S. Corrsin and R. Bolgiano.

While the pioneers' eponymous laws were in many ways highly successful, when applied to the atmosphere they faced two basic obstacles: the atmosphere's extreme intermittency and its strong stratification, which increases systematically at larger and larger scales. As usual in physics, when one faces such a situation there are two choices. Either one abandons the old law and moves on to something different, or else one generalizes the law so that it is able to fully fit the facts. On several occasions during its development, the law of conservation of energy was faced with such a choice: either treat it as no more than a (sometimes) poor approximation, or else extend the notion of energy beyond mechanical energy to heat energy, to chemical energy, to electrical energy and eventually to mass energy. In this book we follow the latter choice with respect to the classical laws of turbulence: we argue that these obstacles of stratification and intermittency can be overcome with appropriate generalizations. For weather, the key generalizations are from isotropic to anisotropic notions of scale and from smooth, quasi-Gaussian variability to strong, cascade-generated multifractal intermittency. Together, this leads to a model of atmospheric dynamics as a system of coupled anisotropic cascade processes.

An initial application of this model takes us up to the limits of the weather domain: in space to the size of the planet, in time to the lifetime of planetary-sized structures ($\tau_w \approx 10$ days), after which there is a drastic change in the behaviour of all the atmospheric fields. It turns out that whereas at shorter time scales fluctuations tend to grow with scale – the weather is perceived as unstable – at longer time scales the average fluctuations tend to decrease, giving on the contrary the impression of stability. It is tempting to identify this new regime with the climate, but we argue that this would be a mistake, that the climate is not just the long-term behaviour of the weather. The reason is that it turns out that the same anisotropic cascade that explains the weather variability can be extended to much lower frequencies. When this is done, there is indeed a "dimensional transition" at τ_w but the model continues to accurately reproduce the lower-frequency variability beyond the transition. It turns out that this is also true of unforced global climate models (GCMs), so that the label "low-frequency weather," or "macroweather," is appropriate. To paraphrase a popular dictum, "macroweather is what you expect, the weather is what you get." From a stochastic perspective, the climate is unexpected in much the same way as the weather, with analogous consequences for its prediction.

In order to find something really new that corresponds to our usual notion of "climate," we have to wait quite a long time – about $\tau_c \approx$ 10–30 years – until we find that the mean fluctuations again start to increase with scale – a behaviour which apparently continues to $\tau_{lc} \approx$ 30–50 kyr. Yet even this true climate regime – where genuinely new processes and/or forcings are dominant – shares features (including scaling) with the weather/macroweather regime. We show how a single overall weather/macroweather/climate process emerges, and we derive its statistics and estimate its exponents and other parameters.

In these pages, we therefore show how to considerably generalize the classical turbulence laws and to obtain emergent laws of atmospheric dynamics. Empirically, we show that they apply from milliseconds to decades to tens of millennia, from millimetres to the size of the planet. In more detail, we argue (a) that the atmosphere is a strongly nonlinear system with a large number of degrees of freedom, (b) that it nevertheless respects an (anisotropic) scale-invariance symmetry, (c) that this leads to new emergent properties, new dynamical laws. These new laws are statistical, and physically they imply that the variability builds up scale by scale in a cascade-like manner. This variability sports many nonclassical statistical characteristics including long-range statistical dependencies and nonclassical extreme values ("heavy"-tailed, algebraic probability distributions). Finally, they can be exploited to understand and to forecast atmospheric fields.

The basic ingredients needed to effect this generalization are multifractals, cascades and generalized scale invariance. The development of these notions was largely motivated by atmospheric applications and arose in the 1980s, a period when nonlinear dynamics was generating excitement in many areas of science. Although a book comprehensively treating multifractals is still lacking (see however Schertzer and Lovejoy, 2011), they are *not* the focus here. For our purposes, they are rather the tools needed to generalize the classical turbulent laws to the atmosphere. In the last five years or so, the scope of these applications has dramatically increased, thanks to the existence and accessibility of massive global-scale databases of all kinds. Whereas only ten years ago we were still speculating on the ranges and types of scaling of atmospheric fields, today we can already be confident about a great deal. This confidence is due partly to the qualitative – and in many cases quantitative – agreement between quite different databases over wide ranges of scale, but also to the surmounting of several obstacles in the interpretation of the data (in particular of aircraft data). We therefore place much emphasis on the empirical underpinnings of the new laws. Although over the meteorological scales we extensively analyze the traditional sources of satellite, lidar, drop-sonde and aircraft data, we also investigate at length reanalysis fields that are hybrid products somewhere between the data and the models, as well as the outputs of the models themselves.

Therefore, although many of the ideas in this book have been around since the 1980s, most (perhaps 90%) of the examples are from research performed in only the last five years. It is thanks to these new global datasets that the original 1980s models of 23/9 D anisotropic scaling dynamics, and of three scaling regimes from weather to climate, can be convincingly validated and a new, comprehensive view of atmospheric variability established. Beyond new results on global spatial scales, there are also new results on the space-time variability, including the emergence of waves (Chapter 9). The last two chapters – the research for which was largely undertaken specifically for this book – include the generalization of the emergent weather laws into the macroweather regime (Chapter 10), i.e. between the $\approx$ 10-day lifetime of planetary structures out to 10–100 years where the true climate regime begins. These longer climate scales are mostly beyond the instrumental range, so in Chapter 11 we analyze various surrogates including multiproxies, paleotemperatures and climate forcings (including solar, volcanic and orbital) as well as GCMs (unforced and forced climate "reconstructions"). We show how the space-time climate variability can be understood by a further extension of the weather/macroweather space-time scaling framework. By quantifying the natural variability as a function of space and time scale, this provides the information necessary to construct statistical tests for assessing anthropogenic influences on the climate. This approach is complementary to the current GCM approach but has the advantage of being largely data rather than model-driven.

Although long in gestation, this book comes at a critical moment for atmospheric science. While ever bigger computers, ever higher resolution devices and ever larger quantities of data have resulted in our present golden age, it has come at a price: they have gobbled up most of our resources. Sometimes, it seems that there are only barely enough left over to support a narrow focus on applications to numerical weather and climate modelling. One can easily get the impression that a basic understanding of the atmosphere's variability in space and in time is a luxury that we cannot afford. Yet today's continuing lack of consensus about these questions is increasingly hampering the development of the numerical models themselves. For example, without knowledge of the effective dimension of atmospheric motions it will not be possible to place the currently ad hoc "stochastic parametrizations" in modern Ensemble Forecasting Systems on a solid theoretical basis. As we argue here, this new synthesis – which is remarkably simple – provides a compelling and consistent picture of atmospheric variability and dynamics from weather through climate scales and suggests numerous ways forward, including the possibility of direct stochastic forecasting (Chapter 9).

From the above, the reader may correctly infer that this book is squarely oriented towards practising atmospheric scientists (especially meteorologists and climatologists) and that it includes a (hopefully) accessible exposition of the necessary nonlinear tools. Occasionally, when a topic is a bit too technical but nevertheless important either for applications or for the theory, details are given in appendices. Similarly, advanced or optional sections are indicated by asterixes. In addition, at the end of each chapter, under the rubric *Summary of emergent laws in Chapter ...*, we give a succinct summary of the developments in the chapter that are important for developing the main theory. These summaries will allow readers to skip details that are unimportant while maintaining the basic thread of the argument. Finally, to highlight them, the more important formulae have been placed in boxes. Let the reader be warned, however, that this is neither a textbook nor a conventional monograph. It is rather a systematic presentation of arguments and evidence for a new framework for understanding atmospheric dynamics.

In order to make the material as accessible as possible, the basic philosophy has been to first present empirical analyses demonstrating the existence of wide-range scaling: an overview in Chapter 1, the horizontal wind in Chapter 2, the state variables and radiances in the horizontal in Chapter 4, in the vertical in Chapter 6, and in time in Chapters 8, 10 and 11. The analyses proceed from the (familiar) Fourier (power) spectra applicable to essentially any field, to trace moments in Chapter 3 needed to analyse cascades, followed by further related analysis techniques (generalized structure functions, wavelets, the probability distribution multiple scaling technique etc.) in Chapter 5. For readers primarily interested in the longer time scales, Chapters 10 and 11 are to some degree independent of the preceding, making only light use of the formalism and relying extensively on the use of Haar fluctuations (wavelets). However, this underexploited technique is actually quite straightforward – even intuitive – and allows systematic comparisons to be made of different types of data and over different and large scale ranges. It gives a far clearer picture of the macroweather and climate variability than is otherwise possible, so that any effort expended to understand this analysis technique will be rewarded.

Following the empirical motivation, the theory is introduced gradually and as needed: first the basic elements of turbulence theory (Chapter 2), then elementary (discrete in scale) cascades (moment statistics, Chapter 3), with the more general treatment of multifractals including probabilities and continuous in scale simulations reserved for Chapter 5. In Chapter 6 we go beyond isotropy, by introducing generalized scale invariance, but only in the simplest self-affine form needed to handle scaling different in two orthogonal directions: atmospheric stratification. Only in Chapter 7 do we treat the more general case needed for cloud and other morphologies whose anisotropies vary both with scale and with position. Going beyond space to space-time involves extra complications, if only because causality must be taken into account, and this is why its introduction is delayed until Chapter 8, where we give both an empirical overview and the basic theory needed to understand the space-time scaling in the weather regime. In Chapter 9 we extend this to an explicit treatment of causality, to turbulence-driven waves as an emergent scaling process, to predictability and (stochastic) forecasting. In Chapter 10 we extend the space-time weather model into the macroweather regime, showing that it not only predicts the observed sharp "dimensional transition" between weather and macroweather at about 10 days (the lifetime of planetary structures), but that it does remarkably well up to scales of decades and centuries. At scales below the transition in the weather regime, fluctuations generally grow with increasing scale, but at larger scales, in the macroweather regime, on the contrary they diminish with scale – the atmosphere appears "stable." However, this is not the full story. In Chapter 11 we show – with the help of instrumental, multiproxy and paleodata – how the macroweather regime eventually gives way to a new climate regime where fluctuations once again grow with scale, and attempt to address the question as to whether or not GCMs predict the climate or merely macroweather.

Acknowledgments

S.L. would like to thank the hospitality of NOAA's Climate Diagnostics Center (CDC), Boulder, Colorado, for hosting a sabbatical (August 2009 to July 2010) during which much of this book was written. CDC director Prashant Sardeshmukh and CDC colleague Cecile Penland are especially thanked for long discussions, critical comments and suggestions (including on early versions of the manuscript), and also for their friendship. Thanks also go to the Cooperative Institute for Research in the Atmosphere for the award of a fellowship during this time. Several granting agencies are thanked for allowing S.L. a quiet, virtually grant-free period without which the writing and research for this book would not have been possible.

DS thanks his students and colleagues, in particular those of his laboratory (LEESU, Laboratory Water, Environment and Water Systems), as well as his project partners, for having kept him busy with many stimulating problems that require new methodologies and paradigm changes.

The authors would also like to thank Adrian Tuck for detailed comments and suggestions on the manuscript, Gil Compo for helpful discussion, and Julien Pinel and Daniel Varon for useful comments and help with the editing and the figures.

We would also like to thank our editors at Cambridge University Press, especially Susan Francis, for exemplary patience as this project evolved over the last decade.

Acronyms and abbreviations

20CR	Twentieth-Century Reanalysis
AMDAR	Aircraft Meteorological Data Relay
AMV	atmospheric motion vectors
AOGCM	atmosphere–ocean general circulation model
ASAT	anisotropic scaling analysis technique
AVHRR	Advanced Very High Resolution Radiometer
BO	Bolgiano–Obukhov
BP	before present
CAM	correlated additive and multiplicative
CDC	Climate Diagnostics Center
CPC	Climate Prediction Center
CRUTEM3	Climate Research Unit Temperature version 3
DEM	digital elevation model
DFA	detrended fluctuation analysis
DIRTH	direction interval retrieval with thresholded nudging
DNS	direct numerical simulation
DO	Dansgaard–Oeschger
DTM	double trace moment
ECHAM4	ECMWF Hamburg model version 4
ECHO-G	ECHAM4 and HOPE-G
ECMWF	European Centre for Medium-range Weather Forecasts
EFS	ensemble forecasting systems
EKE	eddy kinetic energy
ENSO	El Niño–Southern Oscillation
EOF	empirical orthogonal function
EOLE	(French name for the Greek god of the winds)
ER2	Environmental Research 2
ERA40	ECMWF 40-year reanalysis
ETOPO5	Earth Topography 5 (minutes resolution)
FIF	fractionally integrated flux
GASP	Global Assimilation and Prognosis System
GCM	general circulation model or global climate model
GEM	global environment model
GFS	global forecasting system
GISP2	Greenland Ice Sheet Project 2
GISS	Goddard Institute for Space Studies
GOES	Geostationary Operational Environmental Satellite
GPS	global positioning system
GRIP	Greenland Ice Core Project
GSI	generalized scale invariance
GTOPO30	Global Topography 30 (seconds resolution)
HadCRUT	Hadley Climate Research Unit Temperature
HadSST	Hadley Centre Sea Surface Temperature
HOPE-G	Hamburg Ocean Primitive Equation model (global version)
INS	inertial navigation system
IPCC	International Panel on Climate Change

IPSL	Institut Pierre Simon Laplace
IR	infrared
KBO	Kolmogorov–Bolgiano–Obukhov
LANDSAT	Land remote-sensing Satellite
LGM	Last Glacial Maximum
MET	multiplicative ergodic theorem
MFDFA	multifractal detrended fluctuation analysis
MIES	Modernized Imagery Exploitation System
MODIS	Moderate resolution Imaging spectroradiometer
MOZAIC	Measurement of Ozone by Airbus In-service Aircraft
MSM	Markov-switching multifractal
MSU	microwave sounding unit
MTSAT	Multi-function Transport Satellite
NAO	North Atlantic Oscillation
NASA	National Aeronautics and Space Administration
NCDC	National Climatic Data Center
NCEP	National Centers for Environmental Prediction
NOAA	National Oceanographic and Atmospheric Administration
NSSTC	National Space Science and Technology Center
NVAG	Nonlinear Variability in Geophysics
OU	Orenstein–Uhlenbeck
PDE	partial differential equation
PDMS	probability distribution multiple scaling
PDO	Pacific Decadal Oscillation
PR	precipitation radar
PV	potential vorticity
QG	quasi-geostrophic
RMS	root mean square
SCF	second characteristic function
SCT	saturated cascade theory
SIG	scale-invariant generator
SLF	stochastic linear forcing
SLM	stochastic linear modelling
SOC	self-organized criticality
SOI	Southern Oscillation Index
SORCE	Solar Radiation and Climate Experiment
SPOT	Système Pour l'Observation de la Terre
SST	sea surface temperature
TAMDAR	Tropospheric Airborne Meteorological Data Reporting
TIMS	Total Irradiance Monitor Satellite
TMI	TRMM Microwave Imager
TRMM	Tropical Rainfall Measurement Mission
TSI	total solar irradiance
VIRS	visible infrared scanner
WRF	weather research and forecasting

Chapter 1 Introduction

1.1 A new synthesis

1.1.1 Two (irreconcilable?) approaches to understanding the atmosphere

In the last 20 years there has been a quiet revolution in atmospheric modelling. It's not just that computers and numerical algorithms have continued their rapid development, but rather that the very goal of the modelling has profoundly changed. Whereas 20 years ago, the goal was to determine the (supposedly) unique state of the atmosphere, today with the advent in Ensemble Forecasting Systems (EFS), the aim is to determine the possible future atmospheric states including their relative probabilities of occurrence: this new goal is *stochastic*. A *stochastic process* is a set of random variables indexed by time (Kolmogorov, 1933), and this definition includes that of deterministic processes as a special case.

At present, the EFS are really hybrids in the sense that they operate by first generating an initial ensemble of atmospheric states compatible with the observations and then use conventional deterministic forecasting techniques to advance each member in time to produce a distribution of future states. Once the leap was taken to go beyond the forecasting of a unique state to forecasting an ensemble, the next step was to make the subgrid parametrizations themselves stochastic (e.g. Buizza *et al.*, 1999; Palmer, 2001; Palmer and Williams, 2010). This is an attempt to take into account the variability of different possible subgrid circulations. The artificial deterministic/stochastic nature of these hybrids suggests that the development or pure stochastic forecasts would be advantageous, a possibility we explore in Chapter 9.

Interestingly, the tension between determinism and stochasticity has been around pretty much since the beginning, although for most of the (still brief) history of atmospheric science the deterministic approaches have been in the ascendancy and the stochastic ones left in the wings. To see this, let us recall the important developments. Drawing on the classical (deterministic) laws of fluid mechanics, Bjerknes (1904) and Richardson (1922) extended these to the atmosphere in the now familiar form of a closed set of nonlinear partial differential governing equations. From a mathematical point of view, their deterministic character is evident from the absence of probability spaces; from a conceptual point of view, it is associated with classical Newtonian thinking. In physics, Newtonian determinism began to disappear with the advent of statistical mechanics (starting with the "Maxwellian" distribution of molecular velocities: Maxwell, 1890), which showed that physical theories could indeed be stochastic. The break with determinism was consecrated with the development of quantum mechanics, which is a fundamental yet stochastic theory where the key physical variable – the wavefunction – determines probabilities.

At roughly the same time as the basis of modern deterministic numerical weather prediction was being laid, an alternative stochastic "turbulent" approach was being developed by G. I. Taylor, L. F. Richardson, A. N. Kolmogorov and others. Just as in statistical mechanics, where huge numbers of degrees of freedom exist but where only certain "emergent" macroscopic qualities (temperature, pressure etc.) are of interest, in the corresponding turbulent systems the new theories sought to discover new emergent statistical turbulence laws.

The first of these emergent turbulent laws was the Richardson "4/3 law" of atmospheric diffusion: $\nu(L) \approx KL^{4/3}$, where $\nu(L)$ is the effective viscosity at scale L and K is a constant to which we return (Richardson, 1926): see Fig. 1.1. This law is famous not only as the precursor of the Kolmogorov (1941) law of 3D isotropic homogeneous turbulence (the "5/3" law for the spectrum – or, if expressed for the fluctuation $\Delta v(L)$, the "1/3" law: $\Delta v(L) = \varepsilon^{1/3}L^{1/3}$ where Δv is the velocity fluctuation and ε is the energy flux), but it is also celebrated thanks to the

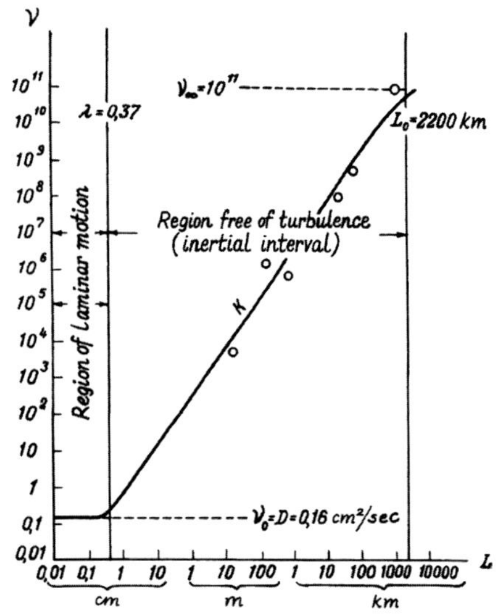

Fig. 1.1 Effective viscosity as a function of scale, reproduced from Monin (1972), adapted from (Richardson, 1926). The text (inserted by Monin) should read "region of free turbulence"(!)

ingenious way that Richardson experimentally confirmed his theory with the help of balloons and later even with parsnips and thistledown (Richardson and Stommel, 1948)! While this attention is all well deserved, the law was perhaps even more remarkable for something else: that Richardson had the audacity to conceive that a unique scaling (power) law – i.e. a law without characteristic length scales – could operate over the range from millimetres to thousands of kilometres, i.e. over essentially the entire meteorologically significant range. In accord with this, Richardson believed that the corresponding diffusing particles had "Weierstrass function-like" (i.e. fractal) trajectories. Nor was the 4/3 law an isolated result. In the very same pioneering book, *Weather Prediction by Numerical Process* (Richardson, 1922), in which he wrote down essentially the modern equations of the atmosphere (Lynch, 2006) and even attempted a manual integration, he slyly inserted:

> Big whirls have little whirls that feed on their velocity,
> and little whirls have lesser whirls and so on to
> viscosity – in the molecular sense.

Thanks to this now iconic poem, Richardson is often considered the grandfather of the modern cascade theories that we discuss at length in this book.

Had Richardson been encumbered by later notions of the meso-scale – or of isotropic turbulence in either two or three dimensions – he might never have discovered his law. Already, 15 years after he proposed it, Kolmogorov (1941) humbly claimed only a relatively small range of validity of the stringent "inertial range" assumptions of statistical isotropy and homogeneity which he believed were required for the operation of his eponymous law (which was also discovered, apparently independently, by Obukhov, 1941b, 1941a; Onsager 1945; Heisenberg, 1948; and von Weizacker, 1948) – and this even though it has strong common roots with Richardson's law. Indeed, it implies that Richardson's proportionality constant depends on the energy flux ε: $\nu(L) = L\Delta v(L) = \varepsilon^{1/3}L^{4/3}$; in this sense Kolmogorov's contribution was to find $K = \varepsilon^{1/3}$. Echoing Kolmogorov's reservations, Batchelor (1953) speculated that the Kolmogorov law should only hold in the atmosphere over the range 100 m to 0.2 cm! Even in Monin's influential book *Weather Forecasting as a Problem in Physics* (1972), the contradiction between the small and wide ranges of validity of the Kolmogorov and Richardson 4/3 laws is pushed surprisingly far, since on the one hand Monin confines the range of validity of the $\Delta v(L) = \varepsilon^{1/3}L^{1/3}$ law to "micrometerological oscillations ... up to $\approx$ 600 m in extent," while on the other hand publishing (on the opposite page!) a reworked copy of Richardson's figure demonstrating the validity of the $\nu(L) \approx L^{4/3}$ up to thousands of kilometres (Fig. 1.1). For the latter, he comments that it "is valid for nearly the entire spectrum of scales of atmospheric motion from millimeters to thousands of kilometres," in accord with Richardson. In Monin and Yaglom (1975), the contradiction is noted with the following mysterious explanation: "in the high frequency region one finds unexpectedly, that relationships similar to those valid in the inertial subrange of the microturbulence spectrum are again valid." In Chapter 6 we argue on the basis of modern reanalyses and other data that the law $\Delta v(L) = \varepsilon^{1/3}L^{1/3}$ does indeed hold up to near planetary scales in the horizontal, but paradoxically that, even at scales as small as 5 m, it does *not* hold in the vertical (and hence 3D isotropic turbulence does not seem to hold anywhere in the atmosphere)! By proposing a theory of anisotropic but scaling turbulence, we attempt to explain how it is possible that Kolmogorov was simultaneously *both* so much *more* accurate (the horizontal) and yet so much *less* accurate (the vertical) than anyone expected. This was achieved with the help of a generalized notion of scaling (Schertzer and Lovejoy, 1985a, 1985b) which ironically led to an effective "in between" dimension of atmospheric turbulence $D = 23/9 = 2.55...$ and enables the Fractal Geometry to at last (!) escape from the Euclidean metric (Schertzer and Lovejoy, 2006).

Facing colossal mathematical difficulties, turbulence theorists, starting with Taylor (1935), concentrated their attentions on the simplest turbulence paradigm: turbulence that is statistically isotropic, first in 3D, and then – following Fjortoft (1953) and Kraichnan (1967) – on the special isotropic 2D case. While Charney did extend Kraichnan's 2D theory to the atmosphere in his seminal paper "Geostrophic turbulence" (1971), meteorologists had already begun focusing on numerical modelling. By the end of the 1970s, there had thus developed a wide divergence between, on the one hand, the turbulence community with its focus on statistical closures and statistical models of intermittency (especially cascades) and, on the other hand, the meteorology community with its focus on practical forecasting and which treated turbulence primarily as a subgrid parametrization problem.

1.1.2 Which chaos for geophysics, for atmospheric science: deterministic or stochastic?

The divergence between statistical and deterministic approaches was brought into sharp relief thanks to advances in the study of nonlinear systems with few degrees of freedom. The new science of "deterministic chaos" can be traced back to the pioneering paper "Deterministic nonperiodic flow" (Lorenz, 1963) (and has antecedents in Poincaré, 1892). Lorenz's 1963 paper caused excitement by showing that three degrees of freedom were sufficient to generate chaotic (random-like) behaviour in a purely deterministic system. At the time, it was widely believed (following Landau, 1944) that on the contrary, random-like behaviour was a consequence of a very large number of degrees of freedom, so that as the nonlinearity increased (e.g. the Reynolds number) a fluid became fully turbulent only after successively going through a very large (even infinite) number of instabilities. By showing that as few as three degrees of freedom were necessary for chaotic behaviour, Lorenz's paper opened the door to the possibility that turbulence could have a relatively low-dimensional "strange attractor" so that effectively only a few degrees of freedom might matter. However, Lorenz's observation did not immediately lead to practical applications because theorists can readily invent nonlinear models, and at the same time it appeared that each model would require its own in-depth study in order to understand its behaviour. The problem of apparent lack of commonality in different nonlinear systems is the now familiar problem of "universality" which Fischer, Kadanoff and Wilson were only then successfully understanding and exploiting in the physics of critical phenomena; we shall revisit universality later in this book (Chapter 3). It is therefore not surprising that the turning point for deterministic chaos was precisely the discovery of "metric" (i.e. quantitative) "universality" by Grossman and Thomae (1977) and Feigenbaum (1978): the famous Feigenbaum constant in period doubling maps. Soon, with the help of theorems such as the extension of the Whitney embedding theorem (Whitney, 1936) and the practical "Grassberger–Procaccia algorithm" (Grassberger and Procaccia, 1983a, 1983b), all manner of time series were subjected to nonlinear analysis in the hope of "reconstructing the attractor" and of determining its dimension, which was interpreted as an upper bound on the number of degrees of freedom needed to reproduce the system's behaviour. In fact – as argued by Schertzer *et al.* (2002), Schertzer and Lovejoy (2003) – the mathematics do not support such a statement: they showed that indeed a stochastic cascade process may yield a finite correlation dimension, whereas the process itself has an infinite dimension! They therefore raised the question "which chaos?" For climate models essentially the same question was asked by Lorenz (1975), and more recently Palmer (2012) has strongly defended stochastic approaches.

Other developments in the 1980s helped to transform the "deterministic chaos revolution" into a more general "nonlinear revolution." Of particular importance for this book was the idea that many geosystems were fractal (scale invariant) (Mandelbrot, 1977, 1983) and later, that they commonly displayed "self organized criticality" (SOC) (Bak *et al.*, 1987; Bak, 1996), implying that many real-world systems could be "avalanche-like." Indeed, SOC is so extreme that even "typical" structures are determined by extreme events (see Chapter 5 for the connection between SOC and turbulent cascades).

The success of the apparently opposed paradigms of deterministic chaos and (stochastic) fractal systems thus sharply posed the question "which chaos for atmospheric science: deterministic or stochastic?" The question was not the philosophical one of whether or not the world is deterministic or stochastic, but rather whether deterministic or stochastic models are the most fruitful: which is the closest to reality (Lovejoy and Schertzer, 1998)? The answer to this question essentially depends on the number of degrees of freedom that are important: since stochastic systems are usually defined on infinite dimensional probability spaces they are good approximations to systems with large numbers of degrees of freedom. As applied to the atmosphere, the classical estimate of that number is essentially the number of dissipation scale fluid elements in the atmosphere, roughly 10^{27}–10^{30} (see Chapter 2 for this estimate). However, at any given moment clearly many of these degrees of freedom are inactive, and indeed we shall see that multifractals (via the codimension function $c(\gamma)$, Chapter 5) provide a precise estimate of the fraction of those at any given level of activity and at any space-time scale.

1.2 The Golden Age, revolution resolution and paradox: an up-to-date empirical tour of atmospheric variability

1.2.1 The basic form of the emergent laws and spectral analysis

Without further mathematical or physical restrictions, the high number of degrees of freedom paradigm of stochastic chaos is too general to be practical. But with the help of a scale-invariant symmetry such that in some generalized sense the dynamics repeat scale after scale, it becomes tractable and even seductive. It turns out that the equations of the atmosphere are indeed formally scale-invariant (Chapter 2), and even fields for which no theoretically "clean" equations exist (such as for precipitation) still apparently respect such scale symmetries. However, even if the equations respect a scaling symmetry, the solutions (i.e. the real atmospheric motions) would not be scaling were it not for the scale invariance of the relevant boundary conditions.

We have briefly mentioned the Kolmogorov law as being an example of an emergent law. Indeed, all the emergent laws discussed in this book are of the form:

$$\textit{Fluctuations} \approx (\textit{turbulent flux})^a \times (\textit{scale})^H \quad (1.1)$$

The Kolmogorov law mentioned in the previous section is recovered as a special case if the velocity difference Δv across a fluid structure of a given scale (L) is used for the fluctuations and we take the scaling exponent $H = 1/3$ and the turbulent flux is ε and $a = 1/3$. The book is structured around a series of generalizations of this basic equation. For example, rather than considering smooth or weakly varying (for example quasi-Gaussian) fluxes, we show in Chapters 3 and 5 how to treat wildly variable fluxes that are the results of multiplicative (and multifractal) cascades (this involves interpreting the equality in Eqn. (1.1) in the sense of random variables). Then in Chapters 6 and 7 we generalize the notion of "scale" to include strong anisotropy – needed in particular for handling atmospheric stratification ("generalized scale invariance"). In Chapters 8 and 9 this is further generalized from anisotropic space to anisotropic space-time (including causality). Finally in Chapter 10 we show how the long-time behaviours of space-time cascades involve "dimensional transitions" and low-frequency weather fluctuations with $H < 0$. According to Eqn. (1.1), since the mean of the turbulent flux is independent of scale this "macroweather" regime is characterized by mean fluctuations that decrease with scale. This contrasts with the higher-frequency "weather" regime in which typically $H > 0$ so that, on the contrary, mean weather fluctuations increase with scale. Box 1.1 (below) discusses the typical types of variability associated with different H values.

We now proceed to give an empirical tour of some of the fields relevant either directly or indirectly to atmospheric dynamics. This overview is not exhaustive, and it partly reflects the availability of relevant analyses and partly the significance of the fields in question. Our aim is to exploit the current "golden age" of geophysical observations so as to demonstrate as simply as possible the ubiquity of wide-range scaling – even up to planetary scales – and hence the fundamental relevance of scaling symmetries for understanding the atmosphere. However, before setting out to empirically test Eqn. (1.1) on atmospheric fields, a word about fluctuations. Often, the definition of a fluctuation as simply a difference is adequate (strictly speaking when $0 < H < 1$), but sometimes other definitions are needed. Indeed, there has arisen an entire field – wavelets – centred essentially around systematic ways of defining and handling fluctuations. For most of the following, thinking of fluctuations as differences is adequate, but some mathematical formalism is developed in Section 5.5, and as a practical matter, differences are not adequate in Chapter 10, where we treat macroweather which has $H < 0$ and requires other definitions of fluctuations (we recommend the simple Haar fluctuation, but others are possible).

In the following scaling overview, it will therefore be convenient to use the Fourier (spectral) domain version of Eqn. (1.1), which avoids these technical issues. In Fourier space, Eqn. (1.1) reads:

$$\left(\frac{Variance_{observables}}{wavenumber}\right) = \left(\frac{Variance_{flux^a}}{wavenumber}\right)(wavenumber)^{-2H} \qquad (1.2)$$

Consider a random field $f(\underline{r})$ where $\underline{r}$ is a position vector. Its "variance/wavenumber" or "spectral density" $E(k)$ is the total contribution to the variance of the process due to structures with wavenumber between k and $k + dk$, i.e. due to structures of size $l = 2\pi/k$ where l is the corresponding spatial scale and $k = |\underline{k}|$ (the modulus of the wavevector); we postpone a more formal definition to Chapter 2. The spectral density thus satisfies:

$$\langle f(\underline{r})^2 \rangle = \int_0^\infty E(k)dk \qquad (1.3)$$

where $\langle f(\underline{r})^2 \rangle$ is the total variance (assumed to be independent of position; the angular brackets "<·>" indicate statistical averaging).

In the following examples we demonstrate the ubiquity of power law spectra:

$$E(k) \approx k^{-\beta} \qquad (1.4)$$

If we now consider the real space (isotropic) reduction in scale by factor λ we obtain: $\underline{r} \rightarrow \lambda^{-1}\underline{r}$ corresponding to a "blow up" in wavenumbers: $k \rightarrow \lambda k$; power law spectra $E(k)$ (Eqn. (1.4)) maintain their form under this transformation: $E \rightarrow \lambda^{-\beta}E$ so that E is "scaling" and the (absolute) "spectral slope" β is "scale-invariant." If empirically we find E of the form Eqn. (1.4), we take this as evidence for the scaling of the field f. For the moment, we consider only scaling and scale invariance under such conventional isotropic scale changes; in Chapter 6 we extend this to anisotropic scale changes.

1.2.2 Atmospheric data in a Golden Age

As little as 25 years ago, few atmospheric datasets spanned more than two orders of magnitude in scale; yet they were challenging even to visualize. Global models had even lower resolutions, yet required heroic computer efforts. The atmosphere was seen through a low-resolution lens. Today, in-situ and remote data routinely span scale ratios of 10^3–10^4 in space and/or time scales, and operational models are not far behind. We are now beginning to perceive the true complexity of atmospheric fields which span ratios of over 10^{10} in spatial scales (the planet scale to the dissipation scale). One of the difficulties in establishing the statistical properties of atmospheric fields is that it is impossible to estimate spatial fields without making important assumptions about their statistical properties. We now survey the main data types, indicating some of their limitations, and briefly discuss the various relevant data sources.

In-situ networks

In-situ measurements have the advantage of directly measuring the quantities of greatest interest, the variables of state: pressure, temperature, wind, humidity etc. However, at the outset, these fields are rarely sampled on uniform grids; more typically they are sampled on sparse fractal networks (see Fig. 3.6a for an example). In addition, standard geostatistical techniques such as Kriging require various regularity and uniformity assumptions which are unlikely to be satisfied by the data (as we shall see, the latter are more accurately densities of measures which are singular with respect to the usual Lebesgue measures). This means that the results will depend in power law ways on their resolutions.

At first sight, an in-situ measurement might appear to be a "point" measurement, but this is misleading since while their spatial extents may be tiny compared to the analysis grids, what is relevant is rather their *space-time resolutions*, and in practice this is never point-like – nontrivial amounts of either spatial or temporal averaging are required. The main exceptions would be measurements simultaneously near 10 kHz in time and at 0.1–1 mm in space, which would allow one to approach the typical viscous dissipation (and hence true homogeneity) space and time scales.

In-situ measurements: aircraft, sondes

In-situ measurement techniques such as aircraft (horizontal) or sondes (vertical) have other problems, some of which we detail in later chapters. Aircraft data are particularly important. In many cases they provide our only direct measurements of the horizontal statistics. Unfortunately aircraft don't fly in perfectly flat straight trajectories; due to the very turbulence that they attempt to measure, the trajectories turn out to be more nearly fractal and – this turns out to be even more important – their average slopes with respect to the vertical are typically nonnegligible. If one assumes that the turbulence is isotropic (or at least has the same statistical exponents in the horizontal as in the vertical), then this issue is of little importance: if one measures a scaling exponent, then by the isotropy assumption it is unique so that the exponent estimate is assumed to be correct. However, it turns out that if the turbulence is strongly anisotropic, with different exponents in the horizontal and vertical directions, then (as we show in Chapter 6) the interpretation of the measurements is fraught with difficulties and one will generally observe a break in the spectrum/scaling. For the smaller scales the statistics are dominated by the horizontal fluctuations, while at the larger scales they are dominated by the vertical fluctuations. In Chapter 2 we see that naive use of isotropy assumptions has commonly led researchers to misinterpret this spurious transition from horizontal to vertical scaling as a signature of a real physical transition from an isotropic 3D turbulence regime at small scales to an isotropic 2D turbulence regime at large scales.

Remote sensing

One way of overcoming the problems of in-situ sampling is to use remotely sensed radiances. There is a long history of using radiances in "inversion algorithms" in an attempt to directly estimate atmospheric parameters (Rodgers, 1976). However, to be useful in numerical weather models, the data extracted from the inversions must generally be of high accuracy. This is because models typically require gradients of wind, temperature, humidity etc., and taking the gradients greatly amplifies errors. The fundamental problem is that classical inversion techniques aim to estimate the traditional numerical model inputs (variables of state) and they rely on unrealistic subsensor resolution homogeneity assumptions to relate these parameters to the measured radiances. Since the heterogeneity is generally very strong (scaling, multifractal) there are systematic power law dependencies on the resolution of the measurements (a consequence of the cascades structure, Section 5.3). Therefore, new resolution-independent algorithms are needed (Lovejoy *et al.*, 2001).

Reanalyses

Having recognized that in-situ measurements have frequent "holes," and that the inversion of remote measurements is error-prone, one can attempt to combine all the available data as well as the theoretical constraints implied by the governing atmospheric equations to obtain an "optimum estimate" of the state of the atmosphere; these are the meteorological "reanalyses." Reanalyses are effectively attempts to provide the most accurate set of fields consistent with the data and with the numerical dynamical models, themselves believed to embody the relevant physical laws. The data are integrated in space with the help of a variational algorithm either at regular intervals ("3D var"); or – in the more sophisticated "4D var" – both in space and time (see e.g. Kalnay, 2003). In these frameworks, remotely sensed data can also be used, but in a *forward* rather than an inverse model: one simply calculates theoretically the radiances from the guess fields of the traditional atmospheric variables. Once all the guess fields are calculated at the observation times and places, then the two are combined by weighting each guess and measurement pair according to pre-established uncertainties. While these sophisticated data assimilation techniques are elegant, one should not forget that they are predicated on various smoothness and regularity assumptions which are in fact not satisfied because of the very singular scaling effects discussed in this book. These resolution effects introduce nonnegligible uncertainties and possible biases on the reanalyzed fields.

1.2.3 The horizontal scaling of atmospheric fields

We start our tour by considering global-scale satellite radiances, since they are quite straightforward to interpret. Fig. 1.2 shows the "along track" 1D spectra from the Visible Infrared Sounder (VIRS) instrument of the Tropical Rainfall Measurement Mission (TRMM) at wavelengths of 0.630, 1.60, 3.75, 10.8, 12.0 μm, i.e. for visible, near infrared and (the last two) thermal infrared. Each channel was recorded at a nominal resolution of 2.2 km and was scanned over a "swath" 780 km wide, and $\approx$ 1000 orbits were used in the analysis. The scaling apparently continues from the largest scales (20 000 km) to the smallest available. At scales

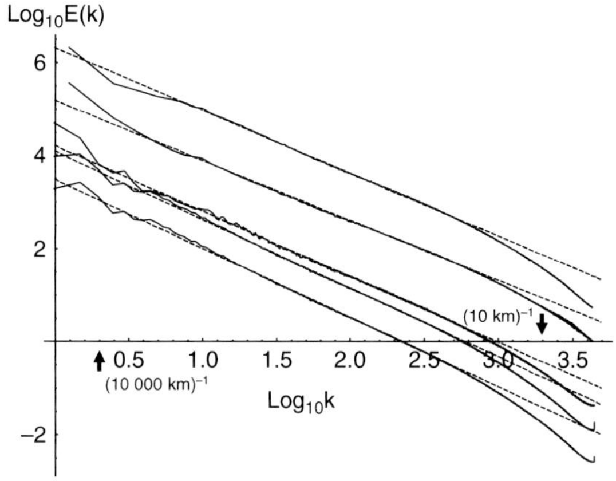

Fig. 1.2 Spectra from ≈ 1000 orbits of the Visible Infrared Sounder (VIRS) instrument on the TRMM satellite channels 1–5 (at wavelengths of 0.630, 1.60, 3.75, 10.8, 12.0 μm from top to bottom, displaced in the vertical for clarity). The data are for the period January through March 1998 and have nominal resolutions of 2.2 km. The straight regression lines have spectral exponents $\beta = 1.35, 1.29, 1.41, 1.47, 1.49$ respectively, close to the value $\beta = 1.53$ corresponding to the spectrum of passive scalars (= 5/3 minus intermittency corrections: see Chapter 3). The units are such that $k = 1$ is the wavenumber corresponding to the size of the planet $(20\ 000\ \text{km})^{-1}$. Channels 1, 2 are reflected solar radiation so that only the 15 600 km sections of orbits with maximum solar radiation were used. The high-wavenumber fall-off is due to the finite resolution of the instruments. To understand the figure we note that the VIRS bands 1, 2 are essentially reflected sunlight (with very little emission and absorption), so that for thin clouds the signal comes from variations in the surface albedo (influenced by the topography and other factors), while for thicker clouds it comes from nearer the cloud top via (multiple) geometric and Mie scattering. As the wavelength increases into the thermal IR, the radiances are increasingly due to black body emission and absorption with very little multiple scattering. Whereas at the visible wavelengths we would expect the signal to be influenced by the statistics of cloud liquid water density, for the thermal IR wavelengths it would rather be dominated by the statistics of temperature variations – themselves also close to those of passive scalars. Adapted from Lovejoy *et al.* (2008).

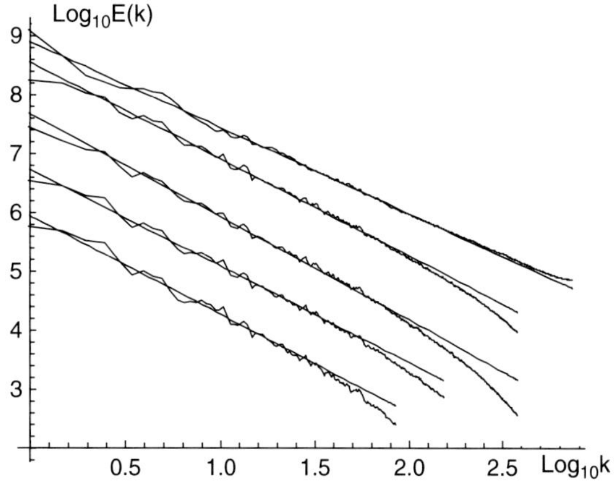

Fig. 1.3 Spectra of radiances from the TRMM Microwave Imager (TMI) from the TRMM satellite, ≈ 1000 orbits from January through March 1998. From bottom to top, the data are from channels 1, 3, 5, 6, 8 (vertical polarizations, 2.8, 1.55, 1.41, 0.81, 0.351 cm) with spectral exponents $\beta = 1.68, 1.65, 1.75, 1.65, 1.46$ respectively at resolutions 117, 65, 26, 26, 13 km (hence the high wavenumber cutoffs), each separated by one order of magnitude for clarity. To understand these thermal microwave results, recall that they have contributions from surface reflectance, water vapour and cloud and rain. Since the particles are smaller than the wavelengths this is the Rayleigh scattering regime and as the wavelength increases from 3.5 mm to 2.8 cm the emissivity/absorbtivity due to cloud and precipitation decreases so that more and more of the signal originates in the lower reaches of clouds and underlying surface. Also, the ratio of scattering to absorption increases with increasing wavelength so that at 2.8 cm multiple scattering can be important in raining regions. The overall result is that the horizontal gradients – which will influence the spectrum – will increasingly reflect large internal liquid water gradients.

below about 10 km, there is a more rapid fall-off but this is likely to be an artefact of the instrument, whose sensitivity starts to drop off at scales a little larger than the nominal resolution. The scaling observed in the visible channel (1) and the thermal IR channels (4, 5) are particularly significant since they are representative respectively of the energy-containing short- and long-wave radiation fields which dominate the earth's energy budget. One sees that thanks to the effects of cloud modulation, the radiances are very accurately scaling. This result is incompatible with classical turbulence cascade models which assume well-defined energy flux sources and sinks with a source and sink-free "inertial" range in between (see Section 2.6.6). Also of interest is the fact that the spectral slope β is close (but a little lower) than the value $\beta = 5/3$ expected for passive scalars in the classical Corrsin–Obukhov theory discussed in Chapter 2. This result is consistent with theoretical studies of radiative transfer through passive scalar clouds (Watson *et al.*, 2009; Lovejoy *et al.*, 2009a). Although we cannot directly interpret the radiance spectra in terms of the wind, humidity or other atmospheric fields, they are strongly nonlinearly coupled to these fields so that the scaling of the radiances are *prima facie* evidence for the scaling of the variables of state. To put it the other way around: if the dynamics were such that it predominantly produced structures at a characteristic scale L, then it is hard to see how this scale would not be clearly visible in the associated cloud radiances.

To bolster this interpretation, we can also consider the corresponding images at microwave channels (corresponding to black body thermal emission with wavelengths in the range 0.351–3.0 cm) (Fig. 1.3). In

(a)

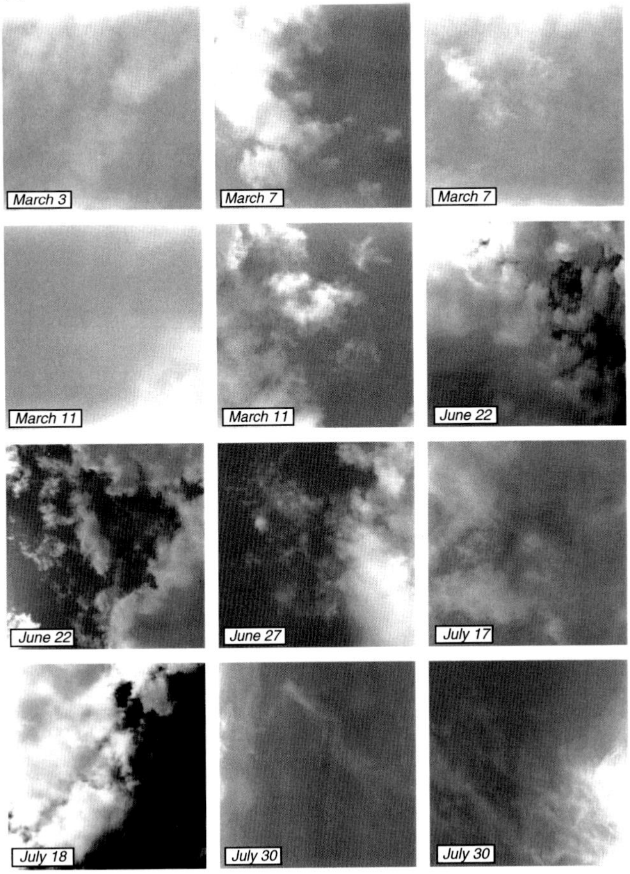

(b)

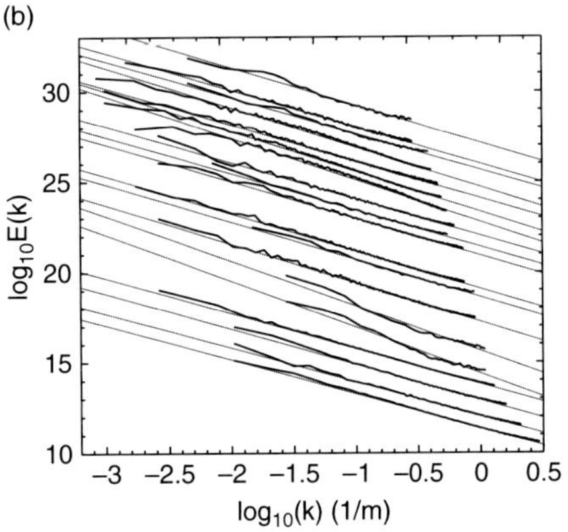

Fig. 1.4 (a) A sample of cloud pictures taken looking upward from the ground near midday, Montreal, Quebec. To get a useful resolution of several thousand pixels on a side, the standard 8 bit imagery of commercial digital cameras is not adequate. In the figure it was necessary to scan black and white negatives (with effectively 13–14 bit dynamical range); the figure shows typical results in the latter case using large-format (60 × 60 mm) negatives to resolutions (for low-lying clouds) down to 50 cm or so. Reproduced from Sachs *et al.* (2002). (b) The spectra of the 19 (of 38) highest-resolution clouds analyzed in with a spectral slope $\beta \approx 2$; see Fig. 1.4a for 12 of the samples. Reproduced from Sachs *et al.* (2002).

order to extend these results to smaller scales, we can use either finer-resolution satellites such MODIS, SPOT or LANDSAT, or we can turn to ground-based photography (Figs. 1.4a, 1.4b). Again, we see no evidence for a scale break. Interestingly, the average exponent $\beta \approx 2$ indicates that the downward radiances captured here (with near-uniform background sky) are smoother (larger β) than for the upward radiances analysed in Figs. 1.2 and 1.3 (the variability falls off more rapidly with wavenumber since β is larger).

The remotely sensed data analyzed above give strong direct evidence of the wide-range scaling of the radiances and hence indirectly for the usual meteorological variables of state. For more direct analyses, we therefore turn our attention to reanalyses. Fig. 1.5a shows representative reanalyses taken from the European Medium Range Weather Forecasting Centre (ECMWF) "interim" reanalysis products, the zonal and meridional wind, the geopotential height, the specific humidity, the temperature, vertical wind. The ECMWF interim reanalyses are the successor products to the ECMWF 40-year reanalysis (ERA40) and are publicly available at 1.5° resolution in the horizontal and at 37 constant pressure surfaces (every 25 mb in the lower atmosphere). At the time of writing, the fields were available every 6 hours from 1989 to the present. The data in Fig. 1.5a were taken from the 700 mb level. The 700 mb level was chosen since it is near the data-rich surface level, but suffers little from the extrapolations necessary to obtain global 1000 mb fields (which is especially problematic in mountainous regions); it gives a better representation of the "free" atmosphere (see Section 4.2.2 for more information and analyses, and Berrisford *et al.*, 2009, for complete reanalysis details).

(a)

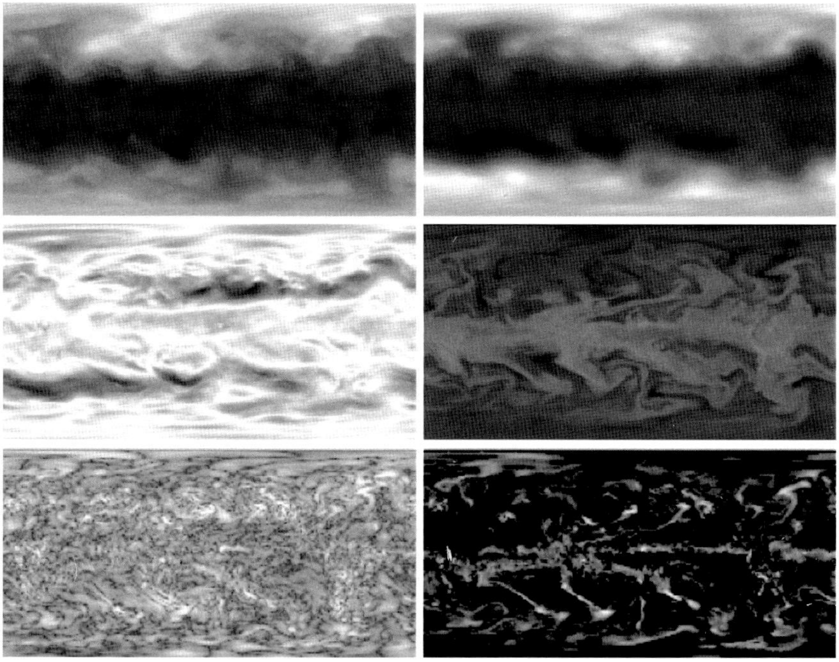

(b)

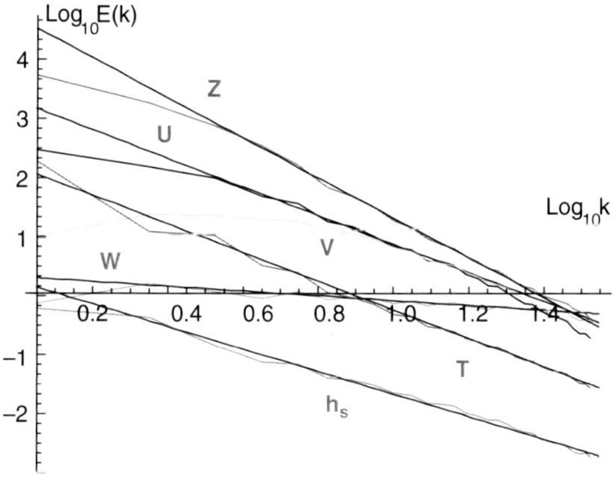

Fig. 1.5 (a) Comparison of various reanalysis fields for January 1 2006, 0Z, ECMWF interim. This shows the specific humidity (top left), temperature (top right), zonal, meridional wind (middle left and right), and vertical wind and geopotential height (bottom left and right). All fields are at 700 mb. Reproduced from Lovejoy and Schertzer (2011). (b) Comparisons of the spectra of different atmospheric fields from the ECMWF interim reanalysis. Top is the geopotential ($\beta = 3.35$), second from the top is the zonal wind ($\beta = 2.40$), third from the top is the meridional wind ($\beta = 2.40$), fourth from the top is the temperature ($\beta = 2.40$), fifth from the top is the vertical wind ($\beta = 0.4$), at the bottom is the specific humidity ($\beta = 1.6$). All are at 700 mb and between $\pm$ 45° latitude, every day in 2006 at 0 GMT. The scale at the far left corresponds to 20 000 km in the east–west direction, at the far right to 660 km. Note that for these 2D spectra, Gaussian white noise would yield $\beta = -1$ (i.e. a positive slope = +1). Reproduced from Lovejoy and Schertzer (2011).

The data analyzed were daily data for the year 2006 with only the band between $\pm$ 45° latitude used (with a cylindrical projection). The reason for this choice was twofold: first, this region is fairly data-rich compared to the more extreme latitudes; second, it allows us to conveniently compare the statistics in the east–west and north–south directions in order to study the statistical anisotropies between the two. In addition, the east–west direction was similarly broken up into two sections, one from 0° to 180° and the other from 180° to 0° longitude. For technical reasons (discussed in Chapter 6), the spectrum was estimated by performing integrals around ellipses with aspect ratios 2 : 1 (EW : NS). The wavenumber scale in Fig. 1.5b indicates the east–west scale; a full discussion of the anisotropy is postponed to Chapter 6.

From the figure we can see that the scaling is convincing (with generally only small deviations at the largest scales, $\geq$ 5000 km), although for the geopotential the deviations begin nearer to 2500 km. In spite of this generally excellent scaling, the values of the exponents are not "classical" in the sense that they do not correspond to the values predicted by any accepted turbulence theory. An exception is the value $\beta \approx 1.6$ for the humidity, which is only a bit bigger than the Corrsin–Obukhov passive scalar value 5/3 (minus intermittency corrections, which for this are of the order of 0.15; see Chapter 3), although in any case classical (isotropic) turbulence theory would certainly not be expected to apply at these scales. We could also mention that classically the atmosphere is "thin" at these scales (since the horizontal resolution $\approx$ 166 km is much greater than the exponential "scale height" $\approx$ 10 km), and hence according to the classical isotropic 3D/2D theory one would expect 2D isotropic turbulence to apply. For the horizontal wind field this leads to the predictions $\beta = 3$ (a downscale enstrophy cascade) and $\beta = 5/3$ (an upscale energy cascade; see Chapter 2). In comparison, we see that the actual value for the zonal wind ($\beta = 2.35$) is in between the two. In Chapter 6 we argue that this is an artefact of using gradually sloping isobars (rather than isoheights) in a strongly anisotropic (stratified) turbulence. These spectra already caution us that in spite of the intentions of their creators, the reanalyses should not be mistaken for real-world fields. Indeed, it is only by comparing the reanalysis statistics (especially the scaling exponents) with those from other (e.g. aircraft) sources that they can be validated through scale-by-scale statistical comparisons.

Satellite imagery and reanalyses are the only sources of gridded global scale fields, and we have mentioned some of the limitations of each. We therefore now turn our attention to in-situ aircraft data. First consider the 12 m resolution data from an experimental campaign over the Sea of China (Figs. 1.6a, 1.6b). We see that the scaling for both the temperature and horizontal wind is excellent. In both cases, the value $\beta \approx 1.7$ (near the Kolmogorov value 5/3) is reasonable, although in the case of the temperature we have added reference slopes with $\beta = 1.9$, which seems closer to those of the more recent data analyzed in Fig. 1.6c over the larger range 560 m to 1140 km. Once again, the scaling is excellent. We have deliberately postponed discussion of the larger-scale wind field to Chapters 2 and 6, since somewhere between $\approx$ 30 and 200 km (i.e. a bit beyond the range of Fig. 1.6b) it displays what is

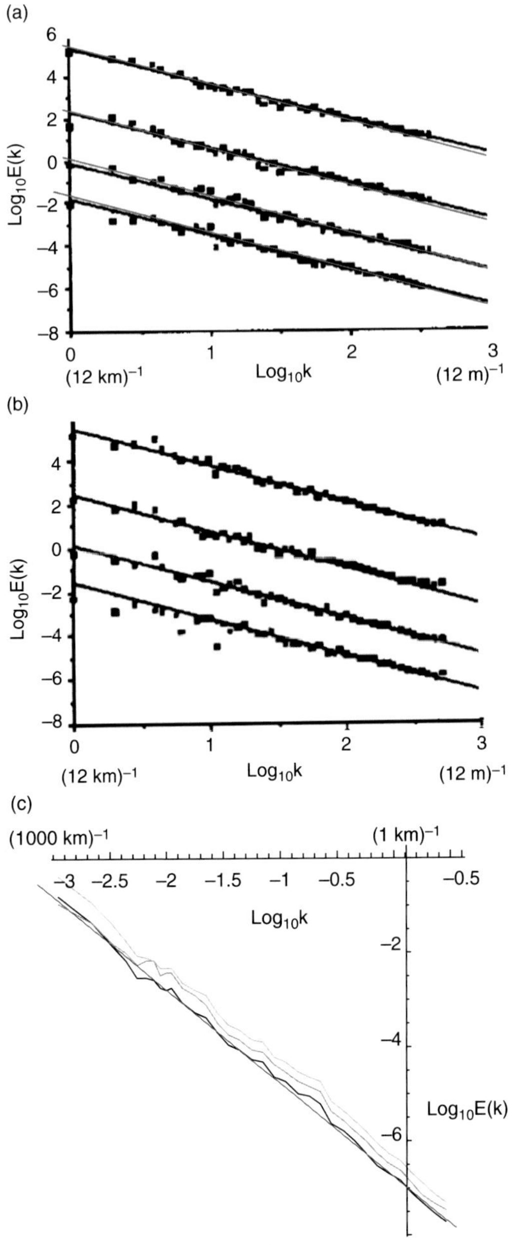

Fig. 1.6 (a) Aircraft temperature spectra. Grey slopes are 1.9, black 1.7. The bottom three curves are averages of 10 samples and each curve is taken at roughly a one-year interval; the top curve is the overall ensemble average. The curves are displaced in the vertical for clarity. Adapted from Chigirinskaya *et al.* (1994). (b) The same as Fig. 1.6a but for the horizontal wind spectrum; slopes of 1.68 are indicated. Adapted from Chigirinskaya *et al.* (1994). (c) Aircraft spectra of temperature (bottom), humidity (middle), log potential temperature (top); reference lines $\beta = 2$. These are averages over 24 isobaric aircraft "legs" near 200 mb taken over the Pacific Ocean during the Pacific Winter Storms 2004 experiment; the resolution was 280 m; Nyquist wavenumber = $(560\text{ m})^{-1}$. Adapted from Lovejoy *et al.* (2010).

apparently a spurious break due to the aircraft flying on isobars at nontrivial slopes.

1.2.4 The atmosphere in the vertical

In spite of the fact that gravity acts strongly at all scales, the classical theories of atmospheric turbulence have all been quasi-isotropic in either two or three dimensions. While a few models tentatively predict possible transitions in the horizontal (for example between $k^{-5/3}$ and k^{-3} spectra for a 3D to 2D transition for the wind), in contrast, in the vertical any 3D/2D "dimensional transition" would be even more drastic (Schertzer and Lovejoy, 1985b). This is also true for passive scalars: if a passive scalar variance flux is injected at wavenumbers k_i, then we expect $E(k) \approx k^{+1}$ if $k < k_i$ and $k^{-5/3}$ for $k > k_i$ (see e.g. Lesieur, 1987). It is therefore significant that wind spectra from relatively low-resolution (50 m) radiosondes (Fig. 1.7a, taken as part of the same experiment as the horizontal data analysed in Figs. 1.6a, 1.6b) are scaling over nearly the entire troposphere (up to 13.3 km). The slope is near (but a little larger than) that predicted by Bolgiano and Obukhov (11/5), a fact we discuss in Chapter 6. In any case, the empirical vertical spectral slope value $\beta_v \approx 2.4$ is greater than the horizontal spectral slope value $\beta_h \approx 5/3$ (Fig. 1.6b). Although it is not obvious, this implies in fact that the atmosphere is more and more stratified at larger and larger scales. Let us mention that the analysis of another radiosonde dataset (Schertzer and Lovejoy, 1985b; see Fig. 5.19a) convinced the authors that the dimensional transition does not occur along the vertical and hence would not occur along the horizontal either; this was evidence for the existence of a new type of scaling. This differential stratification can be observed directly by eye in Figs. 1.7b and 1.7c, which are vertical cross-sections of state-of-the-art lidar aerosol backscatter fields with resolutions down to 3 m in the vertical. Starting at the low-resolution image (Fig. 1.7b), we see that structures are generally highly stratified. However, zooming in closer (Fig. 1.7c), we can already make out waves and other vertically (rather than horizontally) oriented structures. In Fig. 1.7d we confirm – by direct spectral analysis – that the fields are scaling in both the horizontal and the vertical directions, and that the exponents are indeed different in both directions: the critical exponent ratio $(\beta_h - 1)/(\beta_v - 1) = H_z$ is quite near the theoretical value 5/9 discussed in Chapter 6.

1.2.5 The smallest scales

Conventional turbulence theory has primarily been applied at small scales, and there are a great many published spectra showing that they are scaling with various exponents over various ranges. Indeed, classical turbulence theory predicts that viscosity becomes dominant (and breaks the scaling) when the turbulent viscosity from Richardson's 4/3 law equals the molecular viscosity (equivalently, when the turbulent Reynolds number is unity), i.e. at scales $L \approx (\nu^3/\varepsilon)^{1/4}$, where $\nu \approx 10^{-5}$ m^2/s is the kinematic viscosity of air and $\varepsilon \approx 10^{-3}$ m^2/s^3 is the typical energy flux to smaller scales (see discussion in Chapter 8); this leads to the estimate $L \approx 0.1$–1 mm. Also classically, the number of degrees of freedom is roughly the number of these 0.1–1 mm sized cubes contained in the troposphere, therefore a number somewhere around 10^{27}–10^{30} (for an atmosphere of 10^4 km of horizontal and 10 km vertical extent).

Up until now, rather than survey the abundant literature on small-scale scaling of turbulence, we have deliberately concentrated on the far less numerous (and more controversial) large-scale analyses showing the little-known fact that scaling applies not only at the smallest but also to the largest scales. However, an interesting nonclassical exception to this is the case of rain, where the interdrop distances even in fairly heavy rain are of the order of 10 cm and hence much larger than the turbulent dissipation scales. In addition, at large enough scales, rain clearly follows the wind field (except for a superposed mean drop-fall speed), so that it is important to determine the scale where the turbulence and raindrops effectively decouple. Unfortunately, up until now the study of rain and turbulence have been almost entirely divorced from each other, so it is only very recently, with the help of stereophotography of individual drops, that this question can finally be answered. Fig. 1.8a shows a representative 3D "drop reconstruction" in which the positions and sizes of roughly 20 000 drops in a 2 × 2 × 2 m volume were determined (for clarity only the largest 10% are shown). Ninety percent of the drops larger than 0.2 mm in diameter were identified, and the positional accuracy is of the order of ± 4 cm (depth) and ± 2 cm (left to right). The drop liquid water volumes were binned to this accuracy (i.e. on 4 cm cubes) and the 3D isotropic spectrum estimated (using spherical shells in Fourier space, Appendix 2A). The result is shown in Fig. 1.8b for five storms (a total of 18

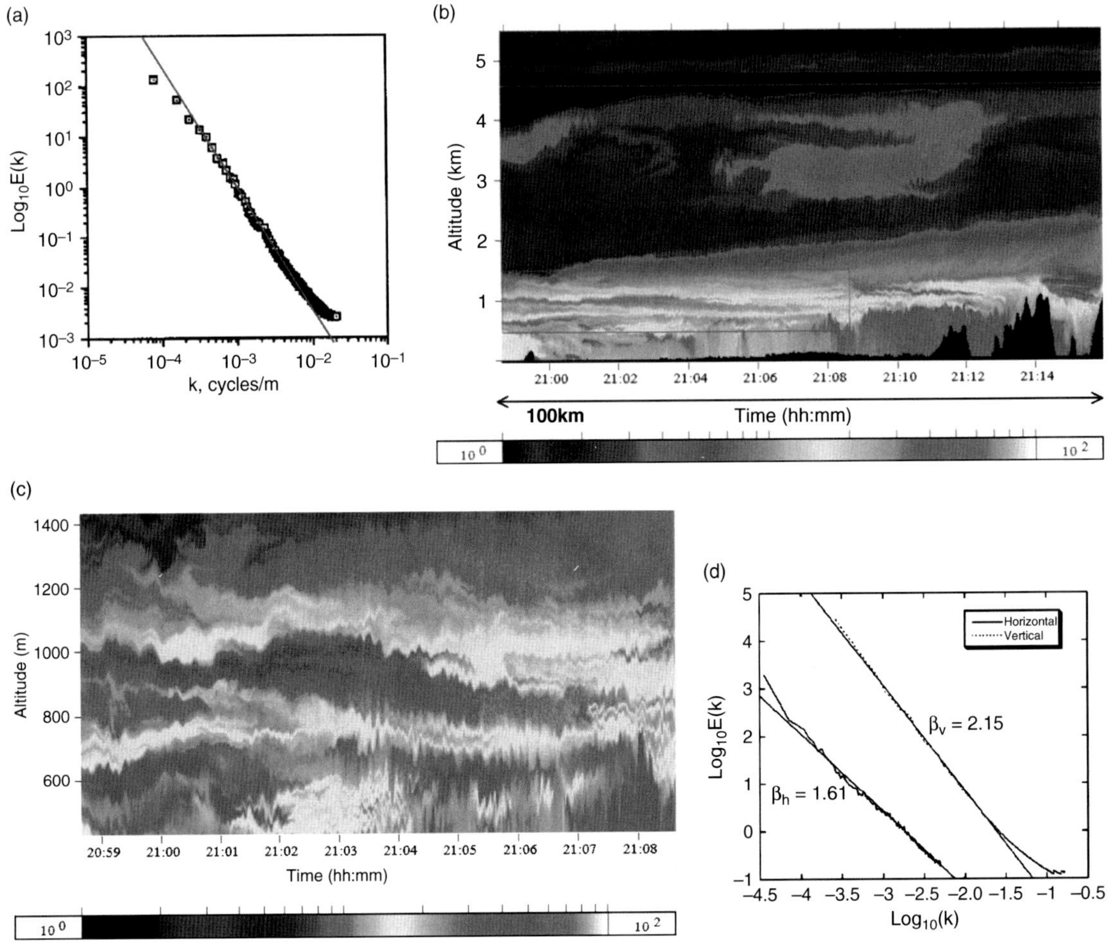

Fig. 1.7 (a) Adapted from Lazarev *et al.* (1994), slope −2.4 indicated (287 radiosondes, 50 m resolution, dropped from 13.3 km altitude). (b) Typical vertical–horizontal lidar cross-section acquired on August 14 2001. The scale (bottom) is logarithmic: darker is for smaller backscatter (aerosol density surrogate), lighter is for larger backscatter. The black shapes along the bottom are mountains in the British Columbia region of Canada. The line at 4.6 km altitude shows the aircraft trajectory. The aspect ratio is 1 : 96. Reproduced from Lilley *et al.* (2004). *See colour plate section.* (c) Zoom of Fig. 1.7b showing that at the small scales, structures are beginning to show vertical (rather than horizontal) "stratification" (even though the visual impression is magnified by the 1 : 40 aspect ratio, the change in stratification at smaller and smaller scales is visually obvious). Reproduced from Lilley *et al.* (2004). *See colour plate section.* (d) The lower curve is the power spectrum for the fluctuations in the lidar backscatter ratio, a surrogate for the aerosol density (*B*) as a function of horizontal wavenumber *k* (in m^{-1}) with a line of best fit with slope $\beta_h = 1.61$. The upper trace is the power spectrum for the fluctuations in *B* as a function of vertical number *k* with a line of best fit with slope $\beta_v = 2.15$. Adapted from Lilley *et al.* (2004).

reconstructions from sets of three images ("triplets") taken a slightly different angles in 3D). We see a clear transition from the white noise spectrum $E(k) \approx k^2$ at small scales corresponding to the usual "homogeneous" assumption (Poisson drop statistics), to the form $E(k) \approx k^{-5/3}$ at scales larger than 30–50 cm (depending somewhat on the rain), which is the spectrum predicted for passive scalars in Corrsin–Obukhov theory. Interestingly, the introduction of cascades for the rainfall (Schertzer and Lovejoy, 1987) was argued on the basis of coupled cascades of dynamics and of passive scalar. As detailed in Lovejoy and Schertzer (2008), the transition occurs where the mean turbulent Stokes number is of order unity; this is effectively the scale at which the turbulence and drops decouple due to the drop inertia. Down to this homogeneous "patch" scale, rain is a thoroughly turbulent field.

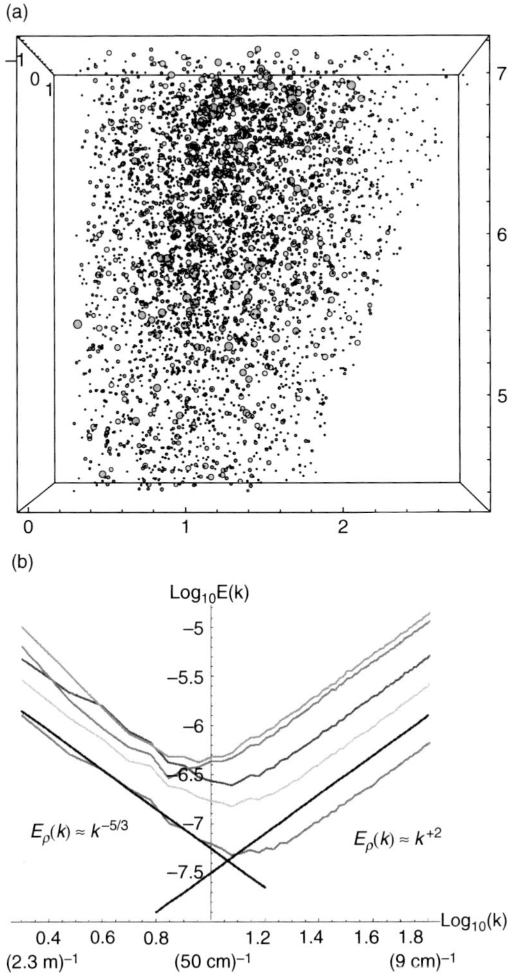

Fig. 1.8 (a) An example of a 3D drop reconstruction. For clarity only the 10% largest drops are shown, only the relative sizes and positions of the drops are correct, the colours code the size of the drops. The boundaries are defined by the flash lamps used for lighting the drops and by the depth of field of the photographs. Adapted from Lovejoy and Schertzer (2008). *See colour plate section.* (b) The angle averaged drop spectra for five storms, 18 image triplets and, for reference, Corsin–Obukhov passive scalar theory lines (rain has statistics like a tracer). This shows the 3D isotropic (angle-integrated) spectrum of the 19 stereophotographic drop reconstructions for ρ, the particle mass density. Each of the five storms had 3–7 "scenes" (from matched stereographic triplets) with 5000–40 000 drops, each taken over a 15–30 minute period. The data were taken from regions roughly 4.4 × 4.4 × 9.2 m^3 in extent (slight changes in the geometry were made between storms). The region was broken into 128^3 cells (3.4 × 3.4 × 7.2 cm^3, geometric mean = 4.4 cm); we used the approximation that the extreme low wavenumber ($\log_{10}k = 0$) corresponds to the geometric mean (i.e. 5.6 m, the minimum in the plot corresponds to about 40–70 cm). The single lowest wavenumbers ($k = 1$) are not shown since the largest scales are nonuniform due to poor lighting and focus on the edges. The reference lines have slopes −5/3, +2, i.e. the theoretical values for the Corrsin–Obukhov ($l^{1/3}$) law and white noise, respectively. Adapted from Lovejoy and Schertzer (2008).

1.2.6 Temporal scaling, weather, macroweather and the climate

If the wind field is scaling in space, then atmospheric fields are likely to be scaling in time – at least up to scales of the order of the lifetime of the largest eddies/structures in the wind field. This is because physically the wind transports the fields and dimensionally a velocity is all that is needed to convert spatial fluctuations into temporal ones. A typical example of temporal scaling is shown in Fig. 1.9a; the mean of hourly temperature spectra from four years and four stations from the northwestern USA from the US Climatological Reference Network. We can already note two key features: the division of the spectrum into two scaling ranges with a transition frequency roughly $\omega_w = (7\ \text{days})^{-1}$, and the very sharp diurnal (and harmonic) "spikes" roughly three orders of magnitude above an otherwise scaling "background."

Fig. 1.9b gives more justification for the existence of a straightforward space-time relation. It is based on two months of hourly 30 km resolution thermal IR data over the west Pacific from 30° S to 40° N from the geostationary MTSAT satellite. One can see that if the time scales are converted to space using a fixed speed of ≈ 900 km/day then the 1D spatial (zonal, meridional) and temporal spectra are nearly identical. Although at the largest scales (corresponding to ≈ 5000 km), the spectrum is slightly curved, the curvature for both space and time are virtually identical, so that even over the full range, time and space are statistically connected by this constant speed. In addition, much of this small curvature can be explained by the spectral anisotropy and the finite range of wavenumbers empirically available. In Chapters 8 and 9 we discuss this in more detail and argue that the same data show evidence that atmospheric waves also display emergent scaling laws.

But what about the transition and the behaviour at frequencies below ω_w? As we discuss in detail in Chapters 8 and 10, this transition scale is roughly the lifetime of planetary-sized structures, and the break is a "dimensional transition" whose mechanism is fairly obvious. The high frequencies where both spatial and temporal interactions are important are statistically quite different from the lower frequencies where (almost) only temporal interactions are important. In the former case, this means interactions between neighbouring structures of all sizes and at their various

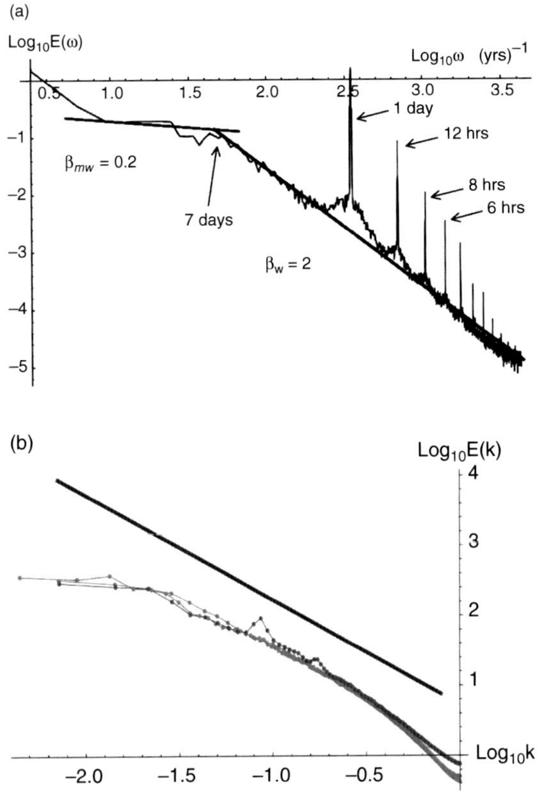

Fig. 1.9 (a) The scaling of hourly surface temperatures from four stations in the northwest USA, for four years (2005–2008) from the US Climate Reference Network. The data are discussed more fully in Section 8.1.2. To reduce noise the data were divided into sections of 112 days and the 48 spectra averaged (the rise at the extreme low frequency is connected with the annual cycle; see Fig. 8.3c for the full four-year spectra and discussion of detrending). One can see that in spite of the strong diurnal cycle (and harmonics) the basic scaling extends to about 7 days. The reference lines (with absolute slopes 0.2, 2) are theoretically motivated: see Chapter 10. (b) 1D spectra from the thermal infrared over the Pacific Ocean (MTSAT). Analyzed in time (with diurnal peak), in the east–west direction (bottom at right), and in the north–south direction. Units are such that the highest wavenumber is $(60\ \text{km})^{-1}$ and highest frequency is $(2\ \text{hours})^{-1}$ (i.e. the Nyquist wavenumber and frequency of data at 30 km and 1 hour resolutions). In Chapter 8 we show that the low-frequency/wavenumber curvature is an artefact of the finite geometry of the MTSAT scene coupled with some horizontal and space-time anisotropy. The reference line has slope = 1.5. Reproduced from Pinel (2012).

stages of development, whereas in the latter case only interactions between very large structures at various stages in their development are important. This break – which is universally observed (see Chapters 8 and 10 for many examples) – provides an objective basis for determining the low-frequency limit of the weather regime.

It turns out that this lower-frequency regime is at least roughly scaling down to the beginning of a third new even lower-frequency $\omega_c \approx (10\ \text{yr})^{-1}$ – $(100\ \text{yr})^{-1}$ regime; see Fig. 1.9c for an instrumental/paleotemperature composite spectrum of the "three scaling regime" model (the figure is a modern update of that originally proposed in Lovejoy and Schertzer, 1986). Box 1.1 discusses the corresponding types of variability.

Although we are used to the idea that "the climate is what you expect," i.e. that the climate is simply the long-term statistics of the weather, Fig. 1.9c shows that this idea is both vague and misleading. To start with, as shown graphically in Fig. 1.9d, the intermediate regime $\omega_w < \omega < \omega_c$ has statistics which are very close to those predicted by simply extending the weather scale models to low frequencies. This includes the stochastic fractionally integrated flux (FIF) model developed below (Chapter 5), which predicts a realistic dimensional transition as well as standard global climate models (GCMs) when these are run without special anthropogenic, solar, orbital or other "climate forcings": i.e. in "control runs." This regime is therefore no more than low-frequency "macroweather," without any new internal dynamical element, or any new forcing mechanism. Although the three scaling regime picture seems quite realistic, the transition frequency ω_c varies from place to place and even from epoch to epoch, with the Greenland Holocene series exceptional in having particularly small values corresponding to very stable (weakly variable) conditions (see Chapter 10).

On the contrary, the lowest frequencies $\omega < \omega_c$, corresponding to multidecadal, multicentennial, multimillennial variability, correspond to our usual ideas about "climate." At these really long time scales – in addition to various "climate forcings," all kinds of complex deep ocean, land, ice and other internal mechanisms become important – and these may also be expected to be scale-invariant: in effect the synergy between nonlinearly interacting parts of the "climate system" result in the emergence of a unique scaling regime; in this case between about 10–100 years and 100 kyr (see Fig. 1.9c). At the really low frequencies below this, the spectrum decreases; this defines the pseudo-periodicity of the interglacials. Note that the spectral spikes corresponding to the Milankovitch (orbital) forcing mechanism are at fairly narrow bands near the precessional ($\approx (19\ \text{kyr})^{-1}$, $(23\ \text{kyr})^{-1}$) and obliquity frequencies $(41\ \text{kyr})^{-1}$, and the "wobbling" of

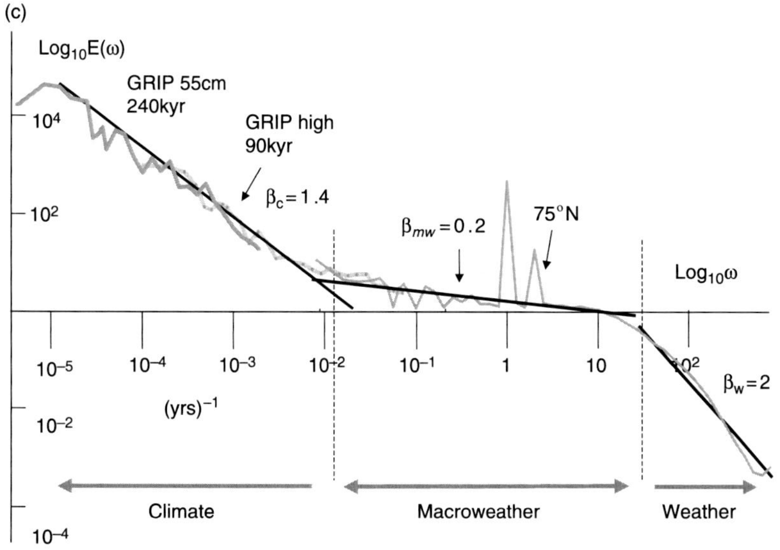

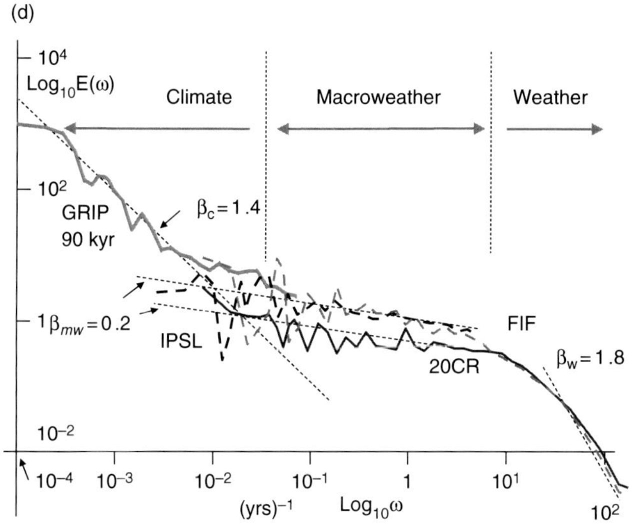

Fig. 1.9 (c) A composite spectrum of the GRIP (summit) ice-core $\delta^{18}O$ (a temperature proxy, low resolutions (and the first 91 kyr at high resolution at left), with the spectrum of the (mean) 75° N Twentieth Century Reanalysis (20CR) temperature spectrum, at 6-hour resolution, from 1871 to 2008, at 700 mb (right). The overlap (from 10-year to 138-year scales) is used for calibrating the former (moving them vertically on the log-log plot). All spectra are averaged over logarithmically spaced bins, 10 per order of magnitude in frequency. Three regimes are shown corresponding to the weather regime (which apparently extends down to turbulent dissipation scales $\approx$ 1 ms, i.e. another seven orders of magnitude to the right), with $\beta_w = 2$; note that the diurnal variation and harmonic at 12 hours are visible at the extreme right. The central macroweather "plateau" is shown along with the theoretically predicted $\beta_{MW} = 0.2 - 0.4$ regime; see Chapter 10. Finally, at longer time scales (left), a new scaling climate regime with exponent $\beta_c \approx 1.4$ continues to about 100 kyr. This composite uses a single instrumental and single paleodata source: data from the 138-year-long 20CR at 75° N. This is roughly the same latitude as the paleotemperatures from $\delta^{18}O$ proxy temperature series from the famous GRIP Greenland summit ice core (Greenland Ice Core Project, 1993). More details on these data are given in Chapters 8 and 10. Reproduced from Lovejoy and Schertzer (2012). (d) Similar to Fig. 1.9c, but showing only the high-resolution paleo spectrum (GRIP 90 kyr, the average of nine consecutive 10 kyr sections at 5.2-year resolution, left), the daily resolution, annually detrended 20CR spectrum at 75° N. These empirical spectra are compared with model spectra: the stochastic (FIF: fractionally integrated flux model, daily resolution) and the control run of the IPSL GCM at monthly resolution, both dashed. Reproduced from Lovejoy and Schertzer (2012).

the eccentricity at around $(100\ \text{kyr})^{-1}$ is apparently quite weak: almost all of the variability is due to the scaling "background," presumably to internal nonlinear (and apparently scaling) variability.

1.2.7 The scaling of the atmospheric boundary conditions

In Chapter 2 we shall see that the basic equations of the atmosphere are scaling, so that solutions can potentially also be scaling. However, for this to be true, the boundary conditions must be scaling. We will therefore now take a quick tour of some of these.

One of the lower boundary conditions is the topography, which is of prime importance for surface hydrology and oceanography, and therefore for hydrosphere–atmosphere interactions. The issue of scaling has an even longer history in topography than in atmospheric science, going back almost 100 years to when Perrin (1913) eloquently argued that the coast of Brittany was nondifferentiable.

Box 1.1 Characterizing the dynamics by the type of statistical variability

Based on the "spectral slope" in Fig. 1.9c (the negative spectral power law exponent β), we divided atmospheric dynamics into weather, macroweather and climate regimes; we see that each regime primarily displays a continuous "background" spectrum. Indeed, the annual cycle and its harmonic at six months are the only nonscaling features that stand out after the spectrum is averaged over 10 logarithmically spaced bins per order of magnitude. Each scaling regime can therefore be considered as the outcome of different dynamical processes, each of which is effectively a (scaling) synergy of nonlinear mechanisms repeating scale after scale over a wide range. Although Fig. 1.9c characterizes the regimes by β, in Chapter 3 we see how a whole hierarchy of exponents are needed for full statistical characterizations, and we also show how the hierarchy can usually be reduced to two additional parameters ("universal multifractals": see Table 11.7 for the parameters for the temperature field).

Another way of looking at this is that the dynamics in each regime generate different types of variability so that a statistical characterization of the variability and dynamics are essentially equivalent (although for this to be fully true, the variability must be characterized in space-time and it must include any relevant anisotropy; furthermore, the joint variability of all the dynamically significant fields must be specified, not just the temperature).

Fig. 1.9e shows typical examples of series of temperature data in each regime. In order to make the comparison clear, each series is shown with both smallest resolutions and largest scales in the corresponding scaling regimes and each has the same number of points (720). Perhaps the most obvious feature is that in the top and bottom series the signal has a tendency to "wander" whereas in the middle (the macroweather regime) successive fluctuations have a tendency to cancel each other out. Indeed, we shall see that temperature fluctuations $\Delta T \approx \Delta t^H$ (at lag Δt; see Eqn. (1.1)), and in the weather and climate regimes $H \approx 0.4$ (growing) whereas in the macroweather $H \approx -0.4$. Another feature displayed by the series is the abruptness of the changes, which are largest in the weather regime, a bit smaller in the climate regime and relatively minor in the macroweather regime; this feature roughly corresponds to the intermittency parameter C_1.

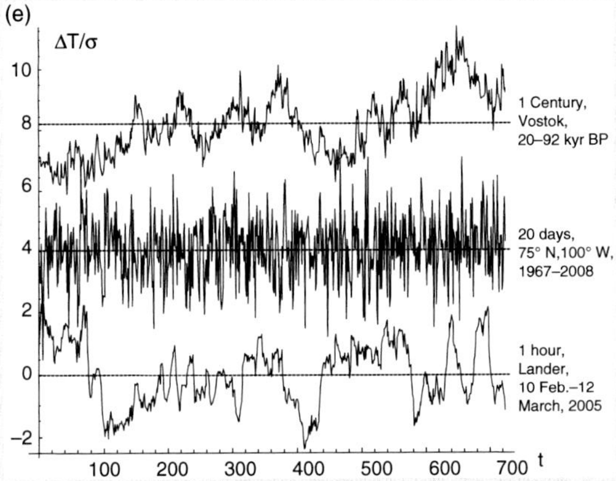

Fig. 1.9 (e) In this book, we associate different regimes with different dynamics on the basis of their types of scaling variability. This figure gives a visual comparison displaying representative temperature series from each of the scaling regimes shown in Fig. 1.9c. Bottom to top: weather, macroweather, climate. To make the comparison as fair as possible, in each case the sample is 720 points long and each series has its mean removed and is normalized by its standard deviation (4.49 ± *K*, 2.59 ± *K*, 1.39 ± *K*, respectively); the two upper series have been displaced in the vertical by four units for clarity. The resolutions are 1 hour, 20 days and 1 century respectively. The data are from a weather station in Lander, Wyoming (one of the stations used in Fig. 1.9a), the Twentieth Century Reanalysis (20CR) and the Vostok Antarctic station, respectively. The scaling exponents characterizing the regimes are discussed in the book and are summarized in Table 11.7. Note the similarity between the type of variability in the weather and climate regimes (reflected in their scaling exponents).

Later, Steinhaus (1954) expounded on the nonintegrability of the River Vistula, while Richardson (1961) quantified both aspects using scaling exponents and Mandelbrot (1967) interpreted the exponents in terms of fractal dimensions. Indeed, scaling in the earth's surface is so prevalent that there are entire scientific specializations such as river hydrology and geomorphology that abound in scaling laws of all types, and these virtually require the topography to be scaling (see Rodriguez-Iturbe and Rinaldo, 1997, for a review; and Tchiguirinskaia *et al.*, 2002, for a comparison of multifractal and fractal analysis of basins).

The first spectrum of the topography was the very-low-resolution one computed by Venig-Meinesz (1951), who noted that it was nearly a power law with $\beta \approx 2$. After this pioneering work, Balmino *et al.* (1973) made similar analyses on more modern data and confirmed Venig-Meinesz's results. Bell (1975) followed, combining various data (including those of

(a)

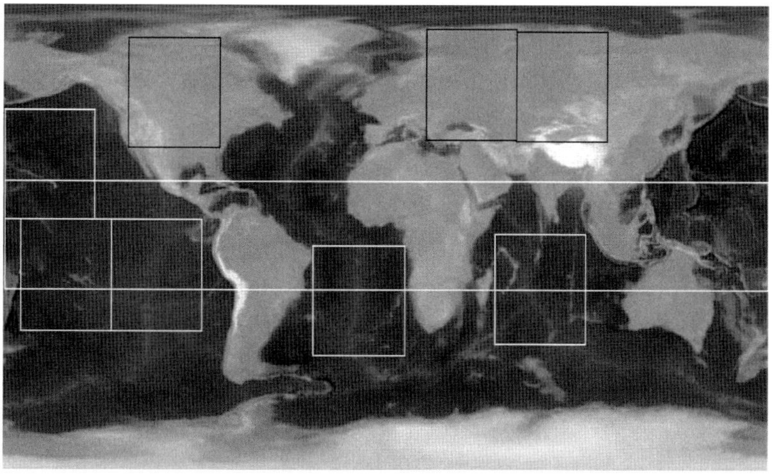

(b)

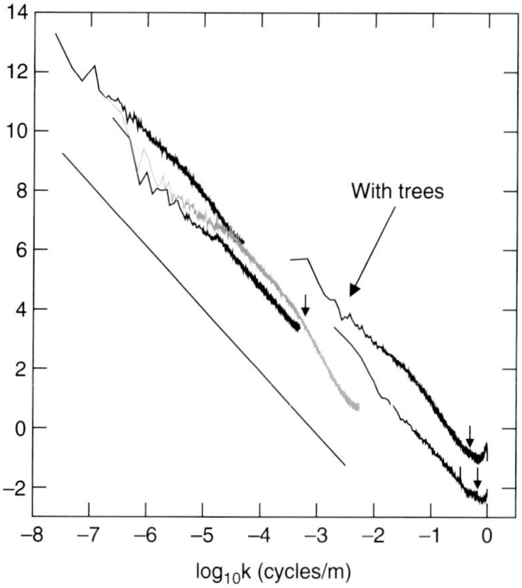

Fig. 1.10 (a) ETOPO5 topography dataset at 5 minutes of arc resolution (roughly 10 km). The squares delineate regions that were subject to a special comparison of continental versus bathymetric/oceanic regions (the *H* exponents were found to be a bit different, $\approx$ 0.4 and 0.7 respectively: see Chapter 5). Reproduced from Gagnon *et al.* (2006). (b) A log-log plot of the spectral power as a function of wavenumber for four digital elevation models (DEMs). From right to left: Lower Saxony (all, i.e. with and without trees, top; a section without trees, bottom), USA (in grey), GTOPO30 and ETOPO5 (top). A reference line of slope −2.10 is shown for comparison. The small arrows show the frequency at which the spectra are not well estimated due to their limited dynamical range (for this and scale-dependent corrections, see Gagnon *et al.*, 2006).

abyssal hills) to produce a composite power spectrum that was scaling over approximately four orders of magnitude in scale (also with $\beta \approx 2$). More recent spectral studies of bathymetry over scale ranges from 0.1 km to 1000 km can be found in Berkson and Matthews (1983) ($\beta \approx 1.6{-}1.8$), (Fox and Hayes (1985) ($\beta \approx 2.5$), Gilbert (1989) ($\beta \approx 2.1{-}2.3$) and Balmino (1993) ($\beta \approx 2$). Attempts were even made to generalize this to many natural and artificial surfaces (Sayles and Thomas, 1978). The resulting spectrum exhibited scaling over eight orders of magnitude with $\beta \approx 2$; see, however, the critique by Berry and Hannay (1978) and Gagnon *et al.* (2006).

In Fig. 1.10a we show a grey-scale rendition of the modern ETOPO5 dataset which is the earth's topography (including bathymetry) at 5 minutes of arc, roughly 10 km, and in Fig. 1.10b we show the corresponding spectrum along with those of other higher-resolution but regional digital elevation models (DEMs). These include GTOPO30 (the continental USA at $\approx$ 1 km) as well as two other DEMs: the USA at 90 m resolution and part of Saxony in Germany at 50 cm

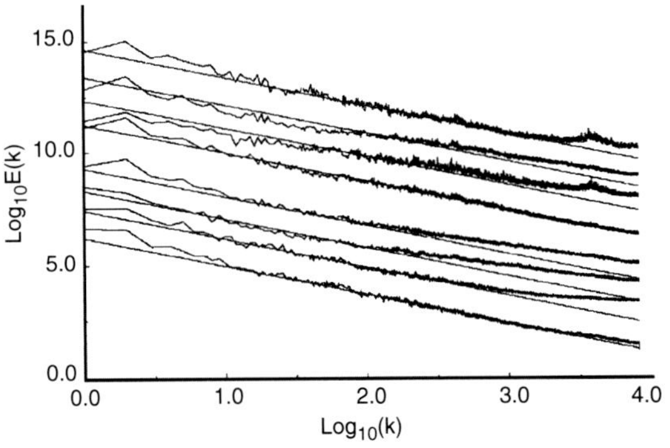

Fig. 1.11 The ocean: channels 1–8 offset for clarity, eight visible channels, 210 km long swath, 28 500 × 1024 pixels, 7 m resolution. The extreme high wavenumber is $(14\ \mathrm{m})^{-1}$. Adapted from Lovejoy *et al.* (2001).

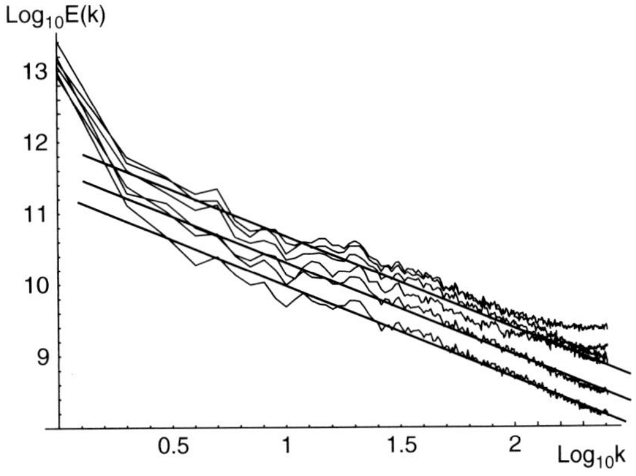

Fig. 1.12 Spectra of six bands of MODIS radiances over a 512 × 512 pixel region of Spain (at 250 m resolution; $k = 1$ corresponds to 128 km): $E(k)$ as a function of the modulus of the wave vector. In order from top to bottom at the point $\log_{10}k = 0.7$, the curves are: band 6, band 1, band 7, band 2, band 4, band 3. Reference lines have slopes −1.3. The band wavelengths are (in nm): channel 1: 620–670; 2: 841–876; 3: 459–479; 4: 545–565; 5: 1230–1250; 6: 1628–1652; 7: 2105–2155. These data are used for determining both vegetation and soil moisture indices. Adapted from Lovejoy *et al.* (2007).

resolution. Overall, the spectrum follows a scaling form with $\beta \approx 2.1$ down to at least ≈ 40 m in scale. The remarkable thing about the spectra is that the only obvious breaks are near the small-scale end of each dataset. In Gagnon *et al.* (2006) it is theoretically shown that starting from the arrows (which are always near the high-wavenumber end of the scaling part), the data are corrupted by an inadequate dynamical range. For example, the DEM at 90 m spatial resolution had an altitude resolution of only 1 m, implying that huge swaths of the country had nominally zero gradients and hence overly smooth spectra. We shall see later that such power law spectra imply fractal isoheight contours, and that the topography itself (the altitude as a function of horizontal position) is multifractal, since each isoheight contour has a different fractal dimension.

The ocean surface is particularly important for its exchanges with the atmosphere, and Fig. 1.11 shows a particularly striking wide-range scaling result: a swath over 200 km long at 7 m resolution over the St. Lawrence estuary in eight different narrow visible wavelength channels from the airborne MIES sensor. The use of different channels allows one to determine "ocean colour," which itself can be used as a proxy for phytoplankton concentration. For example the channels fourth and eighth from the top in this figure exhibit nearly perfect scaling over the entire range: these are the channels which are insensitive to the presence of chlorophyll, and they give us an indication that over the corresponding range ocean turbulence itself is scaling. In comparison, other channels show a break in the neighbourhood of ≈ 200 m in scale: these are sensitive to phytoplankton. The latter are "active scalars" undergoing exponential growth phases ("blooms") as well as being victim to grazing by zooplankton; in Lovejoy *et al.* (2000) a turbulence theory is developed to explain the break with a zooplankton grazing mechanism. Other important ocean surface fields that have been found to be scaling over various ranges include the sea surface temperature field (SST). The scaling of ocean currents and SST is discussed at length in Section 8.1.4.

Finally, many surface fields are scaling over wide ranges, particularly as revealed by remote sensing. Fig. 1.12 shows 6 MODIS channels at 250 m resolution over Spain (a 512 × 512 pixel "scene"). The scaling is again excellent except for the single lowest wavenumber, which is probably an artefact of the contrast enhancement algorithm that was applied to each image before analysis. These channels are used to yield vegetation and surface moisture indices by dividing channel pair differences by their means, so that the scaling is evidence that both vegetation and soil moisture is also scaling. In-situ measurements show not only that underground flows are also scaling, but that the hydraulic conductivity is extremely variable and strongly anisotropic (Tchiguirinskaia, 2002).

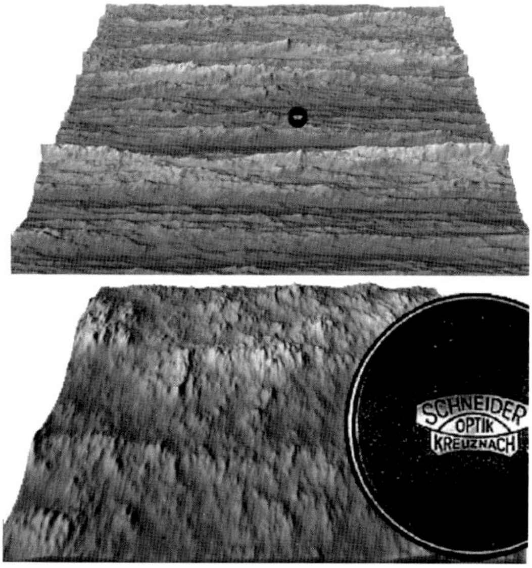

Fig. 1.13 A self-affine simulation illustrating the "phenomenological fallacy": upper and lower images look quite different while having the same generators of the scale-changing operator G (see Chapter 6; G is diagonal with elements 0.8, 1.2) and the same (anisotropic) statistics at scales differing by a factor of 64 (top and bottom blow-up). The figure shows the proverbial geologist's lens cap at two resolutions differing by a factor of 64. Seen from afar (top), the structures seem to be composed of left-to-right ridges, but closer inspection (bottom) shows that in fact this is not the case at the smaller scales. Reproduced from Lovejoy and Schertzer (2007). *See colour plate section.*

1.3 The phenomenological fallacy

We have presented a series of striking wide-range scaling spectra covering most of the meteorologically significant fields as well as for several important atmospheric boundary conditions. In this "tour" of the scaling we have exclusively used a common statistical analysis technique (the power spectrum). The conclusion that scaling is a fundamental symmetry principle of wide applicability is hard to escape, although it is still greeted with deep scepticism by some. Part of the difficulty probably stems from the feeling that "the real world can't be so simple," or that "wide-range scaling must imply that the morphologies of all clouds or all landscapes are basically the same at all scales, and this is absurd" etc. Such reactions illustrate the "phenomenological fallacy" (Lovejoy and Schertzer, 2007), which arises when phenomenological approaches are only based on morphologies, rather than underlying dynamics. This fallacy has two aspects. First, form and mechanism are confounded so that different morphologies are taken as *prima facie* evidence for the existence of different dynamical mechanisms. Second, scaling is reduced to its special isotropic "self-similar" special case in which small and large scales are statistically related by an isotropic "zoom"/"blow-up." In fact, as we explore later in this book, scaling is a much more general symmetry: it suffices for small and large scales to be related in a way that does not introduce a characteristic scale, and the relation between scales can involve differential squashing, rotation etc. so that small and large scales can share the same dynamical mechanism yet nevertheless have quite different appearances.

In order to illustrate how morphologies can change with scale when the scaling is anisotropic, consider Fig. 1.13. This is a multifractal simulation of a rough surface (with the parameters estimated for the topography); its anisotropy is in fact rather simple in the framework of the generalized scale invariance (GSI) that we will discuss in various chapters. More precisely, it is an example of linear GSI (with a diagonal generator) or "self-affine" scaling. The technical complexity with respect to self-similarity is that the exponents are different in orthogonal directions, which are the eigenspaces of the generator, so that structures are systematically "squashed" (stratified) at larger and larger scales. The underlying epistemological difficulty, which was not so simple to overcome and which still puzzles phenomenologists, corresponds to a deep change in the underlying symmetries. The top image in the figure illustrates the morphology at a "geologist's scale" as indicated by the traditional lens-cap reference. If these were the only data available, one might invoke a mechanism capable of producing strong left–right striations. However, if one only had the bottom image available (at a scale 64 times larger), then the explanation (even "model") of this would probably be rather different. In actual fact, we know by construction that there is a unique mechanism responsible for the morphology over the entire range.

Fig. 1.14 gives another example of the phenomenological fallacy, this time with the help of multifractal simulations of clouds. Again (roughly) the observed cascade parameters were used, but each with a vertical "sphero-scale" (this is the scale where structures have roundish vertical cross-sections) decreasing by factors of four, corresponding to zooming out at random locations. One can see from the vertical cross-section (bottom row) that the degree of vertical stratification increases from left to right. These passive scalar cloud simulations

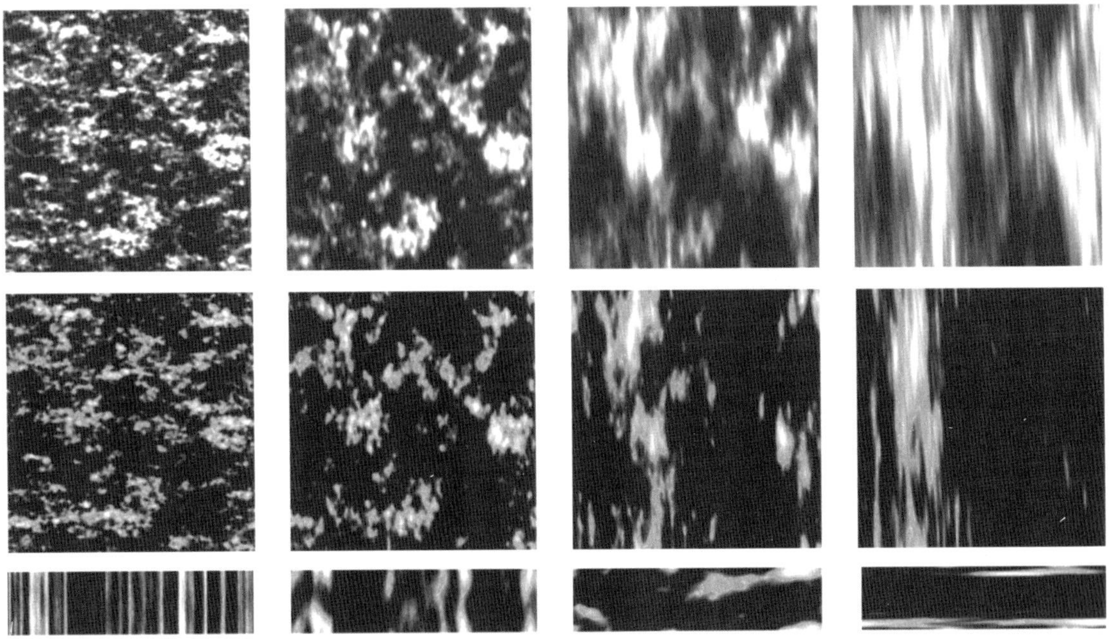

Fig. 1.14 Examples of continuous in scale anisotropic multifractals in 3D (256 × 256 × 64). The effect of changing the sphero-scale (l_s, see Chapter 6) on multifractal models of clouds with $H_z = 5/9$. The cloud statistical parameters are: $\alpha = 1.8$, $C_1 = 0.1$, $H = 1/3$ (see Chapter 3; similar to CloudSat and aerosols, see Chapter 6). From left to right we decrease l_s (corresponding to zooming out by factors of 4) so that we see the initially vertically aligned structures (bottom left) becoming quite flat at scales 64 times larger (right). At the same time, the horizontal structures have anisotropy characterized by the (matrix) generator $G = ((0.8, -0.02), (0.02, 1.2))$ so that they too change orientation, elongation (the horizontal sphero-scale starts at 1 pixel, far left; see Chapters 6 and 7 for this "generalized scale invariance"). The middle row shows a false-colour rendition of the liquid water density field, the bottom row shows the corresponding vertical sections (side view), the top row shows the corresponding single scatter visible radiation; the mean optical thickness is 2, isotropic scattering phase function, sun incident at 45° to the right. Reproduced from Lovejoy *et al.* (2009b).

(liquid water density, bottom two rows; single scattering radiative transfer, top row) show that by zooming out (left to right) diverse morphologies appear. Although a phenomenologist might be tempted to introduce more than one mechanism to explain the morphologies at different scales in the figure, we are simply seeing the consequence of a single underlying mechanism repeating scale after scale. The phenomenological fallacy can undermine many classical ideas. For example, in Lovejoy *et al.* (2009b) (see Box 6.1) it is argued that the classical two-scale theories of convection are incompatible with the data which are scaling, and that the division into qualitatively distinct small and large regimes is unwarranted.

Chapter 2 Classical turbulence, modern evidence

2.1 Complexity or simplicity? Richardson's dreams and the emergence of the laws of turbulence

2.1.1 Numerical weather prediction and statistical theories of turbulence

We have discussed the apparent dichotomy present at the birth of modern meteorology between the brute-force numerical integration of (deterministic) nonlinear partial differential equations, and turbulence approaches seeking emergent (stochastic) turbulent laws. Interestingly, both approaches coexisted within the person of one of its founders, L. F. Richardson. At first both were rather speculative: on the one hand, Richardson's "dream" (Lynch, 2006) of numerical weather prediction was at least in some quarters considered to be no more than wishful thinking. To appreciate how far advanced it was for its time, only a few years earlier, Max Margules had even speculated on the "impossibility" of forecasting (see the discussion in Lynch, 2006). As for his largely forgotten other dream of discovering emergent turbulent laws, for many years this amounted to little more than the empirical Richardson 4/3 law.

In order to understand the atmosphere, both deterministic and stochastic approaches have largely concentrated on analytical approximations. In the absence of computers, meteorologists developed barotropic, quasi-geostrophic, hydrostatic and other approximations. The discipline of "dynamical meteorology" focused on the stability of various strongly idealized prototypical flows such as jets and fronts. In turbulence, with the help of the simplifying paradigms of isotropic turbulence (Taylor, 1935) and statistical stationarity, some progress was initially made, notably through the Karmen-Howarth (1938) equations. These equations are directly or indirectly a cornerstone for a large part of analytic "closure" approximations that occupied theorists through the 1980s and which persist today.

By the end of the 1970s the relative success of the brute-force numerical approach had essentially relegated the use of many simplified sets of equations to the status of research tools, whereas real forecasting had reverted to the numerical integration of "primitive equations" (the name is not quite accurate since they still make approximations – such as the hydrostatic approximation – and these may still have serious consequences). In the parallel turbulence approach, much effort was directed at statistical closure techniques. These were initially appealing due to their ability to reach high Reynolds numbers and their relative simplicity (e.g. the quasi-normal approximation: Millionshtchikov, 1941). Over time, a series of improvements were introduced – for example to obtain statistical models that respected random Galilean invariance (the "test field" and subsequent models: Kraichnan, 1971). They gave some deep insights into the dynamics, e.g. the role of nonlocal interactions (Kraichnan, 1971; Lesieur and Schertzer, 1978) and into the delicate balance between renormalized forcing and viscosity (Schertzer *et al.*, 1998). One may note that the renormalized forcing that emerges from the nonlocal small- to large-scale interactions has since become a practical issue for atmospheric modelling: "backscatter" (Palmer and Williams, 2010). However, these advances only made it more obvious that closures could not address the fundamental question of intermittency (e.g. Frisch and Morf, 1981). In the last decade, the limitations of closures and kindred analytical approaches has led to an increasing use of brute-force "direct numerical simulations" (DNS) in hydrodyamical turbulence.

In comparison, the classical "emergent" turbulent laws – the Kolmogorov law for the wind, the Corrsin–Obukhov law for passive scalar advection, and the Bolgiano–Obukhov law for buoyancy forced turbulence – all assumed a priori the turbulence to be isotropic in three dimensions, and to be relatively homogeneous, calm (e.g. quasi-Gaussian). These limitations implied that their range of applicability was strongly circumscribed. However, as we shall see in the next chapter, thanks to the development of multiplicative cascade models, since the 1980s their generalizations have been quite successful at handling intermittency, and in Chapters 6 and 7 we show how their generalization to anisotropic turbulence has enabled us to apply them up to planetary scales. However, these developments are still not widely known or appreciated, so that for the moment the brute-force approach has triumphed but at the cost of complex codes based on nontrivial numerical truncations, smoothness, and regularity assumptions. In the process the goal of *understanding* has often been sacrificed on the altar of expediency.

Box 2.1 Richardson and Moore's law

To close the circle, we could mention that it is ironic that even in his brute-force approach Richardson was not just a dreamer, but a pioneer having spent – by his own admission – six weeks manually integrating the equations to obtain a 12-hour forecast at two grid points. Indeed, he could perhaps be considered the founder of Moore's law of computing technology (Fig. 2.1)!

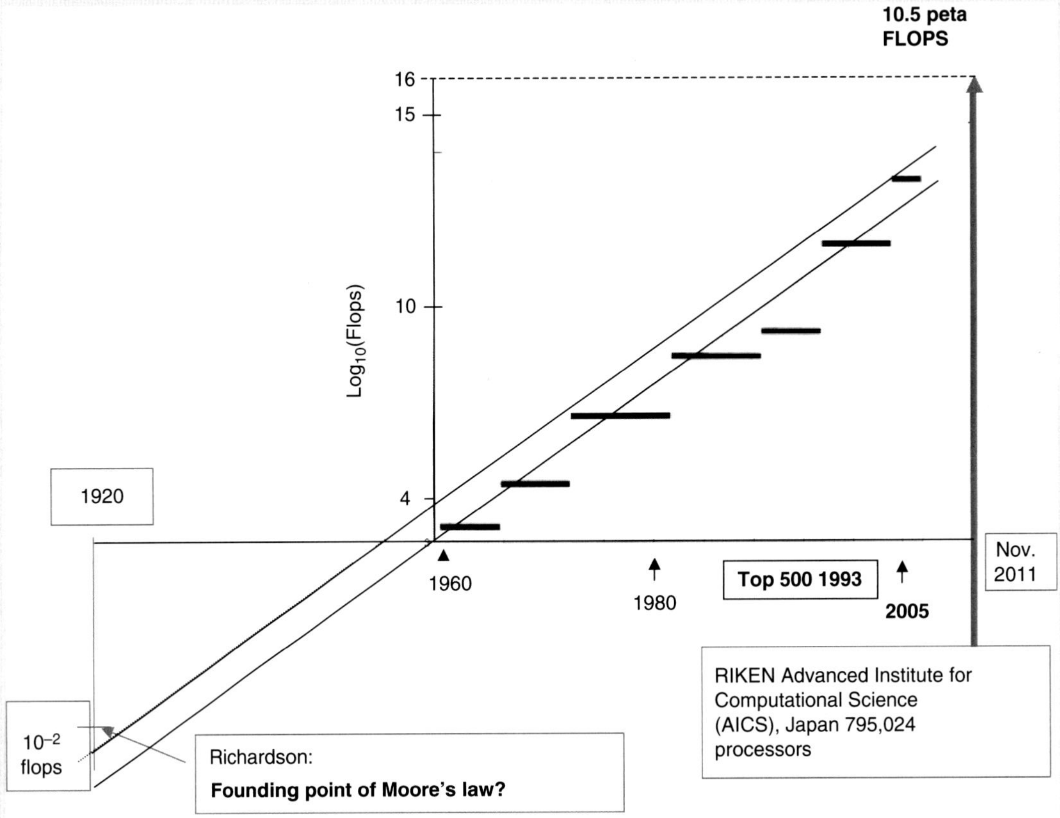

Fig. 2.1 Was Richardson the founding point of Moore's law? The progress of computer power in flops. Two lines are shown: the top is for the fastest computers, the lower for computers used for operational weather forecasting. Extrapolated from Lynch (2006).

2.1.2 The nonlinear revolution: complex or simple?

From this historical overview, it would seem that both numerical and turbulence approaches have increasingly reinforced the idea that the atmosphere is such a highly complex system that it defies simple approaches and frameworks. But the history of science is replete with examples where apparent complexity at one level has given way to simplicity at another. Even continuum mechanics itself could be regarded as an emergent paradigm with respect to more fundamental theories which operate at the atomic level, such as statistical mechanics: the Boltzmann equations. In the same way, classical statistical physics itself consists of a set of macroscopic laws which emerge when Newton's laws are applied to large numbers of particles, and similarly quantum statistical physics emerges from the application of quantum mechanics to the same.

The "nonlinear revolution" in the 1980s resuscitated hope that emergent turbulent laws that would simplify our understanding were within reach. At the time, enthusiasm was sometimes so strong that it waxed lyrical:

> The discovery was that large classes of non-linear systems exhibit transitions to chaos which are *universal* and *quantitatively* measureable . . . But the breakthrough consists not so much in discovering a new set of scaling numbers, as in developing a new way to do physics. Traditionally we use regular motions (harmonic oscilliators, plane waves, free particles, etc.) as zeroth order approximations to physical systems, and account for weak non-linearities perturbatively. We think of a dynamical system as a smooth system whose evolution we can follow by integrating a set of differential equations. The universality theory seems to tell us that the zeroth-order approximations to strongly non-linear systems should be quite different. They show an amazingly rich structure which is not at all apparent in their formulation in terms of differential equations. However, these systems do show self-similar structures which can be encoded by universal equations. . . To put it more succinctly, junk your old equations and look for guidance in clouds' repeating patterns.
>
> (P. Cvitanovic, discussing low-dimensional deterministic chaos in the introduction to his book *Universality in Chaos*, 1984)

Consider Fig. 2.2, an illustration of the Mandelbrot set. While this has been termed "the most complicated object in mathematics" (Dewdney, 1985), it could nevertheless be considered on the contrary to be one of the simplest, since it is generated by the simplest possible nonlinear dynamics: a quadratic map on the complex plane. Indeed, one can define "algorithmic complexity" (e.g. Goldreich and Wigderson, 2008) in terms of the minimal number of bits needed in a computational algorithm. With this definition, the Mandelbrot set is on the contrary exceedingly simple. Of more relevance to our present discussion is the drunkard's walk or "Brownian motion" (Fig 2.3). On the one hand the actual path of the drunkard is highly complex. On the other hand, after enough steps, the statistics become very simple. For example, as long as the variance of each step is finite (and each is independent), and the drunkard uses the same rule to choose the direction and length of each step, the variance of the distance from the starting point is simply proportional to the number of steps. Indeed, there is an even stronger result: under still fairly wide conditions, the limiting probability distribution is Gaussian, i.e. independent of the rule for choosing the steps. The simple behaviour of the random walk illustrates another important aspect of the emergence of simplicity from chaos: it can often be robust, with the same emergent behaviour arising irrespective of many of the otherwise complicated details. As an illustration, Fig. 2.3 shows two different rules for choosing a step – yet after enough steps they lead to the same Gaussian statistics. This is a classical and familiar example of "universality." When generalized from finite variance (Laplace, 1886) to infinite variance (Lévy, 1925), we will see in Chapter 3 that this universality in the form of a generalized additive "central limit theorem" is the basis for a generalization applicable to multiplicative cascade processes, and is relevant to turbulence.

The dichotomy of the simple versus the complex is more relevant to the atmosphere than it might first appear. Consider the 1 s trace of a component of the wind shown in Fig. 2.4 (left). Is this simple or is this complex? In considering data of this sort, Richardson (1926) was led to ask "Does the wind possess a velocity?", continuing, "this question, at first sight foolish, improves upon acquaintance." Indeed, he already proposed that "we may really have to describe the position x of an air particle by something rather like Weierstrass's function": i.e. a nondifferentiable fractal (Fig. 2.4, right). It is significant that one thing that all these examples have in common is their scale invariance: in the case of the Mandelbrot set, it is only approximate but becomes more and more exact at

The Mandelbrot set:

Complex?

...Or simple?

$$Z_{n+1} = Z_n^2 + C$$

Z, C are complex numbers

If $Z_1 = 0$, the Mandelbrot set is those complex C values such that: $\left| Z_\infty \right| < \infty$

Parameters for blow up: −0.75104<Re(C)<−0.7408
0.10511<Im(C)<0.11536

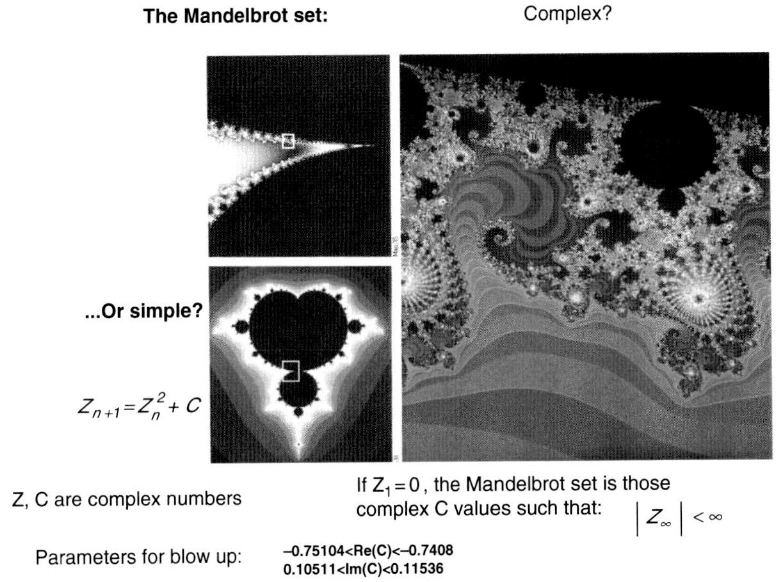

Fig 2.2 The Mandelbrot set: lower left is at lowest resolution; the upper left is an enlargement of the section in the small box. The right-hand side is the enlargement of the small box on the upper left. Adapted from Peitgen and Richter (1986).

Drunkard's walk
(Brownian motion)

Complex?

Or simple?

Statistical average

$$\langle \Delta X^2 \rangle \approx N$$

Distance from initial bar   Number of bars visited

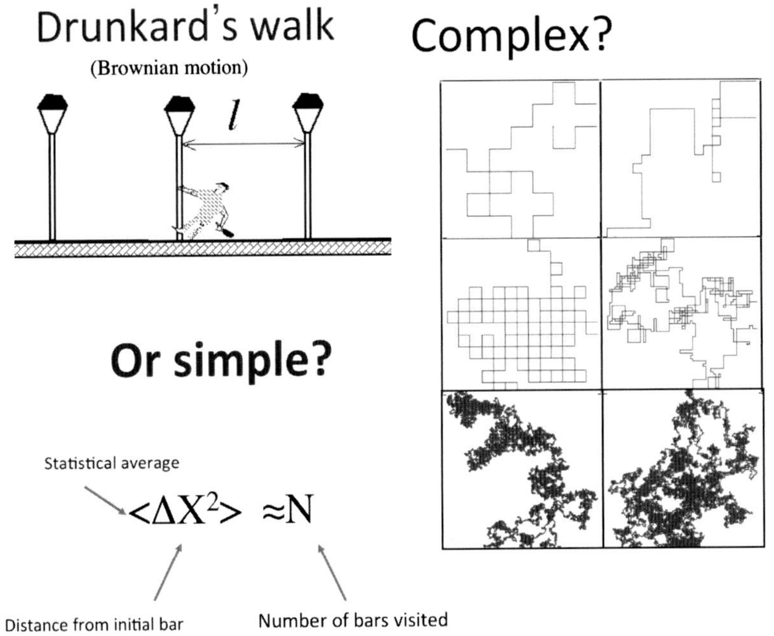

Fig. 2.3 The drunkard's walk: simple or complex? In the boxes on the right, two different walk rules are shown: in the left column, the drunk moves one step, choosing one of the orthogonal directions at random. Moving from top to bottom, the number of steps increases, approaching asymptotic statistics at the bottom. In the right column, the directions are again chosen uniformly at random but some steps are long (5 units) and some are short (1 unit) with probabilities chosen so that the variance is the same as in the left column. This assures that the limiting walk (lower right) has the same statistics as at left. This is an example of central limit theorem universality: since the variance is finite, the distributions are Gaussian in the limit even though the distributions for individual steps are different. Adapted from Schertzer and Lovejoy (1993).

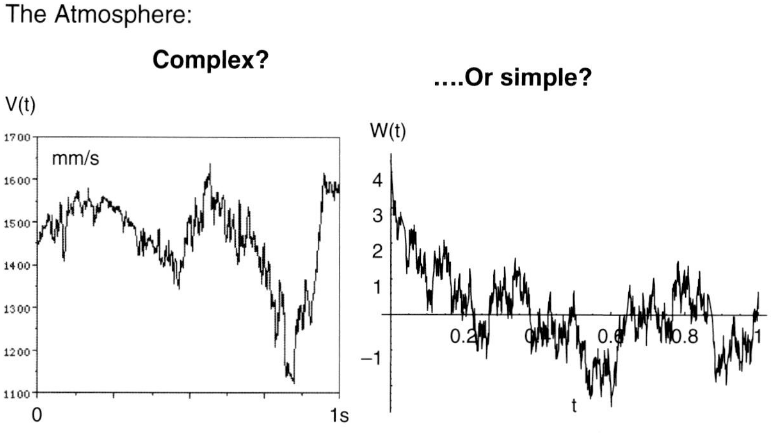

Fig. 2.4 Left: wind data at 2 kHz. Right: the Weierstrass function with parameters $H = 1/3$, $b = 2$. In both cases, the graph is roughly self-affine. For example the turbulent signal, $|\langle\Delta v(\Delta t)\rangle| \propto \Delta t^H$ so that $|\langle\Delta v(\lambda^{-1}\Delta t)\rangle| = \lambda^{-H}|\langle\Delta v(\Delta t)\rangle|$ (empirically, $H \approx 1/3$ here, the Kolmogorov value). In comparison,
$W(t) \propto \sum_{n=0}^{\infty} A_n(t)$ and consecutive terms have the same scaling property: $A_{n+1}(b^{-1}t) = b^{-H}A_n(t)$ where $A_n(t)=b^{-nH}e^{2\pi i b^n t}$ (see Eqn. (2.1)).
However the data are multifractal whereas the Weierstrass function is monofractal.

smaller and smaller scales (see Fig. 2.2 for several magnifications), while in the drunkard's walk, after a large number of steps it is also statistically scale invariant. Similarly, analysis of the turbulent wind trace (Fig. 2.4, left) shows that it appears to follow the (statistically) scale-invariant (Kolmorogov) spectrum $\omega^{-5/3}$ where ω is the frequency. In comparison, the Weierstrass function $W(t)$

$$W(t) = \sum_{n=0}^{\infty} b^{-Hn} e^{2\pi i b^n t}; \quad b > 1; \quad 0 < H < 1 \tag{2.1}$$

is deterministic; by inspection, its Fourier transform has only discrete frequencies b^n decreasing with amplitude b^H. Each term in the expansion is invariant if $t \to t/b$ and the amplitude $\to$ amplitude/b^H. The set of points on the graph $(t, W(t))$ is therefore already an example of anisotropic scaling since unless $H = 1$ the amplitudes must be rescaled by a different factor than the time scales, so that for a small part of the graph to be identical to a larger part the horizontal and vertical axes must be "blown up" by different factors.

Finally, we can easily verify the property for which the Weierstrass function is famous: it is everywhere continuous and nowhere differentiable. This can be checked by noting that for $0 < H < 1$, the moduli of the complex amplitudes of each term in Eqn. (2.1) *decrease* in geometric progression and hence the series unconditionally converges. However on the contrary, the series of the derivatives

$$W'(t) = 2\pi i(1 - b^{-2H})^{-1/2} \sum_{n=0}^{\infty} b^{(1-H)n} e^{2\pi i b^n t} \to \infty; \quad b > 1; \quad 0 < H < 1 \tag{2.2}$$

will have terms geometrically *increasing* in amplitude and will therefore unconditionally diverge, pointing out that the derivative is almost everywhere not defined! A glance at the example in Fig. 2.4 (right) shows graphically what happens: the hierarchical zigzagging structures continue down to infinitely small scales, preventing the small-scale convergence. This highly irregular (singular) small-scale behaviour is indeed a hallmark of fractal processes, although the more interesting cascades lead to something rather different: to multifractals which are the densities of singular measures (Chapter 5).

2.2 The equations of the atmosphere and their scale symmetries

2.2.1 The cascade alternative

Throughout this book, we argue that at high enough levels of turbulence (e.g. large enough Reynolds number, nonlinear variability over a wide enough range of scales) new, simpler statistics emerge: those of multiplicative cascades. Let's see how these statistics emerge from the original equations with the help of three basic cascade properties each satisfied by the equations: (i) scale invariance, (ii) a scale-by-scale conserved quantity and (iii) Fourier "locality": the fact that the interactions are strongest between structures of neighbouring size. Scale invariance implies that the basic dynamical mechanism repeats scale after scale, while the locality in scale property means that large structures do not spontaneously break up into much smaller structures without first passing through intermediate-sized ones. Note that this latter property, while presumably necessary for the existence of multiplicative cascades, may not exclude other types of cascade. For example, 2D enstrophy cascades are presumably nonlocal (Section 2.5), although to our knowledge the existence of multiplicative cascades in them has yet to be investigated.

2.2.2 Scaling

Consider incompressible (and constant-density) hydrodynamic turbulence with energy injection at large scales (the forcing), and with dissipation due to viscosity ν at small scales. In between there is a scaling ("inertial") range. Strictly speaking, the inertial range (so-called because it is dominated by the nonlinear, "inertial" terms) is a scale range with no sources or sinks of energy flux. However in the atmosphere, this is unrealistic – and in any case, as long as the latter are scaling, the assumption is unnecessary.

To see the origin of scaling, let us first consider the equations of incompressible (and dry) hydrodynamics, the Navier–Stokes equations. These can be considered to be the basic equations of the atmosphere and oceans; they are:

$$\frac{\partial \underline{v}}{\partial t} + (\underline{v}\cdot\nabla)\underline{v} = -\frac{\nabla p}{\rho_a} + \nu\nabla^2\underline{v} + \underline{f} \tag{2.3}$$

$$\nabla\cdot\underline{v} = 0 \tag{2.4}$$

where $\underline{v}$ is the velocity, t is the time, p is the pressure, ρ_a is the (fluid) air density, ν is the kinematic viscosity, and $\underline{f}$ represents the body forces (per unit volume) due to stirring, gravity. Eqn. (2.3) expresses conservation of momentum, whereas Eqn. (2.4) expresses conservation of mass in an incompressible fluid: mathematically it can be considered simply as a constraint used to eliminate p.

These equations are known to be formally invariant under isotropic "zooms" $\underline{r} = \lambda\underline{r}'$, as long as one rescales the other variables as:

$$\begin{aligned} \underline{v} &= \lambda^{\gamma_v}\underline{v}' \\ t &= \lambda^{-\gamma_v+1}t' \\ \nu &= \lambda^{\gamma_v+1}\nu' \\ \underline{f} &= \lambda^{2\gamma_v-1}\underline{f}' \end{aligned} \tag{2.5}$$

γ_v is an arbitrary scaling exponent (singularity); hence the possibility of "multiple scaling" discussed below; we do not consider the pressure since, as noted, it is easy to eliminate with Eqn. (2.4). The rescaling of the viscosity may at first glance seem odd, but it may be understood as a rescaling of the eddy-viscosity or renormalized viscosity; similar remarks can be made for the forcing (Schertzer *et al.*, 1998). The rescaling of these equations, although seemingly straightforward, may have various meanings, ranging from deterministic to statistical. A more systematic scaling analysis of differential equations is obtained with the help of the "pullback transforms" (Schertzer *et al.*, 2010, 2012; Schertzer and Lovejoy, 2011). We will return to this later; for the moment, we only note that as far as the scaling constraint is concerned, γ_v can be fairly arbitrary. However, if we impose a condition such as the conservation of the energy *flux* (not the energy: see below) the constraint will be enough to determine its value. Indeed, considering the energy flux $\varepsilon = -\partial v^2/\partial t$, we find:

$$\begin{aligned} x &= \lambda^1 x' \\ \varepsilon &= \lambda^{-1+3\gamma_v}\varepsilon' \end{aligned} \tag{2.6}$$

If it is scale invariant, we obtain $\gamma_v = 1/3$; hence, for fluctuations in the velocity Δv over distances (lags) Δx, we obtain for the mean shear:

$$\begin{aligned} \Delta x &= \lambda^1 \Delta x' \\ \Delta v &= \lambda^{1/3}\Delta v' \end{aligned} \tag{2.7}$$

If we eliminate λ this is perhaps more familiar:

$$\Delta v = \left(\frac{\Delta x}{\Delta x'}\right)^{1/3}\Delta v' \tag{2.8}$$

or in dimensional form:

$$\Delta v \approx \varepsilon^{1/3}\Delta x^{H_v}; \quad H_v = \gamma_v = 1/3 \tag{2.9}$$

which was first derived by Kolmogorov (1941). A similar scaling argument in Fourier space yields the famous $k^{-5/3}$ energy spectrum first derived by Obukhov (1941):

$$E(k) = \varepsilon^{2/3}k^{-5/3} \tag{2.10}$$

Since the two are essentially equivalent, both are sometimes referred to as the "Kolmogorov–Obukhov law" (see Section 3.1 for more discussion on this). Both real and Fourier space results can also be derived by dimensional analyses on $\varepsilon(\mathrm{m^2/s^3})$, and one can pass from one to the other if one squares both sides of Eqn. (2.9), taking ensemble averages followed by Fourier transforms (see Section 2.4.2).

The Kolmogorov law is the prototype for the emergent turbulent laws. In the rest of the book we discuss its limitations as well as its generalizations.

In the following, the exponent H characterizing the scaling of the fluctuations will appear regularly. Although the symbol H was chosen in honour of Edwin Hurst, who was a pioneer in studying long-range statistical dependency, it is not identical to the Hurst exponent (the latter has no simple general expression in multifractal processes).

Although the Kolmogorov law in both real space and Fourier space forms are often presented as though they are almost trivially equivalent, in fact this equivalence is not so obvious. For example, the Fourier space version shown in Eqn. (2.10) is a statement relating the variability (variance/wavenumber, $E(k)$) to the scale ($1/k$), whereas the meaning of the equality in Eqn. (2.9) is less clear. In accord with the Fourier interpretation, it can be read as a relation between a "typical" velocity difference Δv, the scale ("lag" Δx) over which the difference is estimated and the "typical" energy flux ε in the corresponding region. However, the Fourier product in the Obukhov spectral version of the law corresponds to a real space convolution (and vice versa), suggesting that the product $\varepsilon^{1/3}\Delta x^{H_v}$ should be interpreted instead as a convolution between $\varepsilon^{1/3}$ and a power of Δx. More precisely – and this is the basis of the "fractionally integrated flux" model which we describe in Chapter 5 – $\varepsilon^{1/3}\Delta x^{H_v}$ is interpreted as a fractional integration of the highly variable cascade process $\varepsilon^{1/3}$ order $H_v = 1/3$ (integrals of fractional orders are indeed convolutions with power laws).

As with the Weierstrass function, Eqn. (2.1), this already implies nondifferentiability:

$$\frac{\partial v}{\partial x} = \lim_{\Delta x \to 0} \frac{\Delta v}{\Delta x} \approx \Delta x^{-2/3} \to \infty \tag{2.11}$$

Even though at small enough scales Eqn. (2.9) breaks down (due to viscosity) and the derivatives converge, the limit will depend on the dissipation scale details. Similarly, at large scales, the forcing term breaks the scaling symmetry. However, since in the atmosphere the "outer" scale is roughly the size of the planet and the inner "viscous" scale is typically 0.1–1 mm (although it will vary considerably due to intermittency – see below), this leaves a potential scaling range of factor 10^4 km/10^{-3} m $\approx 10^{10}$.

2.2.3 Conservation of turbulent fluxes from one scale to another

We have seen that at least formally, the equations of hydrodynamic turbulence are scaling under isotropic scale changes. We now consider the energy flux whose density is:

$$\varepsilon = -\frac{1}{2}\frac{\partial v^2}{\partial t} \tag{2.12}$$

The term "flux" may appear odd; it refers to an energy per mass per time that passes through a spherical shell in Fourier space – it is a Fourier space flux – see Section 2.4.2. Shortly, we give a (classical) demonstration that it is indeed conserved by the nonlinear "inertial" terms. This is important, because we are interested in strongly nonlinear situations where these terms are dominant. The ratio of the nonlinear term to the dissipative (viscous) term in the Navier–Stokes equation can be estimated using the Reynolds number:

$$\mathrm{Re} \sim \frac{\text{Nonlinear terms}}{\text{Linear damping}} = \frac{|\underline{v}\cdot\nabla\underline{v}|}{\nu|\nabla^2\underline{v}|} \sim \frac{V\cdot L}{\nu} \tag{2.13}$$

where V is a "typical" velocity of the largest-scale motions L (the "outer" scale). In the atmosphere, Re is usually estimated by taking $V \approx 10$ m/s or structures size 10^4 km (see Chapter 8 for more precision and discussion). At standard temperature and pressure the viscosity of air is $\nu = 10^{-5}$ m^2/s, hence $Re \approx 10^{12}$. The Reynolds number is the nonlinear "coupling constant" for the problem – so in this strong coupling limit we may anticipate that many standard methods such as perturbation techniques (which work by solving the easy linear problem and treating the nonlinear term as a perturbation) will not converge. Even so, attempts are still regularly made to find clever "closure" or "renormalization" methods that might succeed in correctly summing the contributions from all the nonlinear effects.

In the atmosphere, we are therefore interested in the limit $\mathrm{Re} \to \infty$. Note that putting $\nu = 0$ reduces the Navier–Stokes equations to the "Euler equations" (discovered 100 years earlier in 1754), but the latter is presumably not the same as taking the limit $\mathrm{Re} \to \infty (\nu \to 0)$, since putting $\nu = 0$ reduces the order of the equation (from 2nd to 1st); hence the limit $\nu \to 0$ is a singular perturbation problem.

We now show that under certain conditions of mathematical regularity, the integral of the energy rate density ε of a fluid parcel is conserved by the nonlinear terms of the Navier–Stokes equation. Starting with the equation of an inviscid fluid (i.e. $\nu = 0$; Euler equation) with no forcing term at constant fluid density:

$$\frac{\partial \underline{v}}{\partial t} = -(\underline{v}\cdot\nabla)\underline{v} - \nabla\left(\frac{p}{\rho_f}\right) \tag{2.14}$$

Multiplying both sides by $\underline{v}$:

$$\varepsilon = -\frac{1}{2}\frac{\partial v^2}{\partial t} = -\underline{v}\cdot(\underline{v}\cdot\nabla)\underline{v} - (\underline{v}\cdot\nabla)\left(\frac{p}{\rho_f}\right) \tag{2.15}$$

Because of incompressibility (i.e. $\nabla\cdot\underline{v} = 0$), Eqn. (2.15) can be written:

$$\varepsilon = -\nabla\cdot\left[\left(\frac{1}{2}v^2 + \frac{p}{\rho_f}\right)\underline{v}\right] \tag{2.16}$$

Integrating over a volume of space V, it yields (due to Gauss's divergence theorem that transforms volume integrals of divergences to surface integrals):

$$\int_V \varepsilon\, dV = -\int_V \nabla\cdot\left[\left(\frac{1}{2}v^2 + \frac{p}{\rho_f}\right)\underline{v}\right] dV = -\oint_S \left(\frac{1}{2}v^2 + \frac{p}{\rho_f}\right)\underline{v}\cdot d\underline{S} \tag{2.17}$$

where the right-hand integral is over the enclosing surface only. The first term in the surface integral represents the transfer of kinetic energy across the surface, the second is the work done by pressure forces; there is no net source or sink of ε inside the volume.

We now consider the dissipation term $\nu\nabla^2\underline{v}$. Multiplying by $\underline{v}$, ignoring the surface term, we obtain:

$$\int_V \varepsilon\, dV = \nu\underline{v}\cdot\int_V \nabla^2\underline{v}\, dV \tag{2.18}$$

Now, using vector identities, we have:

$$\underline{v}\cdot\nabla^2\underline{v} = -\left|\nabla\times\underline{v}\right|^2 - \nabla\cdot[(\nabla\times\underline{v})\times\underline{v}] \tag{2.19}$$

The second term on the right-hand side is a divergence, and when integrated over a volume it can be rewritten as a surface integral (Gauss's theorem):

$$\int_V \varepsilon\, dV = -\nu\int_V \left|\nabla\times\underline{v}\right|^2 dV - \nu\oint_S \left[(\nabla\times\underline{v})\times\underline{v}\right]\cdot d\underline{S} \tag{2.20}$$

Since the surface integral vanishes if S is a current surface ($d\underline{S}\perp\underline{v}$) or a rigid boundary ($\underline{v} = 0$), it can be ignored if we take it to infinity. In these cases, the right-hand side integrand is a positive definite quantity, $\nu > 0$, and hence the viscosity is always dissipative (decreases the total energy). Conversely, if $\nu = 0$, then ε is "conserved" by the nonlinear terms, and even when $\nu > 0$, the dissipation will only be important at small scales where the derivatives $\nabla\times\underline{v}$ (i.e. the vorticity) are important.

2.3 Extensions to passive scalars, to the atmospheric primitive equations

2.3.1 Passive scalars, conservation of passive scalar variance flux

If we include the concentration ρ of a passive scalar quantity (i.e. a quantity such as an inert dye or in atmospheric experiments chaff, which is advected, transported by the wind without influencing the wind), we obtain the additional equation:

$$\frac{\partial\rho}{\partial t} = -\underline{v}\cdot\nabla\rho - \kappa\nabla^2\rho + f_\rho \tag{2.21}$$

where κ is the molecular diffusivity of the fluid.

Equations (2.3), (2.4), (2.21) are also formally invariant under the following scale-changing operations:

$$\begin{aligned} &\underline{x} = \lambda^1\underline{x}'; \quad &&\underline{v} = \lambda^{\gamma_v}\underline{v}'; \quad &&t = \lambda^{1-\gamma_v}t'; \\ &\rho = \lambda^{\gamma_\rho}\rho'; \quad &&f = \lambda^{1+2\gamma_v}f'; \quad &&f_\rho = \lambda^{1+2\gamma_v}f'_\rho; \\ &\nu = \lambda^{1+\gamma_v}\nu'; \quad &&\kappa = \lambda^{1+\gamma_\rho}\kappa' && \end{aligned} \tag{2.22}$$

where γ_v, γ_ρ are arbitrary. This arbitrariness allows the possibility of multiple scaling (i.e. weak and intense turbulent regions which scale differently, and have different fractal dimensions); hence the solutions can in principle be multifractals.

By repeating arguments similar to the above for ρ^2 rather than v^2, one can check that the scalar variance flux:

$$\chi = -\frac{1}{2}\frac{\partial\rho^2}{\partial t} \tag{2.23}$$

analogous to ε is conserved by the nonlinear terms $\underline{v}\cdot\nabla\rho$. Putting $\kappa = 0$ and recalling $\nabla\cdot\underline{v} = 0$:

$$\chi = -\frac{1}{2}\frac{\partial\rho^2}{\partial t} = -\rho\underline{v}\cdot\nabla\rho = -\frac{1}{2}\nabla\cdot(\underline{v}\rho^2) \tag{2.24}$$

hence:

$$\int_V \chi\, dV = -\frac{1}{2}\oint_S \rho^2\underline{v}\cdot\underline{dS} \tag{2.25}$$

This shows that there is no volume contribution to the passive scalar variance; it will be conserved by the nonlinear $\underline{v}\cdot\nabla\rho$ term.

Using the conservation of χ one obtains $\gamma_\rho = (1-\gamma_v)/2$, and since from Eqn. (2.9) (from the conservation of ε) $\gamma_v = 1/3$, we find $\gamma_\rho = 1/3$. This yields the result analogous to the Kolmogorov law: the "Corrsin–Obukhov law of passive scalar advection" (Obukhov, 1949; Corrsin, 1951), which in dimensional form is:

$$\Delta\rho = \chi^{1/2}\varepsilon^{-1/6}\Delta x^{H_\rho}; \quad H_\rho = \gamma_\rho = 1/3 \tag{2.26a}$$

i.e. with the same exponent H as the Kolmogorov law. Similarly, the Fourier space version is:

$$E_\rho(k) = \chi\varepsilon^{-1/3}k^{-5/3} \tag{2.26b}$$

2.3.2* The scale invariance of the equations of the atmosphere: an anisotropic scaling analysis of the "primitive equations"

We can also try the same approach to the scaling of the equations of the atmosphere. However, we will see that the effect of gravity and Coriolis forces make the analysis much more complex. We have seen that the dissipation terms can be either neglected (by considering large enough scales much larger than the dissipation scales), or "rescaled" by changing the values of the viscosity and diffusivity; we therefore neglect them for simplicity. We also neglect water in all of its phases. The basic equations for the velocity, energy, continuity and equation of state are:

$$\frac{\partial\underline{v}}{\partial t} = -(\underline{v}\cdot\nabla\underline{v}) - 2\underline{\Omega}\times\underline{v} - \frac{1}{\rho}\nabla p - \underline{g} + \underline{F} \tag{2.27}$$

$$\frac{\partial T}{\partial t} = -(\underline{v}\cdot\nabla)T - \frac{p}{\rho c_v}\nabla\cdot\underline{v} + \frac{Q}{c_v} \tag{2.28}$$

$$\frac{\partial\rho}{\partial t} = -(\underline{v}\cdot\nabla)\rho - \rho\nabla\cdot\underline{v} \tag{2.29}$$

$$p = \rho RT \tag{2.30}$$

where $\underline{\Omega}$ is the earth rotation vector, $\underline{g}$ is the acceleration of gravity, Q represents sources and sinks of energy, R is the universal gas constant, c_V is the specific heat at constant volume and T is the temperature. To bring out the key differences with the scaling arguments in pure hydrodynamics discussed above, we split the velocity equation into horizontal and vertical components and use the ideal gas law to replace p/ρ by RT in the energy equation. Finally, also for simplicity, we omit the forcing terms whose scaling can be treated in the same way as for the hydrodynamic equations discussed above. The system now becomes:

$$\frac{D}{Dt} = \underline{v}_h\cdot\nabla_h + w\frac{\partial}{\partial z} \tag{2.31}$$

$$\frac{D\underline{v}_h}{Dt} = -2f\hat{z}\times\underline{v}_h - \frac{1}{\rho}\nabla p - g\hat{z} \tag{2.32}$$

$$\frac{DT}{Dt} = -\frac{RT}{c_v}\nabla\cdot\underline{v} \tag{2.33}$$

$$\frac{D\rho}{Dt} = -\rho\nabla\cdot\underline{v} \tag{2.34}$$

where we have introduced the unit vector $\hat{z}$ in the vertical (z) direction and the horizontal components are indicated with the subscript h so that $\underline{v}_h = (u, v)$; $\nabla_h = \left(\frac{\partial}{\partial x}, \frac{\partial}{\partial y}\right)$ with the full 3D wind vector $\underline{v} = (u, v, w)$. We now take into account the vertical stratification by allowing for an anisotropic scale transformation:

$$\underline{r}_h = \lambda^1\underline{r}_h'; \quad z = \lambda^{H_z}z'; \quad \underline{r}_h = (x, y) \tag{2.35}$$

$$\underline{v}_h = \lambda^{\gamma_v}\underline{v}_h'; \quad w = \lambda^{\gamma_w}w_h'; \quad t = \lambda^{1-\gamma_v}t'; \quad \rho = \lambda^{\gamma_\rho}\rho'; \quad T = \lambda^{\gamma_T}T' \tag{2.36}$$

The key point now is that if:

$$\gamma_v - 1 = \gamma_w - H_z \tag{2.37}$$

then the advection operator D/Dt and velocity divergence will have a uniform scaling (i.e. each term will transform the same way with scale):

$$\frac{D}{Dt} = \lambda^{\gamma_v-1}\frac{D'}{Dt'}; \quad \nabla\cdot\underline{v} = \lambda^{\gamma_v-1}\nabla'\cdot\underline{v} \tag{2.38}$$

Hence we require:

$$h = \gamma_v - \gamma_w; \quad h = 1 - H_z \tag{2.39}$$

where the exponent h (equivalently H_z) determines the stratification (Schertzer *et al.*, 2012).

An indication that this is reasonable comes from the empirical estimates discussed in Chapters 4 and 6, where we find for the mean fluctuations: $H_v \approx 1/3$, $H_z \approx 0.44 - 0.56$, $\gamma_w \approx -0.2 - -0.1$ (although the vertical wind is notoriously difficult to measure, so that the parameter γ_w is not well known; see the corresponding statistical parameter H_w in Table 4.1 for an empirical estimate). If this (anisotropic) scaling of the advection operator holds, then the energy and continuity equations above show that γ_T and γ_ρ may be arbitrary.

The analysis of the velocity equations is more involved; we follow Schertzer *et al.* (2012), who find it convenient to consider the vorticity equation obtained by taking the curl of the velocity equation:

$$\frac{D\underline{\omega}}{Dt} = (\underline{\omega}\cdot\nabla)\underline{v} + \underline{b}; \qquad \underline{b} = \frac{1}{\rho^2}\nabla\rho \times \nabla p \tag{2.40}$$

where we have kept the compressibility and the baroclinic term $\underline{b}$ but (temporarily) dropped the Coriolis term. We now find that under anisotropic scale changes (Eqn. (2.35)):

$$\underline{\omega} = \begin{bmatrix} \frac{\partial w}{\partial y} - \frac{\partial v}{\partial z} \\ -\frac{\partial w}{\partial x} + \frac{\partial u}{\partial z} \\ \frac{\partial v}{\partial x} - \frac{\partial u}{\partial y} \end{bmatrix} = \lambda^{1+\gamma_v} \begin{bmatrix} \lambda^h \frac{\partial w'}{\partial y'} - \lambda^{-h} \frac{\partial v'}{\partial z'} \\ -\lambda^h \frac{\partial w'}{\partial x'} + \lambda^{-h} \frac{\partial u'}{\partial z'} \\ \frac{\partial v'}{\partial x'} - \frac{\partial u'}{\partial y'} \end{bmatrix} \tag{2.41}$$

We can see that things are a little more complicated because there are effectively three different scaling exponents ($1 + \gamma_v$, $1 + \gamma_v \pm h$). Finally, consider the vortex stretching term:

while the baroclinic term scales as:

$$\frac{1}{\rho^2}\nabla\rho \times \nabla p = \lambda^{2+\gamma_p-\gamma_\rho} \frac{1}{\rho'^2} \begin{bmatrix} \lambda^{-h}\left(\frac{\partial\rho'}{\partial y'}\frac{\partial p'}{\partial z'} - \frac{\partial\rho'}{\partial z'}\frac{\partial p'}{\partial y'}\right) \\ \lambda^{-h}\left(\frac{\partial\rho'}{\partial z'}\frac{\partial p'}{\partial x'} - \frac{\partial\rho'}{\partial x'}\frac{\partial p'}{\partial z'}\right) \\ \frac{\partial\rho'}{\partial x'}\frac{\partial p'}{\partial y'} - \frac{\partial\rho'}{\partial y'}\frac{\partial p'}{\partial x'} \end{bmatrix} \tag{2.43}$$

We can now see that if $\gamma_p - \gamma_\rho = 2\gamma_v$, then, by equating terms with the same scaling exponents, we obtain three equations.

To understand this most clearly, we drop the baroclinic term and introduce the following notation for the horizontal (h) and vertical (v) wind and gradient operator components: $\underline{v} = \underline{v}_h + \underline{v}_v$; $\nabla = \nabla_h + \nabla_v$. The corresponding decomposition of the vorticity is:

$$\begin{aligned} &\underline{\omega} = \underline{\omega}_v + \underline{\omega}_h; \quad \underline{\omega}_h = \underline{\sigma} + \underline{\tau}; \quad \underline{\omega}_v = \nabla_h \times \underline{v}_h \\ &\underline{\sigma} = \nabla_h \times \underline{v}_v; \qquad \underline{\tau} = \nabla_v \times \underline{v}_h \end{aligned} \tag{2.44}$$

where $\underline{\omega}_h$ and $\underline{\omega}_v$ are the horizontal and vertical components of the vorticity. By assuming that $\underline{\omega}_h$ is negligible at large scales where the (almost vertical) earth rotation vector $\underline{\Omega}$ is assumed to be dominant, then the barotropic vorticity equation (i.e. Eqn. (2.40)) with ($\underline{b} = 0$) then splits into three equations:

$$\begin{aligned} \frac{D\underline{\sigma}}{Dt} &= \underline{\sigma}\cdot\nabla_h \underline{v}_h \\ \frac{D\underline{\tau}}{Dt} &= (\underline{\tau}\cdot\nabla_h + \underline{\omega}_v\cdot\nabla_v)\underline{v}_h \\ \frac{D\underline{\omega}_v}{Dt} &= (\underline{\tau}\cdot\nabla_h + \underline{\omega}_v\cdot\nabla_v)\underline{v}_v \end{aligned} \tag{2.45}$$

These equations, together with the large-scale condition $\underline{\omega}_v \approx \underline{\Omega}$, allow both anisotropic scaling – such as

$$(\underline{\omega}\cdot\nabla)\underline{v} = \lambda^{2(1+\gamma_v)} \begin{bmatrix} \lambda^h\left(\frac{\partial w'}{\partial y'}\frac{\partial u'}{\partial x'} - \frac{\partial w'}{\partial x'}\frac{\partial u'}{\partial y'}\right) - \lambda^{-h}\left(\frac{\partial v'}{\partial x'}\frac{\partial u'}{\partial z'} - \frac{\partial v'}{\partial z'}\frac{\partial u'}{\partial x'}\right) \\ \lambda^h\left(\frac{\partial w'}{\partial y'}\frac{\partial v'}{\partial x'} - \frac{\partial w'}{\partial x'}\frac{\partial v'}{\partial y'}\right) - \lambda^{-h}\left(\frac{\partial u'}{\partial z'}\frac{\partial v'}{\partial y'} - \frac{\partial u'}{\partial y'}\frac{\partial v'}{\partial z'}\right) \\ \frac{\partial v'}{\partial x'}\frac{\partial w'}{\partial z'} - \frac{\partial v'}{\partial z'}\frac{\partial w'}{\partial x'} + \frac{\partial u'}{\partial z'}\frac{\partial w'}{\partial y'} - \frac{\partial u'}{\partial y'}\frac{\partial w'}{\partial z'} \end{bmatrix} \tag{2.42}$$

that displayed in Eqn. (2.41) to (2.43) – and nonlinear growth of the horizontal vorticity. We note that the latter mechanism is absent from the quasi-geostrophic (QG) approximation which corresponds to $\underline{\sigma} = \underline{\tau} = 0$ and the approximation $D\underline{\omega}_v/Dt \approx \underline{\Omega}_v \cdot \nabla_v \underline{v}_v$. Contrary to the QG equations, the fractional vorticity equations (2.45) are not approximations to the vorticity equation because the former are also solutions of the latter. Instead, they correspond to selecting relevant interactions which yield solutions with the anisotropic scaling prescribed by Eqn. (2.35) (all this can be done rigorously using "pullback transforms": see Schertzer *et al.*, 2012). This explains the extreme contrast between Eqn. (2.45), which has 3D nonlinear vorticity stretching, and the linear stretching in the QG approximation: $D\underline{\omega}_v/Dt \approx \underline{\Omega}_v \cdot \nabla_v \underline{v}_v$.

Overall, the solutions of Eqn. (2.45) (and the corresponding equations for the other components) are the solutions of the vorticity equation that statistically respect the anisotropic scaling prescribed by Eqn. (2.35), and statistically break the isotropic scaling of Eqn. (2.40). These solutions correspond to an alternative to QG turbulence: they share the common boundary condition that vorticity is dominated by the earth's rotation at large scales, but the transfer of the vorticity to smaller scales is obtained with the help of a nonlinear stretching term, contrary to the QG approximation. If we include the Coriolis force, then the vorticity equation is the same except for the absolute rather than relative vorticity (Schertzer *et al.*, 2012).

2.4 Classical isotropic 3D turbulence phenomenology: Kolmogorov turbulence and energy cascades

2.4.1 Fourier locality, energy transfer and cascade phenomenology

We have gone through the classical demonstration that the governing equations are formally isotropically scale invariant, that the nonlinear terms conserve the energy and passive scalar variance fluxes, and we have updated the scaling argument to take into account anisotropy. We will now study under which conditions the cascade is local, i.e. the energy transfer is most efficient between neighbouring scales: that it is "local" in Fourier space. To demonstrate this, it is usual (e.g. Rose and Sulem, 1978) to use a discrete hierarchy of eddies, broadly defined as fluid "coherent" structures. The dynamically important quantities for this type of analysis are: v_n (an appropriate characteristic velocity difference, see below); τ_n (the time scale called the "eddy turnover time," which is the typical time necessary for the dynamics to pass energy fluxes from one scale to another); and l_n, the length scale (size of the eddy). It is the shear that is important, because the Navier–Stokes equations are Galilean invariant; it is the difference of velocity across an eddy which intervenes, not the "mean" velocity of an eddy. The subscript n refers to the number of octaves from the largest "outer scale"; thus l_n refers to all values of l in the interval $\left[\frac{l_n}{\sqrt{2}}, \sqrt{2}l_n\right]$. Since we consider energy transfer from one scale to another, only motions which can distort the eddies are dynamically important. Any overall large-scale motion (i.e. translation) will not affect the transfer of energy from one scale to another (by Galilean invariance, we can always move to a reference frame where the mean velocity is zero). Similarly, any very small-scale motions within the eddy will be ineffective at distorting the eddy. This leads us to expect that only velocity gradients over distances approximately l_n will come into play (in Appendix 2A we show this more rigorously on condition that the spectral exponent satisfies $1 < \beta < 3$). We therefore expect the dynamically important velocity v_n at scale l_n to be a typical gradient across the eddy, i.e.:

$$v_n = \sqrt{\langle |\underline{v}(\underline{r}) - \underline{v}(\underline{r} + \underline{l}_n)|^2 \rangle} \tag{2.46}$$

The $\langle \cdots \rangle$ is the ensemble statistical average. Likewise $\tau_n \sim l_n/v_n$, the "eddy turnover time," is the typical time scale of the transfer process. Finally (again using dimensional analysis) the viscous time scale corresponding to the nth octave is:

$$\tau_{n,\,dis} = \frac{l_n^{\,2}}{\nu} \tag{2.47}$$

Viscosity can be ignored if $\tau_{n,\,dis} >> \tau_n$ (i.e. the viscosity is too slow to affect the dynamics).

2.4.2 The Kolmogorov–Obukhov spectrum

Denote by $\prod_n$ the rate at which energy is transferred out of a low wavenumber octave ($k_n/\sqrt{2} \le k \le \sqrt{2}k_n$) into a higher octave ($\sqrt{2}k_n \le k \le 2\sqrt{2}k_n$). This is a Fourier-space energy flux. It is given by the energy per unit mass in the octave (E_n) divided by the typical time scale of the transfer, the eddy turnover time:

$$\Pi_n \approx \frac{E_n}{\tau_n} \qquad (2.48)$$

Now assume that the cascade is local, so that the dominant contribution to E_n comes from the velocity gradient at the same scale, i.e. v_n. This implies $E_n \sim v_n^2$ (recall that due to incompressibility all energies are taken per unit mass) and that $\tau_{vis,n} \gg \tau_n$ so that there is no energy dissipation in this wavenumber band. If the energy injection rate ε (e.g. by stirring) at large scale is balanced by viscous dissipation at small scale then it is possible that the system is stationary (statistically invariant under translations in time) then $\Pi_n \sim$ constant, i.e. there are no viscous losses and no sinks or sources. This is assumed to be a quasi-steady state: energy flows through the nth octave at a rate ε which is on average equal to the large-scale injection rate and to the small-scale dissipation (as we will see, such statistical stationarity is quite compatible with violent fluctuations):

$$\varepsilon = \Pi_n \sim \frac{E_n}{\tau_n} \sim \frac{v_n^2}{\left(\frac{l_n}{v_n}\right)} \sim \frac{v_n^3}{l_n} \sim \text{constant} \qquad (2.49)$$

(assuming that the injection rate is constant). Π_n is therefore a scale-invariant quantity (it is independent of n). This yields Kolmogorov's law (1941):

$$v_n \sim \varepsilon^{1/3} l_n^{1/3} \qquad (2.50)$$

Since the fluctuation v_n is a scaling power law function of size l_n, we expect that the spectrum will also be a power law (see Box 2.2 for more details on Tauberian theorems that relate real space and Fourier space scaling). For wavenumber p, we therefore seek the spectral exponent β:

$$E(p) \sim p^{-\beta} \qquad (2.51)$$

corresponding to the real space exponent $1/3$ in Eqn. 2.50. Assuming $\beta > 1$ we get the following expression for the total variance due to all the wavenumbers in the nth band:

$$v_n^2 \approx l_n^2 \int_{k_n/\sqrt{2}}^{\sqrt{2}k_n} dp\, p^2\, E(p) \qquad (2.52)$$

(since the variance in a spherical shell between p and $p + dp$ is $4\pi\, p^2 dp$, and we ignore the constant factor). We thus obtain:

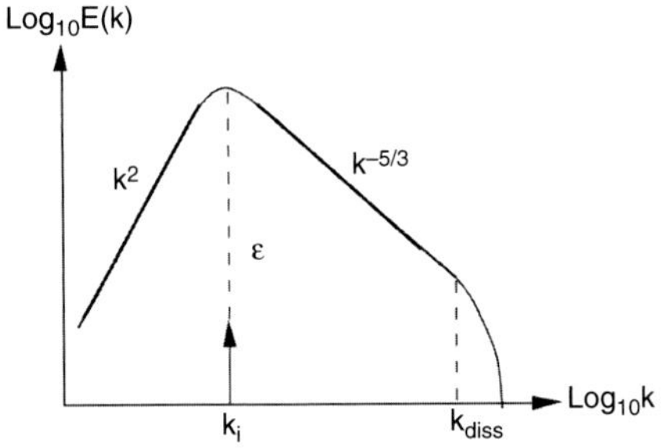

Fig. 2.5 Schematic diagram of 3D energy cascade showing the equipartition ("equilibrium") range at low wavenumbers, the energy flux injection wavenumber k_i, the "inertial" $k^{-5/3}$ range and the dissipation range $k > k_{diss}$ dominated by viscous β dissipation.

$$v_n^2 \approx l_n^2\, k_n^{3-\beta} \sim l_n^{2-3+\beta} \qquad (2.53)$$

(since $l_n \sim k_n^{-1}$). Comparing this with Eqn. (2.50), we obtain $2 - 3 + \beta = 2/3$, or:

$$\beta = \frac{5}{3} \qquad (2.54)$$

The Kolmogorov–Obukhov spectrum is thus derived:

$$E(k) \sim \varepsilon^{2/3} k_n^{-5/3} \qquad (2.55)$$

A schematic diagram of the 3D cascade is shown in Fig. 2.5. The slope of the spectrum on the low-frequency side of the injection wavenumber is of the form $E(k) \sim k^2$. This follows since using statistical mechanical arguments, one expects that there is a low-frequency "equilibrium" range where each mode has roughly the same energy (equipartition). The spectral form $E(k) \sim k^2$ then follows, since there are $k^2 dk$ modes between wavenumbers k and $k + dk$.

2.4.3 Vortex stretching, the break-up of eddies and the cascade direction

It is easy to identify each term in the vorticity equation (2.40): $D\underline{\omega}/Dt$ is the convective (total) derivative of the vorticity (remembering that the total derivative operator is just $\frac{D}{Dt} = \frac{\partial}{\partial t} + \underline{v}\cdot\nabla$, it represents the change in a quantity that moves with the flow; it is also called a Lagrangian derivative), the term $\underline{v}\nabla^2\underline{\omega}$ (ignored in Eqn. (2.40)) represents the molecular dissipation, the term $(\nabla\cdot\underline{v})\underline{\omega}$ is the compressibility term; we consider here the simplest incompressible case, $\nabla\cdot\underline{v} = 0$. The all-important "vortex stretching" contribution $(\underline{\omega}\cdot\nabla)\underline{v}$ is so named because its component is only positive when the gradient of $\underline{v}$ is parallel to $\underline{\omega}$, in

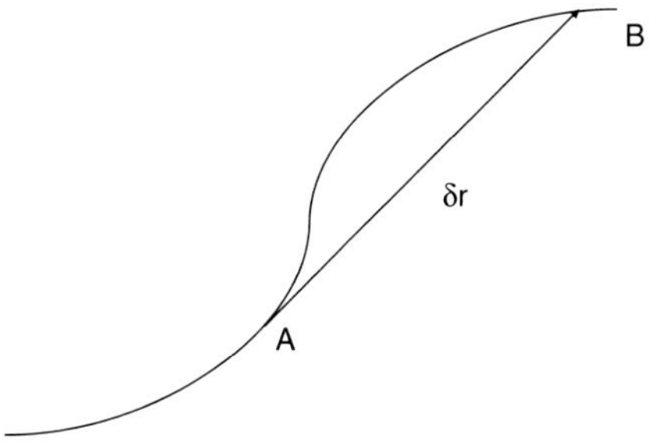

Fig. 2.6 Schematic showing an infinitesimal segment of a vortex line.

which case vortex lines are "stretched" by the velocity field (Fig. 2.6: a vortex line is like a line of electric or magnetic field – its tangent is everywhere parallel to the field lines; the analogous line for the velocity field is called a "streamline").

However, a more important property of vorticity is that – ignoring viscosity – vortex lines are material lines. To see this, let $\delta \underline{r}$ represent the vector between particles A and B (Fig. 2.6). Then the equation of evolution of $\delta \underline{r}$ is:

$$\frac{D(\delta \underline{r})}{Dt} = \frac{D(\underline{r}_A)}{Dt} - \frac{D(\underline{r}_B)}{Dt} = \underline{v}_A - \underline{v}_B = \delta \underline{v} \tag{2.56}$$

and to first order in $\delta \underline{r}$:

$$\delta \underline{v} = (\delta \underline{r} \cdot \nabla)\underline{v} \tag{2.57}$$

then:

$$\frac{D\delta \underline{r}}{Dt} = \delta \underline{v} = (\delta \underline{r} \cdot \nabla) v \tag{2.58}$$

which is identical to the (incompressible) vorticity equation if $\delta \underline{r}$ is taken parallel to $\underline{\omega}$ (recall we are considering negligible viscosity, $\nu = 0$). This shows that if at some initial time a vortex line is composed of a given set of fluid particles then at any later time the (evolved) vortex line will still be composed of the same particles. Vortex lines are therefore material lines.

Now apply this to the evolution of vortex tubes (these are the surfaces bounded by vortex lines): the volumes enclosed by the tubes are constant, since the fluid is incompressible and vortex lines are material lines. As the system evolves, the ends of the tubes move apart on average (this is a statistical effect: in a turbulent fluid, the ends of the tube will execute a convoluted random walk; on average, they will move apart). Since the volumes of the tube are incompressible, this implies that as the lengths of the tubes increase the cross-sectional areas tend to decrease. Hence there will be "pinching" of the tube at certain regions where there is a high stretching, leading locally to extremely high gradients of $\underline{v}$. The $\nu\nabla^2\underline{v}$ term will become large and viscosity will tend to smooth the high gradients and break (smooth out) the vortex tubes. This stretching–pinching mechanism means that a fat (large) vortex tube "slims" (cross-sections become smaller) and then gets broken up, the energy flux being conserved throughout the process, except for the final viscous smoothing/dissipation at very small scales. If we now imagine a complex turbulent flow as a "spaghetti" of vortex tubes evolving in time, we can see that ends of tubes which are far apart will tend to move further apart (just as a drunkard tends to move away from his starting bar), and hence the tubes will be generally stretched and then pinched (Fig. 2.7). Since this causes tubes with initially large cross-sections to tend to evolve into tubes with small cross-sections, this gives a simple explanation for the downscale direction of energy cascades in three-dimensional turbulence, and indeed whenever vortex stretching is important.

2.4.4* The vorticity spectrum

In homogeneous isotropic turbulence $E(k)$ contains a lot (but by no means all!) of the statistical information about the turbulent flow (it is still only a second-order moment depending on only the separation $\underline{r}$ of the two points $\underline{x}$ and $\underline{x} + \underline{r}$; it is a "two-point" statistic). We now derive the relation between $E(k)$ and the spectrum of the vorticity, which will be important in considering two-dimensional turbulence. First we use the vector identity:

$$\underline{A}\cdot(\nabla \times \underline{B}) = \underline{B}\cdot(\nabla \times \underline{A}) - \nabla\cdot(\underline{A} \times \underline{B}) \tag{2.59}$$

If $\underline{A}$ and $\underline{B}$ are functions of $\underline{v}$, and if we assume that the statistical properties of $\underline{v}$ are independent of position (statistical homogeneity) then $\langle \underline{A} \times \underline{B} \rangle$ is a constant and it follows that the expectation of the last term is zero (i.e. $\nabla\cdot\text{constant} \equiv 0$). Now, using $\underline{A} = \nabla \times \underline{v} = \underline{\omega}$ and $\underline{B} = \underline{v}$, we obtain:

$$\langle \omega^2 \rangle = \langle \underline{A}\cdot(\nabla \times \underline{B}) \rangle = \langle \underline{B}\cdot(\nabla \times \underline{A}) \rangle = \langle \underline{v}\cdot(\nabla \times (\nabla \times \underline{v})) \rangle \tag{2.60}$$

Finally, using the following vector identity for incompressible flows:

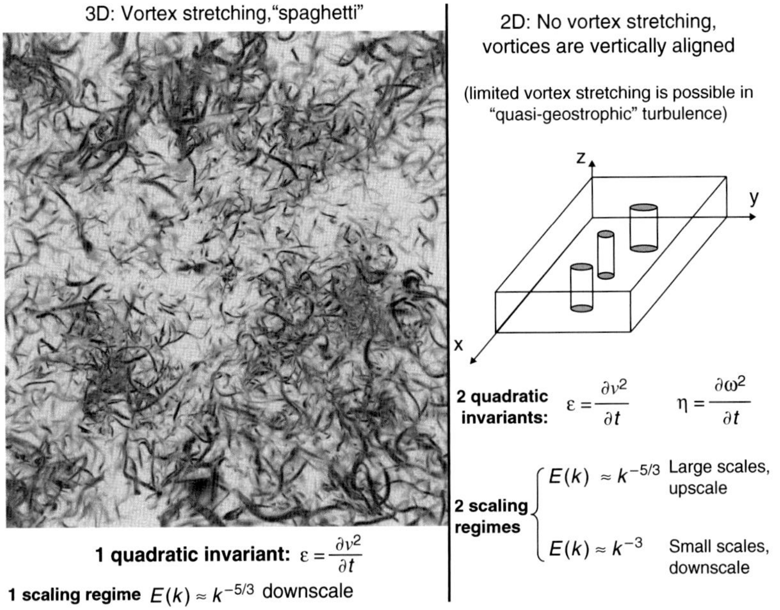
3D: Vortex stretching,"spaghetti"
1 quadratic invariant: $\varepsilon = \frac{\partial v^2}{\partial t}$
1 scaling regime $E(k) \approx k^{-5/3}$ downscale
2D: No vortex stretching, vortices are vertically aligned
(limited vortex stretching is possible in "quasi-geostrophic" turbulence)
z
y
x
2 quadratic invariants: $\varepsilon = \frac{\partial v^2}{\partial t}$ $\eta = \frac{\partial \omega^2}{\partial t}$
2 scaling regimes
$E(k) \approx k^{-5/3}$ Large scales, upscale
$E(k) \approx k^{-3}$ Small scales, downscale

Fig. 2.7 A schematic showing the "spaghetti plate" view of vortices stretching and tangling in 3D turbulence (the left-hand side) compared with the vortex-stretching free dynamics in 2D turbulence (the right-hand side). Spaghetti of vortex tubes thanks to numerical simulations by M. Wilczek, thanks to http://www.vapor.ucar.edu/ software.

$$\nabla \times (\nabla \times \underline{v}) = -\nabla^2 \underline{v} \tag{2.61}$$

we obtain:

$$\langle \omega^2 \rangle = -\langle \underline{v} \cdot \nabla^2 \underline{v} \rangle \tag{2.62}$$

Therefore, since spectra are Fourier transforms of correlations and since the Laplacian corresponds to multiplication by $(ik)^2$ in Fourier space, we have the following relationship between the vorticity spectrum E_ω and velocity spectrum E:

$$E_\omega(k) = k^2 E(k) \tag{2.63}$$

Box 2.2 Scaling and Fourier transforms: correlation functions, structure functions and Tauberian theorems

In the following we will use both real-space and Fourier-space statistics, so it is useful to consider the general relation between real- and Fourier-space scaling. First define the Fourier transform and its inverse (note that ω in this section no longer denotes the vorticity but the angular frequency, i.e. the Fourier conjugate of the time t):

$$\tilde{v}(\omega) = F(v) = \int_{-\infty}^{\infty} dt\, e^{-i\omega t} v(t) \tag{2.64}$$

$$v(t) = F^{-1}(\tilde{v}) = \int_{-\infty}^{\infty} d\omega\, e^{i\omega t} \tilde{v}(\omega) \tag{2.65}$$

We recall two fundamental properties of Fourier transforms:

$$F\left(\frac{d^n v}{dt^n}\right) = (i\omega)^n \tilde{v}(\omega) \tag{2.66}$$

$$F(v * w) = \tilde{v}(\omega)\tilde{w}(\omega) \tag{2.67}$$

where $v * w$ is the convolution of v and w:

Box 2.2 *(cont.)*

$$v * w = \int_{-\infty}^{\infty} dt' v(t - t') w(t') \tag{2.68}$$

We have the "Tauberian theorems" (see Feller, 1972, for the Laplace transform), we have slightly extended it to the Fourier transform (Schertzer and Lovejoy, 1991). For $\lambda >> 1$:

$$v\left(\frac{t}{\lambda}\right) \rightarrow \frac{v(t)}{\lambda^{\sigma}} \Leftrightarrow \tilde{v}(\lambda\omega) \rightarrow \frac{\tilde{v}(\omega)}{\lambda^{\sigma+1}} \tag{2.69}$$

which are rather obvious results of the combined transformations $t \rightarrow t/\lambda^{\sigma}$ and $\omega \rightarrow \lambda\omega$ (hence $e^{i\omega t}$ remains unchanged). This relation will be of fundamental importance in relating scaling behaviours in physical space and Fourier space. This is in particular the case for the autocorrelation function $R(\tau)$ and the energy spectrum $E(\omega)$, since they are Fourier transforms due to the "Wiener–Khinchin theorem" (see Appendix 2A for a demonstration):

$$R(\tau) = \left\langle v(t)v(t+\tau) \right\rangle = F(E), \text{ i.e.} \left\langle v(t)v(t+\tau) \right\rangle = \int_{-\infty}^{\infty} d\omega e^{i\omega\tau} E(\omega) \tag{2.70}$$

Another quantity of interest is the "Δ-variance," "variogram," or (second-order) structure function $S_2(\tau)$ that characterizes the fluctuations Δv:

$$S_2(\tau) = \left\langle \Delta v(\tau)^2 \right\rangle = \left\langle \left(v(t) - v(t-\tau) \right)^2 \right\rangle = 2\left[\left\langle v(t)^2 \right\rangle - \left\langle v(t)v(t-\tau) \right\rangle \right] \tag{2.71}$$

or in terms of the power spectrum:

$$S_2(\tau) = 2\left(R(0) - R(\tau) \right) = 2\int_{-\infty}^{\infty} d\omega\, E(\omega)(1 - e^{i\omega\tau}) \tag{2.72}$$

For scaling spectra $E(\omega) \approx \omega^{-\beta}$ there are low-frequency divergences so that the correlation integral (Eqn. (2.70)) only converges for $\beta < 1$ (a high-frequency cutoff is needed for convergence, but is always present in discretely sampled data). However, for $S_2(\tau)$, at low frequencies the real part of $E(\omega)(1 - e^{i\omega\tau}) \approx \omega^{2-\beta}$, so that the structure function integral in Eqn. (2.71) converges for $\beta < 3$ but $S_2(\tau)$ still satisfies the same Tauberian theorem as $R(\tau)$; Section 5.4.3 considers generalizations to higher-order structure functions. These low-frequency divergences are "infrared catastrophes." (The terms "infrared" and "ultraviolet" catastrophe originate from the theory of black-body radiation. By introducing the quantum hypothesis, Planck inserted a high wavenumber (small wavelength) cutoff which saved the theory from "ultraviolet" divergences, hence the term.)

Hence:

$$E(\omega) \approx \omega^{-\beta} \Leftrightarrow R(\tau) \approx \tau^{\xi(2)}; \quad \beta < 1 \tag{2.73}$$

and/or:

$$S_2(\tau) = \langle \Delta v(\tau)^2 \rangle \approx \tau^{\xi(2)}; \quad 1 < \beta < 3 \tag{2.74}$$

where $\xi(2) = -1 + \beta$ is the autocorrelation function/(second-order) structure function exponent. In terms of $\xi(2)$, we see that the autocorrelation function converges for $\xi(2) < 0$, and the structure function for $0 < \xi(2) < 2$. To extend the range beyond this interval we must use other definitions of fluctuations using wavelets (see Section 5.5). For the weakly variable fluxes considered here, we can simply take the mean square fluctuation from Eqn. (2.9) to obtain $S_2(\tau) \approx \tau^{2H}$ so that $\xi(2) = 2H$ ($H = 1/3$ for the Kolmogorov law). Chapter 5 discusses the generalization to moments of order other than 2 and to situations where intermittency is important so that the scale dependence of the various moments of the turbulent fluxes becomes important.

2.5 The special case of 2D turbulence

2.5.1 Comparing two- and three-dimensional turbulence

The seductive but ultimately naive idea that 2D or quasi-2D turbulence should be relevant for large-scale atmospheric dynamics has been so popular that it is worth realizing how singular it is with respect to 3D turbulence – and indeed with respect to turbulence in any space with $d > 2$ (see the generalizations to stratified "elliptical dimensions" in Chapter 6). Indeed, in two-dimensional turbulence, in addition to the conserved quantity ε, the vorticity is conserved. This leads to the conservation of an additional quadratic quantity analogous to ε: the "enstrophy" $\Omega = \omega^2$ (not to be confused with the earth's rotation vector). The reason is that for a two-dimensional flow the vorticity $\underline{\omega}$ is perpendicular to $\underline{v}$ (i.e. $\underline{\omega} = \omega_z \widehat{z}$, $\omega_z = \partial v_y/\partial x - \partial v_x/\partial x$, $\omega_x = \omega_y = 0$ since $v_z = 0$, $\partial/\partial z = 0$; see the schematic Fig. 2.7), and consequently:

$$(\underline{\omega} \cdot \nabla)\underline{v} \equiv 0 \tag{2.75}$$

i.e. there is no longer any vortex stretching and the incompressible vorticity equation reduces to an advection–dissipation equation for the vorticity:

$$\frac{D\underline{\omega}}{Dt} = \nu \nabla^2 \underline{\omega} \tag{2.76}$$

when the dissipation is negligible, any power of the vorticity is conserved, not only the enstrophy which is its square. We can define the enstrophy flux density:

$$\eta = -\frac{1}{2}\frac{\partial \omega^2}{\partial t} \tag{2.77}$$

Taking the scalar product of the two-dimensional vorticity equation with $\underline{\omega}$ and following exactly the same arguments as for ε, we find that when η is integrated over a volume, it is conserved by the non-linear terms of the Navier–Stokes equation (as is ε). The existence of this second quadratic invariant (η) has a drastic effect on the properties of the associated turbulence, as we show below.

2.5.2 Two-dimensional enstrophy cascades

Two-dimensional turbulence has received a lot of attention for several reasons. It is undeniably much more numerically manageable, and at first sight much simpler, than three-dimensional turbulence. In addition, many geophysical flows are "apparently" two-dimensional at large scales since they are "thin" (e.g. the scale height of the atmosphere $\approx$ 10 km for the pressure). In Section 2.6 and Chapter 6, we examine some of the empirical evidence. In the atmosphere, Charney (1971) introduced a variant of two-dimensional turbulence called "quasi-geostrophic" turbulence. This is based on a number of approximations: the existence of a large- and small-scale separation, the use of the geostrophic wind to determine the material derivative, the assumption of hydrostatic balance and the near uniformity of the static stability. Although this development was historically important, these limitations make it unlikely to be realistic (see more detailed discussion in Schertzer, 2009); it is examined empirically in Section 2.6.

Returning to the ideal case of two-dimensional turbulence, we have seen that both energy and enstrophy are conserved by the nonlinear terms, hence both will be cascaded. Now from Eqn. (2.63) we have:

$$\Omega = \langle \omega^2 \rangle = \int_0^\infty dp\, E_\omega(p) = \int_0^\infty dp\, p^2 E(p) \tag{2.78}$$

where Ω is the enstrophy. The enstrophy in the nth octave is therefore:

$$\begin{aligned}\Omega_n &= \int_{k_n/\sqrt{2}}^{\sqrt{2}k_n} dp\, p^2 E(p) \approx \int_{k_n/\sqrt{2}}^{\sqrt{2}k_n} dp\, p^2 p^{-\beta} \\ &\approx p^{3-\beta}\Big|_{k_n/\sqrt{2}}^{\sqrt{2}k_n} \sim k_n^3 E(k_n)\end{aligned} \tag{2.79}$$

From the spectrum we can estimate the lifetime ("eddy turnover time") of a structure of size $l_n = 1/k_n$ as:

$$\tau_n \sim \left(\int_0^{k_n} dp\, p^2 E(p)\right)^{-1/2} \sim \left(k_n^3 E(k_n)\right)^{-1/2} \tag{2.80}$$

In analogy with the energy cascade (Section 2.4.2), we can also define:

$$\Pi_n^{(\Omega)} = \frac{\Omega_n}{\tau_n} \tag{2.81}$$

as the Fourier-space enstrophy flux (which is constant for a quasi-steady process) through the nth octave.

Finally we obtain the enstrophy flux through the *n*th octave in Fourier space:

$$\Pi_n^{(\Omega)} = \frac{\Omega_n}{\tau_n} \sim \frac{\mathrm{k}_n^3 E(k_n)}{\left(\mathrm{k}_n^3 E(k_n)\right)^{-1/2}} \tag{2.82}$$

If we assume that this is constant in a steady state and independent of *n*, then $\eta \sim \Pi_n^{(\Omega)}$ and we obtain the spectrum in the constant enstrophy flux regime:

$$E(k) \sim \eta^{2/3} k^{-3} \tag{2.83}$$

Using either dimensional analysis or the Tauberian theorems (Box 2.2), we can obtain the corresponding real-space result:

$$\Delta v \approx \eta^{1/3} \Delta x \tag{2.84}$$

These formulae (sometimes called the "Kraichnan" laws; Kraichnan, 1967) need some refinement since the picture of enstrophy being passed mainly from one octave to a neighbouring octave (without significant direct, nonlocal transfer over many octaves) is only strictly valid if the cascade was local, $\beta < 3$. Since we have found $\beta = 3$, we may anticipate that this "marginal" case will involve at least logarithmic corrections. This is indeed the case.`

The result $\beta = 3$ shows that every octave in two dimensional turbulence contributes approximately equally (to within the log corrections) to the nonlinear dynamics, the cascade is on the borderline between local and nonlocal. Each eddy turnover time τ_n is approximately equal. Note that the nonlocalness of two-dimensional cascades is quite serious; for example (Kevlahan and Farge, 1997), using numerical simulations on 1024×1024 grids with the usual Newtonian viscosity dissipation term (i.e. a Laplacian) find $\beta \approx 4$ or larger depending on the boundary conditions, but $\beta \approx 3$ for various higher powers (up to 8th) of a Laplacian (i.e. using "hyperviscosity"). This implies that the spectral exponent β depends on the details of the dissipation term. Due to the effects of nonlocalness, two-dimensional turbulence is thus in many ways more complex than three-dimensional turbulence, and since no clear direct evidence for a two-dimensional cascade has been found in the atmosphere (or in other geophysical

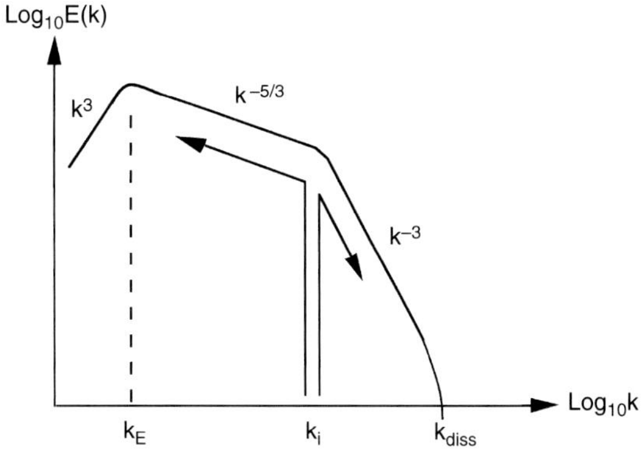

Fig. 2.8 A schematic illustration of the 2D enstrophy cascade with both energy and enstrophy fluxes injected at wavenumber k_i. The energy flux is cascaded to lower wavenumbers while the enstrophy flux is cascaded to higher wavenumbers.

systems), its status as a useful geophysical model is uncertain (it may, however, be relevant in soap films: e.g. Guttenberg and Goldenfeld, 2009).

Before proceeding, note that since $E_\omega(k) = k^2 E(k)$, if the small scales were dominated by an energy flux cascade, we would obtain $E_\omega(k) = k^2 k^{-5/3}$, which would diverge for large wavenumbers, and hence enstrophy could not be conserved. We therefore conclude that enstrophy must be cascaded from large to small scales through a k^{-3} regime, and energy flux from *smaller to larger* scales via a $k^{-5/3}$ regime, an "indirect" cascade. At the lowest wavenumbers, we must either introduce an energy sink, or we obtain a spectral peak (at k_E) that moves to lower and lower wavenumbers in time (the value is determined by dimensional arguments). Fig. 2.8 shows a schematic diagram for the latter case, assuming that injection of both enstrophy and energy fluxes occurs at the same (intermediate) scale. Finally, the dissipation wavenumbers may be estimated by dimensional arguments as in 3D turbulence: we find $k_{diss} = (\eta/v^3)^{1/6}$.

2.6 Atmospheric extensions

2.6.1 Applying isotropic turbulence to the atmosphere: the Gage–Lilly model

Because of the additional conserved enstrophy flux in 2D, the cascades are more complicated than in 3D, depending notably on the (possibly different) injection scales for ε, η. If we follow the classical model which first assumes isotropy, then – due to the $\approx$ 10 km atmospheric scale height – we are forced to introduce at least two isotropic turbulent regimes: a 3D isotropic regime at scales smaller than 10 km, and another 2D isotropic turbulent regime at large scales. From the discussion above, we can see that numerous forcing and dissipation length scales and mechanisms will be required. Indeed, by the early 1980s, theorists had produced a series of complicated ad hoc conceptual models which involved a small-scale 3D direct energy cascade, a larger-scale direct enstrophy cascade and finally a very large-scale indirect energy cascade, the whole system involving three distinct sources of turbulent flux (e.g. Lilly, 1983).

The first major experiment devoted to testing the 2D/3D model was the EOLE experiment. It used the dispersion of constant-density balloons (Morel and Larchevêque, 1974; this is similar in principle to some of Richardson's methods used to obtain Fig. 1.1), and the balloons stayed (nearly) on isopycnals (i.e. surfaces of constant density), not on isobars. Due to the hydrostatic relation $\rho = -g^{-1}\partial p/\partial z$, the vertical spectrum of ρ: $E(k_z) \approx k_z^{-\beta_\rho}$ has exponent $\beta_\rho = \beta_p - 2$ so that a key difference between isopycnals and isobars is that while the latter are gradually sloping ($\beta_p > 3$) the former are highly variable with large-scale average slopes diminishing at larger and larger scales.

The original conclusions of the EOLE analysis (Morel and Larchevêque, 1974) were that the turbulence in the 100–1000 km range was two-dimensional. However, even then discrepancies were noted between the relative diffusivity and the velocity structure function results. Later, and more importantly, the conclusions contradicted those of the GASP and MOZAIC analyses (which found $k^{-5/3}$ out to hundreds of kilometres; see below). This motivated the reanalysis of the original (and still unique) dataset by Lacorta *et al.* (2004), who used velocity structure functions (which they called "finite scale relative velocities") and other techniques to show that, on the contrary, the data followed the $\Delta x^{1/3}$ law (i.e. $\beta = 5/3$), thus vindicating Richardson over this range and invalidating the original conclusions. Fig. 2.9 shows their reanalysis, which supports Richardson over the range of about 200–2000 km.

Interestingly, it seems that to properly understand the behaviour below about 200 km we must revisit their reanalysis! This is because although Lacorta *et al.* (2004) interpolated the EOLE satellite tracked-balloon positions every hour, the actual data were at lower temporal resolution – "mostly" every 2.4 hours – and neither the original nor the recent reanalysis attempted to understand the consequences of this temporal smoothing. The basic effect is straightforward to calculate (the detailed calculations are postponed to Appendix 6A since we haven't yet developed all the necessary theory). If the mean advection velocity is u_0 (estimated by Lacorta *et al.*, 2004, as $\approx$ 100 km/h), and the balloon position is sampled at intervals of Δt, then the estimated velocities are effectively averaged over distances $u_0\Delta t$. This temporal and spatial averaging decreases the variability for distances $<u_0\Delta t$, i.e. on distance scales less than the typical advection distance. Surprisingly neither Lacorta *et al.* (2004) nor Morel and Larchevêque (1974) seem to have noticed this, instead attempting to find physical interpretations for the behaviour down to 50 km even though $u_0\Delta t$ according to their own data was at least 200 km. For example, Lacorta *et al.* (2004) claim that "at distances smaller than 100 km our results suggest an exponential decay with *e* folding time of about 1 day in rough agreement with Morel and Larchevêque (1974)." In Fig. 2.9, we show that even the slope for the range affected by the averaging is roughly as expected theoretically (assuming space-time scaling up to planetary scales: see Appendix 6A). In other words, the re-reanalysis of EOLE is compatible with Richardson's scaling results over the *entire* observed range, not only 200–2000 km.

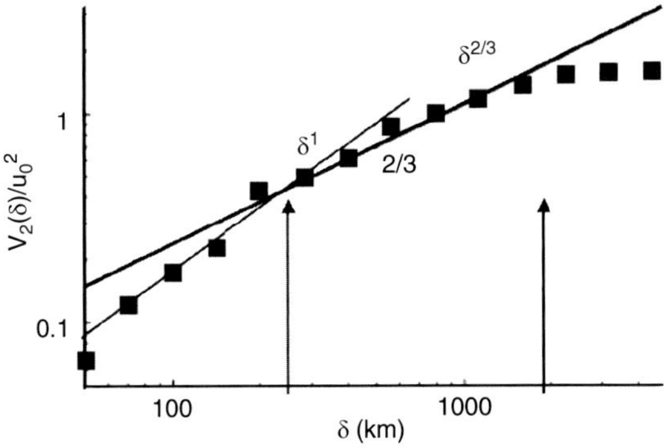

Fig. 2.9 Second-order velocity structure function estimated from the EOLE balloons. The theoretical effect of the low temporal resolution is discussed in Appendix 6A; it predicts the break as indicated on the left by the reference line, slope 1 (added to the original with break point corresponding to $\Delta t = 2.4$ hours and $u_0 =$ 100 km/h). The right-hand arrow shows the true limit ~2000 km. Adapted from Lacorta *et al.* (2004).

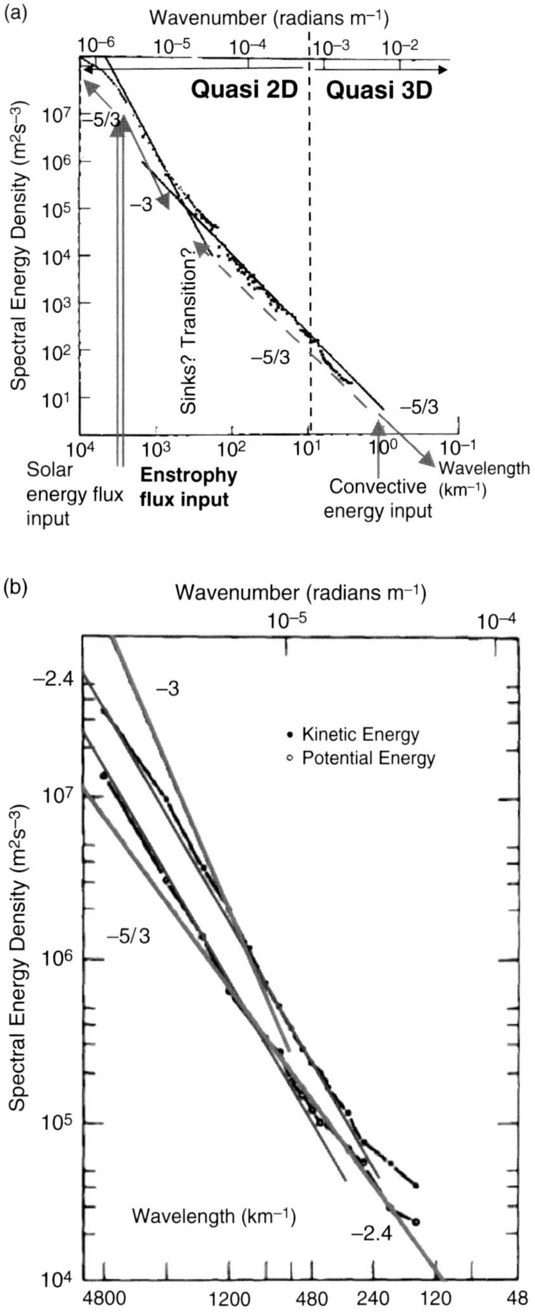

Fig. 2.10 (a) A schematic of the standard model updated to take into account the results of the GASP experiment. The figure is adapted from Lilly (1989) and schematically illustrates the "Gage–Lilly" model. Note that the 2D enstrophy cascade region spans much less than an order of magnitude in scale whereas the speculative inverse energy flux cascade (dashed line) spans over two orders of magnitude. (b) GASP spectrum of long-haul flights (> 4800 km) adapted from Gage and Nastrom (1986) with the reference lines corresponding to the horizontal and vertical behaviour discussed in the text (exponents 5/3, 2.4, i.e. ignoring intermittency corrections corresponding to $H_{h=}$ 1/3, $H_{v=}$ 0.7 as well as to the 2D isotropic turbulence slope −3).

Although at the time the EOLE experiment was influential, it did not include the determination of the spectrum, which was inconvenient to estimate because of the uneven distribution of the balloons. The first serious estimates of the horizontal wind spectrum had to await the GASP experiment (Fig. 2.10a), which was apparently incompatible with the early EOLE interpretations. The key difficulty was that the GASP spectra showed that $k^{-5/3}$ wind spectra extended out to scales much larger than the scale height of 10 km (up to several hundred kilometres). This motivated the development of the more sophisticated "Gage–Lilly" model (Fig. 2.10a: Lilly, 1989). This model suffers from many unsatisfactory ad hoc features, especially the upscale $k^{-5/3}$ energy flux regime from roughly 1 km to ~200 km (dashed line in Fig. 2.10a), which Lilly describes as "escaped" 3D energy transformed to quasi-2D stratified turbulence. The same feature was termed "squeezed 3D isotropic turbulence" by Högström *et al.* (1999). Other difficulties are the unknown flux sinks in the 2D/3D transition region, an unknown large-scale energy flux dissipation mechanism (surface drag?), and speculative energy and enstrophy flux sources at ≈ 2000 km.

2.6.2 The real transition is from $k^{-5/3}$ to $k^{-2.4}$. . . and it is spurious: a review of the classical aircraft campaigns and a new one (TAMDAR)

Even if we accept the plausibility of the various mechanisms invoked in the Gage–Lilly model, the evidence for 2D turbulence is scant: barely an octave in scale of the k^{-3} regime even in the classical (oft-reproduced) GASP spectrum (Fig. 2.10a). Actually, more careful examination of the original GASP analyses proves even more damaging to the k^{-3} hypothesis: while Fig. 2.10a was a composite of all the available data, the more relevant spectrum is the rarely cited Fig. 2.10b,

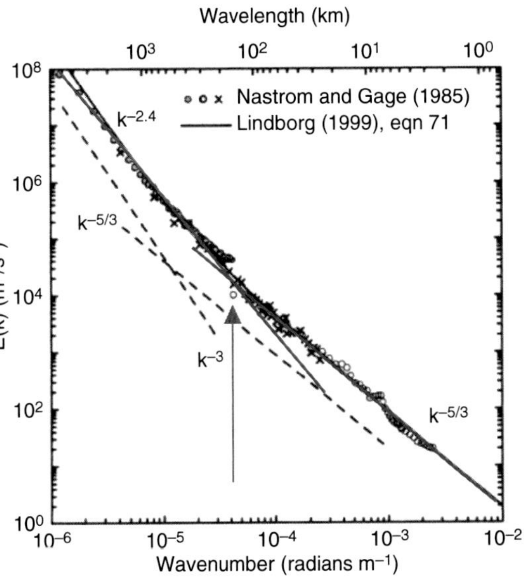

Fig. 2.11 A comparison of the GASP and MOZAIC spectra from commercial aircraft flying on isobars, adapted from Skamarock (2004), reproduced from Lovejoy *et al.* (2010). The thick lines show the behaviour predicted if the atmosphere has a perfect $k^{-5/3}$ horizontal spectrum but estimated from an aircraft following roughly horizontal trajectories until about 100 km (indicated by the arrows) and then following gradually sloping trajectories (either on isobars or gradual changes in altitude due to fuel consumption).

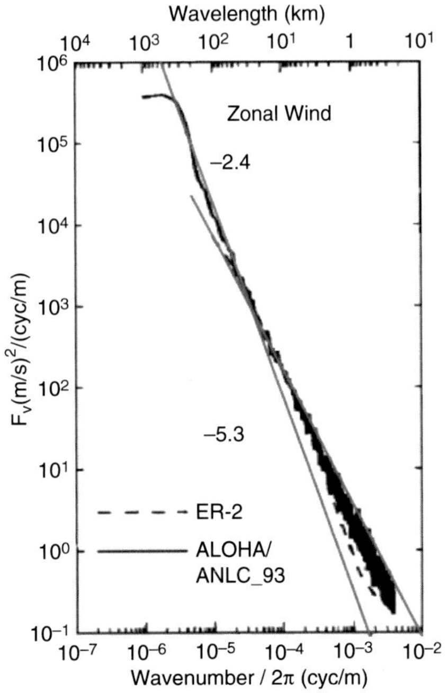

Fig. 2.12 The averaged spectra adapted from Gao and Meriwether (1998) at 6 km altitude with the horizontal and vertical exponents discussed here indicated as reference lines. Reproduced from Lovejoy *et al.* (2010).

which shows only the relevant long-haul flights (> 4800 km). The interpretation of this spectrum is more straightforward than the composite since the composite effectively involves somewhat different ensembles of flights as one moves to larger and larger scales. However, the long-haul spectrum shows no hint whatsoever of a k^{-3} regime; instead, one sees $k^{-5/3}$ at small scales followed by an almost perfect $k^{-2.4}$ spectrum at the larger scales.

A more recent large-scale campaign to estimate spectra has also used instrumented commercial aircraft: the MOZAIC campaign of > 7600 flights between 9.4 and 11.8 km (Cho and Lindborg, 2001; Lindborg and Cho, 2001). Not surprisingly, it is very close to the GASP spectrum, and Fig. 2.11 conveniently summarizes and compares the two. Again it can be seen that any k^{-3} regime must be very narrow, and that in any case $k^{-5/3}$ behaviour at small scales followed by $k^{-2.4}$ at large scales (without any k^{-3} regime) explains the observations quite accurately. By reproducing key figures and adding appropriate reference lines, we can see that the same $k^{-5/3}$ to $k^{-2.4}$ behaviour with similar transition scales (40–200 km) explains other aircraft wind spectra (Gao and Meriwether, 1998: 11 legs of the scientific Electra aircraft, which also flew along isobars but at ≈ 6 km; see Fig. 2.12); for stratospheric spectra, see Fig. 2.13 (Bacmeister *et al.*, 1996). Lovejoy *et al.* (2009) also find similar behaviour in the tropospheric Gulfstream 4 scientific aircraft spectra already discussed in Chapter 1 (Fig. 2.14).

So why is there a break in the spectrum at scales from 40 to 400 km: highly variable yet significantly larger than the atmospheric scale height? And why is it not visible in other spectra of strongly nonlinearly linked fields, such as the radiances (Fig. 1.2) or the temperature or humidity (including from the same aircraft: compare Figs. 1.6c and 2.14)? The answer is surprisingly simple: it suffices that the aircraft have a small but nonzero slope, so that after a critical distance the fluctuations it measures no longer reflect the horizontal statistics, but rather the vertical ones.

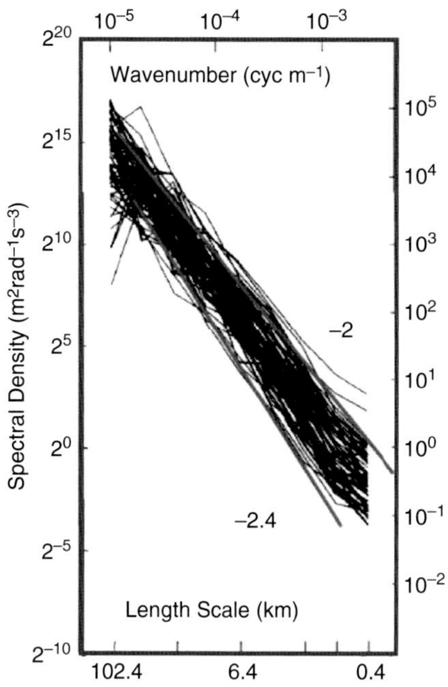

Fig. 2.13 Stratospheric ER2 spectra adapted from Bacmeister *et al.* (1996, Fig. 5). This is a random subset of 1024 s legs, again with reference slopes added. Reproduced from Lovejoy *et al.* (2010).

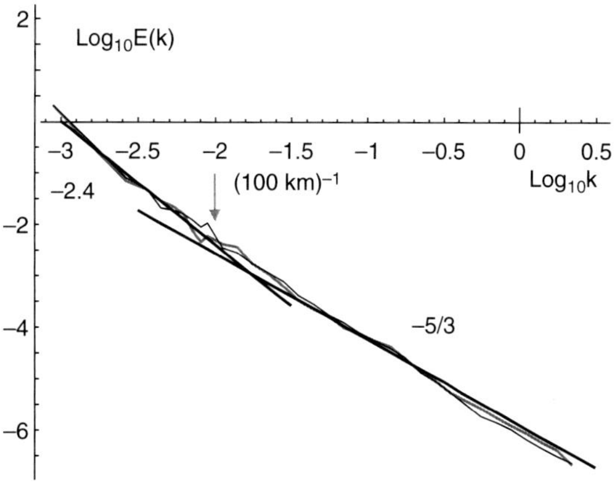

Fig. 2.14 Recent spectra from the Pacific Winter Storm 2004 experiment: 24 aircraft legs, each 1120 km in length (resolution 280 m); units of k: (km)$^{-1}$. For more details on the Winter Storm 2004 data, see Section 6.3. The curves are for the longitudinal and transverse components of the wind (relative to the aircraft direction). The black reference lines show the theoretical Kolmogorov spectrum ($k^{-5/3}$; obeyed roughly up to $\approx$ 40 km) and the $k^{-2.4}$ spectrum at larger scales. Adapted from Lovejoy *et al.* (2009).

To see how this works, consider Fig. 2.15a, which is a contour plot of the mean squared horizontal wind differences $\langle \Delta v^2(\Delta x, \Delta z)\rangle$ for various lags (Δx, Δz) in the vertical plane. It was obtained from data over the year 2009 from a fleet of short-range commercial aircraft flying over the continental USA; the overall sample contained over 14 500 aircraft legs (TAMDAR: Moninger *et al.*, 2003; Mamrosh *et al.*, 2006) sampled somewhat irregularly, but at roughly 20 km resolution in the horizontal. Our investigation required distinguishing statistics on isobars from those on isoheights, and thus required high-accuracy GPS altitude measurements. For our purposes, an essential TAMDAR improvement with respect to the more widespread, older AMDAR equipment was thus that the former included accurate GPS altimetry that enabled altitude differences (Δz) to be estimated to within $\pm$ 4 m, a level of accuracy essential for distinguishing isobars and isoheights. At the same time wind differences are measured to within $\pm$ 2.5 m/s. Although it is possible to estimate Δz, Δv from two different aircraft, here only data from single legs were considered. This eliminates the (relatively poor) *absolute* sensor calibration from the problem, as the wind differences measured from single aircraft only require accurate *relative* calibrations. Using data from single aircraft not only yields much higher-accuracy measurements, but it also greatly simplifies the analysis of the – otherwise extremely complex to analyse – problem of highly nonuniform statistical sampling of Δv^2 in (Δx, Δy, Δz, Δp, Δt) space that results when considering wind differences from two different aircraft with numerous particularities including geographical distributions determined by the commercial flight corridors. More details can be found in Pinel *et al.* (2012).

In Fig. 2.15a one can see that the empirical contours (dark) are nearly of the form theoretically predicted (light) for scaling stratified turbulence discussed in Chapter 6:

$$\langle \Delta v^2(\Delta x, \Delta z)\rangle = C\left(\left|\frac{\Delta x}{l_s}\right| + \left|\frac{\Delta z}{l_s}\right|^{1/H_z}\right)^{\xi(2)} \tag{2.85}$$

where H_z is the ratio of the horizontal to vertical wind exponent and $\xi(2)$ is the exponent of S_2 (Eqns. (2.71), (2.73)), the second-order "structure function exponent," and l_s is the "sphero-scale," which is the scale at which fluctuations have roughly the same vertical and horizontal extents. At scale l_s, we have $\langle \Delta v^2(l_s, 0)\rangle = C = \langle \Delta v^2(0, l_s)\rangle$ so that the

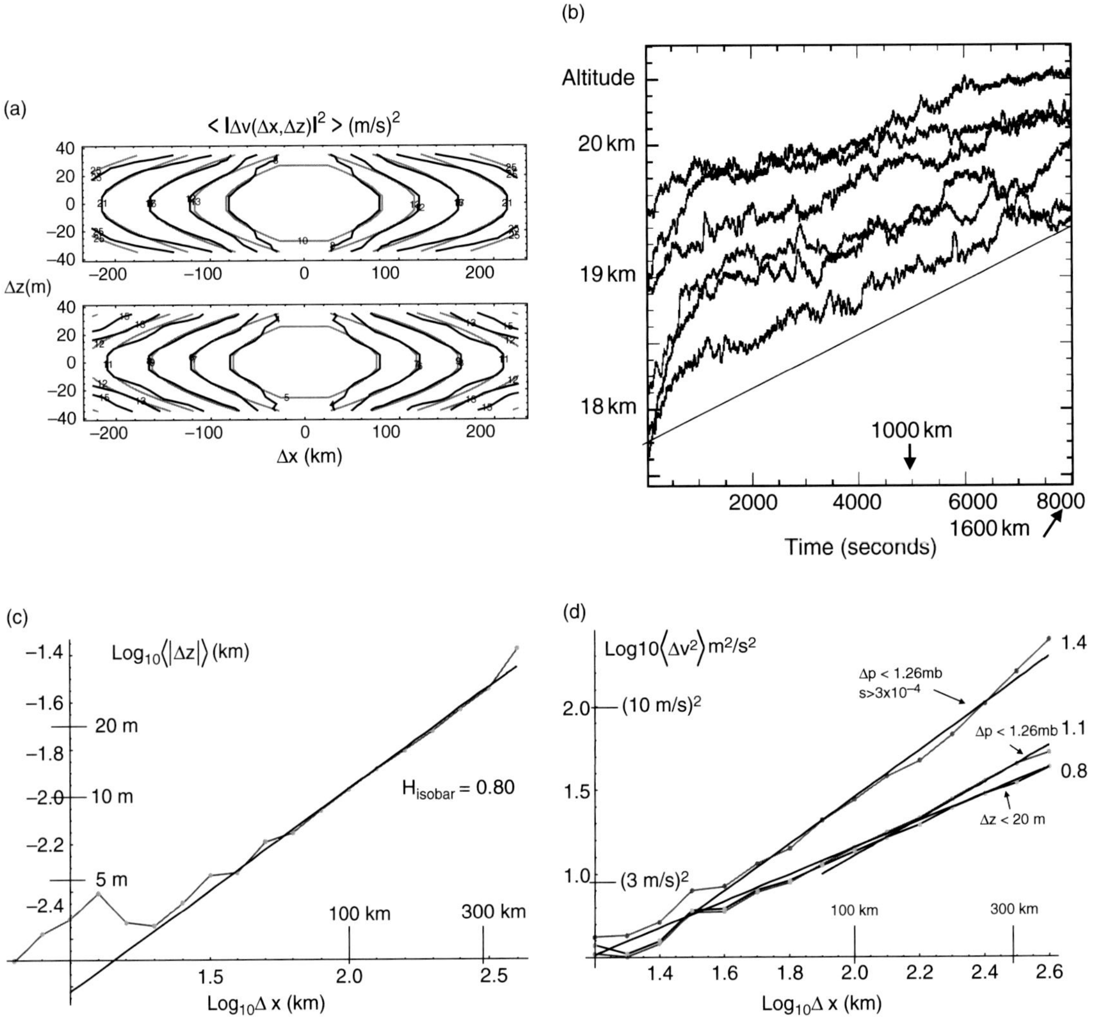

Fig. 2.15 (a) A contour plot of the mean squared transverse (top) and longitudinal (bottom) components of the wind, as estimated by a year's ($\approx$ 14 500) TAMDAR flights, 484 000 wind difference measurements. All the fluctuations were from a single aircraft at different parts of its trajectory, and only trajectories between 5 and 5.5 km were used. Black shows the empirical contours, grey the theoretical contours assuming scaling stratification and the functional form indicated in the text. The numbers next to the contours are the values of the contours (italics is theory, bold is empirical, to improve the statistics, reflection symmetries were used). Reproduced from Pinel *et al.* (2012). (b) A sample of six fractal aircraft trajectories (fractal dimension $\approx$ 1.56) from NASA's ER2 aircraft during missions near Antarctica. The aircraft flew along isomachs but there was a mean vertical "drift" of $\sim$1 m/km (reference line) caused by the lightening of the aircraft due to fuel consumption. Adapted from Lovejoy *et al.* (2004). (c) For the TAMDAR data this shows the mean vertical displacement for points on isobars defined by $\Delta p < 0.126$ mb (373 000 differences were used, flight legs between 5 and 5.5 km). The reference line shows a slope $H_{isobar} = 0.80$ corresponding to a fractal dimension of $D_{isobar} = 1 + 0.80 = 1.80$. Each point is the average of over 1000 measurements. Reproduced from Pinel *et al.* (2012). (d) Comparison of mean squared wind differences of the transverse component of the wind from TAMDAR data sampled from near isobars ($\Delta p <$ 1.26 mb), near isoheights ($\Delta z <$ 20 m) and from sloping isobars i.e. $\Delta p < 1.26$ mb and slope $s > 3.2 \times 10^{-4}$ (the longitudinal components gave very similar behaviour). We see that the latter has the theoretical vertical exponent for a nonfractal vertical section, 1.4, the $\Delta p <$ 1.26 mb curve has (for scales $> \sim$140 km) the exponent 1.1 theoretically predicted for the fractal isobars: $1.1 = H_{isobar}\, 1.4$, whereas the curve for $\Delta z < 20$ m has the theoretical isoheight exponent. Reproduced from Pinel *et al.* (2012).

corresponding structures are "roundish," hence the name. From the regression on $\langle \Delta v^2 \rangle$ in Fig. 2.15a, the best-fit empirical parameters are $H_z \approx 0.57$, $\xi(2) \approx 0.80$ and $l_s \approx 1.0x\ 10^{-3}$ m. The H_z value is very close to the theory value ($H_z = (1/3)/(3/5) = 5/9 = 0.56$); see Chapter 6, although $\xi(2)$ is a little larger than the theory value $2H = 2/3$, and l_s is close to values measured in vertical lidar scans of pollution backscatter and to the somewhat less direct aircraft estimates in Lovejoy *et al.*, 2004, 2009). The constant $C \approx 4.5 \times 10^{-6}$ m^2/s^2 is $\langle \Delta v(L_s^2) \rangle \approx \langle \varepsilon^{2/3} l_s^{2/3} \rangle$; the equality is approximate since there are intermittency corrections due to the highly variable nature of ε, l_s (Lovejoy *et al.*, 2008); ignoring these issues, taking $C \approx \varepsilon^{2/3} l_s^{2/3}$ we obtain $\varepsilon \approx 10^{-5}$m^2/s^3. This low estimate of ε (see Chapter 8) and slightly larger value of $\xi(2)$ could thus be explained by intermittency corrections.

From the functional form in Eqn. (2.85) we can consider two particularly simple extreme cases: pure horizontal and pure vertical displacements. In the former case we easily see that the mean squared horizontal differences vary as $\langle \Delta v^2(\Delta x) \rangle \propto \Delta x^{\xi_h(2)}$ with $\xi_h(2) = \xi(2) \approx 0.8$, whereas in the latter case the mean squared vertical differences $\langle \Delta v^2(\Delta z) \rangle \propto \Delta z^{\xi_v(2)}$ with $\xi_v(2) = \xi(2)/H_z \approx 1.4$. Since the spectral exponent $\beta = 1 + \xi(2)$ (Eqn. (2.73)) this implies different horizontal and vertical spectral exponents: $\beta_h = 1.8$, $\beta_v = 2.4$.

In order to determine the spectrum measured from real aircraft trajectories, we need a model of the latter. For example, in Chapter 6 we consider a simple intermediate model involving a trajectory along a constant slope *s*; i.e. using $\Delta z = s\Delta x$ in Eqn. (2.85), we find a critical value Δx_c such that for $\Delta x < \Delta x_c$ the horizontal behaviour is dominant whereas for $\Delta x > \Delta x_c$ the vertical behaviour is dominant. It is therefore easy to imagine that, depending on how flat the aircraft trajectory is, at small scales $\beta \approx 1.8$, yet at large enough scales one could readily obtain a vertically dominated spectrum with a transition to $\beta \approx 2.4$, as found in the campaigns discussed above.

However, things are potentially more complicated than this simple constant-slope model. As discussed in Chapter 6, both the proportionality constant in Eqn. (2.85) and l_s depend on highly variable turbulent fluxes (energy and buoyancy force variable fluxes), and Fig. 2.15b shows that trajectories are not uniform with constant slope, but can be fractal. However the trajectories in the figure are from aircraft whose autopilot flies on lines of constant Mach number, and this is quite unusual. It is more typical for aircraft to follow isobars; single long, high-resolution (280 m) isobaric aircraft trajectories are discussed in Chapter 6, but here we use the TAMDAR data, which are short (< ~400 km) and sampled somewhat irregularly (on flat legs, typically every ~20–30 km, much more frequently when changing altitude levels). This low sampling rate in individual legs is largely compensated for by the overall high number of TAMDAR legs: for example, 484 000 individual wind differences were used to estimate the contours in Fig. 2.15a. By sampling only wind differences from measurements nearly on the same isobar, we can determine the statistics of vertical isobaric cross-sections. Fig. 2.15c shows that on the isobar the mean vertical displacement $<\Delta z> \approx \Delta x^{H_{isobar}}$ with $H_{isobar} \approx 0.80$, so that the isobar is fractal (fractal dimension $1+H_{isobar} = 1.80$ in vertical sections; see Chapter 3).

We can now combine our information about $<\Delta v^2(\Delta x, \Delta z)>$ with our knowledge of the fractal structure of the isobars to see how $<\Delta v^2>$ varies on various trajectories (Fig. 2.15d). For example, we have already shown from Eqn. (2.85) that on isoheights $\langle \Delta v^2(\Delta x) \rangle \propto \Delta x^{\xi_h(2)}$ with $\xi_h(2) = \xi(2) = 0.8$, and this is confirmed in Fig. 2.15d. Let us now consider the behaviour along isobars. As in the constant-slope model discussed above, in Eqn. (2.85) for small horizontal displacements, the horizontal term dominates and we obtain the same result as for isoheights, i.e. $\langle \Delta v^2(\Delta x) \rangle \propto \Delta x^{\xi_h(2)} \approx \Delta x^{0.8}$. However, for large enough displacements the second term in Eqn. (2.85) becomes dominant so that we expect $\langle \Delta v^2(\Delta x) \rangle \propto \langle |\Delta z(\Delta x)| \rangle^{\xi_v(2)} \approx \Delta x^{H_{isobar}\xi_v(2)} \approx \Delta x^{H_{isobar}\xi(2)/H_z} \approx \Delta x^{1.1}$; this is also confirmed in Fig. 2.15d, although the behaviour is only dominant for scales > ~140 km. As a final test of the model, we can restrict our attention to those isobars which are also steeply sloping (in the figure, with slope > 0.32 m/km). In this case, the fractality of the isobars is no longer important, for a large range in Δx only the vertical displacement is dominant and we expect $\langle \Delta v^2(\Delta x) \rangle \propto \Delta x^{\xi_v(2)} \approx \Delta x^{1.4}$, which is also verified in the figure.

As a final comment we remark that in a recent paper Frehlich and Sharman (2010) used nearly the same TAMDAR data but reached an opposite conclusion: that $<\Delta v^2>$ was the same on isoheights as on isobars. However, detailed analysis in Pinel *et al.* (2012) shows that this conclusion was likely spurious, the reason being the inappropriate use of TAMDAR data over the numerous flight segments where the aircraft changed altitude levels. The corresponding altitude resolution was much lower over these sloping sections than over the roughly flat ones, so much so that the purported isoheight data analyses were not really from isoheights at all – so the data analyses including these low-vertical-resolution sections were unable to adequately distinguish isoheight and isobar data. The TAMDAR system is programmed to sample much more frequently when the aircraft changes altitude levels, so that if care is not taken this low resolution can seriously bias the estimates, making it impossible to distinguish isoheight and isobar statistics. This was graphically demonstrated by Pinel *et al.* (2012), who could almost exactly reproduce the findings of Frehlich and Sharman (2010) by including these low-resolution segments, yet (as Fig. 2.15d shows) when they are removed the difference between isoheight and isobar statistics becomes clear. Additional evidence pointing to bias was the fact that when the low-resolution data points were removed the horizontal scaling of $<\Delta v^2>$ was greatly improved.

This anisotropic scaling model of the vertical structure is developed further in Chapter 6, but it can easily quantitatively reproduce the observations leading to the reinterpretation of the Nastrom–Gage spectrum (Fig. 2.16a). In the new model, rather than having sources and sinks at precisely defined scales separated by wide ranges with no sources or sinks (the classical inertial ranges), the energy flux is instead from solar heating modulated by scaling cloud fields, so that the input is over a wide range of scales in a scaling manner (in accord with the observed scaling of the radiance: Figs. 1.2 and 1.3). In this reinterpretation, the large-scale (low-wavenumber) $k^{-2.4}$ is simply a spurious consequence of the not carefully accounted for effects of anisotropic turbulence on the aircraft motion.

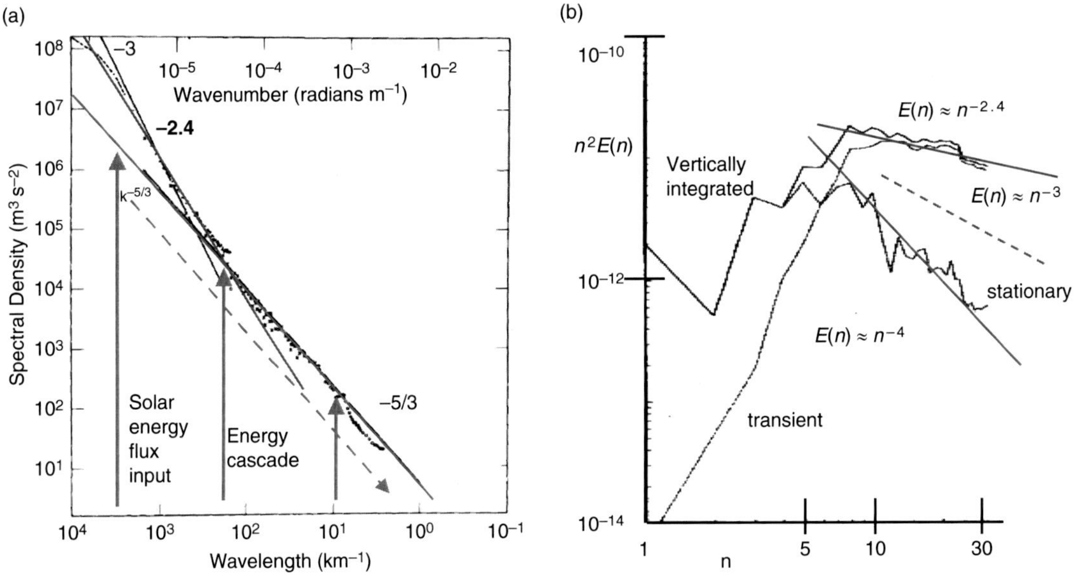

Fig. 2.16 (a) Reinterpretation of the Nastrom–Gage spectrum, as discussed in the text. The transition to large-scale $k^{-2.4}$ behaviour is a spurious consequence of aircraft travelling along gently sloping trajectories, as discussed in detail in Chapter 6. Reproduced from Lovejoy and Schertzer (2010). (b) The enstrophy spectrum ($= n^2E(n)$, where $E(n)$ is the wind spectrum and n is the principal spherical harmonic wavenumber), adapted from Boer and Shepherd (1983). The three curves are from January data; the solid line is for the vertically integrated atmosphere, the lines indicate stationary (spatial spectrum of the monthly average), and the transient is the deviation from the monthly average. Over the range $n \approx$ 5–30 (700–4000 km) the exponents of the spectra of the transient and vertically averaged atmosphere are extremely close to the vertical value $\beta \approx 2.4$, but the stationary spectral exponent is $\beta \approx 4$. No $\beta \approx 3$ regime is observed (dashed line). Reproduced from Lovejoy *et al.* (2010).

2.6.3 The classical approach: conclusions from analyses and reanalyses

The interpretation of aircraft data in terms of large-scale 2D turbulence was very influential, so that other evidence about atmospheric structure and statistics were generally interpreted in the same way. The problem was that the special 2D exponent $\beta = 3$ provided the only theoretical framework for explaining spectra with $\beta > 5/3$. The long absence of a credible alternative theory tempted even early investigators to "shoehorn" their spectra into the k^{-3} mould. For example, by "eyeballing" four spectra over less than an *octave* in scale, Julian *et al.* (1970) concluded that $2.7 < \beta < 3.1$ for the horizontal wind. In the 1980s larger datasets became available, and it was possible to make more direct tests of 2D turbulence theory from atmospheric analyses (Boer and Shepherd, 1983) and later from the ECMWF ERA40 reanalyses (Strauss and Ditlevsen, 1999). Although Boer and Shepherd (1983) gave cautious support to $\beta \approx 3$ and to a 2D interpretation, in hindsight and with the benefit of a simple theory predicting $\beta \approx 2.4$, their conclusions seem unconvincing (Fig. 2.16b). Similarly, when interpreting their reanalyses, Strauss and Ditlevsen (1999) found that "$\beta \approx 2.5$–2.6 ... significantly different than the classical turbulence theory prediction of 3," but again close to the value 2.4.

Today, we can revisit wind spectra using the state-of-the-art successor to the ERA40 reanalysis – the ECMWF interim reanalysis whose spectra were already presented in Fig. 1.5b, and to which we return in Chapter 4 – and calculate the spectrum directly without Strauss and Ditlevsen's complex 2D preprocessing. Fig. 2.17 shows the angle integrated spectrum of the zonal wind at each tropospheric 100 mb level, compensated by the average $k^{-2.4}$ behaviour so as to accentuate the small deviations. Also shown in the figure are straight *reference* lines. These are *not* regressions but rather the predictions of the stratified anisotropic scaling model discussed in Chapter 6: the slopes are those empirically estimated in the *vertical* direction from dropsondes (Fig. 6.2; Lovejoy *et al.*, 2007). Regressions on the reanalysis spectra from $k = 2$ to $k = 30$ (i.e. 5000–330 km) give β differing by less than 0.05 throughout the data-rich lower 4 km, rising to only 0.2 at 10 km (≈ 200 mb). Following the discussion of the previous section, we should not be surprised if

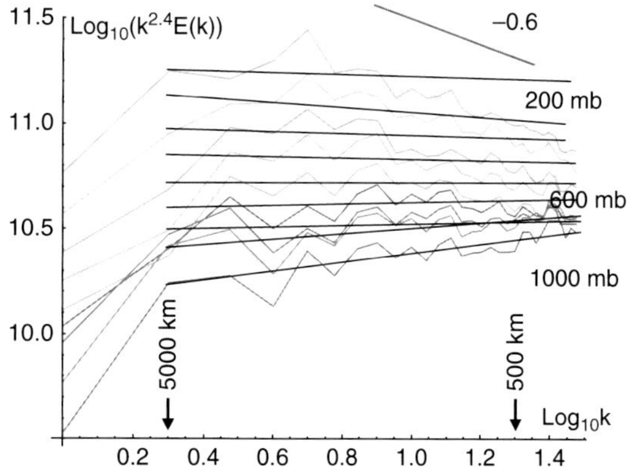

Fig. 2.17 The isotropic spectrum of the zonal component of the wind at 200, 300, 400, ..., 1000 mb from the ECMWF interim reanalysis for January 2006 between ± 45° latitude. The straight lines are *not* regressions, rather they have the slopes of the horizontal wind in the vertical direction as estimated by dropsondes in Lovejoy *et al.* (2007). It can be seen that the isobaric velocity spectra have exponents close to the vertical values, and this is especially true of the data-rich lower levels. The 200 mb spectrum falls off a little too quickly at high wavenumbers, possibly due to poor-resolution data below ~1000 km. The scaling starts at $k = 2$–3, corresponding to $n = 4$–6 in Fig. 2.16b. Reproduced from Lovejoy *et al.* (2010).

these small differences are the consequences of either intermittent aircraft and/or sonde motion (Chapter 6).

2.6.4 Evidence from satellite altimeter winds over the ocean

One way of overcoming the problems and limitations of in-situ wind measurements is to use remote sensing. The most direct remote method is to use the Doppler shift from clear-air radar turbulence measurements. However, existing datasets are over fairly narrow ranges of scales. In addition, the radar measures the radial wind component, which systematically changes direction as the radar scans to build up a three-dimensional field. This makes the interpretations complicated and tempts users to make "products" based on complex-to-analyse assumptions about the statistics in order to correct for these effects.

An alternative is to use ocean surface data from satellite scatterometers to measure wave heights and to correlate these with surface winds over the oceans. Such satellite wind products have been developed since the mid 1980s and rely on measurements of Bragg

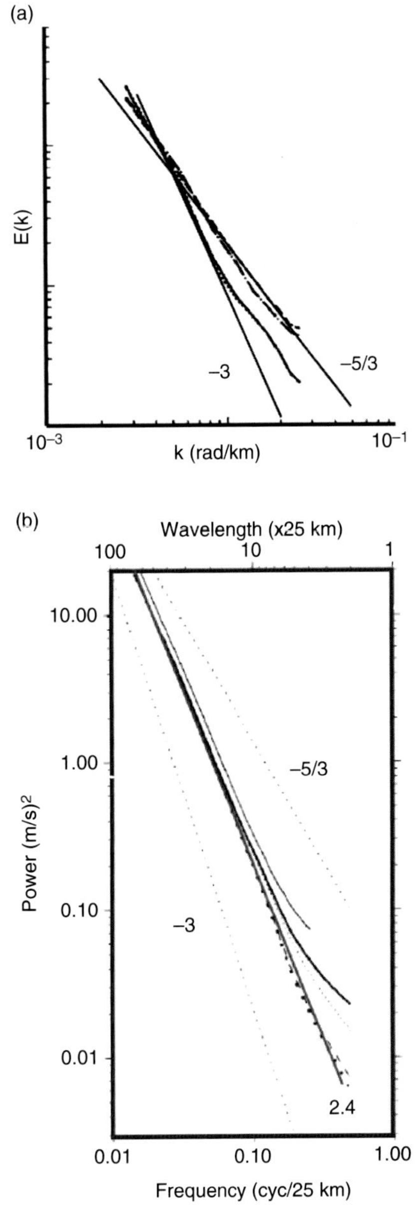

Fig. 2.18 (a) The spectra of sea surface winds as estimated from the SeaSat scatterometer over four regions in the Pacific Ocean (regions 1, 2 are the South Pacific, regions 3, 4 are the north Pacific), adapted from Freilich and Chelton (1986). The smallest wavenumber corresponds to about 2200 km, the reference lines (added) have slopes −3, −5/3. (b) The kinetic energy spectrum estimated over the oceans over a year, using QuikScat satellite altimeter data (adapted from Patoux and Brown, 2001). DIRTH is the recommended product; also shown are reference lines −5/3, −3 (in the original), with −2.4 reference line added. Largest scale is 1700 km, smallest is $2 \times 25 = 50$ km.

scattering amplitudes of ocean waves at the typically centimetric scatterometer wavelength (e.g. 2.1 cm from the SeaSat scatterometer, Fig. 2.18a, and 2.2 cm for the QuikScat data in Fig. 2.18b). At these wavelengths, the scatterometer is sensitive to capillary waves whose amplitudes are only indirectly related to the local winds. Correlating the scatterometer backscatter with the wind speed is only part of the problem; more difficult is determining the wind direction. At the moment, this is currently done by observing the same patch of ocean at different angles and by using meteorological reanalyses to help remove remaining directional ambiguities. The final product is thus dependent in a number of subtle ways on various assumptions about the nature of the turbulence and of the numerical models. Bearing this in mind, we refer the reader to some early scatterometer results that had spectral exponents quite near $\beta = 5/3$ (see the regression lines in Fig. 2.18a; Freilich and Chelton, 1986): over the range 200–2200 km, the regression estimates for the tropics were $\beta \approx 1.9$, and for the mid-latitudes $\beta \approx 2.2$. More recent products use more sophisticated algorithms, but the results are not much different: Fig. 2.18b shows spectra using the DIRTH algorithm (Patoux and Brown, 2001). As can be seen, the spectrum is almost perfectly scaling with $\beta = 2.4$: the (sloping) isobaric value. Although the value of the exponent may well depend on some of the assumptions that went into its derivation, these assumptions would be unlikely to transform an otherwise nonscaling spectrum into a scaling one. In other words, the fact that the spectrum is nearly a perfect power law over the observed range is highly significant and in itself would be difficult to explain with the 2D/3D model. Similar but yet more recent results using 10 years of data (instead of one year; King and Kerr, 2010), are shown in Fig. 2.19, showing that the exponents are apparently even lower, very close to the Kolmogorov value (the isotropic 2D turbulence result $\xi(2) = 2$ corresponding to $\beta = 3$, is completely off the scale!). Also indicated is the Bolgiano–Obukhov value $\xi(2) = 2\ (3/5) - K(2) \approx 1.15$ (the horizontal line) which would be expected for a near-surface layer sloping in the vertical direction ($K(2) \approx 0.05$ is the empirical intermittency correction; see Chapters 5, 6).

Finally we could add that a recent QuickScat paper (Xu *et al.*, 2011) claims that β varies geographically from 1.6 to 2.9, but these estimates are based on little

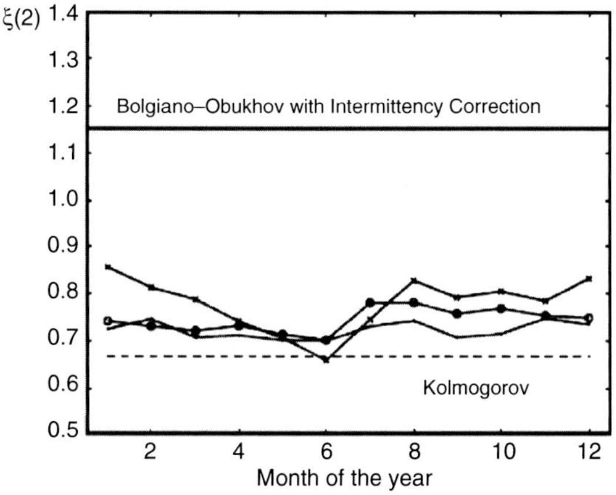

Fig. 2.19 Regression exponents for the second-order wind structure function ξ(2) estimated by regression over the "meso beta scale" (20–200 km), from 10 years of QuikScat sensor data with 1800 km swaths at 25 km resolution, adapted from King and Kerr (2010). The three curves are for somewhat different parts of the Pacific Ocean.

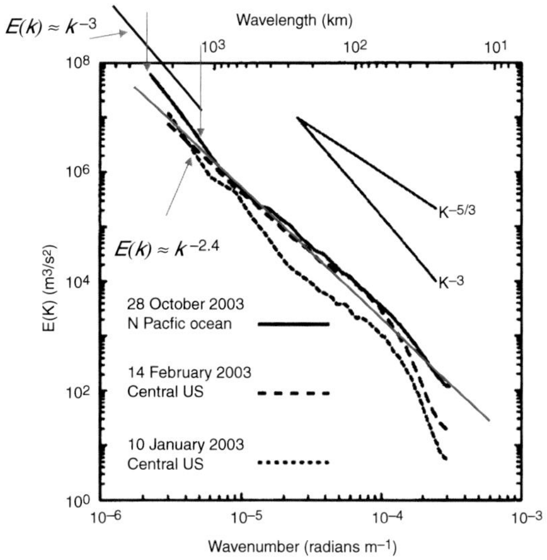

Fig. 2.20 Sample spectra from WRF forecasts of zonal wind averaged over the isobaric surfaces covering roughly the range 3–9 km in altitude, adapted from Skamarock (2004). Although they claimed that this shows a "clear k^{-3} regime" for the solid (oceanic) spectrum it only spans a range of factor 2–3 in scale, and this at the relatively unreliable extreme low wavenumbers (between the downpointing arrows, upper left). Except for the extremes, the spectra again follow the isobaric predictions $k^{-2.4}$ very well over most of the range. Reproduced from Lovejoy *et al.* (2010).

more than an octave in scale. In fact, all the data are very close to $\beta = 1.8$, with high wavenumber deviations from scaling accounting for most of the regional spectral differences.

2.6.5 The continuing difficulties with the classical model and inferences from numerical simulations

In order to improve on these speculative mechanisms needed to combine isotropic 2D and isotropic 3D turublence, efforts have been made to reproduce "realistic" k^{-3} to $k^{-5/3}$ transitions in numerical models. This is not a trivial question, because of the possibility of "three-dimensionalization" of two-dimensional flows discussed in Ngan *et al.* (2004), i.e. the likelihood that three-dimensional turbulence can destabilize an otherwise 2D flow. For the moment, the results are at best equivocal. For example, most numerical weather models do not display the transition (Palmer, 2001), while others may display it although over very small ranges – e.g. the Skamarock (2004) WRF (regional) model spectra, which are in fact very close to $k^{-2.4}$ (Fig. 2.20). To date, the most convincing k^{-3} to $k^{-5/3}$ transitions in numerical models have been produced using the SKYHI model on the earth simulator (Takayashi *et al.*, 2006; Hamilton *et al.*, 2008), yet as pointed out by Lovejoy *et al.* (2010) they have the poorest fit to GASP observations precisely over the range ~400–3000 km which their (painstakingly crafted) $k^{-5/3}$ to k^{-3} transition is supposed to explain. In other words, this model "success" may make them *less* rather than *more* realistic! In addition, Smith (2004) has shown that at least in the case of the quasi-geostrophic simulation by Tung and Orlando (2003), high wavenumber $k^{-5/3}$ regimes are in reality spurious consequences of energy build-up due to unresolved high wavenumbers with respect to an incorrectly "tuned" hyperviscosity.

2.6.6 Empirical determination of the direction of the cascade

Another way to test the classical 2D/3D model and to compare it to the anisotropic scaling model is to examine the spectral energy transfers due to the nonlinear terms. A 2D cascade will display an upscale

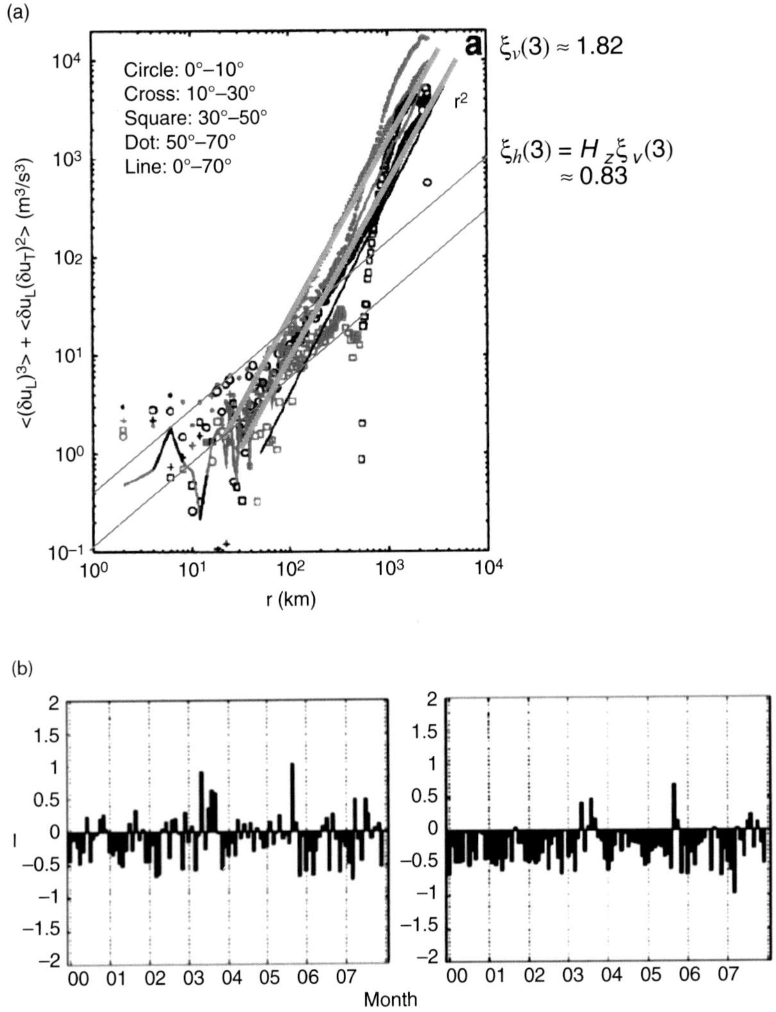

Fig. 2.21 (a) Third-order structure functions (diagonal contributions) adapted from Plate 1 of Cho and Lindborg (2001). Light grey indicates a negative sign, dark grey, positive, indicating large- to small-scale and small- to large-scale transfers, respectively. The theoretical reference lines were added with slopes corresponding to the predictions of the sloping isobaric trajectory model presented here (Section 2.6.2 and Chapter 6) with the third-order vertical structure function $\xi_v(q) = qH - K(q) = 1.82$ (using $q = 3$, $H = 0.77$, $K(3) \approx 0.49$) and horizontal structure function with $\xi_h(q) = \xi_v(q)\, H_z = 0.83$ ($H_z = 0.46$; parameters from Chapter 6 and Lovejoy *et al.*, 2010). The transition is not far from the 40 km found in the Gulfstream 4 analyses (Chapter 6). (b) The third-order structure function of the horizontal wind normalized by the second-order function (the skewness) estimate QuickScat scatterometer data from the N4WP region of the Pacific. These are monthly averages of I = skewness; < 0 means large- to small-scale cascade. The left is averaged over the 250–1000 km ("meso-α") scale range, the right over the 25–250 km ("meso-β") range. Although the transfer is predominantly from large to small, it is highly intermittent in accord with the finding here from the ECMWF reanalysis (Figs. 2.22, 2.23) and with the predictions of the scaling model. Adapted from Fig. 6 of Kerr and King (2009).

energy transfer, a 3D cascade a downscale transfer (Section 2.5.2); furthermore, the usual schematics (e.g. the Gage–Lilly model, Section 2.6.1) also assume that the corresponding ranges are "inertial," i.e. with no sources or sinks, so that the fluxes are roughly constant in Fourier space. In comparison, in the anisotropic scaling model things can be much more complicated, since the sources and sinks (which will be largely solar heating and infrared cooling) will be scaling and nonlinearly coupled to the dynamical fields (via the scaling cloud field).

The empirical determination of the direction of the energy fluxes is very demanding, since in principle all wind components and their derivatives are required. However, with the help of the assumption of statistical isotropy, Lindborg (1999) related the sign of the third-order velocity structure function $<\Delta v^3>$ to the direction of the cascade. The results (using MOZAIC aircraft data) showed mostly negative third-order structure functions (even out to 1000 km) in agreement with a downward (i.e. 3D) cascade. But the signs were not consistent (see Fig. 2.21a, which also shows that their third-order structure functions are close to those theoretically predicted for sloping isobaric aircraft trajectories). An algorithmic correction introduced by Cho and Lindborg (2001), although not clearly explained, surprisingly yielded a more opposite conclusion.

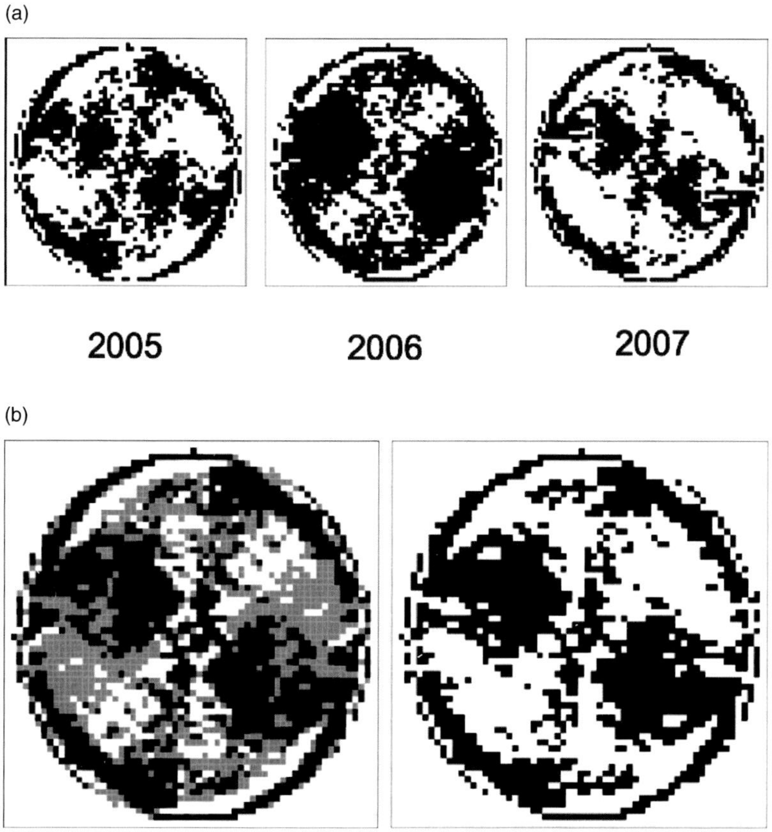

Fig. 2.22 (a) The sign of $T(k)$ for each of the three years analysed. Black indicates $T > 0$, white, $T < 0$. The horizontal axis is k_x (zonal), the vertical axis, k_y (meridional); the largest wavenumber corresponds to 330 km, the smallest to 10 000 km (more accurately, $\pm$ 45° about the equator and 0–90°, 90–180°, 180–270°, 270–0° longitude). (b) The same as Fig. 2.22a, but for the ensemble 2005–2007 (right) and the disagreement areas (in grey, left).

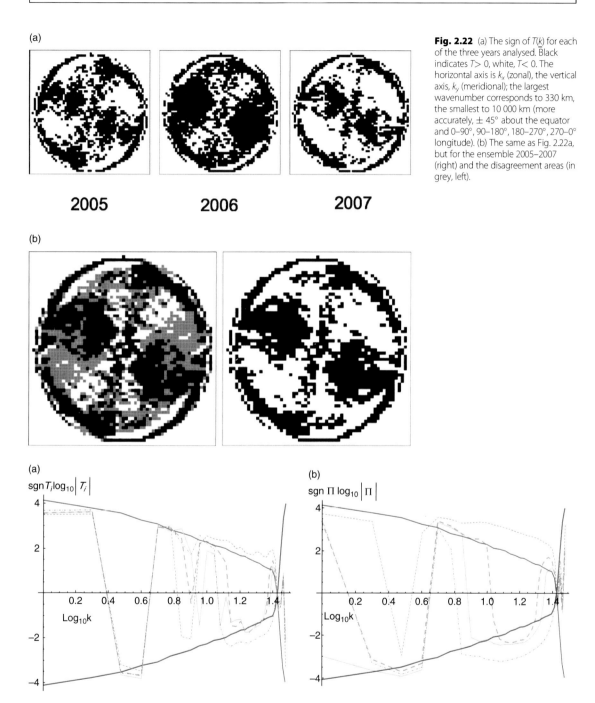

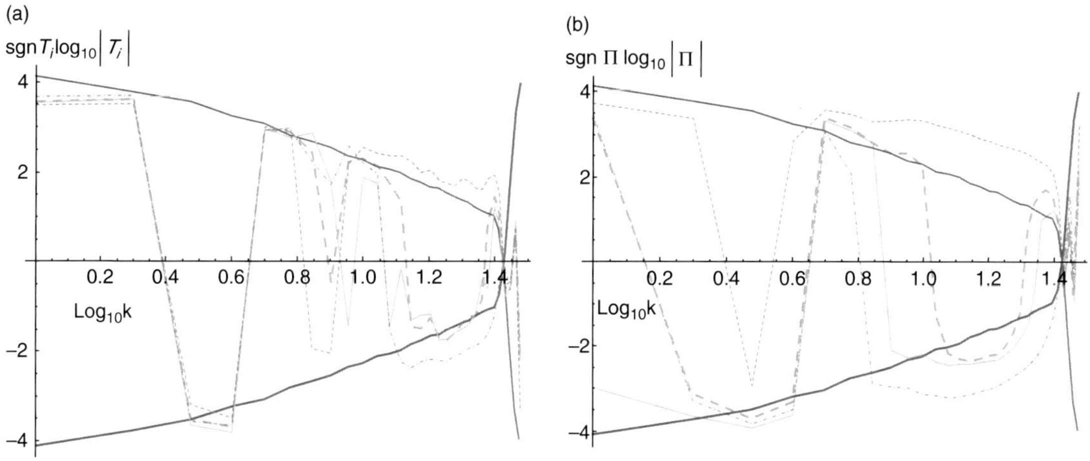

Fig. 2.23 (a)The angular integral of $T(k)$ for each year: 2005, 2006, 2007. The average of the three is the medium grey line which is only clearly distinguishable to the right of the 0.8 position (it's the line that barely drops below the axis at $\log_{10}k = 0.9$). The thick darker lines that roughly define the envelope are the isotropic energy spectrum $E(k)$ (top) and its negative (bottom). (b) The corresponding plots of $\Pi(k)$, which is the integral of $T(k)$, over wavenumbers higher than k for each year: 2005, 2006, 2007. The average of the three (medium grey) is again only clearly distinguishable when its drop below the upper envelope at about $\log_{10}k \approx 1.0$. The thick darker lines that define the envelope are the same as in Fig. 2.23a.

Recently, Kerr and King (2009), using QuickScat scatterometer data, similarly found that the sign of the flux fluctuates as the horizontal lag (scale) changes (Fig. 2.21b).

To test this out in a more direct way (without the third-order structure functions and isotropy assumptions), we can use the ECMWF interim reanalysis products; here the zonal and meridional winds at 700 mb for the years 2005–2007. The main weakness is the hybrid nature of the reanalyses and the neglect of the contribution of the vertical wind terms (the minimum reanalysis scale is 166 km, which is much larger than the atmospheric thickness). Let us recall that the classical (isotropic) spectral energy transfer $T(k, t)$ is defined by the triple velocity correlations coming from nonlinear interactions and which satisfy:

$$\left(\frac{\partial}{\partial t} + 2\nu k^2\right) E(k, t) = T(k, t) \tag{2.86}$$

where $E(k, t)$ is the spectrum as a function of time (see e.g. Lesieur, 1987). The same equation holds before averaging over all wave-vector directions (i.e. without hypothesizing isotropy). In this case consider respectively $\langle v^2 \rangle = u(0)$ and the spectral transfer $T(\underline{k}, t)$ that depends on the vector $\underline{k}(\tilde{u}(\underline{k}, t)$ is the Fourier transform of u: Appendix 2A). The relationship between the two transfers is merely:

$$T(k) = \int_{|\underline{k}'|=k} T(\underline{k}')d\underline{k}' \tag{2.87}$$

(i.e. angle integration). Because the isobaric surfaces are orthogonal to the pressure gradients, the spectral transfer for the "horizontal" velocity along the isobars is somewhat simpler than its more general expression on isoheights (advection is then the unique nonlinear term in the Navier–Stokes equations). One can show (Appendix 2C) that in this case, written out explicitly and considering only the horizontal transfer:

$$\begin{aligned} T(\underline{k}) = 2k_x\, Im\left[\langle \widetilde{u}^*(\widetilde{u^2})\rangle + \langle \widetilde{v}^*(\widetilde{uv})\rangle\right] \\ +2k_y\, Im\left[\langle \widetilde{u}^*(\widetilde{uv})\rangle + \langle \widetilde{v}^*(\widetilde{v^2})\rangle\right] \end{aligned} \tag{2.88}$$

where we have used the notation $u = v_x$, $v = v_y$ (zonal and meridional components). The vertical wind was ignored because in the reanalyses the vertical scales are so much smaller than the smallest horizontal scale (here 1.5°, i.e. 166 km).

Fig. 2.22a shows $sgn(T(\underline{k}))$ when sections of the reanalyses are used in the calculations. Each 700 mb field was broken into sections from ± 45° latitude (this avoids strong distortions from the map projection), and four disjoint 90° longitudinal sections. It can be seen by comparing the results for the different years that the distribution is not at all isotropic, and that much of the details of the anisotropy persist from one year to another (note that standard Hann windowing techniques were used for the numerical Fourier transforms). Fig. 2.22b shows that over half of the wavevectors agree on the sign for each of the three years, but that the region of agreement has a highly complex fractal-like structure. In order to investigate further, we calculated the classical transfer, T (Eqn. (2.87)).

Fig. 2.23a shows the result: the sign oscillates every octave or so in scale, in a largely reproducible way from year to year, yet there is no obvious 2D/3D transition, nor source/sink-free inertial range. At the largest octave or so in scales (smallest k), the transfer is positive (from large to small). Finally, we can calculate the total flux from scales larger than k^{-1} to smaller scales:

$$\Pi(k) = \int_k^\infty T_i(k')dk' \tag{2.89}$$

This is shown in Fig. 2.23b. Although the integration naturally smoothes out some of the oscillations present in $T(k)$, there is still no obvious pattern, with the overall direction/sign changing every factor of 4–5 in scale. Note that here, as for T, at the very high wavenumbers the hyperviscous effects mean that the transfer is poorly estimated and should be ignored. In both Figs. 2.23a and 2.23b we have superposed the envelope defined by the isotropic energy spectrum (see the discussion of this and the slightly different spectrum calculated in Chapter 6 from the 2006 reanalyses). This comparison shows that the magnitude of the transfer closely follows the spectrum itself. Other attempts to test the direction of the cascade using aircraft estimates of the sign of the third-order velocity structure function (Fig. 2.21b) have found similar chaotic/complex variations of cascade direction with scale, much more in accord with a scaling input and output of energy over a wide range.

2.7 Summary of emergent laws in Chapter 2

We derived several scaling laws in dimensional form in both real space and Fourier space (spectra) by using turbulent fluxes linked to the Navier–Stokes equations in three dimensions (the Kolmogorov–Obukhov law, Eqn. (2.90), top) and two dimensions (the Kraichnan law, middle), and linked to the equation of passive scalar advection (the Corrsin–Obukhov law, bottom):

$$\Delta v = \varepsilon^{1/3}|\underline{\Delta r}|^{1/3}; \quad E_v(k) = \varepsilon^{2/3}k^{-5/3}$$
$$\Delta v = \eta^{1/3}|\underline{\Delta r}|; \quad E_v(k) = \eta^{2/3}k^{-3}$$
$$\Delta \rho = \chi^{1/2}\varepsilon^{-1/6}|\underline{\Delta r}|^{1/3}; \quad E_\rho(k) = \chi\varepsilon^{-1/3}k^{-5/3} \tag{2.90}$$

Each equation depends on a turbulent flux: the energy flux (3D, ε), the enstrophy flux (2D, η) and the passive scalar variance flux (χ), respectively. For the moment, each is considered to be classical, i.e. the fluxes are $\approx$ constant (spatially homogeneous, quasi-Gaussian, nonintermittent). To emphasize that the above assumes isotropy, we have replaced the spatial lag Δx used in the earlier sections by the modulus of the vector lag $\underline{\Delta r}$.

The general form of these laws is:

$$\Delta f = \varphi|\underline{\Delta r}|^H \tag{2.91}$$

for fluctuations Δf in a turbulent field f, flux φ and exponent H. This is the prototypical emergent atmospheric law. In this classical form it is valid only for weakly variable fluxes and for statistical isotropy, assumptions which the pioneers doubted would allow them to be valid in the atmosphere over scales much larger than several hundred metres. However, the rest of this book shows how to generalize φ to a highly variable multifractal cascade process and for the isotropic vector norm scale notion$|\underline{\Delta r}|$ replaced by an anisotropic scale function$\|\underline{\Delta r}\|$. Finally, we assumed that the fluctuation was simply a difference, but this is not always adequate – so that, for example, when H is outside the range 0–1, the notion of fluctuation Δf itself can be refined with the help of wavelets (Section 5.5).

The general relations between real-space fluctuations and Fourier-space spectra are obtained using the Wiener–Khinchin theorem (see Appendix 2A for a demonstration), and the specific relations that apply in scaling systems are obtained using Tauberian theorems. The Wiener–Khinchin theorem is valid for statistically stationary processes and relates the autocorrelation $R(\tau)$ to the power spectrum $E(\omega)$:

$$R(\tau) = \left\langle v(t)v(t-\tau)\right\rangle = \int_{-\infty}^{\infty} d\omega e^{i\omega t}E(\omega) = 3(s(x+Dx)$$
$$-3s(x+2Dx/3) + 3s(x+Dx/3) - s(x))/DxE(\omega) \tag{2.92}$$

$R(\tau)$ is also related to the "Δ-variance" of the fluctuation, i.e. the second-order structure function $S_2(\tau)$:

$$S_2(\tau) = \left\langle \Delta v(\tau)^2\right\rangle = \left\langle \left(v(t) - v(t-\tau)\right)^2\right\rangle = 2\left[\left\langle v(t)^2\right\rangle - \left\langle v(t)v(t-\tau)\right\rangle\right] \tag{2.93}$$

or in terms of the power spectrum:

$$S_2(\tau) = 2\left(R(0) - R(\tau)\right) = 2\int_{-\infty}^{\infty} d\omega E(\omega)(1-e^{i\omega t})$$
$$= 3(s(x+Dx) - 3s(x+2Dx/3) + 3s(x+Dx/3) - s(x))/Dx\Big) \tag{2.94}$$

In scaling regimes, we have power law spectra $E(\omega) \approx \omega^{-\beta}$, and we can use a Tauberian theorem to conclude that $R(\tau)$, $S_2(\tau)$ also follow power laws:

$$E(\omega) \approx \omega^{-\beta} \Leftrightarrow R(\tau) \approx \tau^{\xi(2)}; \quad \beta < 1 \tag{2.95}$$

where the condition $\beta < 1$ is needed for low-frequency convergence. A high-frequency (small-scale) cutoff is also needed, but one is always present in discretely sampled data. For the structure function:

$$S_2(\tau) = \left\langle \Delta v(\tau)^2 \right\rangle \approx \tau^{\xi(2)}; 1 < \beta < 3 \qquad (2.96)$$

and in both cases $\beta = -1 + \xi(2)$. For the nonintermittent fields considered in this chapter, $\xi(2) = 2H$, so that $\beta < 1$ corresponds to $\xi(2) < 0$, $H < 0$. The condition $1 < \beta < 3$ for S_2 is for both high- and low-frequency convergence and corresponds to $0 < \xi(2) < 2$, $0 < H < 1$, which is the basic range for atmospheric fields in the weather regime, so that structure functions are generally more useful than autocorrelation functions. However, in the macroweather regime, $\beta < 1$, ($H < 0$), so that it is preferable to use structure functions based on different types of fluctuation defined by wavelets (see Section 5.5).

Appendix 2A: Spectral analysis in arbitrary dimensions

In this appendix we generalize the 1D results of Box 2.2 to higher dimensions, considering space rather than time. Consider the second-order velocity correlation tensor:

$$u_{ij}(\underline{r}) = \langle v_i(\underline{r}')v_j(\underline{r}'+\underline{r})\rangle = \langle v_i(\underline{r}')v_j(\underline{r}'-\underline{r})\rangle \quad (2.97)$$

(the symmetry under inversion follows from translational invariance; this can be a function of time, but we will not denote this explicitly). We will go on assuming statistical homogeneity, i.e. independence of translation when it applies to translation in time and space. Furthermore, we will also assume that the turbulence is statistically isotropic (independent of direction). Then we have $u_{ij}(\underline{r}) = u_{ji}(\underline{r})$ and $u_{ij}(\underline{r}) = u_{ji}(r)$, where $r = |\underline{r}|$. We can define $u(r) \equiv u_{ii}(r)$ the trace of the velocity correlation tensor (using Einstein's notation convention for summing over a repeated index) and the average energy per unit mass is thus:

$$e = \frac{1}{2}\left\langle |\underline{v}(0)|^2 \right\rangle = \frac{1}{2}u(0) \quad (2.98)$$

(by spatial homogeneity, there is no $\underline{r}$ dependence). Introducing the d-dimensional Inverse Fourier transform:

$$u(\underline{r}) = \int d^d\underline{k}\, e^{i\underline{k}\cdot\underline{r}}\tilde{u}(\underline{k}) \quad (2.99)$$

we obtain:

$$\tilde{u}(0) = \int d^d\underline{r}\, u(\underline{r}) \quad (2.100)$$

(using the inverse Fourier transform and setting $\underline{k} = 0$). We now wish to exploit the isotropy by performing the d-dimensional Fourier-space integral above over (d – 1)-dimensional "annuli" or "shells." We obtain:

$$e = \int_0^\infty dk\, E(k) \quad (2.101)$$

where e is the total energy per unit mass, $E(k) \sim k^{d-1}\tilde{u}(k)$ is the (isotropic) "energy spectrum" and $k = |\underline{k}|$ (in one dimension the integral is $\int u dk$, in two dimensions $\int u2\pi k$dk, in three dimensions $\int u4\pi k^2 dk$).

Consider $\tilde{\underline{v}}(\underline{k})$, the Fourier transform of $\underline{v}(\underline{r})$, then the inverse transform gives:

$$\underline{v}(\underline{r}) = \int d^d\underline{k}\, e^{i\underline{k}\cdot\underline{r}}\tilde{\underline{v}}(\underline{k}) \quad (2.102)$$

This implies that $\tilde{v}(k) = \tilde{\underline{v}}^*(-k)$, which follows the fact that $\underline{v}(\underline{r})$ is real, and:

$$\begin{aligned} u(\underline{r}) &= \left\langle \underline{v}(\underline{r}')\cdot\underline{v}(\underline{r}'+\underline{r})\right\rangle \\ &= \int d^d\underline{k}d^d\underline{k}' e^{i\underline{k}\cdot\underline{r}}e^{i(\underline{k}+\underline{k}')\cdot\underline{r}'}\left\langle \tilde{\underline{v}}(\underline{k})\cdot\tilde{\underline{v}}(\underline{k}')\right\rangle \end{aligned} \quad (2.103)$$

Similar expressions are obtained for u_{ij} with the help of the tensor product instead of the scalar product. Now, statistical homogeneity means that the right-hand side is independent of $\underline{r}'$). This implies that the only contribution to the double integral is from $\underline{k}$ = –$\underline{k}'$, hence:

$$\left\langle \tilde{\underline{v}}(\underline{k})\cdot\tilde{\underline{v}}(\underline{k}')\right\rangle = P(k)\delta(\underline{k}+\underline{k}') \quad (2.104)$$

This defines the spectral density of $P(\underline{k})$ and shows that a statistically homogeneous field can be represented as the integral over statistically independent pairs of waves with wavenumber vectors $\underline{k}$, –$\underline{k}$, and with random amplitudes $\tilde{\underline{v}}(\underline{k})$. Using this result we obtain a d-dimensional Wiener–Khinchin theorem:

$$u(\underline{r}) = \int d^d\underline{k}\, e^{i\underline{k}\cdot\underline{r}}P(k) \quad (2.105)$$

relating the autocorrelation function of a stationary process to its harmonic representation via a Fourier transform. Putting $\underline{r} = 0$ shows:

$$u(0) = \int d^d\underline{k}P(k) \quad (2.106)$$

and using isotropy (and ignoring constant factors such as 4π):

$$u(0) = \int_0^\infty dkE(k) = \int_0^\infty dkk^{d-1}P(k) \tag{2.107}$$

Hence if we attribute an energy $\frac{1}{2}P(k)$ to each wavenumber $\underline{k}$ then the total energy in Fourier space equals $\frac{1}{2}u(0)$, which is the energy per unit mass. We also see immediately that:

$$E(k) = k^{d-1}\frac{1}{2}\left\langle |\tilde{\underline{v}}(k)|^2 \right\rangle \tag{2.108}$$

Hence if the d-dimensional spectral density $P(\underline{k})$ is:

$$P(\underline{k}) \approx |\underline{k}|^{-s} \tag{2.109}$$

then:

$$E(k) = k^{-\beta}; \quad k = |\underline{k}|; \beta = s + 1 - d \tag{2.110}$$

This is a relation which will prove useful later. Concerning the enstrophy spectrum, we now repeat the above arguments, but for $\langle \nabla^2 \underline{v} \rangle$ (recalling that in Fourier space $\nabla^2 \to -k^2$ and using Eqn. (2.62)):

$$\langle |\underline{\omega}|^2 \rangle = \langle \underline{v} \cdot \nabla^2 \underline{v} \rangle = \int d^d\underline{k}\, k^2 \langle |\tilde{\underline{v}}(\underline{k})|^2 \rangle \tag{2.111}$$

and integrating as usual over angles in Fourier space:

$$\langle |\underline{\omega}|^2 \rangle = \int_0^\infty dk\, k^2 E(k) \tag{2.112}$$

Hence:

$$E_\omega(k) = k^2 E(k) \tag{2.113}$$

Appendix 2B: Cascade phenomenology and spectral analysis

One of the properties of turbulence to which we appeal to justify the cascade model developed in the following chapters is that the dynamical interactions are strongest between structures whose sizes are nearly the same. This means that for the energy flux to pass from a large to a small eddy/structure it must pass through numerous intermediate steps: large structures don't spontaneously break up into numerous small ones but instead pass energy flux from one scale to another in a cascade-like manner. The development below is close to Rose and Sulem (1978) and shows simply that on condition that $1 < \beta < 3$, the main contribution to the dynamically significant v_n across structures of size l_n is from wavenumbers in the octave near wavenumber $1/l_n$.

Following Section 2.4, consider the dynamically significant velocity gradient. We will express v_n in terms of $E(k)$:

$$v_n{}^2 = \left\langle |\Delta v(\underline{l}_n)|^2 \right\rangle = \left\langle |v(\underline{r}) - v(\underline{r} + \underline{l}_n)|^2 \right\rangle \quad (2.114)$$

$$= 2\left\{ \langle v^2 \rangle - \langle v(\underline{r}) v(\underline{r} + \underline{l}_n) \rangle \right\} \quad (2.115)$$

The first term, $\langle v^2 \rangle = u(0)$ is the total energy, $\int_0^\infty dp\, E(p)$ (we will not worry about constant factors such as π etc.). The second term is just the trace of the velocity correlation tensor $u(l_n) = u_{ii}(l_n)$. Now:

$$u(\underline{l}_n) = \int d\underline{p}\, e^{i\underline{p}\cdot\underline{l}_n}\, \widetilde{u}(\underline{p}) \quad (2.116)$$

but in our case (isotropic turbulence) $l_n = |\underline{l}_n|$ and $\underline{p}\cdot\underline{l}_n = pl_n cos\theta$, where θ is the angle between $\underline{p}$ and $\underline{l}_n$. Hence:

$$u(l_n) = \int_0^\infty dp\, E(p) \int_\Omega d^{d-1}\Omega e^{ipl_n\cos\theta} \quad (2.117)$$

where Ω is the (solid) angle in Fourier space. In spherical polar coordinates (θ, ϕ) $(d=3)$, we have $d^{d-1}\Omega = \cos\theta\, d\theta\, d\phi$. Then we have:

$$v_n{}^2 = \langle |\underline{\Delta} v(l_n)|^2 \rangle = \int_0^\infty dp\, E(p)(1 - \int_\Omega d^{d-1}\Omega e^{ipl_n\cos\theta})$$
$$= \int_0^\infty dp\, E(p) \int_\Omega d^{d-1}\Omega(1 - e^{ipl_n\cos\theta}) \quad (2.118)$$

where we have used the fact that the normalization has been defined so that:

$$\int_\Omega d^{d-1}\Omega = 1 \quad (2.119)$$

To estimate this integral in Eqn. (2.118), we use $k_n = \frac{2\pi}{l_n}$ and divide the range of integration into three parts:

(I) $0 \le p \le \frac{k_n}{\sqrt{2}}$ (low frequency);

(II) $\frac{k_n}{\sqrt{2}} \le p \le \sqrt{2}$ (medium frequency);

(III) $\sqrt{2}k_n \le p < \infty$ (high frequency).

We will now consider each case, starting with the limiting cases I, III.

Term (I)

pl_n is small, i.e. $pl_n \to 0$ and discarding all imaginary parts in first order term of p (since we know a priori that the integral must be real) we are left (ignoring constant factors) with second-order terms $O((pl_n)^2)$:

$$\int_0^{k_n/\sqrt{2}} dp\,(pl_n)^2 E(p) \sim l_n^2 \int_0^{k_n/\sqrt{2}} dp\,p^2\,E(p) \sim l_n^2\langle|\underline{\omega}|^2\rangle \tag{2.120}$$

where we have used $E_\omega(p) = p^2E(p)$. This is the large-eddy contribution to v_n^2. This result can be understood physically in the following way. The effect of large-scale vorticity is to produce a nearly constant velocity gradient across the eddy, and the velocity difference will be approximated by $l_n\omega(p)$ (since $\underline{\omega} = \nabla \times \underline{v}$ and we are interested in a "typical" gradient), hence the mean squared difference will be:

$$l_n^2\langle|\underline{\omega}|^2\rangle = l_n^2 \int_0^{k_n/\sqrt{2}} dp\,p^2\,E(p) \tag{2.121}$$

Term (III)

This is the small-eddy contribution. $pl_n \to \infty$:

$$\int_\Omega d^{d-1}\Omega(1 - e^{ipl_n\cos\theta}) \to 1 \tag{2.122}$$

since the exponential will oscillate very rapidly and will yield zero on average. So the contribution to v_n^2 due to small structures is

$$\int_{\sqrt{2}k_n}^{\infty} dp\,E(p) \tag{2.123}$$

where the contributing wavenumbers are greater than $\sqrt{2}k_n$. The physical interpretation is that the small-scale eddies cause the boundary of l_n-scale eddy to execute a highly convoluted random walk. In the mean, the effect is diffusive. The diffusion constant depends on the mean square velocity of all the contributing eddies, which is:

$$\int_{\sqrt{2}k_n}^{\infty} dp\,E(p) \tag{2.124}$$

Term (II)

Take

$$E_n \equiv \int_{k_n/\sqrt{2}}^{\sqrt{2}\,k_n} dp\,E(p)\int_\Omega d^{d-1}\Omega = (1 - e^{ipl_n\cos\theta}) \tag{2.125}$$

as the definition of energy in band n. We will investigate under which conditions this term is the main contribution to v_n^2. When it dominates terms I and III, the energy spectrum is termed "local," since most of the contribution to the dynamically significant quantity v_n^2 is due to structures with neighbouring wavenumbers; otherwise, it is "nonlocal." The final expression for $v_n^2\big((\mathrm{I}) + (\mathrm{II}) + (\mathrm{III})\big)$ is:

$$v_n^2 \approx l_n^2 \int_0^{k_n/\sqrt{2}} dp\,p^2\,E(p) + E_n + \int_{\sqrt{2}k_n}^{\infty} dp\,E(p) \tag{2.126}$$

Due to the scaling, the dominant behaviour of the spectrum will be a power law. We now consider how the value of the scaling exponent affects the relative value of various terms. Considering $E(p) \sim p^{-\beta}$ (ignoring constant factors) then (I) becomes:

$$l_n^2\int_0^{k_n/\sqrt{2}} dp\,p^2\,p^{-\beta} = l_n^2\,p^{3-\beta}\Big|_0^{k_n/\sqrt{2}} \begin{cases} \sim l_n^2k_n^3 & \text{for } \beta < 3 \\ \to \infty & \text{for } \beta > 3 \end{cases} \tag{2.127}$$

When $\beta \geq 3$ then the term diverges – this low-frequency divergence is called an "infrared catastrophe" and indicates that the spectrum is dominated by low frequencies – it will be nonlocal.

Term (III) becomes:

$$\int_{\sqrt{2}k_n}^{\infty} dp\,p^{-\beta} \sim p^{1-\beta}\Big|_0^{\infty} \begin{cases} \sim k_n^{1-\beta} & \text{for } \beta > 1 \\ \to \infty & \text{for } \beta < 1 \end{cases} \tag{2.128}$$

Hence if $\beta < 1$ the term diverges, we have an "ultraviolet catastrophe," and again the spectrum is nonlocal, this time due to dominance of the higher frequencies.

We can now conclude that if $1 < \beta < 3$, all the terms are dominated by the contributions from wavenumbers near $k_n = \frac{2\pi}{l_n}$, and hence the spectrum will be local. Now as long as $\beta > 1$, term III is negligible and the sum of terms I and II can be approximated by:

$$v_n^2 \approx l_n^2\int_0^{k_n} dp\,p^2\,E(p); \quad \beta > 1 \tag{2.129}$$

(we are interested in an order-of-magnitude estimate only; the angular integration will give a constant correction to the above of order unity).

When viscosity is negligible, the only way to define a quantity with dimensions of time is as follows:

$$\tau_n = \frac{l_n}{v_n} \sim \left(\int_0^{k_n} dp\, p^2 E(p) \right)^{-1/2} \tag{2.130}$$

which is an estimate of the eddy turnover time.

Appendix 2C: Spectral transfers

To calculate the spectral transfers needed for Section 2.6.6, we start with the equations of hydrodynamic turbulence without assuming incompressibility (here we roughly follow Davidson, 2004):

$$\frac{\partial u_i}{\partial t} = -\frac{\partial(u_i u_j)}{\partial x_j} - \frac{1}{\rho}\frac{\partial p}{\partial x_i} + \nu\frac{\partial^2 u_i}{\partial x_j \partial x_j} \tag{2.131}$$

$$\frac{\partial u'_i}{\partial t} = -\frac{\partial(u'_i u'_j)}{\partial x'_j} - \frac{1}{\rho'}\frac{\partial p'}{\partial x'_i} + \nu\frac{\partial^2 u'_i}{\partial x'_i \partial x'_j} \tag{2.132}$$

with:

$$\underline{u} = \underline{u}(\underline{x});\ \underline{u}' = \underline{u}(\underline{x}');\ \underline{x}' = \underline{x} + \underline{r} \tag{2.133}$$

(as usual, we sum over repeated indices). Multiply the first by u_i' and the second by u_i then add and take ensemble averages, we obtain:

$$\begin{aligned}\frac{\partial\langle u_i u_i'\rangle}{\partial t} = &-\left\langle u_i \frac{\partial(u_i' u_j')}{\partial x'_j} + u_i' \frac{\partial(u_i u_j)}{\partial x_j}\right\rangle \\ &-\left\langle \frac{1}{\rho'} u_i \frac{\partial p'}{\partial x_i'} + \frac{1}{\rho} u_i' \frac{\partial p}{\partial x_i}\right\rangle \\ &+\nu\left\langle u_i \frac{\partial^2 u_i'}{\partial x_j' \partial x_j'} + u_i' \frac{\partial^2 u_i}{\partial x_j \partial x_j}\right\rangle\end{aligned} \tag{2.134}$$

Now, the nonlinear transfer (the first term on the right) is the usual term to worry about. The pressure terms vanish in isotropic turbulence; they also vanish along isobars even in nonisotropic turbulence (assuming that we only calculate the "horizontal" isobaric components: $\frac{\partial p}{\partial x_i} = \frac{\partial p'}{\partial x_i} = 0$ for x_i any isobaric component). Finally, the dissipation terms are only important at very small scales.

To estimate the nonlinear term we can use Fourier techniques. For any two real fields A, B:

$$\begin{aligned}A(\underline{x}) &= \int e^{i\underline{k}\cdot\underline{x}}\widetilde{A(\underline{k})}d\underline{k} = \int e^{-i\underline{k}\cdot(\underline{x}'-\underline{r})}\widetilde{A^*(\underline{k})}d\underline{k}; \\ B(\underline{x}') &= \int e^{i\underline{k}'\cdot(\underline{x}+\underline{r})}\widetilde{B(\underline{k}')}d\underline{k}'\end{aligned} \tag{2.135}$$

The far-right top equality is obtained by taking complex conjugates of both sides and using the fact that A is real.

Taking products and averages, we obtain:

$$\langle A(\underline{x})B(\underline{x}')\rangle = \iint e^{i\underline{x}'(-\underline{k}+\underline{k})}e^{i\underline{r}.\underline{k}}\langle\widetilde{A^*(\underline{k})}\widetilde{B(\underline{k}')}\rangle d\underline{k}d\underline{k}' \tag{2.136}$$

Using the assumption of statistical translational invariance, we require that $\underline{k} = \underline{k}'$ so that:

$$\langle A(\underline{x})B(\underline{x}')\rangle = \int e^{i\underline{r}.\underline{k}'}\langle\widetilde{A^*(\underline{k})}\widetilde{B(\underline{k})}\rangle d\underline{k} \tag{2.137}$$

Hence we see that $\langle A(\underline{x})B(x')\rangle$ and $\langle\widetilde{A^*(\underline{k})}\widetilde{B(\underline{k})}\rangle$ are Fourier transform pairs. We can now apply this to the evaluation of the nonlinear terms. In the first, take $A = u_i$; $B = \frac{\partial(u_i u_k)}{\partial x_k}$ so that:

$$\begin{aligned}F.T.\left\langle u_i \frac{\partial(u_i' u_j')}{\partial x_j'}\right\rangle &= \left\langle \widetilde{u_i^*}(ik_j\widetilde{u_i u_j})\right\rangle \\ F.T.\left\langle u_i' \frac{\partial(u_i u_j)}{\partial x_j}\right\rangle &= \left\langle \widetilde{u_i}(ik_j\widetilde{u_i u_j})^*\right\rangle\end{aligned} \tag{2.138}$$

we see that these are complex conjugates, so that the usual transfer is given by:

$$\begin{aligned}T(\underline{k}) &= F.T.\left[-\left\langle u_i \frac{\partial(u'_i u'_j)}{\partial x'_j} + u'_i \frac{\partial(u_i u_j)}{\partial x_j}\right\rangle\right] \\ &= 2k_j Im[\widetilde{u_i^*}(\widetilde{u_i u_j})]\end{aligned} \tag{2.139}$$

Chapter

3 Scale-by-scale simplicity: an introduction to multiplicative cascades

3.1 Cascades as conceptual models

Although the idea of cascades in atmospheric dynamics is conventionally traced back to the famous poem buried in page 66 of Richardson's book on *Weather Prediction by Numerical Process* (cited in Section 1.1.1), for a long time cascade models were mostly inspirational. It was not until the 1960s that explicit multiplicative cascade models were first developed. Even though they almost certainly played a role when Kolmogorov formulated his famous law in 1941, he did not explicitly mention cascades. This is surprising, because in the same year he published a model of rock fragmentation using the allied idea of multiplicative random variables (Kolmogorov, 1941a). The cascade idea was quite explicit, however, in the one-paragraph abstract by Onsager (1945), where he apparently independently discovered the Kolmogorov law:

> The modulation of a given Fourier component of the motion is mostly due to those others which belong to wavenumbers of comparable magnitude ... It has not been pointed out before that the subdivision of the energy must be a stepwise process, such that an n-fold increase of the wavenumber is reached by a number of steps of the order log n. For such a cascade mechanism that part of the energy density which is associated with large wavenumbers should depend on the total volume rate of dissipation ε only. Then dimensional considerations require that the energy per component of wavenumber $k \ldots \approx \varepsilon^{2/3}k^{-11/3} \ldots$
>
> (Onsager, 1945)

(Onsager's spectrum is three-dimensional, i.e. $s = 11/3$ so that $\beta = 5/3$: see Eqn. (2.110)).

Even much later, in the presentation of the log-normal model of turbulent intermittency, cascades are not (quite) explicitly mentioned:

> The hypotheses concerning the local structure of turbulence at high Reynolds number, developed in the years 1939–41 by myself and Obukhov [Kolmogorov, 1941b, 1941c, 1941d; Obukhov, 1941a, 1941b] were based physically on Richardson's idea of the existence in the turbulent flow of vortices on all possible scales $l < r < L$ between the external scale L and the internal scale l and of a certain uniform

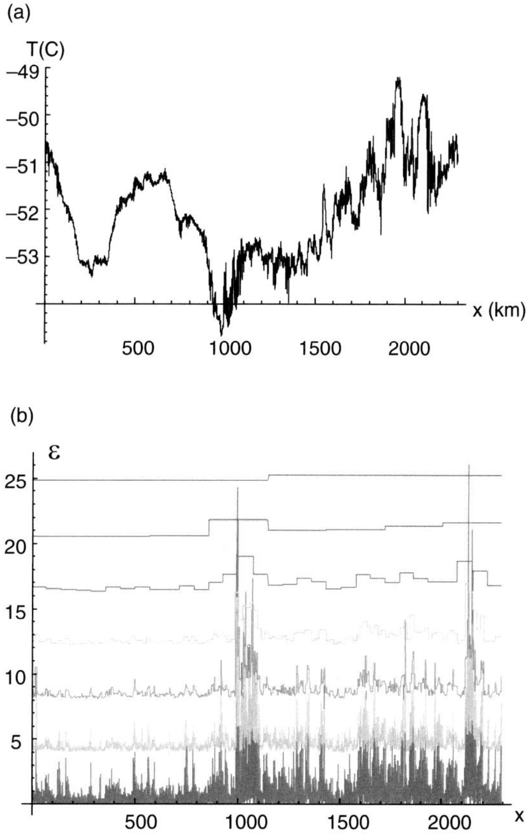

Fig. 3.1 (a) The first 8192 points of the temperature series measured by a Gulfstream 4 flight over the Pacific ocean at 196 mb and at resolution of 1 s corresponding (to within ± 2%) to a spatial resolution of 280 m. (b) The first 8192 points of the flux estimated from the temperature series in Fig. 3.1a by using absolute second differences. The curves starting at the bottom are the normalized energy fluxes. Curves bottom to top are successively degraded by factors of 4 and displaced by 3 units in the vertical for clarity. It turns out that the largest value is nearly the same as that predicted theoretically in Chapter 5 (Section 5.2.3).

(a)

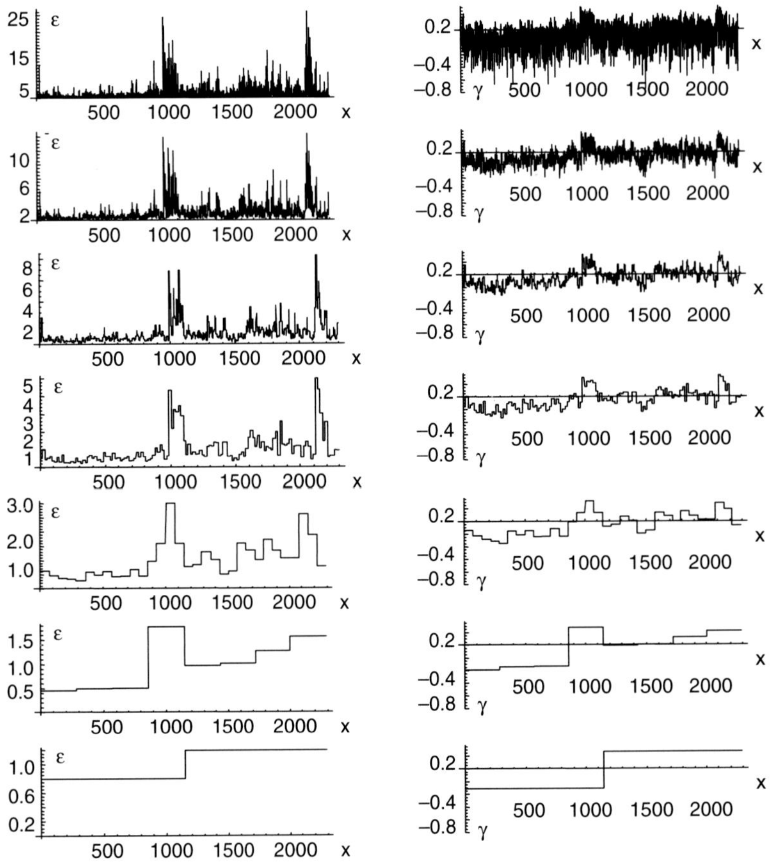

Fig. 3.2 (a) The left column, top to bottom, is the increasingly degraded version of the flux (ε) for the temperature, and the right column is the same but in terms of the singularities: the transformation of variables $\gamma = \log\varepsilon/\log\lambda$ where ε is the normalized temperature flux (estimated from the second finite difference derivative.) The horizontal axis *x* is in km, resolution is 280 m. (The left column is the same as Fig. 3.1b except that the order is top to bottom rather than bottom to top.) Notice that while the range of variation of the ε's rapidly diminishes as the resolution is lowered, on the contrary the amplitude of the fluctuations of the γ's is roughly the same at all scales. A refined analysis of this type of transform and resulting graph can be used to show that this flux is singular (Schertzer *et al.*, 2010).

> mechanism of energy transfer from the coarser-scaled vortices to the finer.
>
> (Kolmogorov, 1962)

In any case, modern explicit cascade models are usually traced back to the "pulse in pulse" model (Novikov and Stewart, 1964), which is close to the β model described below.

The need for explicit cascade models arose because of the problem of turbulent intermittency: the sudden transition from quiescence to chaos which is ubiquitous in turbulent flows, including in the atmosphere. Not only is most of the atmosphere's activity (including the energy, moisture and other transfers) confined to a small fraction of the total volume – mostly storms and other violent events – but even within the latter, they are far from uniform. Fig. 3.1a graphically illustrates this problem with a transect of aircraft temperature data over the distance 280×2^{13} m $\approx$ 2300 km. In Fig. 3.1b, we show the turbulent flux estimated from it by the absolute second finite differences of the temperature (denoted ε; this is the analogue of the energy flux but for the temperature field rather than the velocity field). The flux has been normalized by its mean, and at the bottom (the full resolution) we see that there are occasional large spikes; indeed here the largest normalized flux is $\varepsilon = 26.07$. This corresponds to 16.4 standard deviations, so we immediately see that the transect is far from Gaussian. Moving from bottom to top, the figure illustrates the effect of lowering the resolution by averaging over longer and longer intervals.

The same sequence of degraded ε is shown in Fig. 3.2a (left). This is to be compared to the right-hand column, which is the same but in terms of the singularities γ obtained by the transformation of variables $\gamma = \log\varepsilon/\log\lambda$ where λ is the scale ratio $= L/l$ where L is length of the transect (= 8192 pixels) and l is resolution scale (which increases by factors of 4 from 1 pixel (top) to 4096 pixels (bottom)). Notice

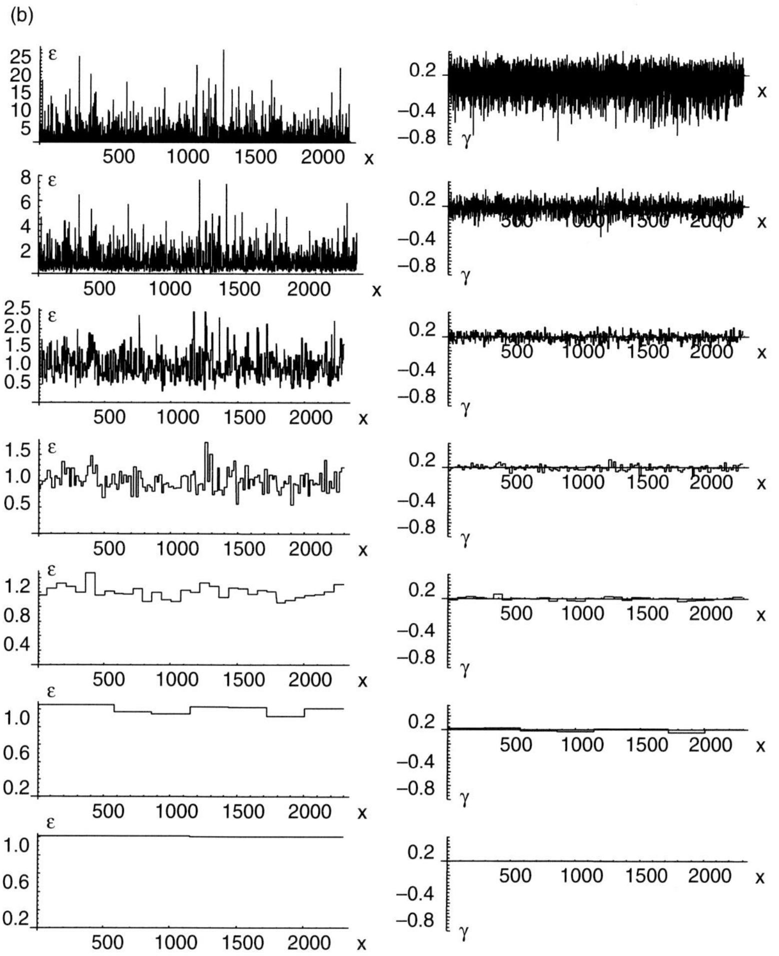

Fig. 3.2 (b) On the left, from top to bottom, the same ε data as in Fig. 3.2a is shown but for the data shuffled/randomized. Note the much more rapid convergence (as compared with Fig. 3.2a) as the series is averaged by factors of 4, top to bottom. On the right, due to the absence of long-range correlations, the range (e.g. the difference between the maximum and minimum values) of the γ's rapidly decreases instead of being roughly constant.

that while the range of variation of the ε's rapidly diminishes as the resolution is lowered, on the contrary the amplitude of the fluctuations of the γ's is roughly the same at all scales, indicating that γ is scale invariant. This result is not trivial; it is a consequence of the long-range correlations in ε. In order to make this clear, in Fig. 3.2b we have shuffled/randomized the same ε transect (top left) and then degraded it (lower left) and compared this with the derived γ transects (right). In this case we see that instead of remaining of roughly constant amplitude the range of variation of the γ's quickly diminishes as we degrade to lower and lower resolutions (bottom right). This visually demonstrates the existence and importance of long-range correlations.

Instead of starting at the bottom of Fig. 3.1b and averaging to lower and lower resolution, we could imagine starting at the top at the lowest resolution (i.e. a line parallel to the x-axis at the constant value 1) and moving downwards by introducing variability at smaller and smaller scales, in this way building up a cascade process. In this chapter we show how to do this by multiplicatively modulating the process at successively smaller and smaller scales.

3.2 Discrete-in-scale multiplicative cascades

3.2.1 The β model: activity on a fractal support

Fig. 3.3 (left-hand side) shows a schematic of a "homogeneous" cascade; a process of this type

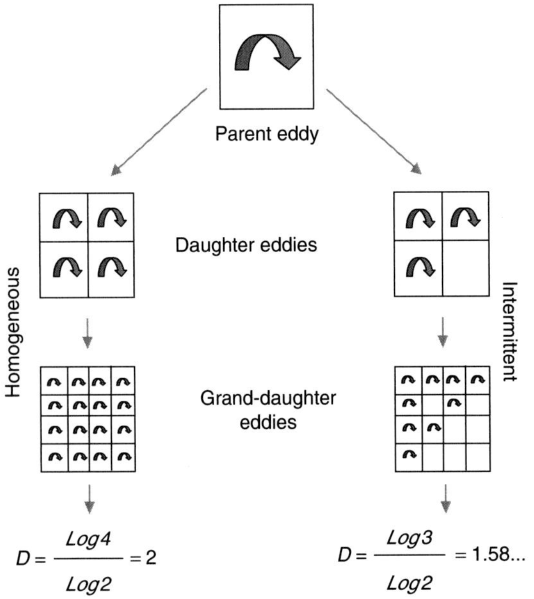

Fig. 3.3 A schematic diagram showing the first few steps in a (discrete in scale) cascade process. At each step, the parent eddy is broken up into 2 × 2 "daughter" eddies, each reduced by a factor of 2 in scale, indicated as squares. The left shows a homogeneous cascade (corresponding to Kolmogorov's 1941 homogeneous turbulence) in which the energy flux is simply redistributed from large to small structures, while keeping its density constant. The right-hand side shows an improvement: "on/off" intermittency is modelled by an "alive/dead" alternative at each step (here only the bottom right sub-eddy becomes dead); the mean conservation of energy flux can be taken into account by boosting the density of the flux in the "active" eddies. For pedagogical reasons, the alternative displayed is purely deterministic, but could be easily randomized (see text). Adapted from Schertzer and Lovejoy (1987).

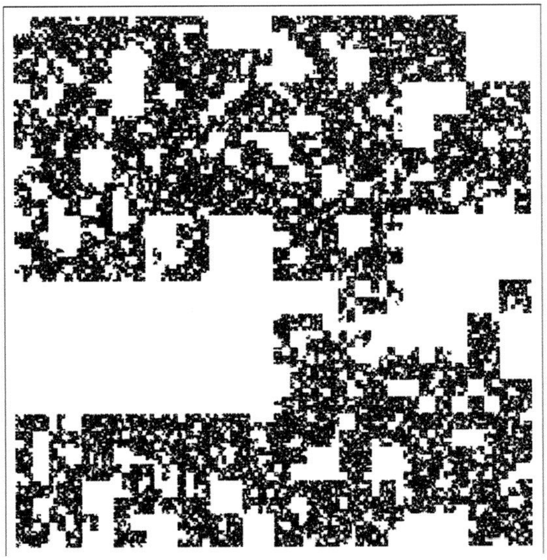

Fig. 3.4 An illustration of four cascade steps of the stochastic "β model." The probability that an eddy will remain alive is $\lambda_0^{-C} = 0.87$ (using the scale ratio at each step $\lambda_0 = 4$ here and the codimension $C = 0.2$). Reproduced from Schertzer and Lovejoy (1987).

may have been Kolmogorov's initial inspiration. One starts with a large-scale structure/"eddy" indicated as square, and this square eddy breaks up, either due to interactions with other eddies or due to internal instability (or both). In this homogeneous version, the conserved quantity (the energy flux density in hydrodynamic turbulence, as discussed in Chapter 2) is simply redistributed from large structures to small; its density remains constant everywhere (i.e. it is homogeneous in the deterministic sense). This homogeneous cascade is clearly a poor model of real turbulence – if only because the latter is highly intermittent; in the words of Batchelor and Townsend (1949), it is "spotty."

An initial attempt to handle intermittency reduces it to the simple notion of "on/off" intermittency, i.e. a cascade with the simple alternative for the offspring of alive/dead. This leads to a confinement of the turbulence to a tiny support, a very small subregion of the flow. The right-hand side in Fig. 3.3 shows the result of such a cascade obtained by multiplying the energy flux of a "mother" eddy to obtain that of the "daughter" eddies either by 0 (dead sub-eddy) or by a positive value (λ_0^c), corresponding to an active sub-eddy, with fixed probability λ_0^{-c}. In the usual, stochastic version of the model, we divide the spatial scales by λ_0 (here $\lambda_0 = 2$) and then flip coins to determine the on or off state; more precisely:

$$\begin{aligned}\Pr(\mu\varepsilon = \lambda_0^c) &= \lambda_0^{-c}\\ \Pr(\mu\varepsilon = 0) &= 1 - \lambda_0^{-c}\end{aligned} \tag{3.1}$$

("Pr" indicates "probability"). The nonzero value is taken as $\mu\varepsilon = \lambda_0^c$ so that the mean $\langle\mu\varepsilon\rangle = 1$; this implies a scale-by-scale conservation of the flux ε.

Since the rule is the same at each level, the process is scale invariant, and in the small-scale limit, the active regions are confined to a fractal set (see e.g. Fig. 3.4). This is essentially the "pulse in pulse" model proposed by Novikov and Stewart (1964), a variant of which was baptized the "β model" in an influential paper by Frisch *et al.*

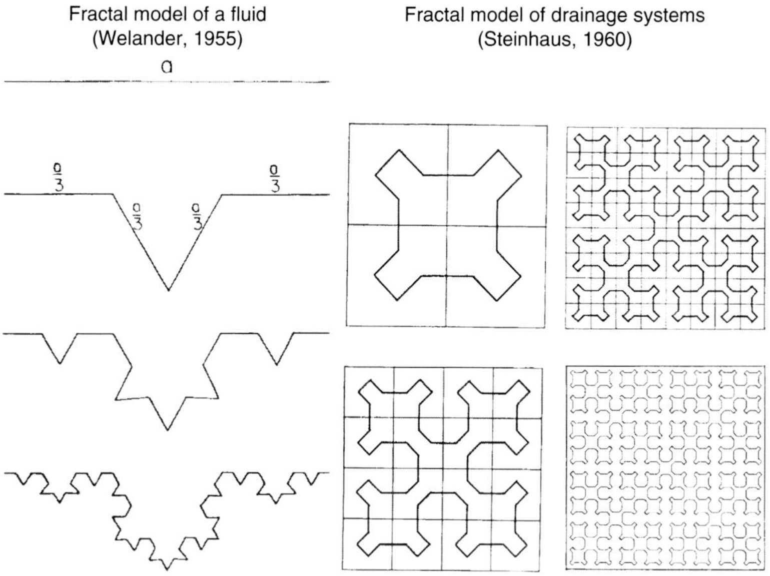

Fig. 3.5 Early examples of geometrical fractal sets applied to geophysical problems, showing the initial stages in construction. Left: the first four steps in the construction of the "Koch" curve (Koch, 1904) reproduced from Welander (1955), who used it as a model of an interface in a turbulent fluid. Each segment is iteratively replaced by a scaled "generator" (shown in the second line). In the limit, the fractal dimension is $D = \log N/\log\lambda_0 \approx 1.25$, where $N = 4$ is the number of segments at each step and $\lambda_0 = 3$ is the scale reduction factor. Right: the first four steps in the construction of a "Peano" curve (Peano, 1890) reproduced from Steinhaus (1960), who gave it as an example of a hydrological drainage network. In this case, each segment is replaced by four others, each reduced by a factor 2 so that $D = \log N/\log\lambda_0 = 2$, implying that the limiting continuous line (topological dimension 1) fills the unit square (dimension 2) so that it ultimately passes through every point in the unit square. Do not be misled by the "necks" which keep the curve from being broken – they do not contribute in the limit.

(1978). The latter paper was actually less explicit, simply specifying a reduction of the volume of active regions by a factor β at each step, hence the name "β model". The active region of the β model is an example of a fractal set since, as we see in the next subsection, its fractal dimension is less than its topological dimension.

3.2.2 Fractal codimensions and dimensions

Let us consider the properties of the β model in more detail. By construction, the set of nonzero points on the β model (the "support") is scale invariant; it is the simplest nontrivial example of scale invariance, a geometric fractal set of points. Again, by construction, at each step in the cascade, the probability that an alive eddy has an alive "daughter" eddy decreases as λ_0^{-c}; after n steps, the total range of scales is $\lambda = \lambda_0{}^n$ and the total probability that a given nth-step eddy is alive is:

$$\Pr(alive) = \left(\lambda_0^{-c}\right)^n = \lambda^{-c} \tag{3.2}$$

Alternatively, we can count the total number of alive eddies. Since at scale ratio λ, λ^{-c} is the fraction of alive eddies, this implies that on average there are:

$$N_{alive} = N_{tot}\Pr = \lambda^d\lambda^{-c} = \lambda^D; \quad D = d - c \tag{3.3}$$

alive daughter eddies, where d = the dimension of the space where the process is developed ($d = 2$ in Fig. 3.4) and $N_{tot} = \lambda^d$ is the total number of intervals ($d = 1$), square boxes ($d = 2$), cubes ($d = 3$) etc. of size λ^{-1}. The exponent $D = d - c$ is the difference between the dimension of the embedding space and the codimension c of the fractal set. In the limit that the process is repeated indefinitely to small scales, and when $D \geq 0$ ($c \leq d$), it is the geometrical fractal dimension of the set of active regions. Note that there is no problem defining the β model cascade process for any $c > 0$, so that when $c > d$ we have $D < 0$ (see Box 3.2 for more discussion of this "latent" dimension "paradox").

The β model effectively approximates turbulence by a geometric fractal set where all the activity is concentrated. We have already seen Richardson's Weierstrass function model of the trajectory of a particle in a turbulent flow (Fig. 2.4); two other early geophysical models using fractal sets are shown in Fig. 3.5: Welander's 1955 model of a turbulent interface, and Steinhaus's 1960 model of a fractal river basin. A little later, Mandelbrot (1967, 1975) proposed geometric fractal models for coastlines and topography.

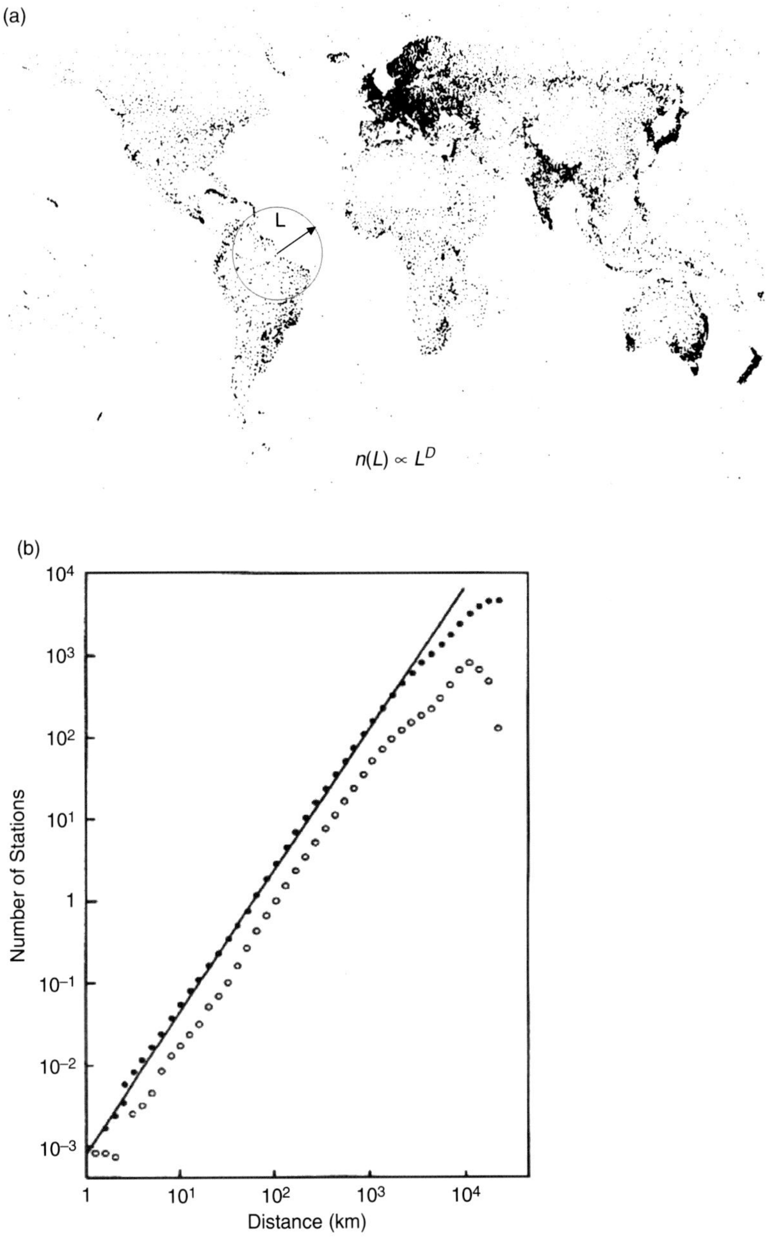

Fig. 3.6 (a) The geographical distribution of the 9962 stations that the World Meteorological Organization listed as giving at least one meteorological measurement per 24 hours (in 1986); it can be seen that it closely follows the distribution of land masses and is concentrated in the rich and populous countries. The main visible artificial feature is the Trans-Siberian Railroad. Also shown is an example of a circle used in the analysis. Adapted from Lovejoy *et al.* (1986). (b) Analysis of the measuring network in Fig. 3.6a: the average number of stations (centred at a station) in a circle radius L. The (logarithmic) slope is the correlation dimension; the line in the plot has $D = 1.75$. The bottom line is the average number between L and $L + \Delta L$, and the top line is the total number within a radius L (ΔL is the spacing between successive annuli; here 10 are taken per order of magnitude in L). The geographical location of the stations was only given to within ~1 km and three pairs were found with nominally the same geographical location, so it is possible that the line extends to even smaller scales. The largest scales are dictated by the finite number of stations available. By inspection, we see that larger and larger circles are clearly more and more filled by holes; this reflects the fact that the number density decreases as L^{-c} with $c = 2 - D = 0.25$. Reproduced from Lovejoy *et al.* (1986).

A fractal set more relevant for atmospheric science is shown in Fig. 3.6a: a meteorological measuring network (Lovejoy *et al.*, 1986; Korvin *et al.*, 1990; Nicolis, 1993; Doswell and Lasher-Trapp, 1997; Mazzarella and Tranfaglia, 2000; Giordano *et al.*, 2006). Although the distribution of measuring stations is a human construct – as can be seen from the figure – it is quite sparse, depending greatly on the distribution of population and wealth, as well as on the distribution of land mass.

In the β model, the probability at a given resolution λ is the same as the fraction or density of points on the fractal at that resolution. In order to empirically demonstrate the fractality of the stations in Fig. 3.6a it suffices to determine the relation between the density of stations and scale. If the density is a power law then its exponent is its codimension; if the latter is > 0, then the set is a fractal set. Alternatively, if the total number of stations as a function of scale is a power law then the exponent is its fractal dimension D. If $D < d$ (the dimension of the embedding space), then $c = d - D > 0$ and again the set is a fractal set (there are also marginal fractal sets when $D = d$, $c = 0$, but we will not pursue this possibility here). This average number is conveniently estimated by determining the average number of stations within a circle radius L (Fig. 3.6a). Fig. 3.6b shows the result using circles centred at given stations. This number is proportional to the number of pairs of stations within a distance L. The resulting exponent (the slope on the log–log plot, Fig. 3.6b) is a "correlation dimension" D_{cor}; in general it is a little smaller than the usual fractal dimension, which is a "box dimension." The box dimension D_{box} would be the result if we performed the same analysis but averaged over circles whose centres were uniformly distributed (rather than being on members of the set, i.e. around stations). D_{box}, D_{cor} are part of a hierarchy of exponents characterizing the density of stations. The relation between them is discussed in Box 5.4; for further analysis of the station density see Tessier *et al.* (1994).

In comparison, the usual method of objective analysis (e.g. "Kriging," "optimal interpolation" or other conventional geostatistical techniques) interpolates atmospheric fields from sparse networks onto regular grids by assuming that the measurements are distributed on a two-dimensional set (the earth's surface) but with "holes" that need filling. For example, taking the area of the earth as $\sim 5 \times 10^8$ km^2, the 9962 stations in Fig. 3.6a would be considered each to represent 5×10^8 km^2/9962 $\approx 5 \times 10^4$ km^2; i.e. the resolution of the network would be taken as the square root: $\approx$ 200 km – but with "holes." Fig. 3.6b shows that this interpretation is incorrect: rather than $d = 2$, the dimension is $D_{cor} \approx 1.75$ with holes at all scales (down to at least 1 km, which was the resolution of the geographical locations of the stations!). Since mathematically two – even large – sets of points ("structures") only intersect each other (with nonzero probability) if the fractal dimension (D) of one of them is greater than the codimension of the other, a new problem arises. For example, active cores of storms (or other regions with $D < 2 - 1.75 = 0.25$) will be so sparse that they will (almost certainly) be missed (will not intersect the network). The reason is that their *dimensional* – not *spatial* – resolution is too low. This intersection theorem is a simple consequence of the fact that codimensions are probability exponents, so the probability of the network (with codimension C_{net}) intersecting a given phenomenon (with C_{phen}) is simply $\lambda^{-C_{net}}\lambda^{-C_{phen}} = \lambda^{-C_\cap}$ and the intersection codimension $C_\cap = C_{net} + C_{phen}$. If $C_\cap > d$, then $D_\cap = d - C_\cap < 0$ and the network is too sparse to measure/intersect the phenomenon (see Box 3.2). (Note added in proof: quantitatively very similar results were obtained by analysing the fractal dimensions of the HadCRUT3 monthly surface temperature climate data set (at $5° \times 5°$ resolution, appendix 10C). The data were also found to be sparse in the time domain with $D_{cor} \approx 0.8$ (i.e. $C_{cor} \approx 0.2$).)

3.2.3 The α model and canonical conservation

By reducing the question of turbulent activity to a geometric "on/off" dichotomy, the β model is not only simple, for most purposes it is simplistic: the more general cascade models described below are necessary to capture the intermediate levels of activity. Indeed, in spite of numerous attempts (many of which, ironically, were inspired by Mandelbrot's seminal works on the *geometry* of fractal sets (1977, 1983)), the last 25 years have shown that most geophysical *fields* must be described statistically, that they *cannot* be reduced to geometry. This is contrary to the case of "monofractal" functions such as the Weierstrass function, or its stochastic generalizations (e.g. fractional Brownian motion), which can be reduced to geometric sets of points by considering the points on their graphs (see Section 5.5).

The β model turns out to be a poor approximation to turbulence, if only because it is unstable under perturbation. As soon as we consider a more realistic alternative to the caricatural dead/alive dichotomy, most of the peculiar properties of the β model are lost. To show this, let us turn to the "α model" (Schertzer and Lovejoy, 1983), so named because of the divergence of moments exponent α that it introduces. In the notation below, we call this exponent q_D where the D emphasizes that it depends on the dimension of space D over which the cascade is averaged: do not confuse this with the Lévy α (used extensively below), nor with the α in the dimension multifractal formalism (Box 5.5).

To see how the α model works, consider constructing it on the unit interval (Fig. 3.8a). At first,

Box 3.1 Box counting and fractal sets, functional box counting and multifractal fields

Eqn. (3.3) relates the number of structures at a given scale to their fractal dimension; it can in fact be taken as the definition of "box-counting dimension," and it can be used to empirically estimate D. Consider the β model (Fig. 3.4): if we distribute disjoint boxes size L (squares in 2D; $L \approx 1/\lambda$) only on those parts which cover at least one point on the set, then the number of the boxes needed for the covering is $N \approx L^{-D}$ where D is the "box-counting" dimension (usually taken as synonymous with the fractal dimension – but more on this in Chapter 5).

Consider what happens if we attempt to characterize a scaling function rather than a (black and white) scaling set of points. Consider a two-dimensional ($d = 2$) field $f(\underline{r})$, and use given thresholds (T) to convert it into (geometrical) exceedance sets S_T, not having in general the same fractal dimension (see Fig. 3.7a for an example with two thresholds): the number of boxes at resolution L, threshold T is $N_T(L)$ and satisfies:

$$N_T(L) \approx L^{-D(T)} \tag{3.4}$$

where $D(T)$ is the box counting dimension for the set S_T.

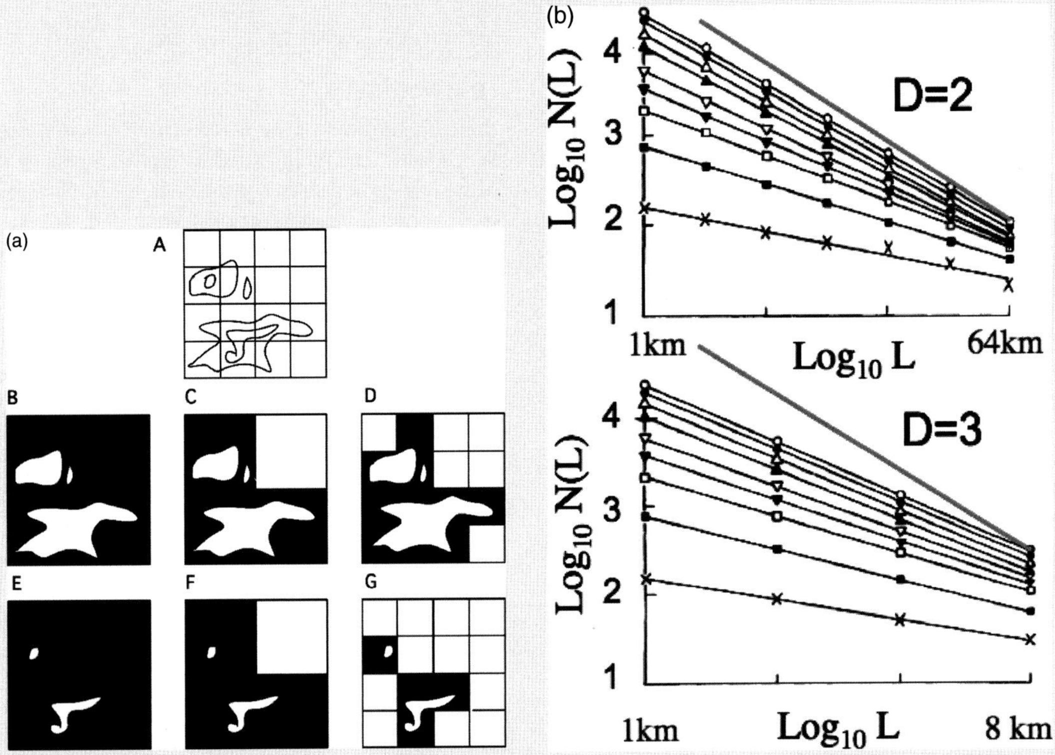

Fig. 3.7 (a) A schematic diagram showing functional box counting. In A (top) the field is covered with a regular grid; two isolines are shown. In B, C, D, a low threshold T_1 is applied so that all the regions $f(\underline{r}) > T_1$ are shown as white; box counting is used to obtain progressively better and better approximations to the set (left to right) and to estimate the box dimension $D(T_1)$ (Eqn. (3.4)). In E, F, G, a higher threshold T_2 and a higher resolution is used and the regions $f(\underline{r}) > T_2$ are used to estimate $D(T_2)$. Reproduced from Lovejoy *et al.* (1987). (b) Functional box counting on radar reflectivity data of rain (data taken from a weather radar in Montreal, Canada). The top graph is for horizontal (2D) sections at 3 km levels, covered with shapes (roughly) $L \times L$; the bottom graph is for 3D sections covered with cube-like shapes. Each line corresponds to a reflectivity factor increasing by a factor of about 2.5 (starting at the top, which is the lowest detectable signal). Although all the different levels are accurately power laws (scaling), the more and more intense rain regions (lower curves) have lower and lower slopes: again, we conclude that rain is multifractal. The geostatistics theory slopes 2, 3 (top and bottom respectively) are shown for reference. Adapted from Lovejoy *et al.* (1987).

Box 3.1 *(cont.)*

Let us now see what happens when this "functional" box counting is applied to radar reflectivities from rain. In Fig. 3.7b, we see that the scaling is excellent: the power law (Eqn. (3.4)) was accurately obeyed for all T, L. However, D_T systematically decreases with threshold: it is not constant (as assumed in the monofractal models such as fractional Brownian motion, Section 5.4.3) but a decreasing function, indicating that the field is multifractal. Similar results (Fig. 3.7c) show that the topography is also multifractal.

Let us now consider some consequences of the scaling of these exceedance sets. If we consider the areas A_T exceeding a given threshold then we find that they systematically decrease as the resolution becomes finer (decreasing L) : $A_T = L^d N_T = L^{C(T)}$; with $C(T) = d - D(T)$. We see that, contrary to standard assumptions (including those of classical geostatistics: Matheron, 1970), unless $C(T) = 0$, the areas depend on the subjective resolution L; Figs. 3.7b and 3.7c show reference lines indicating that for both the rain area and topography, all the regions defined by the thresholds have $C(T) = d - D(T) > 0$ so that they have systematic resolution dependencies.

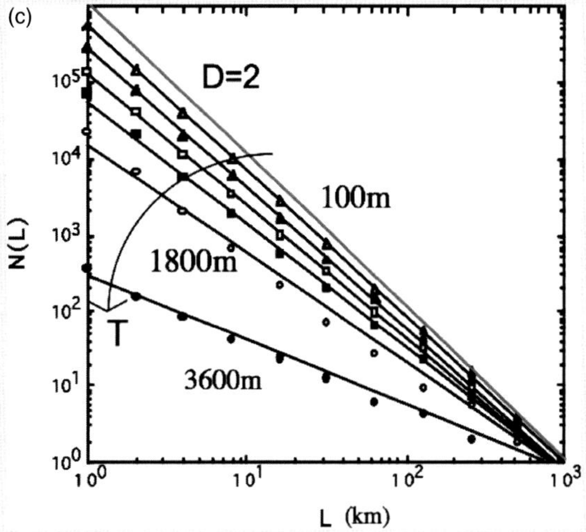

Fig. 3.7 (c) Functional box counting of the topography of France at 1 km resolution. Adapted from Lovejoy and Schertzer (1990).

In meteorology, it is common to use satellite pictures of cloud areas to define "cloud fractions" $F_T = A_T/A \approx L^{C(T)}$ (where A is the total area of the satellite image, independent of L). In this case, the same type of functional box-counting analysis (Gabriel *et al.*, 1988) shows that cloud fraction is indeed a similarly subjective quantity depending on the observer-dependent resolution. Actually, cloud fraction is doubly subjective, since the exact threshold used to separate cloud from no cloud is also somewhat arbitrary, so that both L and T are different when determined by ground observers and by satellite radiance algorithms. It is therefore hardly surprising that there are serious calibration issues (see e.g. Wielicki and Parker, 1992), although even recent attempts to deal with cloud sizes would benefit from a systematic scaling approach (see e.g. Wood and Field, 2011).

the interval is uniform so that the initial energy flux density $\varepsilon_0 = 1$. As in the β model, the cascade proceeds by dividing the unit interval successively into λ_0 subintervals (λ_0 is an integer $= 2$ in Fig. 3.8a) and multiplying the flux density by independent identically distributed random factors $\mu\varepsilon$ (the notation μ indicating "multiplicative increment"; it is analogous to the use of the Δ to denote an additive increment). Therefore after n (discrete) cascade steps, the smallest scale is λ_0^{-n}, the value of the energy flux density at a point $0 \le x \le 1$ is the product:

$$\varepsilon_n = \prod_{j=1}^{n} \mu\varepsilon_j \tag{3.5}$$

In order for the flux to be conserved from scale to scale, we constrain the weights $\mu\varepsilon$ so that $<\mu\varepsilon> = 1$, implying $<\varepsilon_n> = 1$. Each cascade step in the α model is a two-state (binomial) process with $\mu\varepsilon =$ either $\lambda_0^{\gamma+}$ or $\lambda_0^{\gamma-}$ where $\gamma_+ > 0$ corresponds to a boost ($\mu\varepsilon > 1$) and γ_- to a decrease ($\mu\varepsilon < 1$) (Schertzer and Lovejoy, 1985). As in the β model, the corresponding probabilities can be written λ_0^{-c} and $1 - \lambda_0^{-c}$ respectively, where $c > 0$ is a parameter (it corresponds to the maximum codimension of the process, as discussed in Chapter 5). Formally:

$$\begin{aligned} \Pr(\mu\varepsilon = \lambda_0^{\gamma_+}) &= \lambda_0^{-c} \\ \Pr(\mu\varepsilon = \lambda_0^{\gamma_-}) &= 1 - \lambda_0^{-c} \end{aligned} \tag{3.6}$$

An example of the pattern of boosts (+) and decreases (-) is shown in Fig. 3.8b. Although the α model apparently involves three parameters (γ_+, γ_-, c), due to the conservation constraint:

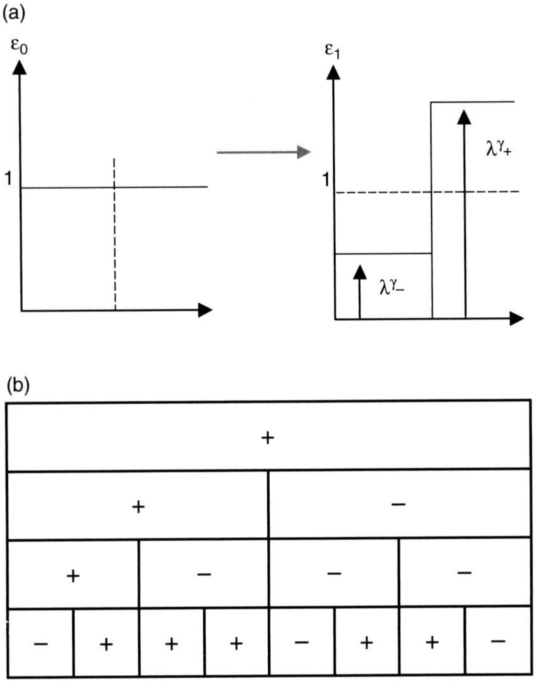

Fig. 3.8 (a) A schematic picture of the α model. At each step, uniform intervals (left) are divided into $\lambda = \lambda_0$ subintervals ($\lambda_0 = 2$ here) and then each is randomly multiplied by either $\lambda_0^{\gamma+}$ of $\lambda_0^{\gamma-}$ (with $\gamma_+ > 0$, a boost, $\gamma_- < 0$, decrease). (b) Schema of tree of boosts (+) or decreases (–) for a one-dimensional β or α model with probabilities as above. Reproduced from Schertzer and Lovejoy (1996).

$$\langle \mu\varepsilon \rangle = \lambda_0^{-c}\lambda_0^{\gamma_+} + (1 - \lambda_0^{-c})\lambda_0^{\gamma_-} = 1 \tag{3.7}$$

only two can be freely chosen. We can see that the β model is recovered in the limit $\gamma_+ \to c$, which is the same as $\gamma_- \to -\infty$. Note that the boost "singularity" γ_+ can in principle take any value $\gamma_+ \geq 0$ and that there is nothing to prevent individual realizations from having "extreme" outcomes such as a cascade step with only boosts or only decreases. The conservation in Eqn. (3.7) is only over an ensemble of processes; in analogy with thermodynamics, it is a "canonical" conservation. Fig. 3.9a shows two examples of the α model in 1D with rather different parameters, showing how the "spikes" – the "singularities" – build up step by step; note the vertical axis scale, which changes considerably as the cascade evolves to smaller scales. Fig. 3.9b shows 2D examples comparing β models (left column) with c increasing from top (0.1) to bottom (0.5); these fractal sets become increasingly sparse as c increases. The second column is a rendition on a linear scale of the corresponding α model with $\gamma_+ = c - 0.09$ with the same c values (and the same random seed so that the difference in the structures that result from $c \to c - 0.09$ can be gauged). The variability is so large that the simulations vary from fairly homogeneous grey (top) to very inhomogeneous when $c = 0.5$ (bottom). In an effort

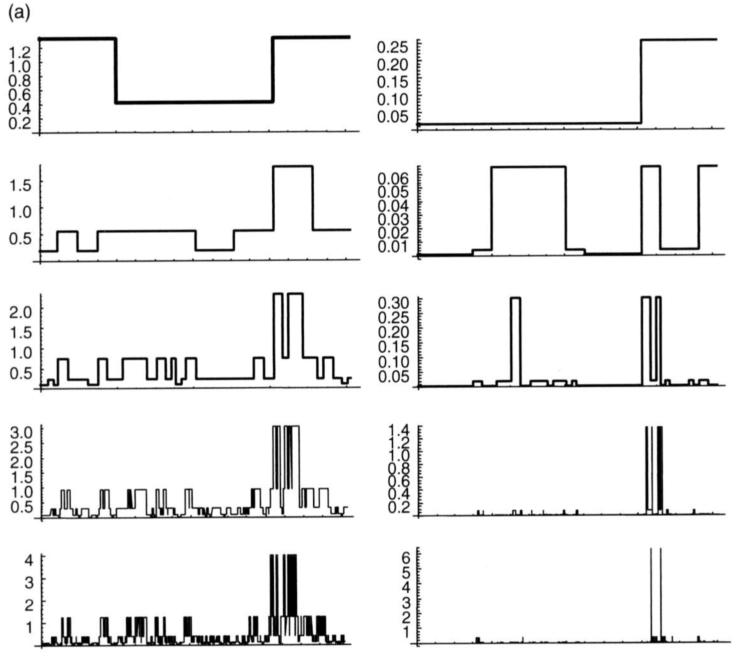

Fig. 3.9 (a) A comparison of two α models with different parameters. Left column: $\gamma_+ = 0.2$, $c = 0.3$ ($C_1 = 0.087$), right column: $\gamma_+ = 1.1$, $c = 1.2$ ($C_1 = 0.82$). From top to bottom every second cascade step (a factor of λ_0^2) is shown, 10 steps in all; the total range of scales is $2^{10} = 1024$). Notice the changing vertical scales.

(b)

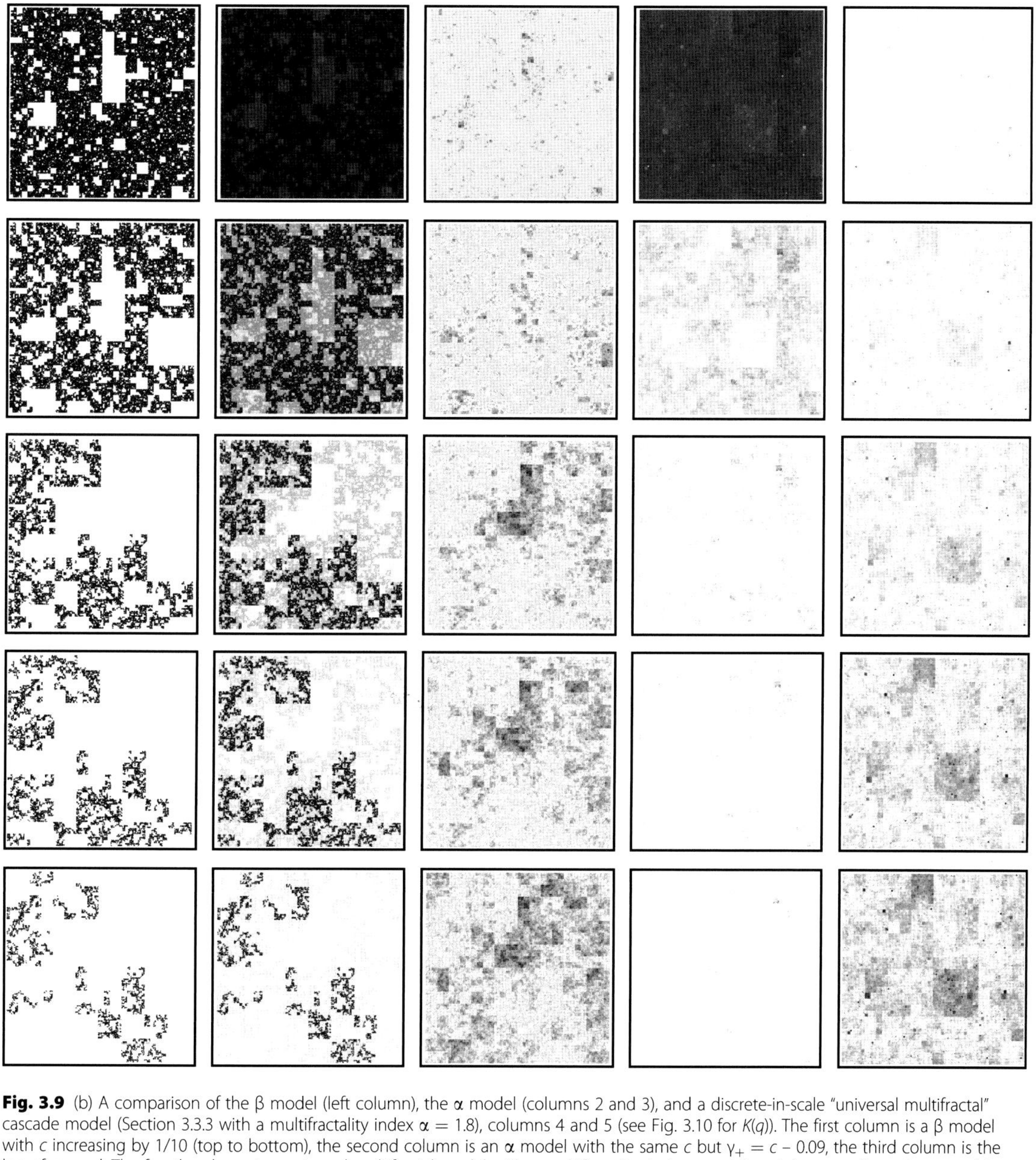

Fig. 3.9 (b) A comparison of the β model (left column), the α model (columns 2 and 3), and a discrete-in-scale "universal multifractal" cascade model (Section 3.3.3 with a multifractality index $\alpha = 1.8$), columns 4 and 5 (see Fig. 3.10 for $K(q)$). The first column is a β model with c increasing by 1/10 (top to bottom), the second column is an α model with the same c but $\gamma_+ = c - 0.09$, the third column is the log of second. The fourth column is a universal multifractal model with a multifractality index $\alpha = 1.8$ and the same C_1 as the α model (0.00050, 0.037, 0.098, 0.170, 0.248 respectively). The fifth column is the log of the fourth. To get an idea of the variations, note that the α model ranges from minima to maxima (0.68, 1.06), (0.028, 1.84), (1.49×10^{-3}, 3.20), (7.52×10^{-5}, 5.58), (2.13×10^{-5},9.71) (top to bottom respectively), while the universal multifractal model has ranges: (0.069, 1.06), (1.99×10^{-13}, 6.36),(1.004×10^{-22}, 19.04), (1.18×10^{-30}, 42.07), (1.08×10^{-37}, 76.80) (top to bottom respectively, all have mean = 1).

to better visualize the intermediate levels of intensity, the third column shows the same simulation but on a log scale. If we converted the α model into a geometric set of points by considering only those regions which are greater than a threshold, then, due to the scaling, the sets are fractal sets and we can see that as we increase the threshold to higher levels the resulting fractal codimension increases (the fractal dimension decreases). This method of converting the multifractal into fractal sets is only approximate; it can only work over a finite range of scales. The "graphs" of the process do not converge: as we increase the resolution, the graph itself does not converge; the small-scale limit is singular. The α model is therefore the simplest example of a multifractal process (to which we return in greater detail in Chapter 5). Finally, columns four and five (linear and log scales respectively) show a discrete-in-scale "universal multifractal" model described below which has not just two possible states at each step but a continuum (but not yet a continuum of spatial scales; there are still ugly construction lines!). Going from left to right, the β and α models share a common parameter c while the Lévy cascade shares the same parameter C_1 (described below) so that their behaviours near the mean are the same.

3.2.4 The *p* model and microcanonical conservation

The α model was developed specifically because it afforded a simple example of nonclassical extreme behaviour: "divergence of statistical moments" – algebraic probabilities – that result when the largest cascade values exceed a critical value (here, when γ_+ exceeds the dimension of space, d: see Section 5.3.2). This interesting behaviour is a consequence of the singular small-scale cascade limit, which turns out to imply that "dressed" cascades – i.e. those cascades completed to infinitely small scales and then spatially integrated over a finite scale – have these nonclassical extremes. The jargon "bare" and "dressed" was borrowed by Schertzer and Lovejoy (1987) from theoretical physics to underline the fact that only in the latter are all the interactions (over the whole hierarchy of scales) taken into account. However, the analysis of the dressed properties is not as straightforward as for those of a cascade developed over a finite range of scale λ and then stopped, i.e. the "bare" properties; we pursue this more advanced topic in Section 5.3.

However, there is a simple way to modify the cascade which avoids at least some of the bare/dressed complications: to change the type of scale-by-scale conservation from canonical (over an ensemble of realizations, as for the α model) to "microcanonical" conservation (at each cascade step, each realization). To understand this, take the example of the 1D α model with $\lambda_0 = 2$: it can be transformed into the microcanonical "*p*" model (Meneveau and Sreenivasan, 1987) by requiring at each step that the two random multipiers $\mu\varepsilon_1$, $\mu\varepsilon_2$ needed to yield a daughter from a parent eddy have exactly an average of 1 so that $\mu\varepsilon_2 = 2 - \mu\varepsilon_1$. In other words, at each step there is exactly one γ_+ and one γ_- satisfying $(\lambda_0^{\gamma_-} + \lambda_0^{\gamma_+})/2 =$ (with $\lambda_0 = 2$). The only choice in the *p* model is thus whether γ_+ is on the right- or left-hand side of each interval. Interestingly, the *p* model was actually first proposed much earlier by de Wijs (1953) as a model for the distribution of minerals and ores in the lithosphere, and it was argued that it led to at least approximately a lognormal distribution.

It turns out that the most important difference between the α and *p* models is in the largest events that they can generate. Whereas we have pointed out that in the α model, any $\gamma_+ \geq 0$ is possible, in the *p* model, the requirement that the multipliers are ≥ 0 (so that γ_- is real) implies an upper limit $\gamma_+ \leq 1$. If we generalize this to other discrete scale ratios $\lambda_0 > 2$ or to higher d-dimensional α models but with the same microcanonical constraint (i.e. that the sum of all the multipliers at each step equals the total number of multipiers $\lambda_0{}^d$) then instead of the canonical conservation condition $<\mu\varepsilon> = 1$ we obtain:

$$\frac{1}{\lambda_0^d}\sum_{i=1}^{\lambda_0^d}\mu\varepsilon_i = 1 \tag{3.8}$$

This microcanonical constraint implies that the most extreme microcanonical model is that in which all the multipliers are zero except for a single one, whose value is thus $\mu\varepsilon = \lambda_0^d$. This implies that in d dimensions, $\gamma_+ \leq d$ is the most extreme microcanonical model possible. It turns out that this is precisely the condition that guarantees that in a space dimension d, the interesting extreme cascade probability tails disappear. By focusing on the special and artificial microcanonical cascades, much of the physics literature has thus missed this interesting "nonclassical route to self-organized criticality" (SOC) discussed in Box 5.2. Similarly, there are also special "Novikov"

Box 3.2 Dimensions and codimensions

Fractal geometry

Fractal geometry (Mandelbrot, 1977, 1983) provides the simplest nontrivial example of scale invariance, and is useful for characterizing fractal sets. Unfortunately in geophysics we are usually much more interested in fields (with values at each point) and rarely interested in geometrical sets. However, over a wide range of scales fractal dimensions can still be useful in "counting the occurrences of a given phenomenon" – as long as this question can properly be posed. If this is the case and the phenomenon is scaling, then the number of occurrences ($N_A(l)$ at resolution scale l in space and/or time of a phenomenon occurring on a set A) follows a power law:

$$N_A(l) \sim \left(\frac{L}{l}\right)^{D_F} \tag{3.9}$$

D_F being the (unique) fractal dimension, generally not an integer, and L the (fixed) largest scale (here and below the sign $\sim$ means equality within slowly varying and constant factors).

Fractal codimensions

The notion of fractal codimension C_F can be defined both statistically and geometrically. While the latter is more popular, we will demonstrate that the former is much more useful and more general, since it applies not only to deterministic but also to stochastic processes.

Definition 1: geometric definition of a fractal codimension
Let $A \subset E$ (the embedding space) with $\dim(E) = D$ and $\dim(A) = D_F(A)$. Then the codimension $C_F(A)$ is defined as:

$$C_F(A) = D - D_F(A) \tag{3.10}$$

This definition corresponds merely to an extension of the (integer) codimension definition for vector subspaces, i.e. E_1 and E_2 being in direct sum (i.e. $E_1 \cap E_2 = \varnothing$):

$$E = E_1 \oplus E_2 \Rightarrow \mathrm{codim}(E_1) = \dim(E_2) \tag{3.11}$$

Definition 2: probabilistic definition of a fractal codimension
In fact the codimension C_F can be considered to be more fundamental than the notion of fractal dimension D_F and should be introduced directly. The probability (Pr) that a ball B_λ (of size $l = L/\lambda$) intersects the set A has the following scaling behaviour:

$$\Pr(B_\lambda \cap A) \sim \lambda^{-C_F(A)} \tag{3.12}$$

and C_F is thus directly defined as an exponent measure of the fraction of the space occupied by the fractal set A (size L) in an embedding space E, which can even be an infinite dimensional space.

Relating the two definitions
Since the probability of the event $(B_\lambda \cap A)$ is defined as:

$$\Pr(B_\lambda \cap A) \sim \frac{N(B_\lambda \cap A)}{N(B_\lambda \subset E)} \approx \frac{\lambda^{D_F(A)}}{\lambda^D}. \tag{3.13}$$

where $N(B_\lambda \cap A)$ refers to for example the number of balls B_λ needed to cover the set A and $N(B_\lambda \subset E)$ is the corresponding number for the entire space, it is easy to check that when $C_F(A) < D = \dim(E) < \infty$ the two definitions (Eqns. (3.10), (3.12), respectively) are equivalent:

$$C_F(A) \leq D < \infty, \{\text{definition 1} \equiv \text{definition 2}\}\ \forall\ D_F \geq 0 \tag{3.14}$$

Rather obviously, the statistical definition does not imply any limitation on C. However, the equivalence between the two definitions does not hold any longer as soon as $C_F(A) > D$, since:

$$\text{for } C_F(A) > D, \{\text{both definition 1 and definition 2}\} \Rightarrow D_F(A) < 0 \tag{3.15}$$

This is the so-called "latent" dimension "paradox" corresponding to the fact that a deterministic geometric definition is no longer possible; indeed, there is no possible definition of a negative Hausdorff dimension! This is not surprising

Box 3.2 (*cont.*)

since definition 2 (Eqn. (3.12)) overcomes many limitations of the Hausdorff dimension, which is defined for compact sets (hence bounded sets): the codimension measures the *relative sparseness* of a phenomenon (the relative frequency of its occurrence), whereas the dimension measures its *absolute sparseness* (the absolute frequency of its occurrence). Obviously, we don't need to know the latter in order to be able to determine the former. However, it turns out historically that the (fractal) dimension was introduced first.

The "intersection theorem"

It is important to point out a direct and important corollary of the probabilistic definition of fractal codimensions. If E_1 and E_2 are statistically independent:

$$C_F(E_1 \cap E_2) = C_F(E_1) + C_F(E_2) \tag{3.16}$$

i.e. statistical codimensions just add for the intersection of independent fractal processes. This is a direct consequence of the fact that the probability of the intersection factorizes $\Pr(E_1 \cap E_1) = \Pr(E_1)\Pr(E_1)$. It is important to note that the derivation and the validity of this "theorem" is not so obvious when using the geometric definition (for discussion, see Falconer, 1990). In Chapter 5 we use statistical codimensions to solve the paradox of apparent negative dimensions by introducing the notion of "sampling dimension."

inequalities which are sometimes invoked but which are only relevant in microcanonical cascades (Novikov, 1994). Finally, the microcanonical assumption is sometimes made implicitly when analyzing data (the "fragmentation ratio" or "multiplier" method), and this will lead to serious biases in estimating the exponents (see Lovejoy, 2010, for a discussion and numerical examples).

3.2.5 Statistical properties of the α model and general cascades

We have seen that in order to respect the scale-by-scale conservation of the mean (its independence of n), we require the canonical conservation $\langle \mu\varepsilon \rangle = 1$, which ensures that $\langle \varepsilon_n \rangle = 1$. However, with an increasing number n of steps, the other statistical moments $\langle \varepsilon_n^q \rangle$ will either "blow up" to infinity (for $q > 1$) or "down" to zero (for $q < 1$). The overall characterization of the statistical properties is conveniently made with the help of the "moment scaling exponent" $K(q)$, which can be defined by the statistics of the distribution of random weights $\mu\varepsilon$:

$$K(q) = log_{\lambda_0}\langle \mu\varepsilon^q \rangle = Log\langle \mu\varepsilon^q \rangle / Log\lambda_0 \tag{3.17}$$

Introducing the (random) cascade "generator" Γ_0, the logarithm of the multiplier:

$$\Gamma_0 = Log_{\lambda_0}(\mu\varepsilon) \tag{3.18}$$

implies that $K(q)$ is actually its (Laplacian, base λ_0) second characteristic function ("cumulant generating function"), because Eqn. (3.17) can be rewritten as:

$$K(q) = log_{\lambda_0}\langle e^{q\Gamma_0} \rangle \tag{3.19}$$

The importance of defining the generator of the cascade implicitly in this manner instead of directly via the weight $\mu\varepsilon$, is that Eqn. (3.19) can be easily generalized for continuous-in-scale cascades, yet it is also useful for discrete cascades (Schertzer and Lovejoy, 1987). Let us first note that in probability theory, characteristic functions are usually introduced as the Fourier transforms of the probability densities of random variables, and their natural logs are the second (Fourier) characteristic functions. For both Laplace and Fourier transforms, due to their invertibility, defining the statistics via the probability density or via the first or second characteristic functions is equivalent. Here, the finiteness of the characteristic function $K(q)$ is equivalent to that of the qth order moment of $\mu\varepsilon$; in fact, conservation of the mean ($q = 1$) requires that at least the positive moments of order $0 \leq q \leq 1$ are finite. Another way of stating this is that only random multipliers $\mu\varepsilon$ with well-defined Laplace characteristic functions can yield physically possible cascade generators.

A trivial but important consequence of the independence of the cascade steps (and of the corresponding weights) is that $K(q)$ is scale invariant, i.e. independent of the number n of steps:

$$\langle \varepsilon_n^q \rangle = \left\langle \prod_{j=1}^{n} \mu\varepsilon_j^q \right\rangle = \prod_{j=1}^{n} \langle \mu\varepsilon_j^q \rangle = \langle \mu\varepsilon^q \rangle^n = \lambda_0^{nK(q)} \tag{3.20}$$

with respect to the overall scale ratio λ, since the cascade started:

$$\lambda = \lambda_0^n \tag{3.21}$$

We can now write the general expression for the statistical properties after a total scale range λ:

$$\langle \varepsilon_\lambda^q \rangle = \lambda^{K(q)} \tag{3.22}$$

This is the basic formula for cascade statistics used throughout this book. As indicated above, this specification of the statistics of $\mu\varepsilon$ (and also of ε_λ) via their statistical moments is equivalent to their specification by their probabilities. We postpone demonstrating the beautiful and simple relation between the two until Chapter 5, concentrating for the moment on the properties of $K(q)$, and in Chapter 4 on the analysis of atmospheric data.

3.2.6 Properties of the moment scaling exponent *K(q)*

In order to see the general shape of the $K(q)$ function, we may first note that conservation from one scale to another $<\varepsilon_n> = 1$ requires $K(1) = 0$. In addition, because any positive number raised to the zero power is one, we have $<1> = 1$, hence $K(0) = 0$. Finally, a basic property of second characteristic functions is that $K(q)$ must be convex, i.e. $K''(q) > 0$; this can be shown directly by doubly differentiating $K(q) = log<e^{q\Gamma}>/log\lambda$ (see Appendix 3A for details). We therefore conclude that the typical $K(q)$ looks something like Fig. 3.10, which shows the $K(q)$ for the α model and the universal multifractal models in the fourth and fifth columns of Fig. 3.9. The models are tangent to each other at $q = 1$ because the derivatives at $q = 1$ were deliberately chosen to be equal to each other. This value:

$$C_1 = K'(1) \tag{3.23}$$

is "the codimension of the mean"; it is a basic characterization of the variability near the mean, to which we return in Chapter 5.

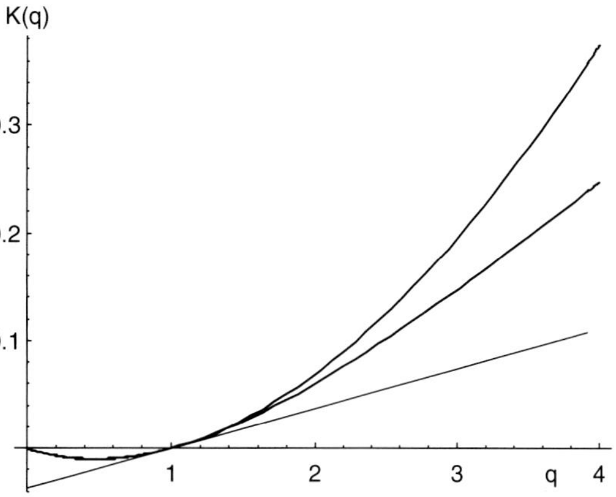

Fig. 3.10 The $K(q)$ functions for the α model (middle curve, $c = 0.2$, $\gamma_+ = 0.11$, from the second from the top of the 5 × 5 grid in Fig. 3.9b), with universal multifractal model (top curve), $\alpha = 1.8$ with same C_1 (= 0.037, same as second from the top of the 5×5 grid in Fig. 3.9b) and β model (bottom, linear) with $c = C_1$.

To illustrate these properties, we can make some explicit calculations for the α model. For example, we have:

$$K_\alpha(q) = log[\lambda_0^{q\gamma_+ - c} + \lambda_0^{q\gamma_-}(1 - \lambda_0^{-c})]/log\lambda_0$$

$$K'_\alpha(q) = \frac{\gamma_+\lambda_0^{q\gamma_+ - c} + \gamma_-\lambda_0^{q\gamma_-}(1 - \lambda_0^{-c})}{\lambda_0^{q\gamma_+ - c} + \lambda_0^{q\gamma_-}(1 - \lambda_0^{-c})} \tag{3.24}$$

From this we see that for the α model:

$$C_1 = K'_\alpha(1) = \gamma_+\lambda_0^{\gamma_+ - c} + \gamma_-\lambda_0^{\gamma_-}(1 - \lambda_0^{-c}) \tag{3.25}$$

Also, we may note that for the α model, there are low and high q asymptotes whose slopes are:

$$\lim_{q \to \pm\infty} K'_\alpha(q) = \gamma_\pm \tag{3.26}$$

Considering now the special case of Eqn. (3.24), when $\gamma_+ = c$, $\gamma_- = -\infty$, we obtain the results for the β model:

$$K_\beta(q) = C_1(q - 1) \tag{3.27}$$

We see that it is linear in q with $c = C_1$. The β model with corresponding C_1 can be said to provide a "monofractal" (on/off) approximation to the mean ($q = 1$) behaviour of the cascade (see Fig. 3.9 for examples), but obviously this approximation is only valid for $q \approx 1$, otherwise it may be misleading. Note that $\lim_{q \to 0} K_\beta(q) = -C_1$ so that in the β model, C_1 is also the codimension of the nonzero regions, of the "support."

It is therefore natural to characterize the multifractality by the deviations of $K(q)$ from the β model, i.e. from its linear tangent $K'(1)$. The deviations can be characterized by the curvature of $K(q)$ near the mean (Schertzer *et al.*, 1991); indeed, we can already use this idea to give a "local" (in q space) definition of the degree of multifractality α:

$$\alpha = K''(1)/K'(1) \tag{3.28}$$

We shall see that this local (C_1, α) description of $K(q)$ can often provide a complete "global" description when the cascade is in the basin of attraction of "universal multifractals" discussed below (note that this α is not the same as the α used in the original formulation of the α model).

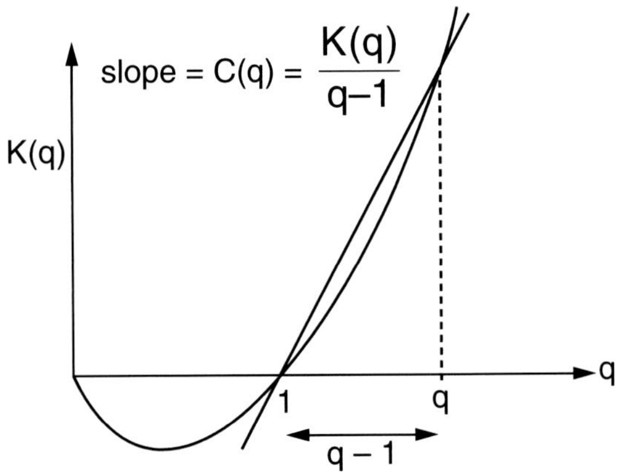

Fig. 3.11 A schematic illustration of the relation between $K(q)$ and $C(q)$. Reproduced from Schertzer and Lovejoy (1996).

3.2.7* The dual codimension function $C(q)$

Functions that will be useful in later analysis are the codimension and dimension functions $C(q)$, $D(q)$:

$$C(q) = \frac{K(q)}{q-1}; \quad D(q) = d - C(q) \tag{3.29}$$

The underlying duality corresponds to the fact that the codimensions $C(q)$ were introduced (Schertzer and Lovejoy, 1983, 1984) with the help of an intersection theorem, as the codimension of sets over which the statistical moments of the field are finite, whereas in order to characterize low-dimensional chaotic systems, Hentschel and Procaccia (1983) and Grassberger (1983) introduced $D(q)$ as "generalized dimensions," but with almost no justification. Both are significant in that they underline the fact that in general a scaling field cannot be defined with only a single dimension, a question that we will pursue in Chapter 5. By the graphical construction (Fig. 3.11) it is clear that $C(q)$ is the slope of the chord between the points (1,0) and (q, $K(q)$), and by the convexity of $K(q)$, $C(q)$ must be an increasing function of q: $C'(q) \geq 0$ for all q, hence the dimension function $D'(q) \leq 0$. Note that using l'Hopital's rule at $q = 1$ we see that:

$$C(1) = K'(1) = C_1 \tag{3.30}$$

This fundamental parameter will be shown to be the codimension of the mean field, i.e. $<\varepsilon>$.

3.2.8 The small-scale cascade limit: a first look

We have seen in Eqn. (3.19) that $K(q)$ is the second characteristic function of the generator Γ. It follows that the small-scale limit ($\lambda \to \infty$) of the cascade is highly singular since logλ diverges implying that $log\langle e^{q\Gamma_\lambda}\rangle$ diverges to ∞ ($q > 0$) or to $-\infty$ ($q < 1$). From the examples in Fig. 3.9a, we see that it is plausible that this corresponds to the limiting density ε_λ becoming either infinite spikes, or zero values, although – because of the conservation property – on average the area under the spikes must be fixed. As $\lambda \to \infty$, we find (Chapter 5) that the spikes are distributed on sparse fractal sets – strictly of measure zero with respect to the usual Lebesgue measure (i.e. in Fig. 3.9a, in the limit, they have zero lengths, zero areas respectively), and that the amplitude of the spikes must diverge so as to maintain the mean fixed. After a large (but finite) λ, if we consider the regions that exceed a given level of activity, then this is a fractal set (truncated at the scale L/λ). If we consider the set of points exceeded by even taller spikes, then we find that the fractal dimension of this different set is smaller. The limiting cascade process is thus an example of a multifractal, a nontrivial hierarchy of fractals.

It turns out that this highly singular multifractal limit is like a Dirac δ function: the limiting cascade density ε_∞ is a generalized function obtained as the limit of the series of functions ε_λ, and is itself only meaningful if integrated over a finite set. The integration over a finite set

Box 3.3 Autocorrelations, spectra and bounded cascades

The moments in Eqn. (3.22) explicitly describe the "one point" cascade statistics (i.e. the statistics at a single point $\underline{r}$ within the cascades) and are the same for all $\underline{r}$ within the region of simulation (in space, the process is "statistically homogeneous"; in time, "statistically stationary"). However, all the statistical properties of the cascade are implicitly defined once $K(q)$ is specified, i.e. the n point joint probabilities. For example, we may wish to estimate the power spectrum of a cascade which is a two-point statistic. Due to the Wiener–Khinchin theorem, it suffices to determine the autocorrelation function $R(\Delta x)$ of the cascade; the spectrum is then obtained as its Fourier transform. From Box 2.2 we have (in 1D):

$$E(k) = \int_{-\infty}^{\infty} R(\Delta x)\varepsilon^{-ik\Delta x}d\Delta x; \quad R(\Delta x) = \langle \varepsilon(x-\Delta x)\varepsilon(x)\rangle \tag{3.31}$$

From the Tauberian theorem (Box 2.2) if $R(\Delta x) \approx \Delta x^{-\delta}$, then $E(k) \approx k^{-\beta}$ with $\beta = 1+\delta$, hence we need only determine δ from the cascade. For the discrete cascade, we can follow the argument from Yaglom (1966) and Monin and Yaglom (1975). Consider a cascade in 1D developed over n steps, for a total scale ratio $\lambda_0{}^n$. Consider next a lag Δx_m such that:

$$\lambda_0^{-(m+1)} < \Delta x_m < \lambda_0^{-m} \tag{3.32}$$

so that $\log\Delta x_m \approx -m\log\lambda_0$. The qth-order autocorrelation is thus:

$$\langle \varepsilon_n^q(x-\Delta x_m)\varepsilon_n^q(x)\rangle = \left\langle \prod_{i=1}^{n}\prod_{j=1}^{n}(\mu\varepsilon_i\mu\varepsilon_j)^q \right\rangle \tag{3.33}$$

where the index i refers to the multipliers at the point $x - \Delta x_m$ and the j to those at the point x. The lag Δx_m is the typical size of the mth-level structures, so that if $m \geq n$, the two points will likely share all the multipliers and $\langle \varepsilon_n^q(x-\Delta x_m)\varepsilon_n^q(x)\rangle = \langle \varepsilon_n^{2q}(x)\rangle = \lambda_0^{nK(2q)}$ (i.e. we discount the unlikely event that they happen to straddle special points at the edge of large structures). If we consider now the case $m < n$ then, typically, we will find that the multipliers at the points x, $x - \Delta x_m$ will be shared up to level m, but will be different for the levels $> m$. This implies:

$$\langle \varepsilon_n^q(x-\Delta x_m)\varepsilon_n^q(x)\rangle \approx \langle \mu\varepsilon^{2q}\rangle^m \left(\langle \mu\varepsilon^q\rangle^2\right)^{(n-m)} = \lambda_0^{mK(2q)-2(n-m)K(q)} \tag{3.34}$$

Using $\lambda_0^{-m} \approx \Delta x_m$ and $\lambda_0^n = \lambda$, we obtain:

$$\langle \varepsilon_\lambda^q(x-\Delta x)\varepsilon_\lambda^q(x)\rangle \approx \Delta x^{-(K(2q)-2K(q))}\lambda^{-2K(q)} \tag{3.35}$$

where we have dropped the subscripts m on the Δx, and indicated the resolution of ε directly by the total scale range λ rather than the number of steps n. Finally, the usual autocorrelation is obtained by taking $q = 1$; using the scale-by-scale conservation condition $K(1) = 0$ we obtain the particularly simple result:

$$\begin{aligned} &\langle \varepsilon_\lambda(x-\Delta x)\varepsilon_\lambda(x)\rangle \approx \Delta x^{-K(2)}; \quad 1 \geq \Delta x > \lambda^{-1} \\ &\langle \varepsilon_\lambda(x-\Delta x)\varepsilon_\lambda(x)\rangle \approx \lambda^{K(2)}; \quad \Delta x < \lambda^{-1} \end{aligned} \tag{3.36}$$

The normalized autocorrelation function $R_\lambda(\Delta x)$ with the property $R_\lambda(0) = 1$ can be obtained by normalizing by the value at $\Delta x = 0$ to obtain:

$$\begin{aligned} &R_\lambda(\Delta x) \approx (\lambda\Delta x)^{-K(2)}; \quad 1 \geq \Delta x > \lambda^{-1} \\ &R_\lambda(\Delta x) \approx 1; \quad \Delta x \leq \lambda^{-1} \end{aligned} \tag{3.37}$$

Finally, using the result from Eqn. (3.31), we obtain:

$$E(k) = \int_{-\infty}^{\infty} R_\lambda(\Delta x)e^{-ik\Delta x}d\Delta x \approx \lambda^{K(2)}k^{-\beta}; \quad \beta = 1 - K(2); \quad K \leq \lambda \tag{3.38}$$

Since $K(1) = 0$, $K'(1) > 0$ and $K'' > 0$, we have $K(2) > 0$ and we see that $\beta < 1$. Fig. 3.12 shows the result on the flux estimated from the temperature series shown in Fig. 3.1a.

Box 3.3 (*cont.*)

The result $\beta < 1$ for a conserved cascade already implies that "observables" with $\beta > 1$ (such as the wind and most of the meteorological fields) cannot be *direct* results of multiplicative cascade processes as defined above. However, this is to be expected since it is only the scale-by-scale conserved quantities that are expected to cascade in the first place. This is worth stressing, since there has been an attempt – the "bounded cascade model" – to directly model atmospheric fields with $\beta > 1$ using multiplicative cascade processes (Cahalan, 1994). For example, in order to model 1D sections of atmospheric turbulence with $\beta = 5/3$ (the Kolmogorov value), at each step in the bounded cascade the multipliers are constrained to be of the form $\mu\varepsilon = (1+spc_n)$ where $0 < p < 1$ is a parameter and $s = \pm 1$ is a random sign. The additional cascade-step *dependent* parameter is $c_n = \lambda_0^{-nH}$ (with $0 < H < 1$) with $\lambda_0 = 2$. Due to this "bounding," instead of being independent of the cascade step, the $\mu\varepsilon$ rapidly approach unity as n becomes large. In fact for large j, λ_0^{-jH} is very small so that at each point:

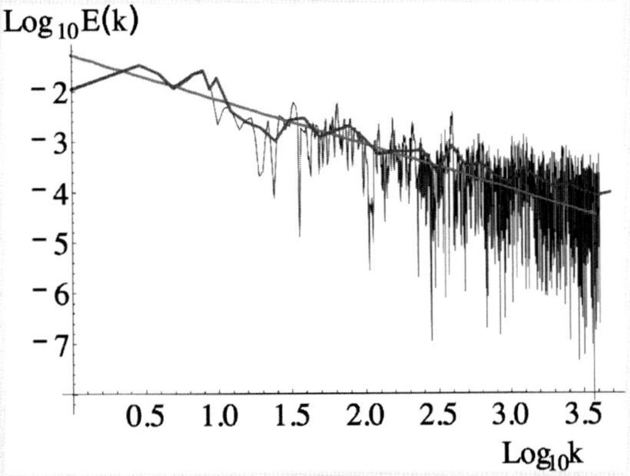

Fig. 3.12 The spectrum (thin line) of the fluxes from the aircraft transect shown in Fig. 3.1a with its average over logarithmically spaced bins (thick line) along with a reference line with slope −0.89 ($K(2) = 0.11$, the value for $C_1 = K'(1) = 0.06$, $\alpha = K''(1)/K'(1) = 1.8$).

$$\varepsilon_n = \prod_{j=1}^{n}(1 + s_j p\lambda_0^{-jH}) \approx 1 + p\sum_{j=1}^{n}\lambda_0^{-jH}s_j \tag{3.39}$$

so that in the small-scale limit, the bounded cascade is actually seen to be a disguised additive process. Indeed, Lovejoy and Schertzer (2006) show that its statistics are essentially those of a truncated, additive fractional Brownian motion, so that the term "cascade" is quite misleading. Since pure multiplicative processes are only sufficient for defining conservative fluxes, in order to deal with processes with $\beta > 1$ a wide generalization is necessary, to fractionally integrated fluxes (Schertzer and Lovejoy, 1987; Schertzer *et al.*, 1997; see Section 5.4.4).

smooths the "bare" (un-integrated) process enough so that a small-scale (large λ) limit does exist for the integrals. Interestingly, the integration does not smooth out the small scales completely, so the statistical variability of such integrated "dressed" cascades is significantly higher than for the cascade constructed down to the same scale and then stopped: the "bare" cascade. This is discussed in further detail in Chapter 5.

3.3 Universal multifractal processes

3.3.1 Universality in physics, in turbulence

As there is only a conservation and convexity constraint on $K(q)$, (i.e. $K(1) = 0$, $K'' > 0$), a priori, an infinity of parameters (the $K(q)$ *function*) is required to specify a cascade process. However, for obvious theoretical and empirical reasons physics abhors infinity; this is the reason why in many different fields of physics the theme of *universality* appears: among the infinity of parameters, it may be possible that only very few of them are relevant. This is especially true as soon as we consider not only ideal systems, but more realistic systems subjected to perturbations or self-interactions. Indeed, such perturbations or interactions may wash out many of the peculiarities of the theoretical model, retaining only some essential features. The system can be expected to converge to some *universal attractor*, in the sense that a whole class of models/processes, belonging to the same domain of attraction, will converge to the same process defined by (far) fewer (relevant) parameters (Fig. 3.13). For example, in critical phenomena most of the many exponents describing phase transitions will depend only on the dimensionality of the system. More precisely –

Table 3.1 How many parameters for turbulence? The idea of universality in turbulence goes back to Kolmogorov (and indeed, to Richardson in his 4/3 diffusion law, which is more or less equivalent to the Kolmogorov law). Universal multifractals reduce the infinity of parameters implicit in $K(q)$ to only two (C_1, α), i.e. back to a finite and hence "manageable" number.

Answer	Date	References	Explanation	Parameters
1	1941	Kolmogorov (homogeneous turbulence)	$\Delta v_\lambda \approx \bar{\varepsilon}^{1/3}\lambda^{-1/3}$	$H = 1/3$
2	1962	Kolmogorov–Obukhov (lognormal model)	$\langle \varepsilon_\lambda^q \rangle = \lambda^{K(q)}$ $K(q) = \frac{\mu}{2}(q^2 - q)$	H, μ
2	1964	Novikov-Stewart, Mandelbrot, Frisch *et al.*, β model	$K(q) = C_1(q-1)$	H, C_1
∞	1974	(Mandelbrot, 1974)	$K(q)$	Any $K(q)$ convex with $K(0) = K(1) = 0$
3	1987	Universal multifractals	$K(q) = \frac{C_1}{\alpha-1}(q^\alpha - q)$	α, H, C_1

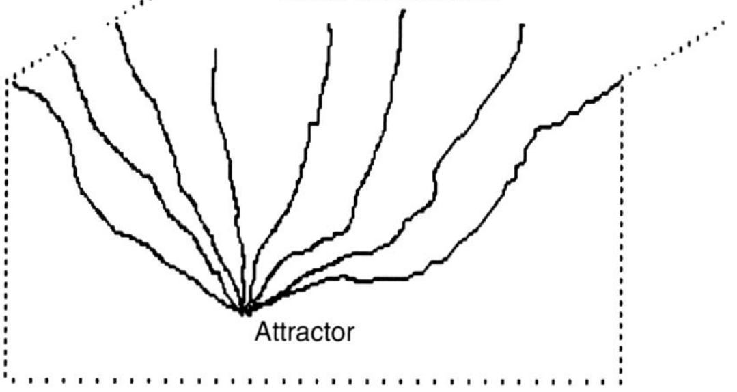

Fig. 3.13 Schematic of a basin of attraction. Reproduced from Schertzer and Lovejoy (1996).

although they could be originally quite different – all the processes belonging to the same "basin of attraction" will converge toward the same limit or "attractor" – hence the notion of universality: the larger the basin, the more universal the attractor. Indeed, it was the realization that low-dimensional systems (such as nonlinear mappings or coupled nonlinear ordinary differential equations) had universal behaviour (leading to the "universal" appearance of the famous Feigenbaum constant) that led to an explosion of interest in deterministic chaos. Universal multifractal processes are an analogous attractor but for a large number of degrees of freedom.

In turbulence, the idea of universality is also quite old, and the early proposals are compared in Table 3.1. It can be seen that in comparison to these one- or two-parameter models, the general cascade process effectively involves an infinity of parameters (the entire convex $K(q)$ function) with the early proposals as special cases. Universality on the basis of a multiplicative central limit theorem is tantamount to reducing the $K(q)$ to two parameters for ε with the third (H) for the observable (e.g. Δv) so that we return from an infinity of parameters to just three: α, C_1, H.

3.3.2 Universality, multiplicative processes and cascades

The study of multiplicative random processes has a long history (see Aitchison and Brown, 1957); it goes back to at least McAlister (1879), who argued that multiplicative combinations of elementary errors would lead to lognormal distributions. Kapteyn (1903) generalized this somewhat and stated what came to be known as the "law of proportional effect," which has been frequently invoked since then, particularly in biology and economics (see also Lopez, 1979, for this law in the context of rain). This law was frequently used to justify the use of lognormal distributions, i.e. it was tacitly assumed that the lognormal was a universal attractor for multiplicative processes. Although Kolmogorov (1962) and Obukhov (1962) did not explicitly give the law of proportional effect as motivation, it was almost certainly the reason that they suggested a lognormal distribution for the energy dissipation in turbulence. Since then, culminating in the multifractal processes, we have seen that there have been many proposals for explicit multiplicative cascade models that would reproduce the strong intermittency in turbulence.

Unfortunately, in the course of development of these models the basic issue of universality was obscured by various technical questions. If we simply iterate the model step by step with a fixed scale ratio λ, we indefinitely increase the overall range of scales

Box 3.4 Scaling gyroscopes cascade (SGC): shell models and beyond ...

The complexity and intractability of the Navier–Stokes equations have lead to the development of some greatly simplified (almost caricatural) models of them, which nevertheless preserve some of their fundamental properties. One well-known example is a 1D turbulence model, the Burgers equation. Although this gives precious hints about the intermittency it has the unfortunate drawback of introducing compressibility. In comparison, the so-called "shell models" involve the conservation of the quadratic interaction and invariance for the flux of energy (Gledzer *et al.*, 1981), a very popular simplification. They are however ultimately simplistic since they are only scalar (not vector) models and they retain only the spatial scale dependence instead of location dependence. Indeed, these models consider the time evolution of the averaged characteristic velocity shear u_n (with corresponding vorticity $k_n u_n$) on the shell defined by the wave-vectors $|k| \approx k_n$, the wavenumber k_n being the inverse of the scale of the corresponding eddies, which is discretized in a algebraic way with $l_n = L/\lambda_0{}^n$, L being the outer scale, with the discrete scale ratio λ_0 (typically = 2), n the number of steps (Section 2.4). Whereas in real flows in d dimensions, the number of degrees of freedom increases as l^{-d}, the shell models keep only a small and fixed number (independent of l) and keep only certain interactions, the typical shell model equation of evolution being: $\left(\frac{d}{dt} + \nu k_n^2\right) v_n = k_n u_n u_{n-1} - k_{n+1} u_{n+1}^2$.

However, as we have seen, the increase in the number of degrees of freedom with scale is crucial for their cascade phenomenology and intermittency (and multifractality). In order to take into account these spatial degrees of freedom, Chigirinskaya and Schertzer (1996), Chigirinskaya *et al.* (1996) and Schertzer *et al.* (1997) proposed a tree structure of eddies, each eddy having $\lambda_0{}^d$ sub-eddies. They show that for both 2D and 3D turbulence the equations of evolution due to the direct interactions of eddies and sub-eddies are analogous to the Euler equations of a gyroscope. The indirect interactions are obtained by coupling an infinite hierarchy of gyroscopes. From a fairly abstract consideration of the structure of the Navier–Stokes equations (its "Lie structure"), they derived space-time models called scaling gyroscope cascades (SGC). The recognition of similarities between the Navier–Stokes equations of hydrodynamic turbulence and the Euler equations of a gyroscope can be traced back to Lamb (1963).

The SGC yields concrete models that can be used to investigate fundamental questions of turbulence, in particular its intermittency. Not only does the SGC in 2D yield the inverse energy cascade sub-range as well as the direct enstrophy sub-range (see Sections 2.4 and 2.5), but the multifractal characteristics of the former are extremely close to those of the SGC direct energy cascade in 3D turbulence. For the energy flux in both cases, in space, $\alpha \approx 1.4 \pm 0.05$, $C_{1,\varepsilon} \approx 0.25 \pm 0.05$ and in time, $\alpha \approx 1.5 \pm 0.05$, $C_{1,\varepsilon} \approx 0.25 \pm 0.05$. This is in excellent agreement with the empirical exponents found in the atmosphere: compare the identical values in time from hot-wire data from Schmitt *et al.* (1993) (Table 8.1) and in space: the aircraft estimate 0.046 (see Table 4.4: note that for the velocity, α is the same and $C_{1v} = C_{1,\varepsilon} 3^{-\alpha} \approx 0.05$; see Eqn. (4.15)). Also in good agreement with experiment is the exponent $q_{D,\varepsilon}$ for the divergence of moments for the energy flux: $q_{D,\varepsilon} \approx 2.3 \pm 0.06$. This implies $q_{D,v} = 3q_{D,\varepsilon} \approx 6.9 \pm 0.2$ for the velocity exponent, which is close to various temporal estimates in the range 7–7.7 (Section 5.3 and Table 5.1a).

Finally, when the SGC is reduced to a shell model (by considering a single path), the intermittency is quite different: $\alpha \approx 0.6 \pm 0.05$, $C_{1,\varepsilon} \approx 0.40 \pm 0.05$, showing the importance of the spatial degrees of freedom.

$\lambda \to \infty$ already posing a nontrivial mathematical problem (the weak limit of random measures: see Kahane, 1985). There were also some hasty claims (e.g. Yaglom, 1966) that iterating the process to smaller scales would lead to the (universal) lognormal model (see also Venugopal *et al.*, 2006, for similar arguments on the smallness of the high-order terms of the cumulant generating function $K(q)$). The claim of universality of the lognormal model was first criticized by Orszag (1970) and then by Mandelbrot (1974), the latter on the grounds that even if the cascade process was lognormal at each finite step, in the small-scale limit the spatial averages of the cascade process would *not* be lognormal. Furthermore, since the particularities of the discrete models (e.g. the form of $K(q)$ for the α model, Eqn. (3.24)) remain as a discrete cascade proceeds to its small-scale limit, the opposite extreme claim has since been made, that multiplicative cascades could not admit *any* universal behaviour: "in the strict sense, there is no universality whatsoever ... this fact about multifractals is very significant in their theory and must be recognized" (Mandelbrot, 1989). In the atmospheric literature, Gupta and Waymire (1993) repeated the same kind of claim. In both cases, their rejection of universality was based on a failure to understand the alternatives discussed by Schertzer and Lovejoy (1987, 1991a); this early debate is discussed in detail in Schertzer and Lovejoy (1997).

On the contrary, keeping the total range of scales fixed and finite, mixing independent processes of the same type (by multiplying them) while preserving certain characteristics (such as the variance or the amplitude of the generator), and *then* seeking the limit $\lambda \to \infty$, a totally different limiting problem is obtained. For instance, this may correspond to *densifying* the excited scales by introducing more and more intermediate scales (Fig. 3.14), and seeking thus the limit of continuous scales of the cascade model. Alternatively, we may also consider the limit of multiplications of independent, identically distributed, discrete cascade models leading also to universal multifractal processes.

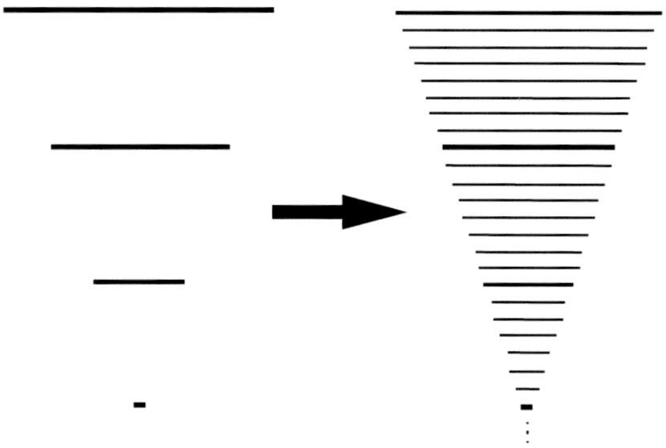

Fig. 3.14 Scheme of densification of scales. Reproduced from Schertzer and Lovejoy (1996).

3.3.3 Universality in cascades: a "multiplicative central limit theorem"

As presented above, the problem with cascades is that we need an entire (nearly arbitrary) convex function $K(q)$ for the specification. Yaglom (1966) sensed the problem and already argued – essentially using the law of proportional effects – for approximate "lognormality" on the basis of the usual central limit theorem applied to $\log\varepsilon$ after a large number of cascade steps. The problem is that the cascade requires a scale-by-scale conservation principle, otherwise there are no well-defined small-scale cascade limits, and it turns out that this normalization is in contradiction with the normalization required for central limit convergence (specifically, the former requires $\langle \mu\varepsilon \rangle = 1$ whereas the latter requires $\langle \Delta\Gamma \rangle = 0$ where $\Delta\Gamma = \log\mu\varepsilon$), and due to the convexity of the logarithm function we have necessarily $\langle \Delta\Gamma \rangle = \langle \log\mu\varepsilon \rangle < 0$ for any probability distribution of $\mu\varepsilon$ which is constrained such that $\langle \mu\varepsilon \rangle = 1$.

However, multifractal universality classes *do* exist. Two different routes to universality have been proposed, and both consider a cascade developed only over a finite range of scales. Only after central limit theorem convergence has been achieved does one consider the small-scale limit. The first route to universality (Schertzer and Lovejoy, 1987) relies on a "densification" of the cascade, adding more and more intermediate scales in a cascade defined over a finite range: an "infinitely divisible" or continuous (in scale) cascade (Fig. 3.14). Another route that was easier to analyze – the nonlinear "mixing" of cascade processes – was proposed by Schertzer *et al.* (1991); indeed, this very practical question of multifractal universality was the subject of debate during the 1990s (Gupta and Waymire, 1993; Schertzer and Lovejoy, 1997); see also Brax and Peschanski (1991) and Kida (1991) for the closely related issue of log-Lévy cascades. We should also mention that a weaker "log-Poisson" universality has also been proposed by She and Leveque (1994), but this is only "infinitely divisible" (continuous in scale); it is neither stable nor attractive (Section 5.2.5 and Eqn. (5.36)).

We now sketch the mathematical argument for multifractal universality (for details, see Schertzer and Lovejoy, 1987, 1991b, 1997). If we assume that the $K_I(q)$ for a single cascade step for each of the interacting processes ε_i are analytic at $q = 0$, then we can make a Taylor expansion about the origin:

$$K_1(q) = \sum_{i=1}^{\infty} A_i q^i \tag{3.40}$$

where the A_i are the expansion coefficients (the sum starts at $i = 1$ since $K(0) = 0$). In order to obtain an exactly lognormal cascade we may consider ε, which is the result of nonlinear (renormalized, multiplicative) interaction of N (generally non-lognormal), statistically independent discrete cascades with a total range of scale λ:

$$\begin{aligned}\varepsilon_\lambda &= \left(\prod_{i=1}^{N} \frac{\varepsilon_i}{a_N}\right)^{1/b_N} \\ K_N(q) &= log_\lambda \langle \varepsilon_\lambda^q \rangle = \frac{Nq}{b_N}(A_1 - Log_\lambda a_N) \\ &\quad + A_2 N \left(\frac{q}{b_N}\right)^2 + A_3 N \left(\frac{q}{b_N}\right)^3 + \ldots\end{aligned} \tag{3.41}$$

Here, i indexes the N independent cascade processes which interact (are multiplied together) and a_N, b_N are recentring and renormalizing constants which must be chosen so that the limit of many interacting processes ($N \to \infty$) is well defined. In the case of analytic $K(q)$ (which turns out to be exceptional!), we can choose to recentre (a_N) and renormalize (b_N) by:

$$b_N = N^{1/2}; \quad A_1 = Log_\lambda a_N \tag{3.42}$$

thus obtaining:

$$K_\infty(q) = \lim_{N\to\infty} K_N(q) = A_2 q^2 \tag{3.43}$$

i.e. the higher-order terms disappear, and $K_\infty(q)$ thus is a pure quadratic function independent of N; it is the moment scaling function of a pure lognormal cascade:

$$\langle \varepsilon_\lambda^q \rangle = e^{A_2 q^2 \log\lambda} = \lambda^{A_2 q^2} \tag{3.44}$$

Once the central limit theorem convergence has been achieved ($N \to \infty$), one then considers the small-scale limit ($\lambda \to \infty$). Here we must normalize the pure lognormal process so that the small-scale cascade limit is well behaved: this is easily performed by noting that an unnormalized ε may be normalized by $\varepsilon \to \varepsilon/\langle\varepsilon\rangle$ so that $K(q) \to K(q) - qK(1)$, so that we obtain:

$$K(q) = C_1(q^2 - q) \tag{3.45}$$

where we have used the notation C_1 for the constant A_2 (since $K'(1) = C_1$, Eqn. (3.23)) and we have dropped the subscript).

However – as pointed out by Lévy (1925) in the context of sums of independent random variables – the (log) Gaussian case does not exhaust the possibilities. Indeed, more generally we must allow for the possibility of nonanalytic single cascade step $K_1(q)$ with the following small q expansion:

$$K_1(q) = A_\alpha q^\alpha + A_1 q + A_2 q^2 + O(q^3) \tag{3.46}$$

If the new nonanalytic term $A_\alpha q^\alpha$ has $\alpha < 2$, then, repeating the above universality argument, with the choice:

$$b_N = N^{1/\alpha}; \quad A_1 = Log_\lambda a_N \tag{3.47}$$

we obtain:

$$K_\infty(q) = A_\alpha q^\alpha; \quad 0 \le \alpha < 2 \tag{3.48}$$

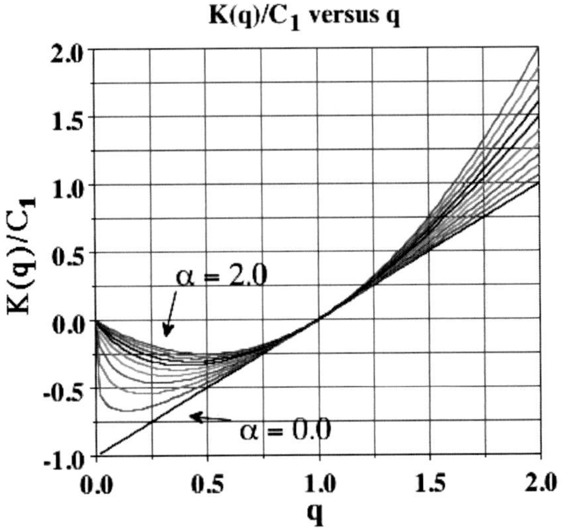

Fig. 3.15 Universal $K(q)$ as a function of q, for different α values from 0 to 2 by increments of $\Delta\alpha = 0.2$. Adapted from Schertzer and Lovejoy (1989).

for $\alpha \ne 1$. When $\alpha = 1$, the nonanalytic term must be taken as $q\log q$ (see below). As a technical point (Schertzer and Lovejoy, 1987, 1991b; Schertzer *et al.*, 1988), note that $K(q) = A_\alpha q^\alpha$ corresponds to a random generator $\Gamma = \log\varepsilon$ that follows an "extremely asymmetric" Lévy distribution, sufficient for cascade processes (see Samorodniitsky and Taqqu, 1994, for the more general Lévys needed for the complete treatment of random sums). The final normalization step needed for small-scale convergence (analogous to the lognormal derivation: $K(q) \to K(q) - qK(1)$) leads to:

$$K(q) = \frac{C_1}{\alpha - 1}(q^\alpha - q); \quad 0 \le \alpha \le 2 \tag{3.49}$$

(Schertzer and Lovejoy, 1987). For $\alpha = 1$, using l'Hopital's rule for the limit $\alpha \to 1$, we have $C_1 q\log q$ (see Fig. 3.15 for the shape of the universal $K(q)$ curves). Once again, the constant has been written this way so that $K'(1) = C_1$. We may also check that the local (near the mean) curvature characterization $\alpha = K''(1)/K'(1)$ is also valid, although here C_1 and α provide a global (complete) characterization of $K(q)$ (Eqn. (3.49)). In the limit $\alpha \to 0$, we recover the (linear) β model (Eqn. (3.27)). Note that when $\alpha < 2$, and $q < 0$, then $K(q) = \infty$; this is a consequence of the extreme Lévy tail on the negative (but not positive) fluctuations of $\log\varepsilon$ (see more discussion of Lévy variables in Section 5.4). The possibility (even likelihood) of $\langle\varepsilon_\lambda^q\rangle \to \infty$ for $q < 0$ means that extreme caution should be used when analyzing negative moments of empirical data. While finite datasets will always have finite negative moments, their values will sensitively depend on the small (near zero) values of the data and would yield spurious scaling properties.

We have already shown an example of a discrete-in-scale universal multifractal (Fig. 3.9b); one of the useful features of the universal multifractals is that they can be obtained as the continuous-in-scale limit of a discrete cascade, and hence they can be used for more realistic continuous-in-scale cascade simulations. This is discussed in more detail in Section 5.4. Finally, although the restriction $\alpha \leq 2$ falls out naturally from the preceding, we could mention that the fundamental reason that α cannot exceed 2 is that the corresponding probability densities would no longer be positive definite everywhere.

3.4 Summary of emergent laws in Chapter 3

The simplest example of intermittency is the "on/off" intermittency exemplified by the β model. The activity is distributed over a fractal set. Fractal sets have the property that if they are observed at resolution l (e.g. by covering them with l-sized "boxes"), then the fraction (= probability, P) of the space covered by the set is:

$$P \propto l^c \propto \lambda^{-c}; \quad \lambda = L_{ext}/l \tag{3.50}$$

where c is the fractal codimension and λ is the ratio of the largest-scale L_{ext} to the box-scale l. The corresponding fractal dimension is $D = d - c$ where d is the dimension of the space in which the set is embedded. More realistic cascades (such as the α model) involve intermediate levels of activity; they are multifractal. In both the β and α models, the turbulent flux ε is the result of a multiplicative cascade process, and it depends on its scale/resolution ratio λ as:

$$\langle\varepsilon_\lambda^q\rangle = \lambda^{K(q)} \tag{3.51}$$

Due to a kind of multiplicative central limit theorem, either via a densification of the cascade, or due to nonlinear interactions of discrete-in-scale cascades, we obtain convergence to "universal multifractal processes" with moment scaling function:

$$K(q) = \frac{C_1}{\alpha - 1}(q^\alpha - q) \tag{3.52}$$

where $0 \leq \alpha \leq 2$ and $0 \leq C_1 \leq d$, ($K(q) = C_1 q \log q$ when $\alpha = 1$).

The normalized autocorrelation function $R_\lambda(\Delta x)$ for the cascade developed over a scale range λ is:

$$\begin{aligned} R_\lambda(\Delta x) &\approx (\lambda\Delta x)^{-K(2)}; \quad 1 \geq \Delta x \geq \lambda^{-1} \\ R_\lambda(\Delta x) &\approx 1; \quad \Delta x \leq \lambda^{-1} \end{aligned} \tag{3.53}$$

so that the power spectrum (the Fourier transform of $R_\lambda(\Delta x)$) is:

$$E(k) = \lambda^{-K(2)} k^{-\beta}; \quad \beta = 1 - K(2) \tag{3.54}$$

Since $K(1) = 0$, $K'(1) > 0$ and $K'' > 0$, we have $K(2) > 0$ so that $\beta < 1$.

Appendix 3A: The convexity of $K(q)$

It turns out that an important property of $K(q)$ is that it is convex: $K'' > 0$. We have already appealed to this property in this chapter, and it is exploited systematically in Chapter 5. In this appendix, we derive this convexity property. Consider:

$$\langle \varepsilon_\lambda^q \rangle = \lambda^{K(q)} = \int \varepsilon_\lambda^q p(\varepsilon_\lambda) d\varepsilon_\lambda \tag{3.55}$$

where $p(\varepsilon_\lambda)$ is the probability density of ε_λ. Then

$$\frac{\partial \langle \varepsilon_\lambda^q \rangle}{\partial q} = \langle \varepsilon_\lambda^q \ln \varepsilon_\lambda \rangle; \quad \frac{\partial^2 \langle \varepsilon_\lambda^q \rangle}{\partial q^2} = \langle \varepsilon_\lambda^q \ln^2 \varepsilon_\lambda \rangle \tag{3.56}$$

Differentiating $K(q) \ln \lambda = \ln \langle \varepsilon_\lambda^q \rangle$, and using the above we obtain:

$$\ln \lambda \frac{\partial^2 K(q)}{\partial q^2} = \frac{1}{\langle \varepsilon_\lambda^q \rangle^2} \left(\langle \varepsilon_\lambda^q \rangle \langle \varepsilon_\lambda^q \ln^2 \varepsilon_\lambda \rangle - \langle \varepsilon_\lambda^q \ln \varepsilon_\lambda \rangle^2 \right) \tag{3.57}$$

To determine the sign of the term in parentheses, we can apply the Schwartz inequality:

$$\left(\int f^2 dx \right) \left(\int g^2 dx \right) \geq \left(\int fg dx \right)^2 \tag{3.58}$$

With $f = \varepsilon_\lambda^{q/2} [p(\varepsilon_\lambda)]^{1/2}$, $g = \varepsilon_\lambda^{q/2} [p(\varepsilon_\lambda)]^{1/2} \ln \varepsilon_\lambda$ and $dx = d\varepsilon_\lambda$, we obtain:

$$\langle \varepsilon_\lambda^q \rangle \langle \varepsilon_\lambda^q \ln^2 \varepsilon_\lambda \rangle \geq \langle \varepsilon_\lambda^q \ln \varepsilon_\lambda \rangle^2 \tag{3.59}$$

from which it follows that

$$\ln \lambda \frac{\partial^2 K(q)}{\partial q^2} > 0 \tag{3.60}$$

and since $\lambda > 1$, we have $K''(q) > 0$, i.e. $K(q)$ is convex (this corresponds to a basic result in probability theory, that second characteristic functions are always convex). Since trivially we have $K(0) = 0$ and via the conservation condition $\langle \varepsilon_\lambda \rangle = 1$ we have $K(1) = 0$, we conclude that $K(q)$ has the general form sketched in Fig. 3.10.

Chapter

4 Empirical analysis of cascades in the horizontal

4.1 The empirical estimation of turbulent fluxes in both dissipation and scaling ranges

4.1.1 Discussion

Before continuing our theoretical analysis of cascade processes, let us pause and survey the territory. In Chapter 1 we used a familiar data analysis technique – Fourier power spectra – to argue that atmospheric fields were generally scaling over potentially most of the dynamically significant ranges. In Chapter 2 we surveyed the main attempts to statistically understand the atmosphere, the elements of turbulence theory, the predictions that at least in 3D there would be cascade processes. We further argued that the unjustified imposition of rotational symmetry (isotropy) leads to the false conclusion that there must be (at least) two qualitatively different regimes: a small-scale 3D isotropic regime and a large-scale 2D isotropic regime. The 2D/3D model predicts drastic differences between the two isotropic regimes for both the statistics of vertical sections, and – at the "dimensional transition" separating the two regimes – for the statistics of passive scalar-like quantities. However, for the horizontal wind field, it predicts a much more mild transition – from $k^{-5/3}$ (3D downscale energy cascade) to k^{-3} (2D downscale enstrophy cascade, upscale energy cascade). In Chapter 2 we argued that by ignoring the literature (especially on the statistics of vertical sections and of remote sensing) and by focusing almost exclusively on the (difficult to measure) horizontal wind in the horizontal direction the 2D/3D model has been kept alive in spite of its serious theoretical and empirical shortcomings.

The decisive argument against the classical picture is quite recent: it comes from the reinterpretion of classical aircraft campaigns. This reinterpretation is based on two findings: (1) that the results of the aircraft campaigns had been misinterpreted: the transition in the horizontal wind spectrum was not that between $k^{-5/3}$ and k^{-3} but the significantly different transition from $k^{-5/3}$ to $k^{-2.4}$; (2) that not only had the $k^{-2.4}$ regime been mis-estimated but the entire regime is in fact a spurious artefact of the vertical aircraft motions. We concluded that in reality the horizontal scaling held over very wide ranges, and that this was possible because the scaling in the vertical was quite different from the horizontal, so that overall the atmosphere respected an anisotropic scale symmetry. The implication of wide-range scaling coupled with the cascade phenomenology was that we should expect emergent turbulent cascade laws to hold from large (perhaps planetary) scales on down.

In Chapter 3 we went beyond "conceptual" cascade models to consider explicit multiplicative cascades, and we examined some of their consequences, notably that the general statistics of the cascades can be specified by their statistical moments via the simple multifractal cascade equation:

$$\langle \varphi_\lambda^q \rangle = \lambda^{K(q)} \tag{4.1}$$

where λ is the ratio of the outer scale where the cascade begins and the resolution of the flux φ. In general, cascades only have the weak convexity ($K'' > 0$) and conservation ($K(1) = 0$) constraints. This is equivalent to an infinite number of parameters and is therefore impractical. However, we showed that due to a kind of multiplicative central limit theorem, under still rather general circumstances, cascades were expected to belong to the basin of attraction of "universal multifractals" in which only two parameters, C_1, α, are necessary: $K(q) = C_1(q^\alpha - q)/(\alpha - 1)$.

We now have enough tools to proceed to test these ideas on atmospheric data. We start with an empirical overview in the horizontal; extensions to the vertical will be discussed in Chapter 6, and to time in Chapters 8, 9 and 10.

4.1.2 Estimating fluxes from the fluctuations

In order to circumvent the isotropy/anisotropy issue and to test the general predictions of multiplicative cascades, we turn to Eqn. (4.1). However, Eqn. (4.1) assumes that both the flux φ and the outer cascade scale are known, whereas they must be estimated empirically without relying on any specific theories of turbulence; we must use an approach that does not require a-priori assumptions about the physical nature of the relevant fluxes, nor of their scale symmetries (isotropic or otherwise), nor of their outer scales.

The empirical determination of the outer scale is fairly straightforward. Consider L_{eff}, the "effective outer scale" where the cascade begins, and use the symbol $\lambda' = L_{eff}/L$ for the (unknown) scale ratio from the beginning of the cascade and the resolution of the flux φ. We will instead use the symbol λ as the ratio of a convenient reference scale to the resolution scale.

Now, starting with Eqn. (4.1), but with λ' in place of λ, the basic prediction of multiplicative cascades applied to a turbulent flux is that the normalized moments φ':

$$M_q = \langle \varphi'^{q}_{\lambda} \rangle; \quad \varphi'_{\lambda} = \frac{\varphi_{\lambda}}{\langle \varphi_1 \rangle} \tag{4.2}$$

obey the generic multiscaling relation:

$$M_q = \lambda'^{K(q)} = \left(\frac{L_{eff}}{L}\right)^{K(q)} = \left(\frac{\lambda}{\lambda_{eff}}\right)^{K(q)};$$
$$\lambda' = \frac{L_{eff}}{L} = \frac{\lambda}{\lambda_{eff}}; \quad \lambda = \frac{L_e}{L}; \quad \lambda_{eff} = \frac{L_e}{L_{eff}} \tag{4.3}$$

where "<.>" indicates statistical (ensemble) averaging. λ is a convenient scale ratio based on the largest great-circle distance on the earth: $L_e =$ 20 000 km and the scale ratio λ/λ_{eff} is the overall ratio from the scale where the cascade started to the resolution scale L; it is determined empirically, although from the foregoing discussion we expect $L_{eff} \approx L_e$ so that $\lambda_{eff} \approx 1$ corresponds to planetary-scale cascades. Since even at planetary scales each field nonlinearly interacts with the other fields, it is possible (and we often find) that L_{eff} is somewhat larger than L_e. We can see that λ_{eff} will be easy to empirically estimate since a plot of $\log M_q$ versus $\log\lambda$ will have lines (one for each q) converging at the outer scale $\lambda = \lambda_{eff}$.

Let us now consider the empirical estimation of the flux φ. There are two basic cases to consider. The first is widely applicable to empirical data, which are nearly always sampled at scales much larger than the dissipation scales (which are typically millimetric): it is the one described above based on the scaling range formula. If instead we have dissipation range data (for example if we estimate fluxes from the outputs of numerical models at the model dissipation scale), then the basic approach still works, but it requires a second interpretation.

Scaling-range flux estimates

If atmospheric dynamics are controlled by scale-invariant turbulent cascades of various (scale-by-scale) conserved fluxes φ, then in a scaling regime the (absolute) fluctuations $\Delta I(\Delta x)$ in an observable I (e.g. wind, temperature or radiance) over a distance Δx are related to the turbulent fluxes by a relation of the form:

$$\Delta I(\Delta x) \approx \varphi \Delta x^H \tag{4.4}$$

(Eqn. (4.4) is a generalization of the Kolmogorov law for velocity fluctuations – the latter has $H = 1/3$ and $\varphi = \varepsilon^{\eta}$, $\eta = 1/3$ where ε is the energy flux to smaller scales and the equality is in a statistical sense; see Eqn. (5.94) and Table 5.2 for a precise definition.) Without knowing η or H – nor even the physical nature of the flux – we can use this to estimate the normalized (nondimensional) flux φ' at the smallest resolution ($\Delta x = l$) of our data:

$$\varphi' = \varphi/{<}\varphi{>} = \Delta I/{<}\Delta I{>} \tag{4.5}$$

where "< >" indicates statistical averaging. Note that if the fluxes are realizations of pure multiplicative cascades then the normalized η powers $\varphi^{\eta}/\langle\varphi\rangle^{\eta}$ are also pure multiplicative cascades, so that $\varphi' = \varphi/{<}\varphi{>}$ is a normalized cascade (${<}\varphi{>}$ is the ensemble mean large-scale flux, i.e. the climatological value; it is independent of scale, hence there is no need for a subscript). The fluctuation, $\Delta I(\Delta x)$, can be estimated in various ways. In 1D a convenient method (which works for the common situation where $0 \leq H \leq 1$) is to use absolute differences: $\Delta I(l) = |I(x+l) - I(x)|$ where l is the smallest reliable resolution and where x is a horizontal coordinate (this is sometimes called "the poor man's wavelet"; other wavelets could be used: see Section 5.5). In 2D, convenient definitions of fluctuations are the (finite difference) Laplacian (estimated as the difference between the value at a grid point and the average of its neighbours), or the modulus of a finite difference estimate of the gradient vector. The resulting high-resolution flux estimates can then be degraded (by averaging) to a lower resolution $L > l$.

Dissipation-scale flux estimates

For data at resolutions high enough for viscous dissipation to be important, the scaling law (Eqn. (4.4)) can no longer be used to estimate the fluxes. In the atmosphere these scales are typically millimetric and such data are rarely encountered. However, in reanalyses and models, the finest resolutions are regularized using artificial "hyper-viscosities," so that their interpretation must be different. To see this, consider the example of the energy flux, recalling that at the dissipation scale the viscous term is dominant:

$$\varepsilon \approx \nu \underline{v} \cdot \nabla^2 \underline{v} \tag{4.6}$$

where ν is the viscosity, $\underline{v}$ the velocity (see Section 2.2.3). Standard manipulations give:

$$\varepsilon \approx \nu \sum_{i,j=1}^{3} \left(\frac{\partial v_i}{\partial x_j} + \frac{\partial v_j}{\partial x_i} \right)^2 \tag{4.7}$$

(the i, j index the velocity components). Therefore if Δx is in the dissipation range (e.g. the finest resolution of the model) then:

$$\Delta v \approx \left(\frac{\varepsilon}{\nu} \right)^{1/2} \Delta x \tag{4.8}$$

Since the meteorological models and reanalyses actually use hyper-viscosities with hyper-viscous coefficient ν^* and a Laplacian taken to the power h (typically $h = 3$ or 4), we have:

$$\varepsilon \approx \nu^* \underline{v} \cdot \nabla^{2h} \underline{v} \tag{4.9}$$

which leads to the estimate:

$$\Delta v \approx \left(\frac{\varepsilon}{\nu^*} \right)^{1/2} \Delta x^h \tag{4.10}$$

In all cases, we therefore have (independently of h):

$$\varphi' = \frac{\Delta v}{\langle \Delta v \rangle} = \frac{\varepsilon^{1/2}}{\langle \varepsilon^{1/2} \rangle} \tag{4.11}$$

We see that this is the same as Eqns. (4.2) and (4.3) with $\varphi = \varepsilon^{1/2}$, the only difference being that for the wind field the exponent $\eta = 1/2$ holds in the dissipation range rather than $\eta = 1/3$, which holds in the scaling regime. If we introduce $K_\eta(q)$, which is the scaling exponent for the normalized η flux φ', then:

$$\langle \varphi'^q_\lambda \rangle = \lambda^{\prime K_\eta(q)}; \quad \varphi' = \frac{\varphi}{\langle \varphi \rangle} = \frac{\varepsilon^\eta}{\langle \varepsilon \rangle^\eta};$$

$$K_\eta(q) = K_\varepsilon(q\eta) - qK_\varepsilon(\eta) \tag{4.12}$$

which for universal multifractals (Eqn. (3.49)) yields:

$$K_\eta(q) = \eta^\alpha K_1(q) \tag{4.13}$$

(note: $K_1(q) = K(q)$), i.e. in obvious notation:

$$C_{1,\eta} = \eta^\alpha C_{1,1} \tag{4.14}$$

so that, comparing the dissipation estimate ($\eta = 1/2$) and the scaling-range estimate ($\eta = 1/3$), we have:

$$C_{1,diss} = \left(\frac{3}{2} \right)^\alpha C_{1,scaling} \tag{4.15}$$

Since we find (for the wind field) that $\alpha \approx 1.8$ we have $C_{1diss}/C_{1scaling} \approx 1.5^{1.8} \approx 2.07$ (most of the fields have fairly similar α values).

The extension of this discussion to passive scalars is also relevant, and shows that the interpretation of the empirically/numerically estimated fluxes in terms of classical theoretical fluxes can be nontrivial. Denoting by ρ the density of the passive scalar, and χ its variance flux, the dissipation range formula analogous to Eqn. (4.6) is $\chi \approx \rho\kappa\nabla^2\rho$ (κ is the molecular diffusivity, assumed constant: see Eqn. (2.21)), leading to $\Delta\rho \approx (\chi/\kappa)^{1/2}\Delta x$, whereas the corresponding formula in the scaling range is $\Delta\rho \approx \chi^{1/2}\varepsilon^{-1/6}\Delta x^{1/3}$ (the Corrsin–Obukhov law). Here, the scaling- and dissipation-range formulae have the same dependency on χ, but the latter also involves the energy flux so that the combined effective flux $\phi \approx \chi^{1/2}\varepsilon^{-1/6}$ measured by the scaling method thus involves two (presumably statistically *dependent*) cascade quantities rather than just one. For example, Schmitt *et al.* (1996) examine various dependency hypotheses in the case of turbulent temperature fluctuations. In summary, although both dissipation and scaling ranges can be used to test for multiplicative cascades and to quantify their variability, the relation between the two normalized fluxes is not necessarily trivial.

Once the fluxes have been estimated at the smallest scale, the next step is to estimate them at the lower-resolution intermediate scales. This is done by spatially averaging the data over disjoint intervals (1D), disjoint boxes (2D) etc. of size L. The last step is to take the statistical (ensemble) average of the qth powers of the spatial averages over the intervals (or boxes). This is equivalent to an ensemble average over a "partition function" of the moments and is called a "trace moment" estimate of M_q (see Chapter 5 for more discussion).

4.2 The scaling properties of reanalyses

4.2.1 Discussion

We start our survey by considering the cascade properties of the ECMWF interim reanalyses whose spectra were analyzed in Fig. 1.5b. Reanalyses provide convenient (and sometimes unique) sources of global-scale state variables (i.e. the wind, temperature, humidity etc.), and today their resolution in space and in time is high enough that their scaling can be reliably tested and their exponents estimated with some confidence.

Recall that a reanalysis is by no means an empirical field; it is rather a highly elaborated "product" obtained by using complex 3D (space: ERA40) or 4D (space-time: ECMWF interim) variational data assimilation techniques, and in the Twentieth Century Reanalysis (20CR) by using an ensemble Kalman filter method. All of these assimilation techniques are based on various smoothness and regularity assumptions: specifically, they don't take the strong resolution dependencies (Eqn. (4.1)) into account, they assume that at scales smaller than one pixel $K(q) = 0$.

In spite of the popularity and importance of reanalyses, the study of their scale-by-scale statistical properties is still in its infancy. The early pioneering studies of Boer and Shepherd (1983) and Strauss and Ditlevsen (1999) attempted to use them to directly test the framework of two-dimensional geostrophic turbulence theory. Rather than studying the scaling properties of the fields themselves, they started by converting them (by vertical integration) into theoretically preordained 2D products; and even these were not studied directly but rather via anomalies with respect to low-frequency means (yielding "stationary" and "transient" components); and these components were further decomposed into rotational and irrotational wind components rather than the more straightforward zonal and meridional components. In Chapter 6 we see that in the spectral exponents there is a systematic (probably spurious) scale-by-scale north–south/east–west anisotropy. By neglecting to perform more basic scaling analyses, the early studies of reanalyses were thus "blind" to what turns out to be a significant source of bias. In contrast, the more recent studies of reanalyses (Stolle *et al.*, 2009, 2012; Lovejoy and Schertzer, 2011) directly focus on the scaling properties of the turbulent fluxes which were expected to be the products of multiplicative cascade processes. They attempted to achieve a wide survey of the statistical properties of the reanalysis fields, comparing many different models, studying the variations in the cascade properties as functions of altitude, of latitude and of forecast horizon.

In order to achieve an overview of the cascade properties of reanalyses we will use the state-of-the-art ECMWF interim reanalysis, chosen because of its high resolution, its ready availability and its recognized overall high quality. We specifically focused on the 700 mb level as being representative of the "free" atmosphere (without too many issues caused by the topography yet low enough to be data-rich), and we studied the daily (0 GMT) products for the year 2006, primarily concentrating on a band between ± 45° latitude. Studies of the other pressure levels ≥ 200 mb showed that the differences were not great (in accord with Stolle *et al.*, 2009). We also made intensive studies of the 20CR reanalysis at 700 mb, since this is particularly long (from 1871 through 2008 and at 6-hour resolution: see Appendix 10C) and will be useful later when we discuss the weather/climate transition and the climate regime.

4.2.2 The reanalysis products

At the time of writing, the ECMWF interim reanalysis products were available from 1989 through 2009. The full analysis is on a T255 spherical harmonic grid corresponding to a resolution of about 0.7° (≈ 79 km), and at 60 pressure levels. The publicly available products used here are on 1.5° latitude/longitude grids (corresponding to 166 km at the equator) and at 37 pressure levels (every 25 mb in the lower troposphere). Although this is slightly lower than the full raw resolution, it has the advantage of being less contaminated by the artificial hyperviscous dissipation used at the smallest scales (although there are still some dissipative effects, as can be seen from the spectra). The reanalyses use a 4D var scheme to assimilate data from both in-situ and remote sources with the help of the ECMWF numerical forecast model (see Berrisford *et al.*, 2009, for more details).

Of the 14 parameters available at the 700 mb level (another 88 are available at the surface), we chose the temperature, specific humidity, zonal wind, meridional wind, vertical wind and geopotential height (respectively: T, h_s, u, v, w, z) as being the most thermodynamically and dynamically important. Fig. 1.5a, shows the corresponding 0 GMT, January 1, 2006 fields, and Fig. 1.5b shows their isotropic (angle-integrated) spectra. At (absolute) latitudes greater than 45°, the pixel size becomes markedly reduced; in addition, the data near the poles are much sparser (hence both the in-situ and satellite data are less reliable); consequently we primarily analyzed the region between ± 45° latitude using a cylindrical projection. This restriction has the advantage that to a reasonable approximation we can ignore the map factor variation in pixel size and treat the data as coming from a Cartesian grid, analysing and comparing exponents in the east–west and north–south directions by using numerically convenient Fourier techniques. This would not be easy using the – otherwise theoretically preferable – spherical harmonic spectral decompositions.

4.2.3 The basic cascade structure of reanalyses

The spatial ECMWF cascade analyses are presented in Fig. 4.1a (zonal), and Fig. 4.1b (meridional) (for the corresponding temporal analyses see Fig. 8.7a). In each case we start with the finite difference absolute Laplacian flux estimate, which was then degraded by spatial averaging in the corresponding direction and then statistically averaged over the other directions (space and/or time). In the figures, one can clearly see the basic cascade structure of lines converging to the external scales; note in particular that the external cascade scales are systematically comparable to the largest great-circle distance (20 000 km), and that the scaling is well respected at all but the largest scales (i.e. for $\log_{10}\lambda > \sim 0.6$, i.e. for scales $< \sim 5000$ km: see below for error estimates). Here, references to "cascade structures" are simply a convenient short-hand to indicate the converging straight lines predicted for the log of the moments versus log of the scale for multiplicative cascades – it does not refer to real-space fluid structures. The moments are only shown up to order $q = 2$ since for large enough q they become dominated by the largest value present in the data sample so that the results spuriously depend on the sample size ($K(q)$ becomes spuriously linear; this is a "multifractal phase transition": see Chapter 5). It was found that in these data the transition always occurred for q somewhat greater than 2, so that the moments shown here are well estimated from the data.

In order to quantify the cascades, we performed nonlinear fits to the universal multifractal form (Eqn. (3.45)). Since the results were found to vary little from zonal to meridional to temporal direction, rather than give separate tables of parameters for each, in Table 4.1 we give the mean parameters. The direction-to-direction differences were sufficiently small (they are apparently less than the systematic and statistical uncertainties in the parameter estimates) that the direction-to-direction mean is given with the corresponding spread indicated by the standard deviation. Two estimates of the C_1 parameter are given, one from the universal multifractal fit (i.e. with $K(q)$ constrained to the form of Eqn. (3.45)) and the other simply from the numerical derivative: $C_1 = K'(1)$; in both cases, the lines were constrained to pass through a single external scale since this is the prediction of the fundamental cascade (Eqn. (4.1)). It can be seen that the two C_1 estimates are very close, although the difference is generally larger than the direction-to-direction spread; we take this as evidence that the direction-to-direction differences in the exponents (corresponding to possible zonal/meridional or space-time scaling anisotropies in the fluxes) are not statistically significant (note that the spread is for the three directions: zonal, meridional and time).

In order to quantify the accuracy of the fits, we estimated the residuals as:

$$\Delta = \overline{|log_{10}(M_q) - K(q)log_{10}(\lambda/\lambda_{eff})|};\ \delta = 100(10^{\Delta} - 1) \tag{4.16}$$

where the overbar represents averaging over all the moments $q \le 2$ and over the scale ratios larger than a critical value, taken as $\log_{10}\lambda = 0.6$ corresponding to 5000 km. The percentage deviation δ over the range is given in Table 4.1 and is estimated as in Eqn. (4.16).

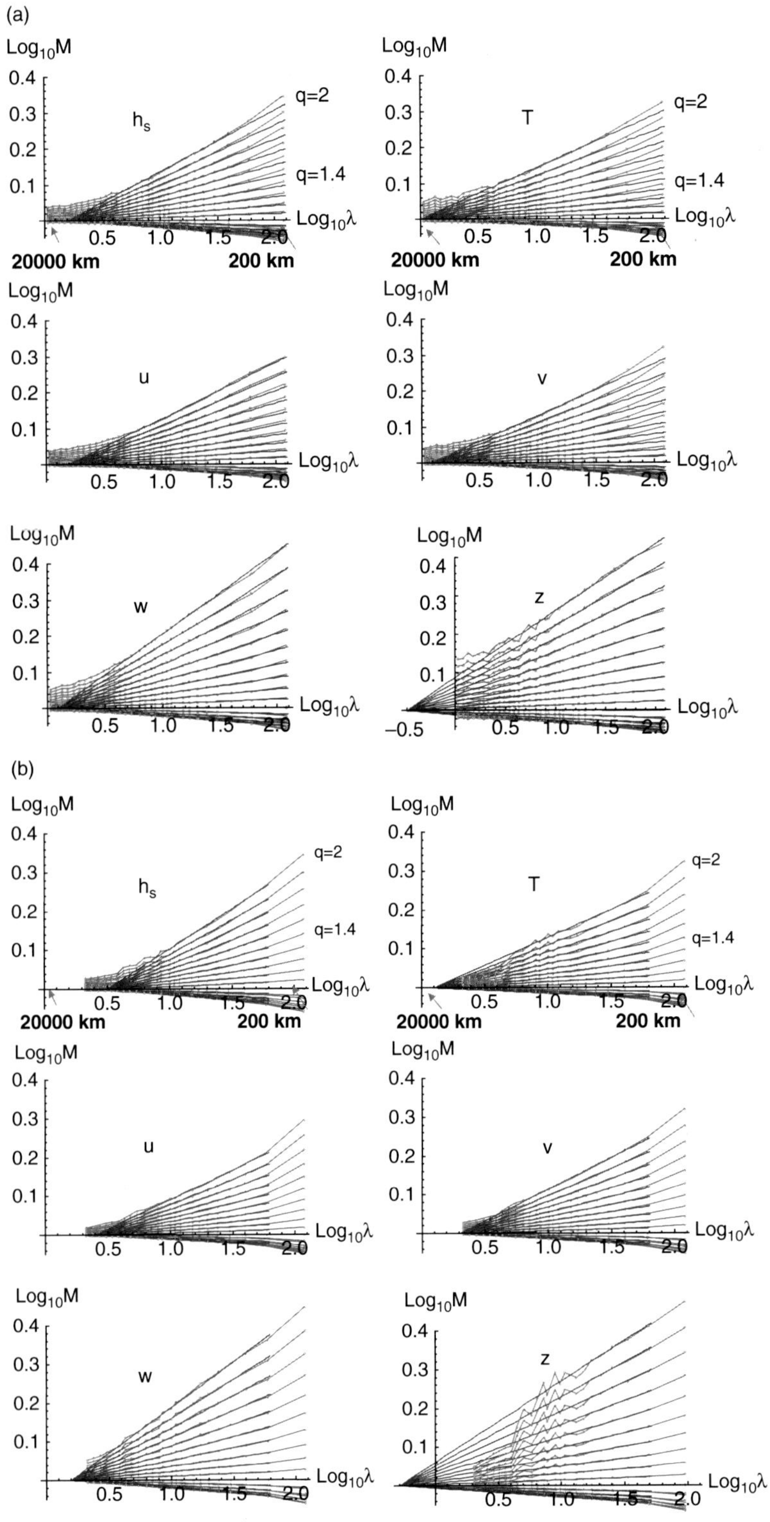

Fig. 4.1 (a) The analysis of the 700 mb ECMWF reanalyses at 0Z for 2006 between latitudes ± 45°. The fluxes were estimated using finite difference Laplacians. The curves are the moments $q = 0, 0.1, 0.2, \ldots, 1.9, 2$ (top). $\lambda = 1$ corresponds to the size of the earth, 20 000 km. Reproduced from (Lovejoy and Schertzer (2011). (b) The same as Fig. 4.1a but for the meridional analysis. The data span 90° in latitude, i.e. 10 000 km, but the reference scale was kept at 20 000 km as in Fig. 4.1a. Reproduced from Lovejoy and Schertzer (2011).

Table 4.1 Fits for C_1, α are for scales < 5000 km, the spreads in the parameters are over east–west and north–south directions, and for time for the ECMWF interim reanalyses. The temporal fits are from 2 to 8 days (these are discussed in Chapter 8). The β and hence H parameter estimate of the v (meridional wind) field is only fit up to 1300 km. The H estimates are for the spatial analyses only, they are from β and $K(2)$: $H = (\beta + 1 - K(2))/2$ with $K(2)$ from the universal multifractal fits and using the β from the isotropic, zonal and meridional horizontal spectral analysis in Chapters 1 and 6; they are for the zonal H values (the only exception is the meridional wind, where it is the north–south value; the value in the orthogonal direction can be obtained by dividing by $H_y \approx 0.80$. See Table 8.1 for temporal H estimates. Note that the parameter estimates are based on dissipation scale fluxes; this implies that the C_1's are different from those based on scaling-range data flux estimates (e.g. aircraft or other typical in-situ data): see Eqn. (4.15).

	h_s	T	u	v	w	Z
β	1.90	2.40	2.40	2.40	0.40	3.35
$C_{1,uni}$	0.102 ± 0.009	0.077 ± 0.005	0.084 ± 0.006	0.087 ± 0.012	0.121 ± 0.007	0.088 ± 0.006
C_1	0.101 ± 0.009	0.072 ± 0.005	0.082 ± 0.007	0.085 ± 0.013	0.115 ± 0.008	0.083 ± 0.005
α	1.77 ± 0.06	1.90 ± 0.006	1.85 ± 0.012	1.85 ± 0.011	1.92 ± 0.009	1.90 ± 0.012
H	0.54	0.77	0.77	0.78	0.14	1.26
$L_{eff,EW}$ (km)	13 000	20 000	13 000	16 000	16 000	63 000
$L_{eff,NS}$ (km)	6300	16 000	8000	10 000	13 000	40 000
$T_{eff,time}$ **(days)**	46	58	29	29	37	290
δ (%)	0.32 ± 0.04	0.35 ± 0.02	0.31 ± 0.09	0.28 ± 0.10	0.33 ± 0.10	0.52 ± 0.30

From the table we see that the typical accuracy is better than ± 0.5% (the worst fit was for the geopotential height in the meridional direction, where it was ± 0.9%). We may also note that all the different fields have very similar intermittency parameters: for the universal multifractal estimates, the C_1's vary only from 0.077 (T) to 0.121 (w), and α only from the lowest 1.77 (h_s) to the highest 1.92 (w). The most significant differences are for the external scales. Whereas the scaling exponent C_1 quantifies the scale-by-scale change of the variability (intermittency), for fixed q, changing the external scale changes M_q at all scales by the same factor. Since the zonal and meridonal $K(q)$'s are very close, there is little scale-by-scale (differential, not absolute) horizontal anisotropy; however, the ratio of the zonal to meridional external scales is 1.6 ± 0.3 (the spread is from one of the six analyzed fields to the other), which indicates a significant "trivial anisotropy" so that typical isolines of flux are (roughly) elongated ellipsoids a factor 1.6 longer in the zonal direction than in the meridional direction. This corresponds to the fact that the gradients of the fluxes (at each scale) are typically about 1.6 times stronger in the north–south direction, and this aspect ratio is roughly independent of scale. We return to this important anisotropy issue later, since for the reanalysis spectra it appears that at least some of it is a spurious artefact of the hyperviscous model dissipation.

For these spatial analyses, the scaling holds over factors of 30 or more, the intermittency is strong enough and the deviation Δ small enough that the results are fairly convincing. However, to gain more confidence in this conclusion, it is worth considering the corresponding M_q graph for quasi-Gaussian processes such as Ornstein–Uhlenbeck processes that have $K(q) = 0$ and are used, for example, as the basis for stochastic linear forcing modelling (e.g. Penland, 1996; Sardeshmukh *et al.*, 2000; for discussion, see Appendix 10B). The resulting "universal" quasi-Gaussian M_q graphs for these processes are discussed in Appendix 4A; they show that convergence to the theoretical $K(q) = 0$ result takes roughly a factor of 10 in scale to achieve. This is more significant for analyses of temporal cascades, where often the data are of sufficiently low resolution (e.g. daily) that the ratio of the outer time scale (≈ 10 days) to the resolution is of the same order as this convergence scale. This makes the establishment of cascade scaling less convincing (see the comparisons in Chapters 8 and 10).

It is of interest to compare the ECMWF interim parameters to those of other models and reanalyses and with those published in Stolle *et al.* (2009). The latter were restricted to the zonal wind (u), temperature (T) and humidity fields (h_s). Considering the spatial (zonal direction) analyses (Table 4.2a) we see that the parameters are very close to those of the other

Table 4.2a The zonal analysis of the fluxes obtained as the Laplacians (the ECMWF interim estimates are from Fig. 4.2; the ERA40, GEM, GFS are from Stolle *et al.*, 2009). All the data are from ± 45° latitude, 700 mb (except the 20CR, which is only between 44° and 46°N but from 1871 to 2008, every 6 hours). The aircraft data are from Lovejoy *et al.* (2010) and have been corrected by the factor $(3/2)^{\alpha} \approx 2.07$, which is a theoretical estimate of the difference between the dissipation-scale flux estimates and the scaling-range flux estimates (using $\alpha = 1.8$; see Eqn. (4.15)). Also, they are for roughly 200 mb flight levels rather than 700 mb levels (but the model parameters did not change too much as functions of altitude). Note that the L_{eff} for the aircraft wind is probably too big due to turbulent intermittency effects of the aircraft trajectory; see Lovejoy *et al.* (2010) for discussion.

		ECMWF interim	ERA40	20CR	GEM	GFS	aircraft
u	α	1.86	1.93	1.87	1.68	1.80	1.94
	C_1	0.081	0.096	0.089	0.104	0.082	0.088
	L_{eff}	12 700	12 000	11 200	11 000	9000	25 000
T	α	1.89	2.11	1.85	1.94	2.00	1.78
	C_1	0.074	0.094	0.088	0.077	0.080	0.107
	L_{eff}	20 000	14 500	11 200	8300	8600	5000
h	α	1.70	1.75	1.73	1.60	1.74	1.81
	C_1	0.095	0.094	0.077	0.100	0.091	0.083
	L_{eff}	12 700	11 000	35 000	11 800	9000	10 000

Table 4.2b A comparison of the 1000 mb fields. The triplets (GEM) represent the parameter estimates for integrations of t = 0, 48, 144 hours, and the pairs (GFS) for t = 0, 144 hours.

	C_1			α			L_{eff} **(km)**			**δ (%)**		
T (GEM)	0.125	0.115	0.112	1.64	1.68	1.69	25 700	20 500	25 700	0.27	0.26	0.80
T (GFS)	0.142	0.138		1.72	1.71		27 900	26 000		0.59	0.60	
u (GEM)	0.121	0.122	0.123	1.68	1.62	1.61	11 000	11 000	12 300	0.32	0.36	1.24
u (GFS)	0.114	0.107		1.80	1.84		12 300	11 200		0.54	0.64	
h (GEM)	0.109	0.106	0.112	1.81	1.80	1.77	15 900	13 800	14 100	0.51	0.49	1.51
h (GFS)	0.128	0.128		1.86	1.81		21 700	20 900		0.46	0.46	

products, although the C_1 parameter is systematically a little smaller while the outer scales are systematically a little larger. At any scale these two effects will tend to cancel each other, since a smaller C_1 indicates that, starting at the external scale, the variability builds up more slowly than for the ECMWF interim; however, since there is an overall increase in the outer scale, this slower variability builds up over a wider range. In Table 4.2a we also show estimates from in-situ aircraft measurements which include various efforts to minimize the biases introduced by the aircraft trajectories (intermittent fractality as well as from the effects of nonzero slopes): see Section 4.3.

Very similar results were found for the forecast products of the Canadian Global Environment Model (GEM; Fig. 4.2a) at both $t = 0$ and $t = 144$, the National Weather Service Global Forecasting System model (GFS; Fig. 4.2b), and the Twentieth Century Reanalysis (20CR; Fig. 4.2c). These are also compared with ERA40, an early version of the ECMWF interim reanalysis; see Table 4.2b for the parameter estimates. We find, for example, that the deviations are of the order ± 0.3% for the reanalyses, ± 0.3% for GEM and ± 0.5% for GFS (the 20CR reanalysis is nearly the same). These small deviations allow us to conclude that the analyses and models accurately have spatial cascade structures. Overall, from the table we can also see that the $K(q)$ "shape parameter" – the difficult-to-estimate multifractal index α – is roughly constant at $\alpha = 1.8 \pm 0.1$. From Table 4.2c, we see that the scale-by-scale characterization of the intermittency near the mean (C_1) has a tendency to decrease with altitude, this effect being somewhat amplified by a decrease in the external scale (which decreases all the moments by

Table 4.2c Comparison of initial ($t = 0$) fields for various fields at 1000, 700, 200 mb. The triplets of values are for, ERA40 (denoted by "ERA"), GEM, GFS respectively. The aircraft estimates are from about 200 mb (the figure in parentheses is from aircraft analyses (Lovejoy *et al.*, 2009c), the second is corrected by the factor $(3/2)^{\alpha}$ needed – at least for the wind field – to estimate the dissipation scale C_1 from the scaling range C_1, see Eqn. 4.15).

	C_1			α			L_{eff} (km)			δ(%)		
	ERA	GEM	GFS	ERA	GEM	GFS	ERA	GEM	GFS	ERA	GEM	GFS
T (1000)	0.113	0.125	0.142	1.94	1.64	1.72	21 900	25 800	28 000	0.31	0.27	0.59
T (700)	0.094	0.077	0.080	2.11	1.94	2.00	14 500	8300	8600	0.29	0.47	1.02
T (200)	0.080	0.080	0.065	1.93	1.88	1.85	12 100	10 700	7800	0.30	0.36	1.17
T (aircraft)	(0.052), 0.107			1.78			5000			0.5		
u (1000)	0.105	0.121	0.114	1.93	1.68	1.80	12 900	11 000	12 300	0.33	0.32	0.54
u (700)	0.096	0.104	0.082	1.93	1.86	1.87	12 000	11 000	9000	0.24	0.29	0.83
u (200)	0.075	0.085	0.073	1.92	1.85	1.89	15 900	16 300	9000	0.267	0.35	0.76
u (aircraft)	(0.040), 0.088			1.94			25 000			0.8		
h_s, h_r (1000)	0.121	0.109	0.128	2.03	1.81	1.86	19800	15 900	21 700	0.33	0.51	0.46
h_s, h_r (700)	0.094	0.100	0.091	1.75	1.60	1.74	11 000	11 800	9000	0.26	0.37	0.46
h_s, h_r (200)	0.085	0.109	0.100	1.73	1.54	1.70	50 000	33 000	9700	0.47	0.56	0.64
h (aircraft)	(0.040), 0.083			1.81			10 000			0.5		

the same factor). Interestingly, the C_1 is very similar for the different fields (it is slightly larger for the humidity), although as expected from our discussion of the difference between dissipation and scaling-range flux estimates the C_1 are quite a bit larger than those measured by aircraft (Section 4.3), also shown in Table 4.2a: the difference is roughly the factor of ~2 estimated in Eqn. (4.15) for the velocity field (i.e. the dissipation versus the scaling-range flux estimate).

In Table 4.2b, we compare the two forecast models (GEM, GFS) in order to see if there are any systematic trends as the model integrations increase (i.e. as the effects of initial conditions become less and less important). No systematic trends are obvious, although for the 144-hour GFS forecast, the scaling is notably poorer (although still quite reasonable) with deviations less than about $\pm$ 1.5%. Note that because even the longest available forecast is still statistically influenced by the analyses, these results do not (quite) establish that the long-time behaviour of the model is cascade-like. In Chapter 8, we examine the cascade behaviour in the time domain.

At this point the reader may be disappointed that the C_1 values are typically "small," and therefore that intermittency apparently is of only minor importance. Indeed, 40 years ago, a common view was to consider intermittency important only for spectral "corrections." In this case, taking the example $C_1 = 0.05$, $\alpha = 1.8$ we find the correction $K(2) \approx 0.09$ to the 5/3 Kolmogorov spectral exponent (Eqn. (3.38)), which is small. However, as soon as we consider fluxes and/or moments away from the mean, the effects can be very important. For example, for the energy flux ($\varepsilon \approx \Delta v^3/\Delta x$) we have $C_1 = 0.053 \times 3^{\alpha} \approx 0.36$ so that the dominant contribution to the mean energy flux comes from a fractal set with codimension 0.36. Taking the dissipation scale as $\approx$ 1 mm and the outer scale as 10^4 km, this implies $\lambda = 10^{10}$, so that the set giving the dominant contribution is the fraction $\lambda^{-C_1} \approx 10^{-4}$, so that 99.99% of the field is too weak to significantly contribute. Similarly, we can easily estimate that the variance (the $q = 2$ moment of ε) is determined by the extremes corresponding to a fraction (probabilities) of 10^{-13}; for quasi-Gaussian processes this would be closer to 10^{-1}. As another example, in Fig. 3.1b, the extreme 16.4 standard deviations event corresponds to a Gaussian probability of $\approx 10^{-118}$, whereas the sample size is $\approx 10^4$ and $C_1 \approx 0.06$. The point is that the variability builds up scale by scale and a "low" exponent simply means that this

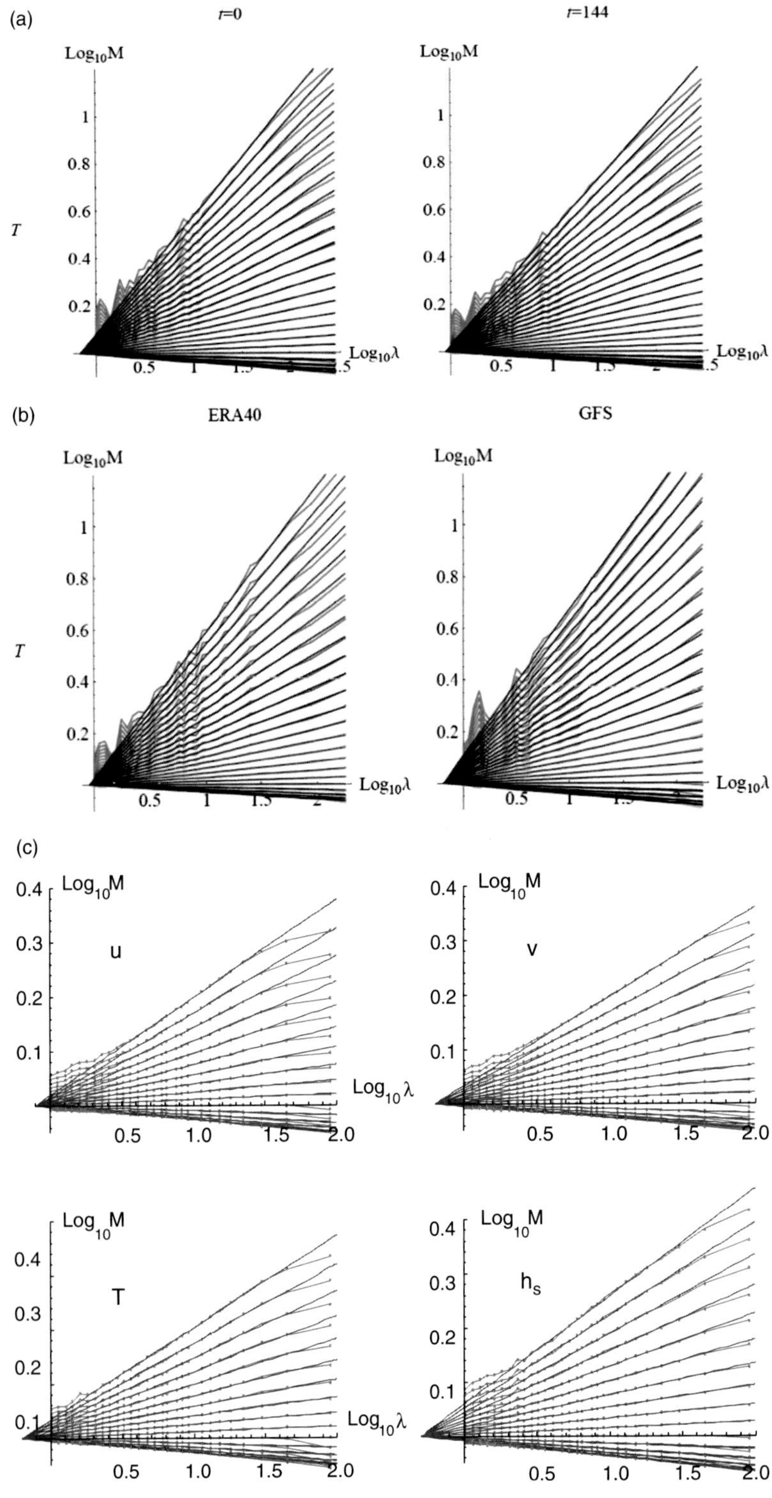

Fig. 4.2 (a) $\lambda = L_e/L$, $L_e = 20\,000$ km, $\pm 30°$ latitude. The left column is $t = 0$, the right column is the GEM analysis ($t = 144$ hours). Moments $q = 0, 0.1, 0.2, \ldots, 2.8, 2.9$. Reproduced from Stolle *et al.* (2009). (b) Same as Fig. 4.2a but for T at 1000 mb for ERA40 (reanalysis, left) and GFS (weather model, right). Reproduced from Stolle *et al.* (2009). (c) The zonal analysis of the spatial Laplacian of the zonal wind, meridional wind, temperature and specific humidity (upper left to lower right), from the 20CR reanalysis, from 1871 to 2008 (at 45° N every 6 hours and for $q = 0, 0.1, 0.2, 0.3, \ldots 2$).

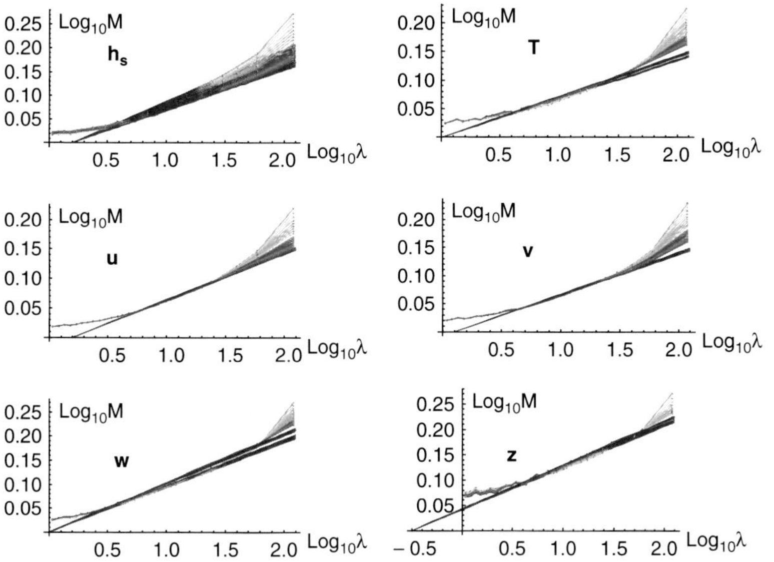

Fig. 4.3 The zonal analysis of the Lévy collapse. The vertical scale is $\log_{10}M'_q$ (Eqn. 4.17). ECMWF interim $q = 0, 0.1, 0.2, 0.3, \ldots 2$). The value of α used in all the collapses was $\alpha = 1.9$. Reproduced from Lovejoy and Schertzer (2011).

happens "slowly" with scale; nevertheless the resulting variability may be enormous!

4.2.4* Lévy collapse: multiplicative processes with and without scaling

We found that the universal multifractal form (Eqn. (3.49)) leads to excellent fits over all the moments (up to $q = 2$) and scales up to 5000–10 000 km, after which the scaling breaks down. However, even at scales so large that they are no longer in the scaling regime, the flux probability distributions may still be roughly of the special log-Lévy form. This would imply that the "reduced moments":

$$M'_q = M_q^{(\alpha-1)/(q^\alpha - q)} \tag{4.17}$$

are independent of q. Fig. 4.3 shows the results in the zonal direction when $\alpha = 1.9$ was used throughout to effect the "collapse" (roughly the mean α value found from the regressions, see Tables 4.1, 4.2). In the scaling regime, all the moments "collapse" onto a single straight line: $M'_q = \lambda^{'C_1}$. However, we see that even in the regime where $M_q{}'$ is no longer a power law (i.e. at large distances), the curves continue to collapse, indicating that they remain of the log-Lévy form (i.e. at a fixed scale λ), the random cascade "generator" $\Gamma_\lambda = \log\varphi_\lambda$ has second characteristic function $\log\langle e^{q\Gamma_\lambda}\rangle \propto (q^\alpha - q)$ so that Γ_λ has a Lévy distribution. The degree of collapse can be simply judged by inspecting the bunching of the lines for different q on top of each other: this can also be quantified scale by scale. For example, even at low λ, in the east–west direction the spreads of the lines for different q values at 20 000 km are: ± 10%, ± 1.5%, ± 6%, ± 6%, ± 10%, ± 10% for h_s, T, u, v, w, z respectively. Note that the spreads are large for the extreme small scale (one pixel in space); this is presumably a finite size effect: the problem of convergence to an accurate flux estimate. In Chapter 8 we reuse this technique in the time domain in the context of the weather/climate transition and climate-scale statistics. Note that while empirically we test the log-Lévy distribution of dressed moments, outside the scaling regime it is quite possible that these are not log-Lévy even though the bare process does have log-Lévy distributions.

4.2.5 The conservation/fluctuation exponent *H*

Up until now, we have concentrated our attention on the spectral properties (especially the exponent β, Chapter 1) and on the cascade properties of the turbulent fluxes (especially the exponents C_1, α). However, we indicated right at the outset that the observables and fluxes are related by the fundamental exponent H, viz: $\Delta v \approx \varphi\Delta x^H$; indeed, we exploited this relationship in order to estimate the fluxes (φ) from

Table 4.3 A comparison of the ECMWF interim multifractal parameters (isobaric) with those estimated for aircraft (horizontal) and dropsonde (vertical). The aircraft C_1's have been multiplied by the factor $(3/2)^{1.8} = 2.07$ in an attempt to take into account the fact that the aircraft flux estimates are in the scaling regime whereas the ECMWF estimates are in the dissipation regime.

	Source	*h*	*T*	*v*
α	ECMWF	1.77 ± 0.06	1.90 ± 0.006	1.85 ± 0.012
	aircraft	1.81	*1.78*	1.94
	dropsonde	1.85	1.70	1.90
C_1	ECMWF	0.102 ± 0.009	0.077 ± 0.005	0.084 ± 0.006
	aircraft	0.083	0.108	0.083
	dropsonde	0.072	0.091	0.088
H	ECMWF	0.54	0.77	0.77
	aircraft	0.51 ± 0.01	0.50 ± 0.01	1/3[a]
	dropsonde	0.78 ± 0.07	1.07 ± 0.18	0.75 ± 0.05

[a] Because of the issue of vertical aircraft movement, this (Kolmogorov) value was inferred, not directly estimated (see Lovejoy *et al.*, 2010).

the fluctuations in the observables (Δv). In the non-intermittent, quasi-Gaussian framework for φ of classical turbulence (e.g. Chapter 2), H is the unique exponent; φ has no scale dependence ($K(q) = 0$) so that it characterizes the variation of the mean fluctuations with size Δx. When, on the contrary, φ is an intermittent multifractal cascade process, the exponent H remains fundamental, but additional exponents are needed for a full characterization of the process. In this case, H characterizes the degree of scale-by-scale conservation of a quantity since $\langle \Delta v(\Delta x)\rangle = \langle\varphi\rangle \Delta x^H$, and $<\varphi>$ is independent of Δx so that the mean fluctuation is proportional to Δx^H; for a pure cascade quantity, $H = 0$. We also see that it has fundamental implications for the scaling of the fluctuations, since for example $H > 0$ implies that they grow with scale whereas $H < 0$ implies that they decrease with scale. This distinction is of fundamental importance in the temporal variability of the atmosphere since the sign changes twice, at about 10–20 days and at 10–100 years: see Chapter 10. Although the symbol H is used in honour of Harold E. Hurst and his pioneering work on long-range statistical dependency (Hurst, 1951), the H here is generally not identical to Hurst's; indeed, for multifractal processes, there appears to be no simple relation between the two.

A simple way to estimate H is to square this relation (Eqn. (4.4) for v) and take ensemble averages to obtain the (second-) order structure function:

$$\left\langle \Delta v(\Delta x)^2 \right\rangle = \left\langle \varphi^2_{\Delta x} \right\rangle \Delta x^{2H} \propto \Delta x^{2H-K(2)} \tag{4.18}$$

where we have used $\langle \varphi^2_\lambda \rangle = \lambda^{K(2)}$ and $\Delta x \propto \lambda^{-1}$. We now use the Wiener–Khinchin and Tauberian theorems (Appendix 2A) to relate the scaling of the spectrum to that of the structure function, to obtain:

$$\beta = 1 + 2H - K(2) \tag{4.19}$$

From this we see that given C_1, α (which determines $K(2)$) and the spectral exponent β, we can easily find H. Methods for estimating H in real space (essentially the direct use of classical different structure functions and their generalizations) will be discussed in Section 5.5 and Appendix 5E. Table 4.1 shows the results when the angle integrated spectrum was used to estimate β. When $E(k)$ is obtained by integrating $P(\underline{k})$ (Eqn. (2.103)) over angles in Fourier space, it is more statistically robust. However, for estimating β, the usual interpretation is only fully justified if the fields are in fact isotropic. We come back to this question and re-evaluate these analyses in Chapter 6.

In order to understand the H values for the wind, humidity and temperature, we have constructed Table 4.3. This compares the ECMWF reanalyses and aircraft (Lovejoy *et al.*, 2010) and dropsonde estimates (Lovejoy *et al.*, 2009b). Starting with the wind, and concentrating on the zonal component, we note that there is excellent agreement between the dropsonde (vertical) value and the ECMWF *isobaric* value, so it is significant that the reanalyses assume hydrostatic equilibrium. As argued in Lovejoy *et al.* (2009b), this agreement is likely because the isobars are gently sloping so that at large enough scales, one obtains the vertical rather than horizontal values. As a consequence, the aircraft wind value given in the table is the theory value 1/3 which was argued to be compatible with the small-scale aircraft statistics

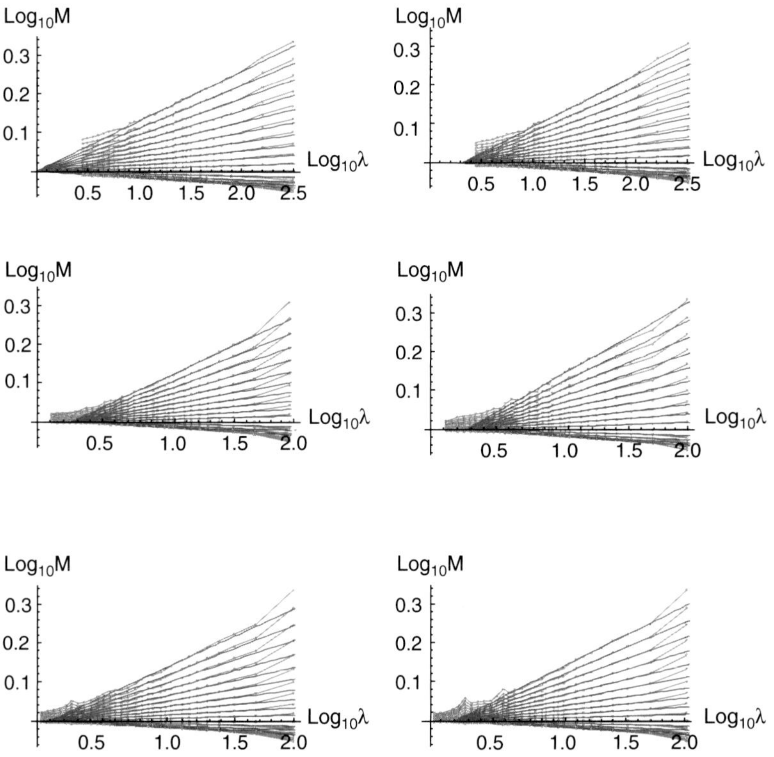

Fig. 4.4 The zonal wind (u) analyzed in the zonal direction for the ECMWF interim reanalysis, 700 mb level for the year 2006. The top row is the band 75–90°, the middle is 30–45°, the bottom is 0–15°; the left-hand column is the northern hemisphere, the right-hand column is the southern hemisphere (spatial finite difference Laplacians were used to estimate the fluxes). The moments are $q = 0.1, 0.2, \ldots 1.9, 2.0$. The outer scale is 20 000 km (corresponding to $\lambda = 1$). Note that at the higher latitudes the largest accessible scales are reduced due to the map projection.

when corrected for intermittent turbulent motions. The H value for the humidity (0.54) is at least close to the in-situ value (0.51), although to our knowledge it is not predicted by any existing theory. Similarly, the temperature value is far from the aircraft value, although it is close to the isobaric wind value, suggesting that at least in the reanalysis it is estimated as a passive scalar – i.e. advection dominates diabatic processes. In this case, it would be an isobaric estimate rather than an isoheight estimate similar to the H for the horizontal wind.

4.2.6 Latitudinal variations in the cascade structure

Up until now, we have taken statistics from $\pm$ 45° latitude in order to concentrate on the basic variation with direction (zonal, meridional, temporal). However, a basic aspect of atmospheric dynamics is its latitudinal dependence, notably due to the Coriolis force. Paradoxically, the fairly limited analysis of latitudinal variation in Stolle *et al.* (2009) found that it was small; this is presumably because the cascade structure is mostly dependent on nonlinear interactions whereas the most important north–south effects involve linear terms and boundary conditions. Let us now investigate this more systematically.

In order to study the latitudinal dependence, we broke up the earth into 15° bands. Fig. 4.4 shows the zonal analysis of several of these bands for the zonal wind (the other fields showed similarly small variations and are not shown, to economize on space); in Chapter 8 we discuss the corresponding temporal cascade structure. The main difference is a small but systematic change in the outer scales. The evolution is pretty small and is confirmed in the other bands. Also, there is generally a good degree of north–south symmetry. In order to quantify this, we turn to Figs. 4.5a and 4.5b, which show (4.5a) the evolution of the exponents C_1, α, and (4.5b) the external spatial scales. We see that for some of the fields (essentially the geopotential and specific humidity near the equator), the α values are a bit larger than the theoretical maximum (2), so that the curvature

(α) estimates (which were made here using $\alpha = K''(1)/K'(1)$) are not too accurate; their excesses above 2 are probably not statistically significant. We notice a slight tendency for the intermittency to increase away from the equator, especially in the southern hemisphere; it is sufficiently systematic that it is probably a real change in C_1.

Even though the exponents show remarkably little latitudinal variation, that does not imply that the cascade structure is nearly independent of latitude. Fig. 4.5c shows the variation of the external spatial scale. The main noteworthy features are: (1) the latitudinal variations (except perhaps the geopotential height) are relatively small in space, almost all between 10 000 and 20 000 km; (2) the external scales have significant north–south asymmetry. The reason for the asymmetry is not clear, and it may be partly due to (spurious) factors such as differences in data density or to (real) factors such as differences in land cover between the two hemispheres (Fig. 8.8a). We return to discuss the physical significance (and comparisons) of parameters at the end of the chapter.

4.3 The cascade structure of in-situ aircraft measurements: wind, temperature and humidity fields

4.3.1 The biases in the wind statistics

In Section 2.6, we discussed the fact that aircraft do not fly on perfectly flat trajectories, that over significant ranges of scales their trajectories are typically fractal; and more importantly that over long distances they typically have small but nonzero mean slopes. This opens up the possibility that their vertical fluctuations significantly influence their measurements. In order to understand this, we need a theory of anisotropic

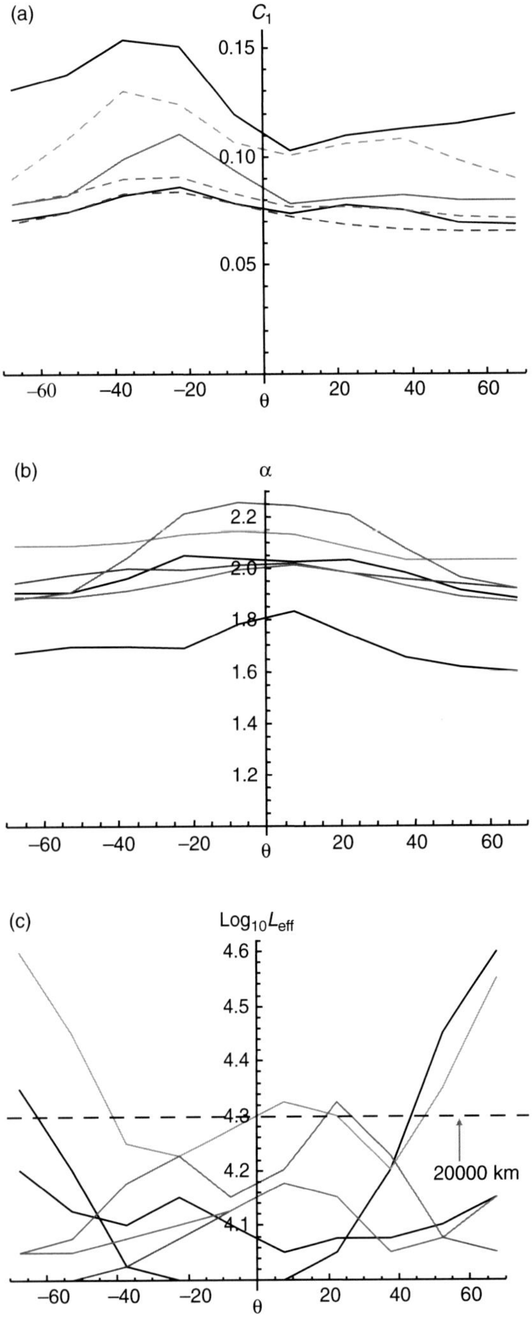

Fig. 4.5 (a) The cascade exponents C_1 from the spatial (zonal) analysis. From bottom to top this shows the zonal wind (u), the meridional wind (v), the temperature (T), the geopotential height (z), the vertical velocity (w), the specific humidity (h_s). The extreme latitude bands ($\pm$ 75–90°) were not used, since the mean map factor is very large and the results were considered unreliable. Reproduced from Lovejoy and Schertzer (2011). (b) Same as Fig. 4.5a except for α. From bottom to top this shows the specific humidity (h_s), the meridional wind (v), the zonal wind (u), the temperature (T), the vertical velocity (w), the geopotential height (z). Since theoretically multifractal indices α must be ≤ 2, the estimates $\alpha > 2$ are presumably unreliable, although error estimates are hard to obtain. Reproduced from Lovejoy and Schertzer (2011). (c) The effective external scales L_{eff} as functions of latitude (in units of km) from the zonal cascade analyses. The dashed line is a convenient reference line, corresponding to the largest great-circle distance on the earth, 20 000 km. From bottom to top (at the extreme left) this shows the zonal wind (u), the meridional wind (v), the geopotential height (z), the temperature (T), the specific humidity (h_s), the vertical velocity (w). Reproduced from Lovejoy and Schertzer (2011).

Table 4.4 Horizontal parameters estimated over the range 100 km down to 2 km, except for z, which is over the range 20 km to 0.5 km. Error estimates are made only for those which are apparently unaffected by aircraft trajectory: they are half the difference of parameters when estimated over the range 200 km to 20 km and 20 km to 2 km. Note that the aircraft α estimates are a bit too big, since the theoretical maximum is $\alpha = 2$. They were estimated with the double trace moment technique (Section 5.5.3), which depends largely on the statistics of the weaker events, and these could be affected by the intermittency of the aircraft altitude. The H parameters were estimated from the spectral exponent β and the value $K(2)$ using the equation $H = (\beta - 1 + K(2))/2$. Since the humidity is very low at the aircraft altitude, the equivalent potential temperature was extremely close to the potential temperature, so that the statistics were indistinguishable and are given in the table. The parameter values for the pressure should be taken with caution, since the aircraft was attempting to follow an isobar.

	T	Logθ	h	v_{long}	v_{trans}	p	z
H	0.50 ± 0.01	0.51 ± 0.01	0.51 ± 0.01	0.46	0.37	0.36	0.43
C_1	0.052 ± 0.012	0.052 ± 0.010	0.040 ± 0.012	0.033	0.046	0.031	0.068
α	*1.78*	1.82	1.81	2.10	2.10	2.2	2.15
L_{eff} **(km)**	5000	10 000	10 000	10^5	25 000	1600	50
δ **(%)**	0.5	2.0	0.5	0.4	0.8	0.5	2.6

turbulence as well as a model of how it affects the aircraft trajectories. Since a detailed discussion of anisotropic scaling is given in Chapter 6, we postpone a full treatment until then. However, we find that it is primarily the wind which is affected by the aircraft motions (especially the sensitive longitudinal or along-track component). Indeed, it is sufficient for the sphero-scales (the scales where typical structures have "roundish" vertical sections) for the temperature and humidity to be sufficiently large that they are relatively unaffected. This is confirmed by an analysis of the spectral coherence between the aircraft altitude and wind, temperature and humidity, which shows that while there is a strong coherence with the wind (and whose phase with respect to the altitude and pressure changes by 180° precisely at the $k^{-5/3}$ to $k^{-2.4}$ transition scale), there are only very low coherences with the temperature and humidity (Appendix 6B). It is therefore worth presenting the cascade analyses for these fields here. The data analyzed are from the same Pacific Winter Storms 2004 experiment described in Chapter 1, where the spectra were shown (Fig. 1.6c). Here we use 24 legs, each with 4000×280 m measurements, i.e. 1120 km long.

4.3.2 Aircraft estimates of horizontal cascade parameters

Figs. 4.6a and 4.6b show the flux analysis results for the longitudinal wind, transverse wind, pressure, temperature, humidity and potential temperature. Lovejoy *et al.* (2010) analyzed this in detail, concluding that as far as estimating horizontal scaling parameters is concerned the range 4–40 km is optimal (between the dashed lines in the figures); at smaller scales the trajectory is too intermittent, while at the longer scales one obtains isobaric rather than isoheight statistics. We nevertheless see a fairly convincing cascade structure for the wind (Fig. 4.6a), while the temperature, humidity and potential temperature show excellent scaling throughout (Fig. 4.6b). Once again, the outer scales are of the order of the size of the earth, although the L_{eff} for the wind is somewhat larger, the variability being presumably increased by the variability of the altitude – which, due to the aircraft response to turbulence, is also cascade-like.

Table 4.4 compares the parameter estimates using the "optimal range" where the statistics are least affected by the aircraft trajectory issues: 4–40 km. We have included the analysis of the vertical coordinate (z) of the aircraft, which was found to follow its own cascade with outer scale very close ($\approx$ 50 km) to the $k^{-5/3}$ to $k^{-2.4}$ transition scale. The interpretation given in Lovejoy *et al.* (2010) is that this scale marked the transition from a turbulence-dominated trajectory to a trajectory dominated by a slowly sloping pressure surface, where the vertical variability in the horizontal wind dominates over its horizontal variability.

We should mention that the aircraft did not fly in particular orientations with respect to meridians so that the exponents and outer scales are "isotropic" estimates. It will be necessary to perform special experiments to properly distinguish the zonal and

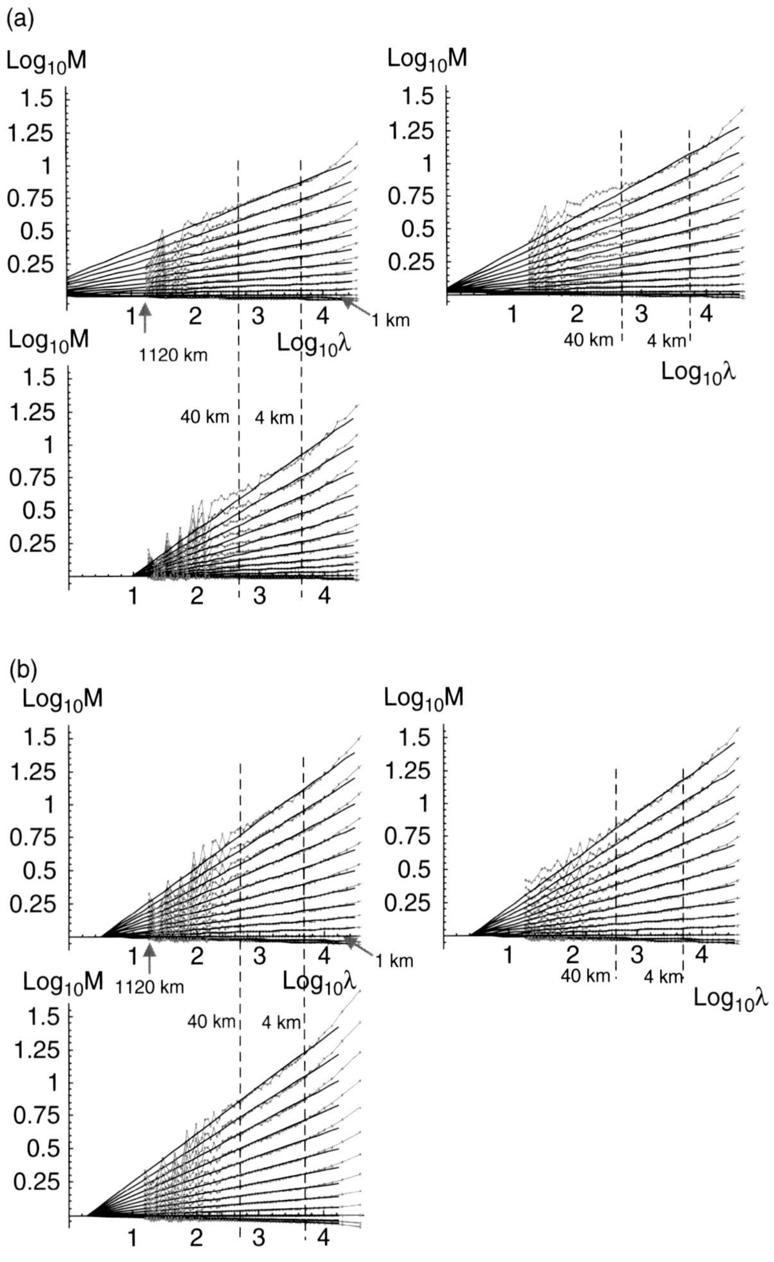

Fig. 4.6 (a) Cascade structures for the fields strongly affected by the trajectories: the longitudinal wind (top left), the transverse wind (top right), pressure (lower left). Reproduced from Lovejoy *et al.* (2010). (b) Same as Fig. 4.6a but for the fields that are relatively unaffected by the trajectory: temperature (top left), relative humidity (top right), log potential temperature (lower left right). Reproduced from Lovejoy *et al.* (2010).

meridional variabilities. For the moment, using the reanalyses (previous section) as a guide, we can assume that this will not affect the C_1, α (and probably H, although see Chapter 6) exponents while it will affect the external scale estimates. In this regard, we can already remark that when comparing with the reanalysis estimates, that for h, the aircraft exponents and external scales are very close while for T both the exponents and external scales are significantly different (due to the problems due to aircraft intermittency, the difference between the horizontal wind external scales is not significant). We further compare the aircraft parameters to those of the reanalyses (and others) in Box 4.1.

Box 4.1 Overview of the horizontal scaling properties of atmospheric fields

We have rapidly surveyed some of the recent studies of the large-scale cascade properties of atmospheric fields and boundary conditions (topography and, in Chapter 1, ocean colour, soil and vegetation indices). These studies profit from the ready availability of massive quantities of high-resolution state-of-the-art data, and most of the analyses would not have been possible more than 10 years ago. Nevertheless, each has its own problems/limitations: nontrivial aircraft trajectories, problems with zero and low rain rates, problems estimating areal rainfall, problems with the hyperviscous smoothing and hydrostatic nature of the reanalyses (see Chapter 6); we must consider that the analyses we have presented are only the first steps towards a full scale-by-scale understanding of the horizontal statistics. Before continuing (in the next chapter) to a more advanced understanding of cascades and multifractals, it is nevertheless worth taking an overview of the basic empirical results presented so far.

Table 4.5 summarizes different categories of values (e.g. radiances) or gives approximate values when there are several sources (e.g. the rain rate but also some of the state variables). The values given for the state variables are those of the in-situ measurements (when available), with parentheses indicating the reanalysis values in those cases where they differ significantly from the former. We can see from the table that the C_1 values are remarkably constant at around 0.1 – the main exception is the very intermittent R field (high C_1) – and that the α values vary in a fairly narrow range around 1.5–1.9.

Table 4.5 A comparison of various horizontal parameter estimates, attempting to summarize categories of values (radiances) or approximate values. When available (and when reliable), the aircraft data were used in precedence over the reanalysis values, with the latter given in parentheses in those cases where there was no comparable in-situ value or when it was significantly different from the in-situ value. The aircraft values were also increased by a factor 2.07 to take into account the difference between dissipation and scaling range estimates: cf. Eqn. (4.15). For L_{eff} where the anisotropy is significant, the geometric means of the north–south and east–west estimates are given; the average ratio is 1.6 : 1 EW/NS (although for the precipitation rate, the along-track TRMM estimate was used). Finally, the topography estimate of L_{eff} is based on a single realization (one earth!), so we only verified that there was no obvious break below planetary scales. The aerosol concentration was estimated from the lidar backscatter ratio as discussed in Section 6.5.1.

		C_1	α	H	β	L_{eff}
State variables	u, v	0.09	1.9	1/3, (0.77)	1.6, (2.4)	(14 000)
	w	(0.12)	(1.9)	(−0.14)	(0.4)	(15 000)
	T	0.11, (0.08)	1.8	0.50, (0.77)	1.9, (2.4)	5000 (19 000)
	h	0.09	1.8	0.51	1.9	10 000
	z	(0.09)	(1.9)	(1.26)	(3.3)	(60 000)
Precipitation	R	0.4	1.5	0.00	0.2	32 000
Passive scalars	Aerosol concentration	0.08	1.8	0.33	1.6	25 000
Radiances	Infrared	0.08	1.5	0.3	1.5	15 000
	Visible	0.08	1.5	0.2	1.5	10 000
	Passive microwave	0.1–0.26	1.5	0.25–0.5	1.3–1.6	5000–15 000
Topography	Altitude	0.12	1.8	0.7	2.1	20 000
Sea surface temperature	SST (see Table 8.2)	0.12	1.9	0.50	1.8	16 000

We should note that the C_1 values for the state variables are from dissipation-scale estimates and will thus be different from the in-situ (aircraft) results, which are for scaling-range estimates (see Eqn. (4.15)); in each case, C_1 effectively characterizes a different flux (e.g. $\varepsilon^{1/2}$ or $\varepsilon^{1/3}$ in the case of the wind). The H values, with the exception of the vertical wind, are > 0, indicating that the fields are smoother than the fluxes and that fluctuations grow with increasing scale Δx. Although the values of H have classically been determined by dimensional analysis, only the value for the horizontal wind – and this is indirectly inferred through data from problematic aircraft trajectories – has a theoretically explained value (1/3). In comparison, Corrsin–Obukhov theory for passive scalars (Chapter 2) also yields $H = 1/3$ (apparently verified for aerosol concentrations: see the table and Section 6.5.1) whereas the Bolgiano–Obukhov buoyancy-driven turbulence (Chapter 6) gives $H = 3/5$ for the wind (close to the observations), but only 1/5 for the temperature. Paradoxically, the value $H \approx 1/2$ (for T, SST, h) corresponds to "normal" diffusion in a solid, but is quite anomalous with respect to the turbulent Corrsin–Obukhov value of 1/3 and we know of no theory that predicts it.

Box 4.1 (*cont.*)

Finally, not shown in the table is the remarkable agreement on the outer scales in both north–south and east–west directions. First, they are all within about a factor of 2 of the planetary scale; second, the north–south and east–west exponents seem to be about the same (although for the precipitation this is not totally clear, and for the aircraft data this has not yet been evaluated); third, even the typical east–west/north–south aspect ratios of structures is nearly constant at about 1.6 : 1.

4.4 The cascade structure of precipitation

4.4.1 The special role and properties of precipitation

Precipitation is highly significant not only for meteorology, where it plays a key role in the earth's energy and water budgets, but also in hydrology, where it is the input field. Yet it has a number of peculiarities that make it somewhat different from most other atmospheric variables. First, from the empirical standpoint, the problem of estimating areal rain-rate fields – whether from (typically sparse) in-situ networks of gauges or from remote sensing (especially radar reflectivities) – is still open. Strictly speaking – due to intermittency, the fact that $K(q) \neq 0$ – this could also be said of any of the other techniques used to produce smooth fields from in-situ data (such as "Kriging," "optimal interpolation," or "3D and 4D var"); however, since rain is much more intermittent (it has a much larger C_1; see Tables 4.5, 4.7.) than the other fields its effect is much stronger and more difficult to ignore. Second, intermittency in rain probably has a significant on/off component: it is apparently zero at most times and places; yet even the exact definition of the (nonzero) "support" – the distinction between rain and no rain – is quite ad hoc, typically depending of the sensitivity of the instrument. Indeed, it may be that the on/off intermittency is a threshold-type effect, in which case it is not scaling at all, since the scale at which the threshold is introduced breaks the scaling. (The alternative – to modulate the rain rate process by another fractal rain/no rain process as proposed by Gupta and Waymire (1993) is also problematic since it would imply, in the small-scale limit, zero rain areas.) Finally, from a modelling viewpoint, there are no theoretically "clean" ways to represent the rain rate in the same way as the main state variables – i.e. as coupled nonlinear partial differential equations – without resorting to strong, unsatisfactory "parametrizations."

The use of scaling in rain – especially in the hydrology literature, where it is mostly implicit as the input field – goes back to at least Hurst (1951), and it includes early contributions on long-range statistical dependencies and extremes (essentially fractional Brownian motion and Lévy processes (Mandelbrot and Wallis, 1968), and wide-range spatial scaling (Lovejoy, 1982). It was also the first field to be used to test multiplicative cascades (Schertzer and Lovejoy, 1987): an early review (Lovejoy and Schertzer, 1995) already included over 50 references. Rather than provide a comprehensive review, which would be too extensive for this book, we will touch upon the main strands of scaling in precipitation and follow this by more in-depth discussion of some recent global-scale analyses.

The scaling precipitation literature is still far from consensual. For example, a key issue is the value of H: is the rain-rate field itself the direct outcome of a cascade process or is it only driven/forced by a turbulent cascade-generated flux? Certainly – although many papers on precipitation simply assume $H = 0$ without discussion – the answer is not self-evident. In rain, it seems likely that $H > 0$ – as theoretically expected for passive scalars in turbulence – at small scales in space $H \approx 1/3$ (as for the liquid water density: Lovejoy and Schertzer, 2008). However, at larger scales – at least for the related radar reflectivity and in the horizontal direction – it was found that $H = 0.00 \pm 0.01$ (Lovejoy *et al.*, 2008). The low H result gives an ex post facto justification for the common assumption in the precipitation literature that H is indeed exactly zero. However, this result could be biased by the satellite radar data, which had a high threshold (only detecting rain in 3.5% of the pixels). Using the huge CPC hourly precipitation database discussed below, we find in time for rain rates $H \approx 0.17 \pm 0.11$ up to 2–3 days (see Serinaldi, 2010, who finds in time series $H \approx 0.19$), a result which itself is somewhat dependent on the rain/no rain threshold of the gauges, notably descending to 0.05 ± 0.10 if the detection threshold is chosen to be so large that it is only exceeded 3.5% of the time. This new value is somewhat larger than several older published results on rain; both Tessier *et al.* (1996) and De Lima (1998) estimate $H \approx -0.1$ in time (see also Larnder, 1995, for theoretical and numerical investigations). Recently there has been some interesting work on this by de Montera *et al.* (2009, 2010) and Verrier *et al.* (2010), who find important differences in exponents for the rain rate over all the data (i.e. dominated by zeroes) or for only the raining part (the support). Using radar data on regions without rain-free "holes," they find $H \approx 0$ for data with holes, but $H \approx 0.4$ for purely raining regions. They also perform numerical simulations that confirm the strong impact of zeroes on the parameter estimates, including C_1. In Verrier *et al.* (2011) it is further shown, using a unique two-year rain series at 15 s resolution, that by using rain-rate moments weighted to compensate for raining fractions the scaling in the rain is much improved. The value $H \approx 0.4$ for rain rates in raining regions is close enough to that of passive scalars ($H = 1/3$), as found for liquid water concentrations (see Figs. 1.8a, 1.8b), that Lovejoy and Schertzer (2008) and de Montera *et al.* (2010) propose turbulent models to explain it.

Other issues where divergences of views persist include the now old debate about multifractal universality (Schertzer and Lovejoy, 1987, 1997; Gupta and Waymire, 1993), which continues today in the guise of (weak) log-Poisson universality (Deidda *et al.*, 1999; Onof and Arnbjerg-Nielsen, 2009; Wang *et al.*, 2010) versus (strong) log-Lévy universality (Olsson and Niemczynowicz, 1996; Olsson, 1998; Olsson and Berndtsson, 1998; De Lima, 1998; Douglas and Barros, 2003; Pathirana *et al.*, 2003; Veneziano *et al.*, 2006; Lovejoy *et al.*, 2008; de Montera *et al.*, 2009, 2010; Verrier *et al.*, 2010, 2011; Serinaldi, 2010; Sun and Barros, 2010). Similarly, the nature of the low and zero rain rates (which are notoriously difficult to accurately measure, and which are prone to spurious breaks and spurious behaviours (Hoang *et al.*, in press) has led to two alternative modelling approaches: that the support (i.e. the region where it is raining) is a fractal set (Over and Gupta, 1994, 1996; Güntner *et al.*, 2001; Pathirana and Herath, 2002; Paulson and Baxter, 2007; Rupp *et al.*, 2009; Wang *et al.*, 2010) or, on the contrary, that the zero rain-rate values are the consequence of a combined physical and instrumental thresholding mechanism wherein the rain below some low value is set to zero (Lovejoy and Schertzer, 2008; Lovejoy *et al.*, 2008; de Montera *et al.*, 2009; Verrier *et al.*, 2010). The former hypothesis is somewhat unsatisfactory – if only because it implies that in the small-scale limit the rain areas are strictly zero, so that finite raining areas depend sensitively on the inner and outer cascade scales. In comparison, the thresholding mechanism implies – as found empirically – that the regions with low (and zero) rates will have poor/broken scaling.

A final important subject of controversy is the nature of the extremes. Although – as discussed in Section 5.3.3 – the general cascade process has "fat" algebraic probability tails, there are several ways in which artificial restrictions on the process can lead to more classical "long-tailed" or simply "thin-tailed" distributions. For example, microcanonical cascades, discussed in Section 3.2.3 and favoured by Carsteanu and Foufoula-Georgiou (1996), Güntner *et al.* (2001), Paulson and Baxter (2007), Rupp *et al.* (2009) and Wang *et al.* (2010), restrict the extremes – as do the log-Poisson models mentioned above. Similarly, the "bounded cascades" (Box 3.3) applied to rain in Menabde (1997) leads to quasi-Gaussian thin tails.

4.4.2 The cascade properties of ground and space radar, gauge networks and reanalysis rain fields

We start our survey of precipitation by considering radar reflectivities of rain which were used for one of the first empirical tests of multiplicative cascade models (Fig. 4.7a: Schertzer and Lovejoy, 1987). This analysis extended over the range 1–128 km, which – due to the curvature of the earth – is about the widest possible range for a single ground-based radar. From both the linearity of log M_q versus log λ shown in the figure and the converging nature of the lines, we see that it gives strong support to the multiplicative cascade idea. The radar reflectivity field is proportional to the sum of the squares of the drop volumes so that it is nontrivially related to the rain rate (which is proportional to the sum over the drops of the products of the volumes with the vertical fall speeds). However – at least above a minimum detectable threshold – the radar reflectivity is an accurately measured atmospheric signal and is strongly coupled

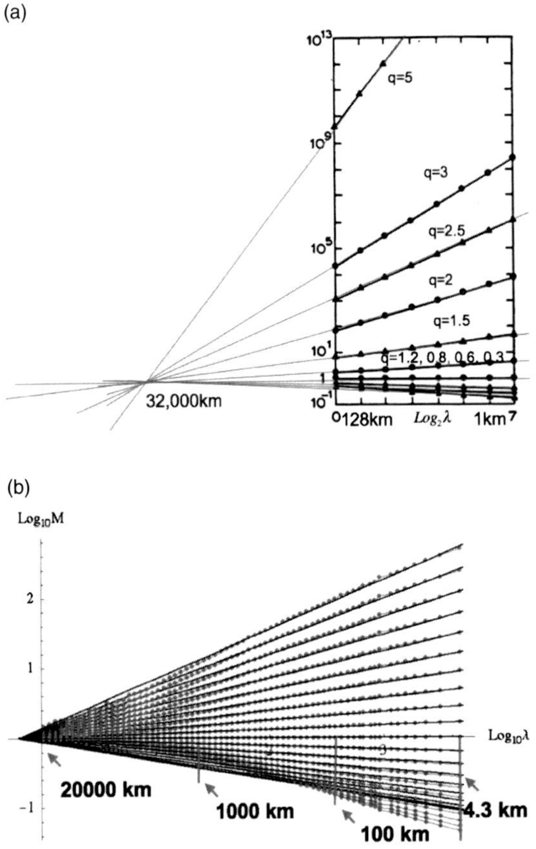

Fig. 4.7 (a) The moments M_q of the normalized radar reflectivity for 70 constant-altitude radar maps at 3 km altitude from the McGill weather radar (10 cm wavelength, 1 km pulse length). The basic figure was adapted from Schertzer and Lovejoy (1987) by Lovejoy *et al.* (2008), who added the straight lines converging to an outer scale at 32 000 km. (b) Same as Fig. 4.7a except for the TRMM reflectivities (4.3 km resolution). The moments are for $q = 0, 0.1, 0.2, \ldots 2$, taken along the satellite track. The poor scaling (curvature) for the low q values can be explained as an artefact of the fairly high minimum detectable signal. $L_{ref} = 20\,000$ km so that $\lambda = 1$ corresponds to 20 000 km, the lines cross at the effective outer scale $\approx 32\,000$ km, $C_1 \approx 0.63$. Reproduced from Lovejoy *et al.* (2009a).

with the rain-*rate* field, so the cascade structure of the reflectivities provides strong evidence in favour of the cascade hypothesis.

In order to achieve estimates at scales > 128 km, networks of ground-based radar can be used to obtain continental-scale reflectivities. These mosaics involve large numbers of radars partially overlapping at their extreme ranges where their resolutions are lowest; the resulting coverage is far from uniform, and in any case it is at most continental in scale. In order to directly verify the cascade behaviour up to planetary scales, we must use satellite data. Fig. 4.7b shows the result using the first orbiting weather radar, the precipitation radar (PR) instrument on the Tropical Rainfall Measuring and Mission (TRMM) satellite (Lovejoy *et al.*, 2008). From the figure we see that, again, the scaling (log-log linearity) is excellent, the main exception being for the low q values. Adopting the convention that any number $x^0 = 1$ if $x \neq 0$ and $x^0 = 0$ if $x = 0$, we find that the $q = 0$ curve corresponds to the scaling of the raining areas (the "support"). However the PR instrument has a very high minimum detectable signal: it is in fact double the mean value, and such thresholding breaks the scaling. In Lovejoy *et al.* (2008), with the help of numerical multifractal simulations, this scale breaking (curved lines for low q) was reproduced as a simple threshold effect. If we use a standard power law Z–R relation, $Z = aR^b$, and assume the statistics follow the universal multifractal form, then Eqn. (4.14) shows that $b^\alpha = C_{1Z}/C_{1R} = 0.63/0.50 = 1.26$. Taking the (spatial) $\alpha = 1.5$ (an estimate sensitive to low and zero rain rates, hence to be taken with caution), then we find $b \approx 1.2$, which is a little lower than the usually cited $b \approx 1.4$ for TRMM reflectivities. We return to this question when considering the corresponding temporal cascades in Chapter 8.

In order to compare the results to in-situ gauge networks and to reanalyses in the zonal, meridional and temporal directions, we gridded 5300 TRMM orbits (the year 1998) onto a 100 × 100 km grid (between ± 40°; the limits were imposed by the orbital parameters) with a four-day temporal resolution (which is roughly the mean time for the satellite to return to a given location). The reflectivities were converted into rain-rate estimates using a power law with the (recommended) value $b = 1.4$. They were then degraded over 100 × 100 km boxes: all the data within a four-day period were averaged (see Table 4.6 for a summary of the characteristics of this and the following precipitation products). Finally, the flux was estimated by the centred differences in time ($|R(t) - (R(t + \Delta t) + R(t - \Delta t))/2|$). This definition of the flux was also used in CPC and ECMWF products discussed below; not much difference was found, so the corresponding spatial flux estimate (the absolute Laplacian) was used instead. Indeed, in space it was found that $H \approx 0$, so the results were nearly the

Table 4.6 The characteristics of the various precipitation datasets discussed in the text.

	Spatial resolution (EW × NS)	Spatial extent EW	Spatial extent NS	Temporal resolution	Length of record
ECMWF (interim reanalysis)	1.5° × 1.5°	360°	180°	3 hours	3 months (1/06–3/06)
CPC (Climate Prediction Center, gauges)	2.5° × 2°	122.5° – 72.5° W (≈ 4000 km)	30° – 54° (≈ 3000 km)	1 hour	29 years (1948–1976)
TRMM (satellite radar)	100 × 100 km	360°	40° S – 40° N	4 days	1 year (1998: 5300 orbits)
20CR (reanalysis)	2° × 2°	360°	2° (only 44–46° analyzed)	6 hours	138 years (1871–2008)

same if no differences were used; the R field being directly taken as a conserved flux.

The results are shown in Figs. 4.8a (east–west) and 4.8b (north–south); these are not too different from the typically NE or SW along-orbit results (Fig. 4.7b) except that a strong NS/EW anisotropy is evident (at least for the outer scales; see Table 4.7 and discussion below), and the fact that the slopes ($K_R(q)$) are much smaller than those for Z due to the power-law transformation.

Since the Z–R transformation can only be theoretically justified using various problematic assumptions (such as the spatial uniformity of the drop size distribution; see e.g. Section 1.2.5), it is important to compare this satellite rain to in-situ (gauge) measurements and to reanalyses. For this purpose, we used NOAA's Climate Prediction Center (CPC) US hourly gridded precipitation (rain-rate) product. This product is unique in its high temporal resolution over a large number of contiguous grid points. We selected a (near complete) subset of the CPC data for the 29 years 1948–1976. The data were detrended annually and daily, which somewhat improves the low q scaling without affecting the high q statistics much. The CPC gridded the station data into 2.5° × 2.0° boxes by using a modified Cressman scheme (a kind of interpolation method); we used the central 13 × 21 point region from −122.5° to −72.5° longitude (every 2.5° ≈ 210 km at these latitudes), and from 30° to 54° latitude (every 2°, about 220 km; each grid point has a near-complete 257 000-point-long hourly series). The analysis results are shown in Figs. 4.8a and 4.8b, showing that the gridded data also have a clear cascade structure in the east–west and north–south directions, and that the extrapolated outer scale is nearly the same as for the ground- and satellite-based radar reflectivities (Figs. 4.7a, 4.7b).

Before making a quantitative satellite/gauge comparison, let us first consider the corresponding analyses on the ECMWF interim "stratiform rain" product (since this analysis was made, another "convective rain" product has been released). We used the first three months of 2006 at the full (three-hour) resolution. The results in the east–west and north–south directions are also shown in Figs. 4.8a and 4.8b, and the parameters and those of the other products are shown in Table 4.7. As a first comment, we could note the much higher values of all the C_1 estimates when compared with any of the meteorological fields considered so far; this confirms our intuition about rain, that it is much more intermittent than the other atmospheric fields. We also note the reasonable agreement between the outer scale estimates: $43 \pm 6 \times 10^3$ km in the east–west direction and $27 \pm 9 \times 10^3$ km in the north–south direction (the spread is the variation from one product to another, and the raw outer scale estimates in the table – and throughout this book – are only quoted to within 1 $dB\lambda$, i.e. to the nearest tenth of an order of magnitude, i.e. to within a factor ≈ 1.25). Taking the ratio of the east–west to north–south outer scales gives us the trivial anisotropy, i.e. the typical aspect ratio in the horizontal (assuming that there is no differential, i.e. scale-by-scale, anisotropy). The ratios of the means gives a factor 1.61, which is almost the same as for the mean of the ECMWF interim fields discussed above.

The variations of the C_1 values in Table 4.7 are not so easy to explain. First, the east–west C_1 values are

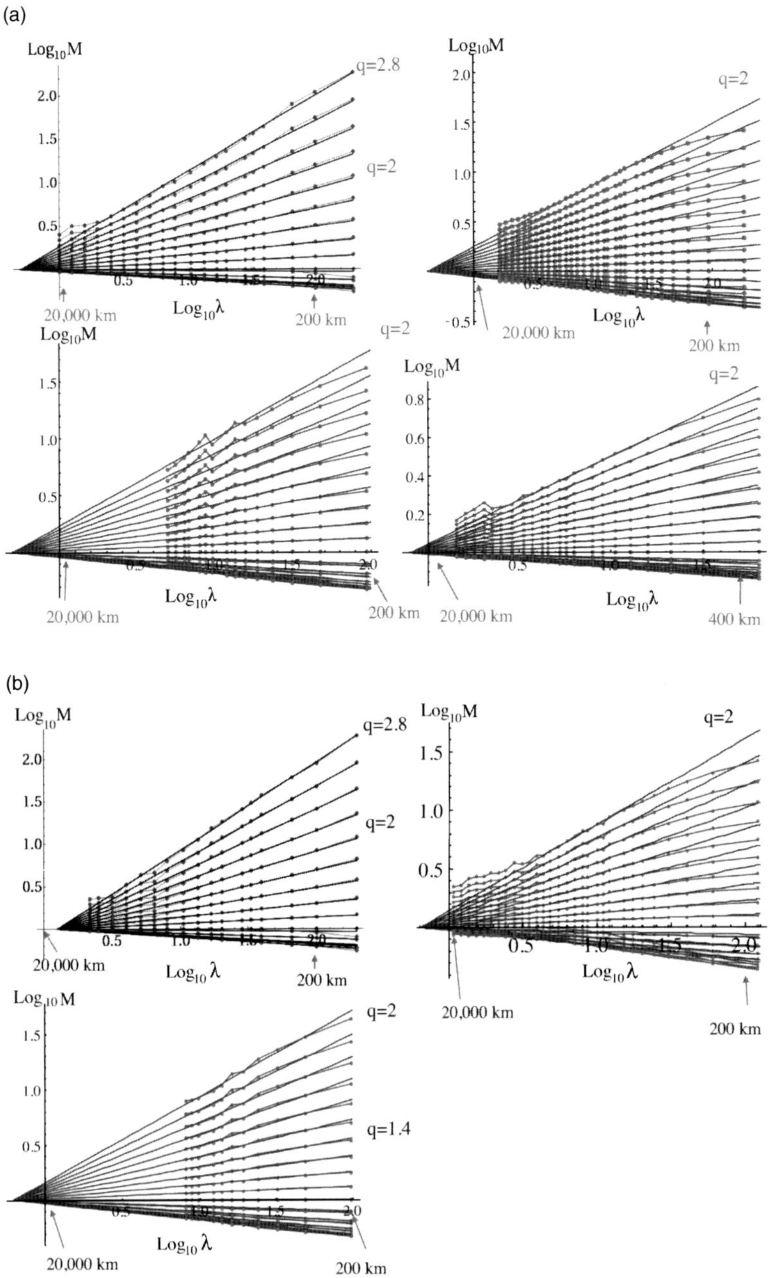

Fig. 4.8 (a) East–west analyses of the gridded precipitation products discussed in the text. Upper left: The TRMM 100 × 100 km, 4-day averaged product. Upper right: The ECMWF interim stratiform rain product (all latitudes were used). Note that the data were degraded in constant-angle bins so that the outer scale is 180°. To compare with the other analyses, a mean map factor of 0.69 has been applied (the mean east–west outer scale was ~14 000 km). Lower left: The CPC hourly gridded rainfall product (USA only). Reproduced from Lovejoy *et al.* (2012). Lower right: 20CR reanalysis at 45°N" (b) Same as Fig. 4.8a but for the north–south analyses (see Fig. 8.7e). Upper left: The TRMM 100×100 km, 4-day averaged product. Upper right: The ECMWF interim stratiform rain product. Lower left: The CPC hourly gridded rainfall product (USA only). Reproduced from Lovejoy *et al.* (2012).

not quite the same as the north–south exponents. The differences are between 4% and 16% in the exponents, and if they are statistically significant it would indicate systematic scale-by-scale scaling anisotropy (as discussed in Chapter 6). However, it is not obvious that they are in fact significant. On the one hand, even with all these data, statistical exponents are very hard to accurately estimate; this is especially true since the large C_1 values indicate huge sample-to-sample variability (intermittency). On the other hand, there are

Table 4.7 A comparison of some of the cascade parameters for the three precipitation products discussed in the text and MTSAT (thermal IR). The α estimates for MTSAT were all 1.5; for the precipitation products, they were not considered reliable since they were sensitive to the low and zero rain rate regions which were poorly determined. The temporal analyses will be discussed in Chapter 8. The zonal/meridional aspect ratios a are: 1.6, 1.3, 2.5, 1.6 for the ECMWF, CPC, TRMM and MTSAT fields respectively. Only the 45° N 20CR data were analyzed (zonally and in time).

	ECMWF		CPC		TRMM $R \propto Z^{1/1.4}$		20CR		MTSAT	
	C_1	L_{eff} or T_{eff}	C_1	L_{eff} or T_{eff}	C_1	L_{eff} or T_{eff}	C_1	L_{eff} or τ_{eff}	C_1	L_{eff} or T_{eff}
East–west	0.41	50 000 km	0.49	40 000 km	0.27	40 000 km	0.26	25 000 km	0.07	50 000 km
North–south	0.45	32 000 km	0.51	32 000 km	0.32	16 000 km	_	_	0.07	32 000 km
Time	0.34	71 days	0.37	42 days	0.30	1 100 days	0.22	50 days	0.07	48 days

fairly large variations from one product to another – larger than the NS/EW differences – and in any case, the east–west values are not even systematically larger than the north–south ones (they are smaller in the ECMWF product, larger in the others). For the moment we conclude that probably the trivial anisotropy is real, but not the scale-by-scale anisotropy (i.e. we may reasonably consider that the C_1's are the same in the east–west and north–south directions). There remains the interesting task of understanding the product-to-product differences, and this should shed light not only on the fundamental nature of rain but also on the optimum way of estimating rain rates from in-situ and remote measurements and of simulating rain in numerical models. Recall that since we are discussing exponents characterizing the scale-by-scale statistical properties, for the products to agree about the rain rate at any particular space-time point it is a necessary – but not a sufficient – condition that they have identical cascade parameters (exponents and outer scales).

In Table 4.7 we did not include estimates of the Lévy index α, since its estimation depends sensitively on the low values of the rain rate, and these are poorly measured. Lovejoy *et al.* (2008) pay much attention to this, and with the help of cascade models conclude that $\alpha \approx 1.5$, but the evidence is still not compelling (see e.g. Lilley *et al.*, 2006, for an review of empirical estimates). Also shown in the table are the MTSAT thermal IR parameters; these are included because the thermal IR field is correlated with high cloud tops and hence precipitation.

The spatial values of C_1 can be compared with others in the literature, notably Tessier *et al.* (1993): $C_1 \approx 0.16$; Olsson and Niemczynowicz (1996): $C_1 \approx$ 0.02 – 0.1; Hubert *et al.* (2002): $C_1 \approx 0.35 \pm 0.2$; de Montera *et al.* (2009) and Verrier *et al.* (2010): $C_1 \approx 0.5$, $H \approx 0$ for data with zeroes, $C_1 \approx 0.15$, $H \approx 0.4$ for data without zeroes. See also Veneziano *et al.* (2006) for estimates obtained by assuming a fractal support (and which are thus not directly comparable). These estimates are (mostly) from gauges (although the latter also consider radar scans); we see that our values are more in accord with the more recent estimate, but there is still much uncertainty in the estimates. The sensitivity to zeroes thus seems to hold the key to a better understanding of the variation in the literature of both H and C_1 parameters.

The temporal estimates of C_1 in Table 4.7 can be compared with those in the literature, notably $C_1 \approx 0.6$ (Tessier *et al.*, 1993); $C_1 \approx 0.2$ (Hubert *et al.*, 1993); $C_1 \approx 0.6$ (Ladoy *et al.*, 1993); $C_1 \approx 0.04$–0.19 (Harris *et al.*, 1996); $C_1 \approx 0.30$–0.51 (De Lima, 1998; De Lima and Grasman, 1999); $C_1 \approx 0.344$, 0.303 (Hubert *et al.*, 2002; Kiely and Ivanova, 1999); $C_1 \approx 0.38 \pm 0.02$, 0.40 ± 0.1 (Hubert *et al.*, 2002); $C_1 \approx 0.345 \pm 0.038$ (Pathirana *et al.*, 2003); $C_1 \approx 0.434 \pm 0.005$ (Garcia-Marin *et al.*, 2008; Pathirana *et al.*, 2003); $C_1 \approx 0.38$ (Serinaldi, 2010); $C_1 \approx 0.3$–0.6 (Sun and Barros, 2010); $C_1 \approx 0.47 \pm 0.08$ (Schertzer *et al.*, unpublished manuscript); $C_1 = 0.59$ for full sample (32-minute resolution), $C_1 = 0.10$ for "rain only" conditional samples (Verrier *et al.*, 2011). These are all from gauges; a radar estimate from Tessier *et al.* (1993) gives $C_1 \approx 0.6$. We see that, as for the spatial C_1 estimates, our values are more in accord with the more recent values but there is still much uncertainty in the estimates, and again the zero/low rain-rate issue is a likely source of this uncertainty (Lovejoy and Schertzer, 2008; Verrier *et al.*, 2010, 2011; Gires *et al.*, 2012).

4.5 The scaling of atmospheric forcings and boundary conditions

4.5.1 The scaling of the earth's energy budget: long- and short-wave radiances, passive microwaves

We have argued that the dynamical equations are compatible with anisotropic scaling over most of their range. However, in order to justify the scaling of the solutions – the state variables – we must also demonstrate the scaling of the forcings and boundary conditions. Basic boundary conditions are the topography (land) and sea surface temperatures (SST, oceans), and the main meteorological forcings are short-wave solar heating and long-wave (thermal infrared) cooling (this implicitly includes the climate forcings discussed in Section 11.3). The scaling and cascade properties of the SST are discussed in Chapter 8; here we consider the forcings and the topography. Other boundary conditions include soil moisture and vegetation (see Fig. 1.12 for relevant spectra and Lovejoy *et al.*, 2007, for cascade analyses).

By estimating the Laplacian of the horizontal wind at the model and reanalysis dissipation scales, we have already given evidence that the energy flux does indeed cascade over the entire available range of scales. By using satellite radiance data (whose spectra were analyzed in Chapter 1), we now show that the corresponding energy forcings and sinks (i.e. the short- and long-wave radiances) are also scaling with corresponding cascade structures whose parameters we estimate. This is in accord with the spectral transfers analyzed in Section 2.7, which did not indicate a clear cascade direction (upscale or downscale). For this analysis, we again used the TRMM satellite which, in addition to the Precipitation Radar instrument had a visible and infrared instrument (VIRS; with 5 wavelengths) as well as a passive microwave instrument (TMI; with 5 wavelengths at 2 polarizations each). The spectra were already shown in Figs. 1.2 and 1.3, and the full analysis is given in Lovejoy *et al.* (2009a). The cascade analyses for the key energy-containing short-wave (visible) and long-wave (thermal IR) wavelengths are shown in Figs. 4.9a and 4.9b. We see once again excellent scaling; Tables 4.7a and 4.7b show the details and comparison with a more limited earlier study (Lovejoy *et al.*, 2001).

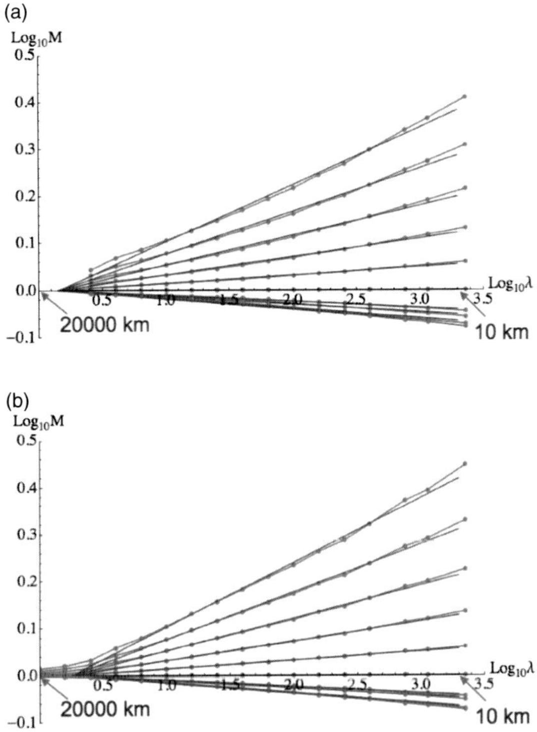

Fig. 4.9 (a) TRMM visible data (0.63 mm) from the VIRS instrument, channel 1 with fluxes estimated at 8.8 km. Only the well-lit 15 000 km orbit sections were used. L_{ref} = 20 000 km so that $\lambda = 1$ corresponds to 20,000 km, the lines cross at $L_{eff} \approx$ 9,800 km. Reproduced from Lovejoy *et al.* (2009a). (b) Same as Fig. 4.9a except for VIRS thermal IR (channel 5, 12.0 μm), $L_{eff} \approx$ 15 800 km (see Table 4.8a for details). Reproduced from Lovejoy *et al.* (2009a).

These TRMM results are suppported by those from thermal infrared data from the geostationary satellite MTSAT (Fig. 4.10; Pinel *et al.*, 2012). Sections from 30° S to 40° N, about 13 000 km in the east–west over the Pacific Ocean were used at 30 km resolution, every hour for two months (1440 images in all). It is interesting to note that the MTSAT analyses were carried out in both east–west and north–south directions; Fig. 4.10 is the mean, presumably closer to the TRMM analyses made along the satellite track which was typically oriented northeast or southeast. The scaling behaviour of these radiances is consistent with the large-scale cascade structure of the wind and temperature fields because it shows that the energy sources and sinks are themselves scaling so that the basic assumptions of the cascade model are still presumably satisfied. In addition, the radiances and cloud fields are strongly nonlinearly coupled, so that the scaling of the radiances is in itself strong evidence

Table 4.8a The statistical characteristics of various sensors in the visible and IR wavelengths. VIRS is the visible-IR instrument on the TRMM satellite. The AVHRR satellite series is an operational NOAA satellite, MTSAT is a Japanese geostationary satellite, and "photography" is ground-based large-format imagery (see Figs. 1.4a and 1.4b; from Sachs *et al.*, 2002). The H estimates are based on structure functions. The mean residues (δ, Eqn. (4.16)) are given both with respect to the restrictive hypothesis that the cascades are universal multifractals (i.e. they respect the cascade, Eqn. (4.1), with the universal form for $K(q)$, Eqn. (3.46) with $\alpha = 1.5$, $C_1 = 0.08$), and for the less restrictive hypothesis, that they only respect Eqn. (4.1).

Channel	Wavelength	Resolution (km)	δ (%) line[a]	δ (%) uni[b]	α	C_1	H	L_{eff} (km)
VIRS 1	0.630 μm	8.8	0.60	0.71	1.35	0.077	0.19	9800
VIRS 2	1.60 μm	8.8	0.83	1.37	1.41	0.079	0.21	5000
VIRS 3	3.75 μm	22.	1.10	1.58	1.99	0.065	0.27	17 800
VIRS 4	10.8 μm	8.8	0.48	0.53	1.56	0.081	0.26	12 600
VIRS 5	12. 0 μm	8.8	0.47	0.81	1.63	0.084	0.33	15 800
AVHRR 14 vis[c]	0.58–0.68 μm	2.2	_	_	1.92	0.075	0.32	18 700
AVHRR 14 IR[c]	11.5–12.5 μm	2.2	_	_	1.91	0.079	0.36	25 200
MTSAT[d]	10.8 μm	30	_	_	1.5	0.74	0.31	40 000
Photography	0.3–0.7 μm	0.5–5 m	_	_	1.77	0.061	0.61	_

[a] This is the residual with respect to pure power-law scaling.
[b] This is the residual with respect to universal multifractal scaling with $\alpha = 1.5$, $C_1 = 0.08$, only the outer scale is fit to each channel.
[c] These were from 153 visible, 214 IR scenes each 280 × 280 km over Oklahoma, from Lovejoy *et al.* (2001), Lovejoy and Schertzer (2006).
[d] This is the average of the north–south and east–west parameters; see Table 4.7.

Table 4.8b The characteristics of the five (TRMM) TMI channels and the Precipitation Radar reflectivity (not rain rate), from Lovejoy *et al.* (2009a). All used vertical polarization. The H estimates are based on structure functions.

Channel	Wavelength	Resolution (km)	δ (%) line[2]	δ (%) uni[3]	α	C_1	H	L_{eff} (km)
TMI1	3.0 cm (10.6 GHz)	111.4	1.40	1.55	1.35	0.255	0.50	15 900
TMI 3	1.58 cm (19.35 GHz)	55.6	1.71	1.93	1.76	0.193	0.331	6900
TMI 5	1.43 cm (22.24 GHz)	27.8	1.62	1.82	1.93	0.157	0.453	5000
TMI 6	8.1 mm (37 GHz)	27.8	1.73	1.95	1.76	0.15	0.377	4400
TMI 8	3.51 mm (85.5 GHz)	13.9	1.40	1.70	1.90	0.102	0.238	6300
TRMM *Z*	2.2 cm (13.2 GHz)	4.3	6.0*	4.6*	1.50	0.63	0.00	32 000

for the scaling of the clouds (i.e. the liquid water field) and hence presumably of the dynamics.

In spite of the variation in the wavelengths from visible through near to thermal IR, the values of the parameters in Table 4.8a are all roughly compatible with $\alpha = 1.5$, $C_1 = 0.08$ – this is indicated directly in the table for the TRMM VIRS data, but the AVHRR and MTSAT results are also quite close. The H values are also in a fairly narrow interval between 0.19 and 0.36, although there seems to be a tendency for H to decrease with wavelength. It is significant that these values are in turn close to those of passive scalars: Table 4.5 indicates the results from lidar backscatter ratio, which is a surrogate for the aerosol concentration (see Figs. 1.7b, 1.7c, 1.7d and Section 6.5.1): $\alpha \approx 1.8$, $C_1 \approx 0.07$, $H \approx 0.33$ (the latter being close the theoretical Corrsin–Obukhov value $H = 1/3$, Section 2.3).

A possible explanation for these results emerges as follows. First consider the visible wavelengths which – ignoring reflection from the surface – are essentially pure scattering (very little emission and absorption).

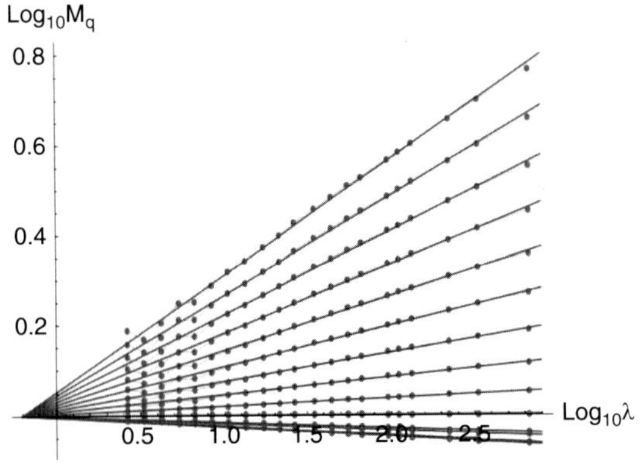

Fig. 4.10 Logs of normalized moments M_q versus $\log_{10}\lambda$ for 2 months (1440 images) of MTSAT, thermal IR, 30 km resolution over the region 40° N to 30° S, 130° east–west over the western Pacific, the average of east–west and north–south analyses. $L_{ref} = 20\ 000$ km so that $\lambda = 1$ corresponds to 20 000 km, the lines cross at the effective outer scale ≈ 32 000 km (from Pinel *et al.*, 2012) and $C_1 \approx 0.074$ (close to the TRMM thermal IR results, Table 4.8a, VIRS 4, 5).

If the clouds are on average sufficiently optically thin and the cloud liquid water density is roughly a passive scalar, then their effect will be roughly a linear modulation of the horizontal radiances, which will then have passive scalar characteristics. However, two effects tend to counter this. First, the surface reflectance, at least over land, is quite "rough," i.e. it has low H values; over relatively cloud-free volcanic terrain, Harvey *et al.* (2002) and Gaonac'h *et al.* (2003) find H in the range 0–0.2 with $C_1 \approx 0.03$–0.08, $\alpha \approx 2$ – so this would tend to lower H below the passive scalar value, perhaps to the observed value ($H \approx 0.19$ for the TRMM visible wavelength result: Table 4.8a). The second effect is that as the cloud gets optically thicker the radiation tends to integrate the optical density, effectively increasing H. The combination of this effect and a smooth (sky) background could explain the visible results from ground-based photography (see Figs. 1.4a and 1.4b), which yield $H \approx 0.6$ (Table 4.8a). Consider now the thermal IR results, which are closer to the passive H values. First, for a given cloud, the optical thickness is much higher in the infrared when compared to the visible but there is very little scattering, so the transfer is dominated by absorption and thermal emission. This makes the clouds less penetrable and makes most of the absorption/emission take place near the cloud "surface." However, the temperature dependence itself is not far from a passive scalar (see Box 4.1, Table 4.5: $H \approx 0.5$, $C_1 \approx 0.1$), so it is plausible that the resulting radiances would have statistics with higher H, thus closer to passive scalars (as observed).

These results are bolstered by the TRMM microwave results shown in Table 4.7b. Recall that the thermal microwave radiation has contributions from surface reflectance, water vapour and cloud and rain. Since the particles are smaller than the wavelengths this is the Rayleigh scattering regime, and as the wavelength increases from 3.5 mm to 3.0 cm the emissivity/absorptivity due to cloud and precipitation decreases so that more and more of the signal originates in the lower reaches of clouds and the underlying surface. Also, the ratio of absorption to scattering decreases so that at 3 cm multiple scattering can be important in raining regions. The overall result is that the horizontal gradients – which we have used to estimate the cascade fluxes – will increasingly reflect large internal liquid-water gradients. We therefore expect the longer wavelengths to give flux statistics close to those of the (2.2 cm) radar reflectivity signal (which is proportional to the second moment of the particle volumes). This explanation is consistent with the trend mentioned above for C_1 to increase sharply at the longest wavelengths towards the reflectivity (Z) value. The relative similarity of the TMI 1 band and Z (and the other bands with the VIRS) is also supported by the fact that the outer scale is in the 5000–7000 km range for the longer wavelengths but is nearly 16 000 km – approaching the reflectivity outer scale – in the TMI 1.

Actually, it is a bit embarrassing that at the moment we cannot improve much on these admittedly hand-waving arguments linking cloud liquid water and radiances. This is indeed frustrating, because work on the interesting statistical physics problem of radiative transfer in fractal (Lovejoy *et al.*, 1990; Gabriel *et al.*, 1990; Davis *et al.*, 1990) and in multifractal clouds (Davis *et al.*, 1991; Lovejoy *et al.*, 1995; Borde and Isaka, 1996; Naud *et al.*, 1996) is hardly new – and as the simulations in this book testify, can give highly realistic-looking cloud fields (see especially Chapters 7 and 9, where the renditions are made using single scatter radiative transfer and a simple linear temperature model for the IR field in Fig. 7.10e). There are also related interesting results for the simpler and different problem of diffusive transfer on multifracals (Meakin, 1987; Weissman and Havlin, 1988; Marguerit *et al.*, 1998; Lovejoy *et al.*, 1998). However, in an effort to understand the basic science issues, the work to date has been largely confined to isotropic conservative ($H = 0$) multifractals, and to the bulk statistical characteristics such as the mean transmission and albedoes (see however Watson *et al.*, 2009; Lovejoy *et al.*, 2009c, for extensions to $H > 0$; see Box 7.2). Before we can confidently apply these results to real clouds, we must extend them to stratified clouds (see Chapter 6), to clouds over reflecting surfaces (themselves having scaling statistics), and to an understanding of the statistical relationship between the cloud and radiation fields. These results promise to lead to more realistic cloud-radiation parametrizations in GCMs, where at the moment clouds are assumed to be plane parallel slabs. Such work is also important in estimating the earth's energy budget, since the multifractal resolution dependencies are currently not taken into account, and this will surely lead to biases in the results.

4.5.2 Atmospheric boundary conditions: the topography

Physically, the TRMM reflectivity signal comes purely from the atmosphere, whereas the visible and infrared radiances depend on the states of both the atmosphere and the surface. Just as various surface features affect the radiances, so they also directly effect the atmosphere; they are important lower boundary conditions. Another important atmospheric boundary condition is the topography; if it had a strong characteristic scale then it could impose this on the atmospheric fields and break the scaling. In Fig. 1.10b we showed the spectral analysis of the largest statistical study of the topography to date, demonstrating that it has accurate spectral scaling (roughly $E(k) \approx k^{-2.1}$) over a range of 10^5 in scale.

Fig. 4.11 shows the cascade structure of the topographic gradients obtained by combining the four different datasets used in Fig. 1.10, spanning the range 20 000 km down to sub-metric scales. As for the spectrum, the scaling holds quite well until around 40 m. Gagnon *et al.* (2006) argue that this break is due to the presence of trees (for the high-resolution dataset used over Germany, 40 m is roughly the horizontal scale at which typical vertical fluctuations in the topography are of the order of the height of a tree). Over the range of planetary scales down to ~40 m, it was estimated that the mean residue of the universal scaling form with parameters $C_1 = 0.12$, $\alpha = 1.79$ (for all moments $q \leq 2$) was ± 45% over this range of

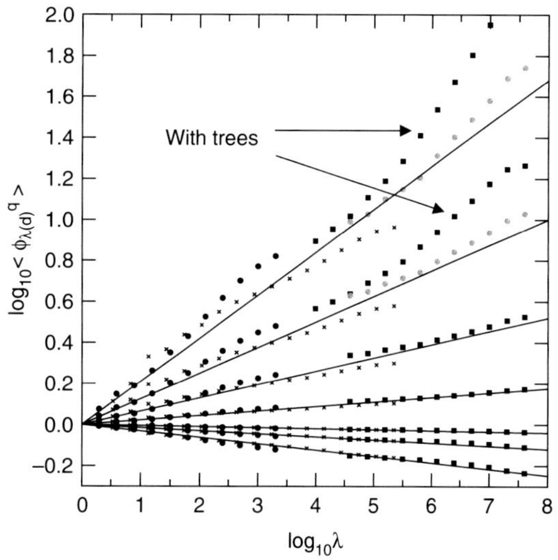

Fig. 4.11 Log-log plot of the normalized trace moments M_q versus the scale ratio $\lambda = L_{ref}/l$ (with $L_{ref} = 20\ 000$ km) for the three DEMs (circles correspond to ETOPO5, X's to the continental USA and squares to Lower Saxony). The solid lines are there to distinguish between each value of q (from top to bottom, $q = 2.18, 1.77, 1.44, 1.17, 0.04, 0.12$ and 0.51). The trace moments of the Lower Saxony DEM with trees for $q = 1.77$ and $q = 2.18$ are on the graph (indicated by arrows). The theoretical lines are computed with the global $K(q)$ function. Figure reproduced from Gagnon *et al.* (2006).

nearly 10^5 in scale (this error estimate was for the "reduced" moments $\langle \varphi^q \rangle^{1/q}$, e.g. for $q = 2$ root mean square moments – i.e. the definition of the errors is a little different than those used here (Eqn. (4.16)), so the values are not directly comparable); the main difference is that Eqn. (4.16) characterizes the errors in logs of the moments whereas for the reduced moments the errors were in the values themselves (hence they are larger).

4.6 Summary of emergent laws in Chapter 4

The basic prediction of multiplicative cascade processes is that there exists a (scale-by-scale) conservative process ("flux") φ' whose statistical moments change with resolution (scale ratio λ' = largest scale of cascade/observation scale = L_{eff}/L) in a power-law way:

$$M_q = <\varphi'^q_\lambda> \approx \lambda'^{K(q)} \quad (4.20)$$

Testing this requires:

(a) Estimating a normalized flux φ'. In the scaling range, this can be done using fluctuations Δf, which are generally related to the underlying flux ε via:

$$\Delta f = \varepsilon^{\eta_s}\Delta x^{H_s} \tag{4.21}$$

If, instead, the finest resolution of the data is in the dissipation range, then we have similar relation but with different exponents η_d, H_d. In both cases, we can estimate normalized fluxes φ'_η as:

$$\varphi'_\eta = \frac{\varepsilon^\eta}{\langle\varepsilon\rangle^\eta} = \frac{\Delta f}{\langle\Delta f\rangle} \tag{4.22}$$

(we temporarily add the subscript η to stress the power with which it is related to the underlying flux ε). For universal multifractals the fluxes φ'_η have the same α but the C_1's vary as: $C_{1,\eta} = \eta^\alpha C_{1,1}$. Comparing scaling-range and dissipation-range cascades, we therefore have:

$$C_{1,d} = \left(\frac{\eta_d}{\eta_s}\right)^\alpha C_{1,s} \tag{4.23}$$

(b) Estimating the ensemble average of φ' using "trace moments," i.e. by (spatial, temporal) averaging the flux over intermediate resolutions L (scale ratio λ) over disjoint "boxes" (intervals in 1D) and then statistically averaging the qth power over all the boxes (intervals) in all the available realizations of the process (the entire sample).

(c) Empirically, the effective outer scale of the process L_{eff} is unknown and must be estimated from the data. Therefore a convenient reference scale L_{ref} and reference scale ratio $\lambda = L_{ref}/L$ is chosen.

(d) A plot of $\log M_q$ versus $\log\lambda$ is made using positive order moments q not too large (in order to avoid possible theoretical divergences for $q < 0$ or for sample-size-limited estimates for large q). The "signature" of multiplicative cascade is the convergence of straight lines to a point $\lambda = L_{ref}/L_{eff}$ (corresponding to $\lambda' = 1$). The slopes of the lines determine $K(q)$, the intercept, $\log L_{ref}/L_{eff}$. And hence the effective outer scale of the cascade.

In this chapter, we applied this trace moment analysis to the empirical verification and parameter estimation of horizontal multiplicative cascades on various fields, including (a) reanalyses: ERA40, ECMWF interim, 20CR; (b) meteorological models: GFS, GEM; (c) aircraft data of T, u, h, $\log\theta$; (d) lidar backscatter data; (e) precipitation data; (f) satellite radiances including the energy-significant visible and thermal IR data; (g) the topography.

Appendix 4A: Trace moments of quasi-Gaussian processes

Classical stochastic processes based on adding rather than multiplying random variables are nonintermittent, with $K(q) = 0$, the prototypical example being the "quasi-Gaussian" processes obtained by filtering Gaussian white noises. In Chapter 8 we note that a common model for the weather/macroweather transition is the Ornstein–Uhlenbeck (OU) process that results from smoothing white noises by integration down to frequency ω_0, yielding a Gaussian white noise with spectrum $E(\omega) \approx 1/(\omega^2 + \omega_0{}^2)$. As discussed in Appendix 10A, multivariate versions of this form the basis of stochastic linear modelling approaches, which are used for example in stochastic forecasting of sea surface temperatures. For relevant empirical spectra, see the SST spectra (Fig. 8.6d) or those of the first principal component of the Pacific SST (called the Pacific Decadal Oscillation, PDO: Fig. 10.8). These show that while there is indeed a transition between two scaling regimes, the exponents are not 0 and 2 as required for OU processes.

Since $K(q) = 0$, applying our analysis technique should theoretically give constant trace moments – indeed, for a large enough range of scales (small λ corresponding to integrating the flux over a wide range of scales) $\log M_q$ should asymptotically converge to 0. However, if the scaling range of the data is not very large (and this is often a problem with meteorological data at daily scales, since the outer cascade scale is only of the order of 10 days: see Chapter 8), then the convergence of the quasi-Gaussian process may be slow enough that it might be hard to convincingly distinguish a quasi-Gaussian process from a scaling process with a short range of scaling.

Let us therefore apply our trace moment analysis procedure (summarized in Section 4.6) to a quasi-Gaussian process; this will be a useful point of comparison. In this example, we analyse an OU process, treating it as an unknown data field. The first step is to estimate the flux by taking the absolute second differences. Since the OU process is essentially a summed Gaussian white noise, this "flux" will be a quasi-Gaussian process (essentially independent of the transition frequency ω_0) and by the central limit theorem, as we degrade the fluxes in order to estimate the moments at lower and lower resolutions, we expect rapid convergence to a flat ($K(q) = 0$) regime. Finally, since the flux is normalized by dividing by its ensemble average, the overall result is a "universal" set of moments valid for any quasi-Gaussian process (at least with $\beta < 4$, the limit is due to the second differencing used to estimate the flux). In Fig. 4A.1, we show these "universal" moments up to order 2.9 obtained from 100 realizations of an OU process 2^{13} points long with $\omega_0 = (128)^{-1}$ (grid points)$^{-1}$. As expected, the result was essentially identical to that of a pure Gaussian white noise. The basic characteristics of the graph that will be useful when comparing with the corresponding model output analyses are (a) the maximum $\log_{10}M \approx 0.46$ ($q = 2.9$; the corresponding value for $q = 2$ is $\log_{10}M \approx 0.20$); (b) the outer scale $\approx$ a factor of 10 larger than the inner scale; (c) the curves start to deviate significantly from the lines at scales larger than about 5 grid points; (d) at scales of a factor of 100 grid points $\log_{10}M$ is already < 0.01; (e) the corresponding universal multifractal parameters (estimated over the range 2–10 grid points) are $C_1 \approx 0.082$, $\alpha \approx 1.79$.

Consider an empirical trace moment $\log M_q$ versus $\log\lambda$ graph with the largest analysis moment $q = q_{max}$. Since the M_q for $q > 0$ is an increasing function with q, the $\log M_q$ versus $\log\lambda$ curve for q_{max} is an envelope bounding all the empirical curves (for $q < q_{max}$). This suggests a straightforward way to use the "universal" quasi-Gaussian curve in Fig. 4A.1 to see how significantly the data deviate from a quasi-Gaussian process.

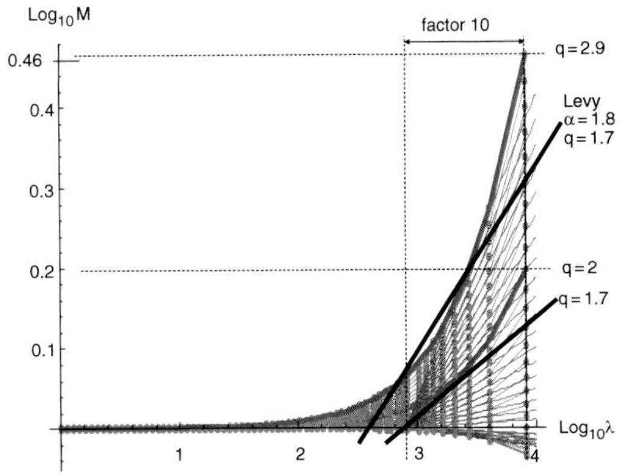

Fig. 4A.1 The "universal" trace moments for quasi-Gaussian processes obtained from 100 realizations of an OU process with $\omega_0 = (128)^{-1}$ (grid points)$^{-1}$ using 100 realizations of process 2^{13} points long. This shows the convergence of the moments up to order 2.9 (at increments of 0.1) to the small λ asymptote $\log M = 1$. Although this is for an OU process, the result is essentially identical to that of a pure Gaussian white noise (and to other quasi-Gaussian processes with $\beta < 4$). The thick curves are for $q = 2, 2.9$. The thin straight lines correspond to $C_1 = 0.082$, $\alpha = 1.79$ and an outer scale of ~10 pixels. The thick black lines compare the log-log linear fit for moments $q = 1.7$ for quasi-Gaussian processes (bottom) and Lévy processes ($\alpha = 1.8$, top; moments for $q > \alpha$ are infinite). Even though the probability distribution is extreme, the outer scale only increases to 20 grid points and the nondimensional M for the $q = 1.7$ moment at the smallest scales is only about 50% larger. In order to obtain significantly stronger variability, strong long-range statistical dependencies are needed.

One need only superpose the envelope of the universal quasi-Gaussian trace moments on the data and compare the corresponding envelopes. Numerous examples of this are given in Chapters 8 and 10 (the spatial analyses in this chapter have such wide ranges of scaling that it is not necessary to make the comparison).

Unfortunately, if the scaling range is of the order of only a factor of 10, then it is hard to come to more definite conclusions, although it isn't easy – without using cascades – to obtain large deviations from the universal quasi-Gaussian curve. For example, we can reproduce the OU process but rather than using random variables with Gaussian distributions we can use (extreme) stable Lévy random variables index α (discussed in Section 5.4.1) with infinite variance (i.e. $\alpha < 2$). In this case, for those finite moments (order $q < \alpha$) it turns out that the trace moments are not very strongly affected. For example, with $\alpha = 1.8$, we obtain an outer scale of 20 pixels (rather than 10 for a Gaussian) and for $q = 1.7$, $\log_{10}M = 0.29$ rather than $\log_{10}M = 0.13$ for a Gaussian OU process (see the thick lines in the figure). This indicates that not only is strong variability at the smallest scales important, but also that strong statistical dependences are necessary to produce moments and outer scales significantly larger than 10–20 grid points.

Chapter 5 Cascades, dimensions and codimensions

5.1 Multifractals and the codimension function

5.1.1 Probabilities and codimensions

We have given evidence that the atmosphere is scaling over wide ranges of scale (Chapter 1), we have argued that the dynamics are also scaling (Chapter 2) and lead to multiplicative cascades (Chapter 3), and finally we have given empirical support for this (Chapter 4). Throughout, our approach has been to provide the minimal theoretical framework necessary for understanding the most straightforward data analyses (e.g. the trace moments method, $\langle \phi_\lambda^q \rangle$). We mentioned that specification of all of the statistical moments is generally a complete statistical characterization of the process and hence this was equivalent to their specification in terms of probabilities. However, we did not go further to specify the exact relation; the moment characterization was convenient and adequate for our purposes. We now turn to the complete formalism needed to understand the probability structure of the cascades. In this chapter, we thus continue to study the properties of cascades, this time emphasizing their probability distributions and their exponents, the codimensions.

5.1.2 Revisiting the β model

We have already described the monofractal stochastic β model in Chapter 3. It is said to be "monofractal" or "unifractal" because it can be defined with the help of a unique codimension; let us examine its probability structure as the cascade develops. Recall from Chapter 3 that it has one parameter $c > 0$ and that two states specify the statistics of the multipliers $\mu\varepsilon$:

$$\begin{aligned} &\Pr(\mu\varepsilon = \lambda_0^c) = \lambda_0^{-c} \quad \text{(alive)} \\ &\Pr(\mu\varepsilon = 0) = 1 - \lambda_0^{-c} \quad \text{(dead)} \end{aligned} \tag{5.1}$$

where λ_0 is the single step (integer) scale ratio. Recall that the magnitude of the boost $\mu\varepsilon = \lambda_0^c > 1$ is chosen so that at each cascade step the ensemble averaged ε is conserved:

$$\langle \mu\varepsilon \rangle = 1 \Leftrightarrow \langle \varepsilon_n \rangle = \langle \varepsilon_0 \rangle \tag{5.2}$$

Indeed, at each step in the cascade the fraction of the alive eddies decreases by the factor $\beta = \lambda_0^{-c}$ (hence the name "β model") and conversely their energy flux density is increased by the factor $1 / \beta$ to assure (average) conservation. Rather than follow Chapter 3 and consider how the moments change with scale, let us now consider how the probabilities evolve as we increase the number of cascade steps. After n steps, the effect of the single-step dichotomy of "dead" or "alive" is amplified by the total (n step) scale ratio $\lambda = \lambda_0{}^n$:

$$\begin{aligned} &\Pr\left(\varepsilon_n = (\lambda_0^n)^c = \lambda^c\right) = (\lambda_0^n)^{-c} = \lambda^{-c} \quad \text{(alive)} \\ &\Pr(\varepsilon_n = 0) = 1 - (\lambda_0^n)^c = 1 - \lambda^{-c} \quad \text{(dead)} \end{aligned} \tag{5.3}$$

Hence the density either diverges ε_n with an (algebraic) order of singularity c, but with an (algebraically) decreasing probability, or is "calmed" down to zero.

Following the discussion (and definitions) given in Section 3.2, c is the codimension of the alive eddies, and hence their corresponding dimension D is:

$$D = d - c \tag{5.4}$$

(d is the dimension of the embedding space, equal to 2 in Figs. 3.4 and 3.9). D is the dimension of the "support" of turbulence, corresponding to the fact that after n steps the average number of alive eddies in the β model is

$$\langle N_n \rangle = \lambda^d \Pr(\varepsilon_\lambda = \lambda^c) = \lambda^{d-c} \tag{5.5}$$

5.1.3 Revisiting the α model

In Chapter 3, we introduced the α model, which more realistically allows eddies to be either "more active" or "less active" according to the following binomial process:

$$\Pr(\mu\varepsilon = \lambda_0^{\gamma_+}) = \lambda_0^{-c} \quad (> 1 \Rightarrow \text{INCREASE})$$
$$\Pr(\mu\varepsilon = \lambda_0^{\gamma_-}) = 1 - \lambda_0^{-c} \quad (< 1 \Rightarrow \text{DECREASE}) \tag{5.6}$$

where γ_+, γ_- correspond to boosts and decreases respectively, the β model being the special case where $\gamma_- = -\infty$ and $\gamma_+ = c$ – due to conservation $<\mu\varepsilon> = 1$, there are only two free parameters, Eqn. (3.6):

$$\lambda_0^{\gamma_+ - c} + \lambda_0^{\gamma_-}(1 - \lambda_0^{-c}) = 1 \tag{5.7}$$

Taking $\gamma_- > -\infty$, the pure orders of singularity γ_- and γ_+ lead to the appearance of mixed orders of singularity, of different orders γ ($\gamma_- \leq \gamma \leq \gamma_+$). These are built up step by step through a complex succession of γ_- and γ_+ values as illustrated in Fig. 3.7b.

What is the behaviour as the number of cascade steps, $n \to \infty$? Consider two steps of the process: the various probabilities and random factors are:

$$\Pr(\mu\varepsilon = \lambda_0^{2\gamma_+}) = \lambda_0^{-2c} \quad \text{(two boosts)}$$
$$\Pr(\mu\varepsilon = \lambda_0^{\gamma_+ + \gamma_-}) = 2\lambda_0^{-c}(1 - \lambda_0^{-c})$$
$$\text{(one boost and one decrease)}$$
$$\Pr(\mu\varepsilon = \lambda_0^{2\gamma_-}) = (1 - \lambda_0^{-c})^2 \quad \text{(two decreases)} \tag{5.8}$$

This process has the same probability and amplification factors as a new three-state α model with a new scale ratio of λ_0^2 defined as:

$$\Pr(\mu\varepsilon = (\lambda_0^2)^{\gamma_+}) = (\lambda_0^2)^{-c} \quad \text{(one large)}$$
$$\Pr(\mu\varepsilon = (\lambda_0^2)^{(\gamma_+ + \gamma_-)/2}) = 2(\lambda_0^2)^{-c/2} - 2(\lambda_0^2)^{-c}$$
$$\text{(intermediate)}$$
$$\Pr(\mu\varepsilon = (\lambda_0^2)^{\gamma_-}) = 1 - 2(\lambda_0^2)^{-c/2} + (\lambda_0^2)^{-c}$$
$$\text{(large decrease)} \tag{5.9}$$

Iterating this procedure, after $n = n_+ + n_-$ steps we find:

$$\gamma_{n_+, n_-} = \frac{n_+\gamma_+ + n_-\gamma_-}{n_+ + n_-}, \qquad n_+ = 1, \ldots, n$$
$$\Pr\left(\mu\varepsilon = (\lambda_0^n)^{\gamma_{n_+, n_-}}\right) = \binom{n}{n_+}(\lambda_0^n)^{-cn_+/n}\left(1 - (\lambda_0^n)^{-c/n}\right)^{n_-} \tag{5.10}$$

where $\binom{n}{n_+}$ is the number of combinations of n objects taken n^+ at a time. The Stirling formula enables us to explicitly compute for $N \to \infty$, $r = n_+/n$ fixed the asymptotic codimensions C_{ij} that define the probability distribution (Schertzer and Lovejoy, 2011):

$$\Pr\left(\varepsilon_{\lambda_0^n} \geq (\lambda_0^n)^{\gamma_i}\right) = \sum_j p_{i,j}(\lambda_0^n)^{-c_{i,j}} \tag{5.11}$$

The p_{ij}'s are the "submultiplicities" (the prefactors in the above) and λ_0^n is the total ratio of scales from the outer scale to the smallest scale. Notice that the requirement that $\langle\mu\varepsilon\rangle = 1$ implies that some of the λ^{γ_i} are > 1 (boosts) and some are < 1 (decreases), that is, some $\gamma_i > 0$ and some $\gamma_i < 0$. Note also that the α model has bounded singularities:

$$\gamma_- \leq \gamma \leq \gamma_+ \tag{5.12}$$

so that the important maximum attainable singularity γ_{max} is equal to γ_+. The final step in "renormalizing" the cascade is to replace the above n-step, 2-state cascade with ratio λ_0 by a single-step cascade with ratio $\lambda = \lambda_0^n$ and with $n + 1$ states. Note that we are not saying that there is absolutely no difference between the n-state α model with ratio λ and the corresponding $(n + 1)$-state model with $\lambda = \lambda_0^n$: their properties will however be identical for integer powers of λ. Finally, doing this and making the replacement $\lambda_0^n \to \lambda$, and taking the limit $\lambda \to \infty$, one of the terms in the sum (Eqn. (5.11)) will dominate (that with the smallest c_{ij}). Hence defining:

$$c_i = \min[c_{ij}] = c(\gamma_i) \tag{5.13}$$

yields for $\lambda \to \infty$:

$$\Pr(\varepsilon_\lambda \geq \lambda^{\gamma_i}) = p_i\lambda^{-c_i} \tag{5.14}$$

where c_i is the codimension and p_i is the corresponding multiplicity. If we now drop the subscripts "i" (this allows for the possibility of a continuum of states, e.g., the original process being defined by a uniform or other continuous distribution) then we obtain:

$$\Pr(\varepsilon_\lambda \geq \lambda^\gamma) = p(\gamma)\lambda^{-c(\gamma)}; \quad \frac{dc}{d\gamma} > 0 \tag{5.15}$$

This is a basic multifractal probability relation for cascades. We now simplify this using the ~ sign, which absorbs the multiplicative (p_i) as well as taking into account the logarithmic number of terms in the sum (which can lead to logarithmic prefactors corresponding

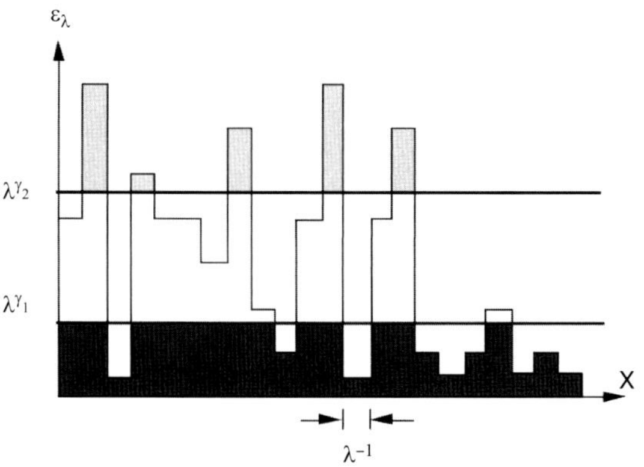

Fig. 5.1 A schematic illustration of a multifractal field analyzed over a scale ratio λ, with two scaling thresholds λ^{γ_1} and. λ^{γ_2}, corresponding to two orders of singularity: $\gamma_2 > \gamma_1$. Reproduced from Schertzer and Lovejoy (1993).

to "subcodimensions"). With this understanding about the equality sign, we may thus write:

$$\Pr(\varepsilon_\lambda \geq \lambda^\gamma) \sim \lambda^{-c(\gamma)} \qquad (5.16)$$

Each value of ε_λ corresponds to a singularity of order γ and codimension $c(\gamma)$ (Fig. 5.1). Note that since the smallest scale $= \lambda^{-1}$, strictly speaking the expression "singularity" applies to $\gamma > 0$ ($\varepsilon_\lambda \to \infty$ for $\lambda \to \infty$), when $\gamma < 0$ it is rather a "regularity." Note that Eqns. (5.11), (5.14)–(5.16) consider probability distributions of events above a given (scaling) threshold; therefore we consider "exceedance probability distributions" $\Pr(\varepsilon_\lambda \geq \lambda^\gamma)$ rather than the standard "cumulative probability distribution function" CDF $= \Pr(\varepsilon_\lambda < \lambda^\gamma)$. However, both are obviously related by $\Pr(\varepsilon_\lambda \geq \lambda^\gamma) = 1 -$ CDF. Here and throughout, this book will always use the term "probability distribution" in the sense of exceedance probability distribution, and therefore for events above a given (scaling) threshold.

5.2 The codimension multifractal formalism

5.2.1 Codimension of singularities $c(\gamma)$ and its relation to $K(q)$

In this section we continue our discussion of multifractal fields in terms of singularities, but also relate this to its dual representation in terms of statistical moments, which we already discussed in Chapter 3. Contrary to the popular dimension $f(\alpha)$ formalism, which was developed for (low-dimensional) deterministic chaos (see Box 5.5), we develop a codimension formalism necessary for stochastic processes; it is therefore more general than the dimension formalism.

The measure of the fraction (at resolution λ with corresponding scale $L = L / \lambda$) of the probability space with singularities higher than γ is given by the probability distribution (Eqn. (5.16)). The previous section underlines this new feature; the exponent $c(\gamma)$ is a function, not a unique value. Rather than dealing with just a scaling geometric set of points, we are dealing with a scaling function (in the limit $\lambda \to \infty$, the density of a measure); from this function we can define an infinite number of sets, e.g. one for each order of singularity γ (Fig. 5.1).

We now derive the basic connection between $c(\gamma)$ and the moment scaling exponent $K(q)$. To relate the two, write the expression for the moments in terms of the probability density of the singularities:

$$p(\gamma) = \frac{d\Pr}{d\gamma} = c'(\gamma)(\log\lambda)\lambda^{-c(\gamma)} \sim \lambda^{-c(\gamma)} \qquad (5.17)$$

(where we have absorbed the $c'(\gamma)\log\lambda$ factor into the "$\sim$" symbol since it is slowly varying, subexponential). This yields:

$$\langle \varepsilon_\lambda^q \rangle = \int d\Pr(\varepsilon_\lambda)\varepsilon_\lambda^q \sim \int d\gamma \lambda^{-c(\gamma)}\lambda^{q\gamma} \qquad (5.18)$$

where we have used $\varepsilon_\lambda = \lambda^\gamma$ (this is just a change of variables ε_λ for γ; λ is a fixed parameter).

Hence:

$$\langle \varepsilon_\lambda^q \rangle = \lambda^{K(q)} = e^{K(q)\log\lambda} \sim \int_{-\infty}^{\infty} d\gamma e^{\xi f(\gamma)};$$
$$\xi = \log\lambda; \quad f(\gamma) = q\gamma - c(\gamma); \quad \lambda >> 1 \qquad (5.19)$$

We see that our problem is to obtain an asymptotic expansion of an integral with integrand of the form $\exp(\xi f(\gamma))$ where $\xi = \log\lambda$ is a large parameter and $f(\gamma) = q\gamma - c(\gamma)$. These expansions can be conveniently performed using the mathematical technique of "steepest descents" (e.g. Bleistein and Handelsman, 1986) which shows that the dominant term in the expansion for the integral is $\exp[\xi \max_\gamma (f(\gamma))]$ (i.e. the integral is dominated by the singularity γ which yields

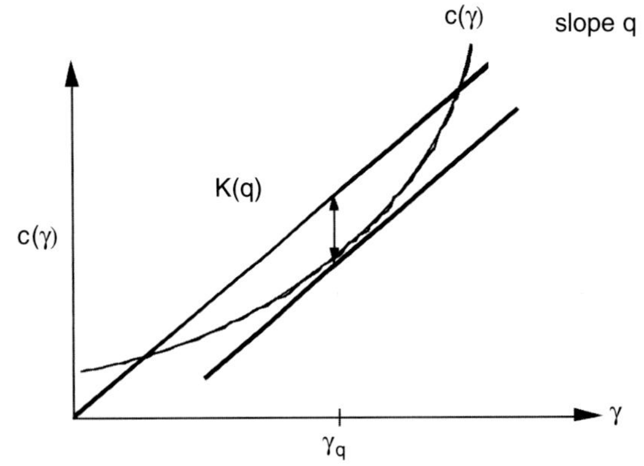

Fig. 5.2 $c(\gamma)$ versus γ showing the tangent line $c'(\gamma_q) = q$ with the corresponding chord γ_q. Note that the equation is the same as $\gamma_q = K'(q)$. Reproduced from Schertzer and Lovejoy (1993).

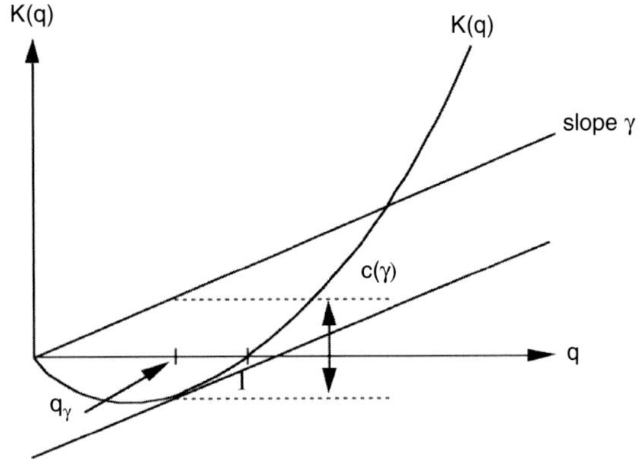

Fig. 5.3 $K(q)$ versus q showing the tangent line $K'(q_\gamma) = \gamma$ with the corresponding chord γ_q. Reproduced from Tessier *et al.* (1993).

the maximum value of the exponent) so that as long as $\xi = \log \lambda >> 1$:

$$K(q) = \max_{\gamma}\left(q\gamma - c(\gamma)\right) \tag{5.20}$$

This relation between $K(q)$ and $c(\gamma)$ is called a "Legendre transform" (Parisi and Frisch, 1985); see Fig. 5.2. We can also invert the relation to obtain $c(\gamma)$ from $K(q)$; just as the inverse Laplace transform used to obtain $K(q)$ from $c(\gamma)$ is another Laplace transform, so the inverse Legendre transform is just another Legendre transform. To show this, consider the twice-iterated Legendre transform $F(q)$ of $K(q)$:

$$\begin{aligned} F(q) &= max_{\gamma}\{q\gamma - (max_{q'}\{q'\gamma - K(q')\})\} \\ &= max_{\gamma;q'}\{\gamma(q - q') + K(q')\} \end{aligned} \tag{5.21}$$

Taking $\partial F/\partial\gamma = 0 \Rightarrow q = q'$, we see that $F(q) = K(q)$. This shows that a Legendre transform is equal to its inverse, and hence we conclude:

$$c(\gamma) = \max_{q}\left(q\gamma - K(q)\right) \tag{5.22}$$

The γ which for a given q maximizes $q\gamma - c(\gamma)$ is γ_q and is the solution of $c'(\gamma_q) = q$ (Fig. 5.2). Similarly, the value of q which for a given γ maximizes $q\gamma - K(q)$ is q_γ so that:

$$\begin{aligned} c'(\gamma_q) &= q \\ K'(q_\gamma) &= \gamma \end{aligned} \tag{5.23}$$

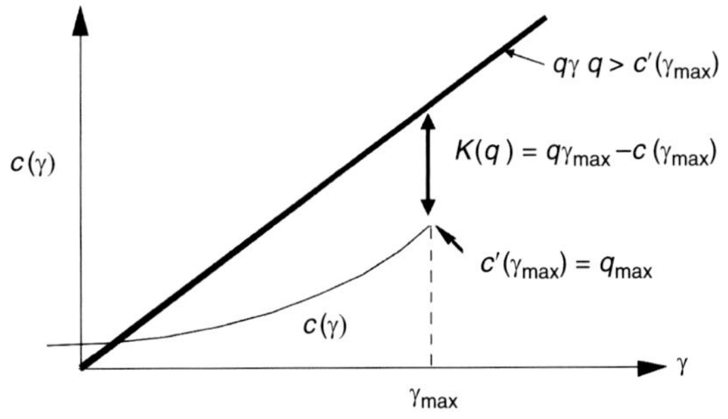

Fig. 5.4 Using the Legendre transformation to derive $K(q)$ when the maximum order of singularity present is γ_{max}; the corresponding moment is $q_{max} = c'(\gamma_{max})$. When $q > q_{max}$ the Legendre transform will have a maximum value for $\gamma = \gamma_{max}$ as shown, which implies that $K(q)$ is linear for $q > q_{max}$. Adapted from Schertzer and Lovejoy (1993).

This is a one-to-one correspondence between moments and orders of singularities (Figs. 5.2 and 5.3). Note that if γ is bounded by γ_{max} (for example in microcanonical cascades, $\gamma \leq d$, Chapter 3, or for the α model, $\gamma \leq \gamma_+$) there is a $q_{max} = c'(\gamma_{max})$ such that for $q > q_{max}$, $K(q) = q\gamma_{max} - c(\gamma_{max})$, i.e. $K(q)$ becomes linear in q (Fig. 5.4).

5.2.2 Properties of codimension functions

We have seen that for each singularity order γ, $c(\gamma)$ is the statistical scaling exponent characterizing how its probability changes with scale. The first obvious property is that due to its very definition (Eqn. (5.16)) $c(\gamma)$ is an increasing function of γ: $c'(\gamma) > 0$. Another fundamental property which follows directly from the

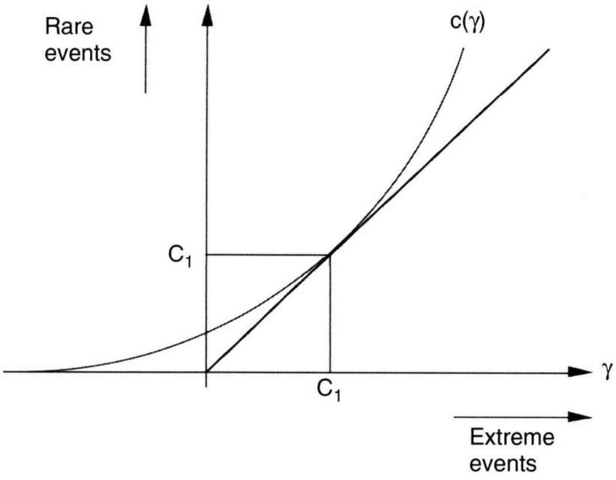

Fig. 5.5 The special properties of the singularity of the mean, C_1. Reproduced from Schertzer and Lovejoy (1993).

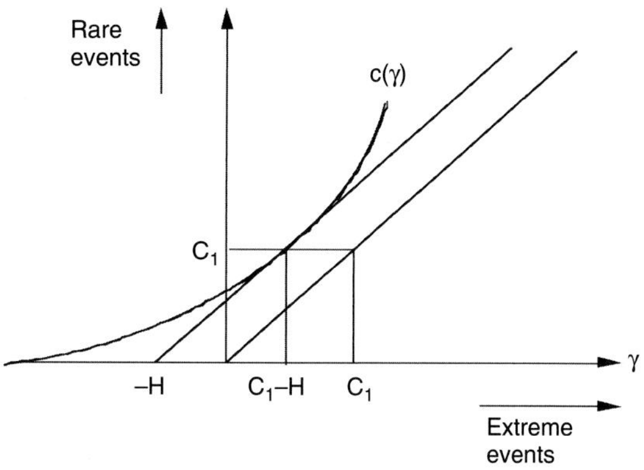

Fig. 5.6 A schematic illustration showing the shift in $c(\gamma)$ of H for nonconserved processes. Reproduced from Schertzer and Lovejoy (1993).

Legendre relation with $K(q)$ is that $c(\gamma)$ must be convex: $c''(\gamma) > 0$.

Many properties of the codimension function can be illustrated graphically. For example, consider the mean, $q = 1$. First, applying Eqn. (5.23) we find $c'(\gamma^1) = 1$, $K'(1) = \gamma_1$, where γ_1 is the singularity giving the dominant contribution to the mean (the $q = 1$ moment). In Chapter 3 we have already defined $C_1 = K'(1)$, so this implies $C_1 = \gamma_1$; the Legendre relation thus justifies the name "codimension of the mean" for C_1. Also at $q = 1$ we have $K(1) = 0$ (due to the scale-by-scale conservation of the flux) so that from Eqn. (5.22) $C_1 = c(C_1)$, as indicated in Fig. 5.5 (this is a fixed point relation). C_1 is thus simultaneously the codimension of the mean of the process and the order of singularity giving the dominant contribution to the mean. Finally, applying $c'(\gamma_q) = q$ (Eqn. (5.23)), we obtain $c'(C_1) = 1$ so that the curve $c(\gamma)$ is also tangent to the line $x = y$ (the bisectrix). If the process is observed on a space of dimension d, it must satisfy $d \geq C_1$, otherwise, following the above, the mean will be so sparse that the process will (almost certainly) be zero everywhere; it will be "degenerate." We will see that when $C_1 > d$ the ensemble mean of the spatial averages (the dressed means) cannot converge.

To understand H, take ensemble averages of Eqn. 4.4. We see that $<\Delta f_\lambda> = <\varepsilon>\Delta x^H \approx \lambda^{-H}$ where we have taken $<\varepsilon> =$ constant and $\Delta x \approx \lambda^{-1}$. H thus determines the deviation from scale-by-scale conservation of $<\Delta f_\lambda>$; it is the basic fluctuation exponent. At the level of random variables, writing $\varepsilon_\lambda = \lambda^\gamma$ we have $\Delta f_\lambda = \lambda^{\gamma-H}$ so that Δf_λ has the statistics of a bare cascade process with a translation of singularities by $-H$

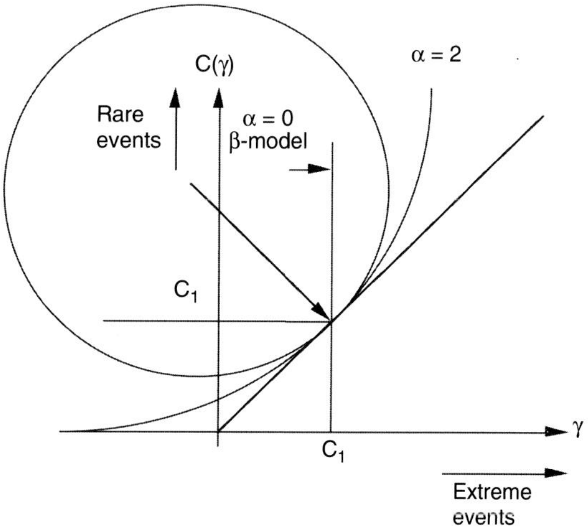

Fig. 5.7 A schematic illustration showing how the $c(\gamma)$ curve can be locally characterized near the mean singularity C_1. Reproduced from Schertzer and Lovejoy (1993).

(Fig. 5.6). This generalizes the Kolmogorov relation ($f = v$, $H = 1/3$), although we shall see that such simple change of cascade normalization ($\gamma \to \gamma - H$) is a poor model of the observable f fields, which are best modelled by fractional integration of order H (see Section 5.4). Finally, since $c(\gamma)$ is convex with fixed point C_1, it is possible (Fig. 5.7) to define the degree of multifractality (α) by the (local) rate of change of slope at C_1 (the singularity corresponding to the mean); its radius of curvature $R_c(C_1)$ is (Schertzer and Lovejoy, 1992):

$$R_c(C_1) = \frac{\left(1 + c'(C_1)\right)^{3/2}}{c''(C_1)} \tag{5.24}$$

Using the general relation $c'(C_1) = 1$ we obtain $R_c(C_1) = 2^{3/2} / c''(C_1)$, and hence we can locally (near the mean $\gamma = C_1$) define a curvature parameter α from either of the equivalent relations:

$$\alpha = \frac{2^{3/2}R_c(C_1)}{C_1} = \frac{1}{C_1 c''(C_1)} \tag{5.25}$$

These local ($q = 1$, $\gamma = C_1$) definitions of α are equivalent to the definition via moments $\alpha = K''(1)/K'(1)$ (Chapter 3). For the universal multifractals (Chapter 3), this becomes global (i.e. is sufficient to describe the entire $c(\gamma)$ function), and we find an upper bound (maximum degree of multifractality) $\alpha = 2$ (a parabola, Eqn. (3.49)). The $\alpha = 0$ case is the monofractal extreme β model whose singularities all have the same fractal dimension.

5.2.3 The sampling dimension, sampling singularity and second-order multifractal phase transitions

The statistical properties we have discussed up until now are valid for (infinite) statistical ensembles. However, real-world data are always finite so that sufficiently rare events will always be missed. In order to understand the effect of finite sample sizes, consider a collection of N_s samples, each d-dimensional and spanning a range of scales $\lambda = L/l$ (= largest/smallest) so that for example there are λ^d pixels from each sample (Fig. 5.8), and introduce the "sampling dimension" D_s:

$$D_s = \frac{\log N_s}{\log \lambda} \tag{5.26}$$

as well as the singularity γ_s corresponding to the largest (and rarest) value $\varepsilon_{\lambda,s}$, the "sampling singularity":

$$\gamma_s = \frac{\log \varepsilon_{\lambda,s}}{\log \lambda} \tag{5.27}$$

The relation between N_s and $\varepsilon_{\lambda,s}$ is thus equivalent to the relation between their base λ logarithms, i.e. between D_s and γ_s (Eqns. (5.26), (5.27)).

In order to relate γ_s and D_s, consider a collection of satellite images ($d = 2$). Our question is thus: What is the rarest event with the most extreme γ_s that we may expect to see on a single picture? On a large enough collection of pictures? The answer to these questions is straightforward: there is a total of λ^{d+D_s} pixels in the sample, so the rarest event has a probability $\approx \lambda^{-(d+D_s)}$ (Fig. 5.9). However, the probability of finding γ_s is simply $\lambda^{-c(\gamma s)}$, so we obtain the following implicit equation for γ_s:

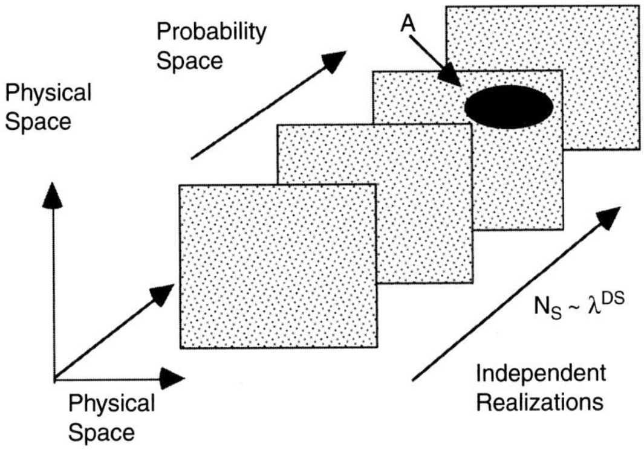

Fig. 5.8 Illustration showing how in random processes the effective dimension of space can be augmented by considering many independent realizations N_s. As $N_s \to \infty$, the entire (infinite-dimensional) probability space is explored. When the process is observed on a low-dimensional cut of dimension d (such as the d = 2-dimensional sketch illustrated by a single rectangle) then we have $N_s = 1$, hence $D_s = 0$. As long as $d > c(\gamma)$, we may introduce the (positive) dimension function $D(\gamma) = d - c(\gamma)$, which is then the geometrical dimension of the set with singularities γ. However, structures with $\gamma > \gamma_s$ ($D(\gamma) < 0$) will be too sparse to be observed (they will almost certainly not be present on a given realization/picture). In order to observe them we must increase the number of samples N_s or equivalently the sampling dimension D_s to reach γ. Reproduced from Schertzer and Lovejoy (1993).

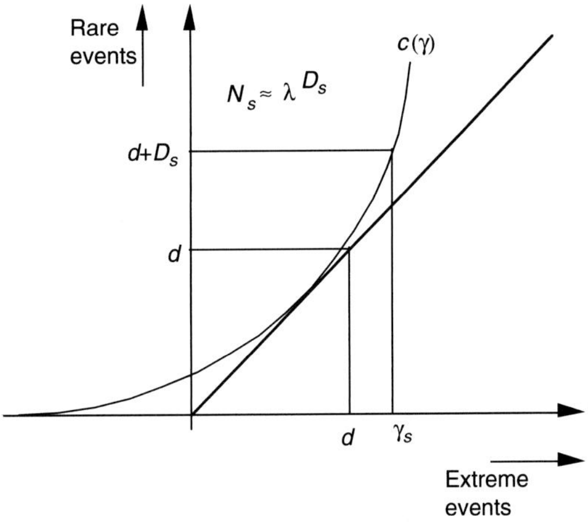

Fig. 5.9 Schematic illustration of sampling dimension and how it imposes a maximum order of singularities γ_s. Reproduced from Tessier (1993).

$$c(\gamma_s) = d + D_s = \Delta_s \tag{5.28}$$

$\Delta_s = d + D_s$ is the corresponding (overall) effective dimension of our sample. More extreme singularities would have codimensions greater than this effective dimension $c > \Delta_s$ and are almost certainly not present in our sample.

As an example, we can use the aircraft data shown in Fig. 3.1 to estimate the largest singularity that we should expect over a single transect 2^{13} points long. We saw that the largest normalized flux value was ≈ 26.5, which corresponds to an order of singularity of γ_s = logε/logλ = log(26.5)/log(2^{13}) = 0.364. Using the estimated multifractal parameters $\alpha = 1.8$, C_1= 0.06 (these are mean C_1, α values for 24 flight legs 1120 km long), we find that in a $d = 1$ section the solution of $c(\gamma_s) = 1.364$ is $\gamma_s = 0.396$, which is very close to the observed maximum.

Let us now calculate the moment exponent $K_s(q)$ for a process with N_s realizations. To do this, we calculate the Legendre transform of $c(\gamma)$ but with the restriction $\gamma \leq \gamma_s$; this is the same type of restriction as discussed earlier (Fig. 5.4, take $\gamma_{max} = \gamma_s$):

$$\begin{aligned} K_s(q) &= \gamma_s(q - q_s) + K(q_s), \quad q \geq q_s \\ K_s(q) &= K(q), \quad q \leq q_s \end{aligned} \tag{5.29}$$

hence at $q = q_s$ there is a jump/discontinuity in the second derivative of K:

$$\Delta K_s'' = -K''(q_s) \tag{5.30}$$

Because of the existence of formal analogies between multifractal processes and classical thermodynamics, this is termed a "second-order multifractal phase transition" (Szépfalusy *et al.*, 1987; Schertzer and Lovejoy, 1992; Box 5.1).

5.2.4 Direct empirical estimation of $c(\gamma)$: the probability distribution multiple scaling (PDMS) technique

In Chapter 3, we saw how to empirically verify the cascade structure and characterize the statistics using the moments, and how to determine their scaling exponent, $K(q)$. In this chapter, we have seen how – via a Legendre transform of $K(q)$ – this information can be used to estimate $c(\gamma)$. However, it is of interest to be able to estimate $c(\gamma)$ directly. To do this, we start from the fundamental defining Eqn. (5.16), take logs of both sides and rewrite it as follows:

$$\mathrm{Log}\,\Pr_\lambda(\varepsilon_\lambda > \lambda^\gamma) = -c(\gamma)\,Log(\lambda) + o(1/\mathrm{Log}(\lambda))O(\gamma) \tag{5.31}$$

$o(1/Log(\lambda))O(\gamma)$ corresponds to the logarithm of the slowly varying factors that are hidden in the ~ sign in Eqn. (5.16) and the subscript λ on the probability has been added to underline the resolution dependence of the cumulative histograms. For each order of singularity γ, this equation expresses the linearity of log probability with the log of the resolution. The singularity itself must be estimated from the fluxes by:

$$\gamma = \frac{\log(\varepsilon_\lambda)}{\log\lambda} \tag{5.32}$$

We now see that things are a little less straightforward than when estimating $K(q)$. First, the term $o(1/Log(\lambda))O(\gamma)$ may not be so negligible, in particular for moderate λ's, so that using the simple approximation $c(\gamma) \approx -\log \Pr_\lambda/\log\lambda$ may not be sufficiently accurate. Second, in Eqn. (5.32), we assumed that ε_λ is normalized such that $<\varepsilon_\lambda> = 1$; if it is not, it can be normalized by dividing by the ensemble mean: $\varepsilon_\lambda \rightarrow \varepsilon_\lambda/<\varepsilon_\lambda>$. However, from small samples, there may be factors of the order 2 in uncertainty over the mean so that even the estimate of γ may involve some uncertainty. In comparison, if one wants to estimate $K(q)$, one needn't worry about either of these issues since (even for the un-normalized ε_λ) the linear relation $\log\langle\varepsilon_\lambda^q\rangle = K(q)\log\lambda$ is exact (at least in the framework of the pure multiplicative cascades): $K(q)$ is simply the slope of the $\log\langle\varepsilon_\lambda^q\rangle$ versus logλ graph (and if the normalization is accurate, the outer scale itself can be estimated from the points where the lines cross: see the examples in Chapter 4). The relative simplicity of the moment method explains why in practice it is the most commonly used. $c(\gamma)$ can then be estimated from $K(q)$ by Legendre transform (either numerically or using a universal multifractal parametrization).

In order to exploit Eqns. (5.31) and (5.32) to directly estimate $c(\gamma)$, we may thus calculate a series of histograms $\Pr_\lambda$ of the lower- and lower-resolution fluxes (obtained from ε_λ by the usual degrading/"coarse-graining"/averaging) and use the transformation of variables in Eqn. (5.32) to write the histograms in terms of γ rather than ε_λ. There are then two variants on this "probability distribution multiple scale" (PDMS) method (Lavallée *et al.*, 1991). The first

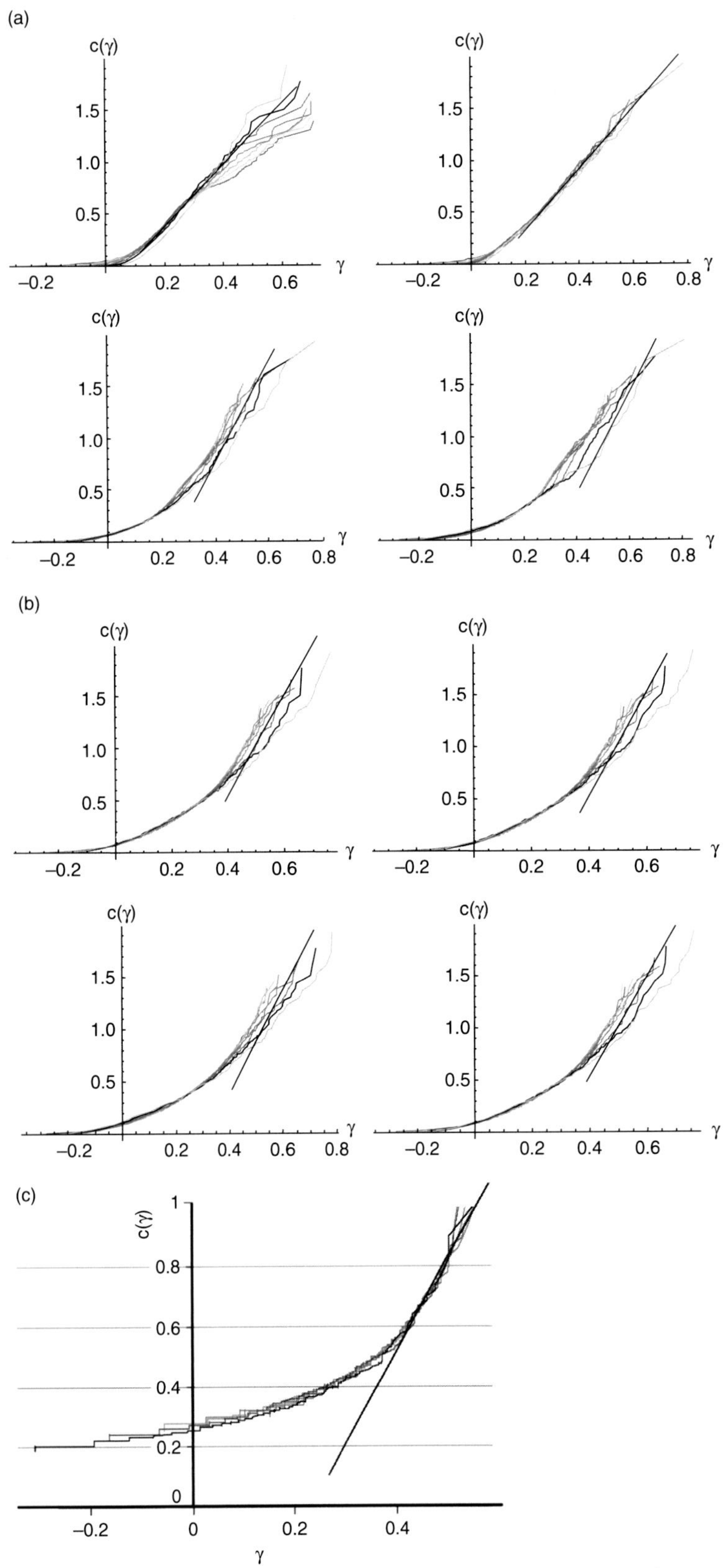

Fig. 5.10 (a) $c(\gamma)$ estimated from the PDMS method, $c(\gamma) \approx -\log Pr/\log\lambda$ (described in the text), shown for resolution degraded by factors of 2 from 280 m to ~36 km (longest to shortest curves). The figure shows aircraft altitude, pressure (upper left, right), longitudinal and transverse wind speed (lower left, right) for 24 flight legs, each 4000 points long, 280 m resolution (i.e. 1120 km). For reference, lines of slope 3 (top row) and 5 (bottom row) are given corresponding to power-law probability distributions with the given exponents. Since the bisectrix touches the curve at the point (C_1, C_1), we can see graphically that C_1's are typically ≈ 0.1 (see Chapter 4 for more precise estimates). (b) Same as Fig. 5.10a except for the thermodynamic variables temperature, potential temperature (upper left, right), humidity and equivalent potential temperature (lower left and right). The reference lines all have slopes of 5. (c) $c(\gamma)$ estimated from hourly rain gauge data at Nîme, France, from 1972–1975 with the resolution degraded by factors of 2–32 hours. The reference line (added) has slope 3. Adapted from Schertzer *et al.* (2010).

and simplest is to ignore the typically small "slow" term in Eqn. (5.31) and simply use $c(\gamma) \approx -\log \Pr_\lambda / \log \lambda$; the second performs (for each of a series of standard γ values) regressions of $\log Pr_\lambda$ versus $\log\lambda$ and then estimates $c(\gamma)$ as the (negative) slope.

In Figs. 5.10a and 5.10b we show some examples of the first method when applied to the aircraft data discussed in Chapter 3: 24 legs, each of 4000 points, were used with a resolution of 280 m (i.e. a total of $24 \times 4000 = 9.6 \times 10^4$ data points for each field), and the absolute differences at 280 m resolution were analyzed. In Chapter 3 we already considered the temperature flux and showed graphically that the transformation from ε_λ's into γ's did indeed result in an apparently stable dynamical range of the γ's (in comparison, the range of the degraded ε_λ's kept diminishing as λ decreased: Figs. 3.2a, 3.2b). Figs. 5.10a and 5.10b simply show the $\log\lambda$-normalized log probability of the histograms at each resolution. The resolution was degraded systematically by factors of 2; only the first seven iterations (i.e. a degradation of resolution factor of 128, corresponding to $\approx$ 36 km) are shown (the corresponding moments are shown in Figs. 4.6a and 4.6b). From the relatively tight bunching of the curves, we see that the method is reasonably successful: we have effectively removed almost all of the scale dependency from the histograms. Although the curves tend to diverge somewhat at large (and rare) values of γ (where the sample size starts to be inadequate), the basic $c(\gamma)$ shape is discernible. To be a bit more quantitative, we can use the estimate of the sampling dimension $D_s = \log N_s/\log\lambda = \log 24/\log 4000 \approx 0.38$ to determine that the estimates are reliable only up to $c = d + D_s = 1.38$. Also shown in the figures are straight reference lines, since we shall see in the next section that one generally expects high-order moments to diverge, implying asymptotically linear $c(\gamma)$'s. Along with both the transverse and longitudinal wind components and the thermodynamic fields, we have also included the essentially aircraft measurement/trajectory-specific fields z, p: the fluxes derived from aircraft altitude and pressure (recall we used absolute differences). These both show extreme tails (compare the reference slopes ≈ 3), and this in spite of the autopilot attempting to enforce an isobaric trajectory! In fact, as discussed in Chapter 6, it seems that due to the aircraft response to wind turbulence the trajectories only begin to be effectively isobaric for scales around 40 km and greater (i.e. just the resolution of the curves is shown in the figures).

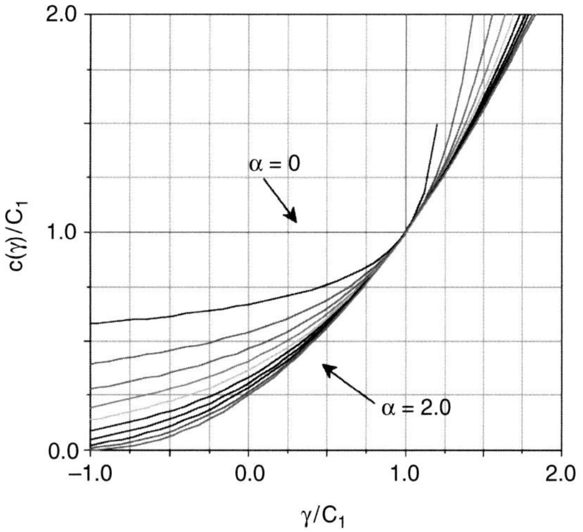

Fig. 5.11 Universal $c(\gamma)$ vs γ, for different $\alpha = 0$ to 2 by increment $\Delta\alpha = 0.2$. Adapted from Schertzer and Lovejoy (1989).

Finally, in Fig. 5.10c, we show a similar analysis of hourly raingauge data from Nîme, France (1972–1975, ~35 000 points). Again we see a good collapse to a unique $c(\gamma)$, and this time there is evidence for an asymptotic linear behaviour with slope $q_D \approx 3$ (see Section 5.3.3).

5.2.5 Codimensions of universal multifractals, cascades

When discussing the moment characterization of the cascades, we have already noted that the two parameters C_1, α are of fundamental significance. C_1 characterizes the order and codimension of the mean singularities of the corresponding conservative flux, it is the local trend of the normalized $K(q)$ near the mean; $K(q) = C_1 (q - 1)$ is the best monofractal "β-model approximation near the mean" ($q \approx 1$). Finally, $\alpha = K''(1)/K'(1)$ characterizes the curvature near the mean. The curvature parameter α can also be defined directly from the probability exponent $c(\gamma)$ by using the local radius of curvature $R_c(C_1)$ of $c(\gamma)$ at the point $\gamma = C_1$, i.e. the corresponding singularity (Eqn. (5.25)). Finally, for the observed field f, there is a third exponent H which characterizes the deviation from conservation of the mean fluctuation $<\Delta f> \approx <\varepsilon>\Delta x^H \approx \Delta x^H$; since $<\varepsilon> =$ constant, it is a "fluctuation" exponent.

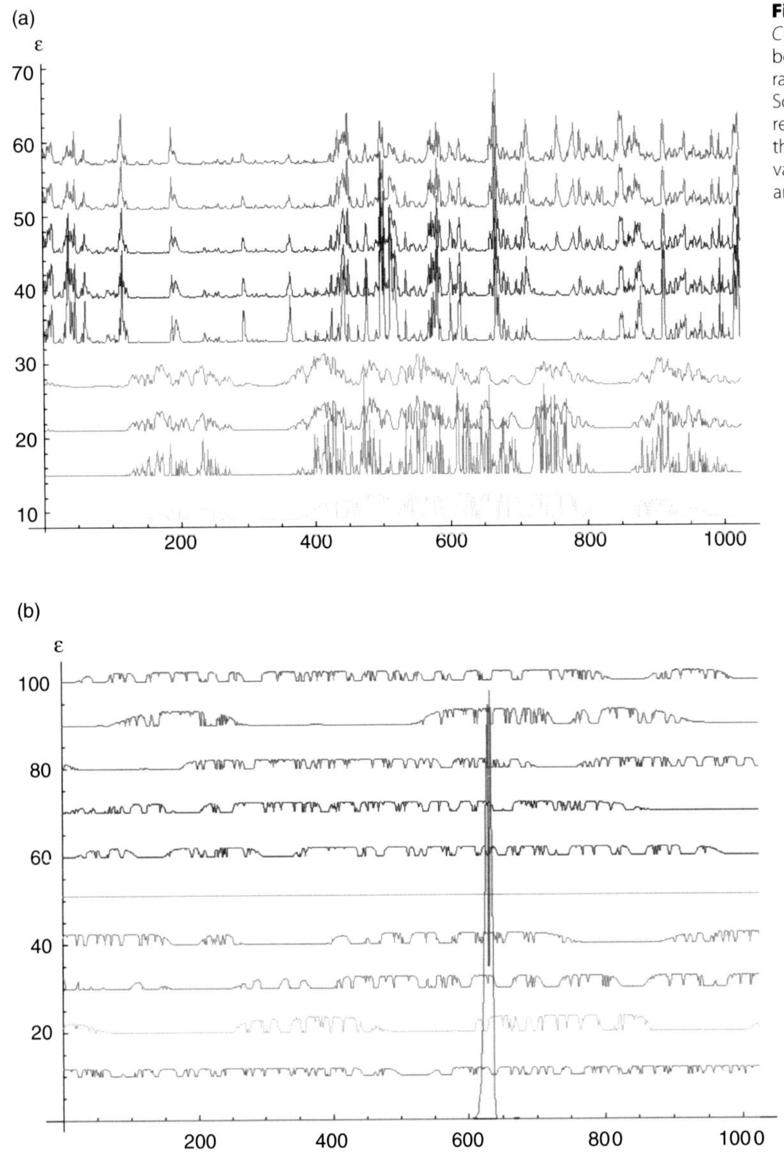

Fig. 5.12 (a) Multifractal simulations $C_1 = 0.1$ and $\alpha = 0.3, 0.5, \ldots 1.9$ from bottom to top, offset for clarity (same random seed). Adapted from Lovejoy and Schertzer (2010a). (b) Eleven independent realizations of $\alpha = 0.2$, $C_1 = 0.1$, indicating the huge realization-to-realization variability: the bottom realization is not an outlier!

In Chapter 3 we discussed how the universal attractors of additive processes can be used to deduce those of the multiplicative processes by studying their "generators", $\Gamma_\lambda = \log \varepsilon_\lambda$. Since multiplying fields ε_λ is equivalent to adding generators Γ_λ (for a fixed scale ratio λ), the generators which are stable and attractive under addition are the Lévy generators discussed in Chapter 3 and to which we return in Section 5.4. The moment exponent $K(q)$ (see Eqn. (3.49)) is given by:

$$K(q) = \frac{C_1}{\alpha - 1}(q^\alpha - q); \quad \alpha \neq 1 \tag{5.33}$$

$$K(q) = C_1 q Log(q); \quad \alpha = 1 \tag{5.34}$$

The top formula is valid for $0 \leq \alpha \leq 2$; however, as discussed in Section 5.4, K diverges for all $q < 0$ except in the special ("lognormal") case $\alpha = 2$. To obtain the

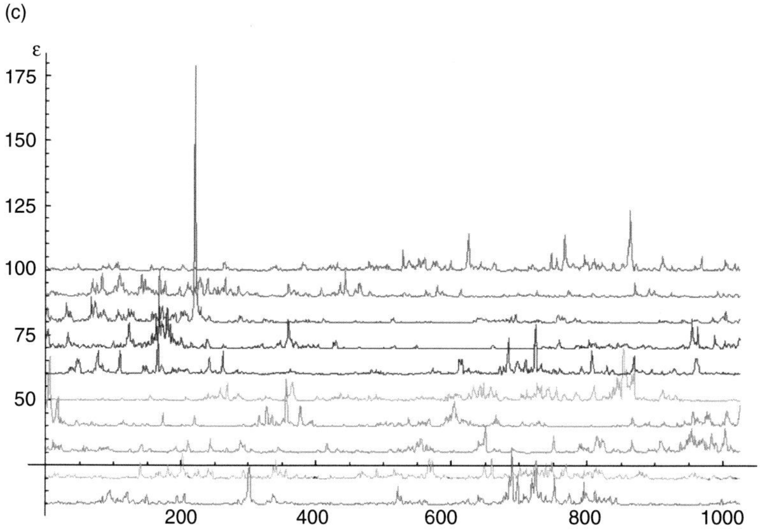

Fig. 5.12 (c) Ten independent realizations of $\alpha = 1.9$, $C_1 = 0.1$. Again, notice the large realization-to-realization variability.

corresponding $c(\gamma)$, one can simply take the Legendre transformation (Eqn. (5.22)) to obtain:

$$c(\gamma) = C_1\left(\frac{\gamma}{C_1\alpha'} + \frac{1}{\alpha}\right)^{\alpha'}; \quad \alpha \neq 1; \quad 1/\alpha' + 1/\alpha = 1 \qquad (5.35)$$

$$c(\gamma) = C_1 e^{\left(\frac{\gamma}{C_1} - 1\right)}; \quad \alpha = 1$$

where α' is the auxiliary variable defined above (Schertzer and Lovejoy, 1987), used to simplify the formula. Figs. 3.15 and 5.11 show respectively the universal $K(q)/C_1$ and $c(\gamma)$ /C_1 curves. Note that since α' changes sign at $\alpha = 1$, for $\alpha < 1$ there is a maximum order of singularity $\gamma_{max} = C_1/(1 - \alpha)$ so that the cascade singularities are "bounded," whereas for $\alpha > 1$ there is on the contrary a minimum order $\gamma_{min} = -C_1/(\alpha - 1)$ below which the prefactors in Eqn. (5.16) dominate ($c(\gamma) = 0$ for $\gamma < \gamma_{min}$) but the singularities are unbounded above.

If we take $H = 0$ (for simplicity) and use Eqn. (5.35) to express the exceedance probability distribution of ε, using $\gamma = \log\varepsilon_\lambda/\log\lambda$, we find for $\varepsilon_\lambda >> \lambda^{C_1/(\alpha-1)}$ that $\Pr(\varepsilon_\lambda' \geq \varepsilon_\lambda) \backsim p(\varepsilon_\lambda) \approx \exp\left(-(\log\varepsilon_\lambda/K)^{\alpha'}\right)$ with constant $K = \alpha'(C_1\log\lambda)^{1/\alpha}$; this is a "stretched exponential" distribution (also known as a "Weibull" distribution) for the *logarithm* of ε. We stress this because eclectic semi-empirical approaches to scaling in turbulence, climate and elsewhere have claimed the existence of the multiscaling of the moments ($\langle\varepsilon_\lambda{}^q\rangle \approx \lambda^{K(q)}$) while simultaneously that the distribution for ε_λ (*not* for $\log\varepsilon_\lambda$) is a stretched exponential. Since, over limited ranges of ε_λ, it may be difficult to empirically distinguish the two behaviours (i.e. stretched exponential for ε_λ or for $\log\varepsilon_\lambda$), theoretical consistency (and clarity!) is particularly important.

Figs. 5.12a, 5.12b and 5.12c show one-dimensional realizations using the continuous cascades described in Section 5.4. In Fig 5.12a, we fix $C_1 = 0.1$ (a typical value for atmospheric fields: Table 4.8) and we demonstrate the effect of increasing α with the same random seed so that the change in the structures with α is apparent. For low α, the series are dominated by "holes" of very low values, while for high α (near the maximum, $\alpha = 2$), it is rather the large singularities that dominate. Fig. 5.13 shows two-dimensional realizations with the same "Lévy hole" phenomenology. In Figs. 5.12b and 5.12c we show the huge realization-to-realization variability by showing 10 realizations with different seeds for $\alpha = 0.2$, 1.9 respectively. Note that due to severe numerical underflow problems, it is difficult to make simulations with α less than 0.2.

Before leaving the topic of universal multifractals, we should mention another, weak form of universality that was proposed by Dubrulle (1994) and She and Leveque (1994): "log-Poisson" cascades. Recall that starting with 2-state binomial processes, one can take

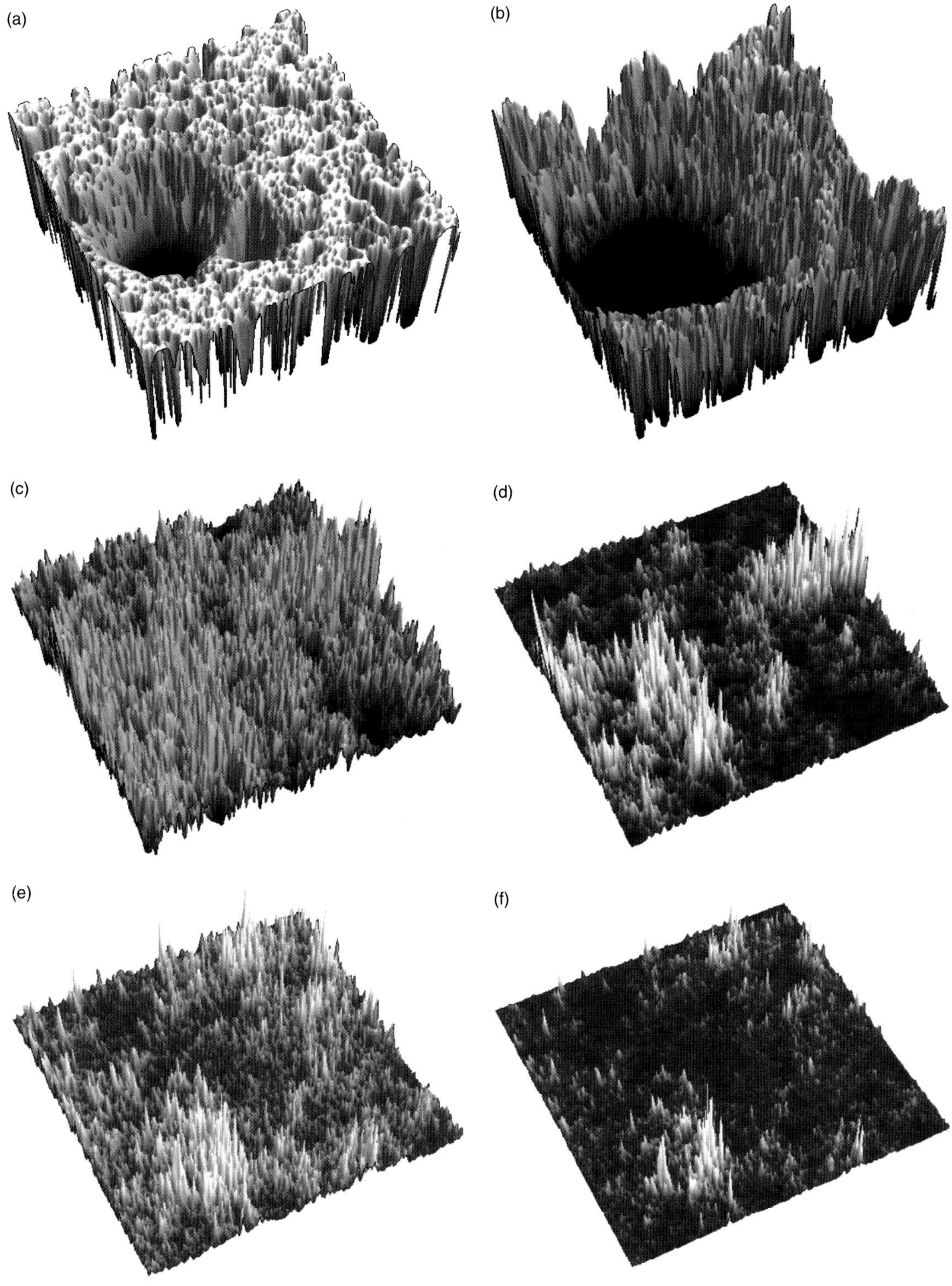

Fig. 5.13 Isotropic realizations in two dimensions with $\alpha = 0.4, 1.2, 2$ (top to bottom) and $C_1 = 0.05, 0.15$ (left to right). The random seed is the same so as to make clear the change in structures as the parameters are changed. The low α simulations are dominated by frequent very low values, the "Lévy holes". The vertical scales are not the same. Reproduced from Lovejoy and Schertzer (2010a). *See colour plate section.*

limits that lead either to Gaussian or to Poisson processes. The corresponding binomial cascade is the α model, and in Schertzer *et al.* (1991) it was shown how to obtain as a limiting case, the $\alpha = 2$ ("log-Gaussian") universal cascade. By taking a different limit of the α model, one obtains "log-Poisson" cascades which have the following form:

$$\begin{aligned} &\gamma_+ = c(1 - \lambda_0^{-\gamma_-}) \\ &K(q) = q\gamma_+ - c + \left(1 - \frac{\gamma_+}{c}\right)^q c \\ &c(\gamma) = c\left(1 - \frac{\gamma_+ - \gamma}{c\gamma_-}\left(1 - \log\frac{\gamma_+ - \gamma}{c\gamma_-}\right)\right); \quad \gamma \le \gamma_+ \\ &\qquad c(\gamma) = \infty; \qquad\qquad\qquad \gamma > \gamma_+ \end{aligned} \tag{5.36}$$

where $\lambda_0 > 1$ is the cascade ratio for a single step, $c > 0$, $\gamma_+ > 0$, and the relation between γ_+ and γ_- (top line) is from the conservation requirement of the α model (Eqn. (5.7); see Schertzer *et al.*, 1995). Clearly, γ_+ is the highest-order singularity and c is the corresponding codimension, so the log-Poisson cascade has intrinsically a maximum singularity that it can produce. The log-Poisson cascade shares with the Lévy generator universal multifractals the possibility of "densifying" the cascade – i.e. it can be made continuous in scale ("infinitely divisible"; Section 5.4, take the limit $\lambda_0 \to 1$), so that it could be said to have "weak" universality properties. However, the Lévy generator cascades could be termed "strongly" universal, since the generator is stable and attractive as well.

The limitations of the log-Poisson cascade can be seen as soon as one considers applications. For example, in hydrodynamic turbulence, She and Leveque (1994) assumed non fractal, $d = 1$ filament-like structures for the highest-order singularity along with homogeneous eddy turnover times, which leads to the parameter choice $c = 2, \gamma_+ = 2/3$ (implying $\lambda_1{}^{\gamma_-} = 3/2$). In magneto-hydrodynamic turbulence one can argue that extreme events occur on current sheets and select $c = 1, \gamma_+ = 1/2$ (Grauer *et al.*, 1994). In both cases, the use of a model with maximum order of singularity seems difficult to justify, as do the particular parameter values. In the atmosphere, Deidda (2000) and Onof and Arnbjerg-Nielsen (2009) have used the log-Poisson model for modelling rainfall, where they found – perhaps unsurprisingly – that the parameters c, γ_+ vary from realization to realization, exactly as predicted by models with unbounded singularities (such as the strongly Lévy generator universal cascades with $\alpha > 1$, where on any realization the maximum singularity is itself simply a random variable rather than a fixed parameter).

5.3 Divergence of statistical moments and extremes

5.3.1 Dressed and bare moments

We now consider the effect of spatially integrating a cascade and then taking the limit $\lambda \to \infty$. This leads to the fundamental difference between the "bare" and "dressed" cascade properties; the former have all moments finite (since by definition, for bare quantities, λ is finite) whereas the latter will generally have divergence for all moments greater than a critical value q_D which depends on the dimension of space over which the process is integrated (see Fig. 5.14 for a schematic).

In order to define the dressed flux, start by defining the Λ resolution flux $\Pi_\Lambda(A)$ over the set A:

$$\Pi_\Lambda(A) = \int_A \varepsilon_\Lambda d^D \underline{r} \tag{5.37}$$

We can now define the "partially dressed" flux density $\varepsilon_{\lambda,\Lambda^{(d)}}$ as:

$$\varepsilon_{\lambda,\Lambda^{(d)}} = \frac{\Pi_\Lambda(B_\lambda)}{vol\,(B_\lambda)} \tag{5.38}$$

where $vol(B_\lambda) = \lambda^{-D}$ is the D-dimensional volume of a ball (interval, square, cube etc.) of size L/λ, and the "(fully) dressed flux density" as:

$$\varepsilon_{\lambda,(d)} = \lim_{\Lambda\to\infty} \varepsilon_{\lambda,\Lambda^{(d)}} \tag{5.39}$$

The terms "bare" and "dressed" are borrowed from renormalization jargon and are justified because the "bare" quantities neglect the small-scale interactions ($<L/\lambda$) whereas the "dressed" quantities take them into account.

Now we can use the factorization property of the cascade (Fig. 5.15): the independence of the large- and small-scale multiplicative factors:

$$\varepsilon_\Lambda = \varepsilon_\lambda T_\lambda(\varepsilon_{\Lambda/\lambda}) \tag{5.40}$$

where the operator T_λ increases scales by a factor λ. This equation should be understood in the following

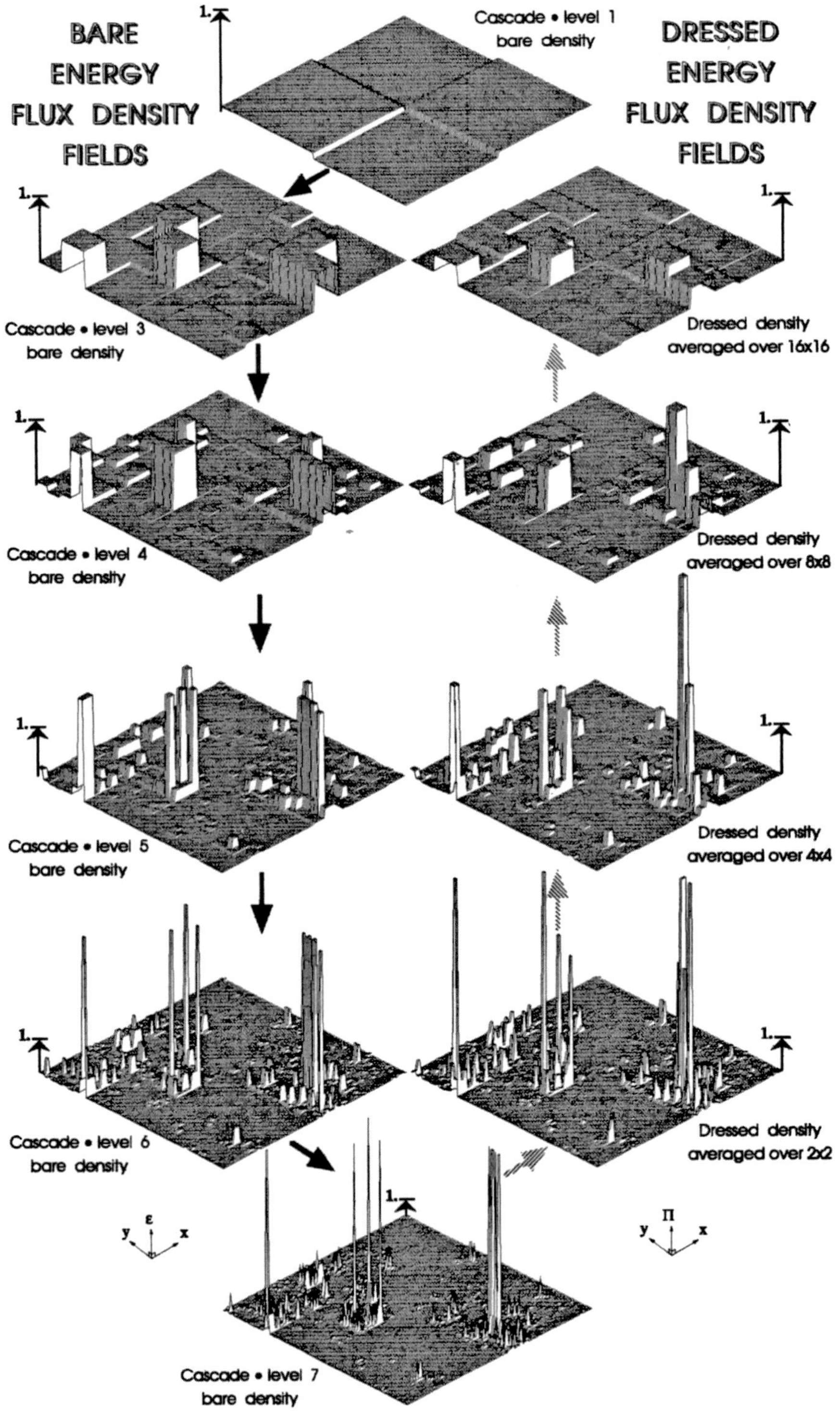

Fig. 5.14 An example of an α-model cascade. The left-hand side shows the step-by-step construction of a ("bare") multifractal cascade starting with an initially uniform unit flux density. The right-hand side shows the "dressed" cascade discussed in the text: the result of spatial averaging (to the same scale as the left image) of the cascade developed over the full range (a factor $\lambda = 2^7$ here, bottom centre) The vertical axis represents the density of energy ε flux to smaller scales which is conserved by the nonlinear terms in the dynamical equations governing fluid turbulence. At each step the horizontal scale is divided by 2, and independent random factors are chosen either < 1 or > 1. Reproduced from Wilson *et al.* (1991).

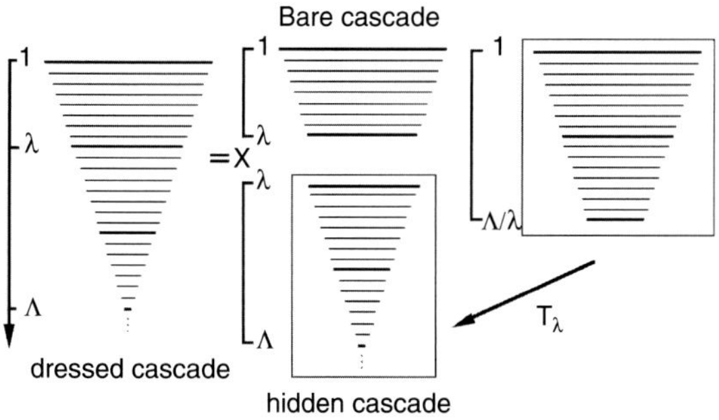

Fig. 5.15 A schematic diagram showing a cascade constructed down to scale ratio Λ, dressed (averaged) up to ratio λ. This is equivalent to a bare cascade constructed over ratio λ, multiplied by a hidden factor obtained by reducing by factor λ, a cascade constructed from 1 to Λ/λ – the action of the scale-changing operator $T_{\Lambda/\lambda}$. Reproduced from Schertzer *et al.*, (1993).

way: to obtain a fine-scale cascade (resolution Λ) we may take a lower-resolution (λ) cascade and multiply each of the λ resolution boxes (balls) by independent cascade processes each developed over a range of scales Λ/λ and reduced in size by factors of λ. This leads to:

$$\varepsilon_{\Lambda,\lambda^{(d)}} = \varepsilon_\lambda \varepsilon_{\Lambda/\lambda^{(h)}} \tag{5.41}$$

where ε_λ is the usual bare density (accounting for variability at scales larger than the observation scale) and the density $\varepsilon_{\Lambda/\lambda^{(h)}}$ (accounting for variability at scales smaller than the observation scale) can be said to be "hidden" since it corresponds to the scales which we average over (see Appendix 5A for more details). We shall see shortly that the small-scale activity is not necessarily smoothed out!

Box 5.1 Flux dynamics and statistical mechanics

Up until now, we have followed a "constructivist" approach. We have constructed specific cascade models and studied their properties. It is also possible to follow a more abstract "flux dynamics" approach (Schertzer and Lovejoy, 1992) because it parallels classical (or quantum) statistical mechanics, but the quantity of interest is the flux of energy whereas in thermodynamics/statistical mechanics the corresponding quantity is just the energy *E*. It turns out that the analogy is not only formal, since it corresponds to mappings from cascade models to Hamiltonian systems.

Analogy with thermodynamics. The correspondences between flux dynamics and thermodynamics with Boltzmann's constant = 1, Σ(β) the Massieu potential, F(β) the Helmholtz free energy. The implications are that just as one can discuss thermodynamic processes without reference to any specific microscopic model of matter, one can similarly discuss multifractal processes without reference to specific models such as cascades.

Flux dynamics	Thermodynamics
probability space	phase space
q	$\beta = \frac{1}{T}$
γ	$-E$
$c(\gamma)$	$-S(E)$
$K(q) = \max_\gamma \left(q\gamma - c(\gamma)\right)$	$\Sigma(\beta) = \max_E \left(S(E) - \beta E\right)$
$C(q) = \frac{K(q)}{q-1}$	$F(\beta) = -\Sigma(\beta)/\beta$

In order to establish the analogy, recall that in thermodynamics (taking Boltzmann's constant =1) we have:

$$\begin{aligned} Z(T) &= \exp\left(\frac{-F(T)}{T}\right) = \left\langle \exp\left(\frac{-E}{T}\right)\right\rangle \\ &= \sum_i \exp\left(\frac{-E_i}{T}\right) \\ &= \int dE \exp\left(S(E) - \frac{E}{T}\right) \end{aligned} \tag{5.42}$$

Box 5.1 *(cont.)*

where $Z(T)$ is the partition function, $S(E) = \log(p(E))$ is the entropy and $p(E)$ is the probability density of states with energy E. In the sum over i we sum over all states, whereas in the sum over j it is only over states with different energy (p_j is the degeneracy associated with the state of energy E_j): the integral form is obtained when the density of states goes to a continuous limit.

In flux dynamics the analogous equations are:

$$Z_\lambda(q) = e^{K_\lambda(q)} = \langle \varepsilon_\lambda^q \rangle = \left\langle e^{q\Gamma_\lambda} \right\rangle = \sum_{\{i\}} \lambda^{q\gamma_i} p(\gamma_i) = \int d\gamma \lambda^{q\gamma - c(\gamma)} \tag{5.43}$$

A summary of the analogies is shown in the table. Such analogies have been discussed in the literature on multifractals, with notably different points of view. For example, our treatment differs somewhat from that of Tél (1988); we rather follow Schuster (1988).

Using the trace moments introduced earlier, we can make parallels with the grand canonical ensemble. The grand canonical ensemble $Z_G(T)$ is obtained by summing not only over all energy states with a fixed number of particles, but also over the number of particles, each weighted by e^{-N}:

$$Z_G(T) = \sum_N \sum_{\{j\}} p_j e^{-E_j/(T-N)} = Tr\left\{ e^{-E_j/(T-N)} \right\} \tag{5.44}$$

where the trace indicates the sum over all states with energy E_j and N particles. In flux dynamics, the sum over energy states is replaced by sums over probability spaces (ensemble averaging, therefore "superaveraging") and the sum over various numbers of particles is replaced by integrals over various observing sets:

$$Z_{G,\lambda}(q) = Tr_{A_\lambda} \varepsilon_\lambda^q \tag{5.45}$$

Hence, formally, we have:

$$e^{-N} = 1_{A_\lambda} d^{D(A)} \underline{r} \tag{5.46}$$

where 1_{A_λ} is the indicator function for the set A_λ.

5.3.2 The divergence of the dressed moments and first-order multifractal phase transitions*

From the above, we see that the bare and dressed densities differ only by the "hidden" factor:

$$\varepsilon^{(h)} = \lim_{\Lambda\to\infty} \varepsilon_{\Lambda/\lambda}^{(h)} = \Pi_\infty(B_1) \tag{5.47}$$

i.e. $\varepsilon^{(h)}$ is a fully developed, fully integrated cascade. In Appendix 5A we derive the following explicit relation for it. From Eqns. (5.39) and (5.41):

$$\varepsilon_\lambda^{(d)} = \varepsilon_\lambda \Pi_\infty(B_1) \tag{5.48}$$

and taking qth moments:

$$\left\langle \varepsilon_\lambda^{(d)q} \right\rangle = \langle \varepsilon_\lambda^q \rangle \langle \Pi_\infty(B_1)^q \rangle \tag{5.49}$$

Since for any q, finite λ, $\langle \varepsilon_\lambda^q \rangle = \lambda^{K(q)}$ is always finite, the finiteness of $\left\langle \varepsilon_\lambda^{(d)q} \right\rangle$ depends on $\langle \Pi_\infty(B_1)^q \rangle$. Using "trace moments" as explained in Appendix 5A, we obtain the result:

$$\langle \Pi_\infty(B_1)^q \rangle \approx \begin{matrix} O(1); & q < q_D \\ \infty; & q \geq q_D \end{matrix} \tag{5.50}$$

where q_D is the solution to the implicit equation:

$$C(q_D) = D \tag{5.51}$$

and $C(q)$ is the dual codimension function (Eqn. (3.29)). This leads to:

$$\left\langle \varepsilon_\lambda^{(d)q} \right\rangle = \begin{cases} \sim \lambda^{K(q)}; & q < q_D \\ \to \infty; & q \geq q_D \end{cases} \tag{5.52}$$

where for $q < q_D$ we have absorbed the proportionality constant $\langle \prod_\infty^q(B_1) \rangle$ into the $\sim$ sign. The overall result is that the dressed and bare densities have identical scaling exponents for $q < q_D$ and only differ for $q > q_D$. The case $q = q_D$ requires some care (Lavallée *et al.* 1991); we obtain divergence. We can thus introduce the dressed exponent scaling function:

$$K_d(q) = \begin{cases} K(q); & q < q_D \\ \infty; & q \geq q_D \end{cases} \tag{5.53}$$

The dressed and bare $K(q)$ are therefore the same except for the extreme fluctuations, which will be much more pronounced for the dressed quantities. To calculate the corresponding dressed codimension $c_d(\gamma)$, we can use the Legendre transform of $K_d(q)$ to obtain:

$$\begin{aligned} c_d(\gamma) &= c(\gamma), \quad \gamma \leq \gamma_D \\ c_d(\gamma) &= q_D(\gamma - \gamma_D) + c(\gamma_D), \quad \gamma > \gamma_D \end{aligned} \tag{5.54}$$

where $\gamma_D = K'(q_D)$ is the critical singularity corresponding to the critical q_D. This transition from convex "bare" behaviour to linear "dressed" behaviour represents a discontinuity in the second derivative of $c(\gamma)$; hence a "second-order multifractal phase transition" for c (for K, see below).

We now note that the microcanonical constraint $\gamma < D$ precludes the possibility of dressed microcanonical cascades having a divergence of moments when averaged over sets of dimension D. To see this, recall that $K(q) = C(q)(q - 1)$ where $C(q)$ is the monotonically increasing dual codimension function (Eqn. (3.29)); i.e. $C'(q) > 0$ and $C(q_D) = D$, Eqn. (5.51). By differentiating $K(q)$ we thus obtain $\gamma_D = K'(q_D) = C'(q_D)\,(q_D - 1) + D$, and hence (since $q_D > 1$), we have $\gamma_D > D$. This shows that microcanonical models (which must have $\gamma_{max} \leq D$) cannot display divergence of moments – at least not on the spaces over which the microcanonical constraint is imposed.

A linear $c_d(\gamma)$ implies a power-law tail on the probability distributions; this is just another way of obtaining the fundamental equivalence between divergence of moments $q \geq q_D$ and "hyperbolic" or "fat-tailed" behaviour of the probability distribution (for large enough thresholds s):

$$\langle \varepsilon_\lambda^{(d)q} \rangle = \infty, q \geq q_D \Leftrightarrow \Pr(\varepsilon_\lambda^{(d)} > s) \sim s^{-q_D}, s >> 1 \tag{5.55}$$

In order to observe the algebraic probability tail however, the sample size must be sufficiently large. Let us consider the effect of varying $D_s = \log N_s/\log\lambda$. Following the argument for Eqn. (5.28), Section 5.2.3, the maximum observable dressed singularity $\gamma_{d,s}$ is given by the solution of $c_d(\gamma_{d,s}) = d + D_s$, and by taking the Legendre transform of $c_d(\gamma)$ with the restriction $\gamma_d < \gamma_{d,s}$ (see Figs. 5.16 and 5.17) we obtain the finite sample dressed $K_{d,s}(q)$:

$$\begin{aligned} K_{d,s}(q) &= \gamma_{d,s}(q - q_D) + K(q_D); \quad q > q_D \\ K_{d,s}(q) &= K(q); \quad q < q_D \end{aligned} \tag{5.56}$$

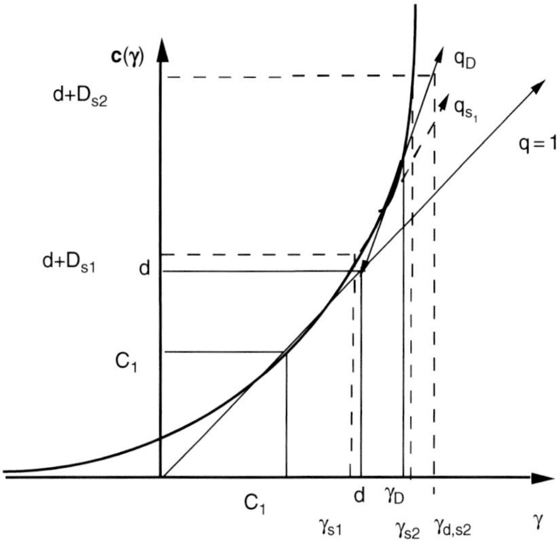

Fig. 5.16 Schematic diagram of $c(\gamma)$, $c_d(\gamma)$ indicating two sampling dimensions D_{S1}, D_{S2} and their corresponding sampling singularities $\gamma_{S1} < \gamma_D < \gamma_{S2} < \gamma_{d,S2}$; the critical tangent (slope q_D) contains the point (D, D). Reproduced from Schertzer *et al.* (1993).

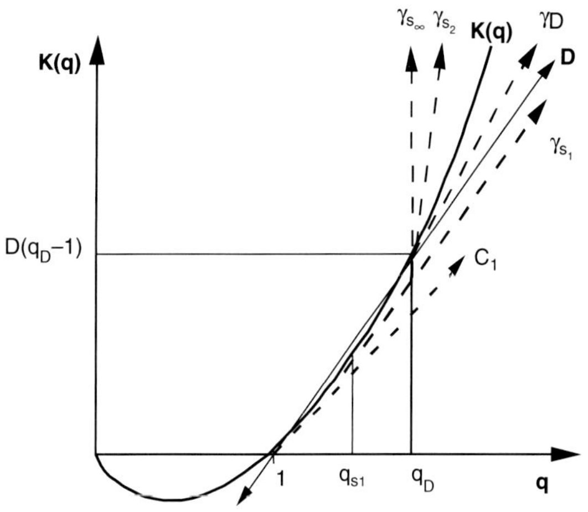

Fig. 5.17 Schematic diagram of $K(q)$, with straight lines of slopes $\gamma_{S1} < \gamma_D < \gamma_{d,S2} < \gamma_{d,S\infty} (= \infty)$ indicating the behaviour for increasing sample size N_S ($N_{S\infty} = \infty$). The line of slope D defining q_D is also shown. Reproduced from Schertzer *et al.* (1993).

In the limit $N_S \to \infty$, $\gamma_{d,s} \to \infty$, and for $q > q_D$, $K_{d,s}(q) \to K_d(q) = \infty$ as expected. This transition corresponds to a jump in the first derivative of the $K(q)$:

$$\Delta K'(q_D) \equiv K'_{d,s}(q_D) - K'(q_D) = \gamma_{d,s} - \gamma_D = \frac{d + D_s - c(\gamma_D)}{q_D} \quad (5.57)$$

Hence, this is a "first-order multifractal phase transition (Box 5.1)."

5.3.3 Empirical tests of divergence of moments

When $\alpha \geq 1$ we have "unconditionally hard multifractal processes"; they are unconditionally hard since $C(q)$ increases without limit, so for any observing space of dimension D there will be a finite q_D satisfying $C(q_D) = D$; this is a consequence of the fact that there is no upper bound on the orders of singularity. Conversely, when $\alpha < 1$, we find that the maximum

Box 5.2 The physical significance of the divergence of moments: the multifractal butterfly effect, self-organized criticality

In Section 5.3.2 we noted the equivalence between the divergence of moments $\langle \varepsilon_\lambda^q \rangle \to \infty$ ($q > q_D$) and "hyperbolic" or "fat-tailed" behaviour of the probability distribution $Pr(\varepsilon_{\lambda(d)} > s) \sim s^{-q_D}$ (Eqn. (5.55)), i.e. for large enough thresholds s, an algebraic fall-off of the probability distribution. In real cascades, viscosity cuts off the cascade process at the viscous scale so that real observables are only partially dressed (Section 5.3.1). This implies that the graphs of log*Pr* versus log*s* will only be linear over a finite interval; for sufficiently large s, they will be truncated. However, it must not be concluded that the divergence of moments is academic – on the contrary, it has a profound physical significance. To see this, denote the inner scale l, the observing scale L, and the ratio $\Lambda = L/l >> 1$. In this case, our previous analysis for dressed moments will be valid, and if we estimate dressed moments with $q > q_D$ (i.e. $C(q) > D$), then the result will be $\sim \Lambda^{(q-1)(C(q)-D)}$ (Appendix 5A), which can be very large and *whose value will depend crucially on the small-scale details*, i.e. the exact value of l, etc. On the contrary, when $q < q_D$ (i.e. $C(q) < D$), the dressed moments will be insensitive to Λ and the small scale. Overall we can say that for $q < q_D$ the moments are *macroscopically* determined whereas for $q > q_D$ the moments will be *microscopically* determined. Because of this dependence on the small-scale details, the divergence of moments is a kind of "multifractal butterfly effect" similar to the "butterfly effect" in deterministic chaos.

It is worth noting that the power-law probability distributions (of dressed quantities) are a basic feature of self-organized criticality (SOC) (Bak *et al.*, 1987). This divergent "hard" statistical behaviour indeed corresponds to the fact that rare and catastrophic events will make dominant contributions. For example, the "mean" shape of a sand pile (the prototypical example of a self-organized critical system: see Fig. 5.18 for a schematic) is maintained due to avalanches of all sizes. In cascades, the flux energy is maintained by instabilities of all intensities. We have also observed that the transition from soft to hard processes (from "thin-tailed" to "fat-tailed" probabilities) corresponds to the fact that microscopic activity cannot be removed to yield a purely macroscopic description of the system. Similarly, we will not be able to understand the "mean" field without understanding its extremes.

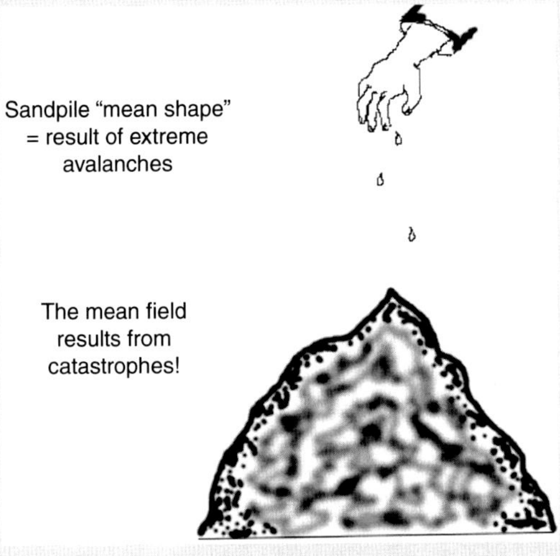
Sandpile "mean shape"
= result of extreme
avalanches
The mean field
results from
catastrophes!

Fig. 5.18 The original paradigm of self-organized criticality. Reproduced from Schertzer and Lovejoy (1996).

The sand pile exemplifies "classical SOC": it produces avalanches of all sizes, and the structure of each avalanche is fractal. However, it is only produced in the "zero flux" limit; in other words, one must add grains of sand so slowly that any avalanches thus induced have the time to stop: if grains are added at a constant rate, then the scaling will be broken. If SOC

Box 5.2 *(cont.)*

is defined as a dynamical system with both fractal structures and power-law probabilities, then we could consider cascade processes to be nonclassical SOC models. While potentially having many of the same statistics, cascades have the advantage that they are generated through a more realistic quasi-constant flux boundary condition.

Another way to generate power-law probability tails is to use correlated additive and multiplicative (CAM) processes (Sardeshmukh and Sura, 2009). These processes are modelled by systems of coupled (linear) stochastic differential equations and have been used as models of atmospheric variability at various scales, the "stochastic linear forcing" (SLF) approach (see Appendix 10C for a detailed comparison of the SLF and scaling approaches). The mechanism is apparently quite close to additive ("dressing") on multiplicative cascade processes.

Finally, we could mention yet another related mechanism yielding power-law probability tails: compound multifractal Poisson processes (Lovejoy and Schertzer, 2006b). This was developed as a model of individual raindrop size distributions in space; it leads to a direct link between the H parameter of the controlling large-scale turbulent fluxes and the individual drop probability exponent q_D.

order of singularity = $\max(C(q)) = C_1/(1-\alpha)$ so that if the "dressing" (averaging/integrating) space has $D > C_1/(1-\alpha)$ there will be no solution $C(q_D) = D$ to the equation and all the moments will converge. Since for small enough D the moments will still diverge, the $\alpha < 1$ multifractals are called "conditionally hard." In Chapter 4, we saw that the value of α is commonly in the range 1.5–2; indeed, empirically, it is almost always found to be > 1, so there are theoretical and empirical reasons to commonly expect power-law probability tails.

In order to empirically test for the divergence of moments, we could either look directly at the probability distributions at a specific resolution, or look for a linear asymptote on a $c(\gamma)$ plot; we have already seen (Figs. 5.10a, 5.10b) some evidence for $q_D \approx 3$, 5 for various aircraft fluxes; Fig. 5.10c for rain rates ($q_D \approx 3$); the $c(\gamma)$ from these figures essentially uses information from a range of scales (the "bunching" of the curves). From the figure, we can already sense a recurring practical problem, that without knowledge of the "effective dimension of dressing" the theory gives no guidance as to at what point in $c(\gamma)$ to expect the linear asymptote to begin, or the value of the asymptotic log-log slope $-q_D$. It may well be at such low probability levels (such high γ) that enormous sample sizes are necessary to observe it. Furthermore, most sensors have difficulty measuring the amplitudes of very rare but extreme events; they are frequently truncated, so that the critical extreme power-law region may be corrupted. This is especially a problem for rain gauges, which tend to saturate at high rain rates. Finally, the sampling properties of such algebraic "fat-tailed" probabilities are quite nontrivial and can even be sensitive to the way the probability densities are estimated from the empirical frequencies of occurrence.

Alternatively, at a single resolution, we can look for asymptotically linear graphs of logPr versus logs. Although this is the approach used in most of the literature, including the examples given below, recent work by Clauset *et al.* (2009) shows that the maximum likelihood estimator is superior, although the problem of determining the region where the power-law tail begins is still nontrivial and requires generalized Hill estimators (Schertzer *et al.*, 2006, Bernadara *et al.*, 2007, 2008).

Figs. 5.19a, 5.19b and 5.19c show the predicted power-law behaviour for velocity differences ($q_D \approx$ 5–7.5 vertical, horizontal, time: see Table 5.1a for an overall comparison). Similarly, Fig. 5.20a shows evidence for $q_D = 5$ in nondimensional raindrop distributions, and Fig. 5.20b, Table 5.1b, $q_D = 3$ for rain rates. We could also add that multifractal rain rates with $q_D \approx 3$ can explain conventional intensity-duration frequency (IDF) relations (Bendjoudi, 1997; Garcia-Marin *et al.*, 2012). Sardeshmukh and Sura (2009) find similar q_D values from reanalysis anomalies of geopotential and vorticity, and give corresponding simulations using correlated additive and multiplicative (CAM) noise (see Table 5.1a, b for values, and Appendix 10B for some theory). The ability of reanalyses (and presumably high resolution meteorological models) to produce these strong probability tails is a consequence of their comparatively large range of scales over which the

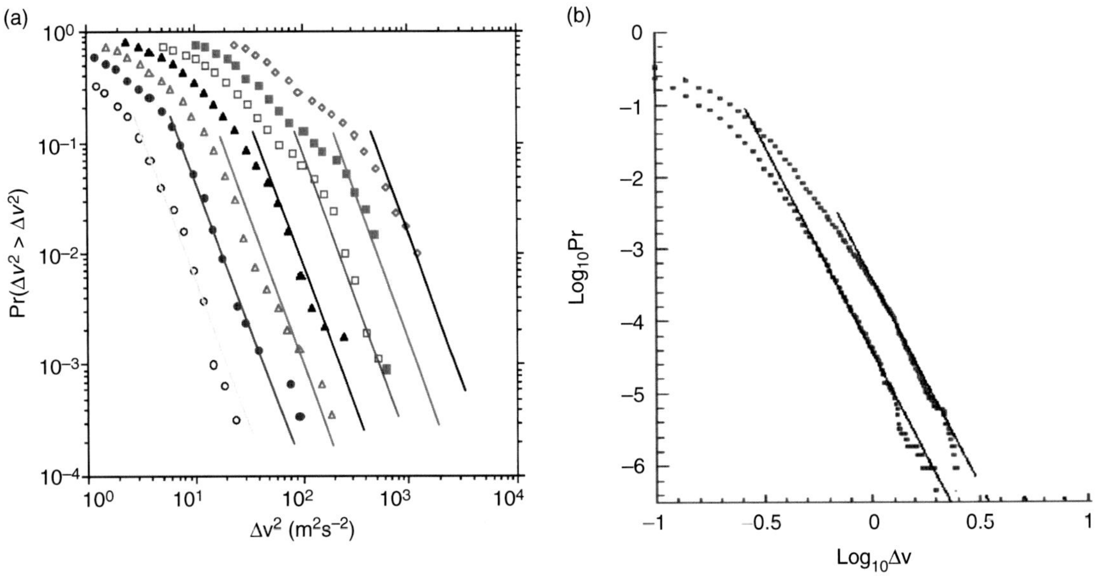

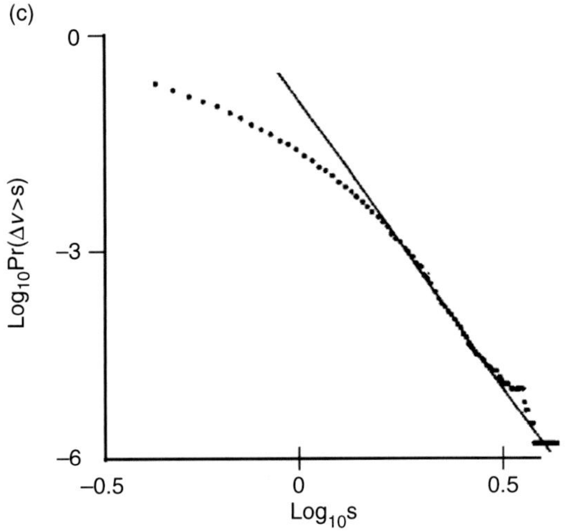

Fig. 5.19 (a) Figure of vertical fluctuations in the quantity $\Delta v^2(\Delta z)$ where v is the horizontal velocity for different layers $\Delta z = 50, 100, 200, \ldots$ 3200 m. The straight lines indicate that the extreme fluctuations follow a power law leading to the divergence of the 5th-order moment; whereas the uniform separation of the curves (corresponding to Δv's for 50, 100, 200, ... 3200 m thick layers) indicates uniform scaling of the extremes; here with $H_h = 3/5$, the Bolgiano–Obukhov value. Reproduced from Schertzer and Lovejoy (1985). (b) The probability distribution (Pr, cumulated from the largest to smallest, i.e. 1 – the cumulative probability distribution) for horizontal wind differences from the stratospheric ER2 aircraft. The horizontal gradients in m/s are given for distances of 40 m, 80 m (left, right), and the reference slopes correspond to $q_D = 5.7$. The data are from 18 aircraft flights, each over paths 1000–2000 km in length. Reproduced from Lovejoy and Schertzer (2007). (c) Probability distribution for the wind fluctuations Δv exceeding a fixed threshold s from a sonic probe at 10 Hz. The line indicates $q_D = 7.5$. Adapted from Schmitt *et al.* (1994).

cascades can develop. In comparison, climate models have narrower ranges, perhaps not enough: this could explain the difference in their respective "regime structures" (spacial intermittencies) noted by Dawson *et al.*, 2012.

Finally, we have included Fig. 5.21, the probability distributions of paleotemperature fluctuations (differences) in the climate regime. When the flucutations are estimated using equal depths (e.g. $\Delta T(\Delta h)$, where h is the depth, here $\Delta h = 55$ cm, 1 m), then a $q_D \approx 5$ regime is clearly visible in both Greenland $\delta^{18}O$ and Antarctic δD proxies (Figs. 11.11, 11.12). Interestingly, exactly the same exponent was obtained on much lower-resolution GRIP data in Lovejoy and Schertzer (1986) (who also found $q_D \approx 5$ for hemispheric temperature fluctuations at annual scales; we have confirmed this result on the monthly global temperature series discussed in Appendix 10C). As we argue in Chapter 11, these power-law tails are the statistical manifestiations of abrupt climate change including (for the GRIP series) "Dansgaard–Oeschger" (DO) events. Indeed,

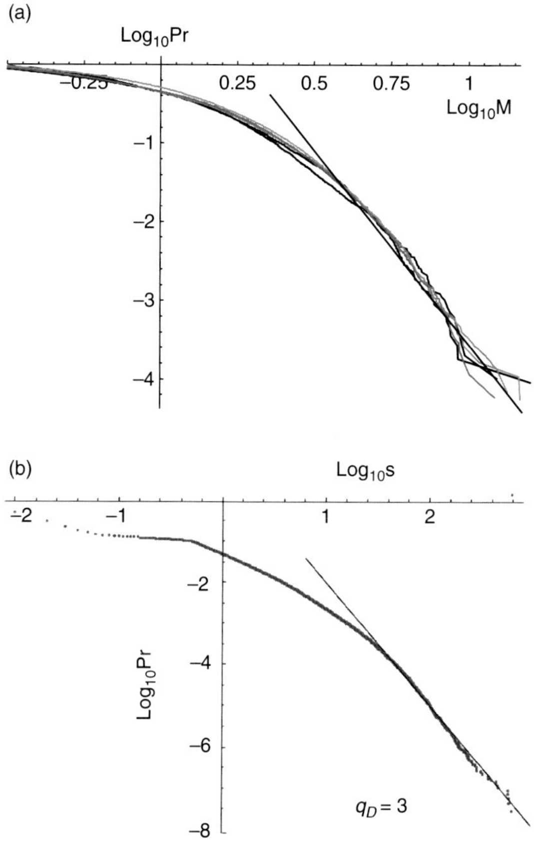

Fig. 5.20 (a) Probability distributions of raindrop volumes nondimensionalized by dividing by the mean mass (each curve is for a different storm). The reference line has absolute slope $q_D = 5$. Note that for the same data, the slightly different statistic, the total liquid water in a given volume in the scaling regime (typically about 30 cm: see Fig. 1.8b), has $q_D \approx 3$, and this value is theoretically predicted from a compound Poisson cascade process obeying Corrsin–Obukhov statistics (Lovejoy and Schertzer, 2006b). The figure is reproduced from Lovejoy and Schertzer (2008), where there are more details. (b) This figure shows the $\log_{10}$ of the probability distribution $Pr(\Delta R > s)$ of the absolute rain rate difference ΔR at times one hour apart for the $13 \times 21 = 273$ CPC series, each 257 000 hours long (a total of 7.0×10^7 points) as a function of the $\log_{10}$ threshold s. The reference line is the theoretically predicted $q_D = 3$ behaviour. This is the most convincing evidence to date for power-law probability tails in rain. This is probably due to the following: (a) the hourly resolution is long enough that many of the high-frequency gauge saturation problems are not so important; (b) the gridded data involve averages from several gauges per grid point, again making the results robust; (c) the dataset is extremely large, so the power law holds over a wide range of scales and the result is clear. The units of s are hundredths of inches per hour. Adapted from Lovejoy *et al.* (2012).

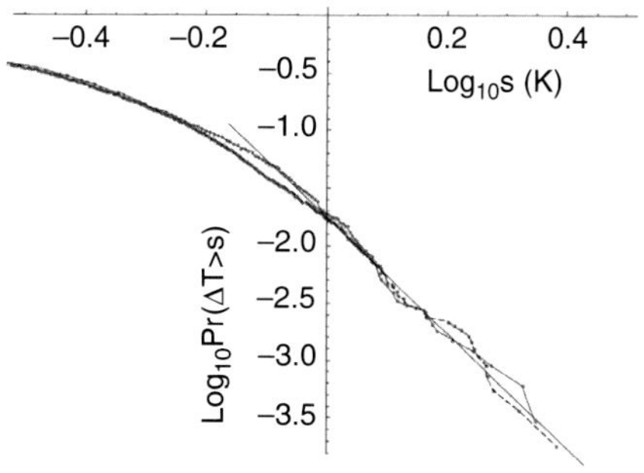

Fig. 5.21 The probability distributions of changes in paleotemperatures for Greenland (GRIP, dashed line, 5425 points last 248 kyr, 55 cm resolution) and Vostok (black line, 3300 points, last 420 kyr, 1 m resolution). The GRIP series is based on $\delta^{18}O$ anomalies with the same calibration as in Fig. 11.11 and a further reduction by a factor 0.13 to account for the difference in sampling thickness and the different mean snow rates. The reference line has slope −5. The implied abrupt changes presumably are the statistical counterparts of "Dansgaard–Oeschger events" at all scales: see in Section 11.2.1.

the Greenland series follows the power law over roughly the extreme 2% of the distribution, corresponding to ~110 "extreme events" over the last 248 kyr, which is a very reasonable estimate of the frequency of DO events (see e.g. Schulz, 2002, although our mechanism is not periodic). Interestingly, when the series is interpolated to series with uniform temporal resolution and we estimate instead the distribution $\Delta T(\Delta t)$, then we find that the extreme power-law behaviour is lost due to the smoothing of extreme events by the interpolation. This illustrates a general problem with empirical data: they often suffer from overly smooth extremes that can mask the power-law tails (see Appendix 11A for further discussion).

These graphs have been shown either for their particular high quality or for their scientific interest. In Table 5.1 we give a more complete summary of the values from the literature. We see that although the values of q_D (notably for the wind) apparently depend on the direction (horizontal, vertical, time; presumably a consequence of anisotropy), the usual variables of state typically have values in the range 5–7.5, with rain rate, potential temperature, geopotential height and vorticity a bit smaller (≈ 3). Since we also have C_1 and α estimates for these fields we could determine the effective dimension of dressing from $C(q_D) = D$ using the empirical C_1, α, q_D to determine $C(q_D)$. When this is done, we find, due to the typically small values of C_1 ($\approx 0.05 - 0.1$), that $C(q_D) < 1$, implying that the effective dressing dimension $D < 1$. It is not clear whether this low

Table 5.1a A summary of various estimates of the critical order of divergence of moments (q_D) for various atmospheric fields.

Field	Data source	Type	q_D	Reference
Horizontal wind	Sonic	10Hz, time	7.5	Schmitt *et al.*, 1994
	Sonic	10 Hz	7.3	Finn *et al.*, 2001
	Hot wire probe	Inertial range	7.7	Fig. 5.22, Radulescu *et al.*, 2002
	Hot wire probe	Dissipation range	5.4	Fig. 5.22, Radulescu *et al.*, 2002
	Anemometer	15 minutes	7	Tchiguirinskaia *et al.*, 2006
	Anemometer	Daily	7	Tchiguirinskaia *et al.*, 2006
	Aircraft, stratosphere	Horizontal, 40 m	5.7	Lovejoy and Schertzer, 2007
	Aircraft, troposphere	Horizontal, 280 m – 36 km	≈ 5	Fig. 5.10
	Aircraft, troposphere	Horizontal, 40 m – 20 km	$\approx 7 \pm 1$	Chigirinskaya *et al.*, 1994
	Aircraft, troposphere	Horizontal, 100 m	≈ 5	Schertzer and Lovejoy, 1985
	Radiosonde	Vertical, 50 m	5	Schertzer and Lovejoy, 1985, Lazarev *et al.*, 1994
	Scaling gyroscopes cascade (SGC) model (Box 3.4)	Time	6.9 ± 0.2	Chigirinskaya and Schertzer, 1996
Potential temperature	Radiosonde	Vertical, 50 m	3.3	Schertzer and Lovejoy, 1985
Humidity	Aircraft, troposphere	Horizontal, 280 m – 36 km	≈ 5	Fig. 5.10
Temperature	Aircraft, troposphere	Horizontal, 280 m – 36 km	≈ 5	Fig. 5.10
	Hemispheric, global	Annual, monthly	$\approx 5, 5$	Lovejoy and Schertzer, 1986, and unpublished analysis respectively
	Daily, stations	Average over 53 stations in France, daily single station (Macon)	4.5, 4.5	Ladoy *et al.*, 1991
Paleotemperatures	Ice cores	350 years (time), 0.55 m, 1 m (depth)	5, 5	Lovejoy and Schertzer, 1986, Fig. 5.21 respectively
Geopotential anomalies	Reanalyses	500 mb, daily	2.7	Sardeshmukh and Sura, 2009
Vorticity anomalies	Reanalyses	300 mb, daily	1.7	Sardeshmukh and Sura, 2009
Visible radiances (ocean surface)	Remote sensing	7 m resolution MIES data	3.6	Lovejoy *et al.*, 2001
Passive scalar (SF_6)	Fast response SF_6 analyzer	1 Hz	4.7	Finn *et al.*, 2001
Vertical CO_2 flux (above a field)	Aircraft new ground	Horizontal ≈ 1 km resolution	5.3	Austin *et al.*, 1991
Seveso pollution	Ground concentrations	In-situ measurements	2.2	Salvadori *et al.*, 1993
Chernobyl fallout	Ground concentrations	In-situ measurements	1.7	Chigirinskaya *et al.*, 1998; Salvadori *et al.*, 1993
Density of meteorological stations	WMO surface network	Geographic location of stations	3.7 ± 0.1	Tessier *et al.*, 1994

Table 5.1b A summary of various estimates of the critical order of divergence of moments (q_D) for various hydrological fields.

Field	Data source	Type	q_D	Reference
Radar reflectivity of rain	Radar reflectivity factor	1 km^3 resolution	1.1	Schertzer and Lovejoy, 1987
Rain rate	Gauges	Daily, Nimes	2.6	Ladoy *et al.*, 1991
	Gauges	Daily, time, France	≈ 3	Ladoy *et al.*, 1993
	Gauges	Daily, USA	1.7–3	Georgakakos *et al.*, 1994
	High-resolution gauges	8 minutes	≈ 2	Olsson, 1995
	High-resolution gauges	15 s	2.8–8.5	Harris *et al.*, 1996
	Gauges	Daily, time	3.6 ± 0.07	Tessier *et al.*, 1996
	Gauges	1–8 days	3.5	De Lima, 1998
	Gauges	Hourly, time	4.0	Kiely and Ivanova, 1999
	Gauges	Daily, four series from 18th century	3.78 ± 0.46	Hubert *et al.*, 2001
	Gauges	Hourly, time	≈ 3	Fig. 5.10c; Schertzer *et al.*, 2010
	Gauges	Hourly, time	≈ 3	Fig. 5.20b; Lovejoy *et al.*, 2012
	High-resolution gauges	15 s, averaged to 30 minutes	2.23	Verrier, 2011
Raindrop volumes	Stereophotography	10 m^3 sampling volume	5	Lovejoy and Schertzer, 2008
Liquid water at turbulent scales	Stereophotography	Total water in 40 cm cubes	3	Lovejoy and Schertzer, 2006b
Stream flow	River gauges (France)	Daily	3.2 ± 0.07	Tessier *et al.*, 1996
	River gauges (USA)	Daily	3.2 ± 0.07	Pandey *et al.*, 1998; Tessier *et al.*, 1996
	River gauges (France)	Daily	2.5–10	Schertzer *et al.*, 2006

Box 5.3 Divergence of moments in laboratory turbulence

Obviously, the cascade theories developed here have implications for laboratory turbulence – at least if the Reynolds number (*Re*) is high enough to allow the cascade to develop over a wide enough range. Aside from its intrinsic interest, the ability to use well-controlled laboratory-scale measurements allows us to resolve the dissipation scale and test our ideas further. To this effect, we analyzed active grid wind tunnel turbulent velocity data at a Taylor microscale Reynolds number of $R_\lambda = 582$, approaching those of the atmosphere (the usual Reynolds number used earlier is roughly the square of R_λ, i.e. $Re \approx 10^5$–10^6; some of the largest *Re* laboratory turbulence to date). The experiments are described by Mydlarski and Warhaft (1998), along with the main characteristics of the turbulent flow. The Kolmogorov characteristic dissipation scale, as estimated from the mean energy dissipation rate, is $L_{diss} = 0.26$ mm. Time series of the velocity fluctuations in the longitudinal direction were obtained via hot-wire anemometry and analyzed in the spatial domain by using Taylor's hypothesis. The inertial scaling range extended more than an order of magnitude in scales, and the increased sampling rate allowed a resolution of the same order of the dissipation scale L_{diss}.

Box 5.3 (*cont.*)

To study the statistics of both the inertial range energy flux and the rate of energy dissipation, we studied the velocity differences of the longitudinal signal. The probability distributions (integral of the probability density functions) of $|\Delta Ur| = |U(x) - U(x + r)|$ normalized by the RMS value are shown in Fig. 5.22 for two typical examples taken at the highest resolution of the data, corresponding to dissipation range (DR) separations $r/L_{diss} \approx 2$, and at a spacing within the inertial (scaling) range (IR) of scales $r/L_{diss} \approx 314$. The tails of both probability distributions display a distinctive power-law distribution, with $q_D \approx 5.4$ and 7.7, respectively. This is confirmed from the compensated probability distributions by the tail exponent $q_{D,V}$, as shown in the inset of Fig. 5.22. A deviation from the power-law dependence is observed for the extreme events measured at inertial range separations, but this is due to the limitation in the observation of events over larger separations for the same length of dataset, making the observation of these extreme events less probable.

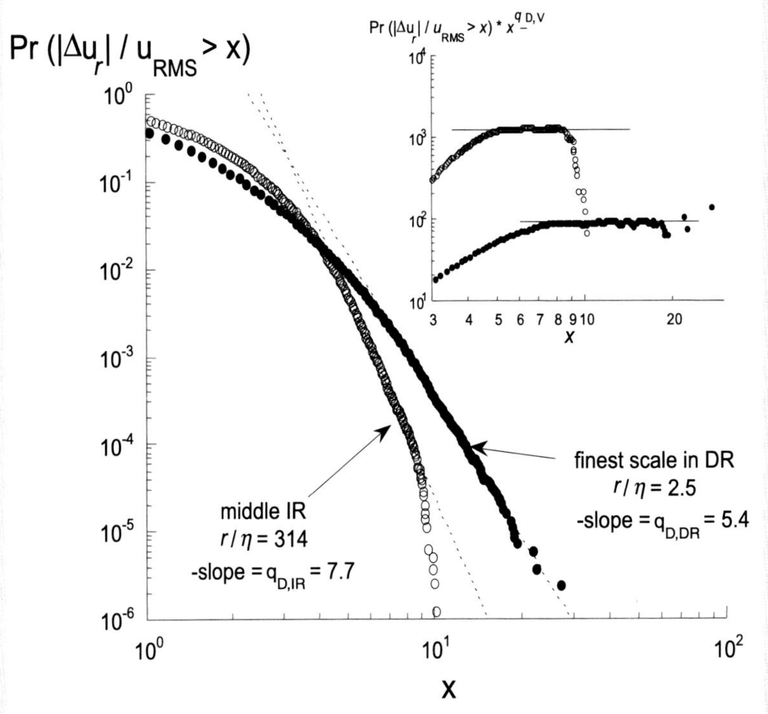

Fig. 5.22 Probability distribution of velocity increments $|\Delta U_r| = |U(x + r) - U(x)|$ estimated at separations r in the inertial scaling range (IR) and in the dissipation range (DR) at $R_\lambda = 582$; compensated distributions in inset (see Radulescu *et al.*, 2002).

According to the cascade model, it is the dressed turbulent fluxes which have the power-law tails (finite q_D's); therefore, following the discussion in Section 4.1.2, the q_D of the observables ought to be different depending on whether we consider dissipation or scaling range values. Recall (Section 4.1.2) that for the velocity field, we had $\Delta U \propto \varepsilon^\eta$ with $\eta = 1/3$ in the scaling range, but $\eta = 1/2$ in the dissipation range. $\Delta U \propto \varepsilon^\eta$ implies $q_{D,\eta} = q_{D,\varepsilon}/\eta$, so we expect $q_{D,IR}/q_{D,DR} = 3/2$.

According to the probability distribution of velocity differences at IR and DR separations in Fig. 5.22 we have $q_{D,IR} \approx 7.7$ and $q_{D,DR} \approx 5.4$ so that $q_{D,IR}/q_{D,DR} = 1.43$, which is close to the theoretical ratio 1.5 and is consistent with atmospheric observations (Table 5.1) and the value $q_{D,\varepsilon} = q_{D,IR}/3 \approx 2.6$. A summary of this study may be found in Radulescu *et al.* (2002).

"effective dimension of dressing" is an indication that it is rather the (fractional) dressing exponent H which is intervening or whether there is some other mechanism at work. Because in practice the origin of the power law is not obvious, for a given field, the usual practice has been simply to regard q_D as an empirical characterization. Finally, we should also mention Tuck *et al.* (2004) and Tuck (2008, 2010), who propose that molecular velocity may also have power-law tails.

5.3.4 The classification of multifractals according to the maximum singularity they can generate

We have seen that the microcanonical constraint places an upper bound d on the orders of singularity (Section 3.2.4). We saw that consequently, for microcanonical cascades, there is no divergence of dressed moments of fluxes averaged over sets with dimension d. However, more restrictive (calm) types of multifractals exist. Here we briefly discuss the geometric multifractals introduced by Parisi and Frisch (1985) which involve neither a probability space nor a cascade process. Parisi and Frisch (1985) coined the term "multifractal" and gave the following quite abstract definition (the following is translated into our notation).

Consider a component multifractal field $v(\underline{r})$ on a space S dimension d; Parisi and Frisch (1985) considered a turbulent velocity field. Define the set of points $S_{\gamma hol} \subset S$ as the set of all points $\underline{r}_{hol}$ for which:

$$\lim_{|\Delta\underline{r}|\to 0} \frac{v(\underline{r}_{\gamma_{hol}} + \Delta\underline{r}) - v(\underline{r}_{\gamma_{hol}})}{|\Delta\underline{r}|^{-\gamma_{hol}}} \tag{5.58}$$

is nonzero and finite. For such points, $\Delta v \approx |\Delta\underline{r}|^{-\gamma_{hol}}$, where γ_{hol} is called the "Hölder" exponent. The set $S_{\gamma hol}$ has a fractal dimension $D(\gamma_{hol})$. If many different singularities γ are present, each distributed over a set dimension $D(\gamma_{hol})$, then $v(\underline{r})$ is a multifractal. Notice that by restricting the multifractal to geometric sets, $D(\gamma_{hol})$ is necessarily nonnegative and $c(\gamma_{hol}) = d - D(\gamma_{hol})$ is bounded above by the embedding dimension d. Since $c(\gamma)$ is an increasing function (see Section 5.2.2) this implies that the orders of singularity are bounded. Just as microcanonical multifractals have maximum orders of singularity $\gamma_{max} = d$, the geometric multifractals have $\gamma_{max} = c^{-1}(d)$, which is less than d due to the convexity of $c(\gamma)$ and the nondegeneracy requirement $C_1 < d$

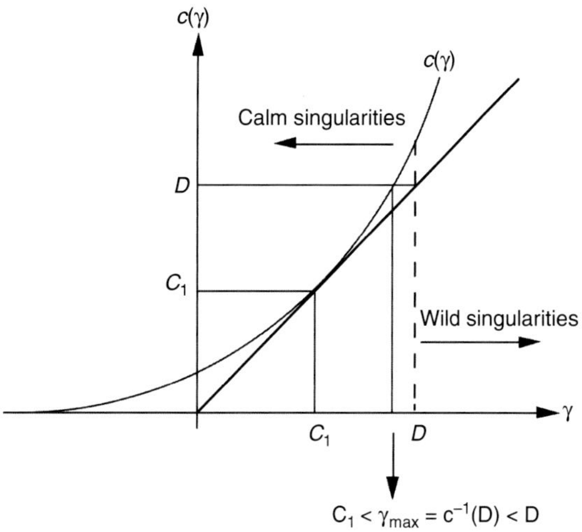

Fig. 5.23 $c(\gamma)$ versus γ. The nondegeneracy of the process requires $D > C_1$ so that γ_{max} is the maximum observable singularity on a single realization of a stochastic process; in geometric multifractals it is the absolute upper limit. Reproduced from Schertzer and Lovejoy, 1996.

(Figs. 5.23, 5.24). It is not obvious how to construct such geometric multifractals, they are simply defined as an abstract (highly nontrivial!) superposition of fractal sets. The basic difficulty is that cascade processes are generally not functions defined at mathematical points; rather they are more like "Dirac δ functions" – where the word "function" is a misnomer because they are not true functions but rather limits of a series of functions. This means that for cascades, at a given point $\underline{r}$, as we increase the resolution of the "incipient singularity" $\gamma_\lambda(\underline{r})$:

$$\gamma_\lambda(\underline{r}) = \frac{\log \varepsilon_\lambda(\underline{r})}{\log \lambda} \tag{5.59}$$

there will generally *not* be a well-defined small-scale limit: $\lim_{\lambda\to\infty} \gamma_\lambda(\underline{r}) \neq \gamma_{hol}(\underline{r})$.

We have noted that geometric multifractals have $\gamma_{max} < c^{-1}(d)$. In comparison, the extreme microcanonical model (with all but one multiplicative factors equal to zero) has $\gamma_{max} = d$ so that generally they have $\gamma_{max} < d$ (Figs. 5.23, 5.24a). Since $c(\gamma)$ is convex, this already means that microcanonical multifractals may have $c(\gamma_{max}) > d$. Since single realizations – even with arbitrarily high resolutions – almost surely do not have these singularities, microcanonical processes with $c(\gamma_{max}) > d$ are not "ergodic." Because this behaviour is already nonclassical, these are termed "wild singularities." However, stronger singularities exist. Indeed, in Section 5.3.2 we showed that

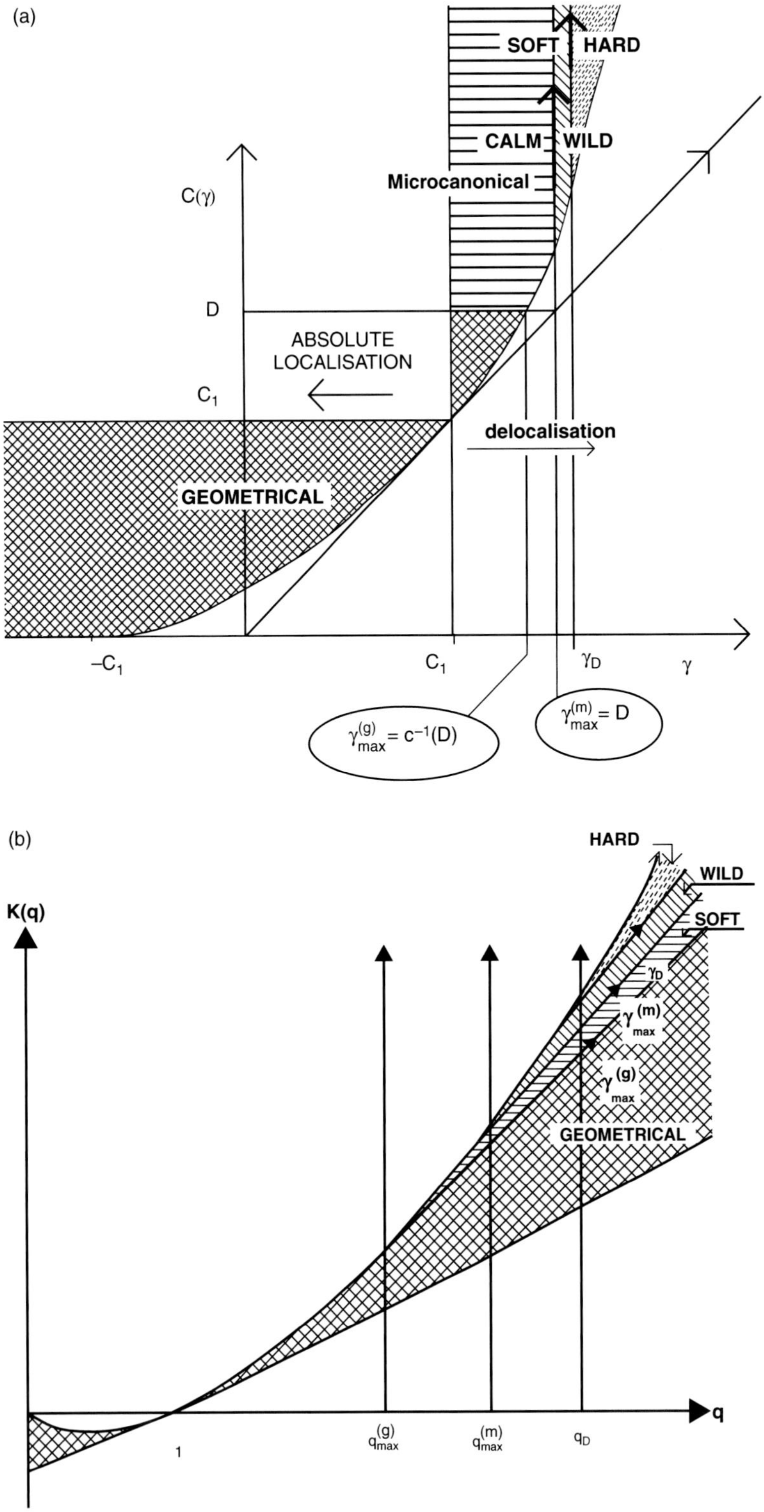

Fig. 5.24 (a) Phase diagram showing the attainable singularities for a (normalized) multifractal, showing the fixed point $C_1 = c(C_1)$, the maximum orders of singularities of geometrical (Parisi–Frisch) multifractal $\gamma_{max}^{(g)}(= c^{-1}(D))$, and microcanonical multifractal $\gamma_{max}^{(m)}(= D)$. Wild singularities are those with $\gamma > \gamma_{max}^{(m)}$ and necessarily involve realizations which do not respect flux conservation. Wild singularities with $\gamma \geq \gamma_D (> \gamma_{max}^{(m)})$ lead to the appearance of hard multifractals with divergent high-order statistical moments. Note that the dimension D refers to the dimension of space in which the various multifractals are defined and that the represented singularities are delimited by the two extreme universal cases, the β model and the lognormal model. Reproduced from Schertzer *et al.* (1991). (b) The dual-phase diagram corresponding to Fig. 5.24a but in moments representation instead of a singularity representation. The (normalized) moment scaling function $K(q)$ corresponds to the codimension function $c(\gamma)$. Reproduced from Schertzer *et al.* (1991).

Box 5.4 Box, information and correlation dimensions

In Box 3.1 we introduced both the box (D_{box}) and correlation (D_{cor}) dimensions of a set of points: the first is the exponent of the average number of disjoint boxes size L/λ needed to cover the set, while the second is the exponent of the number of point pairs separated by a distance $\leq L/\lambda$. Since both dimensions are in common use (D_{cor} particularly for characterizing "chaotic"/"strange" attractors such as the Mandelbrot set: Fig. 2.2) let us now consider the relation between the two. First suppose that the set of interest (denoted A) can be embedded in a d-dimensional "cube" of size L, and cover the cube with a grid of λ^d disjoint boxes each of size $l = L/\lambda$. Denote the number of points in the ith l-sized grid box by $n_{i,\lambda}$ so that the total number of points is:

$$N = \sum_{i=1}^{\lambda^d} n_{i,\lambda} \tag{5.60}$$

If the points are from a strange attractor (such as the Lorenz attractor), then the space is the system's phase space and (with an ergodic hypothesis) $P_{i,\lambda} = n_{i,\lambda}/N$ is an empirical frequency that approximates the probability of finding the system in the ith box at phase space resolution $\lambda = L/l$, this would be its asymptotic limit for an infinite resolution. In order to characterize the scale-by-scale statistics of the attractor, similarly to estimating the "trace moments" (Appendix 5A) we can use a "partition function" approach to introduce the following family of measures indexed by q (Hentschel and Procaccia, 1983; Grassberger, 1983; Halsey *et al.*, 1986):

$$\mu_q(\lambda) = \sum_{i=1}^{\lambda^d} P_{i,\lambda}^q \tag{5.61}$$

and with the corresponding scaling exponents:

$$\mu_q(\lambda) \propto l^{\tau(q)} \propto \lambda^{-\tau(q)} \tag{5.62}$$

To see the meaning of μ_q and its exponent $\tau(q)$, first consider $q = 0$ and adopt the convention that for any x, $x^0 = 1$ if $x > 0$, and $x^0 = 0$ if $x = 0$. In this case, μ_0 is simply the number of boxes needed to cover the set and $\tau(0) = -D_{box}$. Next, consider $q = 2$; in each box, the number of points which are within a distance l of each other is equal to the number of pairs in the box: $n_{i,\lambda}(n_{i,\lambda} - 1)/2 \approx n_{i,\lambda}^2$ (for large n and ignoring constant factors). However, we have $P_{i,\lambda}^2 \propto n_{i,\lambda}^2$ so that we see that μ_2 is proportional to the number of point pairs within a distance l, and hence $\tau(2) = D_{cor}$ (the correlation dimension). The above suggests the definition:

$$D(q) = \frac{\tau(q)}{q-1} = \frac{1}{q-1}\lim_{l\to 0}\left[\frac{\log \mu_q}{\log l}\right]; \quad l = L/\lambda \tag{5.63}$$

This was first proposed by Hentschel and Procaccia (1983) and Grassberger (1983) for strange attractors, and it was acknowledged that such expressions were in fact pointed out by Rényi (1970); they are "Rényi dimensions." In Box 5.5, we relate $D(q)$ to the dual codimension function ($C(q)$; Section 3.2.7) of the multifractal probability p_λ.

What about the value $q = 1$? In this case, since the sum of the probabilities is unity, we have $\mu_1 = \sum P_{i,\lambda} = 1$ so that Eqn. (5.63) reduces to 0/0 and we must use l'Hopital's rule to evaluate the limit $q \to 1$. We find:

$$D(1) = \lim_{l\to 0}\left[\frac{\sum_{i=1}^{\lambda^d} P_{i,\lambda}\log P_{i,\lambda}}{\log l}\right]; \quad l = L/\lambda \tag{5.64}$$

$D(1)$ is thus the exponent of the information I_λ:

$$I_\lambda = \sum_{i=1}^{\lambda^d} P_{i,\lambda}\log P_{i,\lambda} \tag{5.65}$$

so that $I_\lambda \approx l^{D_I}$ where the information dimension $D_I = D(1)$ of the set of points. In Box 5.5 we show that $\tau(q) = d(q-1) - K(q)$ so that the convexity of $K(q)$ (Appendix 3A) implies the concavity of $\tau(q)$ so that $D(q)$ is a monotonically decreasing function of q; we therefore have the hierarchy: $D_{box} \leq D_I \leq D_{cor}$.

Box 5.5 The singular measure ($f(\alpha)$, $\tau(q)$) multifractal formalism

In Box 5.4, we saw that in order to statistically characterize these complex phase space densities, the density can itself be thought of as being a realization of a multifractal probability measure; we saw that this probability density p_λ at resolution λ can be defined by covering the space with λ^{-d}-sized grids (boxes) and using $p_\lambda = n_\lambda/N$ for each box. The resulting measure is a geometric (calm) multifractal, since although it represents the probability of finding the system at a point in the phase space, it is *not itself* random at all!

Halsey *et al.* (1986) wrote an influential paper proposing a notation for dealing with these "geometric attractor" multifractals. Rather than considering the density of the multifractal measure $p_\lambda = P_\lambda / vol(B_\lambda)$ (the nonrandom analogue of the turbulent ε_λ), the measure itself P_λ (i.e. the integral of the density p_λ over a ball (box) size L/λ) was considered.

P_λ defined in this way is really a dressed quantity, but for these multifractals the bare/dressed distinction is irrelevant. They then determined the order of singularity α_d of P_λ:

$$P_\lambda = \int_{B_\lambda} p d^d \underline{x} = p_\lambda \lambda^{-d} \approx \lambda^{-\alpha_d} \tag{5.66}$$

(the subscript d was not used in the original; we have added it to underscore the dependence on the dimension of the system and to distinguish it from the totally different Lévy α). In codimension notation, we may write $p_\lambda = \lambda^\gamma$ so that $P_\lambda = \lambda^{\gamma - d}$; we thus obtain:

$$\alpha_d = d - \gamma \tag{5.67}$$

Each box can thus be indexed according to α_d. The number of boxes at each resolution corresponding to α_d can then be used to define the (box-counting) dimension $f_d(\alpha_d)$:

$$Number[P_\lambda = \lambda^{-\alpha_d}] = \lambda^{f_d(\alpha_d)} \tag{5.68}$$

since *number* $= \lambda^d \times$ *probability*, and probability $\approx \lambda^{-c(\gamma)}$, we obtain:

$$f_d(\alpha_d) = d - c(\gamma), \qquad \alpha_d = d - \gamma \tag{5.69}$$

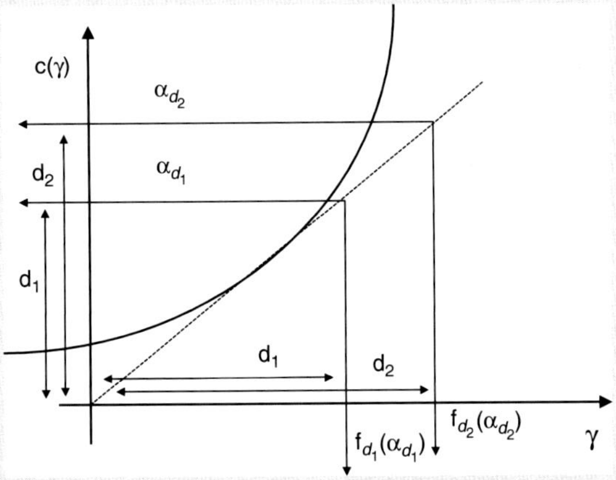

Fig. 5.25 A comparison of codimension (γ, $c(\gamma)$) and dimension (α, $f(\alpha)$) notations. The graph shows the relationship for two different observing dimensions d_1 and d_2; for a given multifractal process with fixed $c(\gamma)$, the corresponding $f(\alpha)$ is obtained by turning the figure upside down and using the axes whose origin slides along the bisectrix (dashed); this is the graphical transformation $\alpha = d - \gamma$ and $f = d - c$. Adapted from Schertzer and Lovejoy (1992).

Finally, we can define the scaling exponents $\tau_d(q)$ for the moments by the partition function:

$$\sum_{i=1}^{\lambda^d} P_{\lambda,i}^q = \lambda^{-\tau_d(q)} \tag{5.70}$$

where the sum is over all the i balls size λ^{-1} needed to cover the d-dimensional phase space region A. Comparison with the moments of the density p_λ shows that they differ only in the ensemble averaging (which gives "trace moments"); since p is nonrandom, we therefore have:

$$\sum_{i=1}^{\lambda^d} P_{\lambda,i}^q = \left\langle \sum_{i=1}^{\lambda^d} \left(p_\lambda \lambda^{-d}\right)^q \right\rangle \approx \lambda^{-\tau_d(q)} = \lambda^{-d(q-1)} \langle p_\lambda^q \rangle = \lambda^{K(q) - d(q-1)} \tag{5.71}$$

This implies:

$$\begin{aligned} \tau_d(q) &= (q-1)d - K(q) = (q-1)D(q) \\ D(q) &= d - C(q) \end{aligned} \tag{5.72}$$

As long as we deal with strange attractors and study the full d-dimensional phase space, the α_d, $f_d(\alpha_d)$, $\tau_d(q)$ and $D(q)$ notation is adequate. However, if we are interested in random multifractals (involving probability spaces) then $d \to \infty$, or if we are interested in looking at subspaces with dimension smaller than d, the d dependence is respectively a fundamental limitation or an unnecessary complication. The turbulent γ, $c(\gamma)$, $K(q)$, $C(q)$ notation (the "codimension" notation for short) has the advantage of being intrinsic to the process. i.e. of being d-independent.

multifractals with canonical conservation have no upper limit on γ, and that it is precisely the singularities $\gamma > \gamma_D > d$ that contribute to the "hard" behaviour associated with the divergence of high-order statistical moments $q > q_D = c'(\gamma_D)$. These various behaviours are shown in Fig. 5.24a (γ, $c(\gamma)$); see Fig. 5.24b for the corresponding (q, $K(q)$) representation.

5.4 Continuous-in-scale multifractal modelling

5.4.1 Review

Up until now, we have discussed discrete-in-scale models constructed by iteratively dividing large structures ("eddies") into disjoint daughter "sub-eddies" each reduced in scale by an integer ratio λ_0. The smaller eddies have intensities which are equal to those of their parents multiplied by weights which are independently and identically distributed. They yield visually weird, highly artificial simulations. Fig. 5.26 shows an example which can be compared to Fig. 5.28, for a continuous-in-scale simulation with the same α, C_1 and scale ranges and to Fig. 5.29 with an additional fractional integration (the FIF model: Section 5.4.3). One way of seeing the problem is that for these processes, the basic cascade equation $\langle\varepsilon_\lambda{}^q\rangle = \lambda^{K(q)}$ only holds exactly for the special scale ratios $\lambda = \lambda_0^n$ where both λ_0 and n are integers, i.e. only a "countable" infinity of scale ratios; in comparison, continuous-in-scale processes satisfy this equation for all $\lambda > 1$. Other key limitations of these "toy model" cascades include (a) isotropy (self-similarity; although limited self-affine extensions are possible:

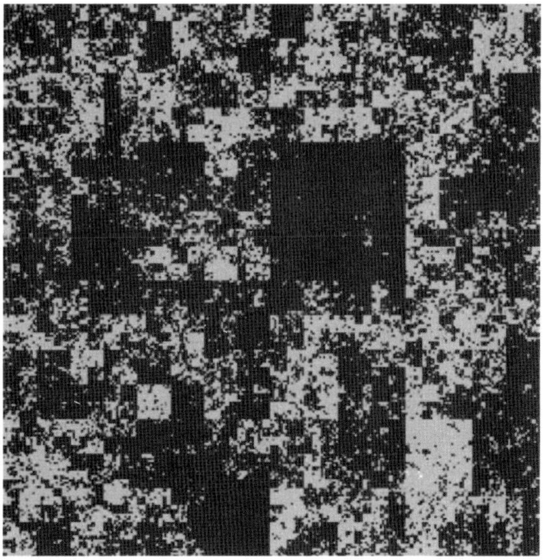

Fig. 5.26 A discrete-in-scale simulation of a universal multifractal with basic scale ratio $\lambda_0 = 2$, $\lambda = 2^9$, $\alpha = 1.8$, $C_1 = 0.1$. The grey scale is proportional to the log of the field, with low values darker. Reproduced from Lovejoy and Schertzer (2010a).

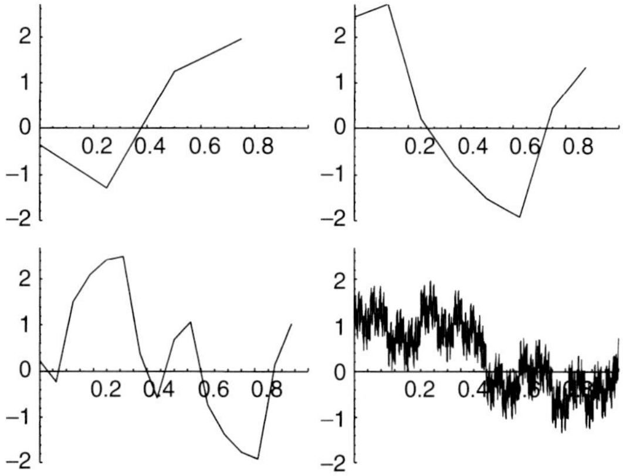

Fig. 5.27 An example of a multfractal produced by a (deterministic) iterated function system. It is based on a basic generator (upper left); each segment is then rescaled and repositioned and the process is iterated. The second, third and ninth iterates are shown (upper right to lower left, to lower right; this example uses the same parameters as in Basu *et al.*, 2004). Only a small and finite range of singularities can be produced in this way (Lovejoy and Schertzer, 2006a).

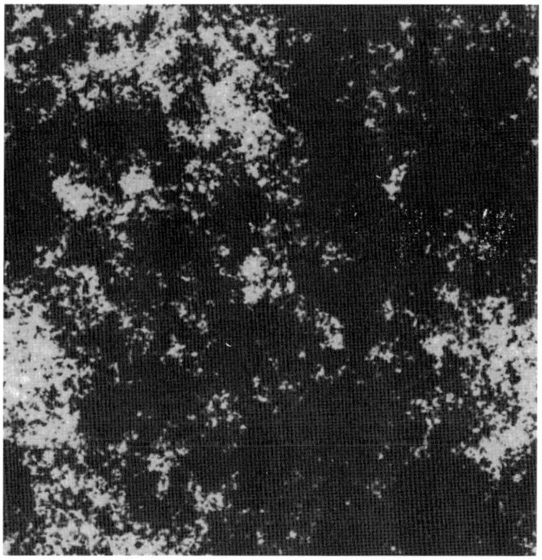

Fig. 5.28 The continuous-in-scale simulation corresponding to Fig. 5.26, using the continuous-in-scale method described in the text (including $\Delta x^{-d/\alpha}$ corrections: see Appendix 5B). Reproduced from Lovejoy and Schertzer (2010a).

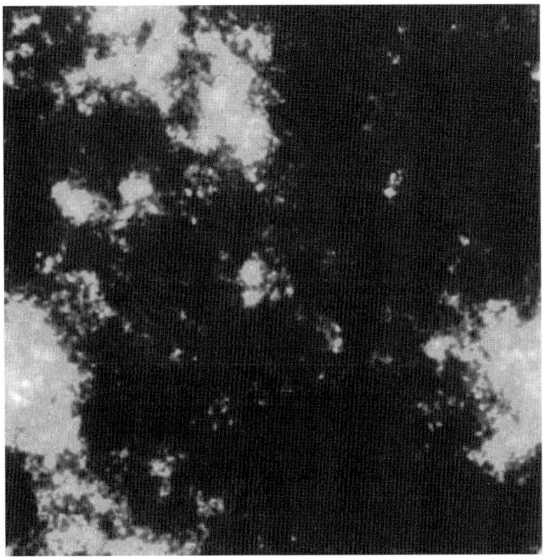

Fig. 5.29 Same as Fig. 5.28 but with an additional fractional integration of order $H = 1/3$ (a scale-invariant smoothing) to simulate a turbulent passive scalar density. Notice that the structures are smoothed. Reproduced from Lovejoy and Schertzer (2010a).

Schertzer and Lovejoy, 1985) and (b) left/right (mirror) symmetry (which precludes causal processes). In addition, most multifractal models are of scale-by-scale conserved cascade quantities, whereas the observables typically have exponents with an extra linear term (e.g. $\Delta v = \varepsilon^{1/3}\Delta x^H$ with $H = 1/3$). A somewhat different but still discrete-in-scale multifractal simulation method is based on iterated function systems (Lévy-Véhel *et al.*, 1994; Basu *et al.*, 2004; see Fig. 5.27 for an example in one dimension). Perhaps the most sophisticated discrete-in-scale approach is the Markov-switching multifractal (MSM) simulation method; it is available only for pure (1D) time series (Calvet and Fisher, 2001, 2008). This model involves causal random transitions but is still based on iterating a finite fundamental scale ratio, so it is not yet truly continuous in scale.

To explain continuous-in-scale cascade processes in a rather elementary manner, let us first review some properties of Lévy variables. First introduce the "unit" (and extremal) Lévy random variable γ_α whose probabilities are implicitly defined by the following characteristic function:

$$\langle e^{q\gamma_\alpha}\rangle = e^{\frac{q^\alpha}{(\alpha-1)}};\quad q \geq 0 \qquad (5.73)$$
$$\langle e^{q\gamma_\alpha}\rangle = \infty;\quad q < 0;\quad \alpha < 2$$

Note that (a) for $\alpha = 2$ we have the familiar Gaussian case and the $q \geq 0$ formula in Eqn. (5.73) is valid for all q; (b) for $\alpha = 1$ we have $\langle e^{q\gamma_\alpha}\rangle = e^{q\log q}$ ($q > 0$, otherwise $= \infty$). An extremal Lévy random variable A with amplitude $a > 0$ and Lévy index α therefore satisfies:

$$A = a\gamma_\alpha$$
$$\langle e^{qA}\rangle = e^{\frac{a^\alpha q^\alpha}{\alpha-1}};\quad q \geq 0 \qquad (5.74)$$
$$\langle e^{qA}\rangle = \infty;\quad q < 0;\quad \alpha < 2$$

(with corresponding exception for $\alpha = 1$). Due to the additivity of second characteristic functions for any independent, identically distributed random variables (Section 3.3.3), this implies that for the sum C of two statistically independent Lévy variables A, B we have:

$$C = A + B;\quad c^\alpha = a^\alpha + b^\alpha \qquad (5.75)$$

where C is also an extremal Lévy with the same α but with amplitude c. Eqn. (5.75) expresses the "stability under addition" property of the Lévy variables, and a, b, c are the corresponding amplitudes. Eqn. (5.75) shows that for Lévy variables $\log\langle e^{qF}\rangle$(the second characteristic function) of the random variables $F = f\gamma_\alpha$ (amplitude f) are "α additive", a property that generalizes to Lévy noises; we use this below.

To understand why only extremals must to be used to generate cascades (Schertzer and Lovejoy, 1987, 1991), and not the more general Lévy variables l_α with $\alpha < 2$, let us recall the general properties of Lévy variables. These have algebraic tails for both positive and negative values:

$$\langle |l_\alpha|^q\rangle \to \infty;\quad q \geq \alpha$$
$$p(l_\alpha) \approx A_+ l_\alpha^{-\alpha-1};\quad l_\alpha >> 1 \qquad (5.76)$$
$$p(l_\alpha) \approx A_-(-l_\alpha)^{-\alpha-1};\quad l_\alpha << -1$$

where p is the probability density and A_+ and A_- are constants that depend on an "asymmetry parameter" (the $\alpha = 2$ case is the qualitatively different Gaussian case, where there is neither power law nor asymmetry). For symmetric Lévys, $A_+ = A_-$, whereas for maximally asymmetric Lévys either A_+ or A_- vanishes. Those without algebraic behaviour for $l >> 0$ have $A_+ = 0$. The latter are required to avoid the divergence of the Laplace transform for $q > 0$, simply because an algebraic fall-off cannot tame an exponential divergence. It turns out that for $\alpha < 1$ the values of extremal Lévy variables all have the same sign as that of the algebraic tail. This is not the case for $\alpha > 1$, although most of the values have the same sign as the tail. The qualitative difference between the $\alpha > 1$ and $\alpha < 1$ processes is more apparent if we consider the corresponding asymptotic forms of the probability densities, which can be seen in fact as a straightforward consequence of the universal multifractal codimension function (Eqn. (5.35)) with $C_1 = 1$:

$$p(\gamma_\alpha) \approx \exp\left[-\left(\frac{\gamma_\alpha}{\alpha'}\right)^{\alpha'}\right]; \quad \left(\frac{\gamma_\alpha}{\alpha'}\right)^{\alpha'} >> 0; \quad \frac{1}{\alpha'} + \frac{1}{\alpha} = 1$$

$$p(\gamma_\alpha) \approx (-\gamma_\alpha)^{-\alpha-1}; \quad \gamma_\alpha << 0; \quad 0 < \alpha < 2 \tag{5.77}$$

where the key point is that the auxiliary variable α' changes sign at $\alpha = 1$. Note that exact closed-form expressions in elementary functions for the probabilities of extremals only exist in the "inverse Gaussian" $\alpha = 1/2$ case; analytic symmetric Lévys exist for the $\alpha = 1$ (Cauchy) and $\alpha = 2$ (Gaussian) cases.

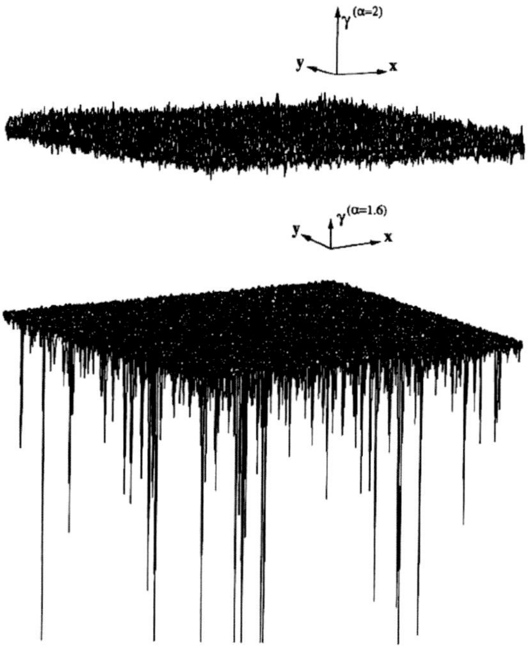

Fig. 5.30 A comparison of the Gaussian ($\alpha = 2$, top) and Lévy ($\alpha = 1.6$, bottom) subgenerators γ_α showing that whereas the former is both positive-negative symmetric with low-amplitude excursions, the latter is asymmetric with huge (algebraic) excursions for negative values. Reproduced from Schertzer and Lovejoy (1987).

5.4.2 Continuous-in-scale cascade processes

An (extremal) white Lévy noise $\gamma_\alpha(\underline{r})$ can be understood as the limit of independent identically distributed extremal Lévy variables on a grid for a mesh size shrinking to zero (with appropriate normalization); this is possible because Lévy distributions are "infinitely divisible" (Feller, 1972). With the help of a suitably normalized convolution kernel $g_\lambda(\underline{r})$, it is then possible to colour this white noise to obtain (Schertzer and Lovejoy, 1987) a generator $\Gamma_\lambda = \log\varepsilon_\lambda$ such that $\langle\varepsilon_\lambda^q\rangle = \langle e^{q\Gamma_\lambda}\rangle = e^{K(q)\log\lambda}$. We now show how to obtain a generator with the appropriate properties (including a mean codimension C_1) by convolving a Lévy noise $\gamma_\alpha(\underline{r})$ with a kernel $g_\lambda(\underline{r})$:

$$\Gamma_\lambda = C_1^{1/\alpha} g_\lambda * \gamma_\alpha = C_1^{1/\alpha}\int g_\lambda(\underline{r} - \underline{r}')\gamma_\alpha(\underline{r})d\underline{r} \tag{5.78}$$

"*" denotes "convolution." We now review, step by step, the different properties that the kernel $g_\lambda(\underline{r})$ and its domain of integration must satisfy in order to obtain the announced result, in particular that the multifractal ε_λ:

$$\varepsilon_\lambda = e^{\Gamma_\lambda} \tag{5.79}$$

will indeed be multiscaling: $\langle\varepsilon_\lambda^q\rangle = \lambda^{K(q)}$. This is the first property to be respected:

$$K_\Gamma(q, \lambda) = Log(< e^{q\Gamma_\lambda} >) = Log(\lambda)K(q) \tag{5.80}$$

i.e. $g_\lambda(\underline{r})$ must be chosen so as to yield a logarithmic divergence of the second characteristic function of the generator Γ_λ (Schertzer and Lovejoy, 1987). Before detailing the corresponding necessary and sufficient conditions, we need to recall that for $\alpha < 2$, g_λ must be nonnegative, otherwise Γ would a (nonextremal) mixture of extremal Lévy variables with opposite signs (see Fig. 5.30 for a graphical illustration). Note that in this chapter we will only consider isotropic d-dimensional multifractals, the only exception being the asymmetric causal processes needed for space-time simulations; in Chapter 7 we consider anisotropic extensions. Since $\gamma_\alpha(\underline{r})$ is statistically homogeneous, from Eqn. (5.78), the statistics of Γ are independent of $\underline{r}$ so that one can take $\underline{r} = 0$ and apply the additivity of the second characteristic function, Eqn. (5.75):

$$K_\Gamma(q) = C_1 \frac{q^\alpha}{\alpha - 1} \left| g_\lambda \right|_\alpha^\alpha; \quad \left| g_\lambda \right|_\alpha^\alpha \equiv \int \left| g_\lambda(\underline{r}) \right|^\alpha d^d \underline{r} \tag{5.81}$$

where the $\alpha - 1$ in the denominator comes from the definition of the unit Lévy variables, Eqn. (5.73).

In order to obtain the desired $\log(\lambda)$ divergence for K_Γ (Eqn. (5.80)), it suffices to choose g_λ to be an isotropic power law with the appropriate power law from the larger-scale L to the resolution L/λ:

$$g_\lambda(\underline{r}) = N_{d,\alpha}^{-1/\alpha} 1_{L/\lambda \le |\underline{r}| \le L} |\underline{r}|^{-1/\alpha} \tag{5.82}$$

where 1_B is the indicator function of the subset B, i.e. $1_B(\underline{r}) = 1$ if $\underline{r} \in B$, $= 0$ otherwise. This implies:

$$\langle \varepsilon_\lambda^q \rangle = \langle e^{q\Gamma_\lambda} \rangle = e^{\frac{C_1}{\alpha - 1} q^\alpha N_d^{-1} \Omega_d \log\lambda}; \quad \Omega_d = \int_{|\underline{r}'|=1} d^d \underline{r}' \tag{5.83}$$

where Ω_d is the integral over all the angles in the d-dimensional space (e.g. $\Omega_1 = 2$, $\Omega_2 = 2\pi$, $\Omega_3 = 4\pi$ etc.). Note that for theoretical work it is more convenient to take $L = 1$ so that the smallest scale is $1/\lambda$, whereas for numerical simulations on discrete grids it is more convenient to take $L = \lambda$ so that the smallest scale is 1.

From Eqn. (5.83), we see that if we choose

$$N_d = \Omega_d \tag{5.84}$$

then we obtain the desired nonlinear part of the multiscaling behaviour:

$$\langle \varepsilon_{\lambda,u}^q \rangle = \lambda^{K_u(q)}; \quad K_u(q) = \frac{C_1}{\alpha - 1} q^\alpha \tag{5.85}$$

where K_u is the unnormalized exponent scaling function corresponding to the fact that ε given by Eqns. (5.78), (5.79) is unnormalized (hence we temporarily add the subscript u). A normalized $\varepsilon_{\lambda,n}$ can now be easily obtained using:

$$\varepsilon_{\lambda,n} = \frac{\varepsilon_{\lambda,u}}{\langle \varepsilon_{\lambda,u} \rangle} \tag{5.86}$$

so that:

$$\langle \varepsilon_{\lambda,n}^q \rangle = \lambda^{K(q)}; \quad K(q) = K_u(q) - qK_u(1) = \frac{C_1}{\alpha - 1}(q^\alpha - q) \tag{5.87}$$

as required (we temporarily add the subscript n to distinguish it from the unnormalized process). The above leads to cascades with the correct statistics at the finest resolution λ. Unfortunately, it turns out that while the above is correct, due to "finite size effects" at both large and mostly at small scales, the internal structure of the realizations is not perfectly scaling. If needed, corrections are not difficult to make (see Appendix 5B and, for software, Appendix 5C). Numerical simulations based on spatially discretizing these continuous-in-scale cascade processes are shown in Figs. 5.12, 5.13, 5.28.

5.4.3 Fractional Brownian and fractional Lévy noises

Up until now, we have concentrated our attention on the underlying turbulent fluxes which are conserved from one scale to another. We have seen that the typical observables such as the wind have fluctuations (Δv) whose statistics are related to the fluxes by a lag Δx raised to a power, the prototypical example being the Kolmogorov law: $\Delta v = \varphi \Delta x^H$ with $\varphi = \varepsilon^{1/3}, H = 1/3$. If we take the qth moments of this equation, we obtain:

$$S_q(\Delta x) \propto \Delta x^{\xi(q)}; \quad S_q(\Delta x) = \langle |\Delta v|^q \rangle; \quad \xi(q) = qH - K(q) \tag{5.88}$$

(we have used $<\varphi_\lambda{}^q> \approx \lambda^{K(q)}$ and $\Delta x \propto \lambda^{-1}$). $S_q(\Delta x)$ is the qth-order structure function and $\xi(q)$ is its scaling exponent; the usual (second)-order structure function (related to the spectrum) was defined in Eqn. (2.72) (see also Box 3.3). Since we now know how to construct fields with the (convex) scaling exponent $K(q)$, our problem is to produce a field with the extra linear exponent qH. Note that if the fluctuations Δv are defined simply as differences (i.e. $\Delta v = v(x + \Delta x) - v(x)$) then the above is only valid for $0 \le H \le 1$; when H is outside this range other definitions of fluctuations must be used (see Section 5.5).

Before the development of cascade processes, the intermittency of the flux φ was considered unimportant, and turbulence was modelled (for the purposes of statistical closure theories, for example) with quasi-Gaussian statistics; this implies that $K(q) = 0$ so that these processes therefore had perfectly linear exponents $\xi(q) = qH$. Before discussing the fractionally integrated flux (FIF) model which gives $\xi(q)$ with convex $K(q)$, let us therefore briefly discuss the simpler processes that give rise to linear $\xi(q)$.

If $\xi(q)$ is linear, we anticipate that the corresponding $v(\underline{r})$ can be modelled using an additive stochastic process. In order to assure that the process can be made continuous, we consider fractional Brownian motion (fBm) and its generalization, fractional Lévy motion (fLm), which are produced by linear combinations of independent Gaussian and Lévy noises with pure singularities (the noises $\varphi_\alpha(\underline{r})$ are of the same type as the $\gamma_\alpha(\underline{r})$ introduced in the previous subsection). More precisely, these are written as convolutions of noises with power laws which are extensions of integration/differentiation to fractional orders; "fractional integrals":

$$v(\underline{r}) = \gamma_\alpha * |\underline{r}|^{-(d-H')} = \int \frac{\gamma_\alpha(\underline{r}')}{|\underline{r} - \underline{r}'|^{d-H'}} d^d \underline{r}';$$
$$H' = H + d/\alpha \tag{5.89}$$

where γ_α is again a Lévy noise made of uncorrelated Lévy random variables (here they need not be extremal) and H' is the order of fractional integration (as usual, the Gaussian case is recovered with $\alpha = 2$). At each point, the resulting v field is a Lévy variable and has fluctuation statistics (of $\Delta v(\Delta r) = v(r + \Delta r) - v(r)$) obeying Eqn. (5.88). When $\alpha < 2$, $\xi(q)$ diverges for $q > \alpha$· These models are monofractal because the fractal codimension of any level set $v(\underline{r}) = T$ has a codimension $c(T) = H$ (i.e. independent of T). Note that for the integral to converge, we require $0 < H < 1$. In order to obtain processes with H outside this range, we may use Eqn. (5.89) for the fractional part and then perform ordinary (integer-valued) differentiation or integration as required.

The power-law convolution (Eqn. (5.89)) is easier to understand if we consider it in Fourier space, by taking Fourier transforms of both sides of Eqn. (5.89). Since the Fourier transform of a singularity is another singularity (Box 2.2; the Tauberian theorem):

$$|\underline{r}|^{-(d-H)} \overset{F.T.}{\rightarrow} |\underline{k}|^{-H} \tag{5.90}$$

We can use the basic property that a convolution is Fourier transformed into a multiplication, to obtain simply:

$$\widetilde{v}(\underline{k}) \propto \widetilde{\gamma_\alpha}(\underline{k})|\underline{k}|^{-H} \tag{5.91}$$

where:

$$v(\underline{r}) \overset{F.T.}{\rightarrow} \widetilde{v}(\underline{k}); \quad \gamma_\alpha(\underline{r}) \overset{F.T.}{\rightarrow} \widetilde{\gamma_\alpha}(\underline{k}) \tag{5.92}$$

We thus see that the convolution with power law $|\underline{r}|^{-(d-H)}$ is the equivalent to a power-law filter $|\underline{k}|^{-H}$. However, such filters are themselves generalizations of differentiation ($H < 0$) or integration ($H > 0$). To see this, recall the Fourier transform of the Laplacian, $(\nabla^2\gamma_\alpha) \overset{F.T.}{\rightarrow} -|\underline{k}|^{-2}\widetilde{\gamma_\alpha}$, so that (ignoring constant factors) $|\underline{k}|^{-H}$ corresponds to real space $(\nabla^2)^{-H/2}$, i.e. for $H > 0$ it corresponds to a negative-order differentiation (i.e. integration) of order H. Let us mention that this definition of fractional derivative with the help of the Laplacian is very convenient but somewhat restrictive. In fact, it is related to an underlying microscopic (isotropic) Brownian motion. As soon as the latter are replaced by Lévy flights (i.e. additive Lévy processes), strongly asymmetric fractional operators are obtained (Schertzer *et al.*, 2001).

5.4.4 The fractionally integrated flux model for nonconservative multifractals

In analogy with the fractional Gaussian/Lévy noises and motions, nonconservative multifractal fields are obtained by a fractional integration of a pure cascade process ε_λ (a "flux," hence the name "fractionally integrated flux," FIF model: Schertzer *et al.*, 1997):

$$v_\lambda(\underline{r}) = \varepsilon_\lambda * |\underline{r}|^{d-H} = \int \frac{\varepsilon_\lambda(\underline{r}')d^d \underline{r}'}{|\underline{r} - \underline{r}'|^{d-H}} \tag{5.93}$$

We already saw that the flux itself can be modelled in the same (power convolution/fractional integration) framework: Eqn. (5.78). Figs. 5.31a and 5.31b show respectively the effect of both fractional integration ($H = +0.333$) and fractional differentiation ($H = -0.333$) on the cascade shown in Fig. 5.12a with $C_1 = 0.1$, varying α. For links between the FIF model and the renormalization of the Navier–Stokes equation, see Schertzer *et al.* (1998a).

The basic structures of the FIF and fBm and fLm processes are compared and contrasted in Table 5.2. We see that the difference is simply the type of noise (Gaussian, Lévy, multifractal) which is (fractionally) integrated. Fig. 5.32a shows a comparison of fBm, fLm and a universal multifractal process with the same H value; we can see that the fBm gives a relatively uninteresting texture. fBm is fairly limited in its possibilities since due to the central limit theorem (the Gaussian special case), a process with the same statistical properties can be produced by using singularities of quite different shapes; the details of the shape get "washed out." In comparison, as one can see in Fig. 5.32a, the fLm has extremes which are on the contrary too strong: several strong mountain peaks stand out. Although far from Gaussian, most atmospheric fields as well as the topography seem to empirically have finite variance (i.e. $q_D > 2$; Table 5.1) so a fractional Lévy motion (fLm) is generally not a relevant

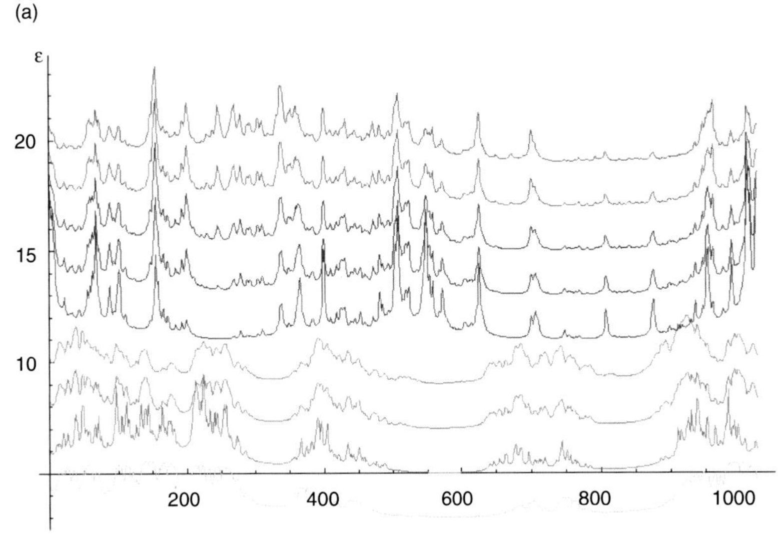

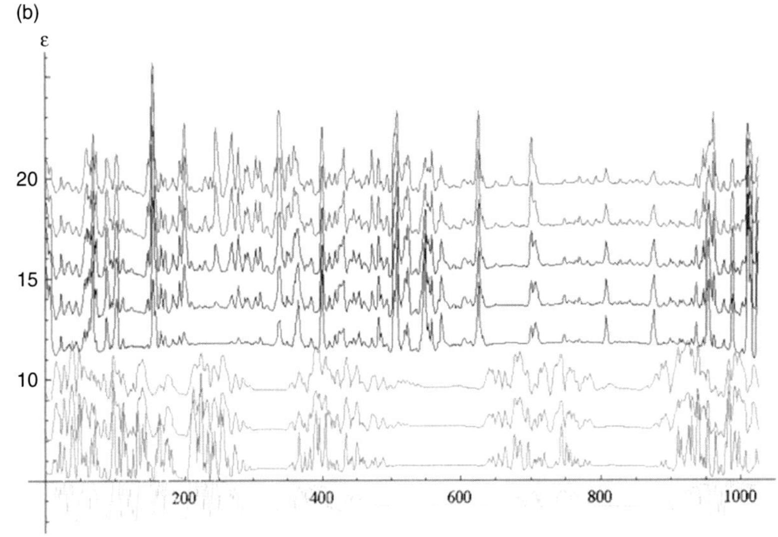

Fig. 5.31 (a) FIF simulations with $\alpha = 0.3, 0.5, \ldots, 1.9$, (bottom to top), $C_1 = 0.1$, $H = 0.333$, each offset for clarity, each with the same random seed. This should be compared to Fig. 5.12a, which shows the corresponding fluxes (i.e. before the fractional integration). (b) The same as Fig. 5.31a but for $H = -0.333$ (i.e. a differentiation, roughening, the same random seeds).

model for those fields. On the contrary, the multifractal simulation has much more interesting structures, but we are still missing the interesting ridges, valleys and other anisotropic features of real geomorphologies, which we deal with in Chapters 6 and 7.

Note that in the case of fBm and fLm, the statistics satisfy the "simple scaling" relation:

$$\Delta v(\lambda \underline{\Delta r}) \overset{d}{=} \lambda^H \Delta v(\underline{\Delta r}) \tag{5.94}$$

where λ is an enlargement factor and $\overset{d}{=}$ means equality in the sense of probability distributions (see Table 5.2 for definitions and statistics). These are monofractal processes, since level sets (i.e. the sets where the sample functions have a given value) are fractal sets with unique dimensions and codimensions (equal to H).

5.4.5 Morphologies of isotropic (self-similar) universal multifractals

Considering the "universal multifractals" v_λ, defined by Eqns. (5.78), (5.79), (5.93), we see that they are isotropic (the singularities have no preferred directions, they depend only on the vector norm). They are therefore "self-similar"; "zoomed" structures will (on

Table 5.2 A comparison of various scaling models for v showing the essential similarities and differences in their mathematical structure, statistical properties. Here $d = 2$ for horizontal planes and the dimension D is the fractal dimension of lines of constant v in the horizontal. The monofractal fractional Brownian motion (fBm) model involves a fractional integration of order H' with a flux $\gamma_2(\underline{r})$ which is simply a ("δ-correlated") Gaussian white noise with variance σ^2. Note that the symbol $a \overset{d}{=} b$ indicates equality in probability distributions, i.e. $a \overset{d}{=} b \Leftrightarrow \Pr(a > s) = \Pr(b > s)$ for all s ("Pr" indicates probability). It results in fluctuations with Gaussian statistics, linear structure function exponent $\xi(q)$ and altitude-independent surface codimension c (or dimension D). The fLm is the generalization obtained by replacing Gaussian variables with stable Lévy variables with index α (fBm is obtained in the case $\alpha = 2$). These have diverging moments q for $q \geq \alpha$. Finally, the multifractal fractionally integrated flux (FIF) model has the same structure, except that the white noises are replaced by multifractal cascades φ_λ where λ is the resolution. The multifractal noise φ is the result of a continuous-in-scale multiplicative cascade. We again find the exponent H, although now there are an infinite number of codimensions $c(\gamma)$ (or dimensions $D(\gamma)$) that depend on the threshold given by λ^γ (do not confuse the singularity γ with the subgenerator γ_α). To generalize fBm, fLm and FIF to anisotropic processes, we must replace the distances in the fractional integration denominators by anisotropic scale functions as discussed in Chapter 6.

Model	Field	Increments	Codimensions of level sets
Monofractal fBm	$v(\underline{r}) = \gamma_2 * \lvert\underline{r}\rvert^{-(d-H')}$ $H' = H + H_2;$ $H_2 = d/2$	$\Delta v \overset{d}{=} \gamma_2 \lvert\Delta \underline{r}\rvert^H$ $\langle\lvert\Delta v\rvert^q\rangle \propto \lvert\Delta \underline{r}\rvert^{\xi(q)}$ $\xi(q) = qH$	$C = H$ $D = d - C$
Monofractal fLm	$v(\underline{r}) = \gamma_\alpha * \lvert\underline{r}\rvert^{-(d-H')}$ $H' = H + H_\alpha;$ $H_\alpha = d/\alpha$	$\Delta v \overset{d}{=} \gamma_\alpha \lvert\Delta \underline{r}\rvert^H$ $\langle\lvert\Delta v\rvert^q\rangle \propto \lvert\Delta \underline{r}\rvert^{\xi(q)}$ $\xi(q) = \begin{matrix} qH; & q < \alpha \\ \infty; & q > \alpha \end{matrix}$	$C = H$ $D = d - C$
Multifractal FIF	$\Gamma_\lambda \propto C_1^{1/\alpha}\gamma_\alpha * \lvert\underline{r}\rvert^{-(d-H_{\alpha'})};$ $\varphi_\lambda = e^{\Gamma_\lambda}$ $v_\lambda(\underline{r}) = \varphi_\lambda * \lvert\underline{r}\rvert^{-(d-H)}$ $H_{\alpha'} = d/\alpha'; \ \frac{1}{\alpha} + \frac{1}{\alpha'} = 1$	$\Delta v = \varphi_\lambda \lvert\Delta \underline{r}\rvert^H$ $\langle\lvert\Delta v\rvert^q\rangle \propto \lvert\Delta \underline{r}\rvert^{\xi(q)}$ $\xi(q) = qH - K(q)$	$c(\gamma) = \max_q \left(q\gamma - K(q)\right)$ $D(\gamma) = d - c(\gamma)$

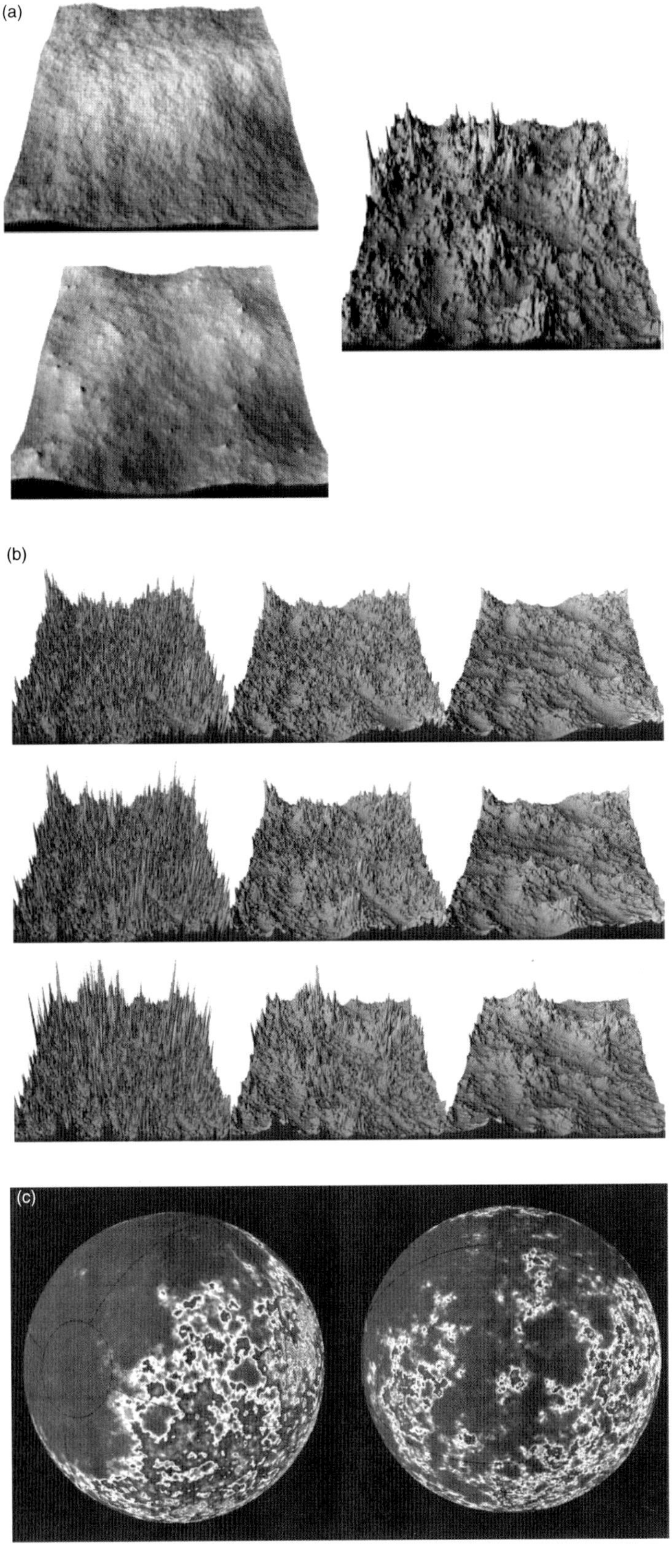

Fig. 5.32 (a) The upper left simulation shows fBm, with $H = 0.7$, lower left fLm with $H = 0.7$, $\alpha = 1.8$, and the right the multifractal FIF with $H = 0.7$, $\alpha = 1.8$, $C_1 = 0.12$ (close to observations for topography, adapted from Gagnon *et al.*, 2006). Note the occasional "spikes" in the fLm which are absent in the fBm; these are due to the extreme power-law tails. (In this fLm positive extremal Lévy variables were used, so there are no corresponding "holes."). (b) Isotropic (i.e. self-similar) multifractal simulations showing the effect of varying the parameters α and H ($C_1 = 0.1$ in all cases). From left to right, $H = 0.2$, 0.5 and 0.8. From top to bottom, $\alpha = 1.1$, 1.5 and 1.8. As H increases the fields become smoother, and as α decreases one notices more and more prominent "holes" (i.e. low smooth regions). The realistic values for topography ($\alpha = 1.79$, $C_1 = 0.12$, $H = 0.7$) correspond to the two lower right-hand simulations. All the simulations have the same random seed. Reproduced from Gagnon *et al.* (2006). (c) A simulation of an (isotropic) multifractal topography on a sphere using the spherical harmonic method discussed in the appendix (both sides of a single simulation are shown, using false colours). The simulation parameters are close to the measured values: $\alpha = 1.8$, $C_1 = 0.1$, $H = 0.7$ (see Chapter 4). The absence of mountain "chains" and other typical geomorphological features are presumably due to the absence of anisotropy. We thank J. Tan for help with this simulation, adapted from Quattrochi and Goodchild (1997). *See colour plate section.*

Box 5.6 The stochastic modelling of vector multifractal processes and atmospheric dynamics: Lie cascades

Up until now, we have exclusively considered individual (scalar) fields such as a wind components or passive scalar concentrations which are derived from scalar turbulent fluxes by fractional integration (the FIF model, Section 5.4). These fluxes are (on average) symmetric with respect to certain scale-changing symmetries (which can readily include anisotropy such as stratification, rotation etc.: see Chapters 7 and 8). However, in order to fully model the atmosphere with many nonlinearly interacting fields (and presumably several nonlinearly coupled cascade processes) we can consider the "state vector" representing the state of the atmosphere (e.g. with components being the usual variables of state such as $\underline{v}, T, p, h, \ldots$). In order to fully model the atmosphere, scalar multifractal cascades must therefore be extended to vector multifractal cascades. Once this is done various additional symmetries (including those representing mass and energy conservation) can be imposed.

The situation is thus paradoxical: classical methods such as those used in GCM modelling deal easily with vectorial interactions but only over a very limited range of scales; in contrast, the scaling models deal easily with (even) infinite ranges of scales but (so far) have mostly avoided treating vectorial interactions. The only partial exception is Schertzer and Lovejoy (1995); we give a short summary below.

Up until now, we have been restricted not only to scalar cascades, but to *positive* scalar cascades. This is already problematic, since turbulence closures have long shown that one expects occasional "backscatter" corresponding to fluctuations with $\varepsilon < 0$ (i.e. transfer from small to large scales even in classical – on average – downscale cascades: see e.g. Lesieur and Schertzer, 1978). The basic problem is that the exponentiation of a real generator Γ yields a positive real result: $\varepsilon = exp(\Gamma)$; to obtain $\varepsilon < 0$ we need to at least use complex generators leading to complex ε.

To see how complex cascades work, consider a large-scale ($\lambda = 1$) complex cascade flux v_1, and complex generator Γ_λ:

$$v_\lambda = v_1 e^{\Gamma_\lambda}$$

The significance of $\Gamma_{R,\lambda} = \mathrm{Re}(\Gamma_\lambda)$ and $\Gamma_{I,\lambda} = \mathrm{Im}(\Gamma_\lambda)$ is obvious: $\Gamma_{R,\lambda}$ generates a nonnegative cascade process which modulates the amplitude of the modulus of v_λ, whereas $\Gamma_{I,\lambda}$ gives the rotation of v_λ, hence the sign of its real part. We may focus on the special case where $\Gamma_{R,\lambda}$ and $i\Gamma_{I,\lambda}$ are independent stochastic processes with corresponding characteristic functions $K_R(q)$, $K_I(q)$:

$$\langle v_\lambda^q \rangle = \langle e^{q\Gamma_{R,\lambda}} \rangle \langle e^{qi\Gamma_{I,\lambda}} \rangle \langle v_1^q \rangle = \lambda^{K_R(q)} \lambda^{K_I(q)} = \lambda^{K(q)} \tag{5.95}$$

The moment scaling function $K(q)$ is therefore simply:

$$K(q) = K_R(q) + K_I(q) \tag{5.96}$$

It is important to note that whereas $K_R(q)$ is real for any real q, $K_I(q)$ is complex, being in general neither real nor pure imaginary. The condition of small-scale cascade convergence ($<v_\lambda> = 1$) still corresponds to $K(1) = 0$, but not to $K_I(1) = 0$, i.e. Γ_R generates a nonconservative process for the vector modulus.

Let us consider the discrete-in-scale complex lognormal case as an example (Gaussian generator, λ_1 being the fixed-step scale ratio). The real and imaginary exponential increments Γ_{R,λ_1} and Γ_{I,λ_1} will be a Gaussian variable of variance and mean σ_R^2, m_R and σ_I^2, m_I, respectively. These lead to a generalization of the scalar universal scaling function (Eqn. (5.33) with $\alpha = 2$):

$$\begin{aligned} &K_R(q) = C_{1,R}(q^2 - q) - H_R q; && C_{1,R} = \sigma_R^2/2; && H_R = C_{1,R} - m_R \\ &K_I(q) = -C_{1,I}(q^2 - q) - H_I q; && C_{1,I} = \sigma_I^2/2; && H_I = C_{1,I} \quad im_I \\ &K(q) = C_1(q^2 - q) - Hq; && C_1 = C_{1,R} - C_{1,I}; && H = H_R + H_I \end{aligned} \tag{5.97}$$

A conservative field is obtained with $m_R = -C_1$ (i.e. $\neq -C_{1,R}$ as required to obtain a conservative cascade of the modulus), $m_I = 0$.

One may note that $K(q)$ remains of the standard universal form even for complex q. Similar properties hold for Lévy processes when Γ_{R,λ_1} and Γ_{I,λ_1} are independent and identically distributed. However, Γ_{R,λ_1} and Γ_{I,λ_1} do not necessarily need to have the same α, and there is no longer the requirement that Γ_{R,λ_1} and Γ_{I,λ_1} should correspond to extremal Lévy processes. For applications to correlated scaling processes where the processes are identified with the real and imaginary parts, see Lovejoy *et al.* (2001).

To generalize this further we may consider nonpositive cascades as being components of more or less straightforward vectorial extensions of positive real processes: $v_\lambda = v_1 e^{\Gamma_\lambda}$, where now the v are d-component vector fields and Γ_λ is a $d \times d$ matrix (tensor). Corresponding to the group property of the transformation for the field, there is a Lie algebra structure for the generators. For more details on these "Lie cascades" see Schertzer and Lovejoy (1995).

average) resemble the unzoomed ones. In addition, they depend on three parameters: the α, C_1 which define the statistics of the generator Γ_λ (Eqns. (5.78)–(5.81)) and the H in Eqn. (5.93). While the parameter H is the order of fractional integration and quantifies the degree of scale-invariant smoothing, the qualitative effects of the codimension of the mean (C_1) and the Lévy index (α) are less easy to see. For $H > 0$, the k^{-H} filter is a kind of scale-invariant smoothing, whereas for $H < 0$ it is scale-invariant roughening. This was already confirmed in the 1D simulations (Figs. 5.31a, 5.31b) To further explore the effect of varying H on the various morphologies, we performed multifractal simulations in 2D (rather than in 1D). Fig. 5.32b systematically shows the morphologies of the structures obtained by varying these three parameters; each has the same initial random "seed" so that the basic generating white noises are the same. For large-scale geo-applications, we require simulations on the sphere (Fig. 5.32c; these are discussed in Appendix 5.D).

To systematically survey the simulation parameter space (α, C_1, H), we use false colours (using a "cloud" palette, Fig. 5.33) and simulations with $\alpha = 0.4$ to $\alpha = 2.0$ in increments of 0.4, as well as C_1, H from 0.05 to 0.80 in increments of 0.15, amply covering the parameter range commonly encountered in the geosciences (this may be compared with Fig. 6.9, showing the same simulations but with scaling stratification). In Fig. 5.34, rather than fixing α and varying C_1, H we show the effect of fixing C_1 and H and varying α. For fixed α, C_1, we see that increasing H systematically smoothes the structures, whereas for fixed α, H, varying C_1 changes the "spottiness," the sparseness of structures. For reference, note that the empirically most common values of α are in the range 1.5–1.9 (the latter being appropriate for topography and cloud radiances, the former, for rain and atmospheric turbulence). The parameter C_1 is often fairly low (e.g. in the range 0.05–0.15 for the wind, cloud radiances, topography; see Table 4.8), although it can be large (0.25–0.7) for rain and turbulent fluxes. While the basic Kolmogorov value of H is 1/3, many fields (such as cloud radiances) are near this; while for rain this parameter is nearly zero, for topography it is in the range 0.45–0.7. From Fig. 5.34 we can see that high values of C_1 lead to fields totally dominated by one or two strong structures, while low α values lead to fields dominated by "Lévy holes": large regions with extremely low values.

5.5 Wavelets and fluctuations: structure functions and other data analysis techniques

5.5.1 Defining fluctuations using wavelets

We have seen that data analyses constantly rely on defining fluctuations at a given scale and location; the simplest definition of fluctuation at position x, scale Δx, being $\Delta v(x,\Delta x) = v(x + \Delta x) - v(x)$. Note that since we typically assume that the statistics of the fluctuations are independent of position, we previously suppressed the x argument. We have already mentioned (Chapter 4) that other definitions of fluctuation are possible and are occasionally necessary. Let us now examine this a bit more closely.

Consider the statistically translationally invariant process $v(x)$ in 1D: the statistics are thus independent of x and this implies that the Fourier components are "δ correlated" (Appendix 2A):

$$\langle \tilde{v}(k)\tilde{v}(k')\rangle = \delta_{k+k'}P(k); \quad \tilde{v}(k) = \int e^{-ikx}v(x)dx \tag{5.98}$$

If it is also scaling then the spectrum $E(k)$ is a power law: $E(k) \approx \langle|\tilde{v}(k)|^2\rangle \approx k^{-\beta}$ (where, here and below, we ignore constant terms such as factors of 2π etc.). In terms of its Fourier components, the fluctuation is thus:

$$\begin{aligned}\Delta v(x,\Delta x) &= v(x+\Delta x) - v(x)\\ &= \int e^{ikx}\tilde{v}(k)(e^{ik\Delta x} - 1)dk\end{aligned} \tag{5.99}$$

so that the Fourier transform of $\Delta v(x,\Delta x)$ is $\tilde{v}(k)(e^{ik\Delta x} - 1)$. We first consider the statistics of quasi-Gaussian processes for which $C_1 = 0$, $\xi(q) = Hq$. Exploiting the statistical translational invariance, we drop the x dependence and obtain the relation to the second-order structure function to the spectrum:

$$\begin{aligned}\langle|\Delta v(\Delta x)|^2\rangle &= 4\int e^{ikx}\langle|\tilde{v}(k)^2|\rangle \sin^2\left(\frac{k\Delta x}{2}\right)dk\\ &\approx \int e^{ikx}k^{-\beta}\sin^2\left(\frac{k\Delta x}{2}\right)dk\end{aligned} \tag{5.100}$$

As long as the integral on the right converges, then the usual Tauberian argument (Box 2.2) shows that:

$$\langle|\Delta v(\Delta x)|^2\rangle \propto \Delta x^{\xi(2)} \approx \int e^{ikx}k^{-\beta}\sin^2\left(\frac{k\Delta x}{2}\right)dk \propto \Delta x^{\beta-1} \tag{5.101}$$

(a)

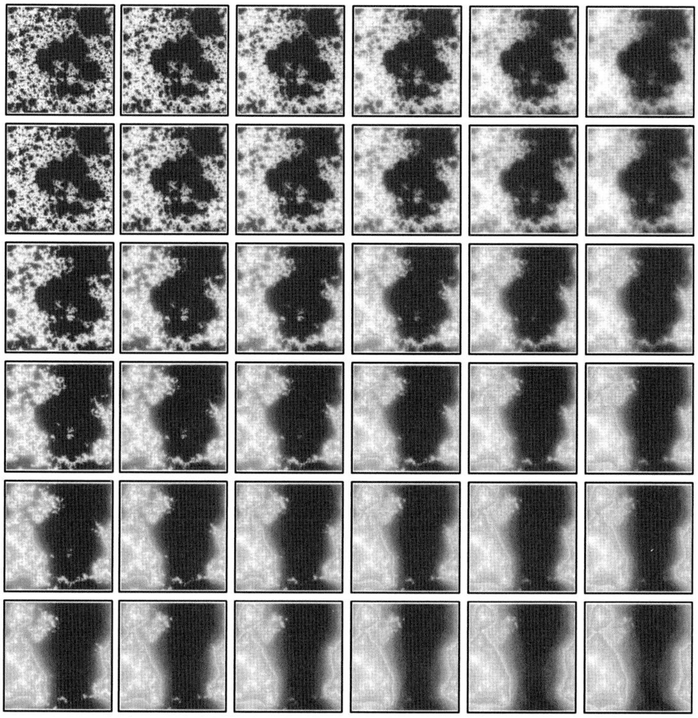

(b)

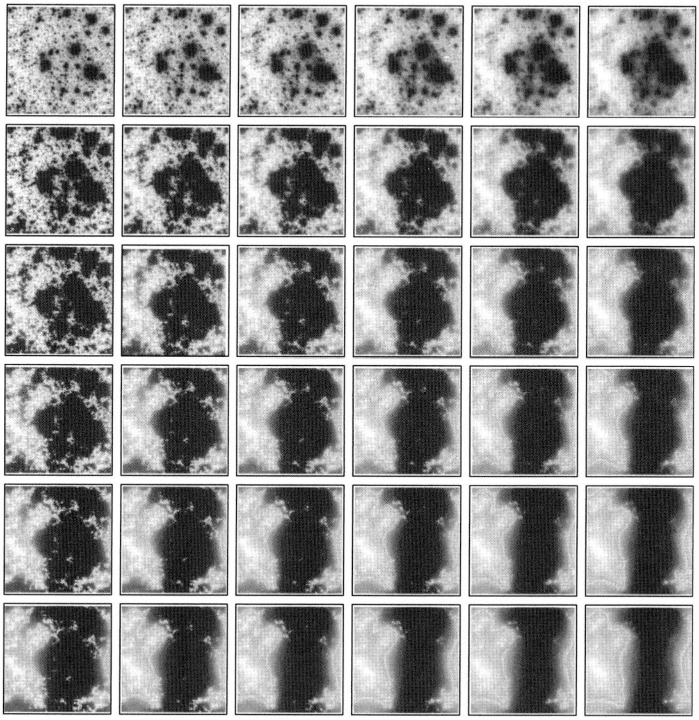

Fig. 5.33 (a) $\alpha = 0.4$, $C_1 = 0.05$ to 0.80, top to bottom row; $H = 0.05$ to 0.80, left to right column. (b) $\alpha = 0.8$, $C_1 = 0.05$ to 0.80, top to bottom row; $H = 0.05$ to 0.80, left to right column.

(c)

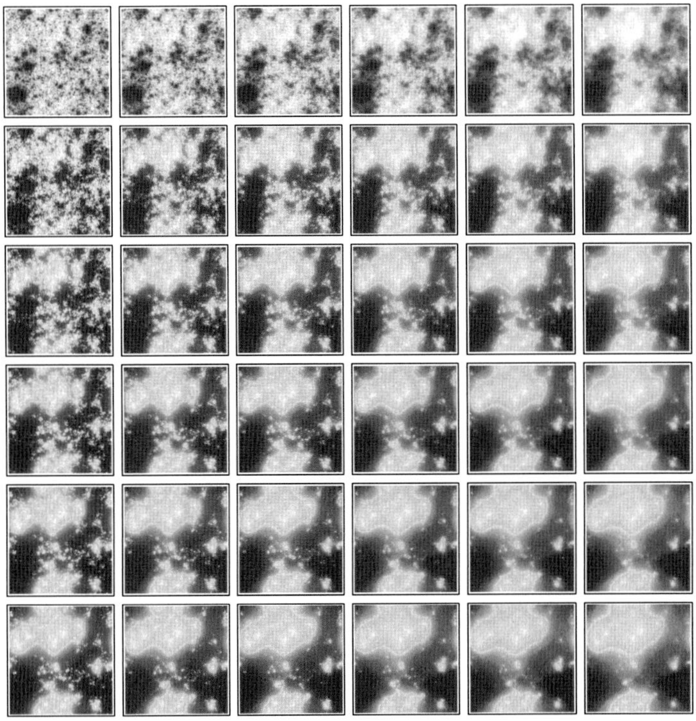

(d)

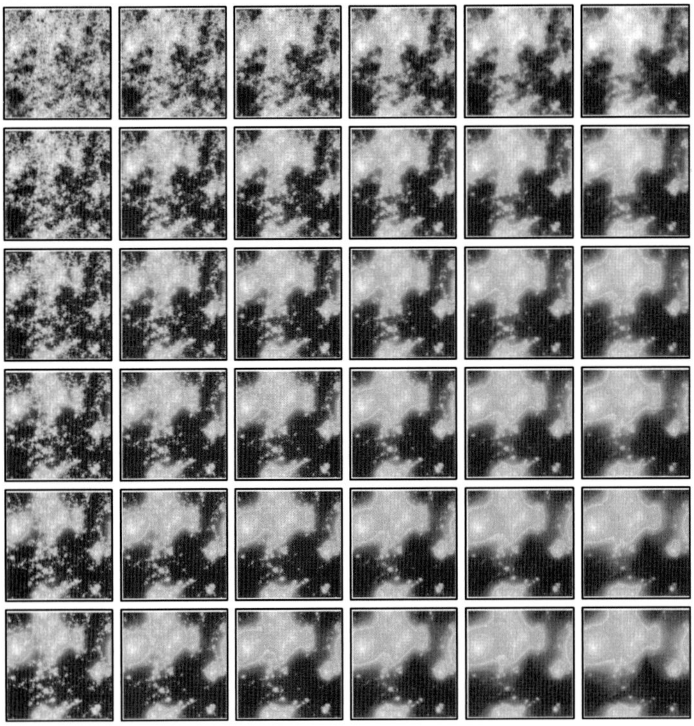

Fig. 5.33 (c) $\alpha = 1.2$, $C_1 = 0.05$ to 0.80, top to bottom row; $H = 0.05$ to 0.80, left to right column. (d) $\alpha = 1.6$, $C_1 = 0.05$ to 0.80, top to bottom row; $H = 0.05$ to 0.80, left to right column.

(e)

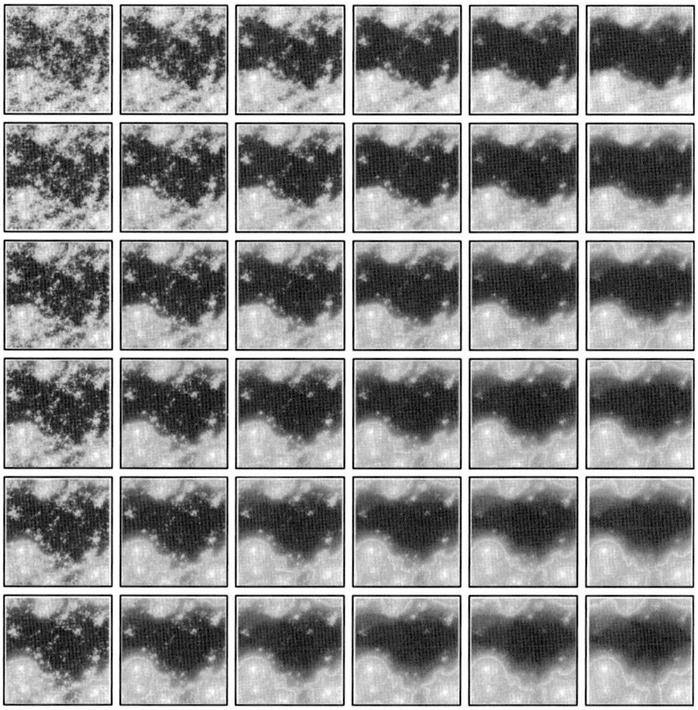

Fig. 5.33 (e) $\alpha = 2$, $C_1 = 0.05$ to 0.80, top to bottom row; $H = 0.05$ to 0.80, left to right column.

so that $\beta = \xi(2) + 1 = 2H + 1$ ($C_1 = 0$ here); see Eqns. (4.18), (4.19). However, for large k, the integrand $\approx k^{-\beta}$, which implies a large wavenumber divergence whenever $\beta \le 1$. However, since for small k $\sin^2(k\Delta x/2) \propto k^2$, there will be a low wavenumber divergence only when $\beta \ge 3$. Although real-world (finite) data will not diverge, in the theoretically divergent cases the structure functions will no longer characterize the local fluctuations, but rather those on either the highest or lowest wavenumbers present in the data. In the quasi-Gaussian case, or when C_1 is small, we have $\xi(2) \approx 2H$ and we conclude that using first-order differences to define the fluctuations leads to second-order structure functions being meaningful in the sense that they adequately characterize the fluctuations whenever $1 < \beta < 3$, i.e. $0 < H < 1$.

Since $0 < H < 1$ is the usual range of geophysical H values, and the difference fluctuations are very simple, they are commonly used. However, we can see that there are limitations; in order to extend the range of H values, one can define fluctuations using finite differences of different orders. To see how this works, consider using second (centred) differences:

$$\Delta v(x, \Delta x) = v(x) - \frac{1}{2}\left(v(x + \Delta x/2) - v(x - \Delta x/2)\right)$$
$$= \int e^{ikx}\widetilde{v}(k)\left[1 - \frac{1}{2}(e^{ik\Delta x/2} + e^{-ik\Delta x/2})\right]dk$$
$$= 2\int e^{ikx}\widetilde{v}(k)\sin\left(\frac{k\Delta x}{4}\right)^2 dk \qquad (5.102)$$

Repeating the above arguments, we can see that the relation $\beta = \xi(2) + 1$ holds now for $1 < \beta < 5$, or (with the same approximation) $0 < H < 2$. Similarly, by replacing the original series by its running sum (a finite difference of order –1) – as done in the Haar wavelet and MFDFA techniques described below – we can extend the range of H values down to –1. Since the "macroweather plateau" (at scales longer than about two weeks, up to decades and centuries) is precisely characterized by $-1 < H < 0$, a corresponding "tendency structure function" technique is indeed useful, and is discussed shortly.

More generally, going beyond Gaussian processes we can consider intermitent FIF processes, which have $\widetilde{v}(k) \approx \widetilde{\varepsilon}(k)|k|^{-H}$, and we see that the Fourier transform of $\Delta v(x,\Delta x)$ is $(e^{ik\Delta x} - 1)|k|^{-H}\widetilde{\varepsilon}(k)$. This implies that for low wavenumbers $\widetilde{v}(k) \approx |k|^{1-H}\widetilde{\varepsilon}(k)$ $(k<<1/\Delta x)$, whereas at high wavenumbers $\widetilde{v}(k) \approx |k|^{-H}\widetilde{\varepsilon}(k)$ $(k>>1/\Delta x)$, hence since the mean of ε is independent of scale for both large and small scales, we see that for $0 < H < 1$ the fluctuations Δv (Δx) are dominated by wavenumbers $k \approx 1/\Delta x$, so that for this range of H, fluctuations defined as differences capture the variability of Δx-sized structures, not structures either much smaller or much larger than Δx. More generally, since the Fourier transform of the nth derivative $d^n v/dx^n$ is $(ik)^n\widetilde{v}(k)$ and the finite derivative is the same for small k but "cut off" at large k, we find that nth-order fluctuations are dominated by structures with $k \approx 1/\Delta x$ as long as $0 < H < n$. This means that $\Delta v(\Delta x)$ does indeed reflect the Δx scale fluctuations.

While finite differences are usually adequate in scaling applications, in the past 20 years there has been a development of systematic ways of defining fluctuations: wavelet analysis. For a meteorological introduction, see Torrence and Compo (1998) and Foufoula-Georgiou and Kumar (1994); for a mathematical introduction, see Holschneider (1995).

In wavelet analysis, one defines fluctuations with the help of a basic "mother wavelet" $\Psi(x)$ and performs the convolution:

$$\Delta v(x, \Delta x) = \int v(x')\Psi\left(\frac{x' - x}{\Delta x}\right)dx' \qquad (5.103)$$

where we have kept the notation Δv to indicate "fluctuation" (technically, Δv is a "wavelet coefficient") The basic "admissibility" condition on $\Psi(x)$ (so that it is a valid wavelet) is that it has zero mean. If we take the fluctuation Δv as the symmetric difference $\Delta v = v(x + \Delta x / 2) - v(x - \Delta x / 2)$ then:

$$\Psi(x) = \delta(x - 1/2) - \delta(x + 1/2) \qquad (5.104)$$

where δ is the Dirac delta function, then we recover the usual first difference fluctuation; see Fig. 5.35b; due to the statistical translational invariance, this is equivalent to the difference: Eqn. (5.99)). This "poor man's" wavelet is usually adequate for our purposes. As the generality of the definition (Eqn. (5.103)) suggests, all kinds of special wavelets can be introduced; for example, special orthogonal wavelets can be used which are convenient if one wishes to "reconstruct" the original function from the fluctuations, or to define power spectra locally in x rather than globally, (averaged over all x). Alternatively, we could consider the second derivative of the Gaussian, which is the popular "Mexican hat" shown in Fig. 5.35a, along with the centred second finite difference wavelet:

$$\Psi(x) = \frac{1}{2}\left(\delta\left(x + \frac{1}{2}\right) + \delta\left(x - \frac{1}{2}\right)\right) - \delta(x) \qquad (5.105)$$

The Mexican hat is thus essentially just a smoothed-out version of the second finite difference. Just as one can use higher- and higher-order finite differences to extend the range of H values, so one can simply use higher- and higher-order derivatives of the Mexican hat. The second finite difference wavelet is easy to implement, it is simply $\Delta v = (v(x + \Delta x) + v(x - \Delta x/2))/2 - v(x)$ and has the advantage that the structure function based on this is valid for $0 < H < 2$.

Because the usual difference structure function is restricted to H in the range $0 < H < 1$, it is useful to introduce variants, in particular to extend the range down to $H = -1$. The idea is simple: perform a running sum (i.e. an integral of order 1, increasing H by 1), and then perform a second finite difference of the running sum, so that with respect to the original function, the range of H is now $-1 < H < 1$. In terms of v, the fluctuation Δv is:

$$\left(\Delta v(\Delta x)\right)_{Haar} = \frac{2}{\Delta x}\left[\int_{x}^{x+\Delta x/2} v(x')dx' - \int_{x-\Delta x/2}^{x} v(x')dx'\right] \qquad (5.106)$$

where we have added the extra $1/\Delta x$ factor so that the scaling is the same as for the poor man's fluctuation, i.e. $\xi_{Haar}(q) = \xi(q)$ for processes with $-1 < H < 1$. In words, to estimate the Haar fluctuation for a lag Δx (Eqn. (5.106)), one simply takes the difference of the mean of the first and second halves of the interval. To within the division by Δx, and a constant factor, the result is the equivalent of using the Haar wavelet as indicated in Fig. 5.35b:

$$\Psi(x) = \begin{matrix} 1; & 0 \le x < 1/2 \\ -1; & -1/2 \le x < 0 \\ 0; & otherwise \end{matrix} \qquad (5.107)$$

It can be checked that the Fourier transform of this wavelet is $\propto k^{-1}\sin^2(k/4)$ so that it

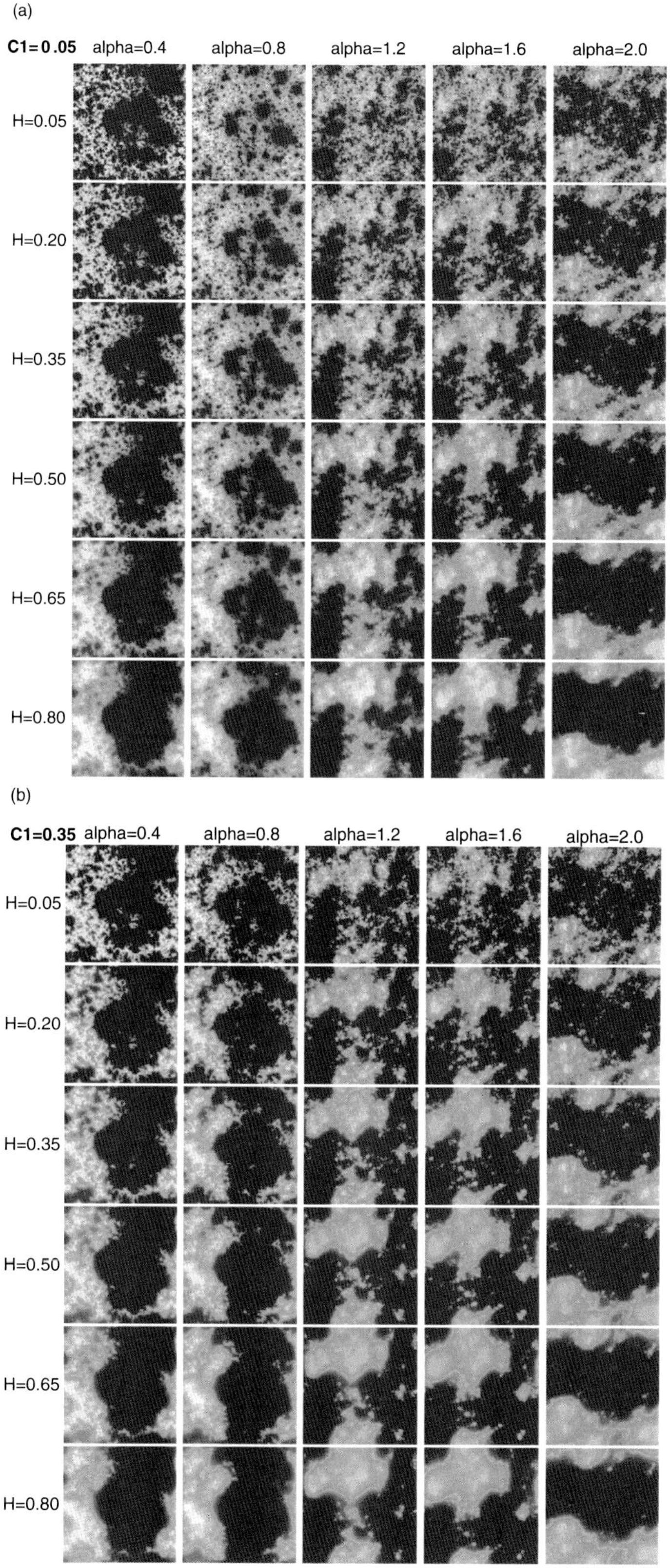

Fig. 5.34 (a) The effect of varying α, H values on multifractal simulations for $C_1 = 0.05$. The upper figures show the effect of increasing α (left to right, 0.4, 0.8, . . ., 2.) and H (top to bottom 0.05, 0.2, . . ., 0.8) with C_1 fixed. Reproduced from Lovejoy and Schertzer (2007). (b) Same as Fig. 5.34a but $C_1 = 0.35$. Reproduced from Lovejoy and Schertzer (2007).

(c)

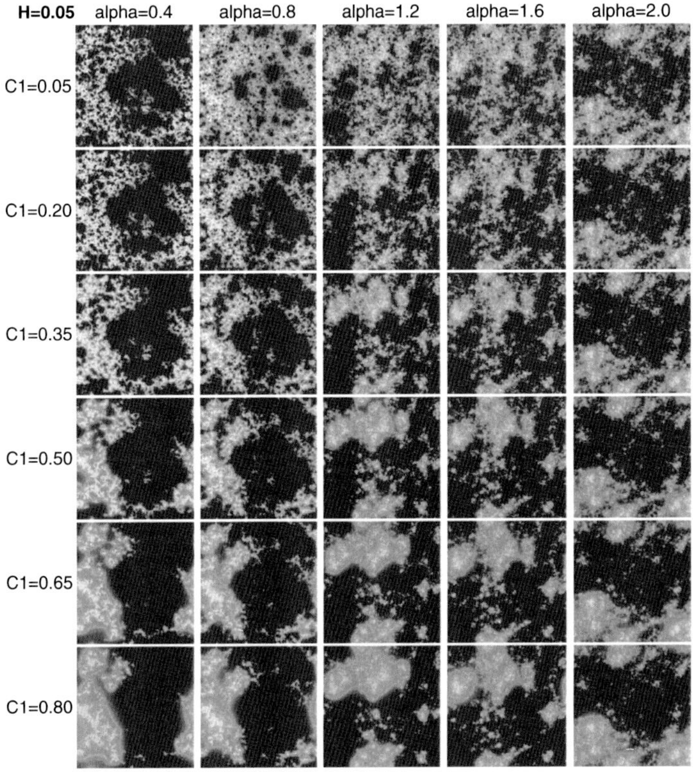

(d)

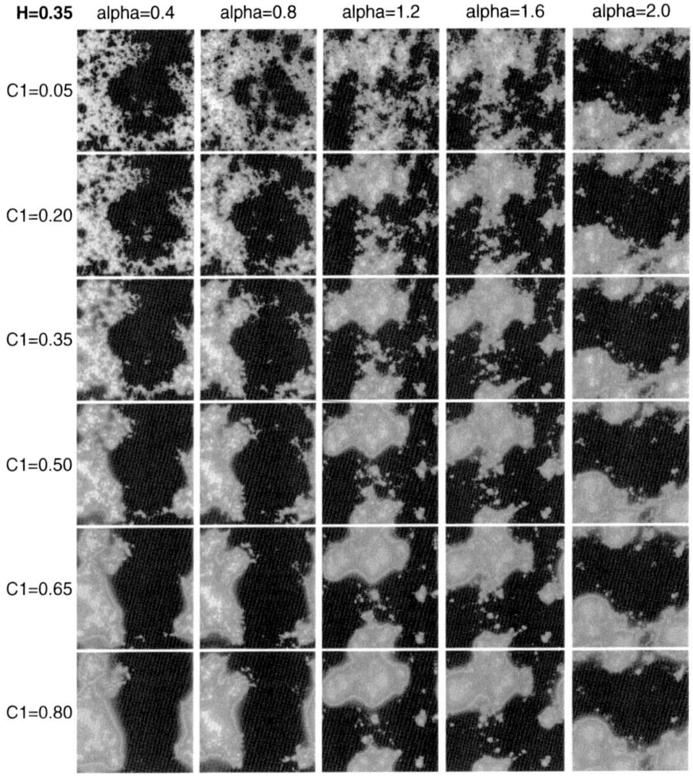

Fig. 5.34 (c) The effect of varying α (left to right, 0.4, 0.8, . . ., 2.) and C_1 (top to bottom 0.05, 0.2, . . ., 0.8) with H fixed = 0.05. Reproduced from Lovejoy and Schertzer (2007). (d) Same as Fig. 5.34c but H = 0.35. Reproduced from Lovejoy and Schertzer (2007).

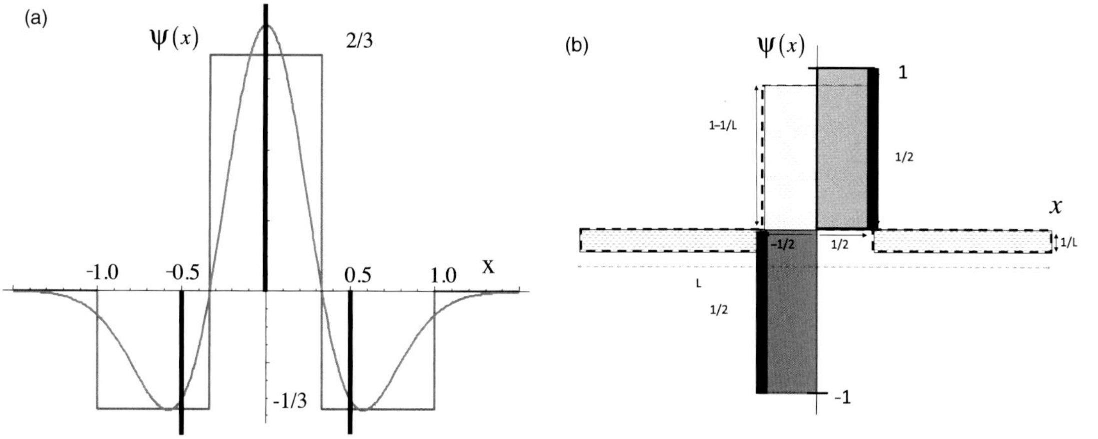

Fig. 5.35 (a) The popular Mexican hat wavelet (the second derivative of the Gaussian (smooth curve) valid for $-1 < H < 2$) compared with the (negative) second finite difference wavelet (solid bars representing the relative weights of δ functions, valid for $0 < H < 2$)), and the second-order "quadratic" Haar wavelet (rectangles) obtained from the third difference of the running sum (i.e.$\Delta v(\Delta x) = [(s(x+\Delta x) - s(x-\Delta x))/3 - (s(x+\Delta x/3) - s(x-\Delta x/3))]/\Delta x$ where $s(x)$ is the running sum valid for $-1 < H < 2$). (b) The "poor man's" wavelet, with solid bars representing the amplitudes of Dirac δ functions (the basis of the usual difference structure function, valid for $0< H< 1$), uniform dark shading showing the Haar wavelet (the basis of the Haar structure function, the second difference of the running sum, valid for $-1< H< 1$), and stippled (light) shading representing the wavelet used for the "tendency" structure function valid for $-1< H< 0$. Reproduced from Lovejoy and Schertzer (2012a)

is better localized in Fourier space with both low and high wavenumber fall-offs ($\approx k$ and $\approx k^{-1}$ respectively). It turns out that for analyzing multifractals, even over their common range of validity ($0 < H < 1$), the Haar structure function does somewhat better than the poor man's wavelet structure function (i.e. the usual structure function). This is demonstrated on numerical examples in Appendix 5E. While the latter has the advantage of being relatively simple to interpret (e.g. the mean of the absolute difference is considered to be the "typical difference"), the corresponding interpretation of the Haar wavelet is less intuitive. To understand it better, including its interpretation for $H < 0$, which is important for macroweather signals which typically have $-1 < H < 0$ (Chapter 10), we introduce a slightly different wavelet based fluctuation that we call the "tendency fluctuation":

$$\left(\Delta v(\Delta x)\right)_{tend} = \overline{\Delta v}(\Delta x) = \frac{1}{\Delta x}\int_{x}^{x+\Delta x} v'(x')dx'; \quad v'(x) = v(x) - \overline{v(x)} \tag{5.108}$$

We can see that the fluctuation $\overline{\Delta v}(\Delta x)$ defined in this way is effectively a spatial average, hence the overbar notation; $v'(x)$ is the transect $v(x)$ with the mean removed: $\overline{v'(x)} = 0$.

In terms of wavelets, this can be seen to be equivalent to using the wavelet:

$$\Psi(x) = I_{[-1/2,\,1/2]}(x) - \frac{I_{[-L/2,\,L/2]}(x)}{L}; \quad L >> 1 \tag{5.109}$$

where I is the indicator function:

$$I_{[a,\,b]}(x) = \begin{matrix} 1 & a \le x \le b \\ 0 & otherwise \end{matrix} \tag{5.110}$$

See Fig. 5.35b for a schematic. Note the closely related Aggregated Standard Deviation (ASD) technique (Koutsoyiannis and Montanari, 2007), which is based on the scaling of the standard deviation of the integrated series (exponent H_{ASD}: their symbol "H"). From Eqn. 5.108, we obtain $H_{ASD} = 1 + \xi(2)/2 = H + 1 - K(2)/2$ with a range of validity $-1 < H < 0$.

The first term in Eqn. (5.109) represents the integral in Eqn. (5.108) whereas the second removes the mean ($L >> 1$ is the overall length of the dataset). The removal of the mean in this way is necessary in order that the wavelet satisfy the admissibility condition that its mean is zero. The only essential difference between wavelet coefficients defined by Eqn. (5.109) and the "tendency fluctuation" defined in Eqn. (5.108) (where the mean is removed beforehand), is the extra normalization by Δx in Eqn. (5.108), which changes the exponent by 1. The term "tendency structure function" is used since for $H < 0$ it gives a direct estimate of the typical tendency of the fluctuations to increase or decrease.

Recall that the Haar fluctuation over an interval is the difference between the means over the first and second halves of the interval; it is in fact the difference between the corresponding tendencies. In Appendix 5E we develop this mathematically and show that it is also equal to the tendency of the differences (the order is immaterial). This observation allows us to straightforwardly interpret the Haar fluctuation. When $0 < H < 1$, the tendencies "saturate" so that the Haar fluctuation is close to the difference between the first and second halves of the interval, whereas when $-1 < H < 0$, the differencing "saturates" and the result is close to the tendencies. The result is that not only does one recover the correct scaling exponents for any $-1 < H < 1$, but the values of the fluctuations themselves are close to the simple differences and tendencies. We could mention the paper by Veneziano and Furcolo (2003), who show how Haar wavelets can be used advantageously for theoretically analyzing multifractal cascades.

From the above we see that if needed, the Haar wavelet can be easily generalized. To see how to do this, introduce the integral $s(x) = \int\limits_{x} v(x')dx'$ (for discrete data, the integral can be replaced by a running sum). Notice that the Haar fluctuation can be written in terms of the second differences of $s(x)$: $(\Delta v(\Delta x))_{Haar} = 2(s(x+\Delta x) - 2s(x+\Delta x/2) + s(x))/\Delta x$. As we saw, generalizations valid for larger and larger H are obtained by taking higher and higher order differences (see also Appendix 5E). For example, the "quadratic Haar" fluctuation can be defined by using the (normalized) third difference of the running sum: $(\Delta v(\Delta x))_{QuadHaar} = 3(s(x+\Delta x) - 3s(x+2\Delta x/3) + 3s(x+\Delta x/3) - s(x))/\Delta x$, which is very close to the Mexican hat wavelet (Fig. 5.35a) but which is numerically much easier to implement. As with the Mexican hat, it is valid for $-1 < H < 2$.

Wavelet mathematics are seductive – and there certainly exist areas such as speech recognition where both frequency and temporal localization of statistics are necessary. However, the use of wavelets in geophysics is often justified by the existence of strong localized structures, which are cited as evidence that the underlying process is statistically nonstationary or statistically inhomogeneous (i.e. in time or in space, respectively). However, we have seen that cascades produce exactly such structures – the singularities – but that their statistics are nevertheless strictly translationally invariant. In this case the structures are simply the result of strong singularities, but nevertheless the local statistics are uninteresting, and they are averaged out in order to improve statistical estimates.

Before continuing to discuss other related methods for defining fluctuations, we should mention that there have been strong claims that wavelets are indispensable for analyzing multifractals (e.g. Arneodo *et al.*, 1999). Inasmuch as the traditional definition of fluctuations as first differences is already a wavelet, this may be true (see however the discussion below of the DFA, MFDFA method). However, for most applications, one does not need wavelet properties or specific wavelet techniques. Indeed, it is ironic that the advantages of the Haar wavelets are due to their combination of summing and differencing: they are not related to their wavelet nature, which is not especially helpful, or even needed! In addition, at a more fundamental level, Arneodo *et al.*'s claim is at least debatable since, mathematically, wavelet analysis is a species of functional analysis, i.e. its objects are mathematical functions defined at mathematical points. On the contrary, we have seen that the generic multifractal processes – cascades – are rather singular measures, they are "delocalized" (See Fig. 5.24; Schertzer and Lovejoy, 1992; Schertzer *et al.*, 2010), so that strictly speaking they are outside the scope of wavelet analysis and therefore may not always be appropriate. A particular example where wavelets may be misleading is the popular "modulus maximum" technique where one attempts to localize the singularities (Bacry *et al.*, 1989; Mallat and Hwang, 1992). If the method is applied to a cascade process, then as one increases the resolution the singularity localization never converges, so the significance of the method is not as obvious as is claimed and its utility is questionable. Similar comments apply to the "wavelet leader" technique (Serrano and Figliola, 2009).

5.5.2 Generalized structure functions and detrended multifractal fluctuation analysis

We have seen (Eqn. (5.88)) that the statistical moments of the fluctuations lead to the exponent $\xi(q) = qH - K(q)$; in the special case where the fluctuations are defined by the first differences, this is the usual qth-order structure function (when $q \neq 2$, this is sometimes called the "generalized structure function"). While these have the advantage that they can be used to directly estimate H ($= \xi(1)$); they are often not optimal for estimating C_1, α. This is because H is often much larger than C_1 so that for low-order moments the qH term in $\xi(q)$ is much larger in magnitude than the $K(q)$ term, and at "large" q (e.g. $q > \sim 3$–4) the moments may become spurious because of the multifractal phase transitions discussed above. In both cases, trying to estimate $K(q)$ using $\xi(q)$ and the equation $K(q) = qH - \xi(q)$ may lead to large errors, since the magnitude of $K(q)$ may be of the same order as the uncertainty in $\xi(q)$. The moment method applied to the turbulent flux (trace moments), as in Chapter 4, gives better accuracy for C_1, α by removing the qH term by taking the absolute fluctuations at the smallest scale. Nonetheless, the universality parameters can be estimated from $\xi(q)$ in a variety of ways; for example using $\xi(1) = H$, $\xi'(1) = H - C_1$ and $\xi''(1) = \alpha C_1$.

An interesting example of the use of structure functions that complement the flux analyses of Chapter 4 (Figs. 4.6a, 4.6b) is shown in Figs. 5.36a and 5.36b. In Fig. 5.36a we show the corresponding results for the longitudinal and transverse wind and pressure (i.e. the wind components parallel and orthogonal to the aircraft direction, respectively). Recall that these are the fields most strongly affected by the aircraft motion, especially the pressure (the aircraft attempted to fly on an isobar). For the reasons discussed in Chapter 2 – the slopes of the isobars – the behaviour for the wind has essentially two scaling regimes with a transition at about 40 km, while, unsurprisingly, the pressure has poor scaling. Contrast this with Fig. 5.36b, where we show the corresponding plot for those fields less affected by the trajectory fluctuations, the temperature, humidity and log potential temperature: here we see that the scaling is indeed very good. These analyses were used to estimate the corresponding H and C_1 values given in Table 4.4, and Fig. 5.36c shows the corresponding $\xi(q)$ functions. Notice that the curvature is small since $H \approx 0.5$ while $C_1 \approx 0.05$. We return to discuss these results in more detail in Chapter 6 (Section 6.4.1), where we compare them to the corresponding vertical analyses.

We now discuss a variant method that defines the fluctuations in a different way but still quantifies the statistics in the manner of the generalized structure function by assuming that the fluctuations are stationary: the "multifractal detrended fluctuation analysis" (MFDFA: Kantelhardt *et al.*, 2001, 2002). Indeed, this is a straightforward generalization of the original "detrended fluctuation analysis" (DFA: Peng *et al.*, 1994) obtained by considering moments other than $q = 2$; below we will use the acronym MFDFA to refer to both techniques. The method only works for 1D sections, so consider the transect $v(x)$ of series on a regular grid with resolution = 1 unit, N units long. The MFDFA starts by replacing the original series by the running sum:

$$s(x) = \sum_{x' \le x} v(x') \tag{5.111}$$

As discussed above, a sum is a finite difference of order −1 so that analyzing s rather than v allows the treatment of transects with H down to −1 (we will discuss the upper bound shortly). We now divide the range into $L_{\Delta x} = int(L/\Delta x)$ disjoint intervals, each indexed by $i = 1, 2, \ldots, L_{\Delta x}$ (int means "integer part"). For each interval starting at $x = i\Delta x$, one defines the "nth-order" fluctuation by the standard deviation σ_s of the difference of s with respect to a polynomial $F_n(x,\Delta x) - \sigma_s$ in the running sum of v as follows:

$$F_n(x, \Delta x) = \sigma_s = \left[\frac{1}{\Delta x}\sum_{j=1}^{\Delta x}\left(s\left((i-1)\Delta x + j\right) - p_{n,i}(j)\right)^2\right]^{1/2}; \quad \left(\Delta v(\Delta x)\right)_{MFDFA} = \frac{F_n}{\Delta x} = \frac{\sigma_s}{\Delta x} \tag{5.112}$$

where $p_{n,i}(x)$ is the nth-order polynomial regression of $s(x)$ over the ith interval of length $x = i\Delta x$. It is the polynomial that "detrends" the running sum s; the fluctuation σ_s is the root mean square deviation from the regression. The F is the standard DFA notation for the "fluctuation function"; as indicated, it is in fact a fluctuation in the integrated series quantified by the standard deviation σ_s. Using F we have $F_n = \sigma_s = \Delta v \Delta x$, so in terms of the usually cited MFDFA fluctuation exponent

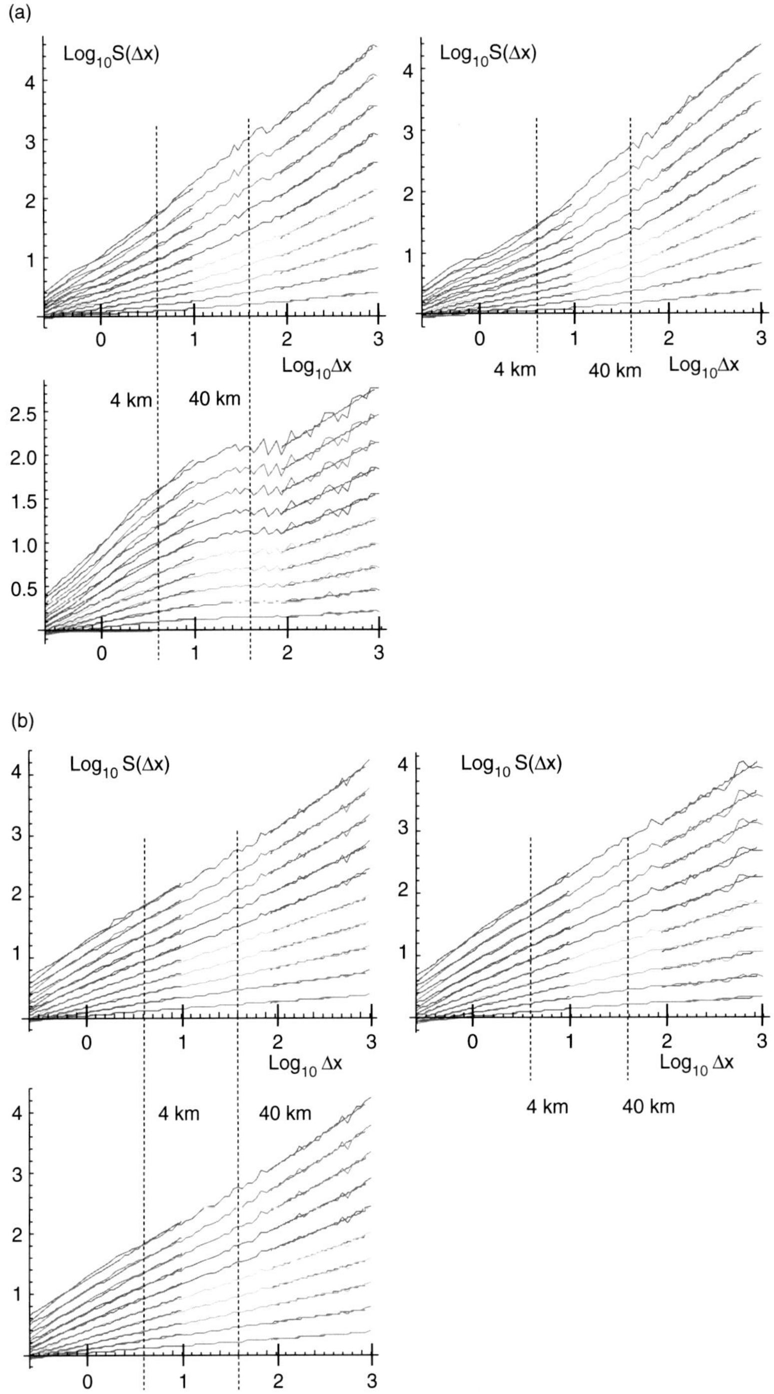

Fig. 5.36 (a) The structure function analyses corresponding to the trace moments of the aircraft data analysed in Fig 4.6 a. Upper left is the longitudinal wind companent; upper right, the transverse component; lower left, the pressure (units of Δx are km). The structure functions of order $q = 0.2, 0.4, \ldots, 1.9, 2.0$ are shown (from bottom to top). All have been nondimensionalized by dividing by the absolute mean first difference at the finest scale (280 m). The dashed lines show the scale range judged to be the least affected by the isobaric trajectory and turbulent aircraft motions (which affect mostly the largest and smallest scales respectively. Reproduced from Lovejoy *et al.* (2010b). Compare these to the vertical structure function analyses in Fig. 6.16a, b.
(b) The same as Fig. 5.36a (corresponding to Fig. 4.6b), but for temperature (upper left), humidity (upper right), log potential temperature (lower left) (fields weakly affected by the trajectory). Reproduced from Lovejoy *et al.* (2010b).

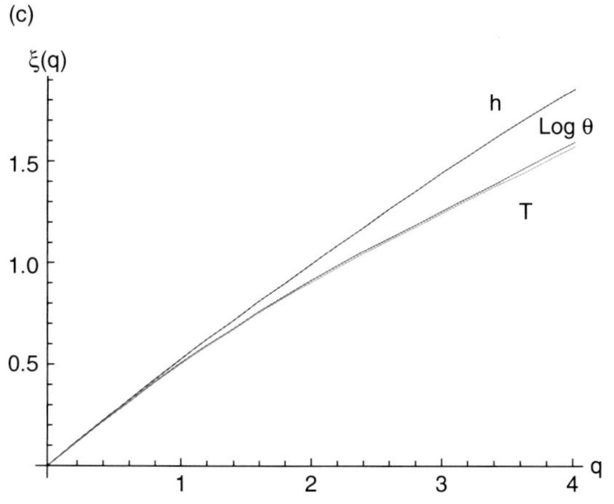

Fig. 5.36 (c) The structure function exponents for T, logθ, h from the aircraft data analysed in Fig. 5.36b. The exponents were estimated by fitting the structure functions over the "optimal" range 4–40 km.

α (not the Lévy α!) we have $F_n \approx \Delta x^{\alpha}$ and $\alpha = 1 + H$, so that the (second-order) spectral exponent $\beta = 2\alpha - 1 - K(2)$ and the oft-cited relation $\beta = 2\alpha - 1$ ignores the intermittency correction $K(2)$. The fluctuation Δv is very close to the nth-order finite difference of v; it is also close to a wavelet-based fluctuation where the wavelet is of nth order, although this – like many MFDFA results – is not completely rigorous (see however Kantelhardt *et al.*, 2001; Taqqu *et al.*, 1995). As a practical matter, in the above sum, we start with intervals of length $\Delta x = n + 1$ and normalize by the number of degrees of freedom of the regression, i.e. by $\Delta x - n$ not by Δx; Eqn (5.112) is the approximation for large Δx.

In the usual presentation of the MFDFA, one considers only a single realization of the process and averages powers of the fluctuations over all the disjoint intervals i. However, more generally, we can average over all the intervals and realizations. If, in addition, we consider moments other than $q = 2$, then we obtain the multifractal DFA (= MFDFA) with the following statistics:

$$\langle|\sigma_s(\Delta x)|^q\rangle^{1/q} = \langle|\Delta x\Big(\Delta v(\Delta x)\Big)_{MFDFA}|^q\rangle^{1/q} = \Delta x^{h(q)} \qquad (5.113)$$

where we have used the exponent $h(q)$ as defined in Davis *et al.* (1996) and Kantelhardt *et al.* (2002). In what follows, we will refer to the method as the MFDFA technique, even though the only difference with respect to the DFA is the consideration of the $q \neq 2$ statistics.

From our analysis and the relation $\langle(\Delta v)^q_{MFDFA}\rangle \approx \Delta x^{\xi(q)}$, we see that the nth-order MFDFA exponent $h(q)$ is related to the usual exponents by:

$$h(q) = 1 + \xi(q)/q = (1+H) - K(q)/q; \quad -1<H<n \qquad (5.114)$$

(the 1 appears because of the initial integration of order 1 in the MFDFA recipe; the method works up to $H = n$ because the fluctuations are defined as the residues with respect to nth-order polynomials of the sum s corresponding to n + 1th-order differences in s and hence n–1th-order differences in v). The basic justification for defining the MFDFA exponent $h(q)$ in this way is that in the absence of intermittency (i.e. if $K(q)$ = constant = $K(0)$ = 0), one obtains $h(q) = 1 + H$ = constant. Unfortunately, contrary to what is often claimed, neither the nonconstancy of $h(q)$ implies that the process is multifractal, nor conversely does its constancy imply a monofractal process. Indeed, it suffices to consider the (possibly fractionally integrated, order H) monofractal β model where $h(q) = 1 + H - C_1 + C_1/q$, which is not only nonconstant but even diverges as q approaches zero. The same divergence occurs for any universal multifractal with $\alpha < 1$!

During the last 10 years, the MFDFA technique has been applied frequently to atmospheric data, and of special note is the work on climate by A. Bunde and coworkers (e.g. Koscielny-Bunde *et al.*, 1998; Kantelhardt *et al.*, 2001; Bunde *et al.*, 2002, 2005; Lennartz and Bunde, 2009; and see Chapter 10). In addition, several numerical studies have compared its performance to various wavelet techniques: for example, for estimating exponents, Oswiecimka *et al.* (2006) find it slightly superior to the modulus maximum method, and Huang *et al.* (2011) superior to the wavelet leader method. However, the exact status of the MFDFA method is not clear, since a completely rigorous mathematical analysis has not been done (especially in the multifractal case), and the literature suffers from unnecessary and misleading claims to the effect that the MFDFA is necessary because by its construction it removes nonstationarities due to trends (linear, quadratic etc. up to polynomials order n - 1) in the data. However, the removal of these trends only accounts for these rather trivial types of nonstationarity: the method continues to make the standard (and strong) stationarity assumptions about the statistics of the deviations which are left over after linear or polynomial detrending. In particular, it does nothing to remove the most common genuine type of statistical nonstationarity in atmospheric science: the diurnal and annual cycles which still strongly break the scaling of the MFDFA statistics. Finally, the corresponding wavelet or finite difference definition of fluctuations can also easily take into account these polynomial trends (see Appendix 5E for the Haar wavelet, structure function and generalizations, and Chapter 10 for many applications); and third, the method removes trends at all scales and locations so that this emphasis on detrending is misleading (recall that that the FIF model is strictly statistically stationary, i.e. it is statistically translationally invariant over the entire range over which it is defined yet it has random trends at all scales and all orders). In other words, the MFDFA is a variant with respect to some of the wavelet methods, in particular with respect to the Haar wavelet and generalizations discussed in Appendix 5E, and has the disadvantage that while it yields accurate exponent estimates, the interpretation of the fluctuations is not straightforward. In comparison, the usual difference and the (new) tendency fluctuations and structure functions have simple physical interpretations in terms of magnitudes of changes (when $1 > H > 0$) and magnitudes of average tendencies (when $-1 < H < 0$).

5.5.3 The Double trace moment technique

Above, we reviewed the results of multifractal analysis techniques which in principle could be applied to arbitrary multifractals. They enjoyed the apparent advantage of making no assumptions about the type of multifractal being analyzed. In practice, however, the techniques are overly ambitious: for a finite (and usually small) number of samples of a process, they attempt to deduce an entire exponent function (an infinite number of parameters), with the result that there is considerable uncertainty in the resulting estimates of $c(\gamma)$ or $K(q)$. With the realization that physically observable multifractals are likely to belong to universality classes, it is natural to develop specific methods to directly estimate the universality parameters (H, C_1, α). These parameters can then be used to determine $c(\gamma)$, $K(q)$ from the expressions for universal multifractals.

The double trace moment (DTM) technique directly exploits universality by generalizing the usual moment method (the trace moments) based on a unique exponent q, which is the order of the moment; it introduces a second exponent η by taking the high-resolution field to the power η. The basic idea is to use the fact that whereas $K(q)$ is not a pure power law in q (due to the extra linear term), the corresponding formula for the normalized η power is a pure power in the exponent η: $K(\eta,q) = K(q\eta) - qK(\eta) = \eta^{\alpha}K(1,q)$ (Eqns. (4.12), (4.13)). The method thus works by taking the fluxes at the finest resolution and then raising them to a series of powers η; these can conveniently be taken as uniformly spaced in their logs. Then η powers of the fields are then degraded in the usual way and at each of a series of scales a fixed qth power is taken ($q = 0.5, 2$ are typical values which are used). We can then estimate $K(\eta,q)$ and repeat for a series of η values, and finally we can estimate α from a plot of $\log\eta$ versus $\log K$. The intercept ($= \log K(1,q)$ with $K(1,q) = C_1(q^{\alpha} - q)/(\alpha - 1)$) can then be used to determine C_1. The whole process can then be repeated with another value of q and the results compared and averaged if needed. Figs. 5.37a, 5.37b and 5.37c show an example of hot-wire anemometer wind data at 2 kHz in the atmosphere yielding the parameters $\alpha \approx 1.50 \pm 0.05$, $C_1 = 0.25 \pm 0.05$ (for the energy flux ε; see Fig. 2.4 for a sample of the wind component data). At the far right of the figure (corresponding to large $q\eta$), the curve flattens; this is because of the multifractal phase transition; for large q, $K(q)$ becomes linear so that $K(\eta,q) = K(q\eta) - qK(\eta)$ becomes constant.

5.6 Summary of emergent laws in Chapter 5

We introduced the codimension function, $c(\gamma)$:

$$\langle \varphi_{\lambda}^{q} \rangle = \lambda^{K(q)} \leftrightarrow \quad \Pr(\varphi_{\lambda} > \lambda^{\gamma}) \approx \lambda^{-c(\gamma)} \qquad (5.115)$$

that characterizes how the probability distributions change with scale ratio λ. It is related to the moment scaling function $K(q)$ introduced in Chapter 3 by a Legendre transformation:

$$\begin{aligned} K(q) &= \max_{\gamma}\left(q\gamma - c(\gamma)\right) \\ c(\gamma) &= \max_{q}\left(q\gamma - K(q)\right) \end{aligned} \qquad (5.116)$$

This implies one-to-one relations between orders of singularities and moments:

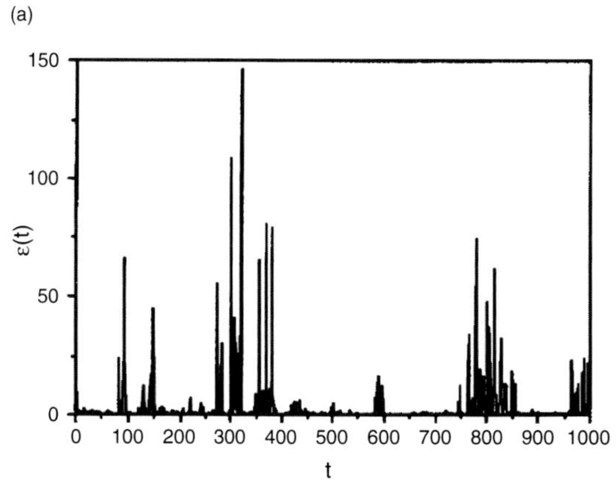

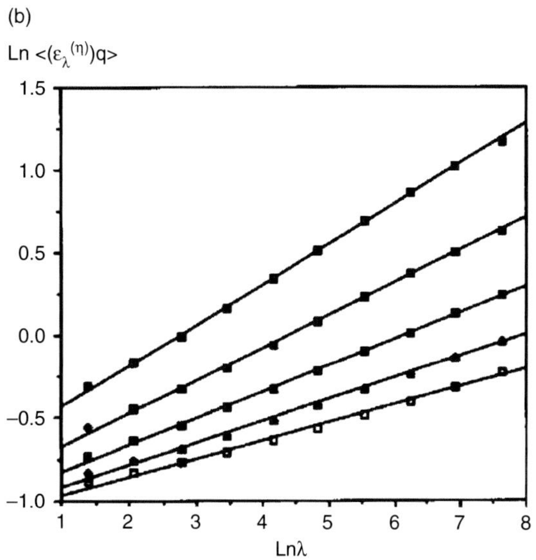

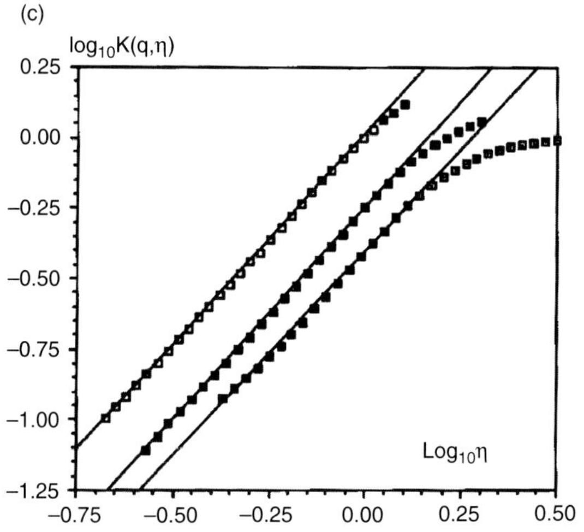

Fig. 5.37 (a) The time series of the (normalized) energy flux estimated from the first 0.5 s of the 2 kHz wind data shown in Fig. 2.4, estimated using the cube of the absolute third power of a fractional integration of order 1/3 of the velocity field (power law filter by $\omega^{1/3}$). This visibly demonstrates the very strong intermittency. Adapted from Schmitt *et al.* (1993). (b) The double trace moments (DTM) for the 2 kHz wind data shown in Figs. 2.4, 5.37a for $q = 2$, $\eta = 0.43, 0.49, 0.56, 0.65, 0.74$ (note that logs are to base e). Reproduced from Schmitt *et al.* (1993). (c) An example of the DTM applied to turbulent wind data at 2 kHz for $q = 2, 2.5, 3$ (bottom to top), yielding slopes $\alpha = 1.50 \pm 0.05$, $C_1 = 0.25 \pm 0.05$. A section of the data is shown in Fig. 2.4. Reproduced from Schmitt *et al.* (1993).

$$
\begin{aligned}
c'(\gamma_q) &= q \\
K'(q_\gamma) &= \gamma
\end{aligned}
\tag{5.117}
$$

For the Lévy generator "universal multifractal" cascade, ($0 \le \alpha < 1$, $1 < \alpha \le 2$, for $\alpha = 1$, take the limit $\alpha \to 1$) this leads to:

$$
\begin{aligned}
K(q) &= \frac{C_1}{\alpha - 1}(q^\alpha - q) \\
c(\gamma) &= C_1\left(\frac{\gamma}{C_1\alpha'} + \frac{1}{\alpha}\right)^{\alpha'}; \quad 1/\alpha' + 1/\alpha = 1
\end{aligned}
\tag{5.118}
$$

The dressed cascade properties (those of a cascade developed over a wide range of scales and then spatially averaged) generally have power-law probability distributions, a kind of self-organized critical (SOC) behaviour:

$$
\langle \varepsilon_\lambda^{(d)q} \rangle = \infty, q \ge q_D \Leftrightarrow \Pr(\varepsilon_{\lambda(d)} > s) \sim s^{-q_D}, s \gg 1
\tag{5.119}
$$

Universal multifractal cascade processes which are continuous in scale ("infinitely" divisible") can be constructed, using the (isotopic) "fractionally integrated flux" (FIF) model:

$$\Gamma = C_1^{1/\alpha} I_{1/\alpha'}(\gamma_\alpha) = C_1^{1/\alpha} \gamma_\alpha * |\underline{r}|^{-(D/\alpha)};$$
$$\varphi = e^{\Gamma} \tag{5.120}$$

$$f = I_H(\varphi) = \varphi * |\underline{r}|^{-(D-H)} \tag{5.121}$$

where γ_α is a unit independent, identically distributed (i.i.d.) Lévy noise with index α and I_H indicates a fractional integration of order H.

The FIF yields f with statistics:

$$S_q(\underline{\Delta r}) = \langle \Delta f^q \rangle = |\underline{\Delta r}|^{\xi(q)}; \xi(q) = qH - K(q);$$
$$K(q) = \frac{C_1}{\alpha - 1}(q^\alpha - q) \tag{5.122}$$

where S_q is the qth order structure function.

Appendix 5A: Divergence of high-order statistical moments

In Section 5.3.2 we gave an outline of the basic argument showing that while the bare moments converge for all moments q, any finite λ, the dressed moments generally diverge for $q \geq q_D$. However, in the interests of simplicity and brevity, we skipped a few nontrivial steps.

First, recall the definition of the λ resolution flux:

$$\Pi_\lambda(A) = \int_A \varepsilon_\lambda d^D \underline{r} \tag{5.123}$$

Since we are interested in the statistics of the dressed and partially dressed density, $\varepsilon_{\lambda,\Lambda(d)} = \Pi_\Lambda(B_\lambda)\ /\mathrm{vol}(B_\lambda)$, we will consider the mean of the qth power of the flux on the set A (dimension D) of the cascade constructed down to the scale L/λ:

$$\langle \Pi_\lambda^q(A) \rangle = \left\langle \left[\int_A d^D \underline{r} \varepsilon_\lambda \right]^q \right\rangle \tag{5.124}$$

when q is an integer ≥ 1:

$$\left\langle \left[\int_A d^D \underline{r} \varepsilon_\lambda \right]^q \right\rangle = \left\langle \int_A \cdots \int_A d^D \underline{r}_1 \cdots d^D \underline{r}_q \varepsilon_\lambda(\underline{r}_1) \cdots \varepsilon_\lambda(\underline{r}_q) \right\rangle \tag{5.125}$$

The complexity of this multiple integral suggests the introduction of "trace moments" which are obtained by integrating over the subset of the integral obtained by taking $\underline{r}_1 = \underline{r}_2 = \underline{r}_3 = \ldots$;:

$$\begin{aligned} Tr_A(\varepsilon_\lambda)^q &= \int_A d^{qD} \underline{r} \langle \varepsilon_\lambda{}^q \rangle \\ &\sim \sum_{A_\lambda} \langle \varepsilon_\lambda{}^q \rangle \lambda^{-qD} = \sum_{A_\lambda} \lambda^{K(q)} \lambda^{-qD} \end{aligned} \tag{5.126}$$

where A_λ is the set A at resolution λ (i.e. obtained by a disjoint covering of A with balls B_λ), ε_λ is the usual (bare) flux density at resolution λ. Since $\varepsilon_\lambda \geq 0$ and we integrate over a subspace (using the fact that, for any x_i, $\left(\sum X_i^q\right)^{1/q}$ is a decreasing function of q), we have that the trace moments are bounds on the usual moments corresponding to a Jensen inequality:

$$\begin{aligned} \langle \Pi_\lambda{}^q(A) \rangle &\geq Tr_A \varepsilon_\lambda{}^q \quad (q > 1) \\ &\leq Tr_A \varepsilon_\lambda{}^q \quad (q < 1) \end{aligned} \tag{5.127}$$

The use of trace moments rather than the usual moments has a number of advantages. First, it is defined for all q (whereas the usual moments can only be expanded as multiple integrals for positive integer q). Second, trace moments are Hausdorff measures since we can use the scaling of $\langle \varepsilon_\lambda{}^q \rangle$ to obtain a Hausdorff measure over a higher dimensional space (for convenience we have left out the inf, etc.). We anticipate that in the limit $\lambda \to \infty$ they will either diverge to ∞ or converge to 0; in fact, they will have two transitions!

To see this, use box counting in the sum ΣA_λ; there will be λ^D terms, each of value $\langle \varepsilon_\lambda{}^q \rangle \lambda^{-qD}$:

$$\begin{aligned} Tr_{A_\lambda} \varepsilon_\lambda{}^q &= \lambda^D . \lambda^{K(q)} . \lambda^{-qD} = \lambda^{K(q)-(q-1)D} \\ &= \lambda^{(q-1)(C(q)-D)} \end{aligned} \tag{5.128}$$

where we have used the dual codimension function $C(q)$ (Section 3.2.7). We recall that $C(q)$ is monotonic in q (Fig. 3.11). Consider first the case where $C(q) > D$ for $q < 1$. Due to the monotonicity of $C(q)$ this is equivalent to $C_1 > D$. In this case:

$$\lim_{\lambda \to \infty} Tr_{A_\lambda} \varepsilon_\lambda{}^q \to 0 \Rightarrow \langle \Pi_\infty{}^q(A) \rangle = 0 \tag{5.129}$$

for all $q < 1$; hence the process is *degenerate* on the space. This implies that when $C_1 > D$, then the mean of the bare process is too sparse to be observed in the space D; in fact, the above shows that if $C_1 > D$ it is impossible to normalize the process so that the dressed mean $\langle \Pi_\infty(A) \rangle$ is finite.

Now consider the nondegenerate case $C_1 < D$. In this case, the trace moments diverge for $q < 1$, but this does not affect the convergence of the dressed moments (the trace moments are upper bounds here). On the other hand, for $q > 1$, we find:

$$\lim_{\lambda \to \infty} Tr_{A_\lambda} \varepsilon_\lambda{}^q \to \infty \Rightarrow \left\langle \Pi_\infty(A) \right\rangle \to \infty \qquad (5.130)$$

for all $C(q) > D$. Using the implicit definition of q_D: $C(q_D) = D$, we thus obtain:

$$\left\langle \Pi_\infty(A) \right\rangle \to \infty; \quad q > q_D \qquad (5.131)$$

i.e. in this case, divergence of the trace moments implies divergence of the corresponding dressed moments. This is the key result needed to complete the demonstration in Section 5.3.2.

Appendix 5B: Continuous-in-scale cascades: the autocorrelation and finite size effects

5B.1 The internal cascade structure

Various numerical details and examples of simulations of continuous in-scale isotropic (self-similar) multifractals were given in Schertzer and Lovejoy (1987) and Wilson *et al.* (1991), and earlier in this chapter. Marsan *et al.* (1996) showed how to extend the framework to causal (space-time) processes, and Pecknold *et al.* (1997a, 1997b) discuss extensions to anisotropic multifractal processes needed, for example, to take into account atmospheric stratification. As we detailed in Section 5.4, the basic method for simulating Γ is to fractionally integrate a Lévy noise, i.e. to convolve it with a singularity. While this method works for large enough scale ranges (i.e. it yields simulations with statistics satisfying $\langle \varepsilon_\lambda^q \rangle = \lambda^{k(q)}$ with $K(q)$ of the universal form (Eqn. (5.33)) for large enough λ), there are significant deviations at small scales ("finite size effects"), especially for $\alpha > 1$, which is the most empirically relevant range (there are also "large-scale effects," which are typically less important). In order to get an idea of the importance of these deviations, consider the theoretical form of the normalized autocorrelation function $R_\lambda(\underline{\Delta r})$ for isotropic cascades as derived on discrete-in-scale cascades in Eqn. (3.37):

$$R_\lambda(\underline{\Delta r}) = \frac{\langle \varepsilon_\lambda(\underline{r})\varepsilon_\lambda(\underline{r} - \underline{\Delta r})\rangle}{\langle \varepsilon_\lambda^2 \rangle} = |\underline{\Delta r}|^{-K(2)}; |\underline{\Delta r}| \geq 1 \tag{5.132}$$

$\underline{r}$, $\underline{\Delta r}$ are D dimensional position vectors and lags respectively. The above uses the convention that the process is developed over the range of scales from λ down to 1 unit so that $R_\lambda(1) = 1$ (Eqn. (3.37)); in this appendix, since we are concerned with the numerical simulation of cascades over a fixed range of scales Λ, we take 1 unit = 1 pixel so that in 1D the Λ resolution cascade is Λ pixels long and the intermediate ratio correction satisfies $1 \leq \lambda \leq \Lambda$ ($K(2)$ appears because the autocorrelation is a $q = 2$ order statistic). Since the spectrum is the Fourier transform of the autocorrelation function we find that for wavenumber $k = |\underline{k}|$, the power spectrum $E(k)$ of ε_λ has the scaling form given by Eqn. (3.38). The discussion here is a summary of key sections of Lovejoy and Schertzer (2010a, 2010b).

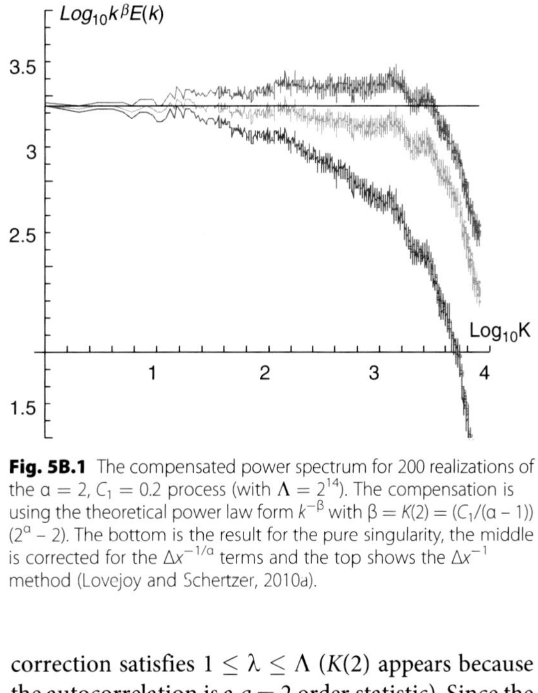

Fig. 5B.1 The compensated power spectrum for 200 realizations of the $\alpha = 2$, $C_1 = 0.2$ process (with $\Lambda = 2^{14}$). The compensation is using the theoretical power law form $k^{-\beta}$ with $\beta = K(2) = (C_1/(\alpha - 1))(2^\alpha - 2)$. The bottom is the result for the pure singularity, the middle is corrected for the $\Delta x^{-1/\alpha}$ terms and the top shows the Δx^{-1} method (Lovejoy and Schertzer, 2010a).

Fig. 5B.1 shows the power spectrum of the process for $\alpha = 2$ compensated by dividing by the theoretical $k^{-\beta}$ so that the theoretically expected spectral scaling leads to horizontal straight lines on the log-log plots. From the figure, we can see that the pure (power-law) fractional integration takes nearly a factor of 10^3 in scale to completely converge (the largest wavenumber for a 2^{14}-point-long series is $k = 2^{13} \approx 10^4$); see Lovejoy and Schertzer (2010b) for similar results for smaller α values. Although the rate of convergence improves as α decreases from 2, deviations are noticeable for all α. Also shown in the figure are the corresponding spectra of the processes obtained using the improved simulation techniques described in Lovejoy and Schertzer (2010a, 2010b) and outlined below. It can be seen that although the spectral results are still not perfect, they are significantly better. Although the biases in the individual realizations are not very visible to the eye, the statistics as revealed by various scale-by-scale analyses are still quite biased at small scales.

A simple way to examine the scaling properties of realizations of cascades developed over a finite scale ratio Λ is to consider two point statistics such as autocorrelation functions, or their Fourier transforms, spectra. To calculate these, we first consider the qth power of the product of the random variables:

$$\left(\varepsilon_\lambda(\underline{r}')\varepsilon_\lambda(\underline{r}'-\Delta\underline{r})\right)^q = e^{q(\Gamma(\underline{r}')+\Gamma(\underline{r}'-\Delta\underline{r}))}$$
$$= \exp\left[qC_1^{1/\alpha}\int_{1\le|\underline{r}|\le\lambda}\left(g(\underline{r}'-\underline{r}'')+g(\underline{r}'-\Delta\underline{r}-\underline{r}'')\right)\gamma_\alpha(\underline{r}'')d^D\underline{r}''\right] \quad (5.133)$$

(cf. Eqns. (5.78), (5.79)); we see that $\Gamma(\underline{r}')+\Gamma(\underline{r}'-\Delta\underline{r})$ is the generator of the autocorrelation and γ_α is the subgenerator. For the statistics we can define the second characteristic function (SCF) of the generator by taking ensemble averages of the above:

$$\log\left\langle\left(\varepsilon_\lambda(\underline{r}')\varepsilon_\lambda(\underline{r}'-\Delta\underline{r})\right)^q\right\rangle = \log\left\langle e^{q(\Gamma(\underline{r}')+\Gamma(\underline{r}'-\Delta\underline{r}))}\right\rangle = \frac{C_1}{\alpha-1}q^\alpha S(\Delta\underline{r}) \quad (5.134)$$

the entire expression is the full SCF of the log of the autocorrelation; the key function $S(\Delta\underline{r})$ defined by Eqn. (5.134) is its spatial part. Note that below we do not need the more complex full two-point SCF $\log\langle\varepsilon_\lambda^{q_1}(\underline{r}_1)\varepsilon_\lambda^{q_2}(\underline{r}_2)\rangle$. Using the statistical translational invariance of the process (by construction the noise subgenerator $\gamma_\alpha(\underline{r})$ is statistically independent of $\underline{r}$), we can take $\underline{r}'=0$ to obtain:

$$S(\Delta\underline{r}) = \int_{1\le|\underline{r}''|\le\lambda}\left(g_\lambda(-\underline{r}'')+g_\lambda(-\underline{r}''-\Delta\underline{r})\right)^\alpha d\underline{r}''$$
$$= \int_{1\le|\underline{r}|\le\lambda}\left(g_\lambda(\underline{r})+g_\lambda(\underline{r}-\Delta\underline{r})\right)^\alpha d\underline{r} \quad (5.135)$$

where in the far right we have taken $\underline{r}=-\underline{r}''$ and used the fact that the domain of integration (but not necessarily g) is invariant under inversion. We have considered isotropic multifractals in a d-dimensional space. For anisotropic multifractals we must replace the vector norm in the above by the corresponding scale function (see Pecknold *et al.*, 1993, 1997a, for anisotropic simulations; the extension of the present technical discussion to anisotropic cascades will be made elsewhere). We note that from its definition (and assumed statistical translational invariance, i.e. independence from $\underline{r}'$ in Eqn. (5.135)), the autocorrelation function (and hence S) is symmetric under inversion: $S(\Delta\underline{r}) = S(-\Delta\underline{r})$. Isotropic processes are symmetric under rotation, in which case S is simply a function of the vector norm $|\Delta\underline{r}|$. In this case, comparing Eqns. (5.132) and (5.134), we see that the theoretical $S(\Delta\underline{r})$ is:

$$S_{theory}(\Delta\underline{r}) = N_d\frac{\alpha-1}{C_1}\log\langle\varepsilon(\underline{r})\varepsilon(\underline{r}-\Delta\underline{r})\rangle$$
$$= -N_d(2^\alpha-2)\log|\Delta\underline{r}| + N_d 2^\alpha\log\lambda \quad (5.136)$$

where we have used $\log\langle\varepsilon_\lambda^2\rangle = \frac{C_1}{(\alpha-1)}2^\alpha\log\lambda$; the corresponding $\log\lambda$ term is absent in the normalized autocorrelation function. We now consider how closely $S_{theory}(\Delta\underline{r})$ (Eqn. (5.136)) is approximated by $S(\Delta\underline{r})$ (Eqn. (5.135)).

5B.2 The general $D = 1$ case

For simplicity, we start with the problem in one dimension. Due to the inversion symmetry of $S(\Delta\underline{r})$ noted above, in 1D it is sufficient to consider $\Delta x > 0$ (x is a coordinate in 1D). The general $d = 1$ case is:

$$S(\Delta x) = \int_{1<|x|<\lambda}\left(g(x)+g(x-\Delta x)\right)^\alpha dx$$
$$= \int_1^\lambda\left[\left(g(x)+g(x-\Delta x)\right)^\alpha + \left(g(-x)+g(-x-\Delta x)\right)^\alpha\right]dx \quad (5.137)$$

The two basic cases of interest are the symmetric acausal case with $g(x) = g(-x)$, and the causal case with $g(x) = 0$ for $x < 0$ (see Chapter 9 and Marsan *et al.*, 1996). The corresponding condition for a space-time process in $d + 1$ dimensions to be causal is $g(\underline{r},t) = 0$, $t < 0$ where $\underline{r}$ is a d-dimensional (spatial) vector (this is equivalent to multiplying an acausal $g(\underline{r},t)$ by a Heaviside function $\Theta(t)$ such that $\Theta(t) = 0$ for $t< 0$, $\Theta(t) = 1$ for $t \geq 0$ (see Lovejoy and Schertzer, 2010b, for numerical implementations).

For the symmetric acausal case we have:

$$S(\Delta x) = \int_1^{\lambda} \left[\left(g(x) + g(x - \Delta x) \right)^{\alpha} + \left(g(x) + g(x + \Delta x) \right)^{\alpha} \right] dx \qquad (5.138)$$

We have already seen that for this case, with g a truncated power law given in Eqn. (5.82), the normalization factor $N_d = 2$.

For the causal case, since $g(x) = 0$ for $x < 0$, we have:

$$S(\Delta x) = \int_1^{\lambda} \left(g(x) + g(x - \Delta x) \right)^{\alpha} dx \qquad (5.139)$$

Here it is easy to see that the corresponding normalization factor is simply $N_d = 1$.

Using these definitions and expansions, Lovejoy and Schertzer (2010a) show:

$$S(\Delta x) = C(\alpha, d) - N_d \left[(2^{\alpha} - 2) \log \Delta x - 2^{\alpha} \log \lambda + 2 \frac{\alpha^2}{d} \Delta x^{-d/\alpha} \right] - O(\Delta x^{-2}) + O(\Delta x^{-2m - nd/\alpha}) + O\left(\frac{\Delta x}{\lambda} \right)^2 \qquad (5.140)$$

where C is constant with respect to Δx, λ and is unimportant; n and m are positive integers and $d = 1$, $N_1 = 2$. The leading order expression for S is thus $= -N_d(2^{\alpha} - 2)(\log\Delta x) + N_d 2^{\alpha} \log\lambda$; i.e. the same as S_{theory} (Eqn. (5.136)) with the leading Δx-dependent correction $-2N_d \frac{\alpha^2}{d} \Delta x^{-d/\alpha}$.

Using the above expansion for $S(\Delta x)$ we can obtain the following expression for the autocorrelation function:

$$\langle \varepsilon(x) \varepsilon(x - \Delta x) \rangle \propto \Delta x^{-\underline{r}(2)} \exp\left(-\frac{2C_1}{d} \frac{\alpha^2}{\alpha - 1} \Delta x^{-d/\alpha} \right) \qquad (5.141)$$

where we have only kept the $(\Delta x^{-d/\alpha})$ correction to the leading power-law term. To gauge the importance of this correction, note that for Δx as large as 100 it can still be a 10% effect.

Having shown the origin of the problem, Lovejoy and Schertzer (2010b) show how to remove the leading order correction; this leads to the improved statistics shown in Fig. 5B.1. Also in the next subsection, we discuss some technical details about the simulations of causal cascades and cascades in higher dimensional spaces. See also Chapter 9 for some space-time examples and discussion of causal versus acausal processes. In Appendix 5C we give a Mathematica code to generate the corrected causal and acausal simulations in 1D and 2D.

5B.3 Some practical (numerical) considerations

Before making numerical simulations, there are a few practical points we should mention. Numerically, it very advantageous to use transform-based convolution routines, and these are periodic. In all our simulations, we therefore used periodic convolutions. Since the noise $\gamma(x)$ is statistically invariant under translations along the x-axis, this will not break the translational invariance, but it does mean that the left half of the simulation is statistically *dependent* on the right half in a way which is artificially strong (due to the constraint that the resulting $\varepsilon(x)$ is periodic). If absolutely needed, zero padding could be used to avoid the periodicity, or – almost equivalently – only one half of the simulation could be used. One consequence is that the periodic $\varepsilon_{\lambda,\lambda(h)}$ is somewhat modified with respect to its non-periodic value. Note that although the Fourier filter corresponding to the convolution with $|\underline{r}|^{-d/\alpha}$ is of the type $|\underline{k}|^{-d/\alpha'}$ (where α' is the auxiliary variable, $1/\alpha' + 1/\alpha = 1$), there is trigonometric prefactor, whose sign is quite important for preserving the extremality of the subgenerator in the physical space. Therefore, this prefractor function cannot be forgotten if one wants to proceed directly from Fourier space. It is therefore simpler to proceed from physical space, i.e. to directly compute convolutions in physical space. While this issue is indeed important for the convolution kernel g of the cascade generator $\Gamma = g * \gamma$, it is not necessarily important for the kernel of the observable $I = |\underline{r}|^{-(d-H)} * e^{\Gamma}$ unless the latter is also a positive definite quantity such as a passive scalar concentration (the FIF: Section 5.4.4). On the contrary, it may be desirable that the observable be symmetric about zero; in this case one may take an antisymmetric convolution kernel for the fractional integration as $|x|^{-(1-H)} sgn(x)$ (in $d = 1$; generalizations to higher d are obvious).

5B.4 Extensions to causal and higher dimensional cascades

In this subsection we proceed to evaluate and compare the statistical accuracy of the causal and acausal simulations, as well as the effect of passing from 1D to 2D. We use the $\Delta x^{-d/\alpha}$ correction method for the small Δx corrections because it is simple to implement, works reasonably well and is mathematically well grounded. Also, to evaluate the simulations, we only consider the (compensated) spectra, i.e. the $q = 2$ statistics rather than statistics of all orders. This is sufficient since we find that: (a) the statistics are more accurate for the causal case than for the acausal case, (b) as expected when going from $d = 1$ to $d = 2$ the convergence is improved, (c) we only consider the case $\alpha = 2$ since the corresponding corrections are the largest.

Consider first the causal 1D simulations. Fig. 5B.2a shows that even without any corrections the causal simulations have significantly smaller deviations from pure power-law behaviour than the uncorrected acausal simulations; indeed they are even comparable to the corrected acausal spectra. With the $\Delta x^{-d/\alpha}$ corrections we see that the causal spectrum is nearly perfect. In comparison, the corrected acausal curve (reproduced from Fig. 5B.1) has significantly larger deviations.

Turning our attention to simulations in 2D, we recall the examples of isotropic acausal realizations of the process in Section 5.4. As discussed there, we expect the corrections to be smaller for $d = 2$ when compared to $d = 1$; Fig. 5B.2b confirms this, and indeed we see that the uncorrected 2D (isotropic, i.e. angle-integrated) spectrum is slightly better than the corrected 1D spectrum, but in all cases the correction method makes the spectra significantly closer to the theoretical pure power-law form.

Since space and time are not symmetric, we made (x,t) causal simulations and evaluated compensated 1D spectra in the spatial and temporal directions separately (see Fig. 5B.3 and Fig. 9.3 for a corresonding realization). We see that the temporal spectra are very close to those of the 1D causal processes (Fig. 5B.2a) – for both the corrected and uncorrected cases. Similarly, the spectra of the corrected and uncorrected cases in the spatial direction have much larger deviations; very close to the acausal 1D results (Fig. 5B.2a), and compare with the 2D acausal spectra (Fig. 5B.2b). We conclude that the correction method

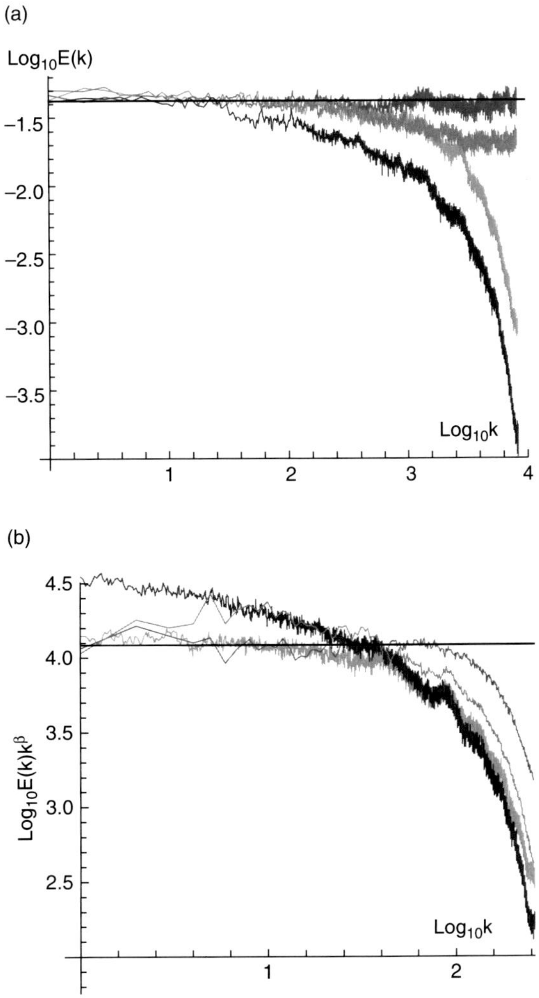

Fig. 5B.2 (a) Compensated spectra for the $d = 1$ causal and acausal cases for $\alpha = 2$, $C_1 = 0.2$ averaged over 200 realizations each 2^{14} long (cf. the comparable Fig. 5B.1) both with and without $\Delta x^{-1/\alpha}$ corrections discussed in the text; using the code in Appendix 5C. From top to bottom: corrected causal spectrum, uncorrected causal, corrected acausal and uncorrected acausal simulations. Reproduced from Lovejoy and Schertzer (2010a). (b) 1D/2D (acausal) comparison for $\alpha = 2$, $C_1 = 0.2$ showing compensated spectra. Numbering the curves 1–4 bottom to top on the far right, nos. 2 and 1 are 1D acausal, compensated spectra (respectively corrected and uncorrected, 200 realizations 2^{14} points). Nos. 4 and 3 are respectively corrected and uncorrected 2D cases (each 10 realizations, $2^9 \times 2^9$ points). They have been shifted horizontally so that for one and two dimensions the highest wavenumbers are the same, and they have been shifted in the vertical so that the corresponding low-wavenumber parts of the spectra roughly overlap (i.e. the corrected with the corrected, and the uncorrected with the uncorrected). The vertical scale is arbitrary and the thick horizontal line is the theoretical pure power law spectrum. Reproduced from Lovejoy and Schertzer (2010a).

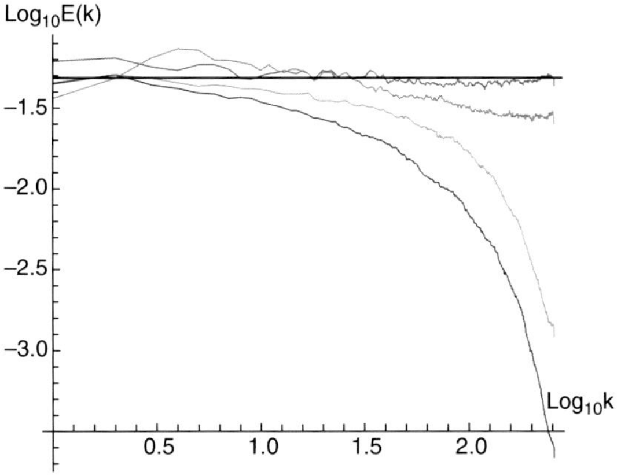

Fig. 5B.3 Compensated spectra for the $d = 2$ causal simulations for $\alpha = 2$, $C_1 = 0.2$ averaged over 10 realizations each $2^9 \times 2^9$ both with and without the $\Delta x^{-d/\alpha}$ corrections; using the code in Appendix 5C. In top to bottom order the curves are: the 1D spectrum of $\Delta x^{-d/\alpha}$ corrected simulations in the t direction; the corresponding spectra of uncorrected simulations; the 1D spectrum of $\Delta x^{-d/\alpha}$ corrected simulations in the x direction; the corresponding 1D spectrum of uncorrected simulations. Reproduced from Lovejoy and Schertzer (2010a).

works well independently of the dimension of the space, and that for the causal extensions the temporal statistics have significantly smaller deviations. This is presumably because of the sharp discontinuity introduced by the Heaviside function, which roughens the simulations along the time axis, thereby somewhat compensating for the otherwise overly smooth behaviour.

Appendix 5C: A Mathematica code for causal and acausal multifractal simulations

The Mathematica code reproduced in Fig. 5C.1 produces conservative ($H = 0$) multifractals for both acausal and causal (Chapter 9) fields in one and two dimensions. It implements the $\Delta x^{-d/\alpha}$ finite size effect correction scheme described in both 1D and 2D for causal and acausal processes (Lovejoy and Schertzer, 2010b). *Lévy* generates random extremal Lévy variables. Attention is paid to avoiding underflows when exponentiating. The basic functions *epsa*, *epsc* generate the corrected multifractal acausal, causal simulations; λ is the length of the resulting vector; α, C_1 are the basic parameters. Note that the built-in function *Convolve* performs a periodic convolution and no zero padding was used (see the text for a discussion).

```
nuni = UniformDistribution[{-Pi/2, Pi/2}]

UniformDistribution[{-π/2, π/2}]
```

this gives a uniform between-Pi/2 and+Pi/2

```
ndist[lam_] := ExponentialDistribution[lam]
RanE[lam_] := Random[ndist[lam]]
Levy[α_] := Module[{ϕ, ϕ0}, {ϕ = Random[nuni], ϕ0 = -(Pi/2) (1 - Abs[1 - α]) / α};
  Sign[(α - 1)] Sin[α (ϕ - ϕ0)] (Cos[ϕ] Abs[α - 1])^-(1/α)
   (Cos[ϕ - α (ϕ - ϕ0)] / RanE[1])^((1 - α) / α)]
```

This gives unit extremal Levy variables index α

```
Fracs[scal_, α_] := Module[{A, a, λ}, {λ = Min[Dimensions[scal]], rat = 2., αp = 1. / (1 - 1 / α),
    lcut = λ / 2., lcut2 = lcut / rat,
    exλ = Chop[E^(-(scal / lcut)^4 /. x_ /; x < -200. → -200.)],
    exλ2 = Chop[E^(-(scal / lcut2)^4 /. x_ /; x < -200. → -200.)],
    sing = (scal)^-(1 / αp), singsmooth = Chop[ sing exλ],
    t1 = Total[Flatten[singsmooth]],
    singsmooth = Chop[ sing exλ2 ],
    t2 = Total[Flatten[singsmooth]],
    A = (rat^-(1. / α) t1 - t2) / (rat^-(1. / α) - 1),
    ff = Chop[E^(-(scal / 3.) /. x_ /; x < -200. → -200.)],
    singsmooth = Chop[sing ff],
    G = Total[Flatten[singsmooth]],
    a = -A / G, sing = Chop[(sing (1 + a ff))]^(1 / (α - 1.))}; sing]
```

this calculates the corrected singularity

```
eps1D[λ_, α_, C1_, switch_] :=
 Module[{ff, xx}, {NDf = If[switch == 0, 1, 0.5],
   Heavi = If[switch == 0, 1., Table[HeavisideTheta[xx], {xx, -(λ - 1), λ - 1, 2}]],
```

Fig. 5C.1 Mathematica code for generating causal and acausal multifractals in one and two dimensions, as described in the text. Reproduced from Lovejoy and Schertzer (2010a).

```
scal = Table[ Abs[xx], {xx, -(λ-1), λ-1, 2}], ggen1 = (C1/NDf) ^(1/α) Table[Levy[α], {λ}],
Fracs[scal, α],
ggen1α = ListConvolve[ggen1, Heavi sing, 1], ggen1α = ggen1α /. x_ /; x < -200. → -200.,
ggen1α = E^ggen1α, ff = Mean[ggen1α], eps1α = ggen1α / ff}; eps1α]
```

1D causal (switch not zero) and acausal (switch = 0)

```
eps2D[λt_, λy_, α_, C1_, switch_] :=
 Module[{ff, tt}, {NDf = Pi/2 If[switch == 0, 1, 0.5],
   Heavi = If[switch == 0, 1., Table[ HeavisideTheta[tt], {tt, -(λt-1), λt-1, 2},
      {yy, -(λy-1), λy-1, 2}]],
   scal = Table[tt^2 + yy^2, {tt, -(λt-1), λt-1, 2}, {yy, -(λy-1), λy-1, 2}],
   ggen1 = (C1/NDf) ^(1/α) Table[Levy[α], {λt}, {λy}], Fracs[scal, α],
   ggen1α = ListConvolve[ggen1, Heavi sing, 1], ggen1α = ggen1α /. x_ /; x < -200. → -200.,
   ggen1α = E^ggen1α, ff = Mean[Flatten[ggen1α]], eps1α = ggen1α / ff}; eps1α]
```

2D causal (switch not zero) and acausal (switch = 0)

Fig. 5C.1 (cont.)

Appendix 5D: Multifractal simulations on a sphere

In this appendix, we outline how to perform isotropic fractional integrals on a sphere using spherical harmonic expansions (Fig. 5.32c; extensions to anisotropic multifractals on a sphere are not as straightforward as in flat space, and we do not consider them here). We require the following addition formula for spherical harmonics:

$$\sum_m Y^*_{lm}(\mu,\phi)Y_{lm}(\mu',\phi') = \frac{2l+1}{4\pi}P_l(\Delta\mu) = \sqrt{\frac{2l+1}{4\pi}}Y_{l0}(\Delta\mu,\phi) \quad (5.142)$$

Where Y_{lm} is the spherical harmonic with principal order l (conjugate to the azimuthal angle θ in spherical polar coordinates) and m is conjugate to the spherical polar angle ϕ, P_l is the l order Legendre polynomial. Note that the right-hand expression is the $m = 0$ harmonic; there is no ϕ dependence, it is indicated purely for notational completeness. The $\Delta\mu = \cos\theta$ is not $\mu - \mu'$, but is the cosine of the angle (θ) between the directions defined by μ, μ' and ϕ, ϕ':

$$\Delta\mu = \mu\mu' + \sqrt{1-\mu^2}\sqrt{1-\mu'^2}\cos(\phi - \phi') \quad (5.143)$$

We are interested in the following convolution:

$$I^H U(\mu,\phi) = \int_0^{2\pi}\int_{-1}^{1} \theta^{-(2-H)}U(\mu',\phi')d\mu' d\phi' \quad (5.144)$$

which is a fractional integration of order H in a space dimension 2 (we have left out normalization factors). In the above, U is the function which we wish to fractionally integrate (i.e. our noise).

We now make the following expansion:

$$\theta^{-(2-H)} = \sum_l \sigma_l Y_{l,0}(\Delta\mu,\phi) \quad (5.145)$$

σ_l is the $m = 0$ component of the full spherical harmonic expansion:

$$\sigma_l\delta_{m0} = \int_0^{2\pi}\int_{-1}^{1} \theta^{-(2-H)}Y_{lm}(\Delta\mu,\phi)d\Delta\mu d\phi \quad (5.146)$$

Combining the expansion of $\theta^{-(2-H)}$ with the addition formula, we obtain the full expansion:

$$\theta^{-(2-H)} = \sum_{l,m}\sigma_l\sqrt{\frac{4\pi}{2l+1}}Y^*_{l,m}(\mu,\phi)Y_{l,m}(\mu',\phi') \quad (5.147)$$

Combining this with the expansion for the noise U:

$$U(\mu,\phi) = \sum_{l,m} u_{l,m}Y_{l,m}(\mu,\phi) \quad (5.148)$$

We can now obtain an expression for the fractional integral:

$$I^H U(\mu,\phi) = \sum_{l,m,l',m'} u_{lm}\sigma_{l'}\sqrt{\frac{4\pi}{2l'+1}}\int_0^{2\pi}\int_{-1}^{1} Y_{lm}(\mu',\phi') Y_{l'm'}(\mu,\phi)Y^*_{l'm'}(\mu',\phi')d\mu' d\phi' \quad (5.149)$$

or:

$$I^H U(\mu,\phi) = \sum_{l,m,l',m'} u_{lm}\sigma_{l'}\sqrt{\frac{4\pi}{2l'+1}}\delta_{ll'}\delta_{mm'}Y_{l'm'}(\mu,\phi) \quad (5.150)$$

hence:

$$I^H U(\mu,\phi) = \sum_{l,m}\sigma_l\sqrt{\frac{4\pi}{2l+1}}u_{lm}Y_{lm}(\mu,\phi) \quad (5.151)$$

We thus see that the required filter is:

$$\sigma_l\sqrt{\frac{4\pi}{2l+1}} \quad (5.152)$$

This is the filter required for the various fractional integrations needed for the FIF model on the sphere. See Fig. 5.32c for an example of a topography simulation.

Appendix 5E: Tendency, poor man's and Haar structure functions and the MFDFA technique

5E.1 A comparison using multifractal simulations

In Section 5.5 we saw that we could characterize the statistics of fluctuations by their moments using structure functions. If the fluctuations at separation Δx are defined by differences (the "poor man's" wavelet), then it is only when $0 < H < 1$ that they are dominated by structures of size Δx (by wavenumbers $\approx \Delta x^{-1}$). When H is outside this range the fluctuations reflect the statistics of much larger ($H > 1$) or much smaller ($H < 0$) structures present in the sample, they are independent of Δx, they "saturate." We have seen that in the weather regime most geophysical H parameters are indeed in the range $0 < H < 1$, justifying the common use of the classical difference structure function. Even more commonly, we find $H > 0$ implying that fluctuations Δf tend to increase with scale $\Delta f \approx \Delta x^H$. The main exception is in the low-frequency macroweather regime (at time scales $> \sim 10$ days), where we generally have $H < 0$ so that fluctuations tend to *decrease* with scale. We noted that the range of H over which fluctuations are usefully defined could be changed by integration and/or differentiation, corresponding to changing the shape of the defining wavelet, changing its real and Fourier space localizations. In the usual wavelet framework, this is done by modifying the wavelet directly, e.g. by choosing the Mexican hat or higher-order derivatives of the Gaussian, or by choosing them to satisfying some special criterion. Following this, the actual convolutions are calculated using with fast Fourier (or equivalent) numerical convolution techniques.

A problem with this usual implementation of wavelets is that not only do these convolutions make the determination of the fluctuations numerically cumbersome, but at the end the physical interpretation of the fluctuations is no longer clear, and in any case we do not need any specific wavelet properties. In contrast, when $0 < H < 1$, the difference structure function gives direct information on the typical difference ($q = 1$) and typical variations around this difference ($q = 2$), and even typical skewness ($q = 3$) or typical kurtosis ($q = 4$) or, if the probability tail is algebraic, of the divergence of high-order moments. Similarly, when $-1 < H < 0$ the tendency structure function directly quantifies the fluctuations's deviation from zero and the exponent, the rate at which the deviation decreases by averaging to larger and larger scales. These poor man's and tendency fluctuations are also very easy to directly estimate from series with uniformly spaced data and – with straightforward modifications – irregularly spaced data (see e.g. Section 6.4.2).

The real drawback of the difference (poor man's) and tendency structure functions is the limited range of H over which they are useful. In this appendix, we show how to extend this range without losing the double advantages of simplicity of implementation and simplicity of interpretation.

Before considering the Haar wavelet (Section 5.5), let us recall the definitions of the difference and tendency fluctuations for series with fixed resolutions. The difference/ poor man's fluctuation is:

$$\left(\Delta v(\Delta x)\right)_{diff} = \delta_{\Delta x} v; \quad \delta_{\Delta x} v = v(x + \Delta x) - v(x) \tag{5.153}$$

where δ is the difference operator. Similarly, in Chapter 10 we use the "tendency fluctuation." The first step is to remove the overall mean $(\overline{v(x)})$ of the series: $v'(x) = v(x) - \overline{v(x)}$ and then take averages over lag Δx. For this purpose, we introduce the operator $\mathcal{T}_{\Delta x}$, defining the tendency fluctuation $(\Delta v(\Delta x))_{tend}$ as:

$$\left(\Delta v(\Delta x)\right)_{tend} = \left|\mathcal{T}_{\Delta x}v\right| = \left|\frac{1}{\Delta x}\sum_{x<x'<x+\Delta x} v'(x')\right| \quad (5.154)$$

Alternatively, with the help of the summation operator S:

$$\left(\Delta v(\Delta x)\right)_{tend} = \left|\frac{1}{\Delta x}\delta_{\Delta x}Sv'\right|; \quad Sv' = \sum_{x'\leq x} v'(x') \quad (5.155)$$

$(\Delta v(\Delta x))_{tend}$ has a straightforward interpretation in terms of the mean tendency of the data but valid for $-1 < H < 0$ (the climate range: Chapter 10). It is also easy to implement: simply remove the *overall* mean and then take the mean over intervals Δx (equivalent to the mean of the differences of the running sum).

We can now define the Haar fluctuation which is a special case of the Daubechies family of wavelets (see e.g. Holschneider, 1995, and, for a recent application, Ashok *et al.*, 2010). This can be done by instead taking the second differences of the mean:

$$\left(\Delta v(\Delta x)\right)_{Haar} = \mathcal{H}_{\Delta x}v = \frac{2}{\Delta x}\delta^2_{\Delta x/2}Sv = \frac{2}{\Delta x}\left(\left(s(x)+s(x+\Delta x)\right) - 2s(x+\Delta x/2)\right)$$
$$= \frac{2}{\Delta x}\left[\sum_{x+\Delta x/2<x'<x+\Delta x} v(x') - \sum_{x<x'<x+\Delta x/2} v(x')\right]; \quad s(x) = Sv \quad (5.156)$$

where $\mathcal{H}_{\Delta x}$ is the Haar operator. Note the use of the shorthand notation $s(x) = Sv$. Although this is still a valid wavelet (Fig. 5.35b), it is almost trivial to calculate, and below we shall see that the technique can be used for $-1 < H < 1$. When applied without the initial running sum to $v(x)$ directly, it can be used for series with $0 < H < 2$ (see Section 6.4.2). The numerical factor (2) is needed so that the Haar fluctuation is close to the usual differences when $0 < H < 1$ (see below).

From the definitions, it is easy to obtain the relation:

$$\mathcal{H}_{\Delta x} = 2\mathcal{T}_{\Delta x/2}\delta_{\Delta x/2} = 2\delta_{\Delta x/2}\mathcal{T}_{\Delta x/2} \quad (5.157)$$

which will be useful for interpreting the Haar fluctuations. Since the difference operator removes any constants, the tendency operator could be replaced by an averaging operator so that Eqn. (5.157) means that the Haar fluctuation is simply the difference of the series that has been degraded in resolution; it is the difference of the mean over the first and second halves of the interval Δx. Note that for a series length L, only a range of scales $L/2$ is accessible; the definition (Eqn. 5.156) identifies the largest lag with L, the smallest with length 2; we could equally well modify the definition so that this would be $L/2$ and 1.

If needed, the Haar fluctuations can easily be generalized to higher order n by using $(n+1)$th-order differences on the sum s:

$$\mathcal{H}^{(n)}_{\Delta x} = \frac{(n+1)}{\Delta x}\delta^{n+1}_{\Delta x/(n+1)}S \quad (5.158)$$

We can note that the nth-order Haar fluctuation is valid over the range $-1 < H < n$; it is insensitive to polynomials of order $n - 1$, although as we see below, for analyzing scaling series, this insensitivity has no particular advantage. Generalizations valid to order $H < -1$ are also possible by iterating the sum operator. Note that $\mathcal{H}^{(1)}_{\Delta x} = \mathcal{H}_{\Delta x}$ and $\mathcal{H}^{(0)}_{\Delta x}v' = \mathcal{T}_{\Delta x}v'$, and hence when applied to a series with zero mean, $\mathcal{H}^{(0)}_{\Delta x} = \mathcal{T}_{\Delta x}$.

Consider for a moment the case $n = 2$ which defines the "quadratic Haar fluctuation":

$$(\Delta v(\Delta x))_{\text{QuadHaar}} = 3(s(x+\Delta x) - 3s(x+2\Delta x/3) + 3s(x+\Delta x/3) - s(x))/\Delta x \quad (5.159)$$

This fluctuation is sensitive to structures of size Δx^{-1}, and hence useful over the range $-1 < H < 2$, and it "filters" out polynomials of order 1 (lines). This is thus essentially equivalent to the quadratic MFDFA technique (Section 5.5) that we numerically investigate below, and computationally it is much faster since there are no time-consuming regression calculations. Below, we only give some limited discussion of this $n = 2$ "quadratic Haar" wavelet since for the most common series studied here, with $H < 1$, it was generally found to be very close to the usual Haar wavelet although with the slight disadvantage of converging a little more slowly for small Δx, and having a less straightforward interpretation. On the plus side, it was a bit more accurate at the very largest factor of two of Δx (see the example in Section 5E.3). We also note that if $H < -1$, fluctuations could be usefully defined using higher-order summation operators, but we do not need them here.

Before proceeding, we could make a general practical remark about these real-space statistics: most lags Δx do not divide the length of the series (L) exactly, so there is a "remainder" part. This is not important for the small $\Delta x << L$, when for each realization there are many disjoint intervals of length Δx, but when $L/3 \approx< \Delta x < L$ the statistics can be sensitive to this since from each realization there will be only one or two segments and hence poor statistics. A simple expedient is to repeat the analysis on the reversed series and average the two results. This can indeed improve the statistics at large Δx when only one or a small number of realizations is available; it has been done in the analyses below.

To have a clearer idea of the limitations of the various fluctuations for determining the statistics of scaling functions, we now numerically compare the performance of these various fluctuations and their corresponding structure functions when applied to the characterization of multifractals. In order to easily compare their performances with those of standard spectral analysis, we will only consider the second-order ($q = 2$) structure functions and will use $\alpha = 1.8$, $C_1 = 0.1$ simulations with H in the range $-7/10 < H < 7/10$ (the most common range for geodata). These parameters yield an intermittency correction $K(2) = 0.18$ for the $q = 2$ moment; 50 simulations were averaged to estimate the ensemble spectrum. The simulations were made using the technique described in Appendix 5B for the flux (i.e. $H = 0$). For the corresponding first fractional integration (for the generator), the $\Delta x^{-d/\alpha}$ correction method was used. For the second fractional integration (for $H \neq 0$ to obtain the field from the flux) we used a Fourier space power-law filter k^{-H}. For this fractional integration, exponential cutoffs were used at high frequencies to avoid small-scale numerical instabilities (this is standard for differentiation, i.e. when filtering by k^{-H} with $H< 0$, it is equivalent to using the "Paul" wavelet; Torrence and Compo, 1998). The series were 2^{16} in length and were then degraded in resolution by factors of 4 to a length of 2^{14} pixels to avoid residual finite size effects in the simulation at the smallest scales. In order to avoid spurious correlations introduced by the periodicity of the simulations, the series were split in half so that the largest scale was 2^{13}.

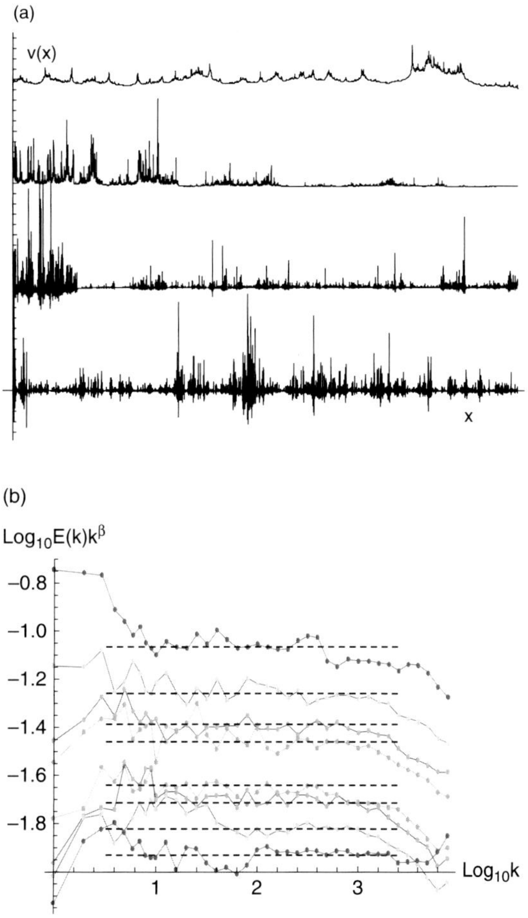

Fig. 5E.1 (a) Samples of the simulations analysed in this appendix. From bottom to top, $H = -7/10, -3/10, 3/10, 7/10$. All have $C_1 = 0.1$, $\alpha = 1.8$; the sections are each 2^{13} points long. One can clearly see the change in character of the series when H changes sign: at the bottom, fluctuations tend to cancel out (fluctuations are "stable"); at the top, they tend to reinforce each other (fluctuations are "unstable"). (b) The compensated spectra for an ensemble of 50 realizations, 2^{14} each, $\alpha = 1.8$, $C_1 = 0.1$, intermittency correction $= K(2) = 0.18$, with H increasing from top to bottom from $-7/10$ to $7/10$. The dashed horizontal line is the theoretical behaviour indicated over the range used to estimate the exponent (i.e. the highest and lowest factor of $10^{0.5}$ in wavenumber has been dropped). Each curve was offset in the vertical for clarity. Reproduced from Lovejoy and Schertzer (2012a).

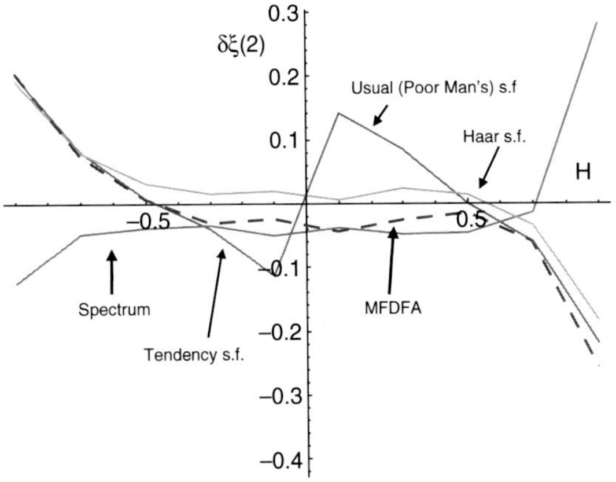

Fig.5E.2 Regression estimates of the compensated exponents for the spectra, Haar structure function ($q = 2$), quadratic, $q = 2$ MFDFA (dashed), the usual difference (poor man's) structure function ($q = 2$, for $H > 0$) and the tendency structure function ($q = 2$, same line, for $H < 0$). Reproduced from Lovejoy and Schertzer (2012a).

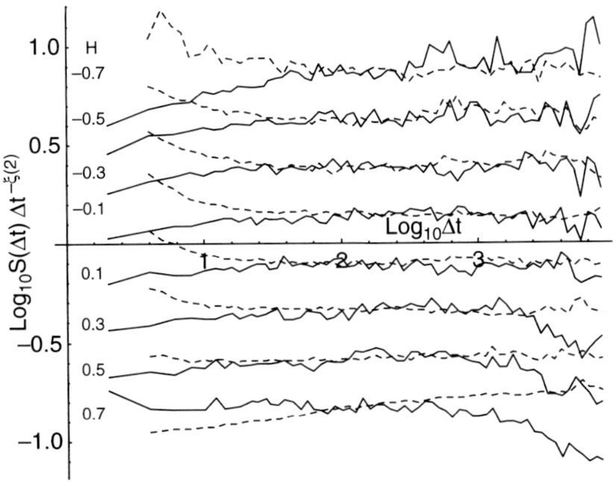

Fig. 5E.3 The compensated Haar structure functions $S_{Haar}(\Delta x) = \langle \Delta v_{Haar}(\Delta x)^2 \rangle^{1/2}$ (solid lines) and the second-order MFDFA technique (dashed lines), again for $H = -7/10$ (top) to $H = 7/10$ (bottom), every 1/5. Each curve was offset in the vertical for clarity by 1/4. Flat regions are the theoretically predicted behaviours. Reproduced from Lovejoy and Schertzer (2012a).

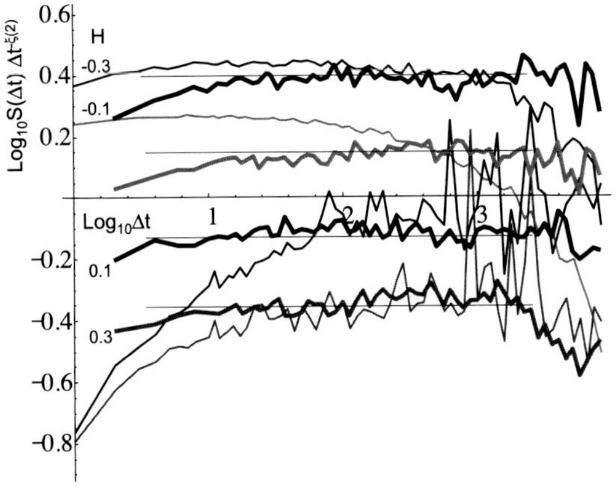

Fig. 5E.4 Comparison of the compensated Haar structure function (thick), the difference structure function (thin, below the axis) and the tendency structure function (thin, above the axis). The pairs of curves, top to bottom, have $H = -3/10$, $H = -1/10$, $H = 1/10$, $H = 3/10$. It can be seen that the standard difference structure function has poor scaling for nearly two orders of magnitude when $H = 1/10$, and one order of magnitude for $H = 3/10$ (see Fig. 5.E.2 for quantitative estimates). Reproduced from Lovejoy and Schertzer (2012a).

In Fig. 5E.1a we show samples of the simulations that visually show the effect of increasing H from -7/10 to 7/10. In Fig. 5E.1b, we see that, as expected, the compensated spectra are nearly flat, although for the lowest and highest factors of about $10^{0.5} \approx 3$ there were more significant deviations; the slopes of the central portions were thus determined by regression, and the corresponding exponents are shown in Fig. 5E.2. In the figure, the spectra were averaged over logarithmically spaced bins, 10 per order of magnitude (with the exception of the lowest factor of 10, where all the values were used). We also performed the regressions using log-log fits using all the Fourier components (rather than just the averaged values) in the same central range, and these gave virtually identical exponent estimates (the mean absolute differences in the estimated spectral exponent β were $\approx \pm\ 0.007$), which are therefore not shown. Similarly, if instead of log-log linear regressions, nonlinear regressions are used, we find that the latter yield identical exponents to within $\pm\ 0.009$, so that these estimates were too similar to the former to be worth showing. From Fig. 5E.2, one can see that there seems to be a residual small bias of about -0.04, whose origin is not clear.

We next consider the Haar structure function, which we compare to the second-order, quadratic MFDFA structure function analogue (based on the MFDFA fluctuation $\Delta v_{MFDFA} = F/\Delta x$ where F is the usual MFDFA scaling function (see Eqn. (5.112)); the former is valid over the range $-1 < H < 1$, the latter over the range $-1 < H < 2$. We can see (Fig. 5E.3) that the Haar structure function does an excellent job with overall bias mostly around + 0.01 to 0.02 (Fig. 5E.2), and that the MFDFA method is nearly as good (overall bias $\approx$ -0.02 to -0.04). It is interesting to compare this in the same figure with the results of the tendency structure function ($H < 0$) and the usual difference (poor man's) structure function ($H > 0$) (Fig. 5E.4). The tendency structure function turns out to be quite accurate, except near the limiting value $H = 0$, with the large scales showing the largest deviations. In comparison, for the difference structure function, the scaling is poorest at the smaller scales, requiring a range factor of ≈ 100 convergence for $H = 1/10$ and a factor ≈ 10 for $H = 3/10$. The overall regression estimates (Fig. 5E.2) show that the biases for both increase near their limiting values $H = 0$ to accuracies $\approx \pm 0.1$ (note that they remain quite accurate if one only makes regressions around the scaling part). This can be understood, since the process can be thought of as a statistical superposition of singularities of various orders, and those singularities whose orders are higher (lower) than the theoretical limit $H = 0$ will have biased difference (tendency) structure functions. The main advantage of the Haar structure function with respect to the usual (tendency or difference) structure functions is precisely that it is valid over the whole range $-1 < H < 1$, so that it is not affected by the $H = 0$ limit.

5E.2 The theoretical relation between poor man's, tendency and Haar fluctuations: hybrid structure functions

We have seen that although the difference and tendency structure functions have the advantage of having simple interpretations in terms respectively of the average changes in the value of the process and its mean value over an interval, this simple interpretation is only valid over a limited range of H values. In comparison, the Haar and MFDFA fluctuations give structure functions with valid scaling exponents over wider ranges of H. But what about their interpretations? We now briefly show how to relate the Haar, tendency and poor man's structure functions, thus giving it a simple interpretation as well.

In order to see the connection between the fluctuations, we use the "saturation" relations:

$$\begin{aligned} \delta_{\Delta x} v &\overset{d}{=} C_{tend} v; \quad H < 0 \\ \mathcal{T}_{\Delta x} v &\overset{d}{=} C_{diff} v; \quad H > 0 \end{aligned} \tag{5.160}$$

where $\overset{d}{=}$ indicates equality in the random variables in the sense of probability distributions and C_{tend}, C_{diff} are proportionality constants. These relations were easily verified numerically and arise for the reasons stated above; they simply mean that for series with $H < 0$ the differences are typically of the same order as the function itself, and that for $H > 0$ the tendencies are of the same order: the fluctuations "saturate." By applying Eqn. (5.154) to Eqn. (5.153) we now obtain:

$$\begin{aligned} \mathcal{H}_{\Delta x} v &= 2\mathcal{T}_{\Delta x/2}\delta_{\Delta x/2} v \overset{d}{=} 2\mathcal{T}_{\Delta x/2} v \overset{d}{=} C'_{tend}\mathcal{T}_{\Delta x} v; \quad H < 0 \\ \mathcal{H}_{\Delta x} v &= 2\delta_{\Delta x/2}\mathcal{T}_{\Delta x/2} v \overset{d}{=} 2\delta_{\Delta x/2} v \overset{d}{=} C'_{diff}\delta_{\Delta x} v; \quad H > 0 \end{aligned} \tag{5.161}$$

where C'_{tend}, C'_{diff} are "calibration" constants (only a little different from the unprimed quantities – they take into account the factor of 2 and the change from $\Delta x/2$ to Δx). Taking the qth moments of both sides of Eqn. (5.161), we obtain the results for the various qth-order structure functions:

$$\begin{aligned} \frac{\left\langle (\Delta v)^q_{Haar} \right\rangle}{\left\langle (\Delta v)^q_{diff} \right\rangle} &= C'^q_{diff}; \quad H > 0 \\ \frac{\left\langle (\Delta v)^q_{Haar} \right\rangle}{\left\langle (\Delta v)^q_{tend} \right\rangle} &= C'^q_{tend}; \quad H < 0 \end{aligned} \tag{5.162}$$

This shows that at least for scaling processes the Haar structure functions will be the same as the difference ($H > 0$) and tendency structure functions ($H < 0$), as long as these are "calibrated" by determining C'_{diff} and C'_{tend}. In other words, by comparing the Haar structure functions with the usual structure functions we can develop useful correction factors which will enable us to deduce the usual fluctuations given the Haar fluctuations. Although the MFDFA fluctuations are not wavelet coefficients at least for scaling processes, the same basic argument applies. Since the MFDFA and usual fluctuations scale with the same exponents, they can only differ in their prefactors, the calibration constants.

To see how this works on a scaling process, we confined ourselves to considering the root mean square (RMS) $S(\Delta t) = <\Delta v^2>^{1/2}$; see Fig. 5E.5, which shows the ratio of the usual $S(\Delta t)$ to those of the RMS Haar and MFDFA fluctuations. We see that at small Δx's, the ratio stabilizes after a range of about a factor 10–20 in scale, and that it is quite constant up to the extreme factor of 3 or so (depending a little on the H value). The deviations are largely due to the slower convergence of the usual RMS fluctuations to their asymptotic scaling form. From Fig. 5E.5 we can see that the ratios

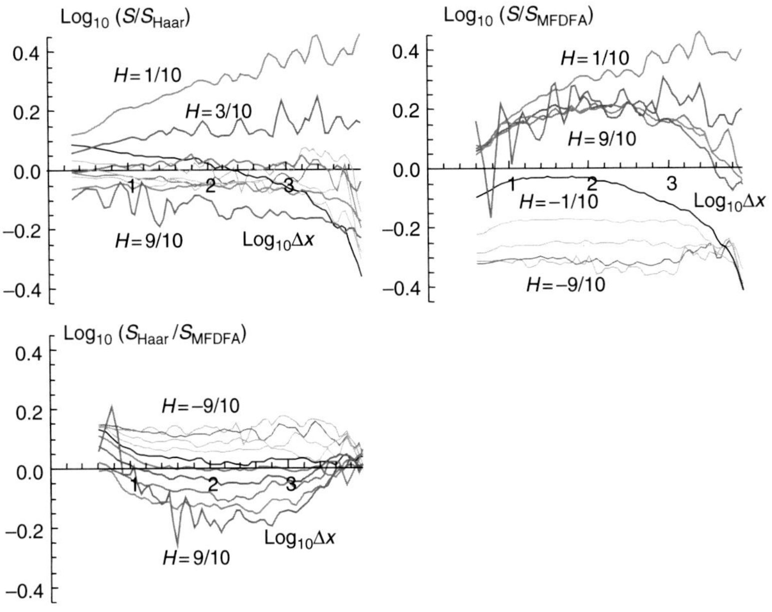

Fig. 5E.5 Comparison of the ratios of RMS structure functions ($S(\Delta x) = \langle \Delta v^2 \rangle^{1/2}$) for the simulations discussed in the text. Upper left shows the ratio of the RMS usual structure function (i.e. difference when $H > 0$, tendency when $H < 0$) to the Haar structure function for $H = 1/10, 3/10, \ldots, 9/10$ (thick, top to bottom), and $-9/10, -7/10, \ldots, -1/10$ (thin, top to bottom). The Haar structure function has been multiplied by $2^{\xi(2)/2}$ to account for the difference in effective resolution. The central flat region is where the scaling is accurate, indicating a constant ratio $C'_{Haar} = S/S_{Haar}$ which is typically less than a factor of 2. Upper right shows same but for the ratio of the usual to MFDFA RMS structure functions after the MFDFA was normalized by a factor of 16 and the resolution correction $4^{\xi(2)/2}$ (its smallest scale is 4 pixels). Bottom left shows the ratio of the MFDFA structure function to the Haar structure function with both the normalization and resolution corrections indicated above (note the monotonic ordering with respect to H, which increases from top to bottom). Reproduced from Lovejoy and Schertzer (2012a).

$C'_{Haar} = S/S_{Haar}$ and corresponding $C'_{MFDFA} = S/S_{MFDFA}$ are well defined in the central region, and are near unity, more precisely in the range $1/2 < C'_{Haar} < 2$ for the most commonly encountered range of H: $-4/10 < H < 4/10$. In comparison (bottom of the figure), over the same range of H, we have $0.23 > C'_{MFDFA} > 0.025$ so that the MFDFA fluctuation is quite far from the usual ones. For applications, one may use the semi-empirical formulae $C'_{Haar} \approx 1.1\, e^{-1.65H}$ and $C'_{MFDFA} \approx 0.075e^{-2.75H}$, which are quite accurate over the entire range $-7/10 < H < 7/10$. Although these factors in principle allow one to deduce the usual RMS structure function statistics from the MFDFA and Haar structure functions, this is only true in the scaling regime. Since the corrections for Haar fluctuations are close to unity, for many purposes they can be used directly.

5E.3 Hybrid fluctuations and structure functions in the case of global monthly surface temperatures: an example with both $H < 0$ and $H > 0$ regimes

For pure scaling functions, the difference ($1 > H > 0$) or tendency ($-1 < H < 0$) structure functions are adequate. The real advantage of the Haar structure function is apparent for functions with two or more scaling regimes, one with $H > 0$ and one with $H < 0$.

Can we "calibrate" the Haar structure function so that the amplitude of typical fluctuations can still be easily interpreted? For these, consider Eqn.(5.161), which motivates the definition:

$$\mathcal{H}_{hybrid,\,\Delta x} v = \max(\delta_{\Delta x} v, \mathcal{T}_{\Delta x} v) \tag{5.163}$$

of a "hybrid" fluctuation as the maximum of the difference and tendency fluctuations; the "hybrid structure function" is thus the maximum of the corresponding difference and tendency structure functions and therefore has a straightforward interpretation. The hybrid fluctuation is useful if a unique calibration constant C_{hybrid} can be found such that:

$$\mathcal{H}_{hybrid,\,\Delta x} v \overset{d}{\approx} C_{hybrid} \mathcal{H}_{\Delta x} v \tag{5.164}$$

To get an idea of how the different methods work on real data, we consider the example of the global monthly averaged surface temperature of the earth (discussed in detail in Section 10.3). This is obviously highly significant for the climate but the instrumental temperature estimates are only known over a small number of years (here, the period 1880–2009; 128 years, three different series; Appendix 10C). Fig. 10.12 shows the comparison of the difference, tendency, hybrid and Haar RMS structure functions, the last increased by a factor $C_{hybrid} = 10^{0.35} \approx 2.2$. It can be seen that the hybrid structure function does extremely well; the deviation of the calibrated Haar structure function from the hybrid one is ± 14% over the entire range of near a factor 10^3 in time scale. This shows that to a good approximation the Haar structure function can preserve the simple interpretation of the difference and tendency structure functions: in regions where the logarithmic slope is between −1 and 0, it approximates the tendency structure function, whereas in regions where the logarithmic slope is between 0 and 1, the calibrated Haar structure function approximates the difference structure function. C_{hybrid} is near the value 2; it was found that for the data studied in this book (primarily with H in the range −0.4 to +0.4) this was usually quite accurate. Therefore, unless otherwise indicated, the Haar analyses presented in this book increased the raw fluctuations by a factor of $C_{hybrid} = 2$.

Before embracing the Haar structure function, let us consider its behaviour in the presence of nonscaling perturbations; it is common to consider the sensitivity of statistical scaling analyses to the presence of nonscaling external trends superposed on the data which therefore break the overall scaling. Even when there is no reason to suspect such trends, the desire to filter them out is commonly invoked to justify the use of quadratic MFDFA or high-order wavelet techniques which eliminate linear or higher-order polynomial trends. However, for this purpose, these techniques are not obviously appropriate since on the one hand they only filter out polynomial trends (and not for example the more geophysically relevant periodic trends), while on the other hand, even for this, they are "overkill" since the trends they filter are filtered at all scales, not just the largest. The drawback is that with these higher-order fluctuations, we lose the simplicity of interpretation of the Haar wavelet while obtaining few advantages. Fig. 5E.6a shows the usual (linear) Haar RMS structure function compared to the quadratic Haar and quadratic MFDFA structure functions. It can be seen that while the latter two are close to each other (after applying different calibration constants: see the figure caption), and while the low- and high-frequency exponents are roughly the same, the transition time scale has shifted by a factor of about 3 so that overall they are quite different from the Haar structure function. It is therefore not possible to simultaneously calibrate the high and low frequencies.

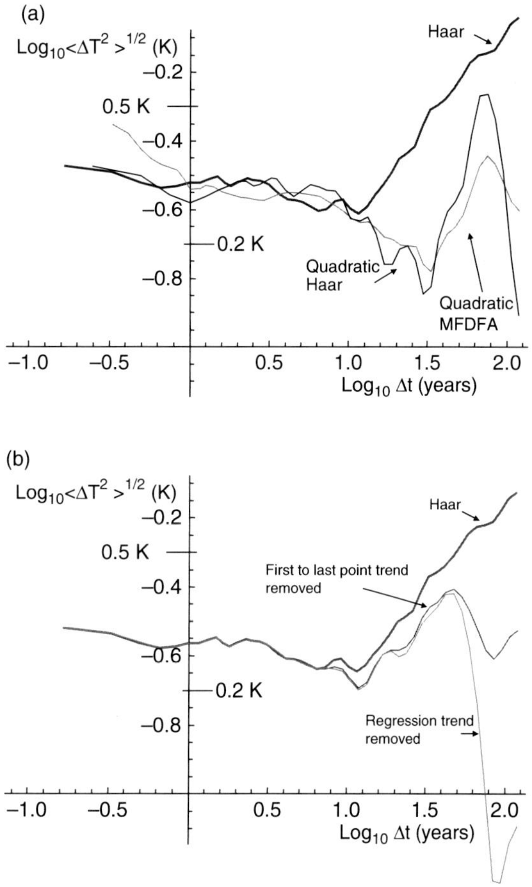

Fig. 5E.6 (a) The same temperature data as Fig. 10.12; a comparison of the RMS Haar structure function (multiplied by $10^{0.35} = 2.2$), the RMS quadratic Haar (multiplied by $10^{0.15} = 1.4$) and the RMS quadratic MFDFA (multiplied by $10^{1.5} = 31.6$). Reproduced from Lovejoy and Schertzer (2012b). (b) The same temperature data as Fig. 5E.6a; a comparison of the RMS Haar structure function applied to the raw data, after removing linear trends in two ways. The first defines the linear trend by the first and last points, while the second uses a linear regression; all were multiplied by $C_{hybrid} = 10^{0.35} = 2.2$. Reproduced from Lovejoy and Schertzer (2012a).

If there are indeed external trends that perturb the scaling, these will only exist at the largest scales; it is sufficient to remove a straight line (or if needed parabola, cubic, polynomial etc.) from the entire dataset, i.e. at the largest scales only. Fig. 5E.6b shows the result when the Haar structure function is applied to (scaling) data that have been detrended in two slightly different ways: by removing a straight line through the first and last points of the series and by removing a regression line. Since these changes essentially affect the low frequencies only, they mostly affect the extreme factor of 2 in scale. If we discount this extreme factor 2, then in the figure we see that there are again two scaling regimes, and that the low-frequency one is slightly displaced with respect to the Haar structure function, but not nearly as much as for the quadratic Haar. In other words, removing the linear trends at all scales, as in the quadratic Haar or the quadratic MFDFA, is too strong to allow a simple interpretation of the result and it is unnecessary if one only wishes to eliminate external trends.

In conclusion, if all that is required are the scaling exponents, the Haar structure function and quadratic MFDFA both seem to be excellent techniques. However, the Haar structure function has the advantage of numerical efficiency, simplicity of implementation, and simplicity of interpretation.

Chapter 6

Vertical stratification and anisotropic scaling

6.1 Models of vertical stratification: local, trivial and scaling anisotropy

6.1.1 The isotropy assumption: historical overview

We have argued that atmospheric scaling holds over a wide range in the horizontal, but this is clearly not possible if the turbulence is isotropic, since it would imply the existence of roundish structures hundreds or even thousands of kilometres thick. Such wide-range scaling is only possible because of the stratification. Following a brief historical overview, in this chapter we discuss the nature of the stratification.

Initially motivated by mathematical convenience, starting with Taylor (1935), the paradigm of isotropic and scaling turbulence was developed for laboratory applications, but following Kolmogorov (1941), three-dimensional isotropic turbulence was progressively applied to the atmosphere. However, there are several features (including gravity, the Coriolis force and stratification) that bring into question the simultaneous relevance of both isotropy *and* scaling. In particular, since the atmosphere is strongly stratified, a model with a single wide range of scaling which is both isotropic and scaling is not possible, so theorists had to immediately choose between the two symmetries: isotropy or scale invariance. Following the development of models of two-dimensional isotropic turbulence (Fjortoft, 1953; Kraichnan, 1967) but especially quasi-geostrophic turbulence (Charney, 1971), which can be seen as quasi-2D turbulence (see Section 2.6), the mainstream choice was first to make the convenient assumption of isotropy and to drop wide-range scale invariance; this could be called the "isotropy primary" paradigm. In Chapter 2 we saw how, starting at the end of the 1970s, this has led to a series of increasingly complex 2D/3D isotropic models of atmospheric dynamics, and we noted that justifications for these approaches have focused almost exclusively on the horizontal statistics of the horizontal wind in both numerical models and analyses and from aircraft campaigns, especially the highly cited GASP (Nastrom and Gage, 1983, 1985; Gage and Nastrom, 1986) and MOZAIC (Cho and Lindborg, 2001) experiments. Since understanding the anisotropy clearly requires comparisons between horizontal and vertical statistics/structures, it is not surprising that this neglect of the vertical has had deleterious consequences.

Over the same 30-year period that 2D/3D isotropic models were being elaborated, evidence slowly accumulated in favour of the opposite theoretical choice: to drop the isotropy assumption but to retain wide-range scaling. This change of paradigm from isotropy primary to scaling primary was explicitly proposed by Schertzer and Lovejoy (1984, 1985a-b, 1987), who considered strongly anisotropic scaling so that vertical sections of structures become increasingly stratified at larger and larger scales, albeit in a power-law manner. Similarly, although not turbulent, anisotropic wave spectra were proposed in the ocean by Garrett and Munk (1972) and an anisotropic buoyancy-driven wave spectrum was proposed in the atmosphere by Van Zandt (1982). Related anisotropic wave approaches may be found in Dewan and Good (1986), Fritts *et al.* (1988), Tsuda *et al.* (1989), Gardner *et al.* (1993), Hostetler and Gardner (1994) and Dewan (1997). These authors used anisotropic scaling models not so much for theoretical reasons, but rather because the data could not be explained without it. In addition, many experiments found nonstandard vertical scaling exponents thus implicitly supporting this position (see the review, and many additional references, in Lilley *et al.*, 2008). Below we shall see that state-of-the-art lidar vertical sections of passive scalars or satellite vertical radar sections of clouds (Section 6.5) give direct evidence for the corresponding scaling (power-law) stratification of structures. These analyses show directly that the standard bearer for isotropic models – 3D isotropic Kolmogorov turbulence – apparently does not exist in the atmosphere at any scale – at least down to 5 m in scale – or at any altitude level within the troposphere (Lovejoy *et al.*, 2007; Section 6.1.5). In Chapters 1 and 4 we used large quantities of high-quality satellite data to directly demonstrate the wide-range horizontal scaling of the atmospheric forcing (long- and short-wave radiances) and showed that reanalyses and atmospheric models display nearly perfect scaling cascade structures over the entire available horizontal ranges. This shows also that the source/sink free "inertial ranges" used in the classical models are at best academic idealizations and at worst unphysical: atmospheric dynamics has multiple energy sources that do not prevent it from being scaling.

At a theoretical level, quasi-geostrophic and similar systems of equations purporting to approximate the synoptic scale structure of the atmosphere are justified by "scale" analysis of various terms in the dynamical equations in which "typical" large-scale values of atmospheric variables (usually their fluctuations) are assumed and terms eliminated if they are deemed too small. However, in Section 2.3.2 we showed that by replacing such *scale* analyses – valid at most over narrow ranges of scales – by *scaling* analysis, focusing on horizontal and vertical exponents valid over wide ranges, the equations were symmetric with respect to anisotropic scale changes. However, just as the classical isotropic scaling analysis of the Navier–Stokes equations is insufficient to determine the Kolmogorov exponent 1/3 – for this we also need dimensional analysis on the scale-by-scale conserved energy flux – so here, the value of the new vertical exponents depends on a new turbulent flux that is the subject of this section.

6.1.2 Empirical status of the isotropy assumption

Kolmorogov's theory of isotropic 3D turbulence is based on the key assumption that there exists an "inertial" range where the turbulence is isotropic and depends only on the energy flux ε across scales (or wavenumber in Fourier space), yielding the $k^{-5/3}$ regime for the energy spectrum. In the atmosphere, the main reason for supposing the existence of *any* isotropic ranges is to do with anisotropic boundary conditions, i.e. it ignores gravity, which acts at all scales throughout the flow. Specifically, the unimportance of boundary conditions is a consequence of the fact that structures at a given scale are mostly coupled with structures at neighbouring scales, so that the effects of large-scale boundary conditions are progressively "forgotten" at small scales. Classically this tendency to "return to isotropy" (Rotta, 1951) has been modelled using second-order statistical closure techniques; however, even within this framework, as soon as buoyancy forces are included, their effect is found to be relatively large (Moeng and Wyngaard, 1986), just as in laboratory flows it is found that even small buoyancy forces readily destroy isotropy (Van Atta, 1991). Even recent theoretical developments (Arad *et al.*, 1998, 1999) assume a priori that fluctuation statistics for points separated by $\underline{\Delta r}$ follow the form:

$$\Delta v(\underline{\Delta r}) \propto \Theta(\psi)|\underline{\Delta r}|^{H} \tag{6.1}$$

where ψ is the angle of $\underline{\Delta r}$ with the vertical, $|\underline{\Delta r}|$ the length of the separation vector $\underline{\Delta r}$. To select only functions of this form is to implicitly assume that the angular variation in the anisotropy is the same at each scale; in the terminology developed below, this mild anisotropy is called "trivial." The main originality of Arad *et al.* (1998) was the proposal that there exists a hierarchy of such isotropic terms each with different H's and Θ's. Since the theory ignores buoyancy, when it was checked in the atmosphere, the data were restricted to the horizontal (Kurien *et al.*, 2000)! Indeed, virtually all empirical surface-layer atmospheric tests of isotropy (i.e. those with the best-quality data) simply assume Eqn. (6.1) with $H = 1/3$ and test the anisotropy at unique scales. It is even common to study the spatial anisotropy of scalars without using any spatial data whatsoever! For example, a common technique involves using *time* series of temporal gradients and converting time to space with "Taylor's hypothesis" of frozen turbulence. Finally, the "anisotropy" is estimated by using a third-order statistical moment (the skewness) to determine the forward/backward trivial anisotropy (Sreenivasan, 1991)! Even in the analysis of laboratory (Rayleigh–Bénard) convection, where there is a debate about whether $H = 1/3$ or 3/5 (respectively energy and buoyancy variance flux dominance: see below), isotropy is still assumed and one still uses data from time series at single points (Ashkenazi and Steinberg, 1999; Shang and Xia, 2001).

6.1.3 The Obukhov and Bolgiano buoyancy subrange 3D isotropic turbulence

Kolmogorov theory was mostly used to understand laboratory hydrodynamic turbulence which is mechanically driven and can be made approximately isotropic (unstratified) by the use of either passive or active grids. In this case, we have seen that fluctuations in Δv for points separated by $\underline{\Delta r}$ can be determined essentially via dimensional analysis using ε (Chapter 2); the latter choice being justified since it is a scale-by-scale conserved turbulent flux. The atmosphere, however, is fundamentally driven by solar energy fluxes which create buoyancy inhomogeneities; in addition to energy fluxes, buoyancy is thus also fundamental. In order to understand atmospheric dynamics we must therefore determine which additional dimensional quantities are introduced by gravity/buoyancy. As discussed in Monin and Yaglom (1975), this is necessary for a more complete dimensional analysis.

Let us start with the thermodynamic energy equation for an ideal gas:

$$\frac{D\log\theta}{Dt} = \frac{\kappa}{T}\nabla^2 T + \frac{\dot{Q}}{c_v T} \tag{6.2}$$

(e.g. Lesieur, 1987) where θ is the potential temperature, κ is the molecular heat diffusivity, T the absolute temperature and $D/Dt \equiv \partial/\partial t + v\cdot\nabla$ is the advective derivative, $\dot{Q}$ is the rate of heat input per unit mass and c_v is the specific heat at constant volume. In the spirit of Section 2.3 we will assume a quasi-steady state where heat sources $\dot{Q}$ create fluctuations/structures at large scales which are transferred by the nonlinear terms to smaller scales where they are eventually smoothed out by the dissipation term. As in our discussion of the energy flux cascade, we argue that at scales much larger than the dissipation scales the right-hand side of Eqn. (6.2) is ≈ 0 so that $\log\theta$ is an advected scalar: $D\log\theta/Dt \approx 0$. Since it determines the buoyancy force and hence modifies the velocity field, it is not passive but it nevertheless defines a scale-by-scale conserved quadratic invariant.

Let us now introduce the Boussinesq approximation (for this we follow Lesieur, 1987). Consider small fluctuations θ' with respect to a time-averaged state θ_0, and express the potential temperature in terms of these:

$$\theta(\underline{r}, t) = \theta_0(\underline{r}) + \theta'(\underline{r}, t) \tag{6.3}$$

Note that for this decomposition to be useful, there must exist a scale separation between "fast" and "slow" processes. The full Boussinesq approximation for the potential temperature fluctuations yields:

$$\frac{D\theta'}{Dt} + \underline{v}\cdot\nabla\theta_0 \approx \kappa\nabla^2\theta' + \frac{\dot{Q}}{c_v} \tag{6.4}$$

We can now consider a vertically stratified fluid in which $\theta_0(\underline{r}) = \theta_0(z)$, and hence $\underline{v}\cdot\nabla\theta_0 \approx w\frac{d\theta_0}{dz}$ where $w = v_z$. If w is small enough, then we have:

$$\frac{D\theta'}{Dt} \approx \kappa\nabla^2\theta' + \frac{\dot{Q}}{c_v} \tag{6.5}$$

Again, assuming that the small-scale dissipation just balances the forcing, we have an equation of scalar advection, only this time directly for the fluctuation θ'. Note that in this model, since the stratification is accounted for in the $\theta_0(z)$ function, the fluctuation θ' is considered isotropic even when the overall fluid is strongly stratified; this is an example of "locally isotropic turbulence" (Kolmogorov, 1941). It has a new term due to the buoyancy force, as does the velocity equation:

$$\frac{D\underline{v}}{Dt} = -\frac{1}{\rho_0}\nabla p' - g\frac{\theta'}{\theta_0}\widehat{k} - 2\underline{\Omega}\times\underline{v} + \nu\nabla^2\underline{v} \tag{6.6}$$

where $\widehat{k}$ is a unit vertical vector and p' is the pressure fluctuation analogous to θ': $p(\underline{r}, t) = p_0(\underline{r}) + p'(\underline{r}, t)$, and $\rho_0(z)$ is the corresponding mean density function. The direct effects of gravity are thus confined to the single term Δf in the equation for the vertical component:

$$\Delta f = g\frac{\theta'}{\theta_0}; \quad f = g\log\theta \tag{6.7}$$

where the fluctuations Δf are thus responsible for the buoyancy effects. We have noted that Eqns. (6.2) and (6.5) are scalar advection equations for $\log\theta$ and θ' respectively, so we may anticipate that – following the arguments in Section 2.4 – the corresponding variance fluxes will be conserved by the nonlinear terms. There are therefore two somewhat different ways to exploit this, one based on θ', the other on $g\log\theta$. Following Obukhov (1959) and Bolgiano (1959) (see e.g. the summary in Monin and Yaglom, 1975, vol. 2), we first consider the original classical approach based on potential temperature variance flux:

$$\chi_\theta = \frac{\partial\theta'^2}{\partial t} \tag{6.8}$$

which is taken as a fundamental cascade quantity along with ε. In order to obtain the scaling laws for the velocity field in a fluid dominated by buoyancy forces (i.e. in a hypothetical isotropic "buoyancy subrange" where the energy flux ε can be neglected) we then argue that only χ_θ (with units K^2/s) and the coupling constant g/θ_0 (with units $m^2/K/s$) between the fluctuation θ′ and the velocity field are dimensionally relevant. Dimensional analysis on χ_θ and g/θ_0 then yields the unique scaling "Bolgiano-Obukhov" (BO) law:

$$\Delta v(\underline{\Delta r}) \approx \chi_\theta^{1/5}\left(\frac{g}{\theta_0}\right)^{2/5}|\underline{\Delta r}|^{3/5} \tag{6.9}$$

corresponding to a $k^{-11/5}$ spectrum (neglecting intermittency; i.e. using $\beta = 1 + 2H$; Section 2.5). In the context of the Boussinesq approximation, this isotropic law applies to the fluctuations in the velocity about a totally stratified anisotropic mean state $\theta_0(z)$.

Staying within this classical framework for isotropic fluctuations, we may now inquire as to over what scale range this new BO law should apply, given that it is in competition with the usual energy-flux-dominated regime. In other words, what happens when we apply a full dimensional analysis to ε, χ_θ and g/θ_0? The answer is that there is a unique "Bolgiano–Obukhov" length scale L_{BO}:

$$L_{BO} = \frac{\varepsilon^{5/4}}{\chi_\theta^{3/4}(g/\theta_0)^{3/2}} \tag{6.10}$$

According to this classical theory, we see that as the effect of gravity is reduced ($g \to 0$), $L_{BO} \to \infty$, so the stratification disappears and we recover the usual isotropic 3D Kolmogorov law (i.e. dominated by ε). Therefore, we interpret the scale L_{BO} as the transition scale from isotropic Kolmogorov turbulence (for scales $L < L_{BO}$) to isotropic BO turbulence for scales $L > L_{BO}$. In order to test this theory we therefore need estimates of L_{BO}, and this requires simultaneous measurements of the difficult-to-measure fluxes χ_θ, ε – observations that Monin and Yaglom (1975) say "will require special observations which one hopes will be carried out in the future." Indeed, at the time, the BO theory did not look at all promising, the only evidence for the BO law being in the vertical direction (from "Jimspheres": Endlich *et al.*, 1969; Adelfang, 1971), and these results were largely ignored in the literature – including by Monin and Yaglom (1975)! Conversely, Adelfang (1971) (but not Endlich *et al.*, 1969) was apparently unaware that his results were close to the BO theory! We shall see in a moment that $L_{BO} \approx 10$ cm so that this classical picture predicts essentially a Kolmogorov regime only at very small scales!

6.1.4 The anisotropic scaling theory: the 23/9D Kolmogorov–Bolgiano–Obukhov model

A second way to approach buoyancy-driven turbulence is to make a more physically based argument (which essentially avoids the Boussinesq and other approximations), noting that the v and θ fields are only coupled by the Δf buoyancy force term ($f = g\log\theta$) so f is the fundamental physical and dimensional quantity rather than θ (in this approach there is no thermodynamic scale separation, so the interpretation of f in terms of buoyancy may be problematic). From Eqn. (6.2) and following the argument for an energy flux cascade, we see that by neglecting dissipation and forcing, $Df/Dt \approx 0$ so that f obeys a scalar advection equation and therefore the corresponding buoyancy force variance flux:

$$\varphi = \frac{\partial f^2}{\partial t} \tag{6.11}$$

is conserved by the nonlinear terms. In this case, the only quantities available for dimensional analysis are ε (units m^2/s^3) and φ (units m^2/s^5), not ε, χ_θ and g/θ_0. In this approach, there is no separation between a stratified "background" state and a possibly isotropic fluctuation field, so there is no rationale for assuming that the φ cascade is associated with any isotropic regime. Indeed, following Schertzer and Lovejoy (1983, 1985), it is more logical to assume that the two basic turbulent fluxes ε, φ can coexist and cascade over a single wide-range regime with the former dominating in the horizontal, the latter in the vertical:

$$\begin{aligned}\Delta v(\Delta x) &= \phi_h \Delta x^{H_h}; \quad \phi_h = \varepsilon^{1/3}; \quad H_h = 1/3\\ \Delta v(\Delta z) &= \phi_v \Delta x^{H_v}; \quad \phi_v = \varphi^{1/5}; \quad H_v = 3/5\end{aligned} \tag{6.12}$$

where Δx is a horizontal and Δz a vertical lag (for the moment we ignore the other horizontal coordinate y). Again, the fluxes ε, φ dimensionally define a unique length scale l_s:

$$l_s = \left(\frac{\phi_h}{\phi_v}\right)^{1/(H_v - H_h)} = \frac{\varepsilon^{5/4}}{\varphi^{3/4}} \tag{6.13}$$

Due to the anisotropy that is characterized by the distinct scaling exponents (H's), Eqns. (6.12) are at odds with the classical notion of scaling, such as that given by Lamperti (1962):

$$\Delta v(\lambda^{-1}\underline{\Delta r}) \overset{d}{=} \lambda^{-H}\Delta v(\underline{\Delta r}); \quad \underline{\Delta r} = (\Delta x, \Delta z) \tag{6.14}$$

(where $\overset{d}{=}$ means equality in the sense of random variables). However, both equations can be recast in a unique scaling equation (Schertzer and Lovejoy, 1985):

$$\boxed{\Delta v(T_\lambda \underline{\Delta r}) \overset{d}{=} \lambda^{-H_h}\Delta v(\underline{\Delta r}); \quad T_\lambda = \lambda^{-G}} \tag{6.15}$$

where T_λ is the "scale changing operator" and:

$$G = \begin{pmatrix} 1 & 0 \\ 0 & H_z \end{pmatrix}; \quad H_z = H_h/H_v = (1/3)/(3/5) = 5/9 \tag{6.16}$$

is the "generator" (this anisotropic "simple scaling" ignores intermittency: Section 5.4.4). As discussed below and in Chapter 7, this is equivalent to introducing a generalized scale function $\|\underline{\Delta r}\|$. Among others, it can be taken under the "canonical" form:

$$\|\underline{\Delta r}\| = l_s\left(\left(\frac{\Delta x}{l_s}\right)^2 + \left(\frac{\Delta z}{l_s}\right)^{2/H_z}\right)^{1/2} \tag{6.17}$$

where we have used l_s to nondimensionalize the coordinates (Eqn. (6.13)). With the help of l_s, and the scale function $\|\underline{\Delta r}\|$, the horizontal and vertical laws (Eqn. (6.12)) can be combined into a single equation:

$$\Delta v(\underline{\Delta r}) = \varepsilon^{1/3}\|\underline{\Delta r}\|^{1/3} \tag{6.18}$$

It is easy to check that if we successively take $\underline{\Delta r} = (\Delta x, 0)$ and $(0, \Delta z)$ in Eqn. (6.17) we recover Eqn. (6.12). Since the resulting scale function $\|\underline{\Delta r}\|$ has the property that $\|(l_s, 0)\| = \|(0, l_s)\|$ we see that l_s is the scale at which moving a vertical distance l_s leads to the same fluctuation as moving a horizontal distance l_s; the contours at scale l_s are thus typically roundish, so l_s is called the "sphero-scale." Since this model combines both Kolmogorov and Bolgiano–Obukhov scaling, it could also be called the Kolmogorov–Bolgiano–Obukhov model (KBO). In Section 6.1.7, we show that structures in turbulence obeying Eqn. (6.14) have volumes proportional to the $D_{el} = 2 + H_z = 23/9$ power of their horizontal extents; they are effectively "23/9 dimensional"; hence the original name "23/9D" for the model (Schertzer and Lovejoy, 1985); D_{el} is the "elliptical dimension" characterizing the volume of nonintermittent structures (see below).

From Eqn. (6.17) we see that in this theory nothing special happens at scale l_s; it is simply the scale at which one power law exceeds another, the scale at which structures are roughly "roundish." Since, dimensionally, both this and the classical isotropic theory predict the existence of a single characteristic scale, these should be equal (at least to within factors of order unity); therefore, the fact that empirically the sphero-scale is typically of the order of 10 cm (although with very large fluctuations) shows that the supposed transition scale $L_{BO} \approx l_s$ is therefore also $\approx \leq 1$ m. This shows that the original isotropic BO law is untenable: it predicts that Kolmogorov 3D isotropic turbulence should be confined to scales below a metre and the larger scales should follow BO scaling (Eqn. (6.9)).

6.1.5 The empirical status of the 23/9D model

If the original isotropic Bolgiano–Obukhov law is untenable, what about the anisotropic alternative? How well do the predictions (Eqns. (6.12), (6.15)) fare empirically? We have already shown spectra (Fig. 1.6b) which are compatible with horizontal exponent $H_h = H_{Kol} = 1/3$ (i.e. Kolmogorov scaling) from ~20 m to ~20 km (Section 1.2.3). Recalling our discussion in Section 2.6, there seems to be little disagreement that $H_h \approx 1/3$ is valid up to horizontal scales which are at least somewhat larger than the atmospheric scale height. The areas where consensus is lacking are for the larger horizontal scales and for the value of the vertical scaling exponent. While the former is discussed in detail in Section 6.3, we now turn to an examination of the latter.

Most of our knowledge of the vertical structure of the atmosphere comes from radiosonde balloons designed to provide inputs to synoptic forecasts rather than for research; they typically have quite coarse vertical resolutions (of the order ± 150 m). In addition to their low spatial resolutions, balloons suffer from swaying payloads and disturbances on ascent caused by the balloon's wake (see however Harrison and Hogan, 2006). In spite of these

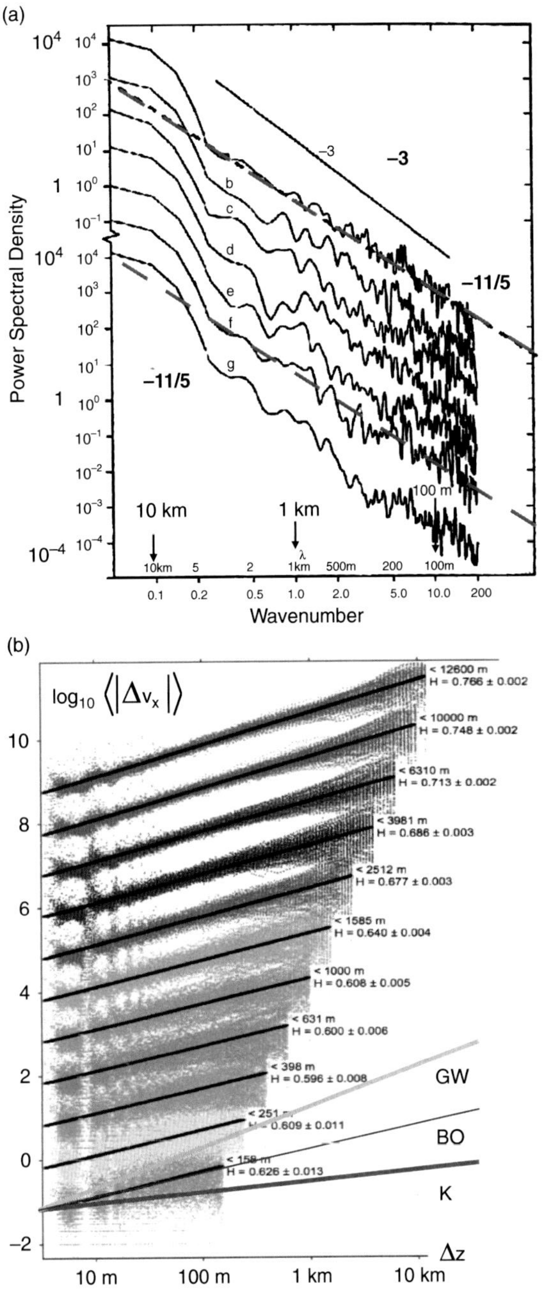

Fig. 6.1 (a) The vertical spectra of seven sondes showing roughly Bolgiano–Obukhov spectra ($\beta = 11/5$; dashed reference lines added); offset in the vertical for clarity. Adapted from Endlich *et al.* (1969). (b) Mean absolute vertical gradients of horizontal wind (first-order structure functions) for layers of thickness increasing logarithmically, with regression lines added. The three reference lines have slopes $H_v = 1/3$ (Kolmogorov, K), $H_v = 3/5$ (Bolgiano–Obukhov, BO), $H_v = 1$ (gravity waves, GW). The regression H_v estimates are given next to the lines. The data for each level are offset by one order of magnitude for clarity, units m/s. Reproduced from Lovejoy *et al.* (2007).

difficulties, experimentalists commonly interpret the vertical spectrum in terms of quasi-linear gravity waves with exponent $H_v = 1$ but with $H_h \approx 1/3$ (see e.g. Allen and Vincent, 1995; Dewan, 1997; Fritts *et al.*, 1988; Gardner, 1994). $H_v = 1$ follows from dimensional analysis if the layers are stable and homogeneous with well-defined Brunt–Väisälä frequencies (Section 6.2). In comparison, the older Lumley–Shur (Lumley, 1964; Shur, 1962) model predicts an *isotropic* $H_h = H_v = 1$ regime, as does the theory of quasi-geostrophic turbulence (Charney, 1971).

The most recent estimate of H_v used state-of-the-art dropsonde data from the NOAA Winter Storms 04 experiment over the Pacific Ocean (ranging over latitudes of about 15° N to 60° N), where 261 sondes were dropped by a NOAA Gulfstream 4 aircraft from roughly 12–13 km altitudes. These GPS sondes had vertical resolutions of ~5 m, temporal resolutions of 0.5 s, horizontal velocity resolutions of ~0.1 m/s and temperature resolutions of ~0.1 K (Hock and Franklin, 1999). While the full analysis of the 2004 experiment is described in Lovejoy *et al.* (2007) and Hovde *et al.* (2011), we concentrate here on analysis of the horizontal velocities. This experiment was in many ways an update of the early work by Endlich *et al.* (1969; Fig. 6.1a) and also the largest previous vertical scaling study to date (Lazarev *et al.*, 1994), which used 287 radiosondes (at 50 m resolution) over the tropical Pacific (Fig. 1.7a). The latter came to conclusions similar to those below but without being able to analyze the fairly thin layers considered here (cf. also the Landes (France) experiment using 80 sondes at about 42° north (Schertzer and Lovejoy, 1985; Fig. 5.19a).

Fig. 6.1b shows the composite analysis of the 235 sondes that are the most complete. For each sonde, the mean absolute shears $\Delta v(\Delta z) = |\underline{v}(z_1) - \underline{v}(z_2)|$ ($\underline{v}$ is the horizontal velocity vector) were calculated using all pairs of points with $\Delta z = |z_1 - z_2|$ in logarithmically spaced intervals, and for all $z_1, z_2 < z_t$ where z_t is the indicated altitude threshold. This method is particularly effective since at 2 Hz there are ~1400 data points per sonde but there are many more pairs of points (roughly 10^6 per sonde); the method also overcomes the problem of irregular vertical spacing of the data without requiring potentially problematic interpolations. Note that for each layer the logarithmic spacing of layers gives predominant weight to the upper part of the range; for results on layers with constant thicknesses, see Fig. 6.2.

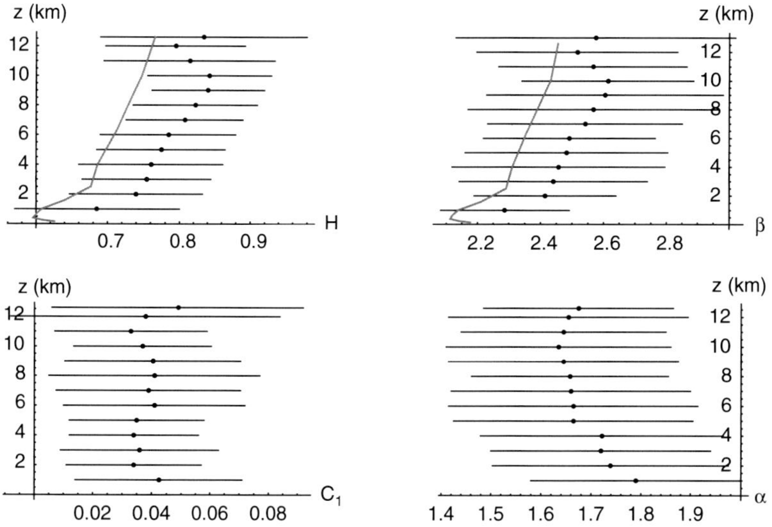

Fig. 6.2 Top left: the means and standard deviations of the H values calculated from the moduli of the vector differences in horizontal winds. The curves to the left of each graph in the top row are from the H values in Fig. 6.1b, i.e. from regressions over all pairs of points below the altitude indicated, estimated over the entire range of scales available (i.e. up to 12.6 km at the highest altitudes). The points are fits from individual sondes, as indicated in Fig. 6.1b. The error bars indicate the sonde-to-sonde variability. Top right: the same, but for the corresponding spectral exponents β: (nonintermittent) Kolmorogov theory yields $\beta = 5/3$, Bolgiano–Obukhov $\beta = 11/5$. The lines are somewhat to the left since they are weighted to be near the indicated altitude whereas the points are from data within a kilometre of the indicated altitude. Bottom left: the C_1 values. Bottom right: the corresponding α values. Reproduced from Lovejoy *et al.* (2007).

Four features of Fig. 6.1b are particularly striking: (a) the overall scaling – even for the thickest layers spanning the entire troposphere – is excellent; the standard errors in the slope (H) estimates are $< \pm 1\%$; (b) the slopes at the lower levels (which are not too affected by the ever-present strong jet streams) are very close to the BO value 3/5, but increase at higher altitudes; (c) there is no evidence for $H_v = 1/3$ (Kolmogorov) behaviour, even at the smallest scales (5 m) and in the lowest layer (< 158 m) which for technical reasons are not accessible to radiosondes; this is especially significant; (d) there is no evidence for $H_v = 1$ (gravity wave, pseudo-potential vorticity) even at the largest vertical scales. However, since this figure pools the data from all the sondes, the result might be an artefact of mixing data from profiles some of which might have $H_v = 1/3$ or $H_v = 1$ scalings. In Lovejoy *et al.* (2007), H_v was estimated for every sonde for every 1 km-thick layer. Of the total 2727 H_v values estimated in this way only a single one at the lowest 1 km level had $H_v \approx 1/3$, only 9 had $H_v > 1$, and only 1 had $H_v > 1.05$. In order to quantitatively characterize the mean and spread of these values, we refer to Fig. 6.2 (top left), which gives the one-standard-deviation spread of values around the mean (the exponents are estimated separately for each sonde). One can see that the Kolmorogov $H_v = 1/3$ value is systematically 2–4 standard deviations below the mean, while $H_v \approx 1$ is about 2 standard deviations above it.

Our results are precise enough to bring to the fore a systematic tendency for the H_v values to increase from the near-surface Bolgiano–Obukhov value 3/5 to values closer to 0.77 in higher layers subject to large (jet) shears. While the exact explanation for this increase is unclear at present, it should be recalled that, like the usual turbulence laws, the 23/9D model presupposes spatial statistical homogeneity, which is violated by the strongly altitude-dependent jets. In Sections 6.4 and 6.5 we investigate the anisotropy of other atmospheric fields. We conclude that the anisotropy of the 23/9D model is a reasonable approximation to the atmosphere but that nevertheless there are notable deviations at higher altitudes.

6.1.6 Local, trivial and scaling anisotropy

The above treatment of atmospheric stratification provides a convenient starting point for discussing

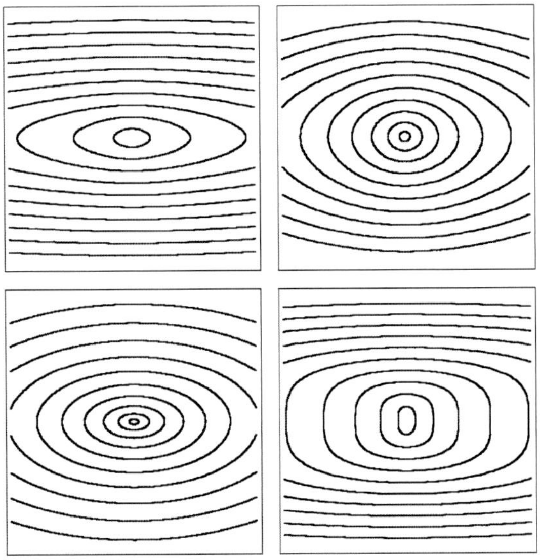

Fig. 6.3 Upper left: local isotropy with broken symmetry (Eqn. (6.20)). The upper right contours are the same but with a factor-25 blow-up showing small-scale "local" isotropy). The bottom left contours show trivial anisotropy with $a = 1.7$, $b = 1$ (Eqn. (6.22)), and the bottom right contours (Eqn. (6.17)) show scaling anisotropy with $H_z = 5/9$ with $l_s = 1$. The range displayed is $-5 < \Delta x < 5$, $-5 < \Delta z < 5$ (except for the 25× blow-up, upper right, in which the range is 25 times smaller).

different types of anisotropy. We can distinguish three cases, which we discuss in turn and illustrate graphically: (a) local isotropy, (b) trivial anisotropy, (c) scaling anisotropy.

Local isotropy

Starting with the classical "isotropic" buoyancy subrange, we can consider the anisotropy of vertical sections by considering the predicted spatial fluctuations:

$$\Delta v(\Delta x,\Delta z)=\Delta v_0(\Delta z)+\Delta v'(\Delta x,\Delta z)\overset{s}{=}\Delta v_0(\Delta z)+\chi_\theta^{1/5}\left(\frac{g}{\theta_0}\right)^{2/5}|(\Delta x,\Delta z)|^{H_{BO}};\quad H_{BO}=3/5 \tag{6.19}$$

where we have made the Boussinesq scale separation for the velocity field, with mean $v_0(z)$ and (isotropic) fluctuating part $v'(x,z)$ (we assume horizontal isotropy, so for simplicity we ignore the y coordinate). We consider fluctuations at a single instant in time, hence time does not enter the equation. Note that here and in the following, the equality $\overset{s}{=}$ means "scales as" so that the quantity on the left has the same scaling law as the quantity on the right. If one prefers, one can take ensemble averages of both sides of the equation and work with (first-) order structure functions.

In the spirit of this classical "locally isotropic" model, we can assume a linear velocity profile $v_0(z)$ so that $\Delta v_0(\Delta z) \propto \Delta z$. To illustrate the implied anisotropy when this is substituted in Eqn. (6.19), we may use the ansatz:

$$\Delta v(\Delta x,\Delta z)\overset{s}{=}\left((a\Delta z)^2+\left(b|(\Delta x,\Delta z)|^2\right)^{H_{BO}}\right)^{1/2} \tag{6.20}$$

where a, b are constants. Note that from Eqn. (6.19), we could have directly used $\Delta v\overset{s}{=}\left(a|\Delta z|+b|(\Delta x,\Delta z)|^{H_{BO}}\right)$, which has the same scaling behaviours, but the resulting iso-lines have unrealistic behaviours due to the absolute value function. In Fig. 6.3 we illustrate this with geometric constants $a = 1.7$, $b = 1$ and we display the large-scale cross-section (left) and small-scale (a 25-times blow-up) on the right. We see that at large scales the velocity is highly stratified, whereas at the small scales it becomes increasingly isotropic (the contours are more circular), as expected.

This model clearly has two scaling regimes, one for small isotropic scales, the other for large strongly stratified scales:

$$\begin{aligned}\Delta v(\lambda^{-1}\underline{\Delta r})&\overset{s}{\approx}\lambda^{-H_{BO}}\Delta v(\underline{\Delta r});\quad |\underline{\Delta r}|<<1\\ \Delta v(\lambda^{-1}\Delta z)&\overset{s}{\approx}\lambda^{-1}\Delta v(\Delta z);\quad \Delta z>>\Delta x>>1\end{aligned} \tag{6.21}$$

where $\underline{\Delta r}=(\Delta x,\Delta z)$. The model is therefore only isotropic in the small $|\underline{\Delta r}|$ limit, since it is only in this case that an isotropic scale reduction of the lag $\underline{\Delta r}$ by the factor λ^{-1} leads to a fluctuation rescaled by $\lambda^{-H_{BO}}$. For large enough Δz, we again have a scale symmetry, although this time it involves reducing only the vertical component of the lag and yields a different rescaling λ^{-1}.

Trivial anisotropy

Local isotropy assumes a scale break and features two different scaling regimes. In this section, we modify this ansatz to restore the overall scaling while maintaining stratification, but with the same aspect ratio at all scales (i.e. in Eqn. (6.16) with $H_z = 1$ so that $G =$ the identity matrix).

A simple modification to Eqn. (6.20) is:

$$\Delta v(\Delta x,\Delta z)\overset{s}{=}\|(\Delta x,\Delta z)\|^{H_{BO}};\quad \|(\Delta x,\Delta z)\|=\left((a\Delta z)^2+|b(\Delta x,\Delta z)|^2\right)^{1/2} \tag{6.22}$$

where we have introduced a scale function $\|(\Delta x, \Delta z)\|$ which is analogous to the vector norm and plays the analogous role in anisotropic scaling systems. In this case it is easy to verify that isotropic scale reductions lead to scaling for all $\underline{\Delta r}$:

$$\Delta v(\lambda^{-1}\underline{\Delta r})\overset{s}{=}\lambda^{-H_{BO}}\Delta v(\underline{\Delta r}) \tag{6.23}$$

The lower left panel of Fig. 6.3 shows the corresponding contours of $\Delta v(\underline{\Delta r})$, and one can see that they are indeed ellipsoids with identical shapes at all scales. The anisotropy implied by the scale function defined in Eqn. (6.22) is an example of "trivial anisotropy" because, while contours are anisotropic, the scale reduction $\underline{\Delta r}\rightarrow\lambda^{-1}\underline{\Delta r}$ needed to relate large and small fluctuations is nevertheless isotropic.

We can now ask, what is the most general form of trivially anisotropic scale functions? By definition they must satisfy the functional scale equation:

$$\|\lambda^{-1}\underline{\Delta r}\|=\lambda^{-1}\|\underline{\Delta r}\| \tag{6.24}$$

By inspection, the general solution is:

$$\|\underline{\Delta r}\|=\Theta(\theta)|\underline{\Delta r}|;\quad \tan\theta=\frac{z}{x} \tag{6.25}$$

where $\Theta(\theta)$ is an arbitrary positive function of the polar angle θ, and $|\underline{\Delta r}|$ is the usual polar radius (the restriction $\Theta>0$ is a physical constraint, so that the notion of scale is always positive definite). If the fluctuations $\Delta v(\Delta x, \Delta z)$ satisfy Eqn. (6.23) then for any exponent H_{BO} they can be written as powers of the trivial scale functions Eqn. (6.21), i.e. $\Delta v(\underline{\Delta r})=\|\underline{\Delta r}\|^{H_{BO}}$.

Scaling anisotropy

We now turn to the anisotropic scaling case, with scale function indicated in Eqn. (6.17). In order to obtain a power-law (scaling) change in the scale function (and hence for the fluctuation $\Delta v(\underline{\Delta r})$) we must now generalize the isotropic scale reduction (Eqn. (6.24)) to make an anisotropic coordinate "blow down":

$$\|T_\lambda\underline{\Delta r}\|=\lambda^{-1}\|\underline{\Delta r}\| \tag{6.26}$$

where T_λ is a scale-changing operator introduced in Eqn. (6.15). If $T_\lambda=\lambda^{-1}\mathbf{1}$, where $\mathbf{1}$ is the identity matrix, then we have the isotropic reduction (Eqn. (6.24)). However, if $\|\underline{\Delta r}\|$ is the scale function defined in Eqn. (6.17) then it satisfies Eqn. (6.26) with T_λ as given in Eqns. (6.15) and (6.16). This can be seen by inspection and by recalling that since G is a diagonal matrix:

$$T_\lambda=\lambda^{-G}=\begin{pmatrix}\lambda^{-1} & 0\\ 0 & \lambda^{-H_z}\end{pmatrix} \tag{6.27}$$

(when G has off-diagonal elements – as in the next chapter – things are more complicated). G is called the "generator" of the scale-changing operator T_λ. The transformations T_λ associated with such diagonal generators are "self-affine." Contour lines of an anisotropic scaling scale function satisfying Eqn. (6.17) (with $l_s=1$) are shown in Fig. 6.3 (lower right), and we see that at scales $<l_s=1$ the structures are ellipsoidal, oriented in the vertical direction, whereas the larger contours are ellipsoidal, oriented in the horizontal direction, and correspond to increasing stratification at larger and larger scales. l_s is called the "sphero-scale" because $\|(l_s,0)\|=\|(0,l_s)\|$, so that typical structures are roundish at this scale. Unlike the previous case of trivial anisotropy, where the shapes of the structures (contours) are independent of scale, here they are scale *dependent*, although this occurs without introducing any characteristic length; l_s is simply the scale at which one power law exceeds another, indeed, from Eqn. (6.17) we can that if the vertical extent of a structure is denoted by Z and its horizontal extent by X, then it is not hard to see that the vertical-to-horizontal aspect ratio of structures $Z/X=(X/l_s)^{H_z-1}$.

We mentioned that the scale function in Eqn. (6.17) is only the simplest, "canonical" scale function which is compatible with the different horizontal (Kolmogorov) and vertical (BO) scaling laws (Eqn. (6.12)). How general can the scale function $\|\underline{\Delta r}\|$ in the equation $\Delta v(\underline{\Delta r})=\varepsilon^{1/3}\|\underline{\Delta r}\|^{1/3}$ be, such that Eqn. (6.12) is recovered as special cases ($\Delta z=0$, $\Delta x=0$, top and bottom of Eqn. (6.12), respectively)? The answer is that it must satisfy the functional Eqns. (6.26) and (6.27); however, since here G is diagonal, we can make a nonlinear coordinate transformation:

$$(\Delta x',\Delta z')=\left(\Delta x, sign(\Delta z)|\Delta z|^{1/H_z}\right) \tag{6.28}$$

that transforms the anisotropic scaling Eqns. (6.26) and (6.27) into the isotropic equation $\|\lambda^{-1}\underline{\Delta r'}\|=\lambda^{-1}\|\underline{\Delta r'}\|$, with $\underline{\Delta r'}=(\Delta x',\Delta z')$. The general solution of this has already been given: Eqn. (6.25). The general solution of Eqns. (6.26) and (6.27) is therefore:

$$\|\underline{\Delta r}\| = \Theta(\theta')r'; \quad r' = (\Delta x^2 + |\Delta z|^{2/H_z})^{1/2};$$
$$\tan\theta' = \frac{\Delta z'}{\Delta x'} = \frac{sign(\Delta z)|\Delta z|^{1/H_z}}{\Delta x} \tag{6.29}$$

where θ', r' are the polar angle angles and radii in the primed system; we have nondimensionalized the distances by l_s so that the unit scale is also the "sphero-scale." The canonical solution is therefore obtained by taking $\Theta = 1$. The function Θ defines the unit ball via the polar coordinate equation $\|\underline{\Delta r}\| = 1$:

$$r' = \Theta(\theta')^{-1} \tag{6.30}$$

so that $\Theta > 0$ is necessary so that the unit ball is closed; this ensures that structures are spatially localized. In Chapter 9 we see how this requirement can be dropped in space-time to treat unlocalized space-time structures such as waves. Of course, one should be cautioned that for sufficiently anisotropic $\Theta(\theta')$ functions, there may not be any scale where structures are "roundish." In the next chapter, we consider generalizations of self-affine transformations to matrix G with off-diagonal elements and then to nonlinear generators.

Before leaving the question of scale functions let us note that a one-parameter model for the unit ball which is sometimes convenient is given by the following scale function:

$$\|\underline{\Delta r}\| = l_s\left(\left|\frac{\Delta x}{l_s}\right|^{2\eta} + \left|\frac{\Delta z}{l_s}\right|^{2\eta/H_z}\right)^{1/(2\eta)} \tag{6.31}$$

This scale function satisfies Eqns. (6.26) and (6.27) for any real η (and with a little effort it can be extended to complex η which gives even more possibilities). An example with $\eta = 1/2$ was already given in Eqn. (2.87) and Fig. 2.15a.

6.1.7 Anisotropic cascades, elliptical dimensions

Physically, scaling anisotropy corresponds to an anisotropic rather than an isotropic cascade. A schematic is shown in Fig. 6.4; compare this to the corresponding isotropic schematic (Fig. 3.3). We see that by subdividing the horizontal direction by ratio λ_0 and the vertical by the different ratio $\lambda_0^{H_z}$, that while the structures start off being horizontally stratified at large scales, at smaller and smaller scales, they are more and more vertically oriented (in the

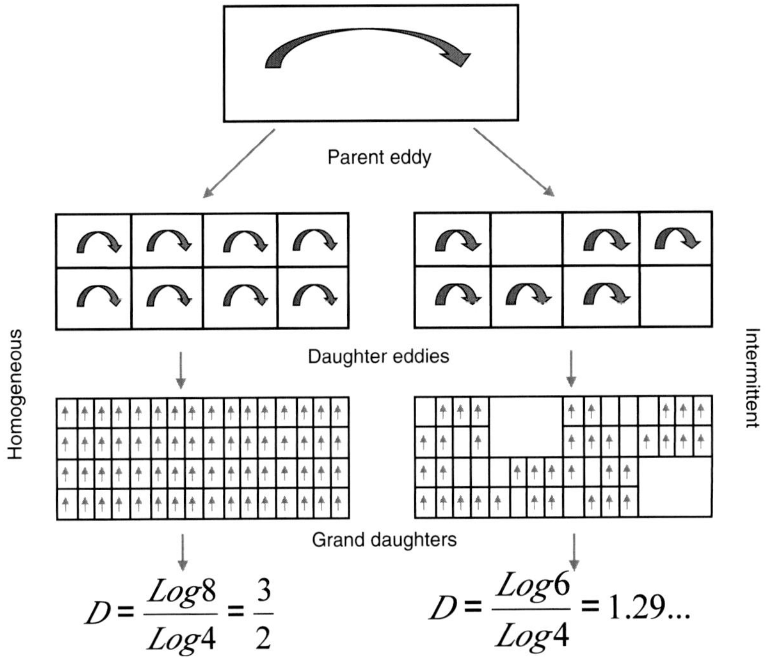

Fig. 6.4 A schematic of an anisotropic cascade; compare with its isotropic counterpart (Fig. 3.3). The exponent governing the decrease in area (equivalently the increase in number) of the sub-eddies with each iteration is $D_{el} = \log8/\log4 = 3/2$. On the right-hand side we illustrate the inhomogeneous (intermittent) anisotropic cascade in which $\lambda_0^C = 6$ of the eight sub-eddies on average survive so the corresponding elliptical (anisotropic) dimension of the active regions of $D - D_{el} - C = \log6/\log2 = 1.29...$ Adapted from Schertzer and Lovejoy (1987).

figure, $\lambda_0 = 4$ and $\lambda_0^{H_z} = 2$ respectively, so $H_z = \log 2/\log 4 = 1/2$). Indeed, the reduction of the areas at each iteration is by the factor:

$$\frac{A_{n-1}}{A_n} = \lambda_0 \lambda_0^{H_z} = \lambda_0^{D_{el}}; \quad D_{el} = 1 + H_z \tag{6.32}$$

where A_n is the area of the anisotropic "eddy" at the nth iteration (rectangular in the figure). The exponent D_{el} which characterizes the change in area of the nonintermittent (homogeneous) cascade (the left-hand path in Fig. 6.4) is called the "elliptical" dimension since typical structures are elliptical (as in Fig. 6.3); indeed, it is easy to see that the contours in the lower left of Fig. 6.3 (corresponding to the anisotropic scaling scale function, Eqn. (6.23)) also have areas proportional to $\lambda^{D_{el}}$ with $D_{el} = 1 + H_z$. Following the discussion in Chapter 3, in Fig. 6.4 the left-hand homogeneous (but anisotropic) cascade can be made intermittent (multifractal) by randomly killing off eddies (i.e. an anisotropic "β model"), a deterministic version of which is shown in Fig. 6.5. If the probabililty of each sub-eddy being "killed off" is λ_0^{-C}, then the average number of eddies at each scale $\lambda_0^{D} = \lambda_0^{D_{el}-C}$. D_{el} is thus purely a characterization of the anisotropy. Extensions to anisotropic multifractal cascades (anisotropic α models etc.) are straightforward; in Section 6.1.8 we illustrate this further with anisotropic continuous-in-scale simulations.

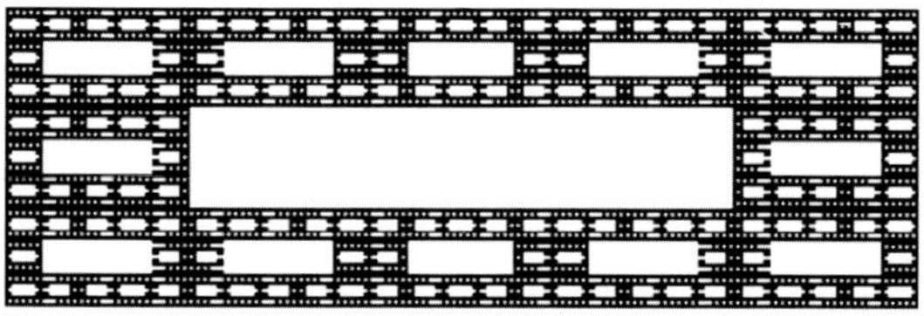

Fig. 6.5 An example of a deterministic β model, an anisotropic "Sierpinski carpet" obtained by dividing the horizontal by factors of 5 and the vertical by factors of 3 at each iteration and removing the three middle rectangles (keeping the 12 outer ones). Reproduced from Schertzer and Lovejoy (1985); $D_{el} = \log 15/\log 5$, $D = \log 12/\log 5$.

The elliptical dimension provides a convenient way to characterize both the atmosphere and turbulence models; Fig. 6.6 shows a schematic with the main isotropic and anisotropic models that have been proposed to date. As expected, the three-dimensional and two-dimensional isotropic models have $D_{el} = 3$, 2 respectively while the anisotropic scaling KBO law has $H_z = 5/9$, $D_{el} = 23/9$. Finally, the quasi-linear gravity wave models based on dimensional analysis on the Brunt–Väisälä frequency have $H_z = 1/3$, hence $D_{el} = 7/3$. Note that if the structures are waves rather than turbulent "eddies" then their characterization in this way by elliptical dimensions is less physically relevant.

Fig. 6.7a shows a zoom sequence illustrating the corresponding stratified structures as functions of scale; it can be compared to the analogous isotropic zoom in Fig. 6.7b. Adding details in this way is a "downscaling" or "disaggregating" of the initial field, and these zoom sequences graphically show how

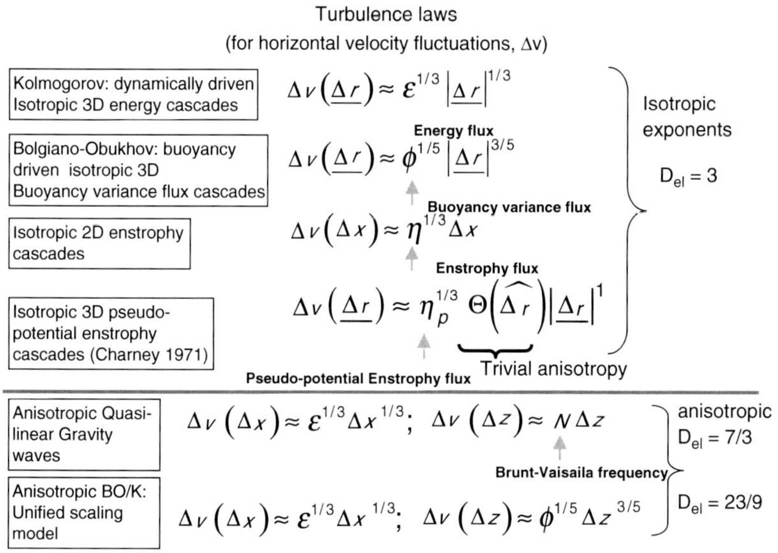

Fig. 6.6 A comparison of various turbulence laws according to their elliptical dimensions.

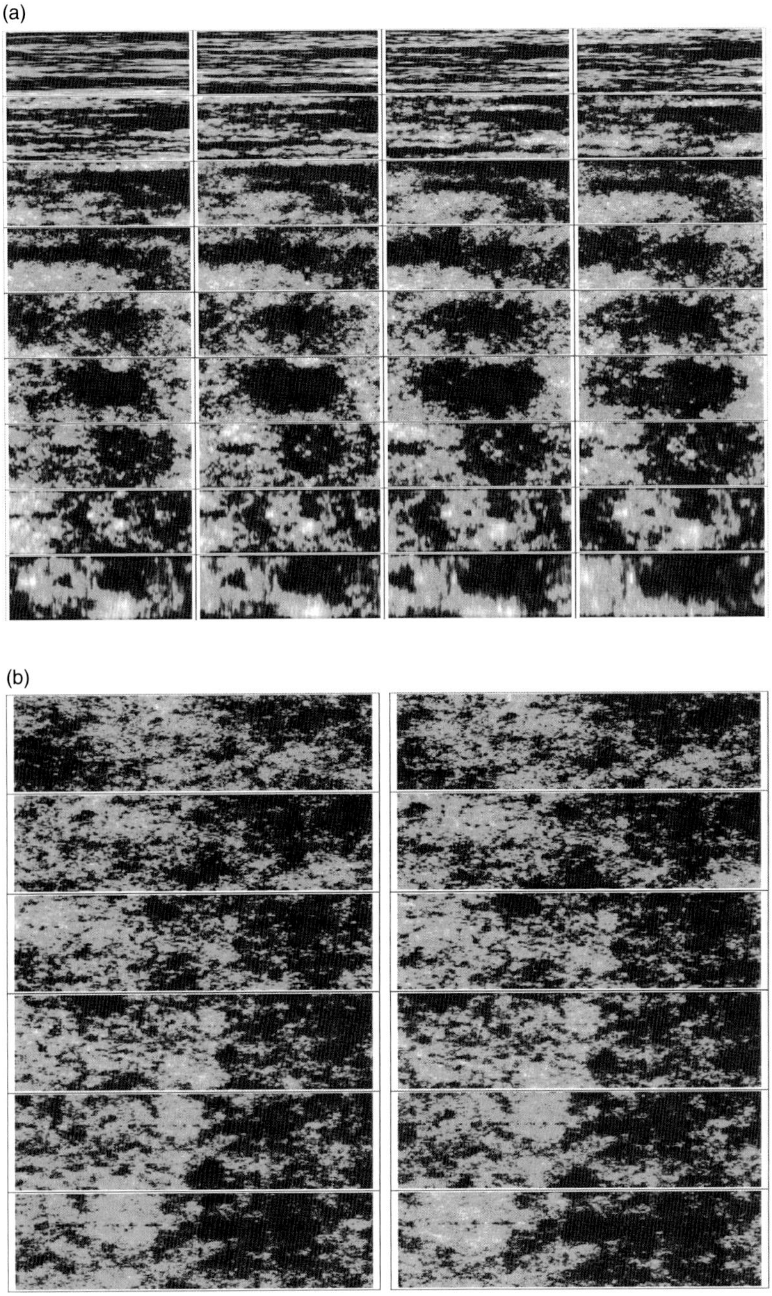

Fig. 6.7 (a) A sequence from a zoom (downscaling, disaggregation) of a stratified universal multifractal cloud model with $\alpha = 1.8$, $C1 = 0.1$, $H = 1/3$, $H_z = 5/9$. From top left to bottom right each successive cross-section represents a blow-up by a factor 1.31 (total blow-up is a factor ≈ 12 000 from beginning to end). If the top left simulation is an atmospheric cross-section 8 km left to right, 4 km thick, then the final (lower right) image is about 60 cm wide by 30 cm high; the sphero-scale is 1 m, as can be roughly visually confirmed since the left–right extent of the simulation second from bottom on the right is 1.02 m, where structures can be seen to be roughly roundish. Reproduced from Lovejoy and Schertzer (2010). *See colour plate section.* (b) A zoom (downscaling, disaggregation) sequence for an isotropic cloud with the same multifractal parameters as for the anisotropic simulation in Fig. 6.7a, from upper left to lower right. Each image is an enlargement by a factor 1.7 of the previous. As in Fig. 6.7a, the grey shades are "renormalized" separately in each image. *See colour plate section.*

multifractal models can be used for downscaling. At present, various essentially ad hoc downscaling techniques (usually simply multivariate regressions) are routinely used in many areas of atmospheric science, for example when (low-resolution) GCM outputs are used to predict the evolution of much smaller-scale "local" atmospheric variables. Multifractal models provide a theoretically well-founded alternative (Deidda, 2000; Paulson and Baxter, 2007; Gires *et al.*, 2011; Nogueira *et al.*, 2012).

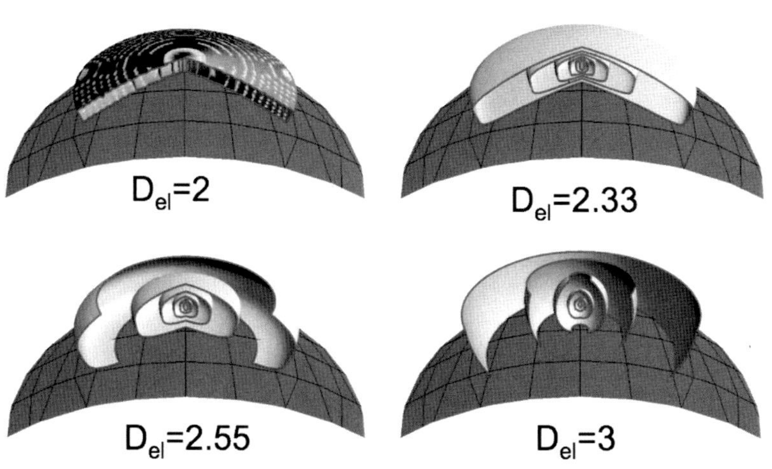

Fig. 6.8 A schematic diagram showing the change in shape of average structures which are isotropic in the horizontal (slightly curved to indicate the earth's surface) but with scaling stratification in the vertical; H_z increases from 0 (upper left) to 1 (lower right); $D_{el} = 2 + H_z$. In order to illustrate the change in structures with scale, the ratio of tropospheric thickness to earth radius has been increased by nearly a factor of 1000. In units of the sphero-scale (also exaggerated for clarity) here, $l_s = 1/10$ the tropospheric thickness (i.e. about 10^3–10^4 times the typical value), the balls shown are ½, 1, 2, 4, 8, 16, 32 times the sphero-scale (so that the smallest is vertically oriented, the second roundish and the rest horizontally stratified). Note that in the $D_{el} = 3$ case, the cross-sections are exactly circles; the small distortion is an effect of perspective due to the mapping of the structures onto the curved surface of the earth. Reproduced from Lovejoy and Schertzer (2010). *See colour plate section.*

Fig. 6.8 shows a schematic with the shapes of structures as functions of scale for the main turbulence models.

6.1.8 Simulations of stratified multifractals: clouds

To see the effect of self-affine stratification on the morphologies of structures, we can make the corresponding continuous-in-scale anisotropic multifractal simulations. The facility with which we can do this underscores both the appropriateness of both the GSI and Universal Multifractal formalisms; we can revisit section Section 5.5.2 on continuous-in-scale cascade processes, and the fractionally integrated flux model for the observables. All that needs to be done is to simply replace the vector norms $|\underline{r}|$ everywhere by scale functions $\|\underline{r}\|$, and the spatial dimensions d by elliptical dimensions D_{el}:

$$\varepsilon_\lambda = e^{\Gamma_\lambda};\quad \Gamma_\lambda(\underline{r}) = C_1^{1/\alpha} N_{D_{el}}^{-1/\alpha} \int\limits_{1\le\|\underline{r}'\|\le\lambda} \frac{\gamma_\alpha(\underline{r}')}{\|\underline{r}-\underline{r}'\|^{D_{el}/\alpha}} d^d\underline{r}' \tag{6.33}$$

where the normalization N_{Del} constant is still given by an angle integral (Eqn. (5.83)):

$$N_{D_{el}} = \Omega_{D_{el}} = \int\limits_{\|\underline{r}'\|=1} d^{D_{el}}\underline{r}' \tag{6.34}$$

(the details of the explicit calculation are somewhat technical and are given in Appendix 7A).

The statistics of the resulting v field will satisfy the anisotropic extensions of the formulae in Section 5.5.3:

$$\left\langle |\Delta v(\underline{\Delta r})|^q \right\rangle = \|\underline{\Delta r}\|^{\xi(q)};\qquad \xi(q) = qH - K(q) \tag{6.35}$$

which for any scale function satisfying Eqn. (6.26) is equivalent to:

$$\left\langle |\Delta v(T_\lambda \underline{\Delta r})|^q \right\rangle = \lambda^{-\xi(q)} \left\langle |\Delta v(\underline{\Delta r})|^q \right\rangle \tag{6.36}$$

In particular, when the generator G of T_λ is diagonal with eigenvalues 1, H_z for the horizontal and vertical respectively (Eqn. (6.16)), and with scale function (Eqn. (6.17)), then we recover the statistics in orthogonal directions:

$$\begin{matrix} \left\langle |\Delta v(\Delta x, 0)|^q \right\rangle = \Delta x^{\xi_x(q)} \\ \left\langle |\Delta v(0, \Delta z)|^q \right\rangle = \Delta z^{\xi_z(q)} \end{matrix};\qquad \begin{matrix} \xi_x(q) = qH - K(q) \\ \xi_z(q) = \xi_x(q)/H_z \end{matrix} \tag{6.37}$$

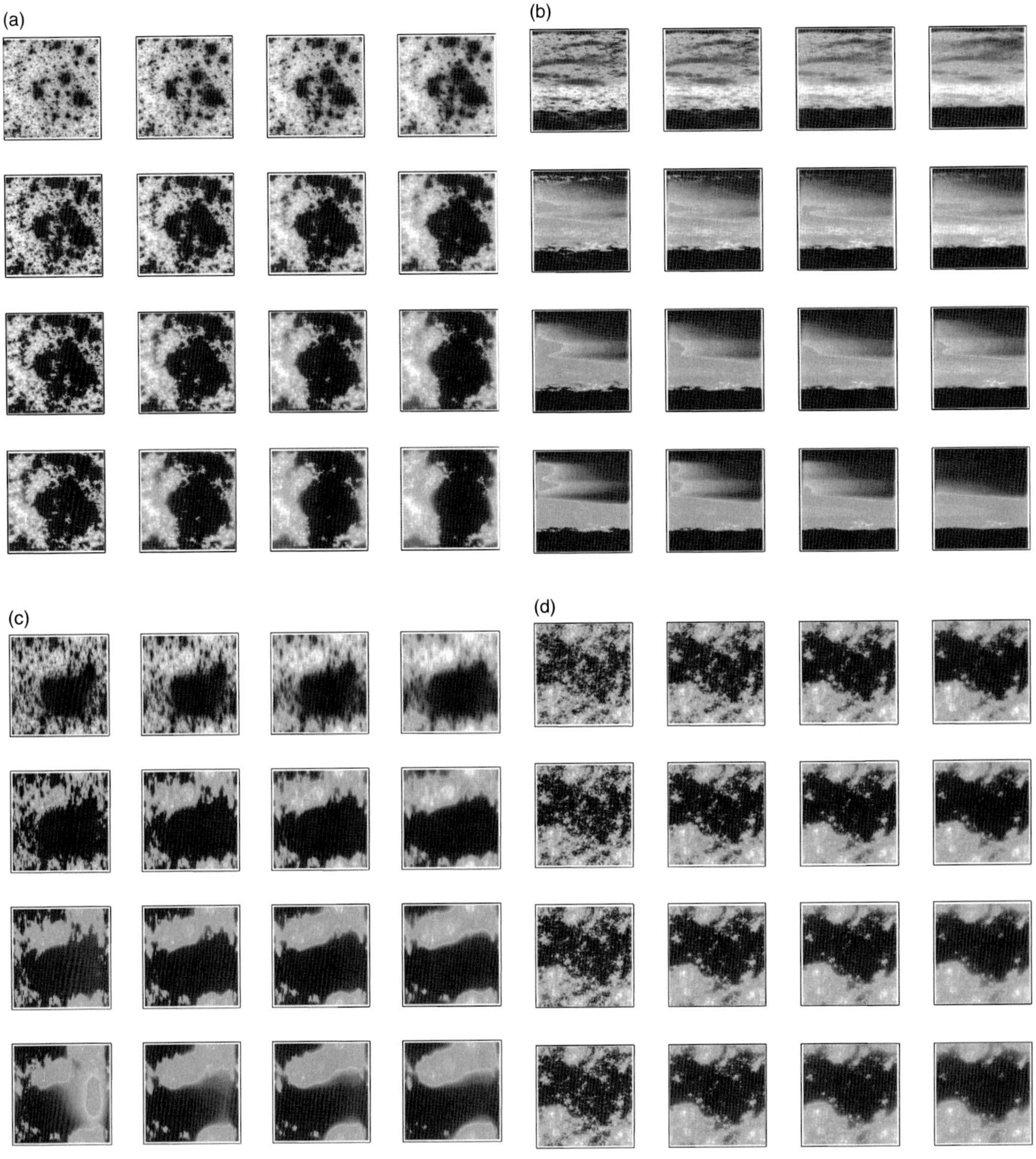

Fig. 6.9 (a) Realizations of self-similar multifractal clouds with the same seed for $\alpha = 0.8$; to be compared with the corresponding stratified (self-affine) simulations in the following figures. From top to bottom, C_1 increases from 0.05 to 0.20, 0.35, 0.50 from left to right, H increases from 0.05 to 0.20, 0.35, 0.50. (b) Same as Fig. 6.9a, except self-affine with $l_s = 1$ pixel; $\alpha = 0.8$. These were obtained by replacing the isotropic vector norm (distance function) by scale functions as described in the text. From top to bottom, C_1 for 1D left–right sections increases from 0.05 to 0.15, 0.25, 0.35 from left to right, H for 1D left–right sections increases from 0.05 to 0.15, 0.25, 0.35. All the simulations have $\alpha = 0.8$ and $H_z = 5/9$. The latter implies that the H, C_1 for the 1D up–down sections are 9/5 times larger (i.e. varying from 0.09 to 0.63). The simulations are 256 × 256 pixels. (c) Same as Fig. 6.9b ($\alpha = 0.8$), except with $l_s = 64$ pixels. (d) Same as Fig. 6.9a (self-similar), except with $\alpha = 2$.

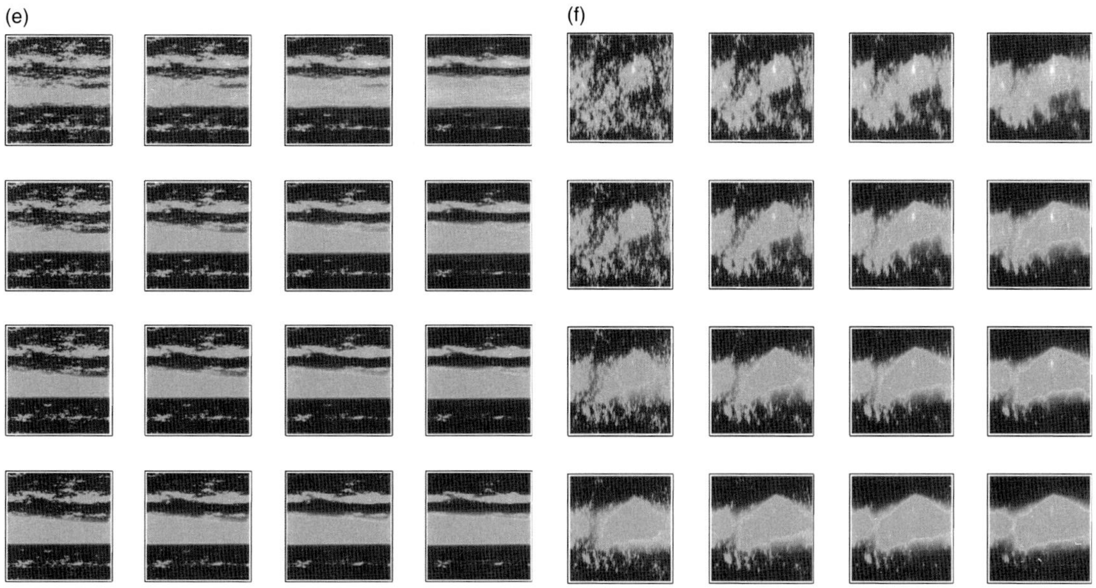

Fig. 6.9 (e) Same as Fig. 6.9d ($\alpha = 2$), except self-affine with $l_s = 1$. (f) Same as Fig. 6.9e, except self-affine with $l_s = 64$.

To see the effect on the resulting morphologies, we turn to Fig. 6.9, which graphically shows the results of simulations with varying C_1, H, and this for $\alpha = 0.8$ (self-similar, and self-affine with $l_s = 1$ and 64; Fig. 6.9a–c, respectively), and $\alpha = 2$ (self-similar, and self-affine with $l_s = 1$ and 64; Fig. 6.9d–f, respectively; compare this to the corresponding isotropic simulations in Fig. 5.34).

6.2 The Brunt–Väisälä frequency and the classical stable layer approach to stratification

6.2.1 Discussion

Up until now, we have exclusively discussed statistical (turbulence) approaches to atmospheric stratification, in the belief that such approaches are mandatory in light of the enormous atmospheric Reynolds numbers ($Re \approx 10^{12}$). However, there also exists a classical nonturbulent "dynamical meteorology" approach to stratification in which the notion of a "stable" atmospheric layer is central. This idealization plays a central role in the use of thermodynamic diagrams and in synoptic meteorology, including in the interpretation of potential vorticity maps (Hoskins *et al.*, 1985). In addition, the notion of stable, smoothly varying layers justifies ubiquitous linear theories. For example, Nappo (2002) states, "Almost all of what we know about the nature of gravity waves is derived from the *linear theory*" (emphasis in the original). Using high-resolution dropsonde data which allow the vertical structure to be measured to 5 m resolution (i.e. 10–20 times better than operational radiosonde data and 100 times better than the standard "significant levels"), we now show that apparently stable layers are punctuated by a fractal hierarchy of unstable layers, making it unlikely that linear theory is appropriate and that the actual stratification is in fact an emergent high Re scaling property.

The explanation for roughly linear temperature fall-off with altitude z (the dry adiabatic lapse rate) is a classic example of dynamical meteorology. Textbooks explain that when a parcel of air is vertically displaced, it expands because of the vertical pressure gradient. The work required lowers the temperature of the parcel; if this process occurs adiabatically without water vapour, then one obtains the dry adiabatic lapse rate ≈ 9.8 K/km. This explanation is at best a first approximation; more interesting is the sign and magnitude of the deviations. Following Väisälä (1925) and Brunt (1927), we can consider an atmosphere with a uniform (i.e. constant) temperature gradient. When a parcel of air is vertically displaced by a small amount, it experiences a restoring force proportional to:

$$N^2 = g\partial\log\theta/\partial z \tag{6.38}$$

where θ is the potential temperature and N is the Brunt–Väisälä frequency. When $N^2 > 0$ the particle will oscillate about its initial position with frequency N; the atmosphere is stable. On the contrary, when $N^2 < 0$, the particle will accelerate away from its equilibrium position; the atmosphere is locally unstable. Since the result neglects the possible destabilizing effect of condensation of water vapour, $N^2 > 0$ implies only "conditional" stability. In a humid atmosphere the same argument can be made: taking into account the latent heat released by condensation of water vapour, the "convective instability" criterion is the same, with the "equivalent" potential temperature θ_E replacing the potential temperature θ and N_E replacing N; both criteria are used below.

The above analysis assumes that the air surrounding the parcel is motionless; $N^2 > 0$, $N_E{}^2 > 0$ are static stability criteria. However, the atmosphere typically has large vertical shears, and we must consider the dynamical stability. Surprisingly, this was actually considered somewhat earlier by Richardson (1920), who noted that buoyancy tended to stabilize shear flows and who quantified this effect by the eponymous dimensionless number:

$$Ri = (N^2/s^2) \tag{6.39}$$

where

$$s = \partial v/\partial z \tag{6.40}$$

is the vertical shear. Layers with Ri exceeding a critical value Ri_c (usually taken $\approx$ 0.25) are considered "dynamically" stable, otherwise they are dynamically unstable (due to the scaling of N^2, s^2, changing Ri_c will only change the fractal exponent characterizing the clustering of the layers). The atmosphere is thus sometimes classified: $Ri < 0$ "unstable stratification", $0 < Ri < Ri_c$ the "stable subcritical regime", $Ri > Ri_c$ the "supercritical regime".

But do stable layers really exist? A long-recognized symptom of problems caused by the strong atmospheric inhomogeneity is that – even at a fixed scale – Ri is an incredibly variable quantity, and its mean barely – if at all – converges (there are empirical and theoretical reasons to suspect that it has a Cauchy probability distribution: Schertzer and Lovejoy, 1985). Indeed, less variable statistically based alternatives such as the "flux" Richardson number (e.g. Garratt, 1992) are frequently used instead. For example, the dropsonde data discussed above show that the mean derivatives defining $\langle s(\Delta z)\rangle \propto \Delta z^{H_v-1}$ with $H_v \approx$ 0.6–0.75 will diverge as the layers become thinner and thinner ($\Delta z \rightarrow 0$), implying that their true values will depend on the turbulent dissipation scale. More recently Dalaudier *et al.* (1994) and Muschinski and Wode (1998) discovered thin (even sub-metric) step-like structures called "sheets" in otherwise supposedly smoothly varying structures. Similar results have been reported in the ocean (Gregg, 1991; Osborne, 1998).

6.2.2 Testing the stable layer notion with high-resolution dropsondes

In order to see if we could define smoothly varying stable layers, we used the dropsondes described in Section 6.1.5, concentrating on the analysis of eight pairs, with the members of each pair dropped within 0.3 s of each other on 2004/02/29. The two sondes within a pair are separated by roughly 50 m and therefore can be used to cross-check each other's accuracy. Such intersonde comparisons put the following upper bounds on the measurement errors: $\pm$ 0.014 K, $\pm$ 1.4 $\times$ 10^{-5} s^{-2}, $\pm$ 7 $\times$ 10^{-5} s^{-2} for temperature, N^2, $N_E{}^2$ respectively. These are sufficiently good that almost all of the layers discussed here are reproduced from one sonde to the other, even at the highest resolution. Fig. 6.10 shows the comparison of N^2 calculated at 5 m resolutions for each sonde in the first pair. That the fluctuations cannot be attributable to instrument noise is clear from their close agreement.

Combining N^2 with velocity data, we can determine the dynamical stability ($Ri >$ 1/4) at various resolutions. At low resolution (320 m, Fig. 6.11), we obtain the usual first-order approximation to the vertical structure familiar from operational radiosonde resolutions: the atmosphere is unstable in the very lowest layer with only a few additional thin unstable layers higher up. At such low resolution, there appear to exist reasonably wide layers which are stable, perhaps allowing the application of quasi-linear gravity wave theories. However, this hope is dashed when we turn to the finer resolutions (80, 20, 5 m; superposed). Upon closer examination, each apparently stable sublayer is found to consist of a hierarchy of unstable subsublayers, themselves embedded with stable subsubsub layers etc. with the same "Russian doll" hierarchical structure holding in reverse for the initially unstable layers; the blow-up on the right-hand side of Fig. 6.11 shows this particularly clearly. Fig. 6.12 shows the same profile using the static stability criterion $N^2 > 0$; we note: (a) dynamical and static criteria are qualitatively similar, (b) both sondes infer almost all the same layers. To show that the unstable layers are indeed fractal subsets of the vertical, we calculated (Fig. 6.13) the (conditional) probability $P(\Delta z)$ of finding a (5 m-thick) unstable layer at a distance Δz from a given (5 m-thick) unstable layer. $P(\Delta z)$ is roughly a power law; its (absolute) exponent is the correlation codimension $C_c = 1 - D_c$ (D_c is the correlation dimension) which characterizes the sparseness of the unstable layers. Fig. 6.13 shows the results for 24 sondes using the "conditional stability" $N^2 > 0$ criterion, the dynamical stability criterion ($Ri > 1/4$) as well as the "convective stability" criterion $N_E^2 > 0$. The bars show the amplitude of the sonde-to-sonde variations. If C_c is estimated on each sonde individually, we obtain: $C_{cN} = 0.36 \pm 0.056$, $C_{cRi} = 0.22 \pm 0.037$, $C_{cNE} = 0.15 \pm 0.016$, based on N, Ri, N_E respectively. This implies an ordering of decreasing sparseness from conditional instability, dynamical instability to convective instability. The deviation of the mean behaviour from perfect power laws is less than 10% over the layers with separations in the range 5 m to 1.5 km. The result $C_{cN} > C_{cRi}$ is a (mathematical) consequence of the fact that the conditionally unstable layers are subsets of the dynamically unstable layers. We should note that vertical scaling laws for N^2, Ri were also

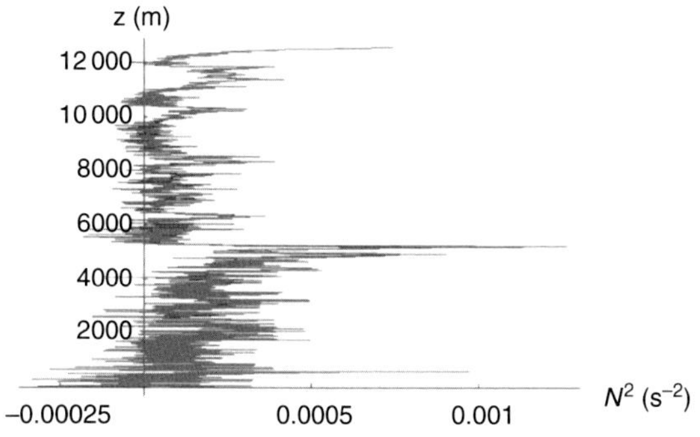

Fig. 6.10 The Brunt–Väisälä frequency squared (N^2) as a function of altitude. The two sondes were released at an interval of 0.3 s, and most of the time their two traces are indistinguishable, indicating that the error in the measurement is less than the width of the lines. Reproduced from Lovejoy *et al.* (2008).

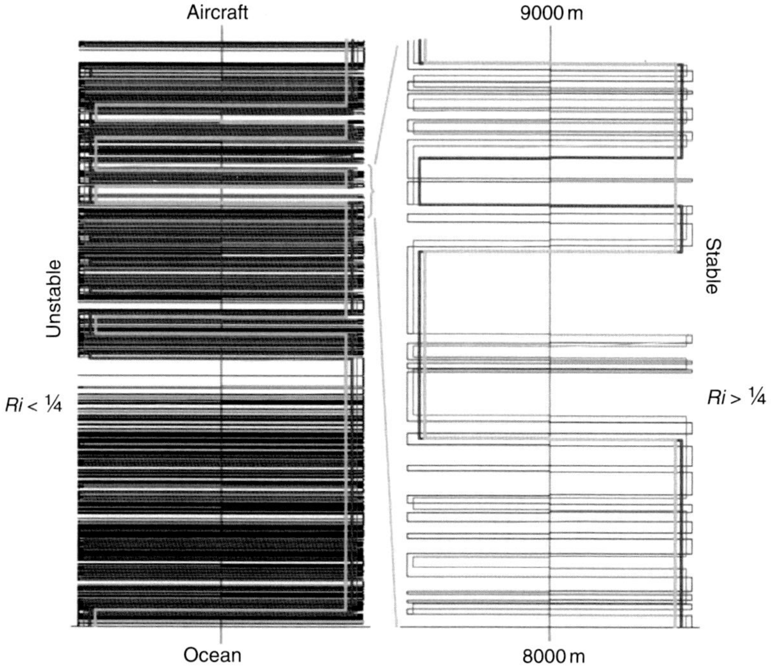

Fig. 6.11 The stability of the atmosphere as determined by a dropsonde using the stability criterion $Ri > 1/4$ where the Richardson number (Ri) is estimated using increasingly thick layers: 5, 20, 80, 320 m thick (black, red, blue, cyan respectively). The figure shows atmospheric columns, the left one from the ocean to 11 520 m (just below the aircraft), while the right is a blow-up from 8000 to 9000 m. The left of each column indicates dynamically unstable conditions ($Ri < 1/4$) whereas the right-hand side indicates dynamically stable conditions ($Ri > 1/4$). The figure reveals a Cantor set-like (fractal) structure of unstable regions. Reproduced from Lovejoy *et al.* (2008). *See colour plate section.*

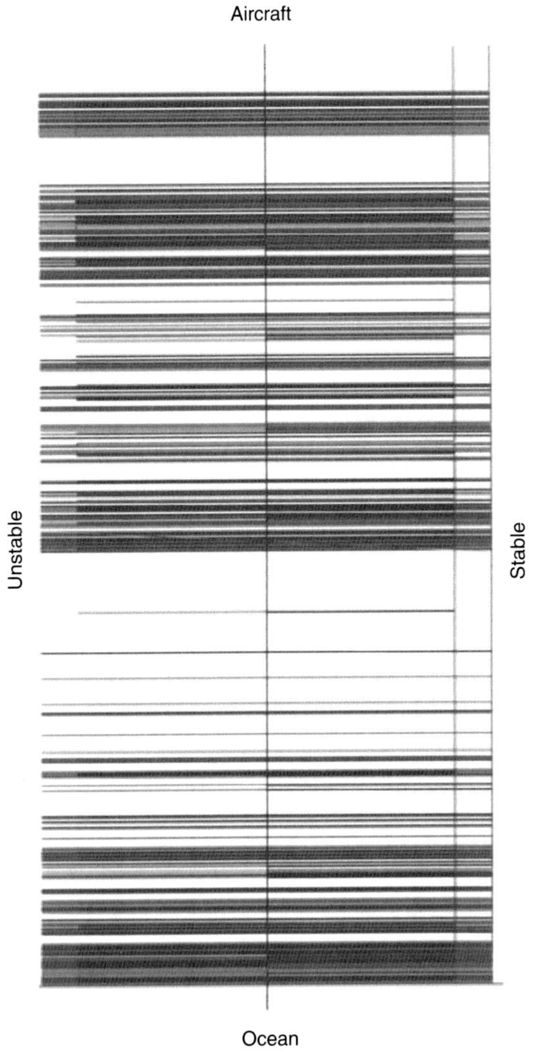

Fig. 6.12 The stability of the atmosphere as determined by two dropsondes dropped about 30 m apart (indicated by darker and lighter transitions), using the stability criterion $N^2 = g\left(\frac{\partial log\theta}{\partial z}\right) > 0$ where N^2 is estimated using layers at 5 m thickness. The transitions from unstable (left) to stable (right) are shown as a function of altitude from the ocean (bottom) to 12 km altitude (top). Nearly the same fractal structure is found in both, showing that the fractality is not an artefact of noise. Reproduced from Lovejoy *et al.* (2008).

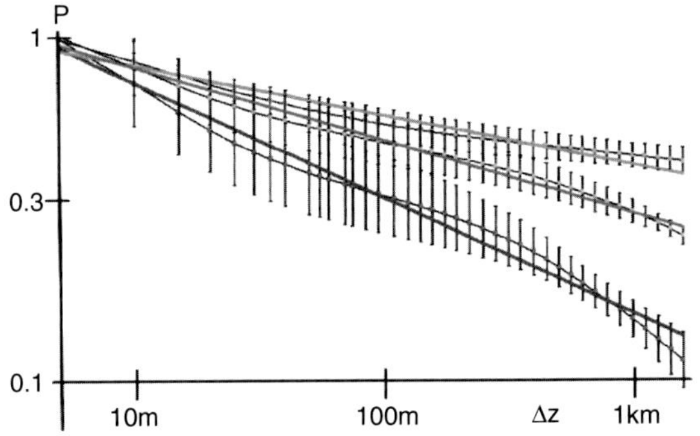

Fig. 6.13 The conditional probability P of finding an unstable layer at a distance Δz from another unstable layer (at 5 m resolution, the average over 24 sondes); $P \propto \Delta z^{-C_c}$ where C_c is the correlation codimension (= $1D_c$ where D_c is the correlation dimension of the unstable layers). The top is for convectively unstable layers, the middle is for dynamically unstable layers and the bottom is for conditionally unstable layers. The best-fit absolute slopes $C_{cN} = 0.36$, $C_{cRi} = 0.22$, $C_{cNE} = 0.15$ implying that the fractal correlation dimensions of the unstable layers are 0.64, 0.78, 0.85, respectively. Reproduced from Lovejoy *ct al.* (2008).

found over land (Schertzer and Lovejoy, 1985), so it is likely that our results are also valid over land. These different instability exponents reflect the importance of water vapour upon the vertical structure.

There are important consequences for the mainstream theories used to interpret vertical sounding data. These are the quasi-linear gravity wave theories notably the saturated cascade theory (Dewan and Good, 1986; Dewan, 1997) and the diffusive filtering theory (Gardner, 1994), which require layers with well-defined, real-valued, smoothly varying Brunt–Väisälä frequencies (N). If the stable propagating gravity waves are broken up by a sparse fractal distribution of unstable layers, it is not obvious that those theories can be saved. However, there is a strongly nonlinear alternative discussed in Chapter 9, where it is shown that one can readily make strongly nonlinear models based on localized turbulence fluxes which have wave-like unlocalized velocity fields, and this respecting the observed horizontal and vertical scaling. This emergent turbulent anisotropic scaling can give rise to (nonlinear) dispersion relations not so different than those predicted by linear theory, so it may be sufficient to reinterpret the empirical studies of waves in this anisotropic scaling framework.

Finally, the concept of a stable layer plays a central role in synoptic meteorology not only through thermodynamic diagrams, but more importantly through the product of N^2 with the absolute vorticity, i.e. the potential vorticity, PV. PV maps are interpreted with the help of balance conditions which are only strictly valid in stable layers (Hoskins *et al.*, 1985). At the moment PV analyses are mostly used in modelling the large scales with vertical resolutions such that layers are stable. However, as the models improve in resolution we may anticipate that the "Russian Matryoshka doll" picture of the fractal embedding of stable and unstable layers will become visible and will have to be taken into account.

6.3 The implications of anisotropic scaling for aircraft turbulence measurements

6.3.1 A simple model for interpreting aircraft measurements in scaling anisotropic turbulence

In Section 2.6 we already reviewed the classical aircraft campaigns and discussed the evidence that they showed transitions from $k^{-5/3}$ to $k^{-2.4}$ spectra with transition scales somewhere in the range 40–200 km. We also reviewed some recent analyses of data from commercial aircraft (TAMDAR), which although they had low horizontal resolutions were sufficiently accurate and numerous so as to allow direct estimates of $< \Delta v^2(\Delta x,\Delta z) >$ in the $(\Delta x,\Delta z)$ (vertical) plane. These structure functions were accurately reproduced by $H_z \approx 0.57$, in agreement with the theory developed earlier in this chapter, which predicts $H_z = 5/9$. We also mentioned that a simple model in which the aircraft fly on constant-slope trajectories could explain the transition. We now take a look at this a little more closely using high-spatial-resolution data from the scientific aircraft campaign Pacific Storms 2004 (for more details see Lovejoy *et al.*, 2009b).

Consider a large-scale section of a trajectory roughly following a sloping isobar with slope s:

$$\Delta v = \varphi_h l_s^{H_h} \left(\left(\frac{\Delta x}{l_s} \right)^2 + \left(\frac{s\Delta x}{l_s} \right)^{2/H_z} \right)^{H_h/2} \quad (6.41)$$

(cf. Eqns. (6.12), (6.17)).

When considering the stratospheric ER2 trajectory, Lovejoy *et al.* (2004) pointed out that in the simplest model of vertical drift where s was constant, then there would exist a critical lag where the two terms in Eqn. (6.41) were of equal magnitude: $\Delta x_c/l_s \approx (s\Delta x_c/l_s)^{1/H_z}$ (i.e. $\Delta x_c = l_s s^{1/(H_z-1)}$) so that for $\Delta x >> \Delta x_c$, the second term would dominate the first and we would obtain:

$$\begin{aligned} \Delta v &= \varphi_h \Delta x^{H_h}; & \Delta x &<< \Delta x_c \\ \Delta v &= \varphi_v s^{H_v} \Delta x^{H_v}; & \Delta x &>> \Delta x_c \end{aligned} \quad (6.42)$$

We would therefore expect a spurious break in the horizontal scaling at Δx_c, after which the aircraft would measure the vertical rather than horizontal statistics with exponent H_v rather than H_h. In the spectra, this transition corresponds to a transition from $k^{-5/3}$ ($H = 1/3$) to $k^{-2.4}$ (i.e. with $H \approx 0.75$ and $K(2) \approx 0.1$: $\beta = 1 + 2H - K(2)$); this simple mechanism thus explains the transition observed in virtually all the horizontal wind spectra, as discussed in Section 2.6.2. Similarly, the sloping nature of the isobars explains how the isobaric wind spectrum can also have a $k^{-2.4}$ spectrum.

6.3.2 Detailed investigation using Pacific Winter Storms 2004 data

In order to test this in more detail, we considered the Pacific Winter Storms 2004 data, which involved 10 aircraft flights over a roughly two-week period over the northern Pacific, each dropping 20–30 dropsondes (these were the data used earlier in this chapter and in Figs. 1.6c and 2.14). The plane flew along either the 162, 178 or 196 mb isobars to within standard deviations of $\pm$ 0.11 mb (i.e. the pressure level was $\approx$ constant to within $\pm$ 0.068%). Each had one or more roughly constant straight and constant-altitude legs more than 400 km long between 11.9 and 13.7 km altitude. The data were sampled every 1 s, and the mean horizontal aircraft speed with respect to the ground was 280 m/s. In addition, we checked that the standard deviation of the distance covered on the ground between consecutive measurements was $\pm$ 2% so that the horizontal velocity was nearly constant (in addition, using interpolation, we repeated the key analyses using the actual ground distance rather than the elapsed time and found only very small differences). Table 6.1 shows some of the characteristics of each of the legs.

The horizontal and vertical winds for the Gulf-Stream 4 aircraft are calculated by solving for the difference of inertial ground speeds in three dimensions and the flow angle measurements from various sensors. In the short term the inertial navigation system (INS) measurements of ground speed are much less noisy than ground speeds taken from GPS positions. While it is possible to smooth the GPS information to gain smooth ground speeds, that has not been done here; the INS data have been used. It is

Table 6.1 Comparison of the various characteristics of the 16 near-straight, flat flight segments (legs) considered in this chapter. The column Max(Δz) is the difference in altitude between the highest and lowest points on the leg, Δx_c is the critical scale beyond which the vertical exponent dominates the horizontal (Eqn. (6.44)), here estimated as the geometric mean between the longitudinal and transverse values. For legs 2 and 7, the transition was not attained over the entire leg, so only a lower bound is given. We also give the energy flux ε, and the sphero-scale l_s are determined by the "scale invariant lag" technique (explained in (Lovejoy *et al.*, 2009b)).

Leg no.	Length (km)	Max(Δz) (m)	Δx_c (km)	$\varepsilon \times 10^4$ (m^2s^{-3})	l_s (m)
1	2100	72	12.4	0.3	0.04
2	1248	83	> 1200	0.2	–
3	2496	69	84	40.	0.14
4	3348	737	108	0.2	0.05
5	2044	631	12.8	0.4	0.07
6	1476	260	7.6	0.2	0.03
7	568	19	> 400	0.3	–
8	1588	206	30.4	1.2	0.08
9	2924	100	384	30.	0.09
10	2272	172	40	40.	0.25
11	2980	899	100	0.4	0.10
12	3292	883	260	0.4	0.13
13	1532	308	52	0.2	0.60
14	3408	597	64	0.5	0.09
15	2780	191	3.6	5.	0.11
16	1844	178	48	2.0	0.05

not clear whether the INS has slower response than the GPS, or whether the latter's high-frequency variability is really "noise" or atmospheric variability. There are no corrections of the position of the inertial platform (near the cockpit) to the aircraft centre of gravity (back toward the trailing edge of the wing, probably 10 m aft of the INS platform).

Since turbulence is highly intermittent, in order to obtain robust estimates of exponents, experimentalists average their velocity fluctuations over as many lags as possible. Since $H_v > H_h$, it is enough that only some lags have a transition from horizontal to vertical behaviour for the spurious vertical scaling to dominate the ensemble statistics for large enough Δx. For each flight segment and for the averages over all the lags Δx, we therefore anticipate (cf. Eqn. (6.41)) that:

$$\langle|\Delta v|\rangle = \left((A\Delta x)^2 + (B\Delta x)^{2H_v/H_h}\right)^{H_h/2} \tag{6.43}$$

for some empirically determined constants A, B.

Fig 6.14a shows the individual structure functions for each flight segment (leg) for the longitudinal components with regressions to the form Eqn. (6.43) constrained to have $H_h = 1/3$ and $H_v = 3/5$ ($H_z = 5/9 = 0.55$, thick line), $H_v = 3/4$ ($H_z = 4/9 = 0.44$, thin line). Alternatively, we can fix $H_h = 1/3$ and find the error in the regressions for the ensemble of legs as function of H_z (Fig. 6.14b), the theoretical fits showing that the optimal $H_z \approx 0.46 \pm 0.05$, 0.45 ± 0.05 for longitudinal and transverse components respectively (with errors indicated for the optimum H_z values $\approx \pm 0.04$); these values are a bit lower than the 5/9 value discussed earlier. Finally (not shown, but see Lovejoy *et al.*, 2009b), we can determine the error-minimizing exponents H_h, H_v simultaneously; we find $(H_h, H_v) = (0.26 \pm 0.07, 0.65 \pm 0.04)$, $(0.27 \pm 0.13, 0.67 \pm 0.09)$ for the transverse and longitudinal components respectively. We see that the values are within a standard deviation of the horizontal (Kolmogorov) value $H_h = 1/3$ whereas the H_v values are a little larger than the Bolgiano–Obukhov values.

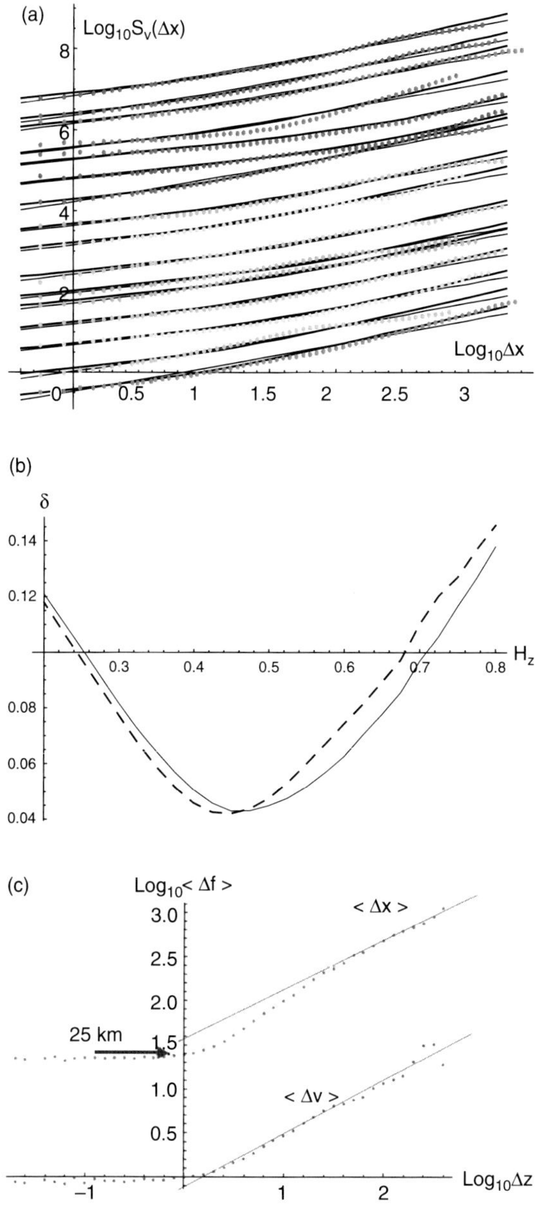

Fig. 6.14 (a) The first-order structure function for the longitudinal component of the horizontal wind for each of the 16 flight segments, each displaced by 0.5 in vertical for clarity. Δx is the horizontal distance in km. The thin line is the regression to the form Eqn. (6.43) with $H_z = 4/9$ while the thick line has $H_z = 5/9$. Reproduced from Lovejoy *et al.* (2009b). (b) The RMS error in estimating $\log_{10}\langle|\Delta v|\rangle$ for longitudinal (far right) and transverse (far left) components respectively, obtained by fixing $H_h = 1/3$. The minima correspond to the estimates: $H_z \approx 0.46 \pm 0.05$, 0.45 ± 0.05 for longitudinal and transverse components respectively. Reproduced from Lovejoy *et al.* (2009b). (c) The first-order structure functions for the longitudinal component of the horizontal wind (<Δv>, bottom, i.e. $f = v$, units m/s), and the horizontal distance (<Δx>, top, i.e. $f = x$, units km) as functions of vertical separations (Δz), estimated for 24 aircraft legs each at 280 m resolution in the horizontal, 1120 km each. For altitude fluctuations less than about 1 m (corresponding to $\approx$ 25 km), the wind fluctuations are independent of the vertical lag; for larger scales they follow almost exactly the Bolgiano–Obukhov $\Delta z^{3/5}$ law (the lower reference line). Similarly, the mean horizontal displacement as a function of vertical lag (<Δx>) closely follows the predictions of the 23/9D model for isobars (reference line slope 5/9). Reproduced from Lovejoy *et al.* (2010a).

We can use the leg-by-leg regression coefficients A, B obtained with $H_h = 1/3$, $H_z = 4/9$ (close to the regression value above) to estimate the critical Δx_c at which the two terms in Eqn. (6.43) are equal:

$$\Delta x_c = A^{H_z/(1-H_z)}/B^{1/(1-H_z)} \tag{6.44}$$

We see that in two cases (leg 2 and leg 7) the $\Delta x^{1/3}$ law holds well over the entire leg so that no transition is observed (the corresponding entry is blank: we are unable to estimate Δx_c and l_s). Detailed examination of the corresponding slopes shows that these are cases with particularly low slopes which, following our preceding analyses, favour the horizontal exponents. If in addition the l_s value is particularly large – and it is highly variable – then there will be no transition over the observed range of lags Δx.

This analysis gives strong support to our conclusions about the effect of anisotropic turbulence. In addition, from the table we can see that the values of both ε and l_s are indeed highly variable (explaining the large leg-to-leg differences in transition scales Δx_c) but the mean values are not unusual. For example the mean $\varepsilon \approx 8 \times 10^{-4}$ m^2s^{-3} is not so far from the troposphere average value $\approx 10^{-3}$ m^2s^{-3} (see e.g. Section 8.1.3) and the estimates of l_s are in the range 3 cm to about 70 cm, which is exactly the range of the direct estimates from lidar (see Section 6.5.1) and a little larger than the estimate from the mean ER2 data ($l_s \approx 4$ cm) (Lovejoy *et al.*, 2004).

The constant-slope model presented here is only an approximation to the rather complex interaction between the aircraft, the wind and the isobars. To get a more in-depth appreciation of the subtleties, we can consider the cross-correlations between the wind fluctuations with both altitude and pressure fluctuations: see Appendix 6B.

We can also re-examine the influence of the sloping trajectories on the wind fluctuations in another way. Recall that the aircraft slopes, combined with the observed sphero-scales and with the observed exponents, lead to transition scales Δx_c of the same order as those observed. Consider now Fig. 6.14c, which includes an analysis of the mean horizontal wind shear along 24 4000-point long legs (1120 km), (a subset of the generally longer legs discussed earlier, some of which were broken into several 4000-point sections). This figure uses all the pairs of points on a trajectory indexed by i, j and calculated as:

$$\langle|\Delta v_k|\rangle = \frac{1}{N_k} \sum_{\lambda_0^k \Delta z_0 < \Delta z' < \lambda_0^{k+1} \Delta z_0} |v(x_i, z_i) - v(x_j, z_j)|; \quad \Delta z' = |z_i - z_j| \tag{6.45}$$

$$\langle|\Delta x_k|\rangle = \frac{1}{N_k} \sum_{\lambda_0^k \Delta z_0 < \Delta z' < \lambda_0^{k+1} \Delta z_0} |x_i - x_j|; \quad \Delta z = |z_i - z_j|; \quad \Delta x = |x_i - x_j| \tag{6.46}$$

where $\lambda_0 = 10^{0.1}$ is a scale factor and N_k is the number of pairs in the kth interval, i.e. between $\Delta z\, \lambda_0{}^k < \Delta z' < \lambda_0{}^{k+1}\, \Delta z$, and Δz_k, Δx_k are the corresponding vertical and horizontal displacements. We can see that for $\Delta z >$ 1 m (corresponding to the mean critical transition scale of about $\langle|\Delta x_k|\rangle$ 25 km (see the upper curve), the wind follows almost exactly the theoretically predicted Bolgiano–Obukhov scaling $\Delta v \approx \Delta z^{3/5}$ (the line is for reference, it is not a regression). At the smaller scales, Δv is independent of Δz, as would be expected if the slopes of the aircraft were unimportant, a consequence of the fact that the smaller-scale fluctuations were dominated by turbulent influences on aircraft drag and lift. By comparing the upper and lower graphs (for $<\Delta v>$ and $<\Delta x>$, respectively) we see that the transition is at ~25 km, i.e. roughly the scale at which the phase relations are reversed and the $k^{-5/3}$ spectrum gives way to a $k^{-2.4}$ spectrum. To put it another way, if the sloping nature of the isobars were irrelevant (so that isoheight and isobaric exponents were the same), then why is there any systematic variation of Δv with Δz – and why does it follow so perfectly the Bolgiano–Obukhov predictions?

6.4 Horizontal and vertical analyses of dynamic and thermodynamic variables

6.4.1 Horizontal structure function analyses

Ideally, in order to properly quantify atmospheric stratification, we should analyse vertical sections. Unfortunately, such data do not exist for any of the usual variables of state, and the best we can do is use remotely sensed data: we turn to this in the next section. For the usual state variables, the best solution is to use (near) simultaneous horizontal aircraft and vertical dropsonde soundings; they are available from the Winter Storms 2004 experiment, and we now give a more complete analysis of those data.

When we discussed spectral and cascade analyses in the horizontal (cf. Figs. 1.6c, 6.14a and Figs. 4.6a, 4.6b) we noted the break induced in the horizontal wind statistics by the gentle slopes of the isobaric aircraft trajectories. In Chapter 5, in order to illustrate the structure function method, we calculated the aircraft structure functions of various orders, but postponed until now a discussion of the results.

In Fig. 5.36a we showed the structure function results for the pressure and the longitudinal and transverse wind components (i.e. those parallel and perpendicular to the direction of the aircraft). The behaviour for the wind has essentially two scaling regimes with a 4–40 km transition regime. In comparison, since the aircraft attempted to follow isobars, the pressure has poor scaling (i.e. the deviations from perfect isobaric trajectories). In Fig. 5.36b we showed the corresponding plot for those fields little affected by the trajectory fluctuations: the temperature, humidity and log potential temperature; we see that the scaling is indeed very good (see also Fig. 1.6c). In Table 6.2 we also give the corresponding C_1 estimates obtained by numerically estimating $K'(1)$ using regressions over the range 4–40 km for the thermodynamic variables with the best scaling. We can see that the various methods (trace moments, spectra, structure functions) give estimates of C_1, H which are very close to each other, with all fields having roughly the same values: $\alpha \approx 1.8$, $C_1 \approx 0.05$, $H \approx 0.50$. In spite of the fact that, classically, we expect H to be given by dimensional analysis of a turbulent flux – and hence to be expressed as a ratio of small integers, here apparently 1/2 – as far as we know there is no theory which plausibly explains these values: passive scalar theory yields $H = 1/3$, and the classical BO theory applied to the temperature gives $H = 1/5$.

Table 6.2 Parameter estimates over the "optimum" range 4–40 km. β is the spectral exponent, H_{st} is from the first-order structure function, $H_\beta = (\beta + K(2) - 1)/2$, and H_{mean} is the average of the two. $C_1 = K'(1)$ from the trace moments and $C_{1,st} = \xi(1) - \xi'(1)$, C_{1mean} is the average.

	T	$\log\theta$	h
α	1.78	1.82	1.81
β	1.89	1.91	1.99
$K(2)$	0.12	0.12	0.11
C_1	0.064	0.063	0.051
$C_{1,st}$	0.040	0.042	0.028
$\mathbf{C_{1mean}}$	**0.052 ± 0.012**	**0.052 ± 0.010**	**0.040 ± 0.012**
H_{st}	0.49	0.50	0.52
H_β	0.51	0.52	0.50
$\mathbf{H_{mean}}$	**0.50 ± 0.01**	**0.51 ± 0.01**	**0.51 ± 0.01**

6.4.2 Intermittent sampling intervals and structure functions in the vertical

A particularity of the vertical dropsonde data discussed in Sections 6.1.5 and 6.2.2 is that the measurements are not uniformly spaced. Whereas estimating fluctuations and fluxes is straightforward enough for data sampled at regular intervals, for data with highly irregular resolutions we must take into account the variability of the resolution. The dropsonde resolution is variable for two reasons: first, even if the sampling was always at the nominal 0.5 s time interval, the variable vertical sonde fall speed would lead to variable vertical sampling intervals. This source of variability is not too large: due to increased air resistance the mean vertical sonde velocity decreases from about 18 m/s to about 9 m/s near the surface; in addition, turbulence-induced fluctuations increase this range of resolutions by another factor of ~2. However, the variability problem is made much, much worse because of data outages – even though these affected only 9.5% of the observations. The problem is that they affected every sonde, that they were highly clustered and that they were sometimes very large (occasionally several kilometres in size). Fig. 6.15 shows the distribution of the distance between consecutive measurements from two near-simultaneous sondes (launched 0.3 s apart); one can see not only that can the outages be large but that they are highly clustered. If such data are interpolated onto uniform grids, then their spectra and other statistics can be significantly biased.

When calculating structure functions these problems are not hard to avoid; in Section 6.1.5 we simply used point pairs to define the fluctuation $\Delta v(\Delta z)$ as the difference between the velocity at two vertical levels (indexed by n, m) separated by altitude Δz:

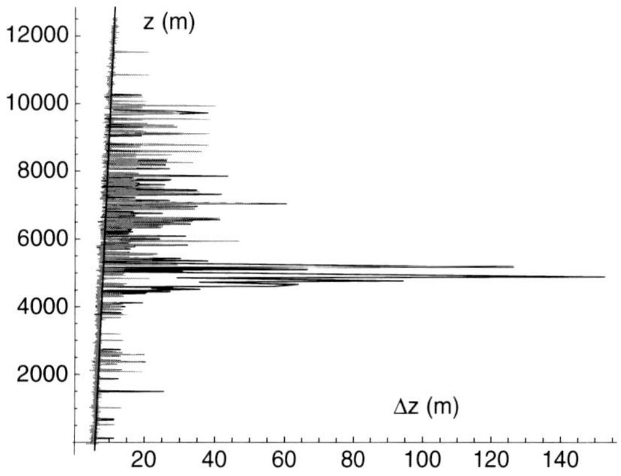

Fig. 6.15 Comparison of vertical sampling intervals of two near-simultaneous sondes. Notice the strong (and typical) clustering of the outages. The mean Δz is larger at high altitudes due to the lower air resistance. Reproduced from Lovejoy *et al.* (2009a).

$$\Delta v(\Delta z) = |v(z_n) - v(z_m)|; \quad \Delta z = |z_n - z_m| \qquad (6.47)$$

By examining all the $N(N - 1)/2$ observation pairs we avoid interpolations; multifractal simulations show that the pair method is quite robust even in the presence of large intermittency in the measurements (Lovejoy *et al.*, 2009a). Defining fluctuations in this way as differences is equivalent to using a "poor man's wavelet" and is only valid for $0 < H < 1$ (Section 5.5). Since the temperature, pressure and potential temperatures are known to be nearly linear with Δz (for the temperature, this is the classical linear adiabatic temperature profile), we must define fluctuations in a way that is valid over a wider range of H values; at least for $2 > H > 0$. In Section 5.5 we saw that the basic trick for doing this is to use the second centred differences that are classically defined for evenly spaced z_n by:

$$\Delta v(\Delta z) = \left| \frac{v(z_{n+m}) + v(z_{n-m})}{2} - v(z_n) \right|; \qquad \Delta z = z_{n+m} - z_n = z_n - z_{n-m} \tag{6.48}$$

where the subscripts are the integer indices of the measurements. Graphically, this definition is equivalent to finding the distance in the v direction between the central point $(z_n, v(z_n))$ and the line joining the points $(z_{n+m}, v(z_{n+m}))$, $(z_{n-m}, v(z_{n-m}))$. To obtain a fluctuation estimate valid for unevenly spaced z_n, we can simply use the distance from the line to obtain the following estimate:

$$\Delta v(\Delta z) = |v(z_{n-m}) + (z_n - z_{n-m})s - v(z_n)|; \quad s = \frac{v(z_{n+m}) - v(z_{n-m})}{z_{n+m} - z_{n-m}} \quad \Delta z = \left((z_n - z_{n-m})(z_{n+m} - z_n)\right)^{1/2} \tag{6.49}$$

Since now there are three points needed to define the fluctuation, there is not a unique choice of scale (lag) Δz with which to associate the fluctuation. The above choice (the geometric mean) evenly weights the logarithm of the scales and is appropriate for scaling fluctuations. In order to implement this definition it is usually not possible to consider all the triplets of points (this is of order N^3 for series of length N). Our choice was based on the fact that the outages affected only 9.5% of the points, so we considered only the triplets with altitudes z_{n-m}, z_n, z_{n+m}, and for each scale we considered all the integer pairs n, m. Note that this choice, along with the use of the geometric mean for Δz, means that there will be few statistics for the largest factor of 2 in scale. Also, since most of the points are regularly spaced, $z_n - z_{n-m}$ is typically not so different from $z_n - z_{n+m}$, so the definition does not "mix" different scales too much.

We may now apply the fluctuation analyses to the atmospheric fields whose fluxes were analyzed above. Using the definitions in Eqns. (6.47) and (6.49) for the fluctuations for estimating the qth-order structure functions (see Eqn. (5.98)), we obtain Figs. 6.16a and 6.16b for the dynamic and thermodynamic fields, respectively. They have been nondimensionalized by dividing by the value $<\Delta f>(\Delta z)$ for $\Delta z = 10$ m, although there is no expectation that the lines converge to a point as for the moments of the fluxes. In the figure, we used definition Eqn. (6.47) for the fluctuations of v, h, w_s (since $H < 1$) and definition Eqn. (6.49) for p, $\log \theta$, $\log \theta_E$, ρ, T (since H is near 1 or larger). In the case of the pressure, we see that $H \approx 2$, so we should perhaps have further generalized the definition of fluctuations so as to obtain a result valid for $H < 3$.

From the figures, the linearity of the log fluctuation moments versus $\log \Delta z$ is quite striking. Indeed, we will see in the next subsection that the scaling of the moments apparently extends to somewhat larger scales than the scaling of the fluxes. In order to assess both the quality of the scaling and to compare the scaling of the different fields we refer the reader to Table 6.3, where we have calculated the exponent $H = \xi(1)$ from the mean of the exponents calculated for $\Delta z < 300$ m and $\Delta z > 300$ m (300 m is the geometric mean of the observed range 10 m to 10 km). The range indicated by the "±" is half the difference. If the spread indicated in this way is small then the scaling is effectively well respected over the whole range; we see that the H values for p, w_s are particularly well defined, whereas for T, $\log\theta$ it is less so.

We made no attempt to investigate the altitude dependence of the exponents from the structure functions, but this was done with the flux moment analyses discussed in the next section, and the parameters for this analysis are also given in the table for comparison. The altitude dependence of the exponents, although small, is partially responsible for the imperfect scaling since the thickest layers necessarily involve points at high altitudes. To estimate C_1 from the structure (fluctuation) exponent $\xi(q)$, we can exploit the equation $\xi(q) = qH - K(q)$ so that $\xi'(1) = H - C_1$ and hence $C_1 = \xi(1) - \xi'(1)$. However, for these fields, H is much larger than C_1; indeed, it is often the order of the error in the estimates of H, hence the C_1 will not be too accurate. Since the fluxes do not depend on H they can be used to directly estimate $K(q)$ and hence C_1; we therefore consider the flux-based estimate of C_1 to be more accurate: see the next section.

6.4.3 Cascades in the vertical

By treating the measurement intervals in both the vertical and in time Lovejoy *et al.* (2009a) obtained a surprising result: the outages had almost exact

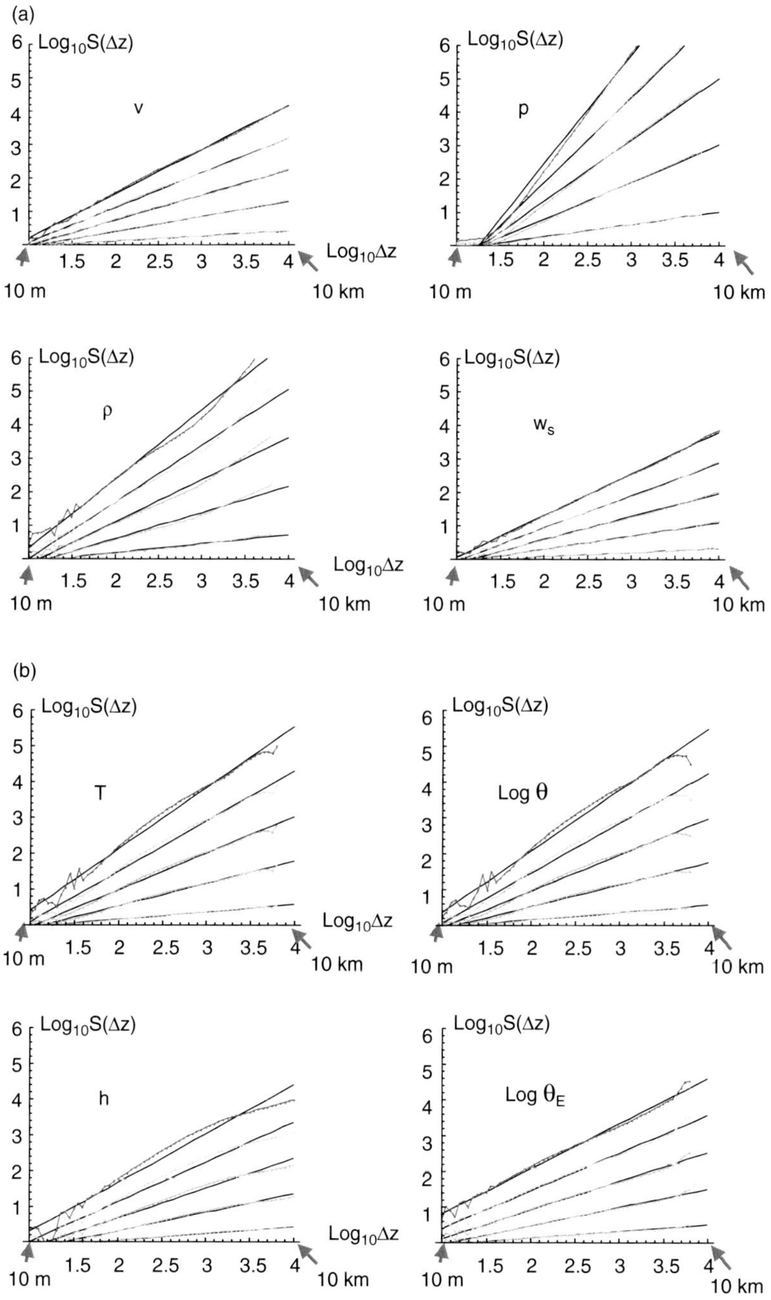

Fig. 6.16 (a) Nondimensionalized dynamical variables, moments of the fluctuations Δf; $S(\Delta z) = \langle \Delta f(\Delta z)^q \rangle$. Clockwise from upper left: $f = v, p, \rho, w_s$. Moments $q = 0.2, 0.6, 1, 1.4, 1.8$. Reproduced from Lovejoy *et al.* (2009a). (b) Same as Fig. 6.16a but for the thermodynamic variables, fluctuation analysis. Clockwise from upper left: $f = T$, $\log\theta$, h, $\log\theta_E$. Moments $q = 0.2, 0.6, 1, 1.4, 1.8$. Reproduced from Lovejoy *et al.* (2009a).

cascade structures with rather large intermittencies ($C_{1time} \approx 0.21$, $C_{1vert} \approx 0.23$) with outer scales near the outer scale of the data ($\approx$ 200 s and $\approx$ 3 km, respectively). In order to overcome this extreme outage problem, two developments are needed. The first is a robust technique to estimate the fluxes, the second a method of statistical correction to the scaling exponents in order to correct for the strongly variable resolutions (Lovejoy *et al.*, 2009a).

In order to estimate the turbulent fluxes from such intermittent data, recall that in a scaling regime for a field v, the fluxes are related to the fluctuations

Table 6.3 Vertical parameter estimates from from dropsondes using both structure functions and moments of the fluxes. The rows H, $C_{1,fluc}$ are from the structure function (fluctuation) analyses over the entire range 0–10 km; they are indicated as the means of the fits from 30–300 m and 300–3000 m; the spread is half the difference (H is from $\xi(1)$, C_1 from $\xi(1)$-$\xi'(1)$). The row C_1 is from the flux analyses in Section 6.4.3 reduced by the factor 1.39 to account for the intermittency of the sonde outages. $C_{1,>6km}$ is the same but for the upper part only (6–10 km); this shows some of the variation with altitude but is also an estimate of the 200 mb values that can be compared to the horizontal aircraft values at roughly this flight level (Table 6.5). The rows α, L_{eff}, δ are from the flux moment analysis over 0–10 km (δ is the residual, Eqn. (4.16))

	T	Logθ	Logθ_E	h	v	p	ρ
H	1.07 ± 0.18	1.07 ± 0.18	0.87 ± 0.10	0.78 ± 0.07	0.75 ± 0.05	1.95 ± 0.02	1.31 ± 0.12
$C_{1,fluc}$	0.066 ± 0.038	0.051 ± 0.027	0.140 ± 0.101	0.144 ± 0.028	0.023 ± 0.016	0.043 ± 0.032	0.123 ± 0.103
C_1	0.049	0.046	0.106	0.103	0.071	0.032	0.065
$C_{1,>6km}$	0.072	0.071	0.069	0.091	0.088	0.072	0.077
α	1.70	1.90	1.90	1.85	1.90	1.85	1.95
L_{eff} (km)	5.0	4.0	25	16	1.3	5.0	13
δ (%)	1.4	1.2	1.9	1.4	2.3	1.1	1.4

through the basic flucuation/flux relation $\Delta v(l) = \varphi_l l^H$. With regularly spaced data, the usual way to estimate the flux is to degrade it starting from the highest resolution in the scaling regime (so that the above law holds), and estimate the fluctuations at the finest scale (e.g. by absolute first differences or by absolute wavelet coefficients). The result is the flux at the finest resolution; one then degrades the result by averaging over larger and larger scales. However, we can also estimate the fluctuations from the local derivatives; for example (for uniformly spaced data) for the nth flux estimate, we can use the data at the (n–1)th and (n+1)th points to yield a "centred difference" estimate:

$$\Delta v_l(z_n) \approx \left|\frac{dv(z_n)}{dz}\right|_l l \approx s_n l; \quad s_n = \left|\frac{v(z_{n+1}) - v(z_{n-1})}{z_{n+1} - z_{n-1}}\right| \tag{6.50}$$

where s_n is the local absolute slope. We saw that in the scaling regime $\varphi_l = \Delta v(l)/<\Delta v(l)>$. At least if $H< 1$, we therefore have the estimate $\varphi_l = s/< s>$. In order to generalize this to intermittent locations z_n, we can simply include the point z_n and estimate s_n by a linear regression of the point triplets: $\left\{(z_{n-1}, v(z_{n-1})), (z_n, v(z_n)), (z_{n+1}, v(z_{n+1}))\right\}$. This method has the advantage of not being too sensitive to the noise in either the data (v) or the position estimates (z) (these factors were found to make interpolations unreliable for derivative estimates; similarly, estimates of spectra obtained from interpolated series will contain serious artefacts due to the interpolation). The resulting series of s_n values determined at the irregular locations z_n can then be interpolated at uniform intervals. Providing that one recognizes that their resolutions are not fixed, the (normalized) slopes can be used as estimates of the normalized fluxes ($\varphi_l = s/< s>$). One can then use the interpolated s series to obtain regularly sampled values, systematically degrading these to obtain a series of lower- and lower-resolution flux estimates as usual. The final step is to statistically correct the result for this variable-resolution effect.

To see how this method works we determined the fluxes for the simultaneous sonde pair analysed in Fig. 6.15. The results for the main dynamic and thermodynamic fields are shown in Fig. 6.17a. We can see that the estimates for the two sondes are very similar (the curves are mostly indistinguishable), even though, as Fig. 6.15 shows, the outages were significantly different. In Fig. 6.17b, we blow up a particularly intermittent section, which shows the enormous variability of the fluxes (especially the humidity and equivalent potential temperature), which is nonetheless well reproduced by both sondes. Finally, in Fig. 6.17c we examine a section of the data that was particularly poorly sampled by the sondes (Fig. 6.15): we see that even here the two sondes give fairly similar flux estimates, although at different resolutions.

When this technique was applied to the dropsonde data, the moments displayed in Figs. 6.18a and 6.18b were obtained. As for the structure function analysis, the quantities that we analysed can be roughly grouped into two categories: dynamic and thermodynamic variables. The dynamic variables (Fig. 6.18a) were the modulus of the horizontal wind v, the pressure p, the total air density (ρ, including that due to humidity), and the sonde vertical velocity w_S. We also separately analyzed the north–south and east–west components of the horizontal wind, but the results were not much different and we will not discuss them further. For the vertical sonde velocity, the fluctuations around a quadratic fit (corresponding to a constant deceleration from 18 m/s to 9 m/s) were used. Due to the parachute drag, the fluctuations in w_s depend on both the vertical and horizontal wind, so it should not be used as a surrogate for the vertical wind.

The cascade structures displayed in Figs. 6.18a and 6.18b are biased due to the outages, so the flux estimates have strongly variable resolutions. It turns out that these can be taken into account and simple statistical corrections can be made (see the somewhat technical details in Lovejoy *et al.*, 2009a); in the case here, they effectively boost the C_1 by a factor of about 1.39. Table 6.3 compares the corrected exponents obtained in this way, both those over the entire range 0–10 km and those from the higher range 6–10 km indicating that the intermittency generally increases with altitude, the exception being the humidity and the derived equivalent potential temperature. The basic regressions were taken over the range 10–300 m. From the figures we can see that with small deviations δ (the residuals Eqn. (4.16); see Table 6.3) all the fields well respect to the predictions of cascade theories over the range 1 km to 10 m.

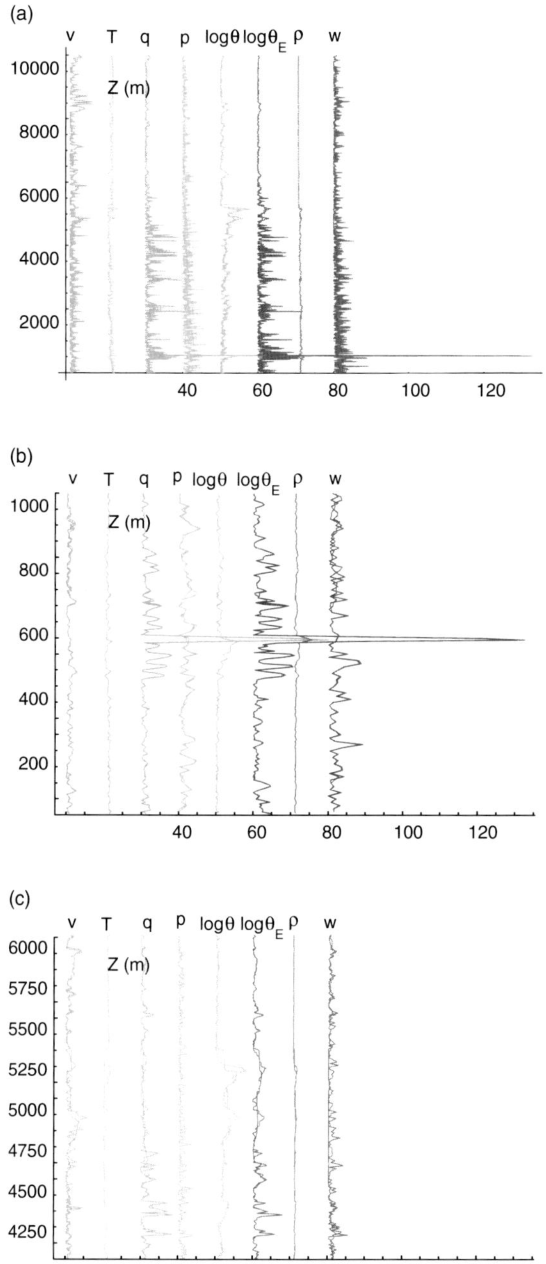

Fig. 6.17 (a) Comparison of normalized (and hence nondimensional) fluxes from a simultaneous sonde pair. For each field, the profiles are so close as to be nearly indistinguishable. The acceleration was not analyzed, since the noise was too large. Reproduced from Lovejoy *et al.* (2009a). (b) Detail of Fig. 6.17a showing that even very intermittent layers can be well reproduced from sonde to sonde. Reproduced from Lovejoy *et al.* (2009a). (c) Comparison of normalized flux estimates from the two sondes for a particularly poorly sampled section from Fig. 6.17a. Reproduced from Lovejoy *et al.* (2009a).

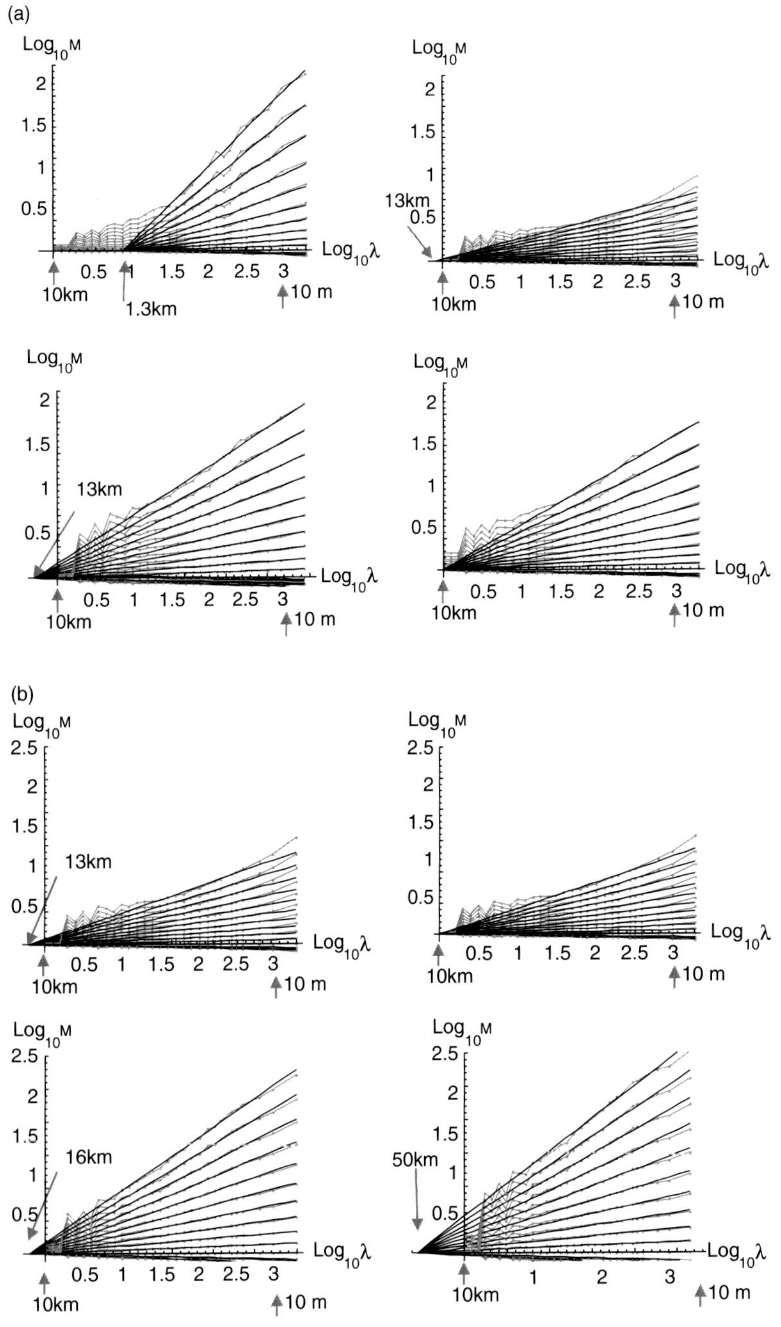

Fig. 6.18 (a) The dynamical fields v, p, ρ, w_s (clockwise from upper left) for $q = 0.2, 0.4, \ldots, 2$. Reproduced from (Lovejoy *et al.*, 2009a). (b) The same as for Fig. 6.18a except for the thermodynamic fields T, $\log\theta$, h, $\log\theta_E$ (clockwise from upper left), reproduced from (Lovejoy *et al.*, 2009a).

6.5 Direct verification of anisotropic cascades using lidar backscatter of aerosols and CloudSat radar reflectivities

6.5.1 Simultaneous horizontal and vertical analyses; structure functions

Lidar and radar are in many ways ideal methods for studying the vertical stratification since they directly yield vertical cross-sections. We therefore return to the unique dataset of airborne lidar backscatter from aerosols already discussed in Section 1.2.4. The data were taken over three afternoons in August 2002 near Vancouver, British Columbia: see Fig. 1.7b, and the zoom (Fig. 1.7c), for an example showing the extremely fine details, including hints that at the small scales the structures are no longer flat, but start to be stretched in the vertical; compare this with the simulation in Fig. 6.7a.

According to the isotropic Corrsin–Obhukhov passive scalar theory (Section 2.3), the mean absolute fluctuations in the aerosol density ρ should scale as:

$$\Delta\rho(\Delta x, \Delta z) \stackrel{s}{=} \chi^{1/2}\varepsilon^{-1/6}|(\Delta x, \Delta z)|^{H_{CO}}; \quad H_{CO} = 1/3 \tag{6.51}$$

where χ and ε are the passive scalar variance and energy fluxes, respectively, and we have used the Corrsin–Obhukhov exponent $H_{CO} = 1/3$. The extension to anisotropic turbulence is effected simply by replacing the (isotropic) vector norm $|(\Delta x, \Delta z)|$ by the scale function $\|(\Delta x, \Delta z)\|$:

$$\begin{aligned} &\Delta\rho(\Delta x, \Delta z) \stackrel{s}{=} \chi^{1/2}\varepsilon^{-1/6}\|(\Delta x, \Delta z)\|^{1/3}; \\ &\|(\Delta x, \Delta z)\| = \Theta(\theta')r' \end{aligned} \tag{6.52}$$

where:

$$\begin{aligned} &r' = l_s\left(\left(\frac{\Delta x}{l_s}\right)^2 + \left|\frac{\Delta z}{l_s}\right|^{2/H_z}\right)^{1/2}; \\ &\tan\theta' = \frac{\Delta z'}{\Delta x'} = \frac{sign(\Delta z)|\Delta z/l_s|^{1/H_z}}{(\Delta x/l_s)} \end{aligned} \tag{6.53}$$

(with $H_z = 5/9$; see Eqn. (6.29) for the nondimensional version).

We first test the special horizontal ($\Delta z = 0$, $\theta' = 0$) and vertical ($\Delta x = 0$, $\theta' = \pi/2$) cases:

$$\begin{aligned} &\Delta\rho(\Delta x, 0) \stackrel{s}{=} \chi^{1/2}\varepsilon^{-1/6}\Theta(0)\Delta x^{1/3} \\ &\Delta\rho(0, \Delta z) \stackrel{s}{=} \chi^{1/2}\varepsilon^{-1/2}\varphi^{1/5}\Theta\left(\frac{\pi}{2}\right)\Delta z^{3/5} \end{aligned} \tag{6.54}$$

The first (horizontal) law is the usual Corrsin–Obhukhov law, but the second (vertical) law (obtained using Eqn. (6.13) for l_s in terms of the buoyancy variance flux φ) is a new prediction of the 23/9D anisotropic scaling theory.

We can now test Eqn. (6.54) by analyzing the first-order ($q = 1$) structure function averaged over the nine available cross-sections (Fig. 6.19a). The first-order structure function is interesting because we expect $K(1)$ to be small enough that the horizontal and vertical H's (H_h and H_v) can be estimated as $\xi_h(1) \approx H_h = 1/3$, $\xi_v(1) \approx H_v = 3/5$. We can see from the figure that not only is the scaling excellent in both horizontal and vertical directions, but in addition the exponents are very close to those expected theoretically. In fact, we find from linear regression: $H_h = 0.33 \pm 0.03$, $H_v = 0.60 \pm 0.04$. Also visible in the figure is the scale at which the functions cross; this is a direct estimate of the sphero-scale, which we find here varies between 2 cm and 80 cm, with an average of 10 cm.

Next, we can empirically verify Eqn. (6.52) for directions other than the coordinate axes; indeed, we can estimate $\Theta(\theta')$. To test this on the data, it suffices to use the nonlinear coordinate transformation $(\Delta x', \Delta z') = \left(\Delta x, sign(\Delta z)|\Delta z|^{1/H_z}\right)$ (Eqn. (6.28)) which makes the structure function $S_1(\Delta x',\Delta z') = \langle\Delta\rho(\Delta x',\Delta z')\rangle$ only trivially anisotropic; this is the anisotropic scaling analysis technique (ASAT; Radkevitch *et al.*, 2007, 2008). Taking constant θ′ rays, we obtain Fig. 6.19b; we see that the lines are parallel with exponent 1/3 in all directions as expected. The difference between the 1D structure functions in different directions means that the unit ball is not a circle (sphere). To clearly see this "trivial" anisotropy $\Theta(\theta')$ we remove the theoretically expected r' dependence by calculating the "compensated" 2D structure function $r'^{-1/3}S_1(r', \theta')$, averaging it over $\log_{10}(r')$: $S_{c,1} = \langle r'^{-1/3}S_1(r', \theta')\rangle_{\log r'} \approx \Theta(\theta')^{1/3}$ as a function of polar angle θ′ – it determines the shape of the unit ball. $S_{c,1}$ is useful because, within rather small intermittency corrections, it is expected to be independent of r'. Fig. 6.19c was determined from $r'^{-1/3}S_1$ values as functions of $\log_{10}(r')$ and θ′ with increments of $\log_{10}(r') = 0.2$. Along rays of fixed angle θ′, the relative error (the ratio of $S_{c,1}$ and standard deviation) does not exceed 10% over wide ranges of scale

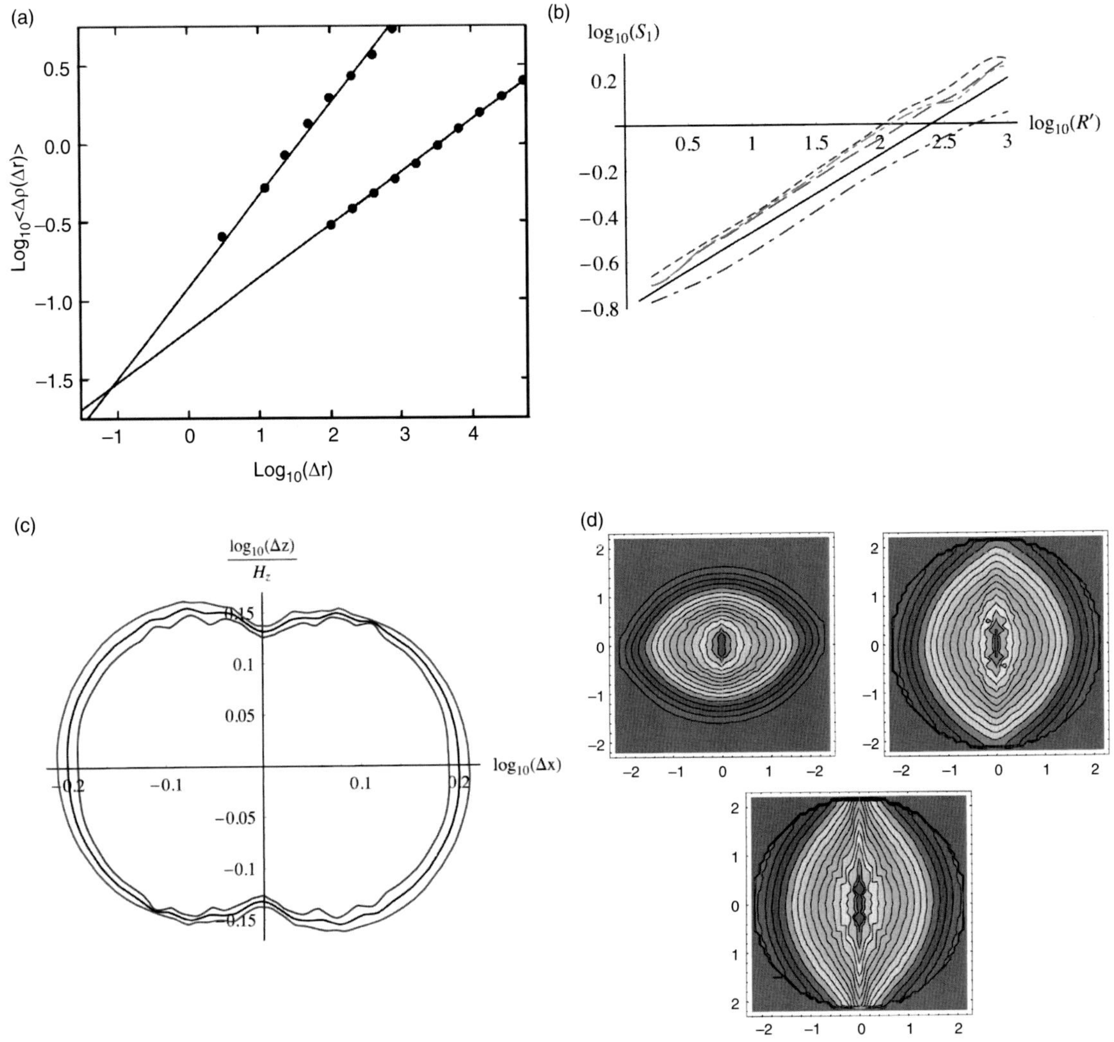

Fig. 6.19 (a) The symbols show the first-order vertical structure function, and first-order horizontal structure function for the ensemble of nine vertical airborne lidar cross-sections. ρ is the dimensionless backscatter ratio, the surrogate for the passive scalar aerosol density; Δr is either the vertical or horizontal distance measured in metres. The lines have the theoretical slopes (*not* regressions!) 3/5, 1/3, and they intersect at the sphero-scale here graphically estimated as ~10 cm. Reproduced from Lilley *et al.* (2008). (b) 2D structure function S_1 as function of r' (indicated R' in the plot) for four directions ($\theta' = 0$ (short dashes, corresponding to $r' = \Delta x'$), $\pi/4$ (long dashes), $\pi/2$ (long, short dashes, corresponding to $r' = \Delta z'$)), $3\pi/4$ (long, short, short dashes) for a single cross-section in the nonlinearly transformed ($\Delta x'$, $\Delta z'$) coordinates. The solid line has the theoretical slope of 1/3. Reproduced from Radkevitch *et al.* (2008). (c) Polar plot of compensated structure function $r'^{-1/3}S_1(\log_{10}(r'),\theta')$ averaged over for the same cross-section as in Fig. 6.19b (black line) with angle representing the direction of $(\Delta x',\Delta z')$ and the distance from the origin representing the size of $S_{1,c}$; the averages are over $0.2 < \log_{10}(r') < 2.8$. The quantities $\log_{10}(\Delta x)$ and $\log_{10}(\Delta z)/H_z$ show distance from the centre of the plot in time and vertical directions, but they are not Cartesian coordinates. The grey lines are compensated averaged 2D structure function ± standard deviation. Reproduced from Radkevitch *et al.* (2008). (d) Contour plot of S_1 with different nonlinear transformations in polar coordinates with radius = $\log_{10}r'$ and the polar angle = θ'. Upper left: isotropic turbulence (no transformation, $H_z = 1$); upper right: 23/9D model ($H_z = 5/9$); bottom: gravity waves ($H_z = 1/3$). Note that although the x' (left–right) and z' (up–down) coordinate directions show the distance from the centre of the plot in the corresponding direction, they are not Cartesian coordinates of the plot. Pacific 2001 0815t6, vertical–horizontal cross-section. If H_z is such that Eqn. (6.50) is satisfied, then the spacing between the contours is constant in all directions (close to the upper right, $H_z = 5/9$ case), this criterion does not imply that the contours have shapes independent of scale. Reproduced from Radkevitch *et al.* (2008). *See colour plate section.*

($r'_{max}/r'_{min} > 10^3$ in many instances); often this variation does not exceed 5% (Fig. 6.19c). The ranges chosen for averaging are indicated in the captions for the figures. Thus, the ASAT technique applied to first order ($q = 1$) 2D structure functions allows us to verify the theory to within about 10% over more than three orders of magnitude of space-space scales, and this at various angles in (x',z') space. Note that from its definition, the structure functions are symmetric with respect to inversion about the origin.

While the near-constancy of the compensated structure functions as functions of angle gives strong support to the correctness of the coordinate transformation and passive scalar theory (the exponent 1/3 in Eqn. (6.52)), one can also use the same technique to test the alternatives, the $D_s = 7/3$ gravity waves, and the $D_s = 3$ isotropic turbulence. This can be easily done by using the corresponding nonlinear coordinate transformations and visually checking for the constancy of the resulting Θ functions.

Fig. 6.19d shows a comparison of contour plots of $\log(S_1(\log(r'/r_i), \theta'))$ with the nonlinear transformation Eqn. (6.28) corresponding to different models: isotropic turbulence, the 23/9D model and quasi-linear gravity waves. If the contours of $\log(S(r', \theta'))$ are invariant under an isotropic scale change (they have the same shapes), then the corresponding contours of $\log(S_1(\log(r'/r_i), \theta'))$ will not have the same shapes but they will be rather equally spaced in all directions (R_i is a nondimensionalizing inner scale; below R_i, the signal is dominated by instrumental noise). The advantage of using a $((\log(r'/r_i),\theta'))$ space representation is that we can visually represent a huge range of scales on a single picture. For the upper left case in Fig. 6.19d there is no transformation of coordinates, and we can see that as we move from contour to contour the spacing between the contours is different in the horizontal and vertical directions. For the upper right case (using the theoretical transformation from the 23/9 model), we can see that the contours are spaced pretty much the same distance apart (i.e. equally spaced in all directions as expected). Finally, in the bottom case, the gravity wave value of $H_z = 1/3$ is used in the transformation, again leading to contours which are not equally spaced – this time they are closer in the vertical than in the horizontal. Note that all the contours are spaced at equal factors of S_1 of 1.12 (for the 23/9 model, i.e. with $H = 1/3$, this corresponds to a factor of 1.41 in scale), the total range of scales is roughly 100. Overall, the 23/9D model gives the best scale invariance of the three tested.

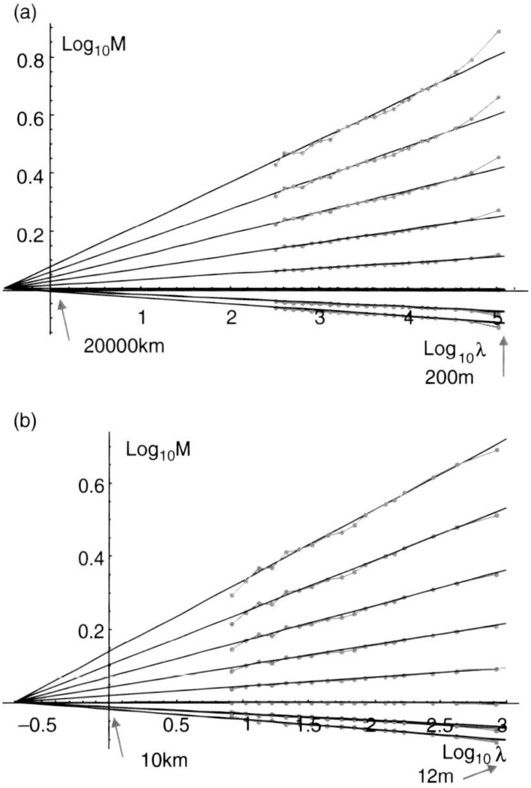

Fig. 6.20 (a) Horizontal analysis of the moments of the normalized lidar backscatter ratio for 10 atmospheric vertical cross-sections (L_{ref} = 20 000 km, corresponding to $\lambda = 1$). The curves are for the moments of order $q = 0.2, 0.4, \ldots, 2$. The largest directly accessible scale is ~100 km, and the lines converge to an effective outer scale of $L_{eff} \approx 25\,000$ km. Reproduced from Lovejoy *et al.* (2009a). (b) The same cross-sections as in Fig. 6.20a but analyzed in the vertical direction (L_{ref} = 10 km corresponding to $\lambda = 1$). The largest directly accessible scale is ~3 km, the point of convergence is $L_{eff} \approx 50$ km: see Table 6.4. Note that the vertical axis is not the same as for the horizontal analysis; this is a consequence of the scaling anisotropy; the exponents are roughly in a constant ratio. Reproduced from Lovejoy *et al.* (2009a).

6.5.2 Simultaneous horizontal and vertical cascades from lidar data

In order to determine the cascade structures, we can estimate the fluxes from the absolute finite difference Laplacian (as discussed in Chapter 4). A direct horizontal/vertical comparison of the normalized fluxes M_q is given in Figs. 6.20a and 6.20b. We see that the cascade structure predicted by Eqn. (4.1) is well

Table 6.4 Statistics derived from backscatter ratios of 10 vertical lidar cross-sections. Table from Lovejoy *et al.* (2009a).

Field	Resolution (m)	$\bar{\delta}$ (%)	C_1	H	α	L_{eff} (km)
B vertical	12 m × 192 m	0.4	0.11	0.60	1.82	50
B horizontal	12 m × 192 m	0.5	0.076	0.33	1.83	25 000

respected: not only are the lines quite straight, they also "point" to the effective outer scale of the process, – i.e. the scale at which a multiplicative cascade would have to start in order to account for the statistics over the range observed. We see that, as before, for the horizontal analysis L_{eff} is a little larger than the physical scales ($\approx$ 25 000 km, 50 km for the horizontal and vertical respectively). Table 6.4 shows some of the parameters characterizing $K(q)$, and shows that they are indeed quite different for the horizontal and vertical directions.

6.5.3 The construction of space-space diagrams from lidar data

If we define a "structure" in a field f as a fluctuation in the value of f of magnitude Δf, then we can use this to statistically define the relation between the horizontal and vertical extents of structures. For example, using the first-order structure function we can equate the horizontal and vertical fluctuations: $\langle|\Delta f(\Delta x)|\rangle = \langle|\Delta f(\Delta z(\Delta x))|\rangle$, which gives an implicit relation $\Delta z(\Delta x)$ between the horizontal and vertical extents (Δx, Δz respectively); i.e. for each horizontal lag Δx, what is the vertical lag Δz which gives the same mean fluctuation? Using the scale function (Eqn. (6.17)) and corresponding relation for the fluctuations in terms of the scale function, we see that this is equivalent to using $\|(\Delta x, 0)\| = \|(0, \Delta z(\Delta x))\|$ or $\Delta z = l_s(\Delta x/l_s)^{H_z}$. The same idea is used in Chapter 8 on space-time cross-section data to produce classical "space-time" ("Stommel") diagrams, so here we use the expression "space-space" diagrams. The existence of spatial vertical lidar cross-section data spanning many orders of magnitude in scale allows us to empirically determine this statistical correspondence directly and accurately.

Fig. 6.21 shows the result on the nine vertical cross-sections used in Fig. 6.20. We see that on a log-log plot the inferred log Δx – logΔz relationship is reasonably linear and that the slope is very near to the theoretical value H_z = 5/9 (shown by reference lines; the scaling is not good at the largest distances where the statistics are poor). The extrapolations of the lines both to larger and to smaller scales are particularly striking and have important implications. First, at smaller scales, we can estimate the sphero-scale (l_s) by the intersection of the extrapolation of the empirical line with the solid black reference line, $\Delta x = \Delta z$. We find that it is in the range 20 cm to 2 m, similar to the other estimates discussed above; it thus seems that the extrapolation is quite reasonable down to metric scales or less. However, equally impressive is the extrapolation to larger scales: we see that extroplation to the planetary scale (20 000 km) implies a corresponding vertical extent of ~10 km, i.e. the thickness of the troposphere. In other words, there is no obvious reason why the scaling stratification

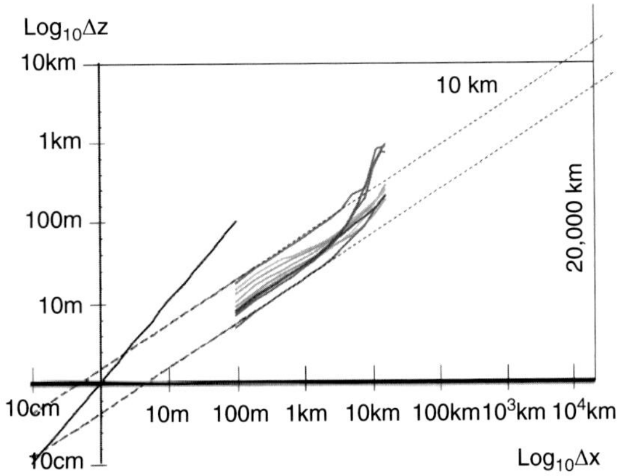

Fig. 6.21 A space-space (horizontal/vertical) diagram from nine vertical lidar sections obtained from first-order structure functions (this is from a slightly different subset of the data analysed in Figs. 6.20a, 6.20b). The dashed lines have theoretical slopes 5/9, the thick black line is the bisectrix ($x = y$). The sphero-scale is the intersection of the empirical lines with the bisectrix (the solid line at the left). It can be seen that the sphero-scales are somewhat variable, but mostly between 20 cm and 2 m. At the larger scales, we see that the earth size (20 000 km) roughly corresponds to the troposphere thickness (10 km). Note that the empirical statistics are not so good at their largest scales (~3–10 km, where there are few structures over which to average; this is the likely explanation for the rise of three of the curves over the extreme factor 2 in scale). Reproduced from Lovejoy and Schertzer (2010).

Box 6.1 CloudSat: reconciling convection with wide-range scaling

On the face of it, our claims of wide-range horizontal and vertical scaling fly in the face of the phenomenology of atmospheric convection which – ever since Riehl and Malkus (1958) – has relied on phenomenological models based on scale separations (although see Lilly, 1986a, 1986b, for attempts to reconcile it with turbulence theory with the help of helicity). The apparent incompatibility of horizontal scaling and convection was debated by Yano (2009) and the appendix of Lovejoy *et al.* (2009b). In Section 1.3, we discussed the phenomenological fallacy, i.e. the dangers of inferring mechanism from form, from phenomenology. In principle, convective phenomenology need not contradict the observation of wide-range anisotropic scaling. To make this more concrete and convincing, we need to show how to get structures traversing the troposphere in height while being only 100 km or so across (the "typical" horizontal scale cited for convection). In terms of generalized scale invariance this implies vertical cross-sections being roundish in shape at around 1–10 km, i.e. with "sphero-scales" being much larger than the range 0.01–1 m observed in passive scalars or in the horizontal wind (Section 6.3). The point is that if we define a "convective cell" as one that spans most of the vertical extent of the troposphere – and which typically has a comparable horizontal extent (e.g. the 100 km cited by Yano, 2009) – then for the argument to work, one would require a cloud liquid water field with a much larger sphero-scale than those which have been observed in the aircraft and lidar data.

In order to empirically examine this question, we can appeal to the CloudSat orbiting radar (1.08 km resolution in the horizontal, 250 m in the vertical from 2006 to the present: see Fig. 6.22 for an example). With wavelength (3 mm, 94 GHz), much smaller than that of the TRMM radar (2.2 cm; see Section 4.4.2), CloudSat can detect signals about 10^4 times weaker (down to $Z \approx 0.01$ mm^6/m^3) and therefore can detect much smaller drops. Since Z is proportional to the sum of the squared drop volumes it is highly correlated with the sum of the drop volumes (i.e. with the liquid water content), and it is thus a good surrogate for convection.

We can use the CloudSat data to determine structure functions (as in Section 6.5.3), and from there determine the corresponding space-space diagrams. Fig. 6.23a shows the result for fluctuations defined from orbit-by-orbit averages

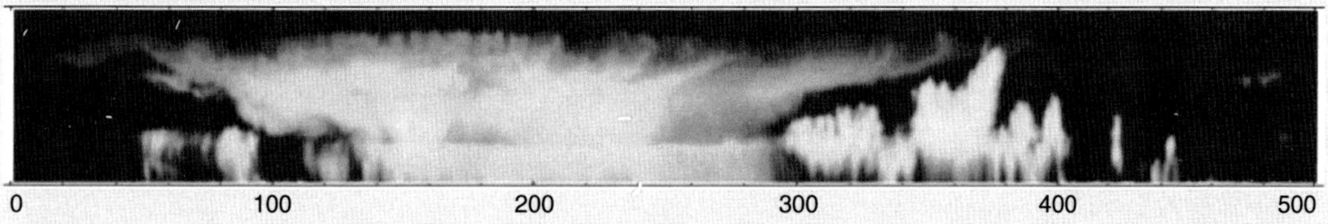

Fig. 6.22 A small sample of the first of the 16 CloudSat orbits analysed here (log density plot). The cross-section is 16 km high, 650 km wide (presented with a 1 : 4 aspect ratio). Notice the large convective cell about 15 km high, 200 km across.

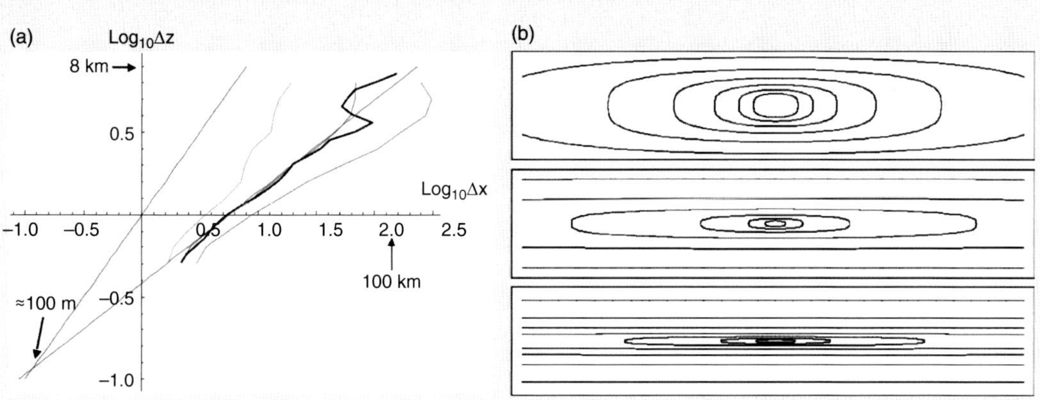

Fig. 6.23 (a) A space (horizontal)–space (vertical) diagram estimated from the absolute reflectivity fluctuations (first-order structure functions) from 16 CloudSat orbits. Reproduced from Lovejoy *et al.* (2009b). (b) The theoretical shapes of average vertical cross-sections using the CloudSat-derived mean parameters from Fig. 6.23a: $H_z = 5/9$, with sphero-scales 1 km (top), 100 m (middle), 10 m (bottom), roughly corresponding to the geometric mean and one-standard-deviation fluctuations. The distance from left to right is 100 km, from top to bottom 20 km. It uses the canonical scale function (Eqn. (6.17)). The top figure in particular shows that structures 100 km wide will be about 10 km thick whenever the sphero-scale is somewhat larger than average. Reproduced from Lovejoy *et al.* (2009b).

Box 6.1 *(cont.)*

as well as from an ensemble average over all the orbits. The mean of the individual orbit-by-orbit $\Delta x(\Delta z)$ curves (obtained from orbit-by-orbit structure functions) and the curve obtained from the ensemble-averaged structure function ensemble are nearly identical; the orbit-by-orbit spread is shown as one-standard-deviation curves above and below (the curves are occasionally double-valued along the Δx axis due to statistical fluctuations). In addition to the empirical curves, we have provided two theoretical reference lines with slopes $H_z = 5/9, 1$, the latter corresponding to isotropy. The basic behaviour is very similar to that of the lidar backscatter (Fig. 6.21). The main difference is the value of the "sphero-scale" determined by the intersection of the two lines, which is about 100 m. Structures at larger scales are flat, while at smaller scales they are elongated in the vertical. Although the exponents for Z and for lidar aerosol backscatter are nearly identical (given by the theoretical anisotropic Corrsin–Obhukov values), the corresponding sphero-scales l_s are about a factor 1000 larger (although, as can be seen from the error, there is a large scatter; the mean of $\log_{10} l_s$ with l_s in km is $\approx -1.6 \pm 0.9$, i.e. one-standard-deviation bars are 5–500 m, geometric mean 50 m). Fig. 6.23b shows the corresponding average contours of cloud reflectivity structures, showing how they very gradually tend to rounder shapes at the larger scales. It thus seems possible that the fundamental distinction between convective as opposed to stratiform dynamics is the much larger sphero-scale in the former. In the 23/9D model, this would be the natural consequence of the much larger buoyancy variance flux φ in convection.

Box 6.2 Comparison of the stratification of different fields: estimating H_z, D_{el}

Through the analyses presented in this book, we have shown that atmospheric fields are scaling over much of the meteorologically significant range in both the horizontal and vertical, so that the dynamics are scaling, turbulent but anisotropic. The simplest anisotropic turbulence model involves a unique scale function for all the fields. This would imply that the ratio of horizontal and vertical components is $H_{hor}/H_{ver} = H_z =$ constant, so that for universal multifractals $\alpha_{hor} = \alpha_{ver}$ and $C_{1hor}/C_{1ver} = H_z$.

Combining the results from the aircraft and the dropsondes and taking into account a small apparent altitude dependence of the sonde exponents (so as to estimate them at the ~200 mb aircraft level), we obtain Table 6.5. Although in the table we give the ratio of the C_1 values, their values are small, so that their relative errors are large and their ratios have large uncertainties. Since the H's are larger, the ratio H_{hor}/H_{ver} to estimate H_z is more reliable than using C_{1hor}/C_{1ver}; indeed, in the latter case the error is very hard to reliably estimate and is not indicated in the table except in the lidar case. The main conclusion is that T, $\log\theta$ and B are within a standard error bar of the 23/9D result ($H_z = 5/9$) whereas h has a somewhat larger value. At the same time, the v, T, $\log\theta$ fields are apparently anomalously low with regard to the 23/9D prediction of 5/9; this was the conclusion of the detailed analysis of aircraft data in Sections 2.6.2 and 6.3.

If the ratios in Table 6.4 are taken at face value then we are led to the conclusion that two or more scale functions are required to specify the scale of atmospheric structures. While this is certainly possible, let us for the moment underline the various difficulties in obtaining the in-situ estimates: the nontrivial vertical sonde outages, the nontrivial aircraft trajectory fluctuations. In addition, detailed analysis of the altitude dependence of the horizontal velocity exponent in Section 6.1.5 indicates that starting with the theoretical Boligano–Obukhov value 3/5 near the surface, the exponent increases somewhat with altitude to the value ≈ 0.75 at 10–12 km. Similarly, the humidity may have both horizontal and vertical variations, which may account for their high H_z values. We should therefore regard these studies as only early attempts to quantify the stratification.

It is interesting to note that the stratification of the earth's crust and mantle also display anisotropic scaling but in the opposite direction to the atmosphere. Rather than having a small sphero-scale with stratification increasing at larger scales, (i.e. $H_z < 1$), the sphero-scales are planetary in size and the stratification increases at *smaller* scales (i.e. $H_z > 1$). For example, the magnetic susceptibility and rock density have $H_z = 2, 3$ respectively (see Lovejoy and Schertzer, 2007).

Table 6.5 Statistics derived from the estimate of the vertical H_v, C_{1v} from sondes (Table 6.3, the $C_{1,>6km}$ value). The horizontal values for H_h for T, $\log\theta$, h are from Table 6.2 (from 4–40 km); for the lidar reflectivity B it is from Table 6.4. Finally, the C_1 for v is for the range 4–40 km using trace moments. For the horizontal v, the transverse wind component was used, since it was not very coherent with the altitude fluctuations and was considered more reliable.

	T	$\mathrm{Log}\theta$	h	v	B
$H_z = H_h/H_v$	0.47 ± 0.09	0.47 ± 0.09	0.65 ± 0.06	0.46 ± 0.05	0.55 ± 0.02
$H_z = C_{1h}/C_{1v}$	0.72	0.71	0.44	0.45	0.69 ± 0.2

should have a break anywhere in the meteorologically significant range of scales.

There are various ways to generalize and extend the method. For example, we could determine the horizontal/vertical relations for weak and strong events by considering structure functions with exponents $q < 1$ or $q > 1$. Alternatively, we could use the statistics of the fluxes (the normalized moments M_q) to establish the relation using $M_q(\Delta x) = M_q(\Delta z(\Delta x))$; this method is used in Chapter 8, where we also use it to determine space-time ("Stommel") diagrams.

6.6 Zonal/meridional anisotropy in reanalyses

6.6.1 Spectral scaling in stratified anisotropic scaling systems

In this chapter, we have only discussed the special case of scaling anisotropy with diagonal G (stratification in orthogonal directions, "self-affine" scaling). While this is at least a good approximation to vertical sections, in the horizontal things will generally be more complicated, with structures (such as cloud morphologies) displaying not only differential "squashing" but also differential rotation with scale; G has off-diagonal elements, and furthermore these will also vary from place to place. A potential exception is the north–south/east–west anisotropies imposed notably by the strong equator-to-pole gradients. We shall see that such gradients are apparently strong enough to lead to differential stratification in the horizontal reanalysis fields, although – significantly – not in the corresponding fluxes, which on the contrary display trivial anisotropies. This EW/NS stratification is the subject of this last section, and it is important since virtually the only statistical scaling technique that has been applied to characterizing reanalyses is to estimate the angle-integrated spectra (below) and then to assume isotropy.

Let us recall the basics (see also Appendix 2A). For a process in a d-dimensional space with wavevector $\underline{k}$, Fourier transform $\widetilde{f(\underline{k})}$ we have the spectral density:

$$\left\langle \widetilde{f(\underline{k})}\widetilde{f(\underline{k}')} \right\rangle = \delta(\underline{k} + \underline{k}')P(\underline{k}) \quad (6.55)$$

In data analysis, f is discrete and over a finite domain, the δ function is simply a proportionality constant (the number of degrees of freedom) so that $P(k) \propto \langle |v(k)|^2 \rangle$. If the system is statistically isotropic, then P only depends on the vector norm of $\underline{k}$: $P(\underline{k}) = P(|\underline{k}|)$. Now perform an isotropic Fourier space "zoom" $\underline{k} \to \lambda \underline{k}$ (i.e. a standard "blow-up") by factor $\lambda > 1$ so that in physical space there is an inverse blow-up: $\underline{r} \to \lambda^{-1}\underline{r}$. If the system is "self-similar," i.e. if it is both isotropic and scaling, then the condition that the smaller scales are related to the larger scales without reference to any characteristic size (i.e. that it is "scaling") is that the spectra follow power-law relations between large wavenumbers $\lambda|\underline{k}|$ and smaller ones $|\underline{k}|$:

$$P(|\lambda \underline{k}|) = P(\lambda|\underline{k}|) = \lambda^{-s}P(|\underline{k}|) \quad (6.56)$$

i.e. that the form of P is independent of scale. Eqn. (6.54) is satisfied by the following scaling law for P:

$$P(\underline{k}) = |\underline{k}|^{-s} \quad (6.57)$$

We can now obtain the power spectrum $E(k)$ (with $k = |\underline{k}|$) by integrating over all the directions:

$$E(k) = \int_{\delta S_k} P\left(\underline{k}'\right) d^d\underline{k}'. \quad (6.58)$$

where S_k is the d-dimensional sphere of radius k and δS_k is its boundary: in $d = 1$ it is the end points of the interval from $-k$ to k, in $d = 2$ it is the circle radius k, in $d = 3$, the spherical shell radius k. If the process is isotropic in 2D, $E(k) = 2\pi\, kP(k)$, and in 3D, $E(k) = 4\pi\, k^2\, P(k)$. In terms of data analysis, where one has a finite rather than an infinite sample size, this angle integration is advantageous because it reduces the noise. In the following examples, we therefore take power-law dependence of the spectrum:

$$E(k) \approx k^{-\beta}; \quad \beta = s - d + 1 \quad (6.59)$$

as evidence for the scaling of the field f, the exponent β being the "spectral slope." Note that in some areas of geophysics, angle averages ($P(k)$) are used rather than angle integrals; this has the disadvantage that the resulting spectral exponents will depend on the dimension of space, so that 1D sections for example will have exponents which differ by one from 2D sections. In contrast, the angle integrations used here yield the same exponent β in spaces of any dimension (i.e. β is independent of d but s is not: see Eqn. (6.59)).

The angle integrals over the spectral densities $P(k_x,k_y)$ are advantageous since it reduces the spectrum to a function of a single variable (the modulus of the wavevector) while simultaneously improving the statistics. While clearly the angle integral (Eqn. (6.58)) can be applied to any field to obtain $E(k)$ from

$P(k_x,k_y)$, if the latter is not isotropic then the classical interpretation of the result can be wrong or misleading. This is especially true if the scaling is anisotropic with different exponents in different directions.

As a first hint that horizontal anisotropy might complicate the estimation of spectral exponents from reanalyses, recall that in Section 4.2.3 we have already examined the horizontal cascade structure in both zonal and meridional directions and concluded that the exponents of the turbulent fluxes were (within statistical error) the same, although typically (depending somewhat on the field in question) there was a trivial anisotropy corresponding to an aspect ratio of about 1.6, with structures in the meridional direction "squashed" by this much with respect to the zonal direction. In order to directly check the isotropy/anisotropy we therefore proceed as follows: first, since the largest east–west distance is 180°, each latitudinal band from -45° to +45° is broken into two longitudinal sections, one from 0° to 180° and the other from 180 to 0° longitude, i.e. each 180° × 90° or 20 000 × 10 000 km or 120 × 60 reanalysis pixels (as usual, all spectra use Hanning windows to reduce spectral leakage). Although it might appear that the use of this cylindrical map projection might in itself lead to a significant statistical anisotropy, the actual bias in the spectral density is only of the order of a few percent (Lovejoy and Schertzer, 2011).

Fig. 6.24 shows the P contours obtained by taking averages over the squared moduli of the 2 × 365 transforms: the spectrum is shown with 2 : 1 aspect ratio such that circular large k contours indicate isotropy. We note that this is more or less the case for the largest wavenumbers corresponding to ~2 pixels (the Nyquist wavenumber) in real space. One sees that at larger and larger scales (smaller and smaller wavenumbers, near the centre), the contours become increasingly elliptical with the ellipses oriented in the k_y direction corresponding to real-space structures extended in the east–west direction; overall the small Fourier space ellipses have aspect ratios ≈ 2. The only – but significant – exception is for the meridional wind, which is also increasingly elliptical (by about the same amount) but elongated in the east–west rather than the north–south direction. In comparison, for the fluxes there was also an EW elongation of structures – but with the key difference that due to the apparent isotropy of the flux exponents their anisotropy was the same at all scales; it was "trivially anisotropic" with about the same factor

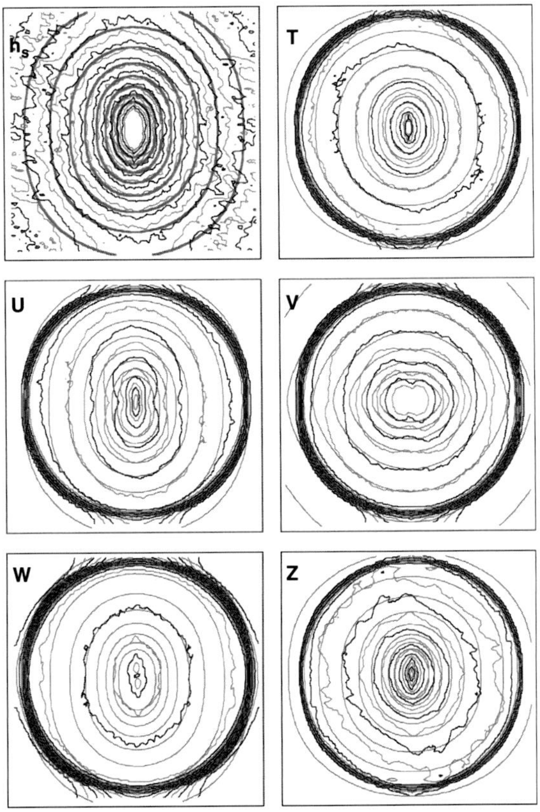

Fig. 6.24 Contour plots of log P. Left–right is k_x, vertical direction is k_y. Upper left is the specific humidity, upper right is the temperature, middle is the zonal (left) and meridional (right) winds. Lower left is the vertical wind, and lower right is the geopotential. Contours of the logarithm of the theoretical canonical scale functions (blue) all have $H_y = 0.8$ (except for v, which has $H_y = 1/0.8$) and the sphero-wavenumbers are $k_s = 60, 30, 60, 30, 60, 30$ respectively for h_s, T, u, v, w, z. Due to the Nyquist frequency, the largest k_y is 30 cycles/90° corresponding to 2 pixels. Due to the 2 : 1 aspect ratio (which compensates for the 2 : 1 change in range of k_x with respect to k_y), a circle the diameter of the square in the figure corresponds to isotropy at a 2-pixel scale. Reproduced from Lovejoy and Schertzer (2011). *See colour plate section.*

(1.6 ± 0.3: Table 4.1). In contrast, for the fields the effect changes markedly with scale, from near isotropy at single-pixel scales to fairly strong stratification elongation at the largest scales.

The scale-by-scale change of the aspect ratio combined with the scaling exhibited by the fluxes suggests that the spectral density respects an anisotropic scaling symmetry:

$$P(\underline{k}) \propto \|\underline{k}\|^{-s} \tag{6.60}$$

where $||\underline{k}||$ is the (nondimensional) "scale function" satisfying the functional scaling equation:

$$||\lambda^{-G}\underline{k}|| = \lambda^{-1}||\underline{k}|| \tag{6.61}$$

where λ is a scale ratio and G is the generator of the anisotropy (e.g. Eqns. (6.26), (6.27)). It is important to note that here we consider the change in the anisotropy of a *scalar* quantity over a range of scales; this is quite different from the more usual approach to anisotropy in the meteorological literature (e.g. Hoskins *et al.*, 1983), which is rather to consider the anisotropy of a *vector* quantity (e.g. the wind) at a unique (e.g. model) scale. Here, in the horizontal (x,y), using a cylindrical projection, we take:

$$G = \begin{pmatrix} 1 & 0 \\ 0 & H_y \end{pmatrix} \tag{6.62}$$

From Section 6.1.6 we see that the "canonical" solution of the scale Eqn. (6.61) with G given by Eqn. (6.62) is:

$$||\underline{k}|| = \left((k_x/k_s)^2 + (k_y/k_s)^{2/H_y}\right)^{1/2} \tag{6.63}$$

where H_y is the ratio of the east–west and north–south scaling exponents and k_s is the "sphero-wavenumber," the size of the roughly isotropic (circular) contours (Eqn. (6.63) is a dimensionless scale function; a dimensional one can be obtained by multiplying by k_s). Before continuing, let us note that the real-space counterpart to Eqns. (6.60) to (6.63) is that the fluctuations in the field Δf follow:

$$\Delta f(\underline{\Delta r}) = \varphi l_s^H ||\underline{\Delta r}||^H; \quad \underline{\Delta r} = (\Delta x, \Delta y);$$
$$||\underline{\Delta r}|| = \left((\Delta x/l_s)^2 + (\Delta y/l_s)^{2/H_y}\right)^{1/2} \tag{6.64}$$

where the real-space "sphero-scale" $l_s \approx 1/k_s$ and again the canonical scale function in Eqn. (6.63) is only the simplest. From Eqn. (6.64) we see that the east–west (Δx) and north–south (Δy) exponents are different:

$$\Delta f(\Delta x, 0) \propto \Delta x^{H_{EW}}; \quad \Delta f(0, \Delta y) \propto \Delta y^{H_{NS}};$$
$$H_{EW} = H; \quad H_{NS} = H_{EW}/H_y \tag{6.65}$$

One way to test Eqns. (6.60) to (6.65) is to use the 1D east–west and north–south spectra $E_{EW}(k_x)$ and $E_{NS}(k_y)$ obtained by integrating the P defined by Eqns. (6.60), (6.63):

$$E_{EW}(k_x) = \int P(k_x, k_y)dk_y = A_x\left(\frac{k_x}{k_s}\right)^{-\beta_{EW}};$$
$$\beta_{EW} = s - H_y; \ s > H_y$$
$$E_{NS}(k_y) = \int P(k_x, k_y)dk_x = A_y\left(\frac{k_y}{k_s}\right)^{-\beta_{NS}};$$
$$\beta_{NS} = \frac{s-1}{H_y}; \quad s > 1 \tag{6.66}$$

where A_x, A_y are dimensionless constants of order unity (which will change somewhat depending on the exact scale function solution (Eqn. (6.61)) – recall that the canonical scale function (Eqn. (6.63)) is only a special case). A useful consequence of Eqn. (6.66) is that it implies the following simple relation between exponents:

$$H_y = \frac{\beta_{EW} - 1}{\beta_{NS} - 1} \tag{6.67}$$

To determine space-space relations (between k_x, k_y) we can use the fact that the contribution to the total variance from all the structures smaller than a given wavenumber k is given by the integral of $E(k)$ from k to infinity. We can therefore exploit this fact to obtain a 1 : 1 relation (an implicit equation) between k_x and k_y:

$$\int_{k_x}^{\infty} E_x(k'_x)dk'_x = \int_{k_y}^{\infty} E_y(k'_y)dk'_y \tag{6.68}$$

If $\beta_x > 1$, $\beta_y > 1$, then the k_x – k_y relation implicit in Eqn. (6.68) (with the help of Eqns. (6.66), (6.67)) reduces to:

$$k_y = k'_s\left(\frac{k_x}{k'_s}\right)^{H_y} \tag{6.69}$$

where $k'_s \approx k_s$ (since $A_x \approx A_y \approx H_y \approx 1$). It is interesting to note that Endlich *et al.* (1969) applied essentially the same idea in order to obtain a space-space relation for vertical sections of the horizontal wind using aircraft and Jimsphere data to estimate the corresponding 1D spectra. However, they mistakenly used the spectral power densities directly ($E_x(k_x)$ and $E_z(k_z)$) to relate k_x and k_z rather than the total power as in Eqn. (6.68). The results will generally be different, since the densities are measured with respect to different spectral resolutions in the k_x, k_y directions whereas the integrated power is independent of the spectral resolutions.

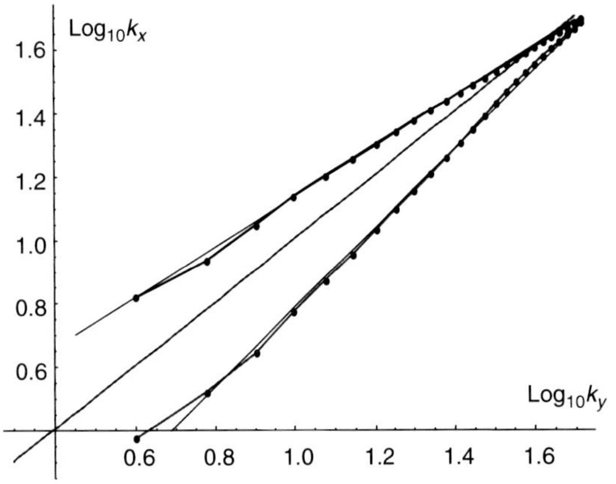

Fig. 6.25 The empirical space-space relations for the zonal wind (*u*, bottom), and meridional wind (*v*, top) calculated using the implicit relation Eqn. (6.68). The reference lines have slopes = $1/H_y = 0.8, 1, 1/0.8$. The sphero-wavenumber is where the bisectrix (middle) intersects the space-space line. Reproduced from Lovejoy and Schertzer (2011).

This gives us a simple way to check if the anisotropy is scaling: we can directly determine the k_x, k_y relation from the 1D spectra (Eqn. (6.68)) and then see if it is roughly a power law (Eqn. (6.69)). For the key zonal and meridonal winds, the corresponding space-space relations are shown in Fig. 6.25, where we have used lines with slope 1.25, which closely fits the *u* data, and slope $1/1.25 = 0.80$, which closely fits the *v* data (in both cases, the slopes $= 1/H_y$). It is noteworthy that the scaling of the space-space relations in Fig. 6.25 is superior to that of the individual zonal and meridional 1D spectra (Figs. 6.26a, 6.26b); this is possible because the latter have some residual deviations to scaling caused by the hyperviscosity. Our result is pleasingly symmetric, since if we reflect the system about a NE/SW line then *u* and *v* components as well as *x* and *y* axes are interchanged and H_y is replaced by $1/H_y$ so that the observed anisotropy respects this basic symmetry. Also shown in the figure is the bisectrix; the point at which this line intersects the space-space $k_y(k_x)$ curve has $k_x = k_y = k'_s$, where k'_s is the "sphero-wavenumber" (Eqn. (6.63)). We see that k'_s is indeed close to the largest wavenumber available (due to the Nyquist wavenumber, for k_y this is $60/2 = 30$ with $k_y = 1$ corresponding to $(10\ 000\ \text{km})^{-1}$). Similarly, the maximum anisotropy is given by the extreme low k_y aspect ratios: it corresponds to a factor $\approx 10^{0.3} \approx 2$.

To test the idea further, we refer the reader to Fig. 6.24, which shows the spectral densities *P* with theoretical contours superposed corresponding to $H_y = 0.8$ (or for *v*, $H_y = 1/0.8 = 1.25$) and with k_s fit to the nearest factor 2. It was found to be always either 30 or 60 cycles/10 000 km, corresponding to either twice the pixel scale of the public data (1.5°) or twice the pixel scale of the raw reanalysis data (0.7°).

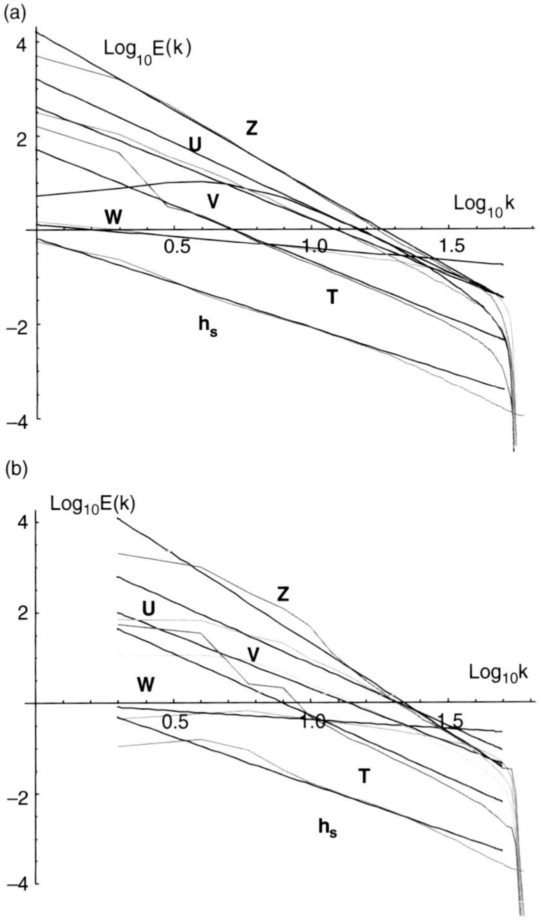

Fig. 6.26 (a) The zonal spectra obtained by integrating *P* in the *y* (east–west) direction averaged over logarithmically spaced wavenumber bins. The slopes of the reference lines are those predicted from the (more accurately estimated) slopes of the angle integrated spectrum using Eqn. (6.71). The reference lines have absolute slopes (β): 1.90, 2.40, 2.40, 2.75, 0.52, 3.35 (for h_s, *T*, *u*, *v*, *w*, *z* respectively, see Table 4.1). The smallest wavenumber ($k = 1$) corresponds to 20 000 km. Reproduced from Lovejoy and Schertzer (2011). (b) The meridional spectra obtained by integrating *P* in the *x* (north–south) direction averaged over logarithmically spaced wavenumber bins. The slopes of the reference lines are those predicted from the (more accurately estimated) isospectral slopes using Eqn. (6.71). The reference lines have absolute slopes (β): 2.12, 2.75, 2.75, 2.40, 0.40, 3.94 (for h_s, *T*, *u*, *v*, *w*, *z* respectively, see Table 4.1). The smallest wavenumber ($k = 1$) corresponds to 20 000 km. Reproduced from Lovejoy and Schertzer (2011).

6.6.2 The implications of scaling horizontal anisotropy for reanalyses

In order to understand this anisotropic spectral behaviour, let us note three key aspects of the results which appear to be robust: (a) the fields appear to have systematic scale-by-scale anisotropies whereas the corresponding turbulent fluxes do not; (b) the anisotropic exponent H_y seems to be nearly the same for all fields (the only exception being v, for which it is the reciprocal, as would be expected by symmetry); (c) all the fields display an apparent isotropy at wavenumbers corresponding to $\sim 1.5°$ resolutions, i.e. very nearly the resolution of the reanalyses. Since physically there is nothing special at $\sim 1.5°$ resolution that would make north–south and east–west fluctuations typically equal in magnitude (and this for all the fields), it would appear that the overall effect is an artefact of the numerics, which have strong isotropic constraints at the (hyper) dissipation scale while at the same time having large-scale anisotropies imposed by the boundary conditions. These large-scale constraints correspond physically to the strong north–south gradients, which are typically much larger than the east–west ones. This suggests that the models/reanalyses could be significantly improved by doubling their north–south resolutions with respect to their east–west resolutions.

Notice that we are making the perhaps surprising suggestion that boundary conditions could change scaling exponents. However, it is difficult to see how this could be otherwise, since for the anisotropic scaling to have more profound causes one would expect the fields to be dominated by physically different turbulent fluxes in the north–south and east–west directions, yet we have shown (Table 4.1) that the fluxes (which are responsible for the nonlinear contribution to the exponents $K(q)$, $\xi(q)$) seem only to display trivial anisotropy (i.e. $H_y = 1$ for the fluxes, only the outer scales are different). It seems that the scaling anisotropy is restricted to the linear part (Δx^H corresponds to a fractional integration of order H of the turbulent flux), while the more fundamental nonlinear part appears to be isotropic. The ability of the boundary conditions to introduce scaling anisotropy has in fact already been noted in radiative transfer on isotropic multifractal clouds when anisotropy is introduced by the boundary conditions, which are cyclic in the horizontal and in the vertical with radiative flux impinging from the cloud top only. The consequence is that the internal cloud radiative fluxes have scaling anisotropies even though the cloud liquid water does not (Lovejoy *et al.*, 2009c).

Is it worth mentioning one last implication of the $H_y \approx 0.8$ differential scaling. If we consider the average area of horizontal structures as functions of their zonal extent, then we find $A \approx \Delta x\, \Delta x^{H_y} \approx \Delta x^{D_{el}}$, where the effective "elliptical dimension" $D_{el} = 1 + H_y = 1.80$. This yields yet another argument against the relevance of two-dimensional isotropic turbulence to the atmosphere: even horizontal cross-sections are not isotropic – they are not even two-dimensional! If the atmospheric models (not just reanalyses) share this feature of having horizontal sections with $D_{el} = 1.80$, then it would seem that attempts such as those by Takayashi *et al.* (2006) to demonstrate the possible coexistence of 2D and 3D isotropic turbulent regimes are doomed to failure – if only because even the horizontal fails to display 2D isotropy.

6.6.3 The power spectrum of anisotropic scaling fields, spurious breaks in the scaling and reanalyses

No matter what the correct explanation for the reanalysis horizontal anisotropy, it has consequences for estimates of the (supposedly) isotropic spectral exponents, and even for the extents of the scaling regimes. To see this, let us estimate the angle-integrated spectrum (Eqn. (6.58)) assuming that the spectral density P has scaling anisotropy (i.e. that it has the form given by Eqns. (6.60) to (6.63)). One obtains:

$$\begin{aligned} E(k) &\approx k^{-\beta_l}; \quad \beta_l = \min(\beta_{EW}, \beta_{NS}); \quad k << k_s \\ E(k) &\approx k^{-\beta_h}; \quad \beta_h = \max(\beta_{EW}, \beta_{NS}); \quad k >> k_s \end{aligned} \tag{6.70}$$

with β_l the low- and β_h the high-wavenumber spectral exponents of the angle-integrated spectrum. We thus see that there is a break in the spectrum at k_s. Note that this break is spurious in the sense that it is a consequence of the isotropic integration; the full 2D spectral density P is perfectly scaling. However, β_{EW} and β_{NS} are related by Eqn. (6.67) so that for $H_y < 1$:

$$\beta_l = \min(\beta_{EW}, \beta_{NS}) = \begin{matrix} \beta_{EW}; & \beta_{EW} > 1 \\ \beta_{NS}; & \beta_{EW} < 1 \end{matrix} \tag{6.71}$$

with a corresponding equation for β_h using the maximum rather than the minimum (and the converse inequalities for $H_y > 1$).

Although the spectrum is not isotropic, if we are careful we can still use the isotropic $E(k)$, which has the advantage of improving the estimates, since the integration reduces the statistical fluctuations. Since we found that k_s is very nearly the wavenumber corresponding to one pixel we always have $k < k_s$, so we do not expect a break: the angle-integrated exponent will always correspond to β_l (Eqns. (6.70), (6.71)). The approximation (Eqn. (6.70)) improves at small k so that good estimates can easily be obtained by numerically estimating P using numerical Fourier transforms over the available 120×60-point grids and then integrating the latter over circular annuli, keeping only the $k < 30$ part. Since there is a $2:1$ aspect ratio, this is equivalent to integrating over ellipses with corresponding $2:1$ aspect ratios; the same result (Eqn. (6.70)) holds but the convergence to the power law $k^{-\beta_l}$ is faster.

Fig. 1.5b shows the results for the six fields. To make the scaling even more evident, we have averaged the spectrum over 10 logarithmically spaced intervals per order of magnitude (except for the lowest decade, where all the wavenumbers are indicated). Due to the $2:1$ aspect ratio, the spatial scale corresponding to a wavenumber $k = 1$ corresponds to 20 000 km in the east–west and 10 000 km in the north–south direction. From the figure, we see that the scaling is generally excellent. The only exception is for the meridional wind, which exhibits a sharp break at wavenumbers corresponding to 1250–2500 km; for the low wavenumbers it follows another scaling regime with $\beta \approx -0.2$ whose origin is obscure – it is far from any theoretically proposed value. Table 4.1 shows the values of the isotropic exponents β_l estimated to the nearest 0.05.

We can now use the relatively accurate angle-integrated exponent estimates to understand the 1D spectra in the zonal and meridional directions (Figs. 6.26a, 6.26b). Since the scaling is not as good as for the angle-integrated spectrum, we have added reference lines with the theoretical exponents (calculated from β_l using formulae (6.66), (6.70), (6.71) above with $H_y = 0.80$). We see that the agreement between the 1D spectra and the line theoretically predicted from the angle=integrated spectra is excellent.

In order to understand the significance of the various H exponents deduced from the spectra, recall that H is the classical nonconservation parameter. In the fractionally integrated flux model, it is also the order of fractional integration needed to obtain the field from a pure cascade. Contrary to α, C_1, which characterize the intermittent cascade processes and which can apparently only be evaluated numerically or empirically, H is a value that traditionally has been estimated by dimensional analysis (e.g. the $H = 1/3$ in the Kolmogorov law). However, scanning the values in Table 4.1, we see that several are problematic. Let us first consider the geopotential height z. Direct empirical estimates are problematic because although aircraft generally attempt to follow isobaric surfaces so that $p \approx$ constant, we have seen that because of turbulence this is only a good approximation at large scales (and the transition point varies considerably from flight to flight); the mean spectral exponent for wavenumbers $< (200\ \text{km})^{-1}$ from the 24 aircraft legs discussed above is $\beta \approx 4$ (see Fig. 3d in Lovejoy *et al.*, 2009b), but because of the small range it may be compatible with the reanalysis estimate $\beta \approx 3.35$. Its large value indicates that it varies very smoothly; the fact that it is greater than unity means that horizontal pressure derivatives are smooth: $\Delta p/\Delta x \approx \Delta x^{H-1}$, which for $H > 1$ is well behaved at small Δx. Turning our attention to w, although the H was too difficult to reliably measure from aircraft, it was indirectly estimated from lidar backscatter in Radkevitch *et al.* (2008) as being in the range -0.1 to -0.2; this analysis thus supports the conclusion that the vertical wind has an exponent H with small negative value. Unlike the other H exponents in the weather regime, which are positive, this implies that vertical wind fluctuations diminish rather than increase with scale (see Chapter 10).

In order to understand the H values for the wind, humidity and temperature, we have constructed Table 6.6. This compares the ECMWF reanalyses, aircraft (Lovejoy *et al.*, 2010b) and dropsonde estimates (Sections 6.2, 6.3). Starting with the wind, and concentrating on the zonal component, we note that there is excellent agreement between the dropsonde (vertical) value and the ECMWF isobaric value. As argued in Section 6.3, this is because the isobars are gently sloping so that at large enough scales one obtains the vertical rather than horizontal values. As a consequence, the aircraft value given in the table is the theory value 1/3 which was argued to be compatible with the small-scale aircraft statistics when corrected for intermittent turbulent motions of the aircraft. The H value for the humidity (0.54) is at least close to the measurements (0.51), although to our knowledge it is not predicted by any existing theory. Similarly, the

Table 6.6 A comparison of the ECMWF interim multifractal parameters (isobaric) with those estimated for aircraft (horizontal) and dropsondes (vertical). The aircraft C_1's have been multiplied by the factor $(3/2)^{1.8} = 2.07$ in an attempt to take into account the fact that the aircraft measure scaling-regime estimates of fluxes whereas the ECMWF estimates are more dissipation-scale fluxes (Eqn. (4.15)).

	Source	h	T	v
α	ECMWF	1.77 ± 0.06	1.90 ± 0.006	1.85 ± 0.012
	aircraft	1.81	1.78	1.94
	dropsonde	1.85	1.70	1.90
C_1	ECMWF	0.102 ± 0.009	0.077 ± 0.005	0.084 ± 0.006
	aircraft	0.083	0.108	0083
	dropsonde	0.072	0.091	0.088
H	ECMWF	0.54	0.77	0.77
	aircraft	0.51 ± 0.01	0.50 ± 0.01	1/3[a]
	dropsonde	0.78 ± 0.07	1.07 ± 0.18	0.75 ± 0.05

[a] Because of the issue of vertical aircraft movement, this (Kolmogorov) value was inferred, not directly estimated (see Lovejoy and Schertzer, 2011).

temperature H is far from the aircraft H although it is close to the isobaric wind H, suggesting that at least in the reanalysis it is estimated as a passive scalar (it would then be an isobaric estimate rather than an isoheight estimate similar to the H for the horizontal wind).

6.7 Summary of emergent laws in Chapter 6

In the simplest "trivial" anisotropy, the shapes of (average) structures are the same at all scales. Self-affinity is the simplest form of nontrivial, scaling anisotropy: structures become more and more stratified as the scale changes. The self-affine scale function $\|\underline{r}\|$ is linear with respect to the contraction parameter λ^{-1}

$$\|T_\lambda \underline{r}\| = \lambda^{-1}\|\underline{r}\| \quad ; \quad T_\lambda = \lambda^{-G} \tag{6.72}$$

where T_λ is the scale-changing operator and the generator G is a diagonal matrix. In the (x,z) (e.g. vertical) plane, the simplest "canonical" self-affine scale function is:

$$\|\underline{r}\| \approx l_s\left(\left(\frac{x}{l_s}\right)^2 + \left(\frac{z}{l_s}\right)^{2/H_z}\right)^{1/2} \quad ; \quad G = \begin{pmatrix} 1 & 0 \\ 0 & H_Z \end{pmatrix} \tag{6.73}$$

which satisfies Eqn. (6.72) with the G indicated. To take account of such stratification, we simply replace the isotropic scale defined by the vector norm $|\underline{r}|$ by the anisotropic scale function $\|\underline{r}\|$, i.e. make everywhere the replacement $|\underline{r}| \to \|\underline{r}\|$ so that for example the anisotropic extension of the basic emergent laws can be written:

$$\Delta f(\underline{\Delta r}) = \varphi_{\|\underline{\Delta r}\|}\|\underline{\Delta r}\|^H \tag{6.74}$$

For horizontal and vertical lags this implies:

$$\begin{aligned} \Delta v(\underline{\Delta r}) &= \varphi_h \Delta x^{H_h}; \quad \underline{\Delta r} = (\Delta x, 0) \\ \Delta v(\underline{\Delta r}) &= \varphi_v \Delta z^{H_v}; \quad \underline{\Delta r} = (0, \Delta z) \end{aligned} \tag{6.75}$$

with:

$$H_z = H_h / H_v \tag{6.76}$$

and where the horizontal and vertical are dominated by the fluxes φ_h, φ_v respectively and satisfy:

$$\begin{aligned} \varphi_h &= \varphi \\ \varphi_v &= \varphi l_s^{1-1/H_z} \end{aligned} \tag{6.77}$$

where l_s is the (roughly) isotropic "sphero-scale" satisfying:

$$\|(l_s, 0)\| = \|(0, l_s)\| \tag{6.78}$$

A specific example is the 23/9D model with:

$$\begin{aligned} \varphi_h &= \varepsilon^{1/3}; \quad H_h = 1/3 \\ \varphi_v &= \phi^{1/5}; \quad H_v = 3/5 \end{aligned} \tag{6.79}$$

where ε is the energy flux and ϕ is the buoyancy variance flux. This leads to:

$$l_s = \frac{\varepsilon^{5/4}}{\phi^{3/4}} \tag{6.80}$$

with $H_z = (1/3)/(3/5) = 5/9$. Since the volumes of nonintermittent structures is $Vol = \Delta x \Delta y \Delta z$, we find:

$$Vol \approx \Delta x^{D_{el}}; \quad D_{el} = TraceG = 2 + H_z \qquad (6.81a)$$

where D_{el} is the "elliptical dimension."

The real-space statistics of the fluctuations in this 2D example are given in terms of the structure function:

$$\langle|\Delta v|^q\rangle = \|\underline{\Delta r}\|^{\xi(q)}; \quad \xi(q) = qH_h - K(q) \qquad (6.81b)$$

with the scale function in Eqn. (6.79) and where $K(q)$, $\xi(q)$ are the horizontal exponents. The corresponding spectral density is:

$$P(\underline{k}) = \|\underline{k}\|^{-s}; \; s = 1 + \xi(2) + H_z \qquad (6.81c)$$

where the canonical Fourier scale function is:

$$\|(k_x, k_z)\| = \left(\left(\frac{k_x}{k_s}\right)^2 + \left(\frac{k_z}{k_s}\right)^{2/H_z}\right)^{1/2} \qquad (6.81d)$$

where k_s is the "sphero-wavenumber" with (generally) $k_s \approx 2\pi/l_s$.

Appendix 6A: Revisiting the revised EOLE experiment: the effect of temporal averaging

In Section 2.6.1 we took a new look at a classical EOLE constant-density balloon experiment designed to test the 2D/3D isotropic turbulence model and revisited by Lacorta *et al.* (2004). The EOLE satellite tracked the balloons ("mostly every 2.4 hours": Lacorta *et al.*, 2004, then interpolated them to an hourly resolution). The structure function is thus effectively averaged over this time Δt. In their notation we have:

$$\overline{S} = \overline{\left\langle \left| \underline{\dot{X}}^{(1)} - \underline{\dot{X}}^{(2)} \right|^2 \right\rangle} \tag{6.82}$$

where the overbar is an averging over Δt and $\underline{X}^{(1)}$, $\underline{X}^{(2)}$ are the coordinates of two different balloons (in the horizontal; ignore the vertical here). Writing this out explicitly, using $\underline{V} = \underline{\dot{X}}$ and assuming translational invariance:

$$\begin{aligned}\overline{S} &= \overline{\left\langle \left| \underline{V}^{(1)} - \underline{V}^{(2)} \right|^2 \right\rangle} = \\ &\frac{1}{\Delta t^2} \int_0^{\Delta t} \int_0^{\Delta t} dt'' dt' \left[S\left(\left| \underline{X}^{(2)}(t'') - \underline{X}^{(1)}(t') \right|, \left| t'' - t' \right| \right) \right. \\ &\left. - S\left(\left| \underline{X}^{(1)}(t'') - \underline{X}^{(1)}(t') \right|, \left| t'' - t' \right| \right) \right]\end{aligned} \tag{6.83}$$

where $\Delta t = t'' - t'$ and $S(|\underline{\Delta X}|, |\Delta t|)$ is the space-time Eulerian structure function, assumed here to be horizontally isotropic:

$$S(|\underline{\Delta X}|, |\Delta t|) = \langle |\underline{\Delta V}|^2 \rangle \tag{6.84}$$

We now follow Lacorta *et al.* (2004) and Morel and Larchevêque (1974) and assume that during the averaging period Δt the separation does not change much:

$$|\underline{X}^{(2)} - \underline{X}^{(1)}| \approx \delta \tag{6.85}$$

Finally, if u_0 is the typical large-scale velocity (estimated as 30 m/s $\approx$ 100 km/hour by Lacorta *et al.*, 2004), then we also have:

$$\left| \underline{X}^{(1)}(t'') - \underline{X}^{(2)}(t') \right| << u_0 \left| t'' - t' \right| \tag{6.86}$$

This just says that the overall advection is much larger than the diffusion during the interval Δt. We can now determine the effect of the temporal averaging, i.e. the low temporal resolution. With these approximations, Eqn. (6.83) becomes:

$$\begin{aligned}\overline{S} &= \frac{2\sqrt{2}}{\Delta t^2} \int_0^{\Delta t/\sqrt{2}} \left(\int_0^{(\Delta t - \tau\sqrt{2})} [S(\delta, \tau) - S(0, \tau)] dt' \right) d\tau \\ &= \frac{2}{\Delta t^2} \int_0^{\Delta t/\sqrt{2}} (\sqrt{2}\Delta t - \tau)[S(\delta, \tau) - S(0, \tau)] d\tau\end{aligned} \tag{6.87}$$

If we now consider the particles separated in the y (e.g. north–south) direction, and the main advection in the x direction (e.g. east–west), then:

$$S(\Delta x, \Delta y, \Delta t) = \frac{u_0^2}{L^{2/3}} \|(\Delta x, \Delta y, \Delta t)\|^{2H} \tag{6.88}$$

with $H = 1/3$. The scale function can be taken as either:

$$\|(\Delta x, \Delta y, \Delta t)\| = |\Delta x - u_0 \Delta t| + |\Delta y| \tag{6.89}$$

or:

$$\|(\Delta x, \Delta y, \Delta t)\| = (|\Delta x - u_0 \Delta t|^2 + |\Delta y|^2)^{1/2} \tag{6.90}$$

(the difference between the two choices is only in the trivial anisotropy). In the former case we have:

$$S(\delta, \tau) \approx \frac{u_0^2}{L^{2/3}} (\delta + u_0 \tau)^{2H} \tag{6.91}$$

while for the latter:

$$S(\delta,\tau) \approx \frac{u_0^2}{L^{2/3}}\left(\delta^2 + (u_0\tau)^2\right)^H \tag{6.92}$$

These different choices correspond to different shapes for the space-time unit balls (others are also possible).

We now use the following approximations:

$$S(\delta,\tau) - S(0,\tau) \approx \frac{u_0^2}{L^{2/3}}\left[(\delta + u_0\tau)^{2H} - (u_0\tau)^{2H}\right]$$

$$\approx \begin{cases} \frac{u_0^2}{L^{2/3}}\delta^{2H}; & \tau << \delta/u_o \\ \frac{u_0^2}{L^{2/3}}(u_0\tau)^{2H}\delta; & \tau >> \delta/u_o \end{cases} \tag{6.93}$$

Plugging these into the integral (Eqn. (6.87)), we find two regimes:

$$\overline{S} \approx \begin{cases} \frac{u_0^2}{L^{2/3}}\delta^{2H}; & \delta >> u_0\Delta t \\ \frac{u_0^2}{L^{2/3}}(u_0\Delta t)^{2H}\frac{\delta}{u_0\Delta t}; & \delta << u_0\Delta t \end{cases} \tag{6.94}$$

(for $H < 1$).

Or, using the other scale function (quadratic form, Eqn. (6.92)), we obtain

$$S(\delta,\tau) - S(0,\tau) \approx \frac{u_0^2}{L^{2/3}}\left[\left(\delta^2 + (u_0\tau)^2\right)^H - (u_0\tau)^{2H}\right]$$

$$\approx \begin{cases} \frac{u_0^2}{L^{2/3}}\delta^{2H}; & \tau << \delta/u_0 \\ \frac{u_0^2}{L^{2/3}}(u_0\tau)^{2H}\frac{\delta^2}{(u_0\tau)^2}; & \tau >> \delta/u_0 \end{cases} \tag{6.95}$$

Plugging this into Eqn. (6.87), we obtain:

$$\overline{S} \approx \begin{cases} \frac{u_0^2}{L^{2/3}}\delta^{2H}; & \delta >> u_0\Delta t \\ \frac{u_0^2}{L^{2/3}}\frac{\delta^{2H+1}}{(u_0\Delta t)^{2H+1}}; & \delta << u_0\Delta t \end{cases} \tag{6.96}$$

(for $H < 1/2$; see Fig. 2.9).

The exact behaviour when the averaging is important ($\delta << u_0\Delta t$) thus depends on various detailed assumptions. However, the qualitative behaviour and the transition scale is robust.

In all cases, the interpretation is straightforward: the temporal averaging decreases the variability for distances $< u_0\Delta t$, i.e. on distance scales less than the typical advection distance. It is hard to understand how both Lacorta *et al.* (2004) and Morel and Larchevêque (1974) failed to notice this; instead they attempted to find physical interpretations for the behaviour down to 50 km even though $u_0\Delta t$ according to their own data was at least 200 km. For example, Lacorta *et al.* (2004) claim "At distances smaller than 100 km our results suggest an exponential decay with *e* folding time of about 1 day in rough agreement with Morel and Larchevêque (1974)".

Appendix 6B: Cross-spectral analysis between wind, altitude and pressure

In Sections 2.6.2 and 6.3 we discussed a simple model for the aircraft trajectory: that it was straight but gently sloping. At the same time, various analyses indicated that it was more fractal than straight, so that this model is at best only a rough approximation and it is necessary to further pursue the relationship between aircraft altitude, pressure and wind in order to understand the scale-by-scale statistical relations. A convenient way to do this is to use the spectral coherence. Consider the cross-spectrum S_{hg} and normalized (complex) cross-spectrum σ_{hg} of two (1D) functions h, g:

$$\sigma_{hg} = \frac{S_{hg}}{(S_{gg}S_{hh})^{1/2}}; \quad S_{hg} = \langle \tilde{h}(k)\tilde{g}^*(k)\rangle$$
$$\tilde{h}(k) = \int_{-\infty}^{\infty} e^{ikx}h(x)dx; \quad \tilde{g}(k) = \int_{-\infty}^{\infty} e^{ikx}g(x)dx \tag{6.97}$$

where the ensemble average is estimated from the 24 disjoint legs each 4000 points (1120 km) long and the tilde (~) indicates Fourier transform. We can define the coherence C_{hg} and the phase θ_{hg} as the modulus and phase of σ_{hg}:

$$\sigma_{hg}(k) = C_{hg}(k)e^{i\theta_{hg}(k)} \tag{6.98}$$

(see e.g. Landahl and Mollo-Christensen, 1986). In Fig. 6B.1, averaging over all the legs, we show these both for h = altitude and g = longitudinal wind and for h taken as the pressure and g as the longitudinal wind. Recall that because of the normalization $0 \leq C \leq 1$, C is a kind of wavenumber-by-wavenumber correlation coefficient, with the important difference that it is positive definite. For identical functions ($h = g$) $C = 1$, while for statistically independent functions $C(k) \approx 1/\sqrt{n}$, where n is the number of independent samples used to estimate the ensemble average. Here we considered the first 4000 points of each sufficiently long flight segment (so that $n = 24$), and hence the coherence for statistically independent wind and

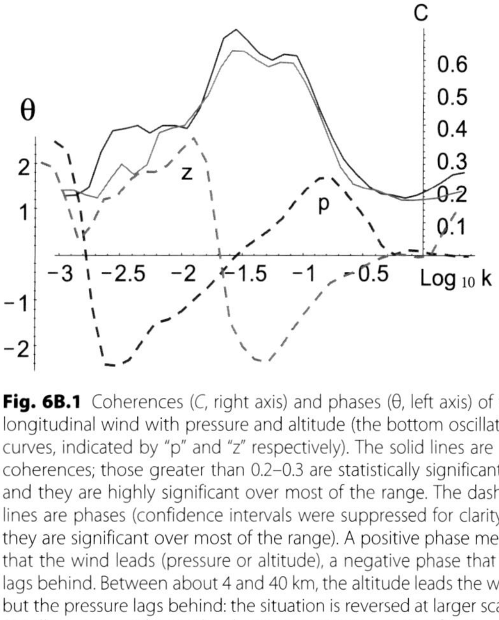

Fig. 6B.1 Coherences (*C*, right axis) and phases (θ, left axis) of the longitudinal wind with pressure and altitude (the bottom oscillating curves, indicated by "p" and "z" respectively). The solid lines are coherences; those greater than 0.2–0.3 are statistically significant, and they are highly significant over most of the range. The dashed lines are phases (confidence intervals were suppressed for clarity; they are significant over most of the range). A positive phase means that the wind leads (pressure or altitude), a negative phase that it lags behind. Between about 4 and 40 km, the altitude leads the wind but the pressure lags behind: the situation is reversed at larger scales (smaller wavenumbers). The direct interpretation is that for the higher wavenumbers ($(4\text{ km})^{-1} > k > (40\text{ km})^{-1}$, corresponding to time scales of 10–150 s) the aircraft autopilot and inertia cause the change in altitude, with the pressure then following the altitude. For the smaller wavenumbers ($k < (40\text{ km})^{-1}$), the situation is reversed, with the pressure changes causing the change in wind and altitude; this is presumably the regime where the aircraft tightly follows the isobars. Comparing with Fig. 2.14, we see that the spectra of the longtitudinal and transverse curves, which have slopes −5/3 and −2.4, have transitions at wavenumbers approximately at the phase change scale. Adapted from Lovejoy *et al.* (2009b).

altitudes is $C(k) \approx 0.20$ (see Lovejoy *et al.*, 2009b, for more refined "bootstrap" estimates of uncertainties).

The coherence is only the modulus; we therefore also considered the phases: $\theta = \theta_{zv}, \theta_{pv}$ (i.e. with $h = z$ and p respectively and $g = v$ in Eqn. (6.98), see Fig. 6B.1). With this choice, $\theta > 0$ indicates that the altitude (pressure) fluctuations lag behind the wind fluctuations while $\theta < 0$ indicates the converse. From Fig. 6B.1 we consider the various regimes.

(a) $k > (3\ \mathrm{km})^{-1}$ ($\log_{10}k > -0.5$)

Starting the analysis at the small scales (large wavenumbers), we see that, as expected, due to the inertia of the aircraft, which prevents it from rapidly responding to changes in wind, the coherence and phase with respect to the altitude is not statistically signficiant (left column).

(b) $(40\ \mathrm{km})^{-1} < k < (3\ \mathrm{km})^{-1}$ ($-1.5 < \log_{10}k < -0.5$)

Moving to lower wavenumbers, we first remark that there are apparent significant and even very strong coherences and phase relations for essentially all the larger scales (although the statistics are poor below about about $k < (500\ \mathrm{km})^{-1}$) with the relation between pressure and wind a bit stronger than that between altitude and wind. When we consider the phases, we see that whereas the pressure continues to lag behind the wind ($\theta_{pv} > 0$), the wind lags behind the altitude changes ($\theta_{zv} < 0$). This could be a consequence of the autopilot (on a time scale of 10–100 s) adjusting the level due to the smaller-scale turbulent trajectory fluctuations (typical time constants for aircraft roll modes are of the order of several seconds, and response to rudder and aileron commands are also of this order). In this range the wind follows the classical $k^{-5/3}$ spectrum so that the fractal aircraft trajectories do not significantly affect the spectrum (they only affect the intermittency corrections).

(c) $k < (40\ \mathrm{km})^{-1} - (60\ \mathrm{km})^{-1}$ ($\log_{10}k < -1.5$)

Finally, at the larger scales, we see that the phases of both the altitude and pressure with respect to the longitudinal component reverse sign. In this regime, the pressure leads the wind fluctuations while the altitude lags behind. This is presumably the regime in which the aircraft closely follows the isobars. From Figs. 6B.1 and 2.14 we see that this is also the regime where the wind spectrum follows the $k^{-2.4}$ rather than $k^{-5/3}$ law; in Section 6.3.1 we argue that it is this "imposed" vertical displacement that leads to the spurious appearance of the vertical exponent 2.4. This regime is consistent with the aircraft closely following isobars with the latter causing the wind and altitude fluctuations. More details and evidence for these interpretations can be found in Lovejoy *et al.* (2009b).

Chapter 7

Generalized scale invariance and cloud morphology

7.1 Beyond self-similarity and self-affinity

7.1.1 The basic elements of GSI

We have seen that the usual approach to scaling is first to posit (statistical) isotropy and only then scaling, the two together yielding "self-similarity." Indeed, the combination of the two is so prevalent that the terms "scaling" and "self-similarity" are often used interchangeably! The generalized scale invariance (GSI) approach that we have presented in this book is rather the converse: it first posits scale invariance, and then studies the remaining nontrivial symmetries. In the previous chapter we discussed simple models of scaling stratification in which the relation between large and small structures was different in two orthogonal directions; to distinguish it from "self-similarity" this is often called "self-affinity." More generally, a scale-invariant system will be one in which the small and large scales are related by a scale-changing operation that involves only the scale ratios; the system has no characteristic size. In what follows we outline the basic elements necessary for defining such a system; we follow the development initiated in Schertzer and Lovejoy (1985). To be completely defined, GSI needs not only a rule that determines how to change scale; it also requires a definition of how to measure the scale. This can be done in a constructive manner by starting from a unit scale (see below for a more abstract definition). Although this is not indispensable, it is closer to our usual way of thinking, i.e. to refer to a standard of measurement.

In this approach, the general idea of GSI is to build up a family of balls B_λ defining vectors of generalized scale L/λ, where L is a given (usual) scale, which can be taken as unity. This can be done with the help of rather arbitrary unit ball B_1 and a generalized scale transform T_λ that transforms the unit ball B_1 into a ball B_λ reduced by a factor λ in scale. The main question is to find the minimal requirements on T_λ (and a given measure on B_1) so as to obtain something corresponding to a (generalized) notion of scale. Keeping this general idea in mind, we can now consider these requirements.

Following Schertzer *et al.* (1999, 2002) and Schertzer and Lovejoy (2011), we start by giving a compact axiomatic definition of GSI. Following this terse presentation, we flesh it out with a detailed discussion of the 2D case.

The elements of GSI

T_λ is a generalized contraction on a vector space E; it is a one-parameter (semi-) group for the positive real-scale ratio λ ($\lambda \geq 1$ for a semi-group), i.e.:

$$\forall \lambda, \lambda' \in R^+ : T_{\lambda'} \circ T_\lambda = T_{\lambda'\lambda} \qquad (7.1)$$

and admits a generalized scale denoted $\|\underline{r}\|$ (double lines to distinguish it from the usual Euclidean metric $|\underline{r}|$), which in addition to being nonnegative, satisfies the following three properties:

(i) *nondegeneracy*, i.e.:

$$\|\underline{r}\| = 0 \Leftrightarrow \underline{r} = \underline{0} \qquad (7.2)$$

(ii) *linearity* with the contraction parameter $1/\lambda$, i.e.:

$$\forall \underline{r} \in E, \forall \lambda \in R^+: \ T_\lambda \|\underline{r}\| \equiv \|T_\lambda \underline{r}\| = \lambda^{-1} \|\underline{r}\| \qquad (7.3)$$

(iii) *strictly decreasing balls*: the balls defined by this scale, i.e.:

$$B_\ell = \{\underline{r} | \ \|\underline{r}\| \leq \ell\} \qquad (7.4)$$

must be strictly decreasing with the contraction T_λ:

$$\forall L \in R^+, \forall \lambda > 1 : \quad B_{L/\lambda} \equiv T_\lambda(B_L) \subset B_L \qquad (7.5)$$

and therefore:

$$\forall L \in R^+, \forall \lambda' \geq \lambda \geq 1 : \quad B_{L/\lambda'} \subset B_{L/\lambda} \qquad (7.6)$$

The usual Euclidean norm $|\underline{r}|$ of a metric space is the scale associated with the isotropic contraction $T_\lambda \underline{r} = \underline{r}/\lambda$. Properties (i), (ii) are rather identical to those of a norm, whereas (iii) is weaker than the triangle inequality, which is required for a norm. As for norms, unicity of generalized scale is not expected for a given T_λ.

Using a function to define the unit scale and unit ball B_1

From the above, we can see that if we define a unit scale (the vectors such that $\|\underline{r}\| = 1$), this defines the borders of a unit ball B_1, and using the scale-changing operator, this defines the other (nonunit) scales and balls. In general, the unit ball B_1 will be defined by an implicit equation:

$$B_1 = \{\underline{r} : f_1(\underline{r}) < 1\}; \quad \partial B_1 = \{\underline{r} : f_1(\underline{r}) = 1\} \tag{7.7}$$

where ∂B_1 is the "frontier" of the unit ball, and f_1 is a function of position ($\underline{r}$) (the use of open balls has the advantage that they generate a topology of the space (Lovejoy and Schertzer, 1985; Schertzer and Lovejoy, 1985)). Comparing Eqn. (7.7) with Eqn. (7.4), we see that $f_1(\underline{r}) = \|\underline{r}\|$.

The ball B_1 defining the unit scale vectors can be arbitrary, but if among all the balls B_λ's there is an isotropic ball (e.g., circle or sphere), we call the corresponding scale the "sphero-scale," and it is often convenient to take the corresponding ball as the unit ball B_1. When it exists, this sphero-scale may only correspond to the scale where structures are roughly spherical, vaguely roundish, approximately isotropic – and we use the expression for these cases too (cf. the self-affine canonical scale function when $l_s = 1$, Eqn. (6.17)). For simplicity we often assume the existence of a sphero-scale.

The scale-changing operator T_λ which transforms the scale of vectors by scale ratio λ

In order to define the nonunit scales and balls, we can exploit the scale changing operator T_λ (Eqn. (7.3)). T_λ is the rule relating the statistical properties at one scale to another and involves only the scale ratio. This implies that T_λ has certain properties. In particular, if and only if $\lambda_1\lambda_2 = \lambda$, then:

$$B_\lambda = T_\lambda B_1 = T_{\lambda_1\lambda_2} B_1 = T_{\lambda_1} B_{\lambda_2} = T_{\lambda_2} B_{\lambda_1} \tag{7.8}$$

i.e. T_λ has the same group properties as multiplication by the scale ratio λ and it is also commutative:

$$T_\lambda = T_{\lambda_2} T_{\lambda_1} = T_{\lambda_1} T_{\lambda_2} \tag{7.9}$$

This implies that T_λ is a one-parameter multiplicative group with parameter λ:

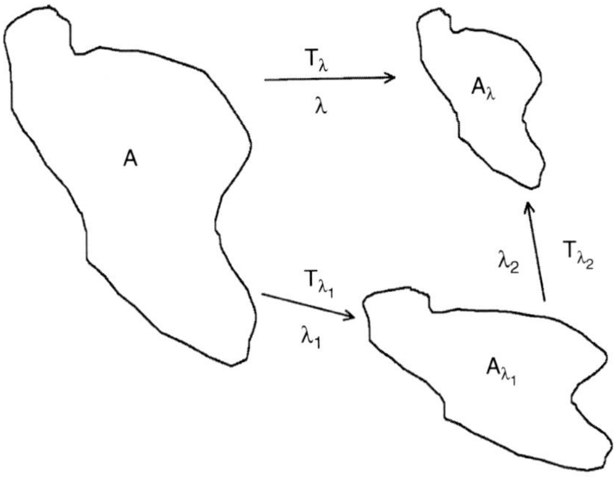

Fig. 7.1 Illustration of the group property of the scale-changing operator. Reproduced from Schertzer and Lovejoy (1996).

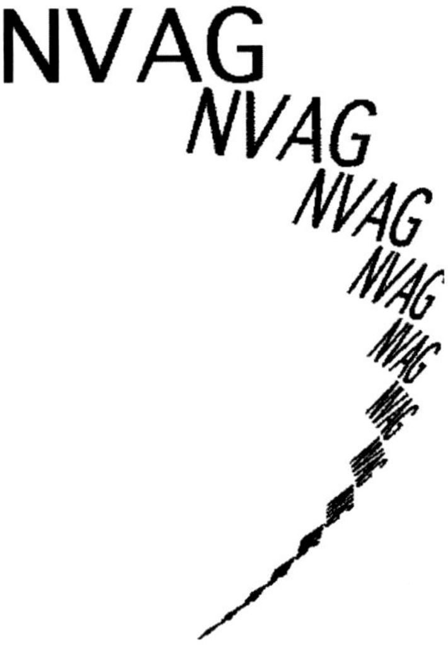

Fig. 7.2 A generalized blow-down with increasing λ of the acronym "NVAG." If $G = I$, we would have obtained a standard reduction, with all the copies uniformly reduced converging to the centre of the reduction. Here the parameters determining G are $c = 0.3$, $f = -0.5$, $e = 0.8$ (see Section 7.1.4), and each successive reduction is by 28%. Reproduced from Schertzer and Lovejoy (1996).

$$T_\lambda = \lambda^{-G} \tag{7.10}$$

where G is the generator of the group (Fig. 7.1). We use the negative sign since in turbulent cascades energy flux is typically transfered from large to small scales, so that we are usually interested in reductions by factor λ; T_λ *reduce* sizes by factor λ. We will not require that the inverse operators $T_\lambda^{-1} = T_{\lambda^{-1}}$ exist, hence we really only have a semi-group (the inverse, however, usually does exist if G is matrix). We saw in Chapter 6 that the action of G is straightforward when G is diagonal – it leads to stratification along one of the axes. However, when G has off-diagonal elements, the effect is more complex; it involves rotation: see Fig. 7.2, for an example showing a generalized "blow down" of the acronym "NVAG" showing how the reduction is combined with both stretching and rotation; this is discussed further below.

A technical consideration in constructing a viable GSI system is that the corresponding balls B_λ must be decreasing $(B_{\lambda_1} \subset B_{\lambda_2}; \lambda_1 > \lambda_2)$: this is necessary to insure that the vectors $\underline{r}_\lambda$ are unique. Let us consider this question in more detail by introducing the function f_λ to define B_λ. From Eqns. (7.3) and (7.7) we can define B_λ from the function f_λ defined as:

$$\begin{aligned} &\forall \underline{r} : f_\lambda(\underline{r}) = f_1(T_\lambda^{-1}\underline{r}), \\ &B_\lambda = \{\underline{r} : f_\lambda(\underline{r}) < 1\} \\ &\partial B_\lambda = \{\underline{r} : f_\lambda(\underline{r}) = 1\} \end{aligned} \tag{7.11}$$

Alternatively, in terms of the scale function, we have: $f_\lambda(\underline{r}) = \|T_\lambda^{-1}\underline{r}\|$. In order to ensure that the frontiers of the balls do not cross and that the scale is thus uniquely defined, consider the balls defined by λ and $\lambda + d\lambda$; it is easy to see that since a crossing point satisfies $\partial f_\lambda / \partial \lambda = 0$ we must have:

$$\frac{\partial f_\lambda}{\partial \lambda} > 0 \tag{7.12}$$

for all $\underline{r}$. This positivity (rather than negativity) requirement is necessary to ensure that T_λ corresponds to a scale reduction rather than enlargement.

A simple example is when $f(\underline{r})$ is a quadratic form:

$$f_1(\underline{r}) = (\underline{r}^T A \underline{r})^{1/2} \tag{7.13}$$

where A is a matrix and T indicates "transpose" (see Box 7.1) For this case, there is an exact result obtained for the decrease condition (Schertzer and Lovejoy, 1985). Indeed, the necessary and sufficient condition that the balls $T_\lambda(B_L) \equiv B_{L/\lambda}$ are strictly decreasing with the contraction group T_λ is:

$$Spec(sym(AG)) > 0 \tag{7.14}$$

where $sym(.)$ denotes the symmetric part of a linear application and where $Spec(.)$ denotes the set of eigenvalues. When A is furthermore positive and symmetric, i.e. the ball B_L is an ellipsoid, this condition reduces to:

$$Spec(sym(G)) > 0 \Leftrightarrow \mathrm{Re}(Spec(G)) > 0 \tag{7.15}$$

We now show step by step how to construct a GSI system respecting the axiomatic definition above.

If G is an $n \times n$ matrix and $\underline{r}$ is an n-dimensional vector and (the usual case) G is diagonalizable with transformation matrix Ω:

$$G' = \Omega^{-1} G \Omega \tag{7.16}$$

so that G' is diagonal (with eigenvalues Λ_i, with $i = 1\ldots, n$), then we obtain:

$$\lambda^{-G} = \Omega \lambda^{-G'} \Omega^{-1} = \Omega \begin{pmatrix} \lambda^{-\Lambda_1} & 0 & \ldots & 0 \\ 0 & \lambda^{-\Lambda_2} & \ldots & 0 \\ \ldots & \ldots & \ldots & 0 \\ 0 & 0 & 0 & \lambda^{-\Lambda_n} \end{pmatrix} \Omega^{-1} \tag{7.17}$$

The most general case of linear GSI requires Jordan matrices, which can be considered as almost diagonalizable and introduces more complex terms (Schertzer *et al.*, 1999). When T_λ is a more general, i.e. no longer linear, generator, we must define it with the help of differential equations (Section 7.3) – equivalently by using the corresponding infinitesimal generator that defines infinitesimal scale transformations.

In order to get a feeling for these transformations, consider the 2D case where G is a 2×2 matrix. When G is diagonizable, there are two cases of interest: the first when Λ_1, Λ_2 are both real and the second when they form a complex conjugate pair. Consider first the real case: we see that in the space $\underline{r}' = \Omega^{-1}\underline{r}$ where G' is diagonal; that the situation is identical to the cases of stratification discussed in Chapter 6: at large λ the structures are stratified along one of the axes, while at small λ they are stratified (squashed) along the other. In the original $\underline{r}$ space, this corresponds to squashing along (generally nonorthogonal) eigenvectors, (Fig. 7.3, lower left). However, when the eigenvalues are complex, the behaviour is different. Writing $\Lambda_1 = d - a$, $\Lambda_2 = d + a$ where d is real and a is pure imaginary, we see that in addition to (isotropic) scale reductions associated with λ^{-d}, there will be rotations associated with the factors $\lambda^{\pm a} = e^{\pm i|a|\log\lambda}$. As $\lambda \to \infty$ the phase $i|a|\log\lambda$ goes through an infinite range corresponding to an infinite number of rotations of structures (see Fig. 7.3 lower right for an example).

Defining measures/integrals

For a single point with position vector $\underline{r}$, the scale/size is determined by the scale ratio λ such that $\|T_\lambda^{-1}\underline{r}\| = 1$, but this is not enough to define the "size" of sets of points. For smooth sets, size is defined with the help of metric properties, such as the diameter of the set. However, for convoluted sets, this is not always manageable. It was therefore useful to use a more general approach based on the mathematical measure/integral of these sets (Schertzer and Lovejoy,

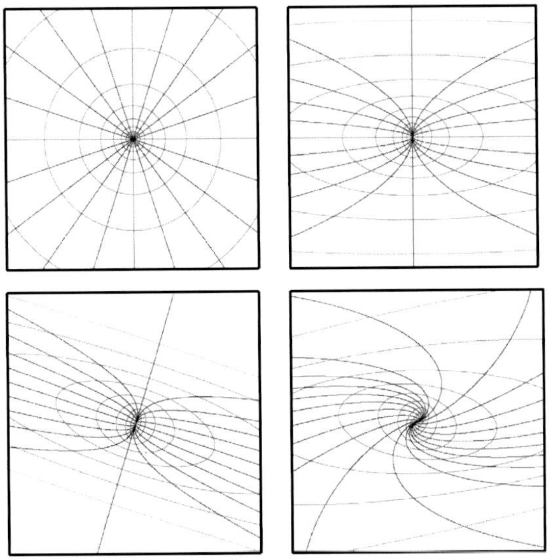

Fig 7.3 Examples of balls and trajectories for linear GSI with sphero-scale isotropic case: $c = 0, f = 0, e = 0$, (top left); self-affine case: $c = 0.35, f = 0, e = 0$ (top right); stratification-dominant case $a^2 > 0$ with no rotation: $c = 0.35, f = 0.25, e = 0$ (bottom right); rotation-dominant case $a^2 < 0$: $c = 0.35, f = 0.25, e = 0.6$; $d = 1$ for all cases with sphero-scale at 30 units (pixels) out of a total of 512. The parameters are defined in Eqns. (7.23) and (7.24). Reproduced from Lewis *et al.* (1999).

1985). This corresponds to first defining a family of elementary sets (here, the B_λ's), and then the measure/integral of each of these. Using these elementary sets, the measure (for example an anisotropic Hausdorff measure) of an arbitrary compact set can then be defined by covering the arbitrary set by the B_λ's.

There are various ways to define the generalized scale ϕ of the B_λ's; for example, we may take ϕ^D as a usual D-dimensional (mathematical) measure, e.g. the ordinary volume integral in a space of dimension D:

$$\phi^D(B_\lambda) = \int_{B_\lambda} d^D \underline{r} \tag{7.18}$$

so that for example in a two-dimensional space, $D = 2$, ϕ will be proportional to the square root of the area of B_λ. We can rewrite this as:

$$\phi^D(B_\lambda) = \int_{f_\lambda(r)<1} d^D \underline{r} = \int_{f_1(T_\lambda^{-1} r)<1} d^D \underline{r}$$

$$= \lambda^{-D_{el}} \int_{B_1} d^D \underline{r'} = \lambda^{-D_{el}} \phi^D(B_1); \quad \underline{r} = T_\lambda \underline{r'} \tag{7.19}$$

where we have used the transformation of variables indicated and the Jacobean of the transformation of variables:

$$d^D \underline{r} = \det(T_\lambda) d^D \underline{r'} = \lambda^{-Tr(G)} d^D \underline{r'} = \lambda^{-D_{el}} d^D \underline{r'} \tag{7.20}$$

If the "elliptical dimension" $D_{el} = TrG$ (the trace of G) does not correspond to the dimension of the embedding space D, then a convenient "elliptical" scale ϕ_{el} is defined by the following relation:

$$\phi_{el}^{D_{el}}(B_\lambda) = \phi^D(B_\lambda) \tag{7.21}$$

so that:

$$\phi_{el}(B_\lambda) = \phi^{D/D_{el}}(B_\lambda) = \lambda^{-1} \phi_{el}(B_1) \tag{7.22}$$

We could also use some (positive) power $\phi_{el} \to (\phi_{el})^d \phi'_{el}$ (e.g. the area of B_λ rather than the square root of the area) which would be equivalent to $\lambda \to \lambda^d = \lambda'$; this amounts to using a different generator $G \to G/d = G'$, $(d > 0)$ and hence a different elliptical dimension $D_{el} \to D_{el}/d = D'_{el}$. The resulting triplet G', ϕ'_{el} and D'_{el} is an equally good GSI system; the actual choice is a matter of convenience (see the next section for an example). In terms of the examples in Chapter 6, where we used the horizontal extent of the ball as our definition of size, we could for example alternatively take the area or the square root of the area of the balls.

7.1.2 Some properties of linear GSI

In order to illustrate the basic properties of the scale-changing operators $T_\lambda = \lambda^{-G}$ we shall again only explicitly consider the two-dimensional case; see Section 7.1.6 for three dimensions. However, even in 2D, G is a 2×2 matrix which depends on four parameters; fortunately not all the parameters lead to qualitatively different families of balls B_λ, and we can therefore establish a hierarchy of their influence.

First, it is convenient to decompose G into "pseudo-quaternions" (or equivalently, Pauli matrices):

$$G = d\mathbf{1} + e\mathbf{I} + f\mathbf{J} + c\mathbf{K} \tag{7.23}$$

where:

$$\mathbf{1} = \begin{bmatrix} 1 & 0 \\ 0 & 1 \end{bmatrix}, \quad \mathbf{I} = \begin{bmatrix} 0 & -1 \\ 1 & 0 \end{bmatrix}$$
$$\mathbf{J} = \begin{bmatrix} 0 & 1 \\ 1 & 0 \end{bmatrix}, \quad \mathbf{K} = \begin{bmatrix} 1 & 0 \\ 0 & -1 \end{bmatrix} \tag{7.24}$$

These matrices satisfy the following anticommutation relations:

$$\{\mathbf{I},\mathbf{J}\}=0,\quad \{\mathbf{I},\mathbf{K}\}=0,\quad \{\mathbf{J},\mathbf{K}\}=0 \tag{7.25}$$

(e.g. $\{I,J\} = IJ + JI = 0$).
A fundamental parameter for the description of the overall type of anisotropy present in the system will be given by the discriminant of the characteristic polynomial of G:

$$a^2 = (Tr(G)/2)^2 - Det(G) = c^2 + f^2 - e^2 \tag{7.26}$$

This is equivalent to the determinant of the traceless part of G and determines whether the eigenvalues:

$$\Lambda_x = d + a;\quad \Lambda_y = d - a \tag{7.27}$$

are real or complex. For clarity we have used the notation $\Lambda_x = \Lambda_1$, $\Lambda_y = \Lambda_2$.

The condition of decrease of the balls (Eqn. (7.5)) for G is merely:

$$Tr(G) > 0 \; and \; \det(G) > 0 \Leftrightarrow d > 0 \; and \; d^2 > a^2 \tag{7.28}$$

In Box 7.1, we explore the effect of changing the parameters on the balls defined by quadratic forms (ellipses). However, to more generally understand the various possible morphologies of linear GSI in two dimensions, it is very helpful to first reduce the number of parameters. Clearly parameters which simply rotate structures or which give them isotropic dilations do not change the morphologies, so of the four parameters c, d, e, f therefore only two need to be considered. For example, we are interested in characteristics of G which are independent of absolute orientation. Therefore, consider the primed coordinate system rotated by angle θ:

$$\begin{aligned} G' &= R^{-1}GR \\ R &= 1\cos\theta + I\sin\theta \end{aligned} \tag{7.29}$$

where R is the rotation matrix. Clearly, the 1 and I components of G commute with R so that they are unaffected by the rotation; hence d, e are rotationally invariant. In addition, both the trace ($= 2d$) and determinant ($= d^2 - a^2$) are rotational invariants (alternatively, and equivalently, the eigenvalues $d \pm a$ are invariant, Eqn. (7.27)), so that we have d, e, a as rotational invariants. We therefore may conclude that since $a^2 + e^2 = f^2 + c^2$ the latter is also invariant; let us define $r = \sqrt{c^2 + f^2}$ (not to be confused with the polar coordinate).

It is useful to now consider the interpretation of the above in terms of eigenvectors. First, the eigenvalues of G are $d \pm a$ and the (unnormalized) eigenvectors are:

$$\left(\frac{c \pm a}{f + e}, 1\right) \tag{7.30}$$

The angle between the two eigenvectors ($\Delta\theta$) is given by:

$$\cos(\Delta\theta) = \cos(\theta_+ - \theta_-) = \frac{e}{r} \tag{7.31}$$

which is rotationally invariant as expected. When $e < r$ and as e approaches r, the eigenvalues are real and the eigenvectors become more and more parallel; for $e > r$, they become complex. The average eigenvector orientation angle ($\overline{\theta}$) satisfies:

$$\cos(2\overline{\theta}) = \cos(\theta_+ + \theta_-) = \frac{-f}{\sqrt{c^2 + f^2}} = \frac{-f}{\sqrt{a^2 + e^2}} \tag{7.32}$$

c, f thus determine the absolute orientation of the balls but not the shapes. This means that we can consider only the case $r = f$, $c = 0$ without loss of generality (alternatively, as in Fig. 7.6, taking $r = c$, $f = 0$ is the same but in a frame rotated by $\theta = \pi/4$). This means we need only consider the following matrix:

$$G = \begin{pmatrix} d & r - e \\ r + e & d \end{pmatrix} \tag{7.33}$$

A further restriction on the parameter space is a consequence of the fact that interchanging the x and y axes (i.e. a reflection of structures about the bisectrix, the line $x = y$) is equivalent to changing the sign of e (this follows since $e \rightarrow -e$ implies $G \rightarrow G^T$ (the transpose). Therefore one need only consider $e > 0$.

Finally, overall "blow-ups" do not change the morphologies of structures. For example, if we normalize G by (half) its trace:

$$G' = G/d \tag{7.34}$$

then a scale function $\|\underline{r}\|'$ satisfying the scale equation for G':

$$\|\lambda^{-G'}\underline{r}\|' = \lambda^{-1}\|\underline{r}\|' \tag{7.35}$$

is obtained from the original scale function (satisfying $\|\lambda^{-G}\underline{r}\| = \lambda^{-1}\|\underline{r}\|$) by:

$$\|\underline{r}\|' = \|\underline{r}\|^d \tag{7.36}$$

so that in 2D we can always take Trace $G = 2$ (i.e. $d = 1$). Therefore, if we are only interested in exploring the various morphologies in 2D linear GSI, it suffices to consider $d = 1$, $r = f$, $c = 0$, i.e. to only consider the matrix:

$$G = \begin{pmatrix} 1 & r-e \\ r+e & 1 \end{pmatrix} \tag{7.37}$$

the scale functions corresponding to arbitrary 2×2 matrices can be obtained from the above by rotations and dilations (isotropic blow-ups). Finally, this simplified G can be diagonalized by:

$$G' = \Omega^{-1} \cdot G \cdot \Omega = \begin{pmatrix} 1-a & 0 \\ 0 & 1+a \end{pmatrix} \tag{7.38}$$

where $a^2 = r^2 - e^2$ (note that if in the above a is real, then it is always > 0) and the diagonalizing matrix Ω is:

$$\Omega = \frac{1}{2}\begin{pmatrix} -(r+e)^{-1/2} & (r+e)^{-1/2} \\ (r-e)^{-1/2} & (r-e)^{-1/2} \end{pmatrix};$$
$$\Omega^{-1} = \begin{pmatrix} -(r+e)^{1/2} & (r-e)^{1/2} \\ (r+e)^{1/2} & (r-e)^{1/2} \end{pmatrix} \tag{7.39}$$

Box 7.1 The example of quadratic balls

While the nonlinear transformations described in Section 7.1.4 are the most convenient for numerical simulations, further insight into the operation of λ^{-G} with nondiagonal G can be obtained by considering the effect of the scale-changing operator on the shapes of the balls in a particularly simple family: those defined by quadratic forms. Consider the unit ball defined by f_1:

$$f_1(\underline{r}) = \underline{r}^T A_1 \underline{r} = 1 \tag{7.40}$$

in two dimensions where A_1 is a 2×2 matrix and $\underline{r} = (x,y)$ is a positive vector on the frontier of the unit ball, and A_1 is a symmetric 2×2 matrix describing the unit ball (A_1 must be symmetric so that the eigenvalues are positive, so the balls are closed). The lack of a subscript on the position vectors will henceforth be taken to mean vectors on the unit ball unless otherwise specified.

Defining A_λ implicitly from the equation $f_\lambda(\underline{r}) = \underline{r}^T A_\lambda \underline{r}$, we have:

$$A_\lambda = (T_\lambda^{-1})^T A_1 T_\lambda^{-1} = \lambda^{G^T} A_1 \lambda^G \tag{7.41}$$

The no-crossing conditions (Eqns. (7.5), (7.12)) now reduce to:

$$\underline{r}^T sym(A_\lambda G)\underline{r} > 0 \tag{7.42}$$

where *sym* indicates the symmetric part (i.e. $symA_\lambda G = ((A_\lambda G)^T + (A_\lambda G))/2$). The above condition is satisfied as long as the eigenvalues of $sym(A_\lambda G)$ are > 0 and in fact for sym $(A_1 G)$, i.e. Eq. 7.14, with the help of the mapping λ^G (Schertzer and Lovejoy, 1985). In the case where a sphero-scale exists, then A_1 can be taken as the identity, and we require only the positivity of the eigenvalues of *symG*. In two dimensions, this is equivalent to Trace $G > 0$, $det(symG) > 0$. Pecknold *et al.* (1997) show how to extend this result to the case of quartic (and more general polynomial) balls which have various qualitative differences with quadratics, notably that they can be closed and nonconvex (e.g. Fig. 7.4).

To obtain an explicit expression for $\lambda^{-G} = e^{-G \ln\lambda}$, we can use the series expansion of the exponential function with pseudo-quaternions (Eqn. (7.20)) combined with the following identities:

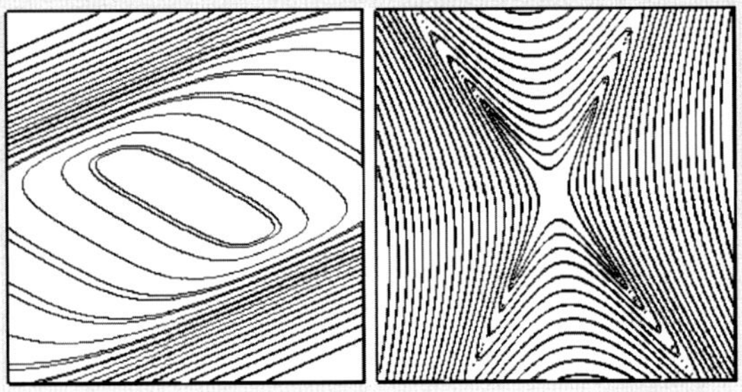

Fig. 7.4 Fourth order polynomial ball with $f_1(\underline{r}) = (\underline{r}^T A_1 \underline{r})^2 + (\underline{r}^T B_1 \underline{r})^2$ where A_1, B_1 are 2×2 matrices, showing nonelliptical contours. On the left is a convex stratification-dominant example, with $G = \begin{bmatrix} 1.5 & 0.13 \\ 1.27 & 0.5 \end{bmatrix}$; on the right is a rotation-dominant example showing highly nonconvex shapes, with $G = \begin{bmatrix} 1.1 & -0.1 \\ 0.3 & 0.9 \end{bmatrix}$. Adapted from Pecknold *et al.* (1997).

Box 7.1 *(cont.)*

$$(G - d\mathbf{1})^{2n} = a^2\mathbf{1} \tag{7.43}$$

where n is an integer and $a^2 = c^2 + f^2 - e^2$. Using this and writing $u = \log\lambda$ we therefore obtain:

$$T_\lambda = \lambda^{-G} = \lambda^{-d}\lambda^{-(G-d\mathbf{1})} = \lambda^{-d}\left[\mathbf{1}\cosh(au) - ((G - d\mathbf{1})\frac{\sinh(au)}{a}\right] \tag{7.44}$$

When $a^2 < 0$ the above formula holds, but with $|a|$ replacing a and ordinary trigonometric functions rather than hyperbolic functions. Examples of both balls and trajectories (the locus of points $\underline{r}_\lambda = T_\lambda \underline{r}_1$, obtained by λ varying with $\underline{r}_1$ fixed) are shown in Fig. 7.3.

We can now consider the effect of T_λ on quadratic balls, recalling the two basic cases depending on whether the eigenvalues of G are real or complex ($a^2 > 0$, $a^2 < 0$ respectively) corresponding to domination by stratification or by rotation. To see this explicitly, decompose the matrix T_λ as follows:

$$T_\lambda = R_{\theta_2} S_{AB} R_{\theta_1} \tag{7.45}$$

where R_θ is a rotation matrix which rotates by an angle θ, and S_{AB} is a "stretch" matrix:

$$R_\theta = \begin{bmatrix} \cos\theta & -\sin\theta \\ \sin\theta & \cos\theta \end{bmatrix}, \quad S_{AB} = \begin{bmatrix} A & 0 \\ 0 & B \end{bmatrix} \tag{7.46}$$

If we apply this to a circular unit ball B_1, we obtain:

$$B_\lambda = T_\lambda B_1 = R_{\theta_2} S_{AB} B_1 \tag{7.47}$$

where we have used the fact that $R_{\theta 2} B_1 = B_1$ (a circle is invariant under rotation). Since $S_{AB}B_1$ is an ellipse with axes A, B, we have therefore have the simple interpretation that B_λ is an ellipse with axes A, B rotated by angle θ_2. In order to understand the effect of T_λ on B_1 it therefore suffices to determine how A, B, θ_2 vary with scale ratio λ. A first step is to write:

$$R_{\theta_2} S_{AB} R_{\theta_1} = \frac{1}{2}\begin{bmatrix} (A+B)\cos\theta_+ + (A-B)\cos\theta_- & (A-B)\sin\theta_+ + (A-B)\sin\theta_- \\ (A+B)\sin\theta_+ + (A-B)\sin\theta_- & (A+B)\cos\theta_+ - (A-B)\cos\theta_- \end{bmatrix} \tag{7.48}$$

where $\theta_+ = \theta_2 + \theta_1$ and $\theta_- = \theta_2 - \theta_1$. Equating this element by element to our expression for λ^{-G} (Eqn. (7.43)) we obtain:

$$\lambda = (AB)^{-\frac{d}{2}} \tag{7.49}$$

Recalling that we can always choose $d = 1$, we see that this is equivalent to:

$$\lambda^{-1} = \sqrt{\frac{\text{area}}{\pi}} \tag{7.50}$$

With this we also find:

$$\frac{\varepsilon}{\sqrt{\varepsilon + 1}} = \sqrt{\frac{B}{A}} - \sqrt{\frac{A}{B}} = 2\sqrt{\frac{c^2 + f^2}{a^2}}\sinh^2(au); \quad \varepsilon = \frac{B}{A} - 1 \tag{7.51}$$

where ε is the "ellipticity". For the angle θ_2, we find:

$$\theta_2 = \frac{1}{2}\tan^{-1}\left(\frac{f}{c}\right) - \frac{1}{2}\tan^{-1}\left(\frac{e}{a}\tanh(au)\right) \tag{7.52}$$

Eqns. (7.47) and (7.48) tell us how an initial circle at $\lambda = 1$ ($u = 0$) changes its ellipticity ε and orientation θ_2 with scale λ.

We now consider the two qualitatively different cases, $a^2 > 0$ and $a^2 < 0$.

Box 7.1 (cont.)

Stratification dominance, $a^2 > 0$

In this case, as $u \to \infty$ ($\lambda \to \infty$), $\Rightarrow B/A \to \infty$; as $u \to -\infty$ ($\lambda \to 0$), $A/B \to \infty$, i.e. we have extreme stratification. Considering the rotation, we have:

$$\theta_2 \to \frac{1}{2}\tan^{-1}\left(\frac{f}{c}\right) - \frac{1}{2}\tan^{-1}\left(\frac{e}{a}\right); \quad u \to \pm\infty \tag{7.53}$$

i.e. a total rotation of $\tan^{-1}(e/a)$ (Note that at $u = 0$ the major and minor axes are exchanged, hence there appears to be an extra $\pi/2$). The total rotation is thus bounded.

Rotation dominance, $a^2 < 0$

In Eqns. (7.51), (7.52), we replace the hyperbolic trigonometric functions by the usual trigonometric functions and use $|a|$ to represent the modulus of a. From the equation for θ_2, we now find that there are an infinite number of rotations as $u \to \infty$ (the logarithm "wavelength" $= 2\pi/|a|$) and the ellipticity oscillates, with maximum ratio:

$$\left(\frac{B}{A}\right)_{max} = 2\left(\frac{e}{|a|}\right)^2\left(1 + \sqrt{1 - \frac{|a|^2}{e^2}}\right) - 1 \tag{7.54}$$

From this, we can conclude that if the unit ball is sufficiently elliptical, there will be no circular balls at any scale.

7.1.3* Scale functions in linear GSI in 2D

In Section 6.1.4 we saw how to find a solution of the scaling Eqn. (6.26) when $T_\lambda = \lambda^{-G}$ and G is a diagonal matrix. It is therefore straightforward to apply the same method to any diagonalizable G; it suffices to diagonalize it before applying a (nonlinear) transformation of variables. Consider for simplicity the two-dimensional case, with the two eigenvalues Λ_x, Λ_y. When these are real we have the following coordinate transformations:

$$\begin{aligned} \underline{r}' &= \Omega^{-1}\underline{r} \\ \underline{r}'' &= (x'', y'') = \left(sign(x')|x'|^{1/\Lambda_x}, sign(y')|y'|^{1/\Lambda_y}\right) \end{aligned} \tag{7.55}$$

From the above discussion we see that:

$$\|\lambda^{-G}\underline{r}\| = \lambda^{-1}\|\underline{r}\| \leftrightarrow \|\lambda^{-1}\underline{r}''\| = \lambda^{-1}\|\underline{r}''\| \tag{7.56}$$

so that the solution of the functional scale (Eqn. (7.56)) is the same as for the self-affine diagonal G case (Section 6.1.4) except for the doubly primed variables:

$$\begin{aligned} \|\underline{r}\| &= \Theta(\theta'')r''; \quad r'' = |\underline{r}''| = (x'^{2/\Lambda_x} + |y'|^{2/\Lambda_y})^{1/2}; \\ \tan\theta'' &= \frac{y''}{x''} = \frac{sign(y')|y'|^{1/\Lambda_y}}{sign(x')|x'|^{1/\Lambda_x}} \end{aligned} \tag{7.57}$$

Once again, the condition that the balls are decreasing with λ (no crossing of balls) is that $\Theta(\theta'') > 0$ and the choice of the otherwise arbitrary Θ determines the shape of the unit ball: in polar coordinates its equation is $r^n = 1/\Theta(\theta'')$.

The self-affine case and the above have real eigenvalues. Let us now consider the case where the eigenvalues form a complex conjugate pair: $\Lambda_x = d - a$, $\Lambda_y = d + a$ where $d = Re(\Lambda_x)$, and $a = -i\ \mathrm{Im}(\Lambda_x)$. Once again, we diagonalize by (the now complex) Ω (Eqn. (7.39)), and obtain:

$$\underline{r}' = \Omega^{-1}\underline{r} = (z', -z'^*) \tag{7.58}$$

where z' is complex (the real and imaginary parts of the x' and y' coordinates are independent, but x' and y' are no longer independent of each other) and "$*$" is the complex conjugate. If we now make the nonlinear transformation of variables:

$$z'' = |z'|^{(1+a)/d}\frac{z'}{|z'|} \tag{7.59}$$

then a scale function satisfying the scale function (Eqn. (7.54)) can be obtained by taking:

$$\|\underline{r}\| = \Theta\left(arg(z'')\right)|z''| \tag{7.60}$$

where once again $\Theta > 0$ defines the unit ball ($\arg(z'')$ denotes the argument (phase) of the complex variable z'').

While Eqns. (7.57) and (7.60) are indeed solutions of the scale function (Eqn. (7.56)), they are not optimal for simulations because we see from Eqn. (7.39) that the diagonalization matrix Ω is singular when $r = e$ (i.e. when $a = 0$). This means that for a fixed Θ function, if we slowly vary the parameters r, e the morphologies will change in a correspondingly singular manner near the $r = e$ part of the parameter space where a becomes complex. It is, however, easy to avoid this by a further transformation of variables: multiplying by Ω (instead of Ω^{-1}). For real eigenvalues:

$$\underline{r}^{(3)} = \underline{r}^{(2)}\Theta\left(\theta^{(2)}\right); \quad \underline{r}^{(4)} = \Omega\underline{r}^{(3)} \tag{7.61}$$

where the superscript (i) indicates the ith transformation of variables (so that $\underline{r}^{(0)} = \underline{r}, \underline{r}^{(1)} = \underline{r}', \underline{r}^{(2)} = \underline{r}''$), and $\theta^{(i)}$ denotes the polar angle in $\underline{r}^{(i)}$ space. The final transformation Ω removes the singular transformation Ω^{-1} in the initial diagonalization step so that the overall transformation $\underline{r}^{(0)} \rightarrow \underline{r}^{(4)}$ becomes continuous in the parameter a. Finally, the scale function is simply the usual vector norm of $\underline{r}^{(4)}$:

$$\|\underline{r}\| = |\underline{r}^{(4)}| \tag{7.62}$$

For the case of complex eigenvalues, we first define the complex coordinate $z^{(1)}$:

$$\underline{r}^{(1)} = \Omega^{-1}\underline{r}^{(0)} = \left(z^{(1)}, -z^{(1)*}\right) \tag{7.63}$$

From this we make the following transformations:

$$\begin{aligned} z^{(2)} &= z^{(1)}|z^{(1)}|^{a} \\ z^{(3)} &= z^{(2)}\Theta\left(\arg\left(z^{(2)}\right)\right) \\ \underline{r}^{(4)} &= \Omega\left(z^{(3)}, -z^{(3)*}\right) \end{aligned} \tag{7.64}$$

where $\arg z$ indicates the argument of the complex z and once again the scale function is just the norm or $\underline{r}^{(4)}$: $\|\underline{r}\| = |\underline{r}^{(4)}|$; and once again, the condition that the balls do not cross is: $\Theta > 0$.

In addition to studying the effect of the two G parameters r, e which determine the scale-changing operator, we can also investigate the effect of varying the shape of the unit ball defined by Eqns. (7.57) and (7.60) for $\|\underline{r}\| = 1$, i.e. the polar equation in the doubly primed coordinates:

$$r^{(2)} = 1/\Theta(\theta^{(2)}) \tag{7.65}$$

where $r^{(2)}$, $\theta^{(2)}$ are the polar coordinates in the $\underline{r}^{(2)}$ space. The "canonical" scale function is obtained by taking $\Theta = 1$ so that the unit ball is a circle in the $\underline{r}^{(2)}$ space; this corresponds to a roundish but not exactly circular ball in the original $\underline{r}$ space.

7.1.4 Illustrating the effect of varying G and the unit ball with multifractal simulations

We now would like to see the effect of introducing scaling anisotropy on multifractal simulations. In Section 7.1.2, we showed how to reduce the number of exponents in 2D from four (c, d, e, f) to two (r, e), and in Section 7.1.3 we showed how to easily create scale functions with quite general unit balls and respecting the anisotropic scaling symmetries while being continuous in the parameters. In order to explore the possible morphologies, the last element we need is therefore a specification of the unit ball. A convenient one-parameter parametrization is:

$$\Theta(\theta'') = 1 + \frac{1 - 2^{-k}}{1 + 2^{-k}}\cos\theta'' \tag{7.66}$$

The parameter k allows us to examine the effect of possibly very nonroundish unit scales. Since $k = \log_2(r^{(2)}_{max}/r^{(2)}_{min})$ where $r^{(2)}_{max}$, $r^{(2)}_{min}$ are the maximum and minimum radii of the sphero-scale, k has a simple interpretation (the ratio in $\underline{r}^{(2)}$ space will be close to the ratio in the original $\underline{r}$ space). For example, with this definition, we see $k = 10$ implies a unit scale which "mixes" conventional scales over a factor of more than 1000 ($= 2^k$).

Once again, multifractal simulations proceed as in the stratified case (Section 6.1.8), i.e. by following the same procedure as in the isotropic multifractals but with $|\underline{r}| \rightarrow \|\underline{r}\|$; $d \rightarrow D_{el}$. In addition to the somewhat more complicated calculation of the scale function, the calculation of the angular integral (needed for normalizing the generator, Ω_{Del}; Eqn. (6.34)) is also more involved and is given in Appendix 7A.

An important practical complication is that the simulation region will generally not have the same shapes as any of the balls; i.e. the limits of the simulation region will in general not coincide (even approximately) with lines of constant $\|\underline{r}\|$. This means that a possibly significant range of scales will not be totally resolved by the convolution integrals. Fig. 7.5a graphically shows the problem. There are three scales to consider: the smallest scale that is totally resolved by the rectangular 1×1 pixel grid ($\|\underline{r}_1\|$; i.e. which completely encloses the central pixels), the largest scale that is totally resolved within the simulation region ($\|\underline{r}_2\|$), and the largest scale that crosses (and hence influences) the integration region ($\|\underline{r}_3\|$).

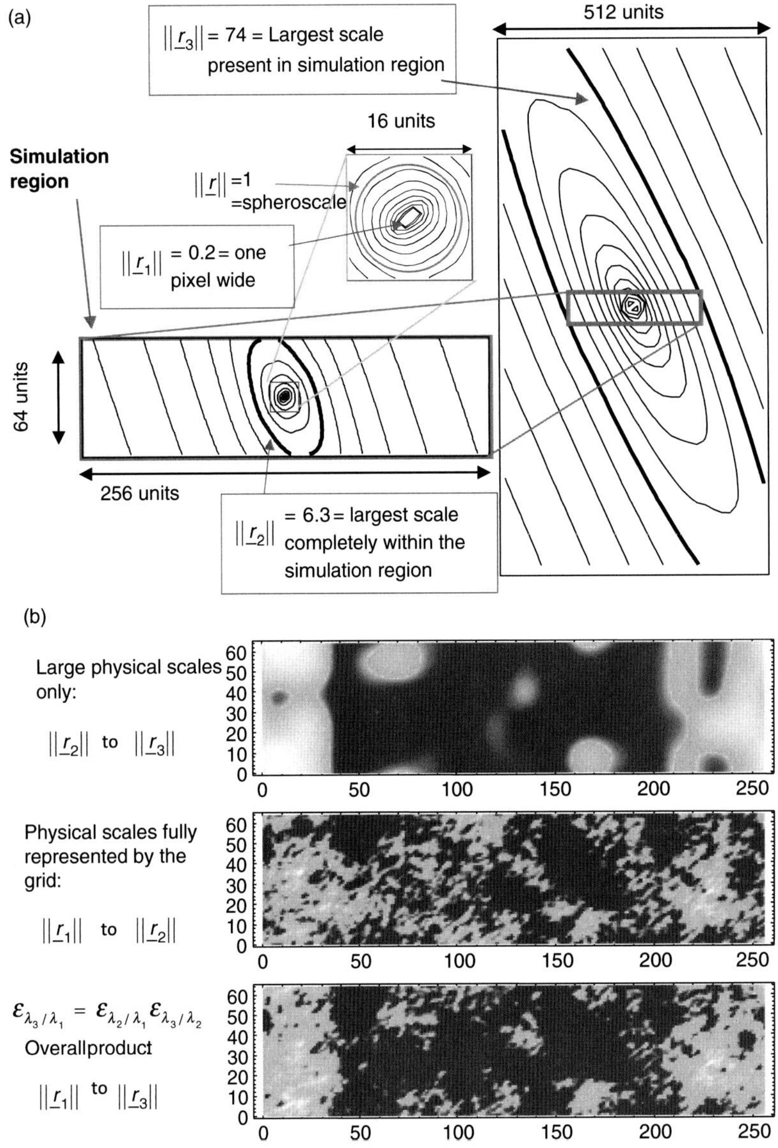

Fig. 7.5 (a) A schematic diagram showing the balls associated with the canonical system with $l_s = 8$ and: $d = 1$, $c = -0.1$, $f = -0.2$, $e = 0.1$. $\|\underline{r}_1\| = 0.2$ is the smallest scale which is completely resolved by the 1×1 pixel grid, $\|\underline{r}_2\| = 6.3$ is the largest scale completely resolved by the 64×256 pixel simulation region; $\|\underline{r}_2\| = 74$ is the largest scale that influences the simulation region. (b) This shows the contributions from the fully resolved band (scales $\|\underline{r}_1\|$ to $\|\underline{r}_2\|$) and the partially resolved band $\|\underline{r}_2\|$ to $\|\underline{r}_3\|$ to the total simulation; $\alpha = 1.6$, $C_1 = 0.1$ (same G as in Fig. 7.5a).

Fig. 7.5b shows the contribution of the structures from each of the ranges $\|\underline{r}_1\|$ to $\|\underline{r}_2\|$ and $\|\underline{r}_2\|$ to $\|\underline{r}_3\|$, showing what is missed if the contribution of the larger scales is not taken into account. The solution adopted for the simulations shown here was to "nest" simulations by using low-resolution versions of scale functions over the range $\|\underline{r}_2\|$ to $\|\underline{r}_3\|$ and then to use the multiplicative property of the cascade (Fig. 7.5b). Finally, we note that, as discussed in Appendix 5B, there will be "finite size effects" at the smallest scales. It turns out that the method outlined in Appendix 5B to alleviate some of these effects still works (but again, with $|\underline{r}|$ replaced by $\|\underline{r}\|$), although when the scale function is too variable near the pixel scale even this is problematic (the method only takes care of the lowest-order correction and this may no longer be adequate).

With these technical issues behind us, we can now consider the effect of varying the parameters c, e, k as shown in the multifractal simulations

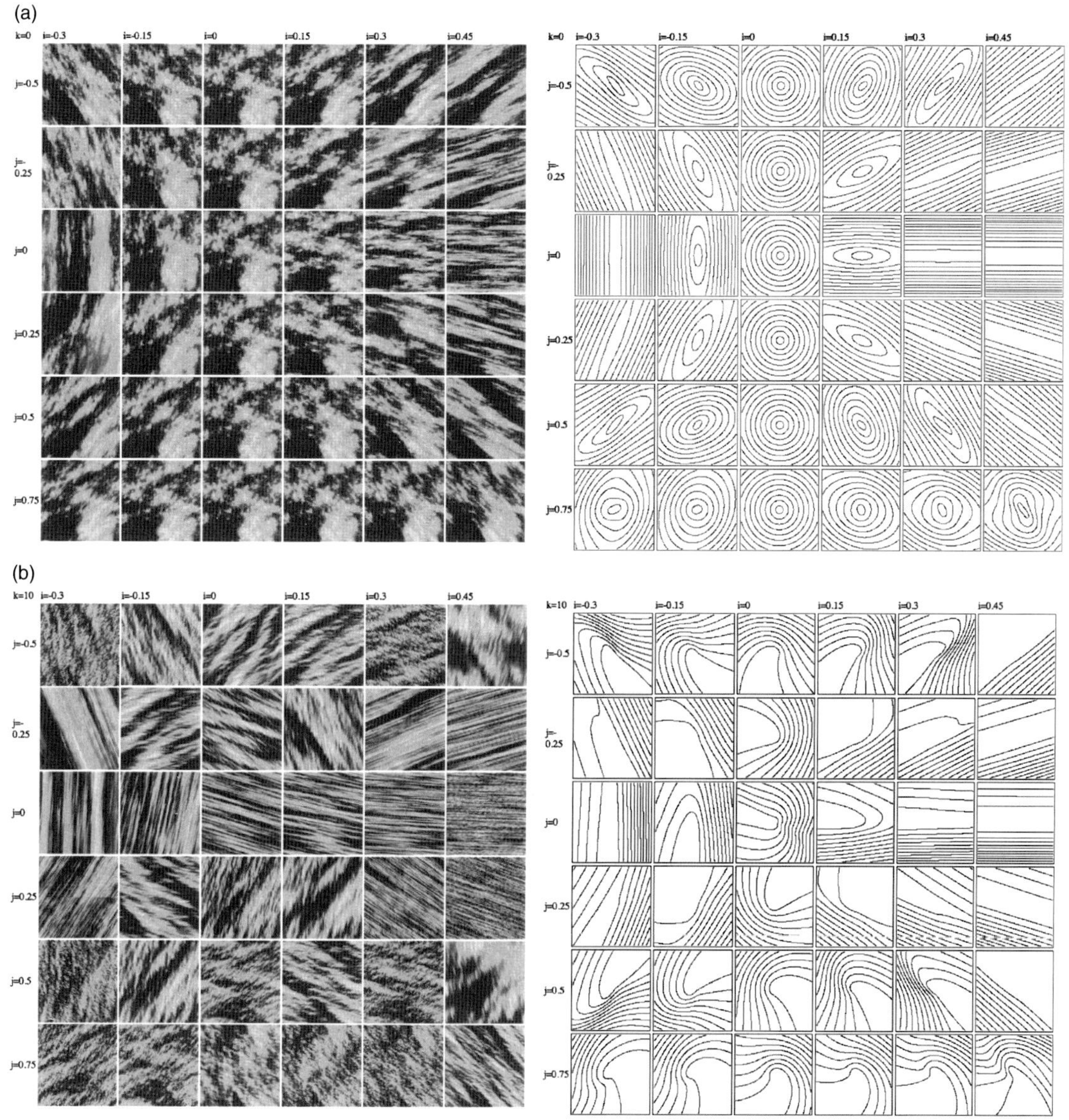

Fig. 7.6 (a) $k = 0$: we vary c (denoted i) from −0.3, −0.15, . . ., 0.45 left to right and e (denoted j) from −0.5, −0.25, . . ., 0.75 top to bottom. On the right we show the contours of the corresponding scale functions. For these and other simulations, see the "Multifractal Explorer": www.physics.mcgill.ca/~gang/multifrac/index.htm. Reproduced from Lovejoy and Schertzer (2007). (b) Same as Fig. 7.6a except that $k = 10$. Reproduced from Lovejoy and Schertzer (2007).

(Figs. 7.6a–7.6f). All the simulations have $\alpha = 1.8$, $C_1 = 0.1$, $H = 0.33$ (roughly the empirical parameters for clouds), and are simulated on 256×256 grids with the same starting seed so that the differences are due only to the anisotropy (the colours go from dark to light, indicating values low to high). For isotropic unit scales ($k = 0$, Fig. 7.6a, top row) we see the effect of varying c. On the right we display the contours of the corresponding scale functions.

Moving on to Fig. 7.6b, we take $k = 10$ so that the unit ball has a range of scales of 2^{10}; the structures are more filamentary. In the bottom row we take $e = 0$, displaying the effect of varying c: G is thus diagonal,

(c)

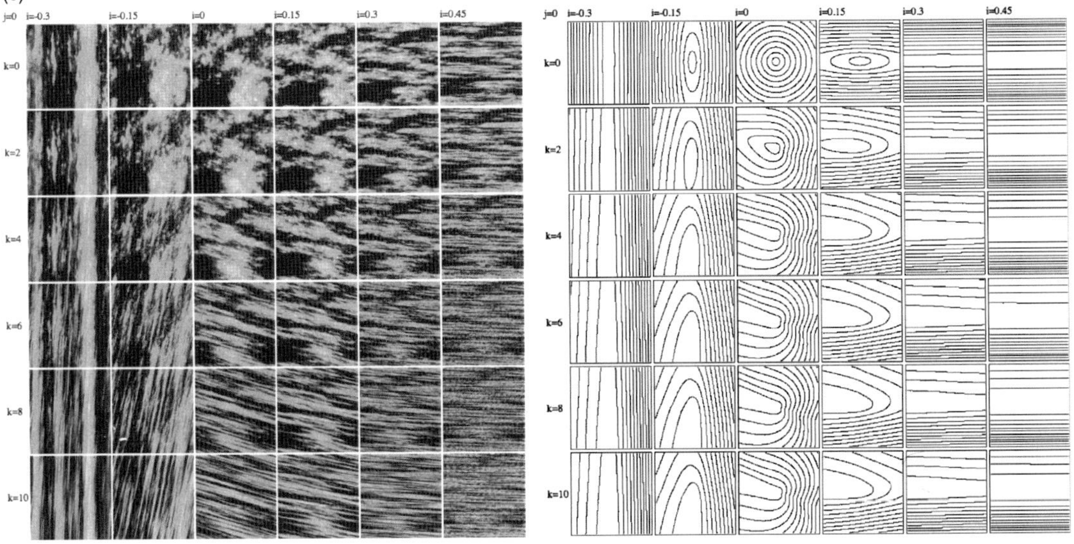

(d)

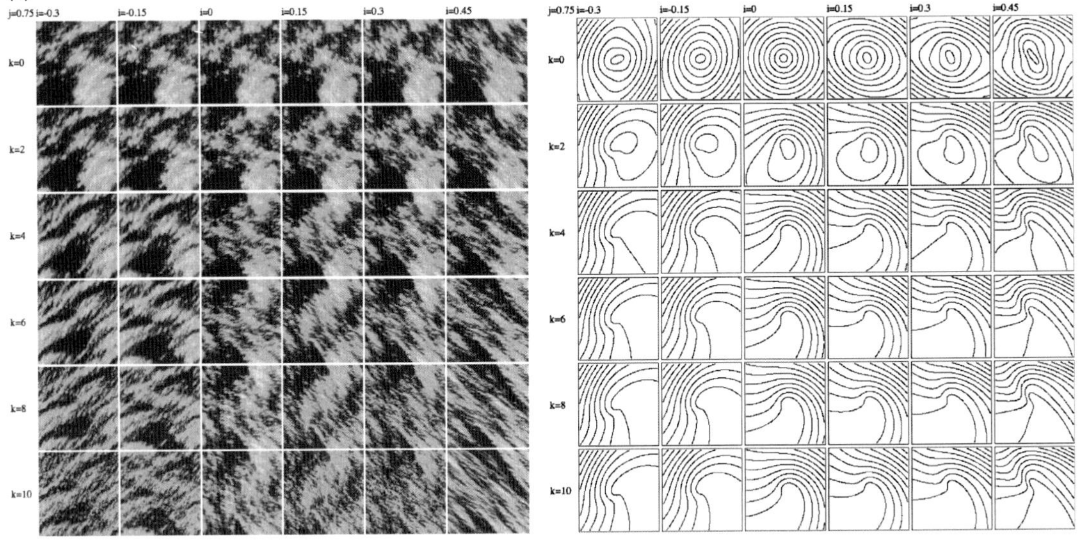

Fig. 7.6 (c) $e = 0$; c is increased from −0.3, −0.15, . . ., 0.45 left to right, from top to bottom, k is increased from 0, 2, 4, . . ., 10. See text for more details. Reproduced from Lovejoy and Schertzer (2007). (d) The same as Fig. 7.6c except that $e = 0.75$. Reproduced from Lovejoy and Schertzer (2007).

the structures are "self-affine" (no rotation). Fig. 7.6d is the same as the bottom row of Fig. 7.6c except that $e = 0.75$, showing the effect of rotation. Since $a^2=c^2-e^2<0$ here ($f = 0$), the eigenvectors of λ^{-G} rotate continuously with scale (this is discussed in more detail in Box 7.1). In Fig. 7.6e we fix $c = 0$ and vary e, and Fig. 7.6f is the same except that $c = 0.15$ so that there is both the effect of stratification (c), and rotation (e). Here the eigenvalues are again complex except in the third column with $e = 0$.

More examples of two-dimensional simulations are shown in Fig. 7.7a and 7.7b; all except the upper left are rotation-dominant with large e values. In the simulations, the unit ball was parameterized by the

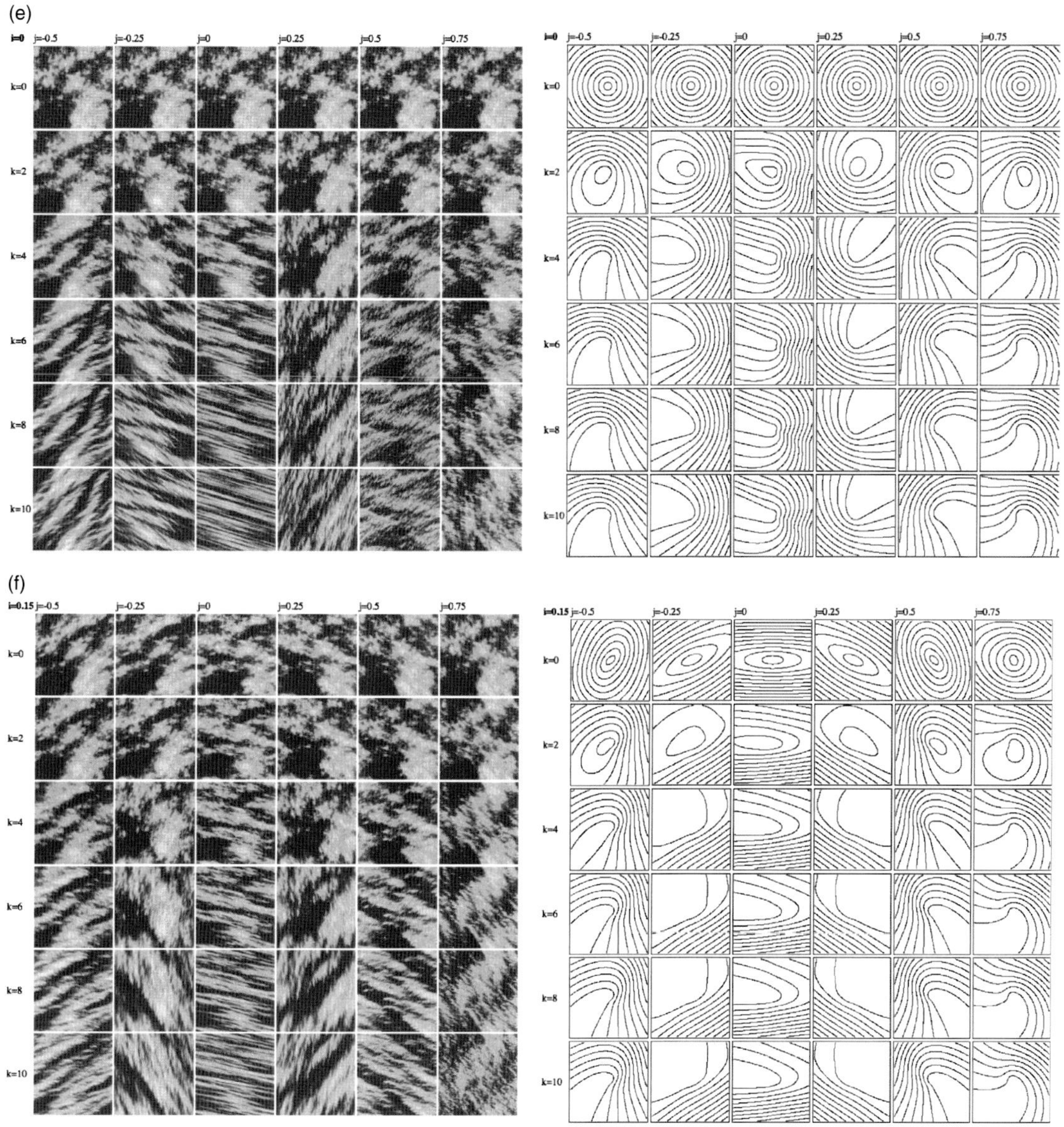

Fig. 7.6 (e) $c = 0$; e left to right is $-0.5, -0.25, \ldots, 0.75$. Reproduced from Lovejoy and Schertzer (2007). (f) Same as Fig. 7.6e except that $c = 0.15$. In all rows, from top to bottom, k is increased $(0, 2, 4, \ldots, 10)$, and the right-hand side shows the corresponding scale functions. Reproduced from Lovejoy and Schertzer (2007).

integer n (here $n = 1$) and by the parameter ξ in the range $-\pi/2 < \xi < \pi/2$:

$$\Theta(\theta'') = 1 + \sin(\xi)\cos(n\theta'') \tag{7.67}$$

When $n = 1$, this is an equivalent parametrization to Eqn. (7.66) using k. The relation between the two can be seen since:

$$r_{max}/r_{min} = \cot^2[(\pi/2 - \xi)/2] = 2^k \tag{7.68}$$

so that the structures are particularly "wispy" for the ξ near $\pi/2 = 1.57\ldots$ (see especially the bottom row with $\xi = 1.5$: $r_{max}/r_{min} \approx 800$). In these large ξ cases, we can still make out the "signature" of the spiral-shaped scale function (Fig. 7.7b). While it may seem artificial

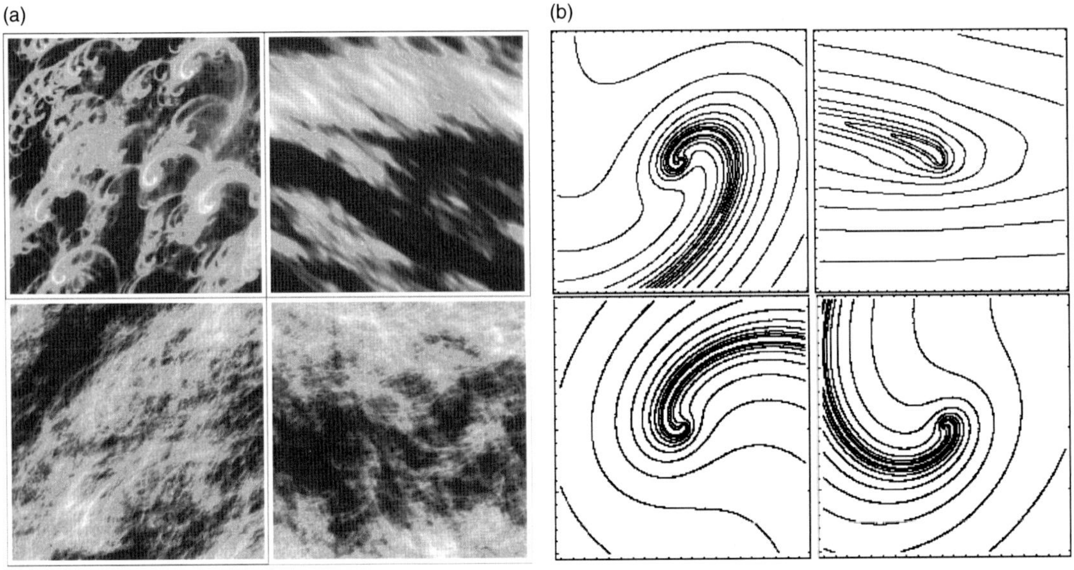

Fig. 7.7 (a) Examples of 2D simulations on 512 × 512 pixel grids with $\alpha = 1.8$, $C_1 = 0.1$, $H = 0.333$, $d = 1$, $f = 0$. Upper left: $c = 0.8$, $e = 2$, $l_s = 512$, $\xi = 1.3$ ($2^k = r_{max}/r_{min} \approx 54$). Upper right: $c = -2/7$, $e = 0.1$, $l_s = 32$, $\xi = 0.7$ ($2^k \approx 5$). Lower left: $c = 0.3$, $e = 1.2$, $l_s = 32$, $\xi = 1.5$ ($2^k \approx 800$). Lower right: $c = 0.3$, $e = 1.2$, $l_s = 1$, $\xi = 1.5$ ($2^k \approx 800$). (b) The scale functions corresponding to Fig. 7.7a.

to use such large ratios, remember that in the real atmosphere a cloud 100 m across has already had a range of (isotropic) scales of 10^5; the wisps of a cirrus would be the result of an even larger range of scales of anisotropic stretching. Starting with a highly anisotropic unit ball and modelling a 1 km structure could be considered as an attempt to take into account this large range of scales that have operated down from a presumably more isotropic planetary scale.

In Chapter 5 we discussed the fact that in general the cascades are "unbounded" in the sense that they produce singularities of arbitrarily high order (this is the case of the universal multifractals for $\alpha \geq 1$, for example). In this case, a sufficiently large number of realizations of the process will almost certainly have singularities which are almost certainly absent on any single realization. However, some of these very strong singularities (structures) are only seen very rarely. In order to simulate these very rare realizations, we can "help" the cascade by artificially boosting the values of the subgenerator $\gamma(\underline{r})$ (Eqn. (5.78)). The simulations in Fig. 7.8a have had their single central subgenerator value $\gamma(\underline{r}_c)$ boosted by a factor N; in the figure N increases from 8 to 64 by factors of 2 (left to right). We see that as N increases, the basic scale function shape appears as a dominant structure out of the chaos.

7.1.5 Simulations in three dimensions, rendering with simulated radiative transfer

Moving from two to three dimensions allows us to obtain much more realistic renditions of clouds by allowing radiative transfer to be modelled instead of simply using false colours. The simplest way to extend the results of the previous sections to three dimensions is to allow for vertical stratification, but without rotation in the vertical plane, i.e. to use matrices of the form:

$$G = \begin{pmatrix} d & r-e & 0 \\ r+e & d & 0 \\ 0 & 0 & H_z \end{pmatrix} \tag{7.69}$$

In any case, 3 × 3 matrices can in general be brought into this form by appropriate similarity transformations. In this case, the diagonalizing matrices are:

$$\Omega = \frac{1}{2}\begin{pmatrix} -(r+e)^{-1/2} & (r+e)^{-1/2} & 0 \\ (r-e)^{-1/2} & (r-e)^{-1/2} & 0 \\ 0 & 0 & 2 \end{pmatrix}$$

$$\Omega^{-1} = \begin{pmatrix} -(r+e)^{1/2} & (r-e)^{1/2} & 0 \\ (r+e)^{1/2} & (r-e)^{1/2} & 0 \\ 0 & 0 & 1 \end{pmatrix} \tag{7.70}$$

and again we perform a series of coordinate transformations starting with the case of real eigenvalues, $a^2 > 0$:

(a)

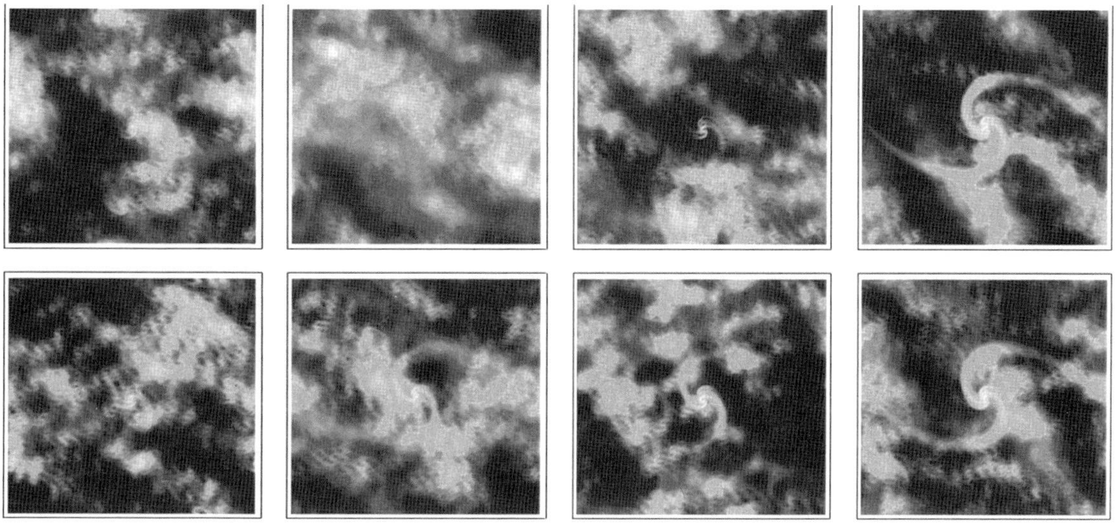

(b)

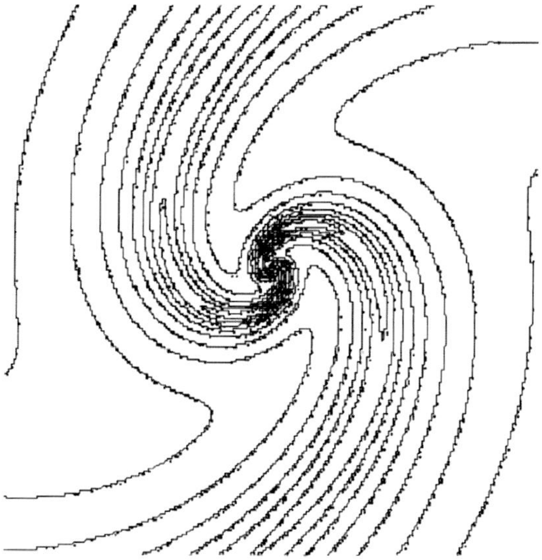

Fig. 7.8 (a) Each row shows a different realization of a random multifractal process with a single value of of the subgenerator $\gamma(\underline{r})$ (Eqn. 5.78) at the centre of a 512×512 grid replaced by the maximum of $\gamma(\underline{r})$ over the field boosted by factors of N increasing by 2 from left to right (from 8 to 64) in order to simulate very rare events ($\alpha = 1.8$, $C_1 = 0.1$, $H = 0.333$). The scaling is anisotropic with complex eigenvalues of G, the scale function is shown in Fig. 7.8b. (b) Contours of the (rotation dominant) scale function used in the simulations (Fig. 7.8a).

$$\underline{r}^{(1)} = \Omega^{-1}\underline{r}^{(0)} \qquad \begin{pmatrix} x^{(2)} \\ y^{(2)} \\ z^{(2)} \end{pmatrix} = \begin{pmatrix} |x^{(1)}|^{1/(1-a)}\left(\dfrac{x^{(1)}}{|x^{(1)}|}\right) \\ |y^{(1)}|^{1/(1-a)}\left(\dfrac{y^{(1)}}{|y^{(1)}|}\right) \\ |z^{(1)}|^{1/H_z}\left(\dfrac{z^{(1)}}{|z^{(1)}|}\right) \end{pmatrix} \tag{7.71}$$

with $a^2 = r^2 - e^2$ as usual. We now use a spherical polar coordinate representation:

$$\begin{aligned} x &= R\sin\phi\cos\theta \\ y &= R\sin\phi\sin\theta \\ z &= R\cos\phi \end{aligned} \tag{7.72}$$

where θ, ϕ are the polar and azimuthal spherical angles. We can express the next transformation as:

$$\begin{pmatrix} R^{(3)} \\ \phi^{(3)} \\ \theta^{(3)} \end{pmatrix} = \begin{pmatrix} R^{(2)}\Theta\left(\phi^{(2)},\theta^{(2)}\right) \\ \phi^{(2)} \\ \theta^{(2)} \end{pmatrix} \tag{7.73}$$

where $\Theta\left(\phi^{(2)},\theta^{(2)}\right) > 0$ is the function specifying the unit ball; in three dimensions, the unit ball is a function of θ, ϕ. Finally, to get the scale function $\|\underline{r}\| = |\underline{r}^{(4)}| = R^{(4)}$, we take:

$$\underline{r}^{(4)} = \Omega \underline{r}^{(3)} \tag{7.74}$$

The case $a^2 < 0$ is the straightforward extension of the corresponding 2D result. Again, the normalization constant $N_{Del} = \Omega_{Del}$ is a nontrivial calculation and is given in Appendix 7A.2.

In order to be able to understand and control the shape of the unit ball, we can parametrize it (cf. Eqn. (7.67) in 2D); a convenient way is:

$$\Theta\left(\phi^{(3)},\theta^{(3)}\right) = \left(1+\sin\xi_\theta \cos\left(n_\theta \theta^{(3)}\right)\right)\left(1+\sin\xi_\phi \cos\left(n_\phi \phi^{(3)}\right)\right) \tag{7.75}$$

where n_θ, n_ϕ are integers, and the two parameters ξ_θ, ξ_ϕ are chosen to vary the shape of the unit ball. If needed, extra parameters θ_0, ϕ_0 can be introduced to rotate the unit ball: replace θ by $\theta - \theta_0$ and ϕ by $\phi - \phi_0$; for an example of some of the convoluted shapes that are possible, see Fig. 7.9. Recall that the spherical polar radius of the unit balls = Θ^{-1} so that as the ξ approach $\pi/2$, the unit ball becomes divergent (there is a wider and wider range of scales in the $\underline{r}^{(0)}$ space: see Eqn. (7.68) and the examples in Section 7.1.4). As a final comment about the unit ball parametrization, in three dimensions, we can separately specify both the scale of roughly isotropic horizontal sections (the horizontal sphero-scale, l_s) and the scale of roughly vertical sections, the vertical sphero-scale, l_{sz}.

In order to visually appreciate the effect of varying some of the (now numerous!) parameters, we refer the reader to Fig. 7.10 for a rotationally dominant ($a^2 < 0$) series showing top (Fig. 7.10a) and side views (Fig. 7.10b) of a multifractal simulation of a cloud liquid water density, along with simulations of thermal infrared (wholly absorbing and emitting radiative transfer) and visible radiance fields (top view, Fig. 7.10c, and bottom view, Fig. 7.10d; totally scattering atmospheres, single scattering only; see also the cover simulation). In Fig. 7.10e we simulate a thermal infrared field (emitting and absorbing radiative transfer) obtained by assuming a linear temperature profile and constant extinction coefficients (ratios of optical to cloud liquid water densities). We see that quite realistic morphologies are possible. In Watson *et al.* (2009) and Lovejoy *et al.* (2009) we give results on radiative transfer in multifractal clouds with full multiple scattering (see Box 7.2 for an overview).

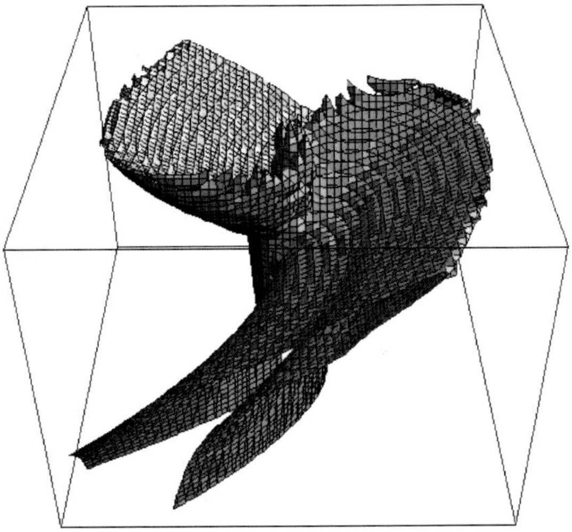

Fig. 7.9 A contour of the scale function corresponding to a single scale; this is a strongly rotationally dominant case with $n_\theta = 2$, $n_\phi = 1$, $\xi_\theta = \xi_\phi = 1.4$, $d = 1$, $c = 0.5$, $e = 1$, $f = 0$, $H_z = 0.8$, $l_s = 64$.

In Fig. 7.11 we use more rotationally dominant examples to simply show the effect of changing the unit ball on the density field; as the scale functions become more and more anisotropic (ξ_θ approaches $\pi/2$), the structures become more and more filamentary, wispy. In Figs. 7.12a–7.12f we show for selected parameters cases of (horizontally) rotationally dominant simulations with only 16 pixels in the vertical direction, also showing the corresponding thermal IR and visible fields. As the parameters of the unit ball (ξ_θ, ξ_ϕ) become larger, the contours of the scale functions become thinner, more stretched, and this tends to break up the structures more and more. In Figs. 7.13a–7.13d we show another series with stratification dominant, also showing the effect of changing the vertical versus horizontal sphero-scale.

7.1.6 Implications of anisotropic multifractals for the interpretation of data

Contrary to the extreme case of deterministic scale invariance, statistical scale invariance is almost certainly broken on every single realization; it is the

(a)

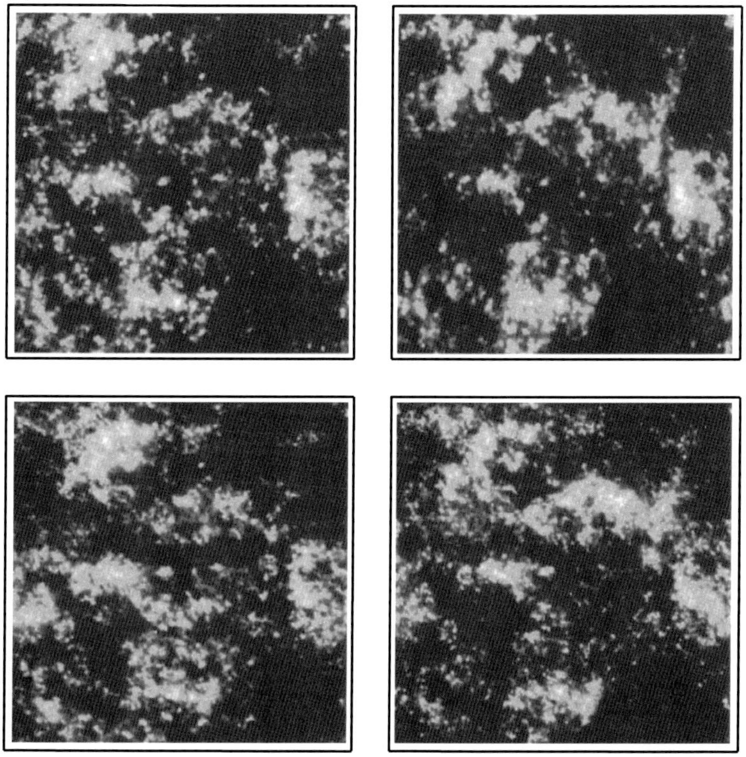

(b)

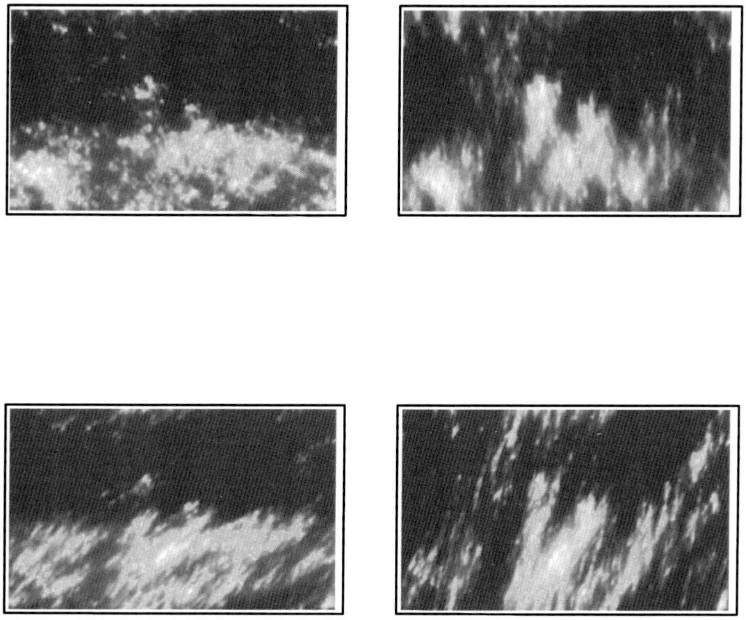

Fig. 7.10 (a) The top layers of three-dimensional cloud liquid water density simulations (false colours), All have $d = 1$, $c = 0.05$, $e = 0.02$, $f = 0$, $H_z = 0.555$, $\alpha = 1.8$, $C_1 = 0.1$, $H = 0.333$ and $\xi_0 = 0.25$ and are simulated on a $256 \times 256 \times 128$ point grid ($a^2 > 0$; stratification dominant in the horizontal). The simulations in the top row have $l_s = 8$ pixels (left column), 64 pixels (right column), $\xi_\theta = 0$ (top row), $\xi_\theta = 3/4$ (bottom row). Note that in these simulations the $l_s = 8, 64$ applies to both vertical and horizontal cross-sections (i.e. $l_s = l_{sz}$). (b) A side view of Fig. 7.10a.

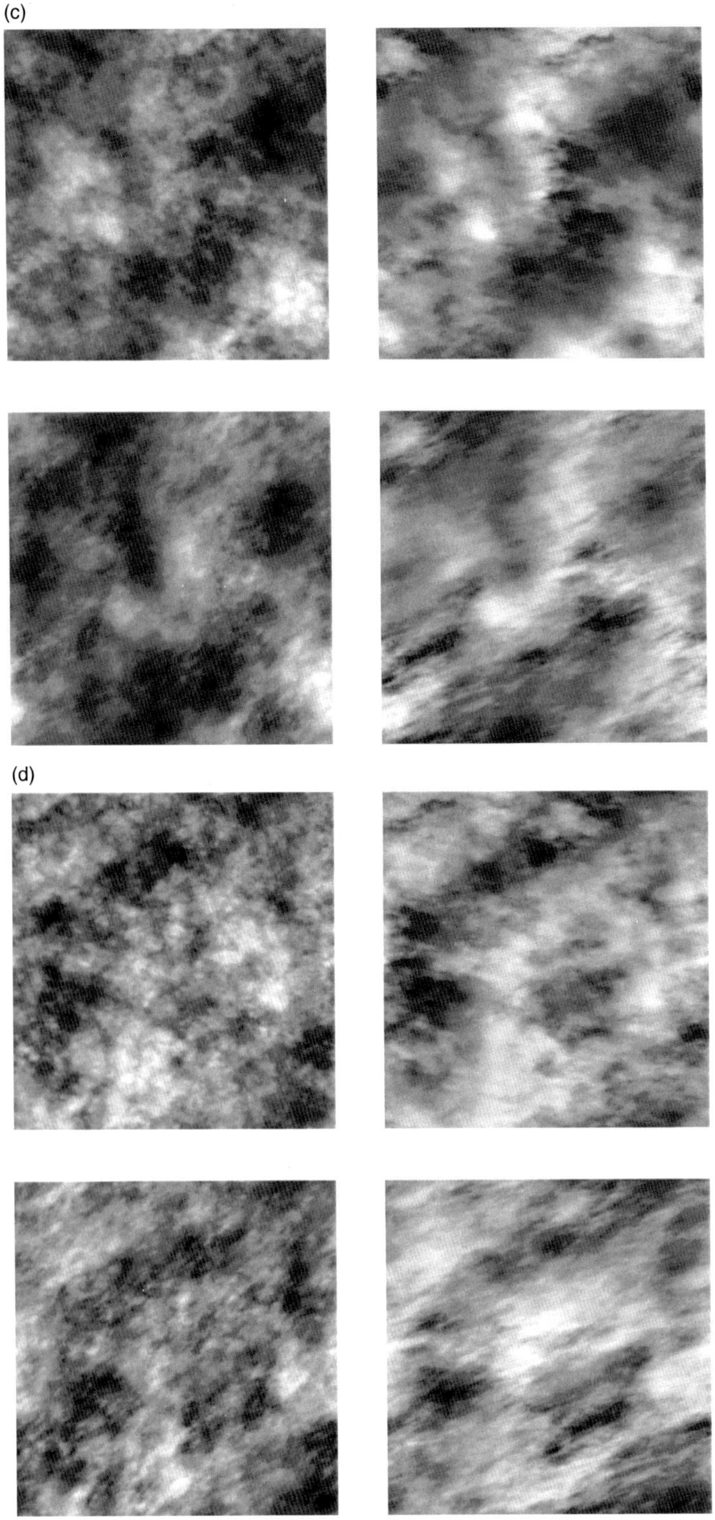

Fig. 7.10 (c) The top view of the same clouds with single scattering radiative transfer; incident solar radiation at 45° from the right, mean vertical optical thickness = 50. *See colour plate section.*(d) Same as Fig. 7.10c except viewed from the bottom. *See colour plate section.*

(e)

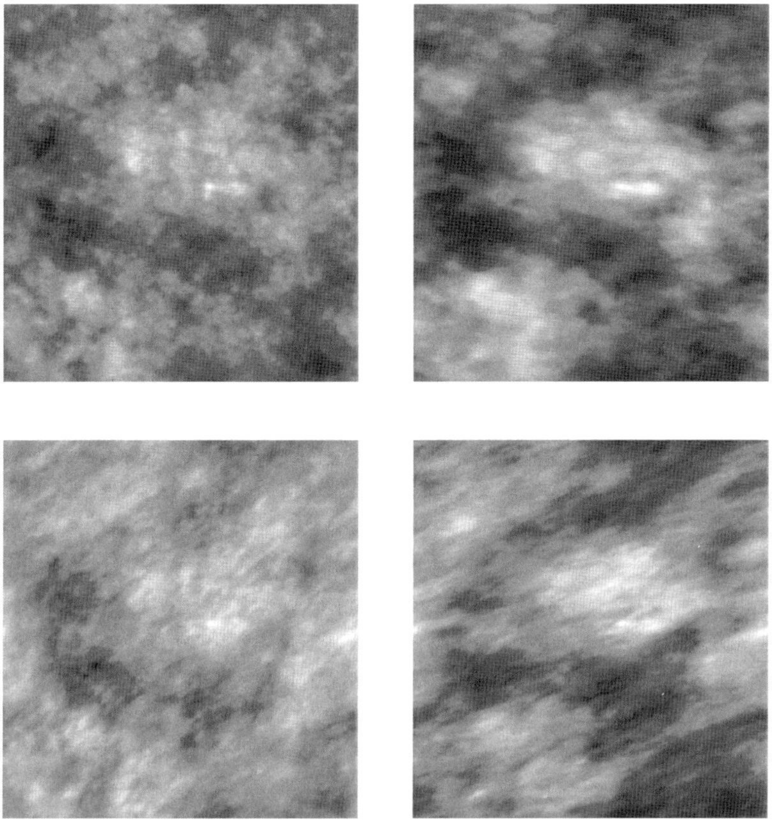

Fig. 7.10 (e) Same as Fig. 7.10d except for a false-colour rendition of a thermal infrared field (assuming a constant extinction coefficient and a linear vertical temperature profile). *See colour plate section.*

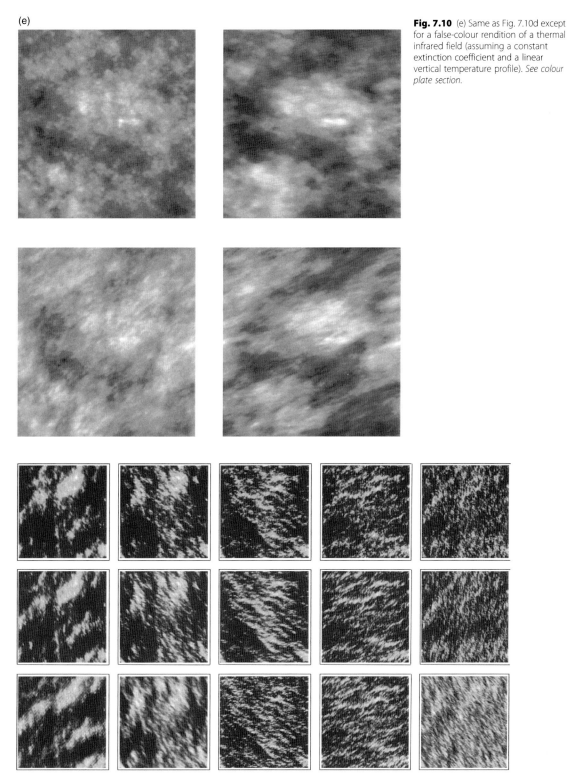

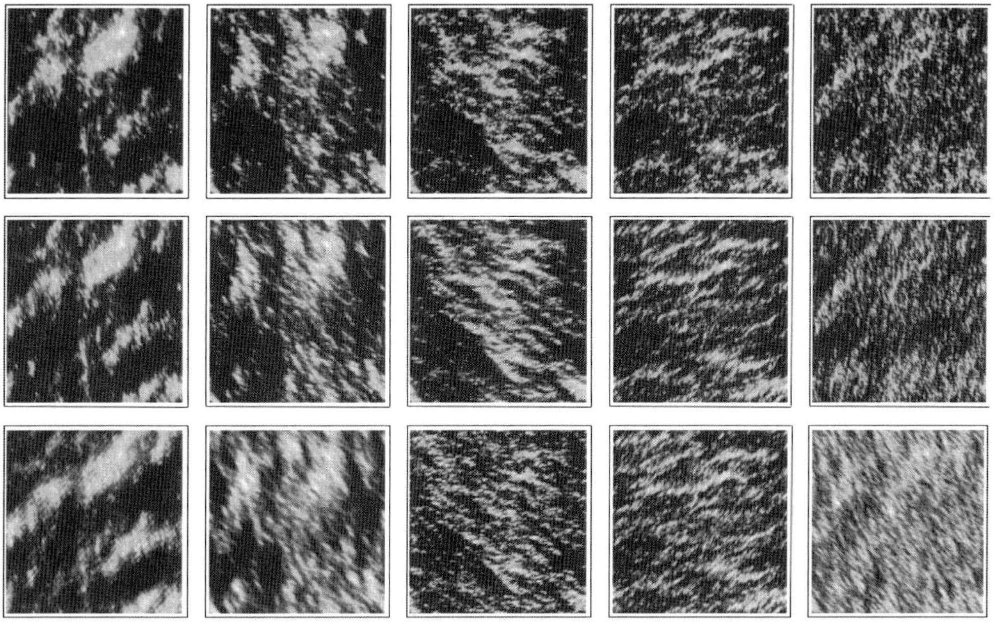

Fig. 7.11 Simulations of cloud liquid water density with l_s increasing by factors of 4 from 1/4 to 64 (right) with $\xi_\theta = 1/2, 1, 3/2$ top to bottom row. All simulations have the same random seed and parameters $\alpha = 1.8$, $C_1 = 0.1$, $H = 0.333$, $\xi_\phi = ½$, $n = 2$, $d = 1$, $c = 0.05$, $e = 0.5$, $f = 0$ (rotation dominant) the grid is 512×512×16 pixels. The parametrization of the unit ball is from Eqn. (7. 60).

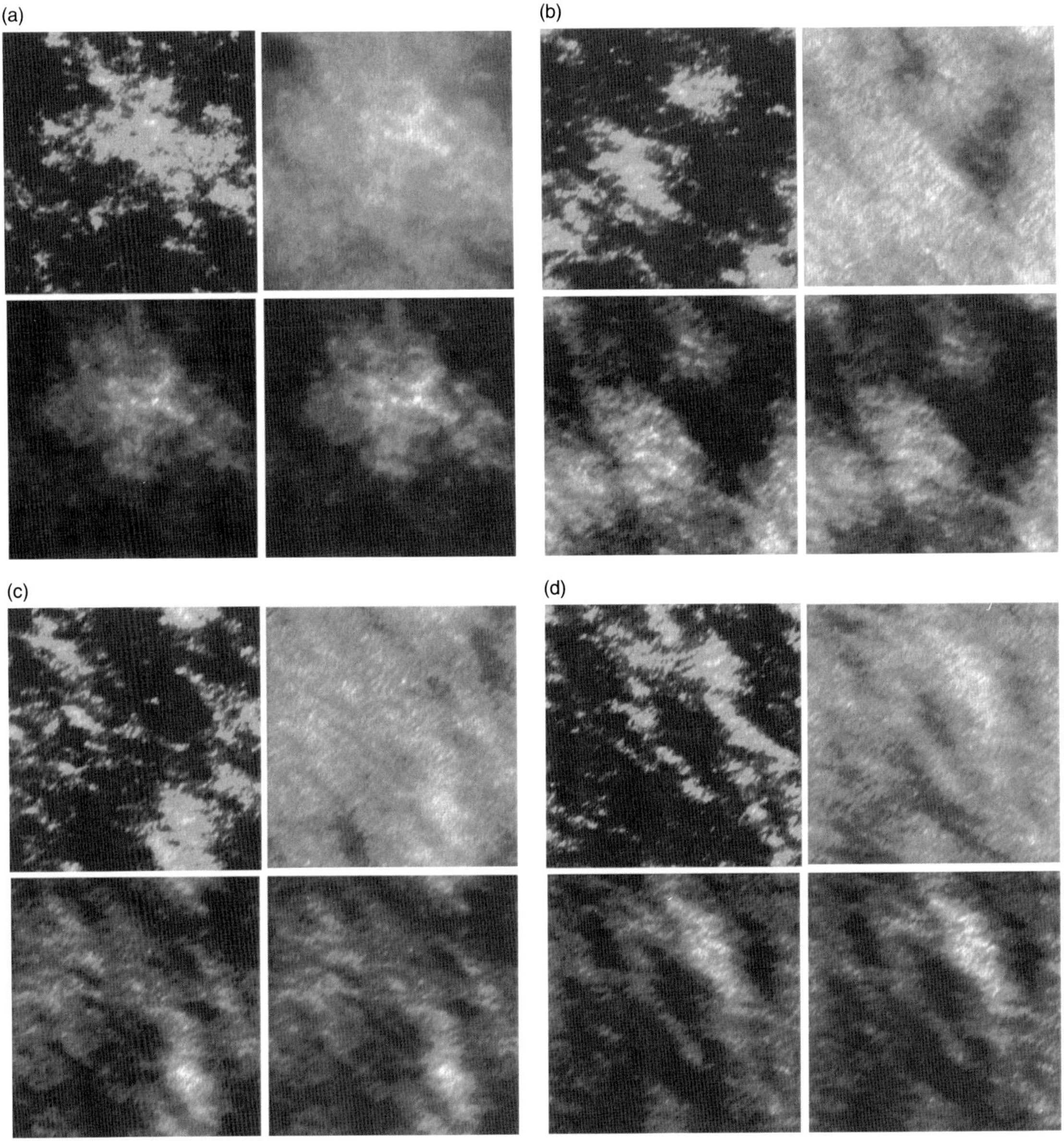

Fig. 7.12 Simulations of cloud liquid water density (top left). Top right is the corresponding thermal IR field, the bottom left is the top view, the bottom right is the bottom view. The parameters are $\alpha = 1.8$, $C_1 = 0.1$, $H = 0.333$, $\xi_\Phi = 1/2$, $n = 2$, $d = 1$, $c = 0.05$, $e = 0.5$, $f = 0$ (rotation dominant). The grid is 512 × 512 × 16 pixels and $\xi_\theta = 1/4$, $l_s = 1/4$. (b) Same as previous but different random seed and $\xi_\theta = 1/2$, $l_s = 1/4$. (c) Same as previous but $\xi_\theta = 1$, $l_s = 1/4$. (d) Same as previous but $\xi_\theta = 1/4$, $l_s = 1$.

statistical ensemble that is symmetric. The enormous variability, intermittency, of multifracals makes it particularly important to have a large database with a large range of scales and many realizations so as to average fluctuations and approximate the theoretically predicted ensemble scaling. In fact, due to the singularities of all orders (see Chapter 5) the variability of multifractals is much greater than that of classical stochastic processes; for example, rare (extreme) singularities are produced by the process

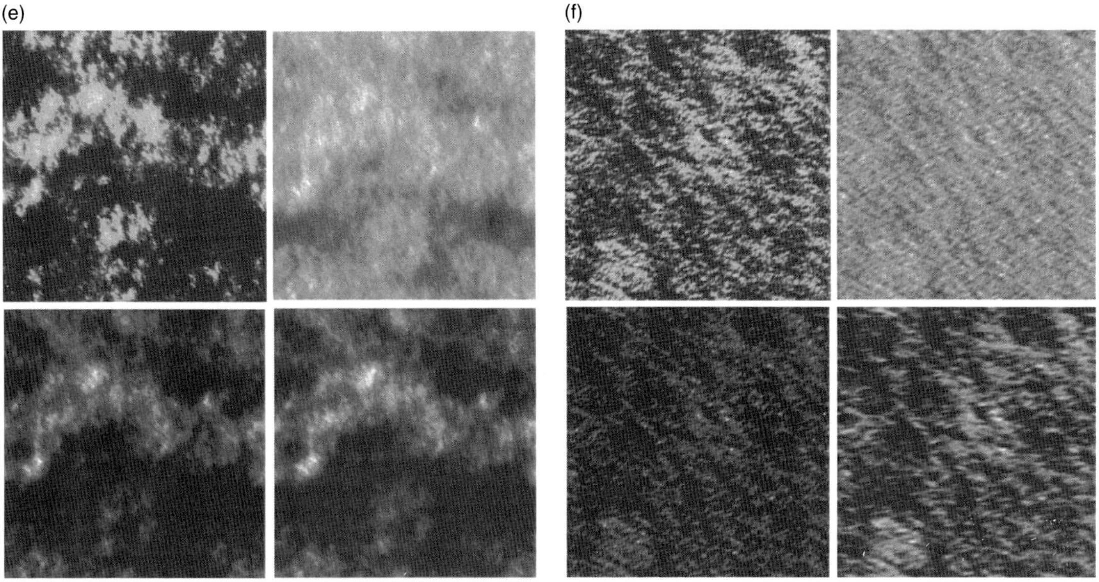

Fig. 7.12 (e) Same as previous but $\xi_\theta = 3/4$, $l_s = 1$. (f) Same as previous but $\xi_\theta = 3/4$, $l_s = 64$.

yet they are almost certainly absent on any given realization (the second-order multifractal transitions discussed in Chapter 5). What may be nothing more than normal multifractal statistical variability can thus easily be interpreted as breaks in the scaling.

Anisotropy introduces further difficulties for analysis and parameter estimation. For example, In Section 6.6 we saw how scaling anisotropy can lead to spurious scale breaks and exponent estimates. Now that we have a more general appreciation of anisotropic scaling, we can return to its effects on data analysis, showing how – unless it is carefully taken into consideration – it can readily lead to spurious conclusions. Rather than giving a purely theoretical discussion, let us consider the analysis of a 1D transect (Fig. 7.16a). Although this example was initially developed in the context of bathymetry transects (Gagnon *et al.*, 2006), it explicitly demonstrates several issues that occur in atmospheric analyses.

Fig. 7.16a compares the energy spectra of two individual transects as well as the ensemble over all the transects. One of the transects passes through "Mt. Multi," the highest peak in the simulated range, another through a randomly chosen transect not far away. One can see that the Mt. Multi scaling is pretty poor; a naive analysis would indicate two ranges with a break at about 10 pixels with high-frequency exponent $\beta = 2.5$, low-frequency $\beta = 1.5$ (this is a large change in spectral slope). Clearly this break has nothing to do with the scaling of the process (which is perfect except – due to finite grid-size effects – for the highest factor of two or so in resolution).

In comparison, the randomly chosen transect has better scaling, but with $\beta \approx 2$ whereas the angle-integrated spectrum averaged over an infinite ensemble of realizations has $\beta = 2.17$. Even the average over the transects shows signs of a spurious break at around 16 pixels (the scale where the north–south and east–west fluctuations are roughly equal in magnitude, the "sphero-scale"); this explains why the theory line does not pass perfectly through the curve corresponding to the average of the transects. Clearly, since a priori the physically relevant notion of scale is not known, the first task should be to determine it (with the matrix G). However, this is still a difficult problem (see however Section 7.2). Obviously, even for fixed parameters, had we chosen a different random seed, the results for the individual transects would have been somewhat different (even the average over the transects would have been a bit different): see the example in the next section.

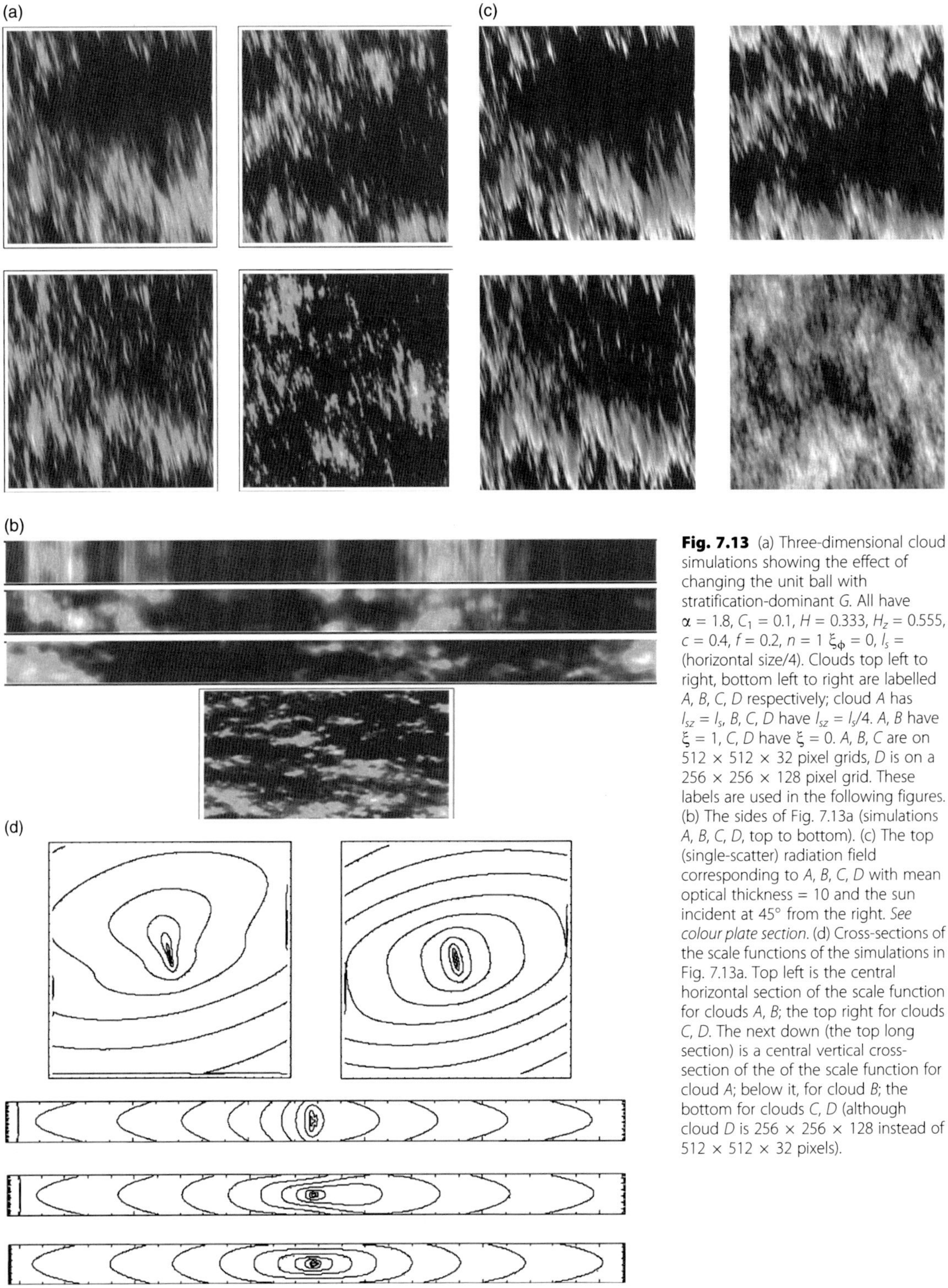

Fig. 7.13 (a) Three-dimensional cloud simulations showing the effect of changing the unit ball with stratification-dominant G. All have $\alpha = 1.8$, $C_1 = 0.1$, $H = 0.333$, $H_z = 0.555$, $c = 0.4$, $f = 0.2$, $n = 1$ $\xi_\phi = 0$, $l_s =$ (horizontal size/4). Clouds top left to right, bottom left to right are labelled A, B, C, D respectively; cloud A has $l_{sz} = l_s$, B, C, D have $l_{sz} = l_s/4$. A, B have $\xi = 1$, C, D have $\xi = 0$. A, B, C are on $512 \times 512 \times 32$ pixel grids, D is on a $256 \times 256 \times 128$ pixel grid. These labels are used in the following figures. (b) The sides of Fig. 7.13a (simulations A, B, C, D, top to bottom). (c) The top (single-scatter) radiation field corresponding to A, B, C, D with mean optical thickness = 10 and the sun incident at 45° from the right. *See colour plate section*. (d) Cross-sections of the scale functions of the simulations in Fig. 7.13a. Top left is the central horizontal section of the scale function for clouds A, B; the top right for clouds C, D. The next down (the top long section) is a central vertical cross-section of the of the scale function for cloud A; below it, for cloud B; the bottom for clouds C, D (although cloud D is $256 \times 256 \times 128$ instead of $512 \times 512 \times 32$ pixels).

(a)

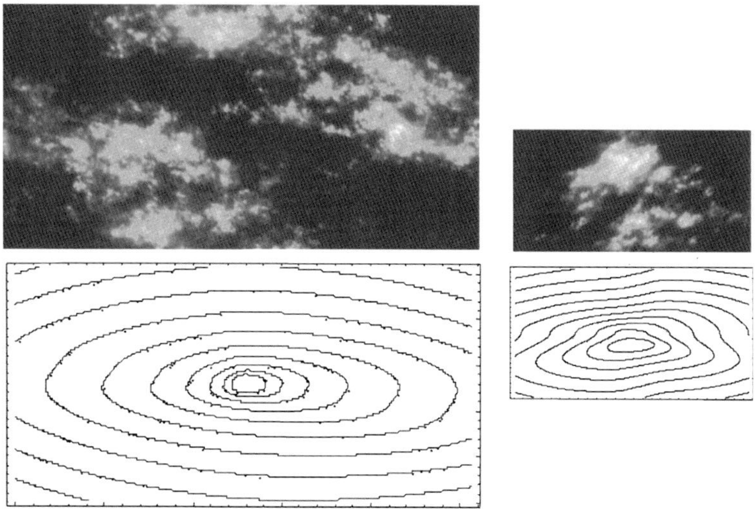

(b)

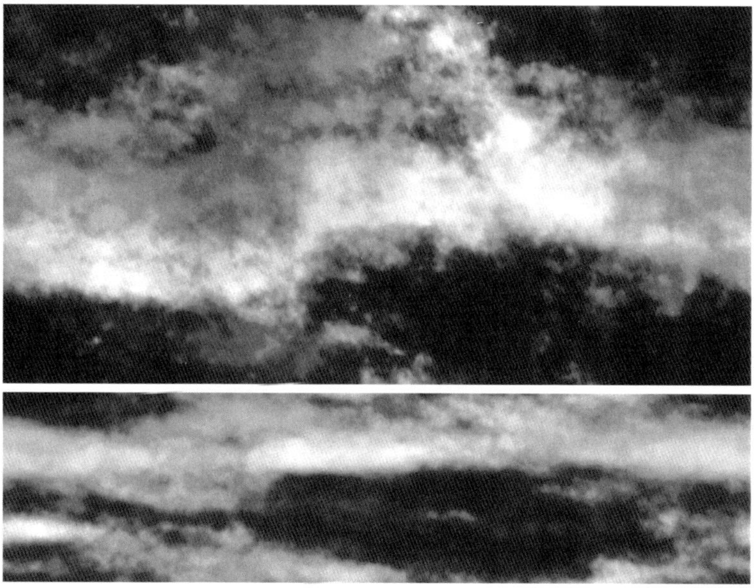

Fig. 7.14 (a) An example with $\alpha = 1.8$, $C_1 = 0.1$, $H = 0.333$, on a 512×256×128 pixel grid (128 is the thickness). The parameters are $n_\theta = n_\phi = 1$, $\xi_\theta = 0.3$, $\xi_\phi = 0.8$, $c = 0.2$, $e = 0.1$, $f = 0.05$, $H_z = 0.555$ with $l_s = 16$, $l_{sz} = 4$. The upper left is the liquid water density field, top horizontal section, to the right is the corresponding central horizontal cross-section of the scale function. The bottom row shows one of the sides (256×128 pixels) with corresponding central part of the vertical cross- section. (b) The top is the visible radiation field (corresponding to Fig. 7.14a) looking up (sun at 45° from the right); the bottom is a side radiation fields (one of the 512 × 128 pixel sides), average optical thickness = 10, single scattering only. *See colour plate section.*

Conclusions about broken scaling in Fig. 7.16a are therefore erroneous. The reason is that the assumption that the scaling is isotropic is false, and therefore breaks in statistics on 1D subspaces (transects) do not imply that the full process is not scaling (this is the same break that was discussed in Section 6.6, Eqn. 6.70). A second reason, discussed a bit further below, is that there can be systematic biases due to the use of conditional statistics – such as studying the transect that happens to pass through a special feature such as Mt. Multi, rather than a randomly chosen transect.

The strong singularities in multifractals can lead to quite different morphologies being in close proximity. This is often interpreted in terms of nonstationarities corresponding to different processes at work in different regions, or, at the very least,

(c)

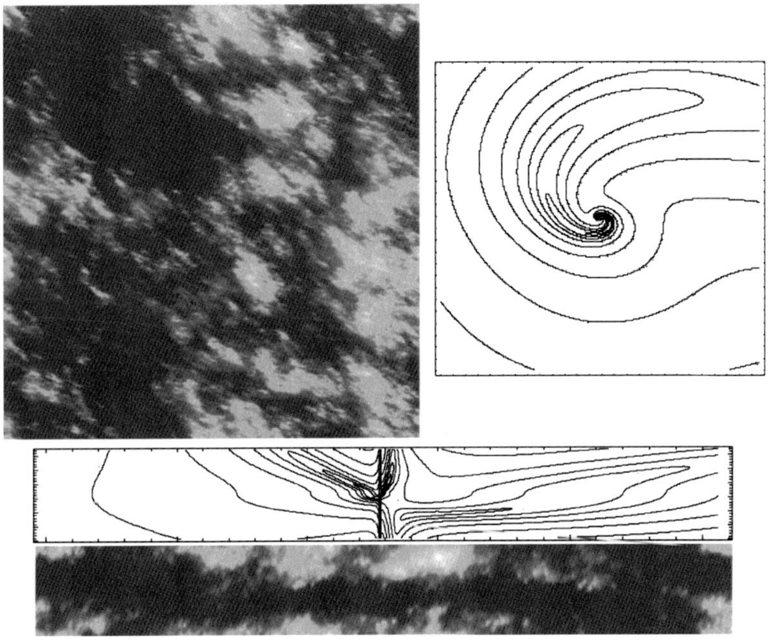

(d)

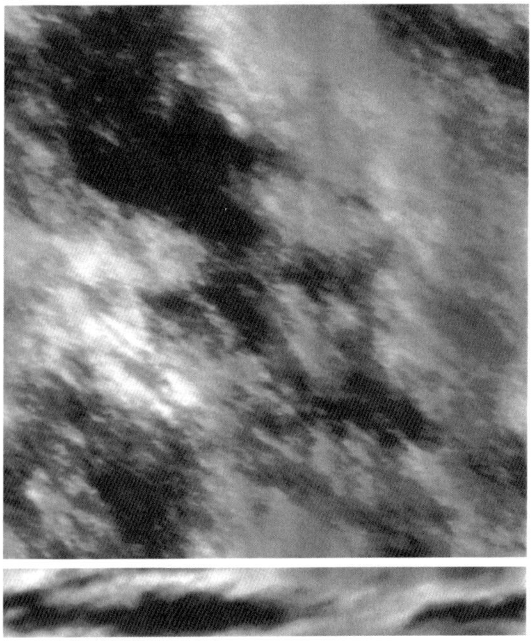

Fig. 7.14 (c) An example with α = 1.8, C_1 = 0.1, H = 0.333, on a 512×512×64 grid (the latter is the thickness). The parameters are $n_\theta = 1$, $n_\phi = 2$, $\xi_\theta = 0.3$, $\xi_\phi = 0.8$, $c = 0.2$, $e = 0.5$, $f = 0.2$ (rotation dominant), $H_z = 0.555$ with $l_s = 128$, $l_{sz} = 32$. The upper left is the liquid water density field, top horizontal section, to the right is the corresponding central horizontal cross-section of the scale function. The bottom row shows one of the sides (512×64 pixels) with corresponding central part of the vertical cross-section. (d) The top is the visible radiation field (corresponding to Fig. 7.14c), looking up (sun at 45° from the right); the bottom is a side radiation fields (one of the 512 × 64 pixel sides), the average optical thickness = 5, single scattering only. *See colour plate section.*

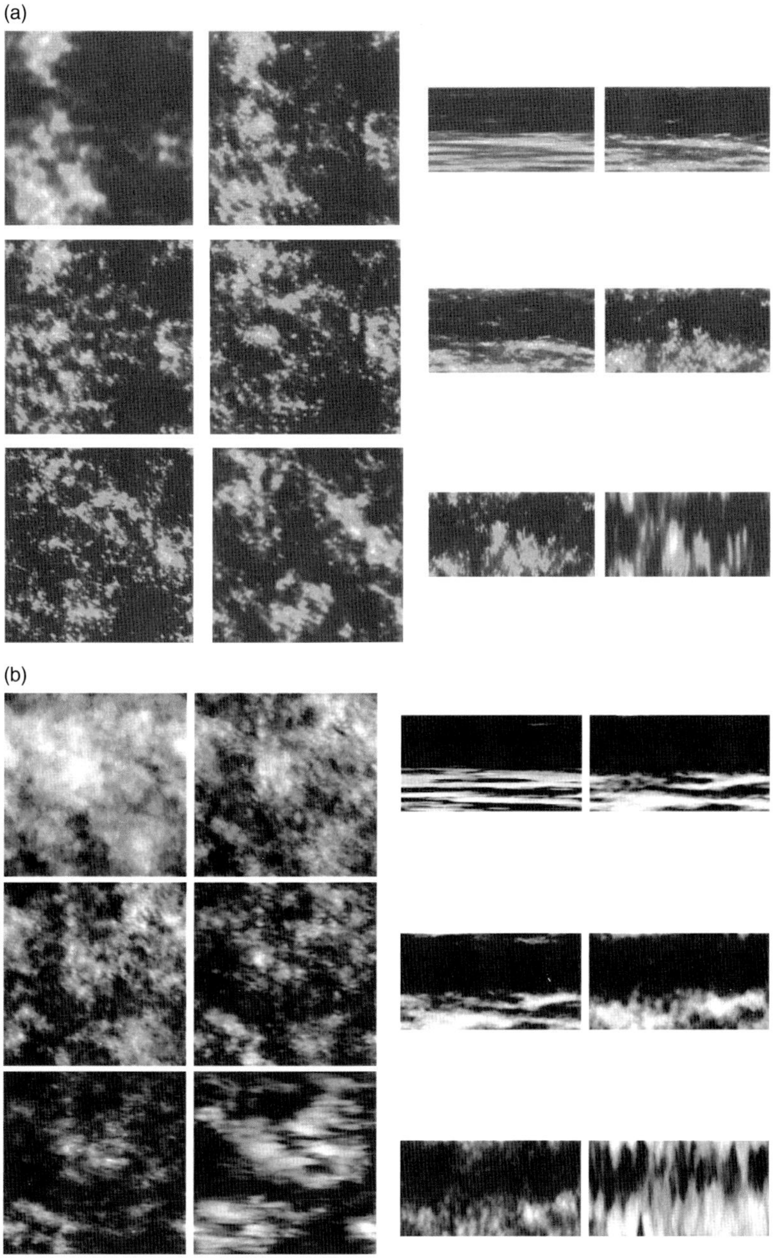

Fig. 7.15 (a) In this sequence, we see the effect of varying the horizontal and vertical sphero-scales: the left column two columns are views of the top cloud liquid water density simulation with (1st column) $l_s = 1$, (2nd column), $l_s = 8$, columns 3, 4 are corresponding side sections, the clouds are 256×256×128 pixels in size. As we move from top to bottom the vertical sphero-scale increases from $l_{sz} = l_s/4$ (top) to $l_{sz} = l_s$ to $l_{sz} = 4\ l_s$ (bottom). All the simulations have $\alpha = 1.8$, $C_1 = 0.1$, $H = 0.333$, $d = 1$, $c = 0.5$, $e = 2$, $f = 0$, $H_z = 0.555$, $n_\theta = n_\phi = 2$, $\xi_\theta = 0.4$, $\xi_\phi = 0.75$. Note here that the vertical structure of the unit ball is strongly anisotropic. (b) The downward (left two columns) and side (right two columns) radiative transfer fields corresponding to Fig. 7.15a with the sun from the right at 45°.

variations in the parameters of a single basic model. However, with multifractals such interpretations may well be unwarranted: the basic multifractal processes are statistically stationary/homogeneous in the strict sense that over the region over which they are defined (which is necessarily finite), the ensemble multifractal statistical properties are independent of the (space/time) location (and this for any spectral slope β). Rather than discussing this at an abstract level, let us see what happens when we analyse a self-similar 1024 × 1024 multifractal simulation (Fig. 7.16b).

Box 7.2 Radiative transfer in fractal and multifractal clouds

An understanding of cloud and radiation variability and their interrelations over wide ranges of scales is of fundamental importance in meteorology and climatology. It is also a challenging problem in the physics of disordered media.

The classical theory of radiative transfer (Chandrasekhar, 1950) is elegant but is only relevant in 1D ("plane parallel," horizontally homogeneous) media. Yet the use of 1D models has long dominated the field; this is because when we turn to horizontally inhomogeneous media, there is no consensus on the appropriate model of heterogeneity, nor is the transport problem analytically tractable. As a consequence, the effect of horizontal variability was underestimated, usually reduced to the problem of inhomogeneity of the external cloud/medium boundaries (e.g. cubes, spheres, cylinders: Busygin *et al.*, 1973; McKee and Cox, 1976; Preisendorfer and Stephens, 1984), with the internal cloud and radiance fields still being considered smoothly varying if not completely homogeneous. When stronger internal horizontal inhomogeneity was considered it was typically confined to narrow ranges of scale so that various transfer approximations could be justified (Weinman and Swartzrauber, 1968; Welch *et al.*, 1980).

When the problem of transfer in inhomogeneous media finally came to the fore, the mainstream approaches were heavily technical (see Gabriel *et al.*, 1993, for a review), with emphasis on comparisons of general-purpose numerical radiative transfer codes (see the C3 initiative: Cahalan *et al.*, 2005) and the application to huge large eddy simulation cloud models (Mechem *et al.*, 2002). At a more theoretical level, the general problem of the consequences of small-scale cloud variability on the large-scale radiation field has been considered using wavelets (Ferlay and Isaka, 2006) but has only been applied to numerical modelling. As a consequence, these "3D radiative transfer approaches" have generally shed little light on the scale-by-scale statistical relations between cloud and radiation fields in realistic scaling clouds (e.g. Marshak and Davis, 2005). Overall, there has been far too much emphasis on techniques and applications, with little regard for understanding the basic scientific issues.

The simplest interesting transport model is diffusion; on fractals see the reviews by Bouchaud and Georges (1990) and Havlin and Ben-Avraham (1987), and on multifractals see Meakin (1987), Weissman (1988), Lovejoy *et al.* (1993, 1998) and Marguerit *et al.* (1998). However, except in 1D (Lovejoy *et al.*, 1993, 1995), diffusion is not in the same universality class as radiative transport (Lovejoy *et al.*, 1990).

The first studies of radiative transport on fractal clouds (with a constant optical density on the support) were those by Gabriel *et al.* (1986, 1990), Cahalan (1989, 1994), Lovejoy and Schertzer (1989), Davis *et al.* (1990), Lovejoy *et al.* (1990), Barker and Davies (1992) and Cahalan *et al.* (1994). These works used various essentially academic fractal models and focused on the (spatial) mean (i.e. bulk) transmission and reflectance. They clearly showed that (1) fractality generally leads to nonclassical ("anomalous") thick cloud scaling exponents, (2) the latter were strongly dependent on the type of scaling of the medium, and (3) the exponents are generally independent of the phase function (Lovejoy *et al.*, 1990). Some general results applicable to multifractal clouds may be found in Naud *et al.* (1997) and Schertzer *et al.* (1997).

Some general theoretical results exist for conservative cascades ($H = 0$), both for single scattering for log-normal clouds ($\alpha = 2$, Lovejoy *et al.*, 1995) and for more general ($\alpha < 2$) "universal" multifractal clouds which are dominated by low-density "Lévy holes" where most of the transport occurs (Watson *et al.*, 2009). The latter show how to "renormalize" the cloud density, to relate the mean transmission statistics to those of an equivalent homogeneous cloud. In Lovejoy *et al.* (2009) these are extended (numerically) to $H > 0$ and with multiple scattering including the case of very thick clouds. By considering the (fractal) path of the multiply scattered photons, it was found that due to long-range correlations in the cloud, the photon paths are "subdiffusive," and that the corresponding fractal dimensions of the paths tend to increase slowly with mean optical thickness. Reasonably accurate statistical relations between N scatter statistics in thick clouds and single scatter statistics in thin clouds were developed, showing that the renormalized single-scatter result is remarkably effective. This is because of two complicating effects acting in contrary directions: the "holes" which lead to increased single-scatter transmission and the tendency for multiply scattered photons to become "trapped" in optically dense regions, thus decreasing the overall transmission.

All results to date are for statistically isotropic media. For more realism, future work must consider scaling stratification as well as the statistical properties of the radiation fields and their (scaling) interrelations with cloud density fluctuations.

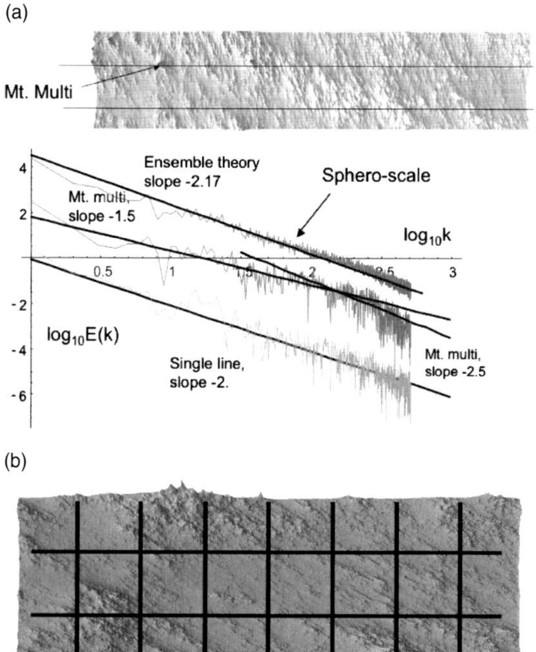

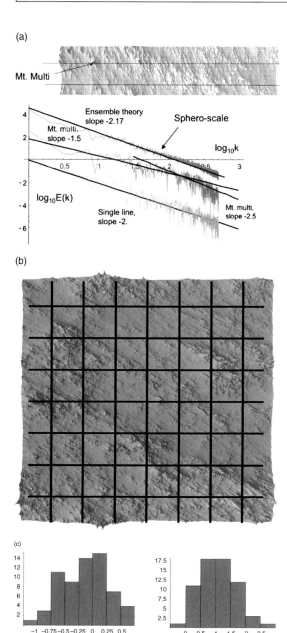

Fig. 7.16 (a) A simulated bathymetry transect. The parameters are $G = \begin{pmatrix} 0.7 & -0.02 \\ 0.02 & 1.3 \end{pmatrix}$, $\alpha = 1.9$, $C_1 = 0.12$, $H = 0.7$. The energy spectra of the transect passing through "Mt. Multi" (the highest peak in the simulation) and through another (randomly chosen) transect are shown as well as the ensemble over all the transects. Reproduced from Gagnon *et al.* (2006). (b) A self-similar multifractal (with some trivial anisotropy) simulated on a 1024 × 1024 point grid with observed universal multifractal parameters ($H = 0.7$, $C_1 = 0.12$, $\alpha = 1.9$); the spectral exponent is $\beta = 1 + 2H - K(2) = 2.17$. Adapted from Gagnon *et al.* (2006). (c) Left: after dividing Fig. 7.16a into 64 128 × 128 squares, we calculated the isotropic spectrum in each, and fit the slope to the lowest factor 16 in scale (we remove the highest factor 4 due to numerical artefacts at the highest wavenumbers). The resulting $\Delta\beta$ is given in the left; it is twice the ΔH, showing that H can vary by 0.5 over a single region. (right): A histogram of the $\log_{10}E_1$ (E_1 is the spectral prefactor: $E(k)=E_1k^{-\beta}$) showing variation of a factor of 1000 from the smoothest to roughest subregion. Adapted from Gagnon *et al.* (2006).

We can now consider the "regional" variability in the spectral exponent β by dividing the simulation into 8 × 8 squares, each with 128 × 128 pixels. Fig. 7.16c (left) shows the histogram of the 64 regression estimates of the compensated spectral exponent $\Delta\beta$: the mean is close to zero as expected, but we see a large scatter, implying that there are some individual regions having $\Delta\beta$ as low as 1.2, some as high as 2.7; the standard deviation is ± 0.3. As we shall see later, this would imply a random variation in local estimates of ΔH of ± 0.3/2 = ± 0.15 (which is of the order of the difference in the H of the topography observed between continents and under oceans, although with this spread in $\Delta\beta$, ΔH will decrease as the size of the dataset increases). In Fig. 7.16c, we can also see the large variations in the log prefactors ($\log_{10}E_1$; $E(k) = E_1k^{-\beta}$). If this is interpreted in terms of roughness, the roughest of the 64 regions has about 10^3 times the variance of the smoothest. While it would obviously be tempting to give different physical interpretations to the parameters in each region, this would be a mistake. On the other hand, this does not imply that the roughest and the smoothest would be associated with identical erosional, orographic or other processes; the point is that in a full coupled model these processes would be also be scaling and would have correlated variations. Fig. 7.16 also demonstrates the fact that if data from special locations (such as near mountains) are analysed we may expect systematic biases in our statistics and parameter estimates.

7.2 GSI data analysis

7.2.1 GSI in Fourier space: determining *G*

In this section, we consider how to empirically estimate G. In the cases considered up until now (stratification), the eigenvectors were orthogonal and their directions known (horizontal, vertical) and it sufficed to make standard 1D analyses along these axes. Alternatively, in some cases, the horizontal direction was sufficiently isotropic that angle integration/averaging (for examples see Figs. 1.8b, 1.10b, 1.12) was sufficient to reduce the system to a nearly self-similar form. However, assumptions of isotropy can be dangerous as shown by the spectral analyses in Section 6.6, where such a reduction led to misinterpretations due to the horizontal stratification!

We now consider the more general case where the directions are not known. Recall the qth-order structure functions ($S_q(\underline{\Delta r})$) of anisotropic but scale invariant multifractal fields ($f(\underline{r})$):

$$S_q(\underline{\Delta r}) = \langle |f(\underline{r}) - f(\underline{\Delta r} + \underline{r})|^q \rangle \quad (7.76)$$

This is the spatial version of Eqn. (2.71). The structure function satisfies:

$$S_q(T_\lambda \underline{\Delta r}) - S_q(\lambda^{-G}\underline{\Delta r}) = \lambda^{-\xi(q)} S_q(\underline{\Delta r}) \quad (7.77)$$

where $\xi(q) = qH - K(q)$ is the structure function exponent; this is a generalization of the isotropic and self-affine statistics (Section 6.1.7). From Eqn. (7.77), we can see clearly the basic difficulty in empirically testing for scaling and estimating the parameters: in order to estimate the statistical parameters $\xi(q)$ (i.e. H, $K(q)$, or for universal multifractals, H, C_1, α), we need to know G; in order to determine G, we need to know $\xi(q)$. In the simplest case, we can assume that G is a matrix so that the generator of the anistropy is constant over our system; in the next section, we show that this may be overly restrictive: G is generally nonlinear, at best admitting linear approximations over limited spatial domains.

Consider the problem of quantitatively understanding different cloud types, or land surface morphologies. These fields are rife with anisotropic structures, textures, and the latter vary over huge ranges of scale. In both cases, isotropic spectra yield excellent scaling; for example, Fig. 7.17a shows an example of a cloud radiance field; Fig. 7.17b shows its power spectral density ($P(\underline{k})$, Eqns. (2.109), (6.55)), which displays somewhat elliptical structures, aligned (as expected) roughly perpendicular to the real-space structures in Fig. 7.17a. In spite of this clear anisotropy, the angle-integrated (isotropic) spectrum (Eqns. (2.109), (6.55)) is quite accurately scaling (Fig. 7.17c). This indicates on the one hand that we can use isotropic analyses to estimate $\xi(q)$; it also suggests using spectral densities as a convenient way to investigate the anisotropies and estimate G.

Before continuing, we first note that by comparing the functional Eqn. (6.62) with the scale function Eqn. (6.26), we see that the general solution is:

$$S_q(\underline{\Delta r}) \propto \|\underline{\Delta r}\|^{\xi(q)} \quad (7.78)$$

where the scale function is symmetric with respect to the G: $\|\lambda^{-G}\underline{\Delta r}\| = \lambda^{-1}\|\underline{\Delta r}\|$. If we consider only $q = 2$, then we have the following relations:

$$S_2(\underline{\Delta r}) = \langle [f(\underline{r}) - f(\underline{\Delta r} + \underline{r})]^2 \rangle = 2[R(0) - R(\underline{\Delta r})] \quad (7.79)$$

where $R(\underline{\Delta r}) = \langle f(0) f(\underline{\Delta r}) \rangle$ is the autocorrelation function and we have assumed translational invariance, i.e. $S_2(\underline{\Delta r})$ is independent of $\underline{r}$ and set $\underline{r} = 0$. We therefore have:

$$S_2(\underline{\Delta r}) = 2\int d\underline{k}' \left(1 - e^{i\underline{k}'\cdot\underline{\Delta r}}\right) P(\underline{k}') \quad (7.80)$$

(see Eqn. (2.72) with $\langle \tilde{f}(\underline{k}) \tilde{f}(\underline{k}') \rangle = P(\underline{k})\delta(\underline{k} + \underline{k}')$, see Eqn. (2.104)). We now introduce $\widetilde{T}_\lambda = \lambda^{\widetilde{G}}$, which is the dual Fourier-space scaling operator and the dual generator $\tilde{G}$ that corresponds to T_λ and G. It is defined such that if $\underline{\Delta r}' = T_\lambda \underline{\Delta r}$ and $\underline{k}' = \widetilde{T}_\lambda \underline{k}$, the scalar product $\underline{k}\cdot\underline{\Delta r} = \underline{k}'\cdot\underline{\Delta r}'$ is invariant under scale changes. This implies that the Fourier-space generator $\tilde{G}$ is:

$$\tilde{G} = G^T \quad (7.81)$$

and with the help of Eqns. (7.77) and (7.78) one can rewrite Eqn. (7.80):

$$\begin{aligned} S_2(\underline{\Delta r}) &= \lambda^{\xi(2)} S_2(T_\lambda \underline{\Delta r}) \\ &= 2\lambda^{\xi(2)}\lambda^{D_{el}} \int d\underline{k}\left(1 - e^{i\widetilde{T}_\lambda \underline{k}\cdot T_\lambda \underline{\Delta r}}\right) P(\widetilde{T_\lambda \underline{k}}); \quad \underline{k}' = \tilde{T}_\lambda \underline{k} \end{aligned} \quad (7.82)$$

where we have used the fact that:

$$d\underline{k}' = (\det \tilde{T}_\lambda) d\underline{k} = \lambda^{Tr\tilde{G}^T} d\underline{k} = \lambda^{D_{el}} d\underline{k} \quad (7.83)$$

Comparing the right-hand side of Eqn. (7.82) with Eqn. (7.80), and using $\tilde{T}_\lambda \underline{k}\cdot T_\lambda \underline{\Delta r} = \underline{k}\cdot\underline{\Delta r}$, we see that:

$$\begin{aligned} &\lambda^{\xi(2)+D_{el}} \int d\underline{k}\left(1 - e^{i\underline{k}\cdot\underline{\Delta r}}\right) P(\tilde{T}_\lambda \underline{k}) \\ &= \int d\underline{k}'\left(1 - e^{i\underline{k}'\cdot\underline{\Delta r}}\right) P(\underline{k}') \end{aligned} \quad (7.84)$$

since the $\underline{k}'$ on the right is simply a variable of integration, we can drop the primes and obtain:

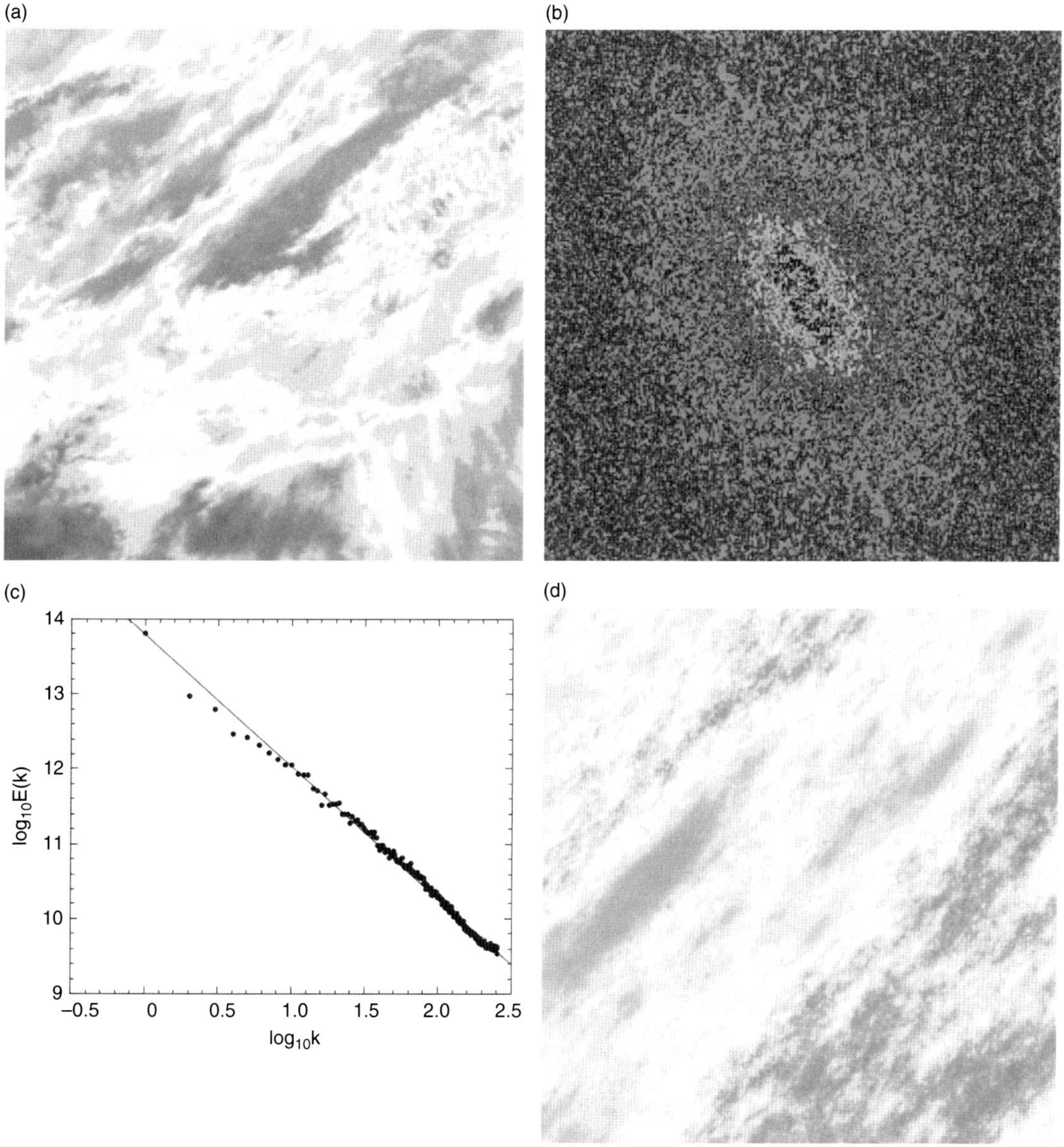

Fig. 7.17 (a) A cloud radiance field in the thermal IR from the NOAA AVHRR satellite off the coast of Florida (512 × 512 pixels, 1.1 km resolution). Various cloud morphologies are visible, although there is a dominant southwest/northeast alignment of large-scale structures. (b) The power spectrum $P(\underline{k})$ of Fig. 7.17a, showing Fourier-space structures roughly perpendicular to the real-space ones in Fig. 7.17a. Reproduced from Schertzer and Lovejoy (1996). (c) The isotropic power spectrum of the radiances displayed in Fig. 7.17a (the spctral density field, Fig. 7.17b, integrated around circular annuli, cf. Eqn. (6.58)). The spectral slope determined by linear regression is given by $\beta = 1.77$. (d) A multifractal FIF model simulation with $c = -0.1$, $f = -0.15$, $e = 0.15$, $d = 1$. Compare this with Fig. 7.17a and notice that much of the NE/SW texture/morophology is reproduced, but that there are clearly regions of Fig. 7.17a where the morphology is different and where the simulation (which assumes the type of anisotropy is constant across the region) does a poor job. All reproduced from Schertzer and Lovejoy (1996). *See colour plate section.*

$$P(\tilde{T}_\lambda \underline{k}) = \lambda^{-s} P(\underline{k}); \quad s = \xi(2) + D_{el} \tag{7.85}$$

The solution of this functional equation is:

$$P(\underline{k}) \propto \|\underline{k}\|^{-s} \tag{7.86}$$

The precise relation between the Fourier scale function (Eqn. (7.86)) and the real-space scale function (Eqn. (7.78)) is thus:

$$\|\underline{\Delta r}\|_R^{\xi(2)} = 2\int d\underline{k}\left(1 - e^{i\underline{k}\cdot\underline{\Delta r}}\right)\|\underline{k}\|_F^{-s} \tag{7.87}$$

where we have temporarily added the subscripts R and F for real space and Fourier space respectively, to underline the fact that the Fourier and real-space scale functions are generally different.

7.2.2* Estimating *G* with the scale invariant generator technique

We now present a technique to test the hypothesis of generalized scale invariance and determine the parameters of the scale-invariant generator and the balls that best describe the scaling field. This technique is an improvement on those described in Pflug *et al.* (1993), but it is more complex to implement than the more geometric one discussed in Beaulieu *et al.* (2007). It is important to remember that the scaling will not hold exactly on any individual realization, but only when averaged over an ensemble of realizations with the same generator and family of balls; there will be random variability about the theoretical ensemble mean spectral density $P(\underline{k})$ (Fig. 7.17b). Also, the anisotropy will change from place to place and from time to time, and thus G itself presumably varies stochastically from realization to realization (e.g. scene to scene, for satellite imagery). An attempt to empirically estimate the ensemble average by averaging many arbitrary realizations would therefore result in a smearing of the parameters (since each realization would have a different generator, see Ch. 8, 9 for analysis of the ensemble statistics). Therefore, we only analyze one realization at a time, and fluctuations about the ensemble average contours of the spectral energy density are taken into account by using statistical regression techniques.

This statistical regression is nonlinear and involves fitting the theoretical spectral density, $P_t(\mathrm{k})$ (the ensemble average spectral energy density generated from the GSI parameters), to the N data points, $P(\underline{k}_i)$ (the spectral energy density of the real-space data estimated at N discrete wavenumbers, $\underline{k}_i$). Here, $P_t(\underline{k}) = P_1\|\underline{k}\|^{-s}$, where P_1 is the value of the spectral energy density on the unit ball. Usual statistical procedures involve the method of least squares, where the parameter estimates are determined by minimizing an error function E^2 (not to be mistaken with the energy spectrum $E(k)$), which we take here as:

$$E^2(G, B_1, P_1, s) = \frac{1}{N}\sum_i \left[\log P(\underline{k}_i) - \log P_t(\underline{k}_i, G, B_1, P_1, s)\right]^2 \tag{7.88}$$

where the full functional dependence of P has been included. This common quadratic error function is somewhat arbitrary, and others may be defined. After some experimentation, we chose the above definition based on the difference of logarithms since for power laws this uniformly weights the different octaves of scale. G is a function of c, f and e (since by convention we can take $d = 1$) and B_1 is a function of the unit ball parameters. The parameter space therefore involves c, f, e, s, P_1, as well as the parameters determining B_1. Searching for the absolute minimum of E^2 in such a large space is computationally prohibitive. Therefore, it is necessary to make some approximation to the error function. The Monte Carlo differential rotation method (Pflug *et al.*, 1993) attempted to do this by estimating the ball parameters before searching the parameter space of G (c, f, e). However, this is not an ideal solution since the statistical scatter of P led to errors in the estimates of B_1, which introduced important biases in the estimates of G.

The scale-invariant generator technique (SIG: Lewis *et al.*, 1999) has the significant advantage of being able to estimate the more fundamental G without prior knowledge of B_1. It reduces the parameter space to four dimensions (c, f, e, s) without introducing errors due to the prior estimation of other parameters. If the anisotropy is not extreme, it is possible to obtain a good prior estimate of s from the isotropic energy spectrum ($s = \beta + 1$; see Pflug *et al.*, 1993) and thus reduce the dimension of the parameter space only to (c, f, e). This method was used in the analysis below, but in general the full four-dimensional parameter space c, f, e, s must be considered.

To see how SIG eliminates any reference to B_1, denote by $k_1(\theta)$ a unit vector parametrized by θ (the polar angle is convenient). Then we can parametrize all vectors, $\underline{k}$, using λ_1 and θ instead of the usual Cartesian coordinates. This can be seen by writing $\underline{k}(\lambda_1, \theta) = \tilde{T}_\lambda \underline{k}_1(\theta)$ and noting that all the

$\underline{k}$ lie on one and only one trajectory which originates from a single unit vector (θ parametrizes the trajectory and λ_1 the points along the trajectory). Further dilations by a factor λ_2 (i.e. $\underline{k}(\lambda_2\lambda_1, \theta) = \widetilde{T}_{\lambda_2}\underline{k}_1(\lambda_1, \theta) = \lambda_2^{\widetilde{G}}\underline{k}_1(\lambda_1, \theta)$) obey:

$$P\left(\lambda_2{}^{\widetilde{G}}\underline{k}(\lambda_1, \theta)\right) = \lambda_2{}^{-s}P\left(\underline{k}(\lambda_1, \theta)\right) \tag{7.89}$$

(from Eqn. (7.85)). Since λ_2 and $\underline{k}(\lambda_1, \theta)$ are arbitrary, this equation shows (with taking logarithms) that for all λ and $\underline{k}$

$$\log P(\lambda^{\widetilde{G}}\underline{k}) + s\log\lambda - \log P(\underline{k}) = 0 \tag{7.90}$$

must be satisfied, i.e. all pairs of points along trajectories will (on average) satisfy Eqn. (7.90). The basic approximation we make is simply to replace ensemble averages by averages over trajectories. That is, we find $\widetilde{G}$ such that the quantity in Eqn. (7.90) averaged over trajectories is as close to zero as possible. The SIG error function used in the analysis is thus defined as:

$$E^2(\widetilde{G}, s) = \frac{1}{n}\sum_{i,j} \left[\log P(\lambda_i{}^{\widetilde{G}}\underline{k}_j) + s\log\lambda_i - \log P(\underline{k}_j)\right]^2 \tag{7.91}$$

The sum is over all the data points, $P(\underline{k}_j)$, and all the possible (discrete) scale ratios, λ_i, which form the n unique pairs $[P(\lambda_i{}^{\widetilde{G}}\underline{k}_j), (\underline{k}_j)]$, i.e. E^2 compares all possible pairs of data points which lie along common trajectories. The power of the SIG error function can be seen in two ways. First, since there is no reference to the unit ball in Eqn. (7.91), no information concerning it is necessary to compute $\widetilde{G}$. Second, since it is expected that we will not need all the pairs to obtain adequate statistics, we can simply choose λ_i and $\underline{k}_j$, from which $\lambda_2{}^{\widetilde{G}}\underline{k}_j$ can be easily computed, and thus a transcendental equation need not be solved.

Since it is not possible to analytically solve for the minimum of E^2, it is necessary to consider E^2 as a continuous function of the four parameters that describes a four-dimensional hypersurface. The parameter space must be searched for the appropriate minimum (i.e. E^2 must be found numerically at intervals in parameter space to trace out the behaviour of the hypersurface). In general, E^2 can be a complicated function with multiple extrema. Therefore, if the absolute minimum is to be found, the intervals must be fine enough such that the estimate of the hypersurface exhibits the same extrema. The hypersurface, however, is expected to be continuous only when an infinite number of independent pairs is used. Since this would require knowledge of P over an infinite range of scales, the actual explicit values of E^2 are expected to be statistically scattered around the theoretical continuous hypersurface. Due to these high-frequency fluctuations, a function is fit to the explicit values of E^2 in an attempt to estimate the continuous hypersurface. The estimated minimum of E^2 can then be found by calculating the minimum of the fitted function.

It is important to note that $P(\lambda_2{}^{\widetilde{G}}\underline{k}_j)$ and $P(\underline{k}_j)$ are data points (i.e. random variables) and therefore both will fluctuate about their average values. This will cause the minimum variance (the minimum value of E^2) to be larger than the case when only one data point is involved. However, it should be possible to compensate for this by use of the greater number of pairs that is available. There is a complication in that the fluctuations of the data points will be more variable than those described by multivariate Gaussian distributions. Therefore, there is no rigorous theoretical justification for using the method of least squares. However, it is still plausible to assume that the behaviour of the hypersurface near the minimum will not be substantially altered if the fluctuations are not too violent. Logarithms of P were used in E^2 since numerical tests showed that taking the logarithm has the effect of decreasing the variability. The results shown below justify the use of the method of least squares and indicate that the bias due to taking the logarithm was small.

$\widetilde{G}$ can then be estimated by searching a three-dimensional parameter space without introducing errors due to the prior estimation of other parameters (with the exception of s, which can usually be accurately estimated using β). The minimization of the error function is still challenging because the param eter space is still large (even if only for c, e, f), and each evaluation of the error function is numerically costly. Lewis (1993) and Lewis *et al.* (1999) give technical details

7.2.3 Estimation of the GSI ball parameters

The next step in the scale-invariant generator technique is the estimation of a family of balls that corresponds to a generalized scale compatible with the generalized dilation/contraction defined by the estimated generator. As already mentioned, uniqueness of such a family is not expected. Since the generator has already been determined, it suffices for any one member of the family of balls to be found, after which the whole family can be generated. Thus, the estimation procedure consists of finding the parameters that describe a unit ball. For simplicity, the balls will be approximated by the second- or fourth-order bivariate polynomials, respectively. (Actually this parametrization seems adequate for clouds, but other (perhaps singular) parametrizations may be needed for other fields such as ice or earthquakes which are characterized by straight-line-like structures.) The relevant parameter space will therefore be three- or five-dimensional.

Unlike the generator parameters, the ball parameters may be found using an analytic method. The ball parameters may be found by fitting a curve of the appropriate form to a level-set of the spectral energy density. It was stated that the large fluctuations about the ensemble average contours cause undesirable errors in the parameter estimates. Ideally, the spectral power density, P, could be smoothed before the fitting procedure. However, conventional smoothing (e.g. averaging adjacent data points) causes nonuniform spreading of the contours of P, and consequently the smoothed field will not be described by the same GSI parameters as the actual P.

Assuming that the estimates of the generator parameters, found using the error function E^2, are reasonably accurate, they can be used to "enhance" the contours of P without affecting the scaling of the field. Regardless of this assumption, fitting a curve to a level-set of the enhanced P will find the best estimate of a unit ball given the estimated generator parameters.

The enhancing technique consists of applying a running average to the data points that lie on the same trajectory. The same principle that was used for E^2 is implemented again, i.e. the fact that the amplitude of any two data points on the same trajectory will be on average related by the scaling:

$$P\left(\lambda^{\widetilde{G}}\underline{k}(\theta)\right) = \lambda^{-s}P\left(\underline{k}(\theta)\right) \tag{7.92}$$

As before, an approximation is made such that only M data points are used in the running average. That is, to generate the enhanced P, $P_{en}(\underline{k}_j)$, the amplitude of each data point, $P(\underline{k}_j)$, is replaced by

$$P_{en}(\underline{k}_j) = \frac{1}{M}\sum_{i=1}^{M}\lambda_i^s\, P(\lambda_i^{\widetilde{G}}\underline{k}_j) \tag{7.93}$$

where λ_i are a series of dilation factors. It can be seen from the results that the enhancing technique has a substantial smoothing effect (see bottom left of Figs. 7.18, 7.19). The parameters of a unit ball, B_1, at some chosen P_1, were found by fitting the appropriate polynomial to a levelset of $P_{en}(\underline{k}_j)$.

Here, the parameters of a unit ball, B_1, at some chosen P_1, were found by fitting a quadratic or quartic polynomial curve to a level-set of $P(\underline{k}_j)$. If increased statistics are required, the parameters of several such balls (by choosing several P_i) can be estimated and transformed (with the known generator parameters) to some arbitrary scale, where they can be averaged. Fig. 7.18 shows the result on a simulation with known parameters, and Fig. 7.19 shows the result on an AVHRR cloud picture (see Lewis *et al.*, 1999, for more examples).

7.2.4 GSI, cloud texture, morphology and type

The multifractal nature of clouds has implications for cloud formation, the earth's radiation budget, climate and weather. From a meteorological perspective, the possibility of classifying cloud types from within a quantitative framework such as that provided by GSI, rather than by visual observation and phenomenology, is seductive. Accordingly, a set of GSI analyses were performed on a series of visible-light AVHRR satellite cloud images. These images were taken by the NOAA-9 satellite off the coast of Florida, longitude 70° W and latitude 27.5° N, during February 1986. The images analyzed were 512 × 512 pixel AVHRR channel 1, visible-light images with wavelength 0.5–0.7 μm. The range of scales was 1.1–560 km.

These images were previously analyzed (Tessier *et al.*, 1993) and were shown to have good isotropic spectral scaling. An example of the scaling of one of these images is given in Fig. 7.17c. We reduced these images to sixteen 256 × 256 pixel images with fairly homogeneous cloud types (according to a professional meteorologist) and analyzed their GSI parameters. The results are shown in Table 7.1 (see Pecknold *et al.*, 1996, for more details).

Some amount of clustering of the parameters by cloud type may be noted, with cumulus having $a^2 < 0$

Table 7.1 GSI parameters of different cloud images with dominant cloud type indicated

Cloud type	β (slope)	c	f	e	a
Cirrus	1.92	0.12	0.0	0.0	0.12
Cirrus	1.77	0.67	0.0	0.0	0.67
Cirrus	1.57	−0.25	0.0	0.0	0.25
Altocumulus	1.88	0.08	0.0	0.0	0.08
Cumulus	1.04	−0.09	0.05	−0.2	0.17i
Cumulus	1.21	0.0	0.0	0.05	0.05i
Cumulus	1.04	−0.15	−0.18	0.33	0.23i
Altostratus	1.91	−0.12	0.0	0.0	0.12
Stratus	1.79	0.04	0.0	−0.05	0.03i
Stratus	1.68	0.05	0.0	0.0	0.05
Stratus	1.65	0.06	0.0	0.0	0.06
Stratus	1.60	−0.15	0.0	0.0	0.15
Stratus	1.84	0.26	0.0	0.0	0.26
Nimbostratus	1.88	0.24	0.0	0.0	0.24
Mesoscale convective complex	1.56		$\approx$ isotropic		
Mesoscale convective complex	1.83		$\approx$ isotropic		

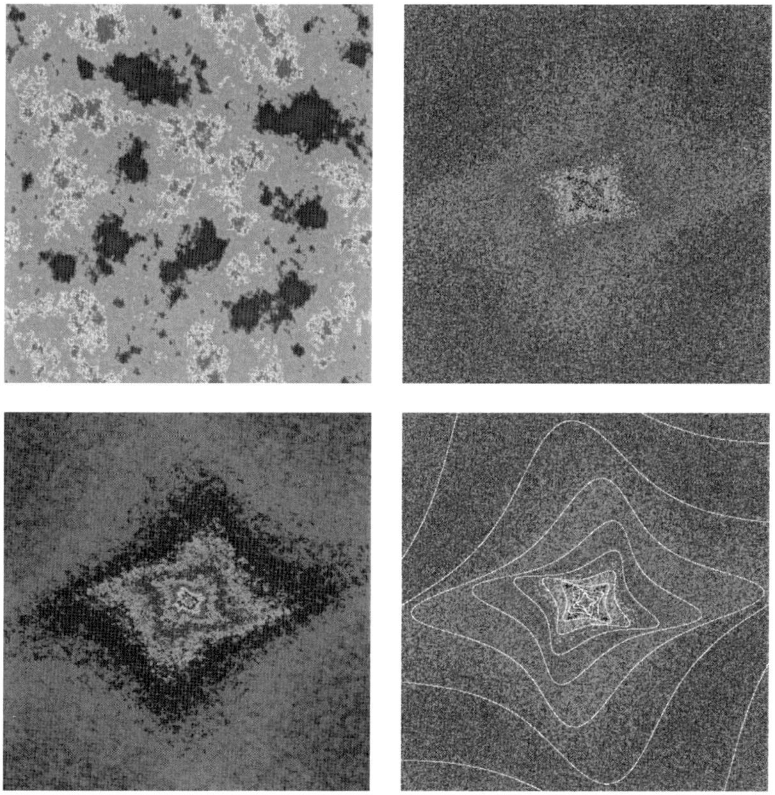

Fig. 7.18 Real-space (top left), spectral energy density (top right), enhanced spectral energy density (bottom left) and spectral energy density with estimated GSI contours (bottom right). This is for a simulation with parameters $s = 2.64$, $c = 0.1$, $f = 0.1$, $e = 0.5$; the measured parameters are $s = 2.63$, $c = 0.05$, $f = 0.08$, $e = 0.51$. Reproduced from Lewis (1993). *See colour plate section.*

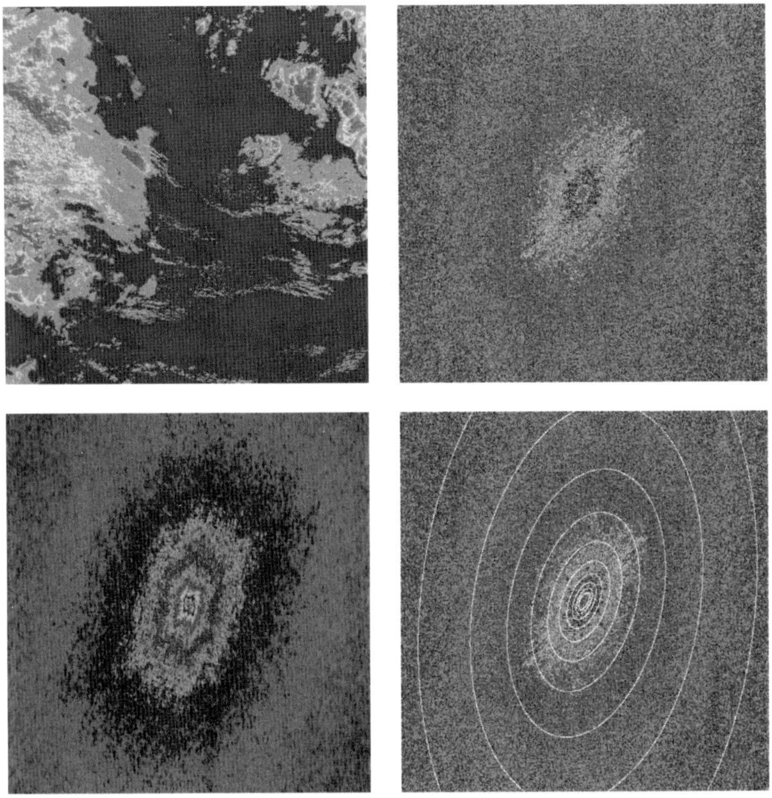

Fig. 7.19 In real space (top left), spectral energy density (top right), enhanced spectral energy density (bottom left) and spectral energy density with estimated GSI contours (bottom right). From AVHRR cloud pictures taken at visible wavelengths; estimated parameters are $s = 2.34$, $c = -0.05$, $f = 0.12$, $e = -0.12$. Reproduced from Lewis (1993). *See colour plate section.*

(rotation dominant) and cirrus and stratus generally having little rotation, stratification dominant. Nevertheless, it is obvious that the large parameter space involved in GSI requires that a large number of images of varying cloud types be analyzed to draw further conclusions. However, before embarking on a massive analysis project, as we discuss in the next section, the extension of linear GSI to nonlinear GSI is probably necessary, since different types of anisotropy appear to frequently coexist in different regions of individual cloud radiance fields.

7.3 Spatially varying anisotropies, morphologies: some elements of nonlinear GSI

7.3.1 The limitations of linear GSI

Linear GSI allows the anisotropy to vary as a function of scale; this already opens up rich possibilities for simulating cloud and other geophysics textures, morphologies. However, given the great diversity of morphologies, it seems obvious that a single G cannot be valid over more than a local region of the earth. Furthermore, the latter and its atmosphere are not vector spaces, but only manifolds, i.e. local vector spaces. A visual confirmation of that can be seen by comparing Fig. 7.17a with Fig. 7.17d; the simulation made with the best-fit G and unit ball parameters. While the basic "texture" and large-scale orientation and elongation of structures in the NE/SW direction is plausibly reproduced (remember that at best they are different realizations of the same random process with the same G, B_1, α, C_1, H), there are nevertheless significant differences which appear to correspond to other G's.

One way to attempt to quantitatively examine the spatial uniformity of G is to estimate $S_2(\underline{\Delta r})$ locally from the data. In Fig. 7.20a we show a satellite infrared image; in Fig. 7.20b we show the corresponding contours of log $S_2(\underline{\Delta r})$ calculated over sub-areas each with 64×64 pixels. It is obvious that the orientations as functions of scale are not constant (although a fair bit of the variability is statistical, i.e. it is due to the fact that we have only a single realization, not an ensemble average). Fig. 7.20c shows an analogous analysis of a topography from a 1 km resolution digital elevation model; we come to the same conclusion: the G matrix varies as a function of spatial location. This conclusion explains why the SIG technique has not been applied more widely on cloud or topographic data: while the technique works quite well for numerical simulations with fixed G, it does not always give consistent results when applied to the data because G is not always uniform enough across the image. If one attempts to avoid the problem by considering G to be constant only over small section, then the statistics become too poor to accurately estimate G. It seems that the spatial variability of G must be taken into account to understand real-world morphologies. Presumably, realistic GSI models will be based on stochastic generators (see Schertzer and Lovejoy, 1991, 2011, for examples).

7.3.2 The generator of the infinitesimal scale change $g(x)$ and nonlinear GSI

To go beyond linear GSI whose generator G is a fixed matrix, one first considers infinitesimal scale transformations; we will consider reductions of scale by a finite $\Delta\lambda$ and then take the small-scale limit.

Consider the vector r_λ obtained by reducing the unit vector by a scale ratio λ:

$$\underline{r}_\lambda = \lambda^{-G}\underline{r}_1 \tag{7.94}$$

In order to change the scale of the vector $\underline{r}_\lambda$ by $\Delta\lambda$, we need to reduce it by a scale ratio $1 + \Delta\lambda/\lambda$:

$$\underline{r}_\lambda + \underline{\Delta r}_\lambda = \left(1 + \frac{\Delta\lambda}{\lambda}\right)^{-G}\underline{r}_\lambda \tag{7.95}$$

Hence dropping the indices and taking the limit $\Delta\lambda \to d\lambda$ we obtain:

$$d\underline{r} = -\frac{d\lambda}{\lambda}G\cdot\underline{r} \tag{7.96}$$

The nonlinear generalization of this is obtained by introducing the infinitesimal (generally nonlinear) generator $\underline{g}(\underline{r})$:

$$d\underline{r} = -\frac{d\lambda}{\lambda}\underline{g}(\underline{r}) \tag{7.97}$$

Linear GSI is the special case where $\underline{g}(\underline{r})$ is linear and G is therefore the (fixed) Jacobian matrix of $\underline{g}$:

$$G_{ij} = \frac{\partial g_i}{\partial x_j} \tag{7.98}$$

where, as usual, $\underline{r} = (x_1, x_2, x_3)$. To keep closer links to the linear case, this can be written in terms of the infinitesimal operator G_{op}, defined as:

$$G_{op}\underline{r} = \underline{g}(\underline{r}) \tag{7.99}$$

so that:

$$d\underline{r} = -\frac{d\lambda}{\lambda}G_{op}\underline{r} \tag{7.100}$$

This can (at least formally) be integrated to obtain:

$$\underline{r}_\lambda = \lambda^{-G_{op}}\underline{r}_1 \tag{7.101}$$

($\underline{r}_1$ is a unit vector, $\underline{r}_\lambda$ is a unit vector reduced by a factor λ). In this way we can keep the power-law notation for the scale-change operator T_λ:

$$T_\lambda = \lambda^{-G_{op}} \tag{7.102}$$

For any vector, T_λ increases scale by a factor λ, and therefore, as usual, the scale function has the basic property:

$$\|T_\lambda \underline{r}\| = \lambda^{-1}\|\underline{r}\| \tag{7.103}$$

We can now obtain the basic equation for the scale function. Consider the scale of a vector reduced from

(a)

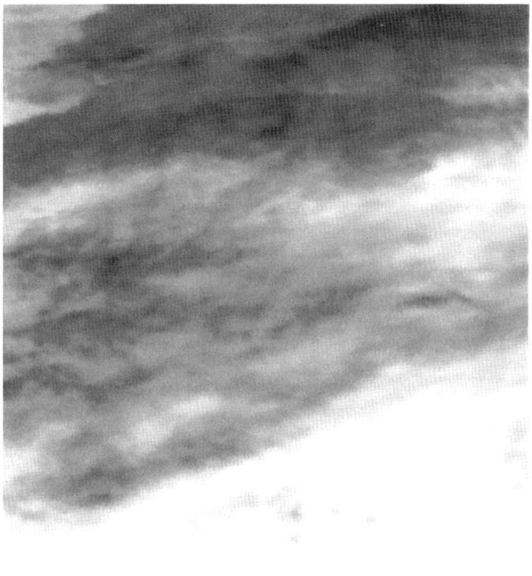

Fig. 7.20 (a) An infrared satellite image from a NOAA AVHRR satellite at 1.1 km resolution, 512 × 512 pixels.

(b)

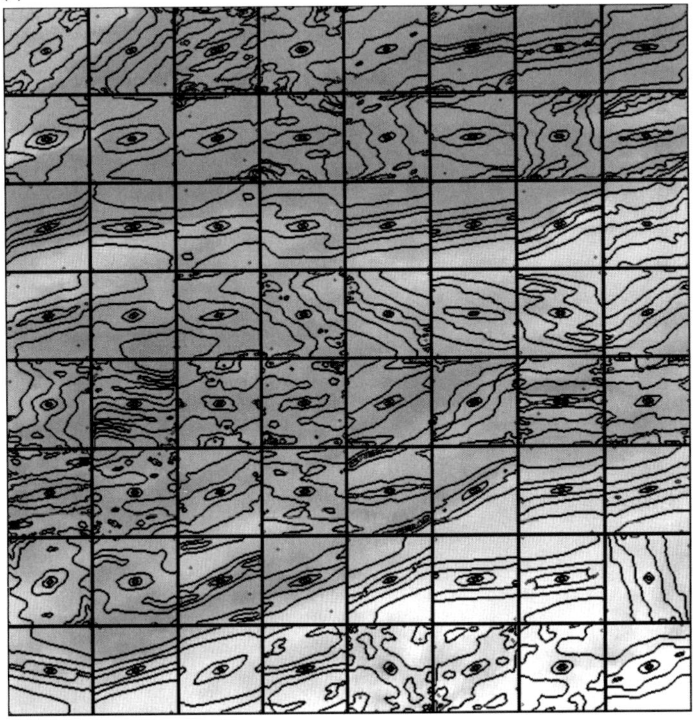

(c)

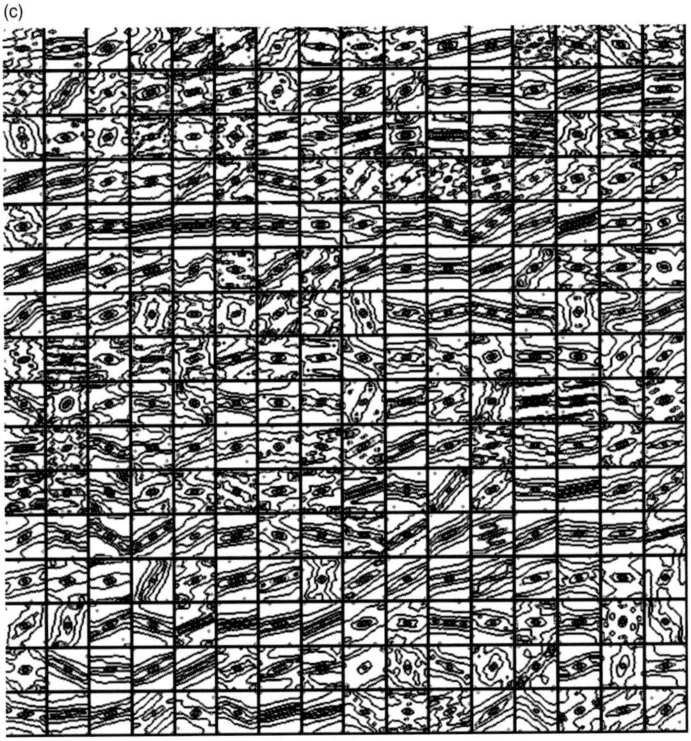

Fig. 7.20 (b) Contours of $S_2(\underline{\Delta r})$ estimated for each 64 × 64 pixel box from the image in Fig. 7.20a. (c) $S_2(\underline{\Delta r})$ for each 64 × 64 pixel box of a 1024 × 1024 digital elevation map of the topography over part of the continental US at 1 km resolution.

scale λ to scale $\lambda + \Delta\lambda$, as above by the reduction factor $(1 + \Delta\lambda/\lambda)$. The basic scale function equation $\|T_\lambda \underline{r}\| = \lambda^{-1}\|\underline{r}\|$ becomes:

$$\left\| \left(1+\frac{\Delta\lambda}{\lambda}\right)^{-G_{op}} \underline{r} \right\| = \left(1+\frac{\Delta\lambda}{\lambda}\right)^{-1} \|\underline{r}\| \tag{7.104}$$

If we now perform Taylor series expansions and take the limit $\Delta\lambda \to 0$, and using $G_{op} = \underline{g}(\underline{r})$, we obtain the basic equation for the scale function:

$$g_i \frac{\partial}{\partial x_i} \|\underline{r}\| = \|\underline{r}\| \tag{7.105}$$

(summing over the indices i), or in vector form:

$$\left(\underline{g}(\underline{r})\cdot\nabla\right)\|\underline{r}\| = \|\underline{r}\| \tag{7.106}$$

In the special case of linear GSI this yields:

$$\underline{r}^T \cdot G^T \cdot \nabla \|\underline{r}\| = \|\underline{r}\| \tag{7.107}$$

As expected, to solve this partial differential equation for the scale function, we can use the same series of transformations of variables (e.g. Eqns. (7.55), (7.57) for $a^2 > 0$ etc.; Section 7.1.3) to reduce Eqn. (7.107) to:

$$\frac{\partial}{\partial \log R^{(2)}} \log\|\underline{r}\| = 1 \tag{7.108}$$

whose general solution is:

$$\|\underline{r}\| = R^{(2)}\Theta\left(\theta^{(2)}\right) \tag{7.109}$$

where $R^{(2)}$ is the polar coordinate representation of $(x^{(2)}, y^{(2)})$ and where Θ (an arbitrary function of angle) here appears as a function of integration.

7.3.3 An example of local scales in quadratic GSI

In order to obtain the scale function for vectors $\underline{\Delta r}$ centred at a point $\underline{r}_c$: $\underline{\Delta r} = \underline{r} - \underline{r}_c$, we must solve the scale function equation for $\underline{\Delta r}$ instead of $\underline{r}$, as well as to use $\underline{g}_c(\underline{\Delta r}) = \underline{g}(\underline{r}_c + \underline{\Delta r}) - \underline{g}(\underline{r}_c)$ instead of $\underline{g}(\underline{r})$. Following the procedure in Eqns. (7.94)–(7.106), the equation corresponding to Eqn. (7.106) but for the "local" vectors $\underline{\Delta r}$ is:

$$\left(\underline{g}_c(\underline{\Delta r})\cdot\nabla\right)\|\underline{\Delta r}\| = \|\underline{\Delta r}\|;$$
$$\underline{\Delta r} = \underline{r} - \underline{r}_c; \underline{g}_c(\underline{\Delta r}) = \underline{g}(\underline{r} + \underline{\Delta r}) - \underline{g}(\underline{r}_c) \tag{7.110}$$

A quasi-linear study would correspond to using a second-order Taylor expansion of $\underline{g}_c$ about the point $\underline{r}_c = (x_c, y_c)$:

$$g_{c,i}(\underline{\Delta r}) = \left(G(\underline{r}_c)\cdot\underline{\Delta r}\right)_i + \underline{\Delta r}^T \cdot A^{(i)} \cdot \underline{\Delta r} + O(\underline{\Delta r})^3 \tag{7.111}$$

where:

$$G(\underline{r}_c)_{i,\alpha} = \left.\frac{\partial g_i}{\partial x_\alpha}\right|_{\underline{r}=\underline{r}_c}; \quad A^{(i)}_{\alpha\beta} = \left.\frac{1}{2}\frac{\partial^2 g_i}{\partial x_\alpha \partial x_\beta}\right|_{\underline{r}=\underline{r}_c} \tag{7.112}$$

The basic behaviour of the local scale functions (near the point $\underline{r}_c$) for small enough scales is thus given by linear GSI with $G(\underline{r}_c)$ (Eqn. (7.112)) combined with quadratic and higher-order corrections as indicated.

It turns out, however, that thanks to a theorem by Poincaré we can do somewhat better than this. If $\underline{g}$ is a nonlinear polynomial of order r, then, around a fixed point, a nonlinear polynomial generator of degree r can be reduced (to within errors of degree $> r$) to a linear generator by a (nonlinear) transformation of variables:

$$\underline{r}' = \underline{r} + \underline{H}(\underline{r}) \tag{7.113}$$

where $\underline{H}$ is a polynomial function of degree r satisfying a "homological" equation. Given a nonlinear GSI system, we can thus effectively reduce it to a linear system by the above nonlinear transformation of variables; we need only determine the function $\underline{H}$ by solving the "homological equation":

$$\frac{\partial H_i}{\partial x_j} G_{jk} x_k - G_{ik} H_k = g_i \tag{7.114}$$

(this is the Poisson bracket for the fields $\underline{H}$, $\underline{Gx}$). Except in certain special cases ("resonances") this equation has a unique solution.

As an example of how this works, consider the nonlinear function:

$$\begin{aligned} g_x(x,y) &= x + \alpha y^2 \\ g_y(x,y) &= y + \alpha x^2 \end{aligned} \tag{7.115}$$

chosen because it has separate regions where the linearization has real and complex values; we obtain:

$$G(\underline{r}_c) = \begin{pmatrix} 1 & 2\alpha y_c \\ 2\alpha x_c & 1 \end{pmatrix} \tag{7.116}$$

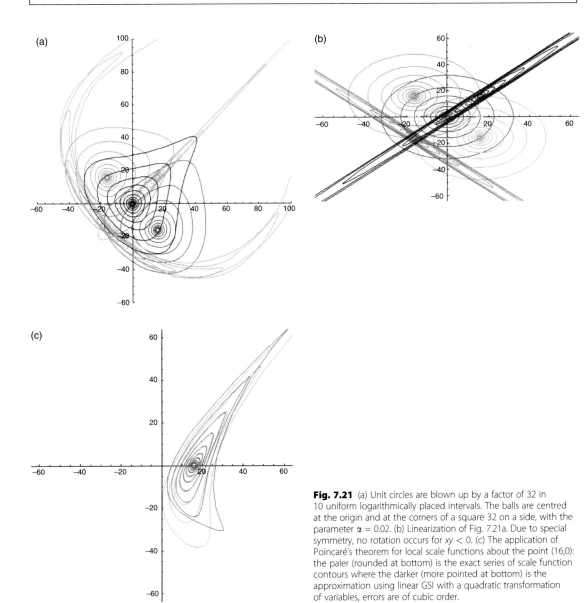

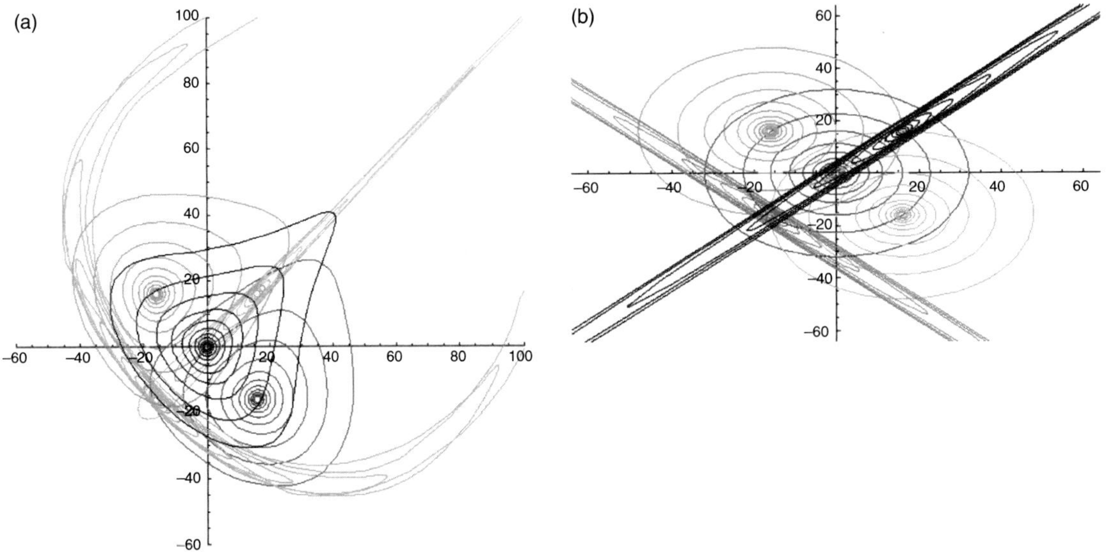

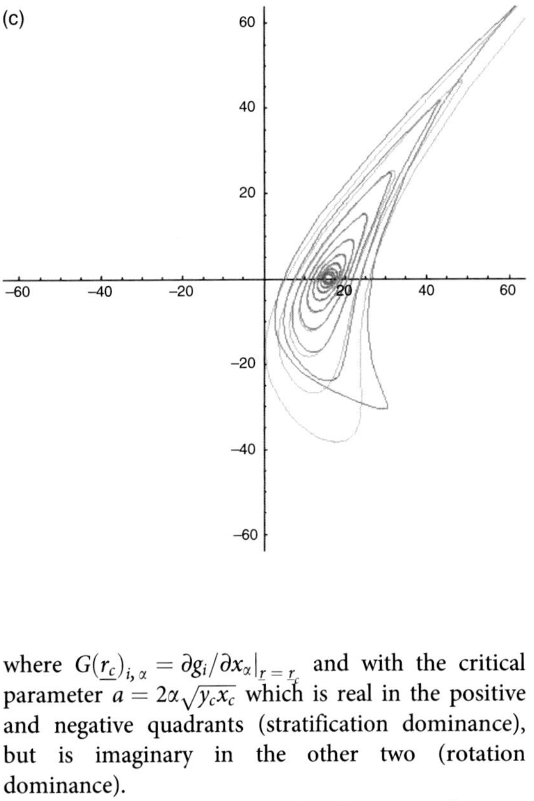

Fig. 7.21 (a) Unit circles are blown up by a factor of 32 in 10 uniform logarithmically placed intervals. The balls are centred at the origin and at the corners of a square 32 on a side, with the parameter $\alpha = 0.02$. (b) Linearization of Fig. 7.21a. Due to special symmetry, no rotation occurs for $xy < 0$. (c) The application of Poincaré's theorem for local scale functions about the point (16,0): the paler (rounded at bottom) is the exact series of scale function contours where the darker (more pointed at bottom) is the approximation using linear GSI with a quadratic transformation of variables, errors are of cubic order.

where $G(\underline{r_c})_{i,\alpha} = \partial g_i/\partial x_\alpha|_{\underline{r}=\underline{r_c}}$ and with the critical parameter $a = 2\alpha\sqrt{y_c x_c}$ which is real in the positive and negative quadrants (stratification dominance), but is imaginary in the other two (rotation dominance).

Fig. 7.21a shows the result of numerically integrating Eqn. (7.115) for five unit balls symmetrically placed on the corners of a square centred at the origin, with an additional ball centred at the origin, and Fig. 7.21b shows the corresponding linear GSI approximation. While the smaller balls are well approximated by linear GSI, the larger ones are quite nonlinear. Fig. 7.21c now shows the result of applying Poincaré's theorem to the scale function around the point (16,0): linear GSI combined with a quadratic change in variables; the result is valid to third order. Note that at large scales the balls "cross" so that the scale function is no longer unique; the

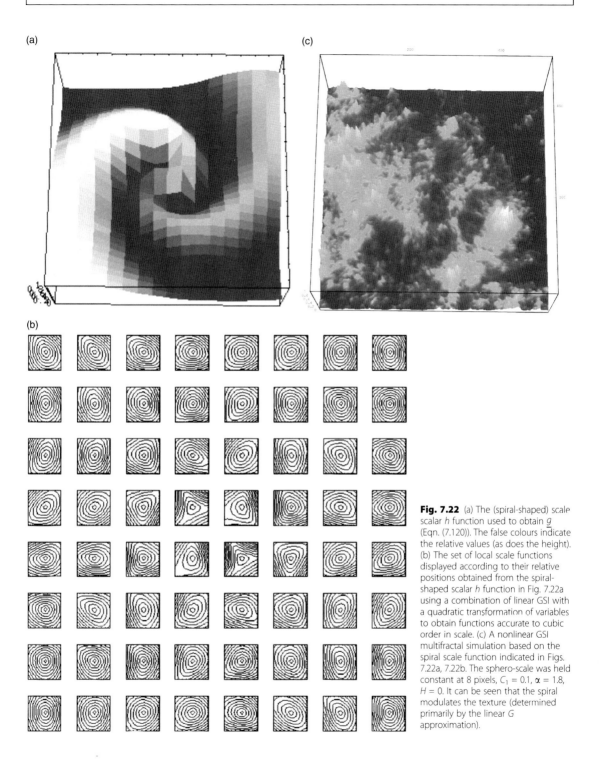

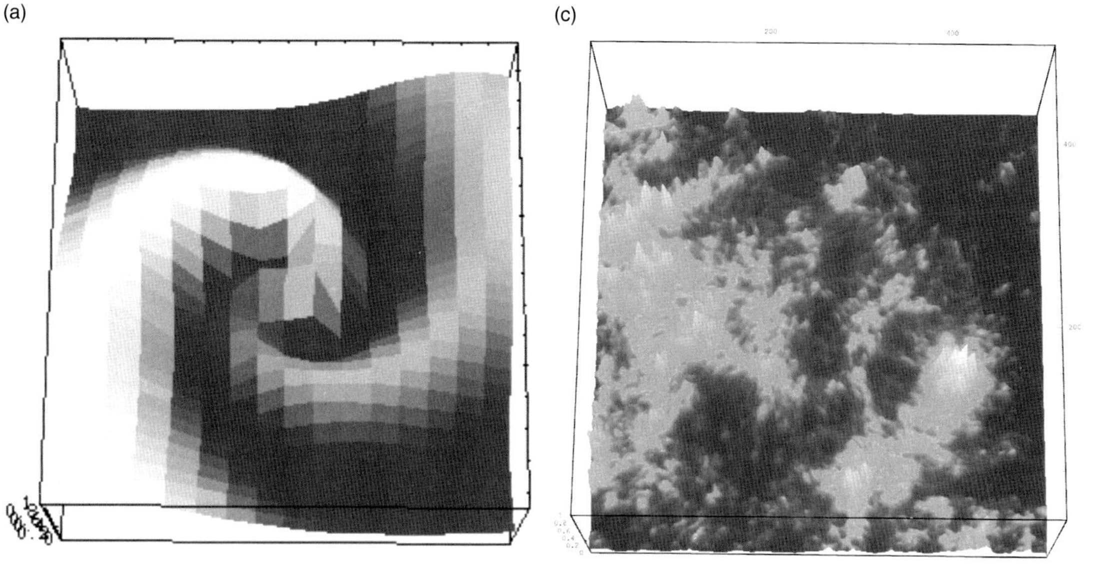

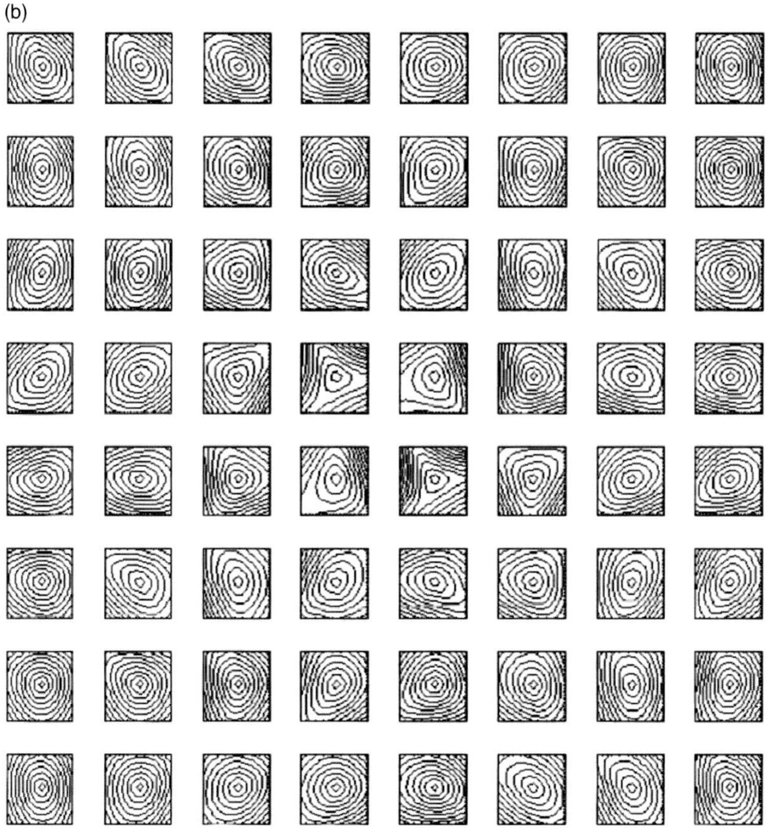

Fig. 7.22 (a) The (spiral-shaped) scale scalar h function used to obtain $\underline{g}$ (Eqn. (7.120)). The false colours indicate the relative values (as does the height). (b) The set of local scale functions displayed according to their relative positions obtained from the spiral-shaped scalar h function in Fig. 7.22a using a combination of linear GSI with a quadratic transformation of variables to obtain functions accurate to cubic order in scale. (c) A nonlinear GSI multifractal simulation based on the spiral scale function indicated in Figs. 7.22a, 7.22b. The sphero-scale was held constant at 8 pixels, $C_1 = 0.1$, $\alpha = 1.8$, $H = 0$. It can be seen that the spiral modulates the texture (determined primarily by the linear G approximation).

generator is only a valid GSI generator for a limited range of scales (and the range can depend on location).

We can use the idea of combining linear GSI with a quadratic transformation of variables to obtain scale functions accurate to third order in systems with more complicated g functions (presumably, even random g functions: see Schertzer and Lovejoy, 2011). Fig. 7.22 gives an example where the scalar function h (from which g is derived, see Eqn. (7.103)) varies in a spiral pattern (Fig. 7.22a); Fig. 7.22b gives the corresponding spatial variation of the local scale functions calculated with this approximation, and Fig. 7.22c shows the corresponding multifractal simulation.

7.3.4 Multifractal simulations in nonlinear GSI

In order to make mulitfractal simulations, we can consider the case in 2D where there is no overall stratification so that $TrG = 2$ everywhere; then:

$$TrG = \nabla \cdot \underline{g} = 2 \tag{7.117}$$

and we can express the vector g through the scalar h, which expresses the deviation from isotropy:

$$\begin{pmatrix} g_x \\ g_y \end{pmatrix} = \begin{pmatrix} x - 2\dfrac{\partial h}{\partial y} \\ y + 2\dfrac{\partial h}{\partial x} \end{pmatrix} \tag{7.118}$$

(the factor 2 is added for convenience). Now that we have considered linear GSI, we can turn to the next level of complication, quadratic GSI:

$$g_i(\underline{x}) = A^i_{jk} x_j x_k \tag{7.119}$$

where we sum over coordinate indices j, k (the summation convention); quadratic g thus corresponds to linearly varying G and to cubic h:

$$G = \begin{pmatrix} 1 & 0 \\ 0 & 1 \end{pmatrix} + 2 \begin{pmatrix} -\dfrac{\partial^2 h}{\partial x \partial y} & -\dfrac{\partial^2 h}{\partial y^2} \\ \dfrac{\partial^2 h}{\partial x^2} & \dfrac{\partial^2 h}{\partial x \partial y} \end{pmatrix} \tag{7.120}$$

Equivalently with $G = \begin{pmatrix} 1-c & f-e \\ f+e & 1+c \end{pmatrix}$, we have:

$$c = 2\frac{\partial^2 h}{\partial x \partial y}; \quad e = \nabla^2 h; \quad f = \frac{\partial^2 h}{\partial x^2} - \frac{\partial^2 h}{\partial y^2} \tag{7.121}$$

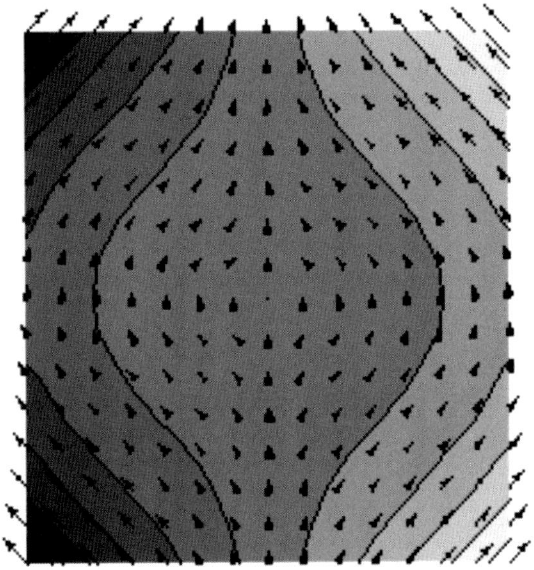

Fig. 7.23 The contours of the field $h(x, y)$ (Eqn. (7.122)) with arrows indicating the corresponding quadratic vector field $\underline{g}$ (Eqn. (7.118)).

To illustrate how this works, we can consider the example:

$$h(x, y) = \frac{x^3}{6} + \frac{xy^2}{2} \tag{7.122}$$

(Fig. 7.23). This yields:

$$G = \begin{pmatrix} 1 & 0 \\ 0 & 1 \end{pmatrix} + 2 \begin{pmatrix} -y & -x \\ x & y \end{pmatrix} \tag{7.123}$$

This is an interesting example since a (the eigenvalue of the traceless part of G) is:

$$a^2 = 4(y^2 - x^2) \tag{7.124}$$

which has both regions of real and imaginary values corresponding to stratification and rotational dominance. Figs. 7.24a, 7.24b, 7.24c show examples of multifractal simulations using canonical scale functions (but with different sphero-scales). One can see that the effect of the spatially varying G is quite subtle; it is certainly not trivial to deduce the underlying generator of the scale-changing group from the observed fields.

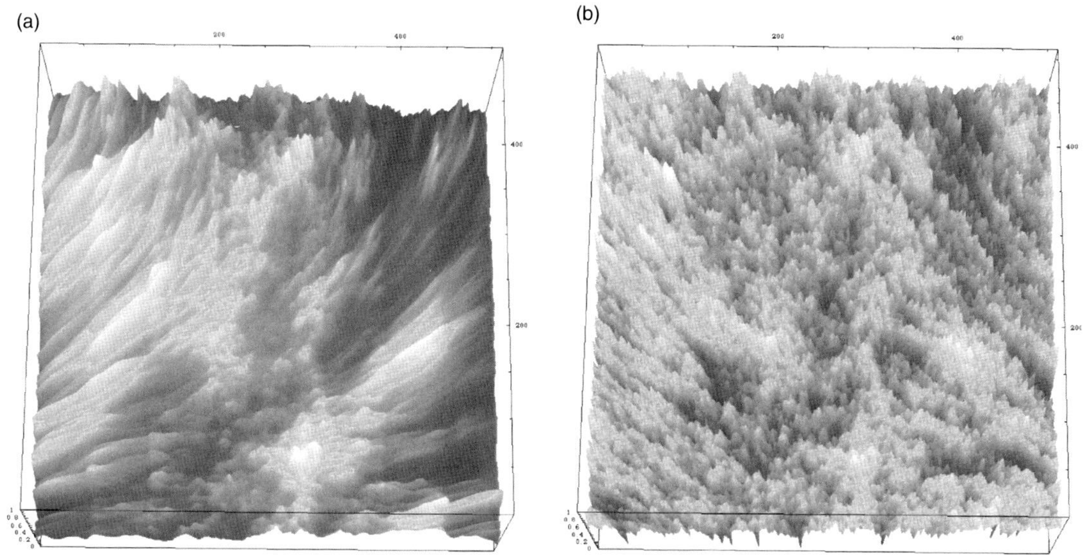

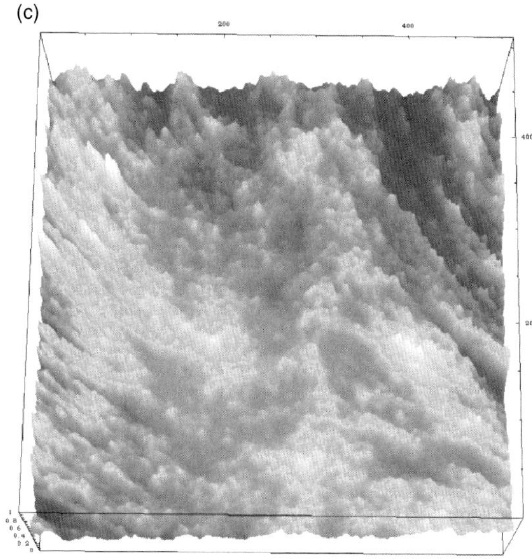

Fig. 7.24 (a) A multifractal simulation of quadratic GSI (with $\underline{g}$ given by the cubic h, Eqn. (7.122)) with $\alpha = 1.8$, $C_1 = 0.1$, $H = 0.33$ and sphero-scale = 256 pixels (the simulation is 512 × 512 pixels). The effect of the varying G is quite subtle. (b) Same as Fig. 7.24a but for $l_s = 1$, $H = 0$. (c) Same as Fig. 7.24a but for $l_s = 1$. *See colour plate section.*

7.4 Summary of emergent laws in Chapter 7

General scale functions $\|\Delta \underline{r}\|$ are linear with respect to the contraction parameter λ^{-1}:

$$\|T_\lambda \underline{\Delta r}\| = \lambda^{-1}\|\underline{\Delta r}\|; \quad T_\lambda = \lambda^{-G} \tag{7.125}$$

Linear GSI

G is a matrix; the self-affine case is when the matrix G is diagonal.

The qth-order structure functions $S_q(\underline{\Delta r})$ and spectral densities $P(\underline{k})$ are related by:

$$S_2(\underline{\Delta r}) = 2\int (1 - e^{i\underline{k}\cdot\underline{\Delta r}})P(\underline{k})d\underline{k} \tag{7.126}$$

For scaling processes they satisfy:

$$S_q(\underline{\Delta r}) = \|\underline{\Delta r}\|_R^{\xi(q)}; \quad \|\lambda^{-G}\underline{\Delta r}\|_R = \lambda^{-1}\|\underline{\Delta r}\|_R$$
$$P(\underline{k}) = \|\underline{k}\|_F^{-s}; \|\lambda^{-\tilde{G}}\underline{k}\|_F = \lambda^{-1}\|\underline{k}\|_F \quad (7.127)$$

where the subscript R and F for "real" and "Fourier" have been used to underscore the fact that the corresponding scale functions are not the same, and the Fourier and real-space generators are related by:

$$\tilde{G} = G^T; \quad s = D_{el} + \xi(2); \quad D_{el} = Trace\,(G) \quad (7.128)$$

Some simple properties of linear GSI can be understood with the help of the pseudo-quaternion decomposition:

$$G = \begin{pmatrix} d+c & f-e \\ f+e & d-c \end{pmatrix} \quad (7.129)$$

(Eqns. (7.23), (7.24)). The eigenvalues are $d \pm a$, where $a^2 = c^2 + f^2 - e^2$, so are real or complex depending on the sign of a^2. When they are real, structures rotate by less than $\pi/2$ as the scale changes from $\lambda = 1$ to $\lambda \to \infty$, when they are complex, they rotate once with every factor $e^{2\pi|a|}$ of λ. The overall change of volumes of structures (areas in 2D) with scale is:

$$Volume \propto \det(\lambda^G) = \lambda^{D_{el}}; \quad D_{el} = Trace(G) \quad (7.130)$$

where D_{el} is the elliptical dimension = $2d$ in a two-dimensional system, where d is the parameter in Eqn. (7.128).

Nonlinear GSI

The basic equation for the scale function is:

$$\underline{g}(\underline{r})\cdot\nabla\|\underline{r}\| = \|\underline{r}\| \quad (7.131)$$

$\underline{g}$ is the infinitesimal generator. Locally, near the point $\underline{r}_c$, we have:

$$G(\underline{r}_c)_{i\alpha} = \frac{\partial g_i}{\partial x_\alpha}\Big|_{\underline{r}=\underline{r}_c} \quad (7.132)$$

for the matrix with elements $G_{i\alpha}$ which defines the local linear GSI approximation.

Appendix 7A: The normalization constant in anisotropic continuous-in-scale multifractal simulations

7A.1 In two dimensions

We saw in Section 7.1.5 that in order to produce multifractal simulations respecting anisotropic GSI, the basic change required with respect to isotropic GSI was to replace the usual norm by the scale function and the usual dimension of space by the elliptical dimension. Beyond that, there was the problem that the simulation region does not generally coincide with contours of constant scale, so that some special methods are required to model the scales that are not totally contained within the simulation region (a "nested" model). The main additional technical point is the determination of the normalization constant:

$$N_{D_{el}} = \Omega_{D_{el}} = \int\limits_{\|\underline{r}\|=1} dx\, dy\,; \quad \underline{r} = (x, y) \tag{7.133}$$

(Section 6.1.7). In this appendix, we show how N_{Del} can be calculated for scale functions obtained via the sequence of transformations of variables outlined in Section 7.1.3. The key is to systematically transform the area element $dxdy$. To see how this works, consider first the case where the eigenvalues are real ($a^2 > 0$), and use the notation $x = x^{(0)}$, $y = y^{(0)}$ and with successive transformations as indicated in Section 7.1.3 (Eqns. (7.55), (7.57)):

$$dx^{(0)}dy^{(0)} = \det \Omega dx^{(1)}dy^{(1)} = \frac{1}{2a}dx^{(1)}dy^{(1)} \tag{7.134}$$

for the second transformation:

$$dx^{(1)}dy^{(1)} = (1 - a^2)\left|\frac{y^{(2)}}{x^{(2)}}\right|^a dx^{(2)}dy^{(2)} \tag{7.135}$$

so that:

$$dx^{(0)}dy^{(0)} = \frac{(1 - a^2)}{2a}\left|\frac{y^{(2)}}{x^{(2)}}\right|^a dx^{(2)}dy^{(2)} \tag{7.136}$$

converting to polar coordinates (R, θ), we have:

$$dx^{(0)}dy^{(0)} = \frac{(1 - a^2)}{2a}\left|\tan\theta^{(2)}\right|^a R^{(2)}dR^{(2)}d\theta^{(2)}$$

We now take:

$$\begin{aligned} R^{(3)} &= R^{(2)}\Theta\left(\theta^{(2)}\right) \\ \theta^{(3)} &= \theta^{(2)} \end{aligned} \tag{7.137}$$

where $\Theta > 0$ is the function specifying the unit ball. With this transformation of variables we obtain:

$$dx^{(0)}dy^{(0)} = \frac{(1 - a^2)}{2a}\left|\tan\theta^{(3)}\right|^a \frac{R^{(3)}}{\Theta\left(\theta^{(3)}\right)^2} dR^{(3)}d\theta^{(3)} \tag{7.138}$$

Finally, since $dx^{(3)}\, dy^{(3)} = adx^{(4)}dy^{(4)}$, and $\|\underline{r}\| = |\underline{r}^{(4)}|$ we have:

$$R^{(3)}dR^{(3)}d\theta^{(3)} = 2a\|\underline{r}\|d\|\underline{r}\|d\varphi \tag{7.139}$$

where we have written:

$$dx^{(4)}dy^{(4)} = \|\underline{r}\|d\|\underline{r}\|d\varphi \tag{7.140}$$

where $d\varphi = \theta^{(4)}$ is the "generalized polar angle" which – along with the scale function (a "generalized" polar coordinate) – parameterizes the points on the plane. The final transformation to GSI angle and scale is:

$$dx^{(0)}dy^{(0)} = \omega(\varphi)d\varphi\|\underline{r}\|d\|\underline{r}\|;$$
$$\omega(\varphi) = \frac{(1-a^2)|\tan\theta^{(3)}|^a}{\Theta\left(\theta^{(3)}\right)^2} \quad (7.141)$$

The normalization constant is thus:

$$N_{D_{el}} = \Omega_{D_{el}} = \int_0^{2\pi} \omega(\varphi)d\varphi \quad (7.142)$$

In order to evaluate this, we can use the relation between φ and $\theta^{(3)}$:

$$\tan\varphi = \sqrt{\frac{r+e}{r-e}}\left(\frac{\tan\theta^{(3)}+1}{\tan\theta^{(3)}-1}\right) \quad (7.143)$$

which implies:

$$\frac{d\varphi}{d\theta^{(3)}} = \frac{a}{r+e\sin 2\theta^{(3)}} \quad (7.144)$$

(this is always > 0 since for $a^2 > 0$, $r > e$ and $r > 0$). As φ varies between 0 and 2π, $\theta^{(3)}$ varies between 0 and 2π so that we finally obtain:

$$N_{D_{el}} = \Omega_{D_{el}} = a(1-a^2)\int_0^{2\pi} \frac{|\tan\theta^{(3)}|^a}{\left(\Theta\left(\theta^{(3)}\right)\right)^2}\frac{d\theta^{(3)}}{r+e\sin 2\theta^{(3)}} \quad (7.145)$$

This integral is straightforward to evaluate numerically.

The case of complex eigenvalues ($a^2 < 0$) follows the same series of steps; we simply give the final result as:

$$N_{D_{el}} = \Omega_{D_{el}} = |a|\int_0^{2\pi} \frac{d\theta^{(3)}}{|e-r\cos 2\theta^{(3)}|\left(\Theta\left(\theta^{(3)}\right)\right)^2} \quad (7.146)$$

which takes into account both positive and negative values of e. Note that when $\Theta = 1$, we have $N_{D_{el}} = \Omega_{D_{el}} = 2\pi$

7A.2 In three dimensions

Following the two-dimensional approach, we need to calculate the transformation of the volume element $dxdydz$ given by the Jacobian of the corresponding transformation of variables. Following the transformations given in Section 7.1.5, and considering first real eigenvalues ($a^2 > 0$), we obtain:

$$d\underline{r}^{(0)} = (2a)^{-1}d\underline{r}^{(1)}$$
$$d\underline{r}^{(1)} = H_z(1-a^2)|\tan\theta^{(2)}|^a|\cos\phi|^{H_z-1}\left(R^{(2)}\right)^{H_z-1}d\underline{r}^{(2)} \quad (7.147)$$

(cf. transformations Eqns. (7.71), (7.72)) using spherical polar coordinates, we have $d\underline{r}^{(2)} = \left(R^{(2)}\right)^2 dR^{(2)}d\cos\phi^{(2)}d\theta^{(2)}$, which yields:

$$d\underline{r}^{(0)} = \frac{(1-a^2)}{2a}H_z|\tan\theta^{(2)}|^a|\cos\phi^{(2)}|^{H_z-1}\left(R^{(2)}\right)^{H_z+1}dR^{(2)}d\cos\phi^{(2)}d\theta^{(2)} \quad (7.148)$$

Now, using the transformation Eqn. (7.73) we obtain:

$$dR^{(2)}d\cos\phi^{(2)}d\theta^{(2)} = \frac{dR^{(3)}d\cos\phi^{(3)}d\theta^{(3)}}{\Theta} \quad (7.149)$$

and hence:

$$d\underline{r}^{(0)} = \frac{(1-a^2)}{2a}H_z\frac{\left(R^{(3)}\right)^{1+H_z}|\tan\theta^{(3)}|^a|\cos\phi^{(3)}|^{H_z-1}}{\Theta\left(\phi^{(3)},\theta^{(3)}\right)^{2+H_z}}dR^{(3)}d\cos\phi^{(3)}d\theta^{(3)} \quad (7.150)$$

Finally, to get the scale function $\|\underline{r}\| = R^{(4)}$, we take $\underline{r}^{(4)} = \Omega\underline{r}^{(3)}$, expressing $\underline{r}^{(4)}$ in spherical polar coordinates, writing:

$$d\underline{r}^{(0)} = d\omega\|\underline{r}\|^{1+H_z}d\|\underline{r}\| \quad (7.151)$$

(ω is the solid angle in the $\underline{r}^{(4)}$ space) and using change of variable $\mu = \cos\phi$ we eventually obtain:

$$d\omega = H_z|a|^{H_z}\frac{|\mu^{(3)}|^{H_z-1}}{\Theta\left(\mu^{(3)},\theta^{(3)}\right)^{2+H_z}\left(|a|^2\left(\mu^{(3)}\right)^2+\left(e-r\cos 2\theta^{(3)}\right)\left(1-\left(\mu^{(3)}\right)^2\right)\right)^{(1+H_z)/2}}d\mu^{(3)}d\theta^{(3)} \quad (7.152)$$

We can check that in the case where Θ is independent of μ, we can integrate out the μ variable and as expected obtain double the 2D result (the total angle $= 4\pi$ in 3D, not 2π). As usual, the normalization constant is the total solid angle in the $\underline{r}^{(4)}$ coordinate system:

$$N_{D_{el}} = \Omega_{D_{el}} = \int_{-1}^{1}\int_{0}^{2\pi} d\omega\left(\mu^{(3)}, \theta^{(3)}\right) \tag{7.153}$$

A technical point is that for $H_z < 1$, there is a singularity $|\mu^{(3)}|^{H_z-1}$ in Eqn. (7.147); it can be removed with the change of variables $\mu' = \mu^{H_z}$; this is advantageous in numerical integrations.

The analogous formula for $a^2 < 0$ is:

$$d\omega = H_z|a|^{H_z}\frac{|\mu^{(3)}|^{H_z-1}}{\Theta\left(\mu^{(3)}, \theta^{(3)}\right)^{2+H_z}\left(|a|^2\left(\mu^{(3)}\right)^2 + \left(e - r\cos 2\theta^{(3)}\right)\left(1 - \left(\mu^{(3)}\right)^2\right)\right)^{(1+H_z)/2}} d\mu^{(3)} d\theta^{(3)} \tag{7.154}$$

followed by the integration Eqn. (7.151).

Chapter 8

Space-time cascades and the emergent laws of the weather

8.1 Basic considerations and empirical evidence

8.1.1 Spatial versus temporal scaling

In atmospheric science – as in physics – it is usual to specify the dynamics with prognostic (partial) differential equations where the time differentiation operator plays the crucial role in specifying the evolution of the system. Consequently, the usual textbook practice is to derive the dynamical laws early on. In our scale-symmetry-based approach, time could be presented in an analogous way; however, for technical reasons discussed in the next chapter, this is not the simplest method of exposition. We therefore introduce time in a different manner, exploring first the consequences of scale symmetries for emergent turbulent laws. In this framework, the scaling is in space-time and such scaling is in many respects a straightforward generalization of spatial scaling. It is therefore simpler to start with space and only later to consider time.

In fluid systems, the key physical process relating time and space is advection. When there is a scale separation, this may justify the common practice of transforming time series measurements into spatial measurements using a constant (non random) advection velocity. This hypothesis of "frozen turbulence" (Taylor, 1938) is justified if small-scale turbulent structures are literally "blown past" an observer at speed V so that the spatial coordinate x is the same as the rescaled time coordinate Vt; the transformation is deterministic. Unsurprisingly, this idea was initially made to interpret the results of laboratory wind tunnel experiments. In our scaling case, it does not strictly apply because there is no scale separation. However, even if the turbulence is not frozen, the statistics in space and time may nevertheless be related by a parameter with the dimensions of velocity. Indeed, in Chapter 2 we saw that mathematically, if the velocity field is spatially scaling, then the advection operator D/Dt can also be scaling, allowing the whole system to be scaling in space-time. Since the boundary conditions cannot be scaling at space-time scales larger than the size of the planet, any such relationship must eventually break down; this turns out to provide the natural distinction between weather and macroweather that we discuss at length in Chapter 10. The slight complication is that the ocean is also a fluid with similar behaviour, although with lower velocities and hence longer critical time scales (Section 8.1.4).

In any case, the existence of space-time relations in fluid mechanics is well known in a different guise: the relatively well-defined lifetime of structures, their "eddy turnover times." These statistical size/duration relations are the basic physics behind the "space-time" or "Stommel" diagrams presented in meteorology textbooks as conceptual tools, but which are in practice rarely empirically calculated (see e.g. Fig. 8.9b). Likewise, although space-time relations are in fact used all the time in meteorological measurements, they are usually implicit rather than explicit, in the form of "rules of thumb." For example, many automatic digital weather stations average measurements at the fairly arbitrary period of 15 minutes. If a scale separation existed, this might have had some justification, but if there is no separation, how long should the averaging be made? Alternatively, how often should a weather radar scan if the spatial resolution is 1 km? If it is 4 km? Conversely, if only "climate" time-scale (say monthly) estimates are needed, what should be the spatial scale of the corresponding maps? In the same vein, in-situ measurements are often considered to be "point measurements," i.e. with infinite (or very high) resolutions, but this is misleading since even if they are at points in *space*, they are never also instantaneous, i.e. they are not points in *space-time*, and it is their *space-time* resolutions that are important for their statistics.

This is the first of four chapters dealing with the space-time structure of the atmosphere. We first focus on the basic empirical evidence and theoretical basis of space-time scaling from dissipation (millisecond) to the beginning of macroweather (10^6 second) scales, considering only the turbulent aspects. In Chapter 9, we consider the important issue of causality and space-time delocalization: i.e. turbulence-driven waves as an emergent scaling phenomenon. Also in Chapter 9 we consider predictability and stochastic forecasting of cascades, multifractal processes. In Chapter 10 we turn to the transition from weather to macroweather, and then in Chapter 11 to scaling in the even lower-frequency climate regime.

8.1.2 A survey of temporal spectral scaling in the weather regime

Before developing the theory, let us first consider evidence for the ubiquity of temporal scaling. As with the discussion of spatial scaling, we'll start with straightforward and familiar spectral analyses which are particularly sensitive to breaks in the scaling. One of the earliest, and certainly the most influential, temporal spectrum was published by Van der Hoven (1957; see also Panofsky and Van der Hoven, 1955). Fig. 8.1a reproduces the graph at the origin of the famous "meso-scale gap," the supposedly energy-poor spectral region between roughly 10–20 minutes and (excluding the diurnal cycle) the spectral "bump" at 4–5 day periods. Even until fairly recently, textbooks regularly reproduced the spectrum (often redrawing it on different axes or introducing other adaptations), citing it as convincing empirical justification for the neat separation between low-frequency isotropic 2D turbulence – identified with the weather – and high-frequency isotropic 3D "turbulence." If the gap were real, it would certainly be convenient, since turbulence would be no more than an annoying source of perturbation to the (2D) weather processes.

Retrospectively, it seems clear that the success of this meso-scale gap was indeed more due to wishful thinking than to hard science. Within barely 10 years it had been subjected to strong criticism by Goldman (1968), Pinus (1968), Vinnichenko (1969), Vinnichenko and Dutton (1969), Robinson (1971) and indirectly by Hwang (1970). For instance, on the basis of much more extensive measurements, Vinnichenko (1969) commented that even if there were occassionally mesoscale gaps they would only be for less than 5% of the time. This may be seen in Fig. 8.1b, which shows how the meso-scale gap might arise under "severe" turbulence with energy fluxes 200 times the large-scale mean. He then went on to note that Van der Hoven's spectrum was actually the superposition of four spectra (roughly indicated by the ellipses which we have added to Fig. 8.1a; the original figure has tiny symbols which are very hard to decipher). Indeed, he noted that the extreme set of high-frequency measurements was taken during a single one-hour period during an episode of "near-hurricane" conditions and that these were entirely responsible for the high-frequency "bump." Ironically, in spite of the gap's prediction of a barren, uninteresting (energy-poor) meso-scale, belief in the gap persisted well after the birth of meso-scale meteorology in the 1970s and 1980s.

The extensive spatial spectra and cascade analyses discussed in Chapters 1 and 4 include the key dynamical velocity field; thus they already give indirect evidence of temporal scaling. We now briefly present more direct evidence using modern temporal spectra. Anticipating the discussion in Chapter 10, we can divide the frequency range into a high-frequency weather regime $\omega > \omega_f$ where $\omega_f \approx (5\ \text{days})^{-1}$ – $(20\ \text{days})^{-1}$, denoted with *w*, a low-frequency macroweather regime $\omega_c < \omega < \omega_f$ (when needed, denoted *mw*) where $\omega_c \approx (10\ \text{years})^{-1}$ – $(30\ \text{years})^{-1}$, and finally a climate regime $\omega < \omega_c$, denoted *c*. We discuss the physical origin of ω_f in Section 8.1.3 and of ω_c in Chapter 10. Fig. 8.2a shows spectra of the horizontal wind from scales ranging from near dissipation scales (5×10^{-4} seconds up to about 1 minute), while Fig. 8.2b covers the range 1 minute to nearly a day and Fig. 8.2c shows daily station data analyzed up to 60 years. In all cases, reference lines with absolute slopes $\beta_w = 1.6$ are indicated. This is roughly the value of β if the *horizontal* exponents $H = 1/3$, $C_1 \approx 0.046$ and $\alpha \approx 2$ are used (see Table 4.4, and recall that $\beta = 1 + 2H - K(2)$); applying these parameters in time, it corresponds to an isotropic (x, y, t) space, a hypothesis to which we return below. According to these spectra, it is plausible that the scaling in the wind holds from dissipation scales out to scales of ~5 days, where we see a transition (Fig. 8.2c). This transition is essentially the same as the low-frequency "bump" in Fig. 8.1a; its appearance only differs because in the latter $\omega E(\omega)$ rather than $\log E(\omega)$ is plotted; we return to this transition momentarily, although notice that the same feature is well reproduced in the older wind data shown in Fig. 8.1b.

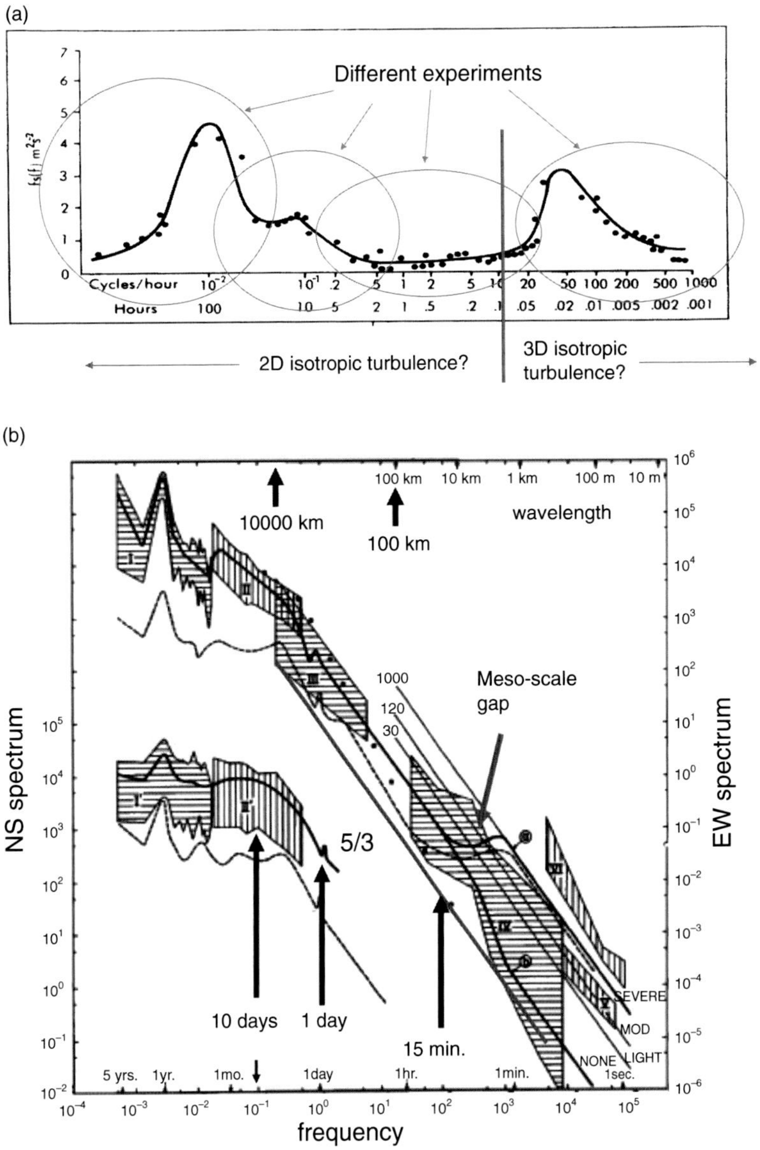

Fig. 8.1 (a) The "meso-scale gap" between the right-most bump and the "synoptic maximum" (left-most bump), adapted from Van der Hoven (1957). The ellipses show the rough ranges of the four experiments which were combined to give the composite spectrum (the actual data points were already replotted from the original); the vertical line which bounds the right-most bump corresponds to about 6 minutes and was often superposed to indicate the limit of 3D isotropic turbulence.
(b) A composite spectrum of horizontal wind (both zonal and meridional components) with bold arrows, reference line and text added (adapted from Vinnichenko and Dutton, 1969). The shading is an attempt to represent the variability, the reference lines labelled in the middle of the graph 1000, 100, 30 are ε in units of cm^2s^{-3} and represent light, moderate and severe turbulence (the mean was given as $5\ cm^2s^{-3}$ ($= 5 \times 10^{-4}\ m^2s^{-3}$), which is very close to the mean from the reanalyses in Section 8.1.3, which is about twice as much). The mean scaling seems to continue up to about 5 days, corresponding to about 5000 km. The meso-scale gap is indicated by the arrow; the top line thus appears only under conditions of severe turbulence with ε about 200 times the mean. Notice the fairly flat spectral "plateau" for frequencies below about (10 days)$^{-1}$ and also the annual spike.

We could mention that here and in the following we do not discuss the difficult question of boundary-layer effects which arise when data are taken close to the surface and which can display scaling breaks and other complications due to their proximity to the "wall." Section 8.2 and Appendix 8A give some indications of the complexities of temporal spectra when compared to spatial spectra; at larger scales, things are simpler because of the dominance of the horizontal wind.

The corresponding spectra for the temperature are shown in Fig. 8.3a (from scales of 0.4 second to ~1 hour), Fig. 8.3b (from 6 minutes to 24 hours), Fig. 8.3c (hourly surface temperatures), Fig. 8.3d (from daily series over 10 years); and in Fig. 8.6d from daily series 6 years long. In all cases multiple series were averaged to reduce the noise. The reference lines have slopes corresponding (roughly) to the horizontally predicted values, (from Table 4.4, with $H \approx 0.50$, $C_1 \approx 0.052$, $\alpha = 1.78$, we obtain $\beta_w \approx 1.9$).

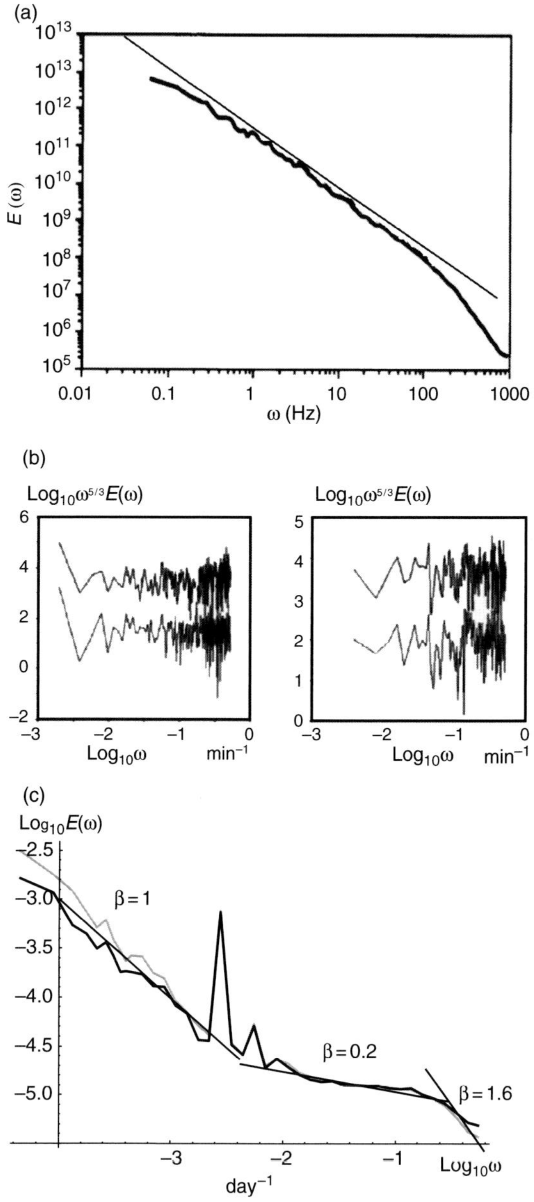

Fig. 8.2 (a) The wind spectrum from 2 kHz data (a sample of which was shown in Fig. 2.4 and which was already partially analyzed in Section 5.5: see Fig. 5.37), from Montreal, Quebec. The lowest frequency represented is about (1 minute)$^{-1}$; the spectra of many samples were averaged to reduce the noise. The reference line has an absolute slope 1.6, which is the Kolmogorov value ($\beta = 5/3 = 1 + 2H$; $H = 1/3$) minus the (horizontally determined) intermittency corrections ($K(2) \approx 0.07$, Table 4.4). Reproduced from Schmitt *et al.* (1993). (b) Spectra of the zonal (top) and meridional (bottom) components of the wind from two French stations: Hayange (left), and St. Menard d'Aunis (right), at 1-minute resolutions (offset in the vertical for clarity), adapted from Tchiguirinskaia *et al.* (2006). The spectra are compensated by $\omega^{5/3}$ so that a flat curve corresponds to $E(\omega) \approx \omega^{-5/3}$. There is no sign of the meso-scale gap (which in the units of the figure would be between $\log_{10}\omega \approx -1$ and $\log_{10}\omega \approx -2$). (c) Wind spectra from 36 near-complete 60-year series of daily data from the continental USA taken from the stations lying nearest to 2° × 2° degree grid points from 30–50° N, 105–71° W. The curve that is lower at left is the daily maximum wind speed and the other curve is daily average (normalized so that the annual peaks coincide). The spectra have been also averaged in the frequency domain; in bins logarithmically spaced, 10 per order of magnitude (except for the lowest factor of 10, where no spectral averaging was performed). As discussed in Appendix 5E, this is an unbiased averaging procedure and leads to much smoother spectra. The reference lines have absolute slopes $\beta = 1$ (low frequencies), $\beta = 0.2$ (the "spectral plateau" which is not totally flat) and $\beta = 1.6$ (high frequencies).

Again, the horizontal values are seen to work well over the entire range up to the low-frequency transition, which in this case starts at around a 7–20-day scale (Figs. 8.3c, 8.3d; see also Fig. 8.6d). Note that the in-situ measurements generally agree that the value of the temporal β for the temperature is about 0.2–0.3 larger than for the wind and that the temporal H values are roughly equal to the horizontal H values of 1/3, 0.50 for wind and temperature respectively (the temperature is thus not a good passive scalar; see also Schmitt *et al.*, 1996; Finn *et al.*, 2001, for the estimates $H \approx 0.38, 0.44$ respectively).

The use of hourly surface data such as in Fig. 8.3c raises practical questions about scaling analyses, since, as can be seen, the diurnal variation and its harmonics are extremely strong (the spike corresponding to

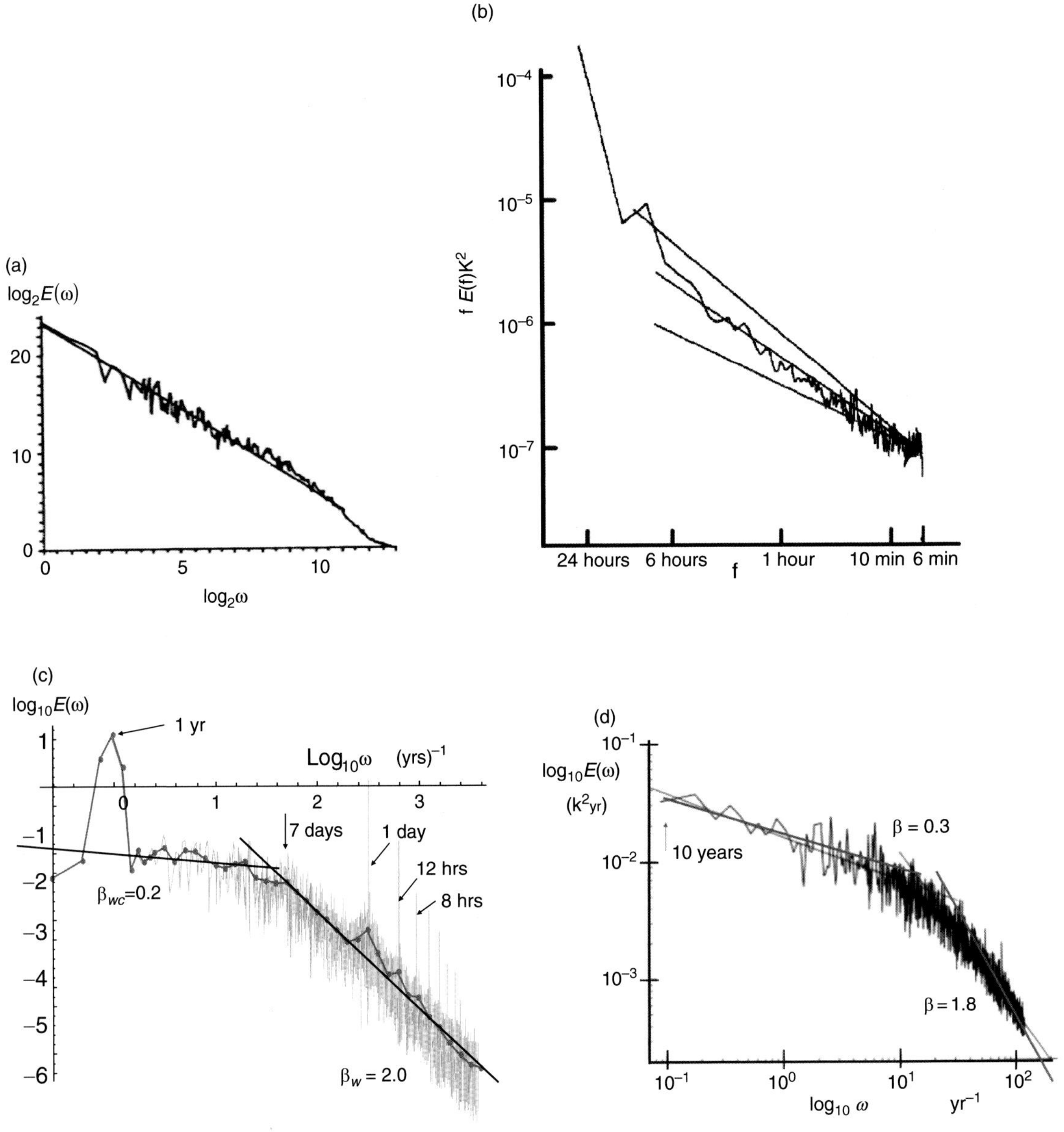

Fig. 8.3 (a) Temperature spectrum from a series of 2^{14} points of 5 Hz data ($\approx$ 1 hour), from Montreal, Québec. Reference line has slope −1.8. Adapted from Schmitt *et al.* (1992). (b) Temperature spectrum from the average of 36 spectra, each 24 hours at 3-minute resolution, from France. Note the diurnal low-frequency peak. Reference lines correspond to β_w = 1.6, 1.8 (closest fit), 2. Reproduced from Lovejoy and Schertzer (1986). (c) The spectra of hourly surface temperature data from four nearly colinear stations running northwest/southeast in the USA (Lander WY, Harrison NE, Whitman NE, Lincoln NE), from the US Climate Reference Network, 2005–2008. See Fig. 8.7h for the corresponding cascade analysis. The thick line is the spectrum of the periodically (diurnally) detrended spectrum, averaged over logarithmically spaced bins as in Fig. 8.2c. Reproduced from Lovejoy and Schertzer (2012). (d) The temperature spectrum from the average of 370 continental daily temperature series from the National Climatic Data Center (NCDC) as a function of frequency (in year^{-1}), adapted from Pelletier and Turcotte (1997), having thick reference lines with absolute slopes β = 0.3, 1.8; the regression values given in the original paper had β = 0.37 (the thin line, left) and 1.37 (right) but depend somewhat on the exact frequency range chosen.

24 hours is about 10^4 times stronger than the background, the 12-hour spike is several thousand times stronger, etc.). Since diurnal effects are essentially periodic, Fourier analysis is very convenient, and shows that except for the spikes the scaling is not much perturbed. However, if instead we want to perform real-space analyses (including trace moment/cascade analyses), the periodicities must be removed as much as possible. In Fig. 8.3c we show the effect of removing the diurnal cycle by calculating the average temperature at each hour of the day and removing it. The curve in the figure shows the spectrum of the periodically detrended temperatures averaged over logarithmically spaced frequency bins, 10 per order of magnitude (except for the lowest 10 frequencies, which are all shown). It can be seen that the detrending is not perfect, but does restore the spectrum to a reasonably power-law form. Cascade analyses of this detrended data are discussed in Section 8.1.5. Note that diurnal cycles are mostly important for surface and stratospheric fields, although cloud convection also displays it and there are tidal as well as specific chemical processes which have prominent diurnal cycles.

Excluding the annual cycle, roughly (but not completely) flat low-frequency spectra are qualitatively reproduced in all the standard meteorological fields, and the transition scale is relatively constant at typically 5–10 days. Using the same database as in Fig. 8.2c, this may be seen in the instrumental spectra of 60-year-long series from daily US station data spectra for precipitation and pressure (Fig. 8.4a), for temperature, dew-point temperature, humidity (Fig. 8.4b) and for wind speed, "gusts" and inverse visibility (an "effective" extinction coefficient) (Fig. 8.4c). Fig. 8.4d shows very similar spectra of precipitation rates from the 20CR, CPC and ECMWF interim reanalyses (all discussed earlier). In Chapter 10, where we focus on the lower frequencies, we show further relevant spectra including 20CR spectra for T, u, v, h (Figs. 10.5a, 10.5b) and discuss the direct extension of the FIF weather model into the macroweather domain. We see that it predicts $\beta_{mw} \approx$ 0.2–0.4 for essentially all the fields (at least those over land; over the ocean it can yield $\beta_{mw} \approx 0.6$ for $\omega < \omega_{eff}^{-1}$); hence we also systematically show this theoretically predicted reference line.

In order to obtain systematic estimates of the weather regime exponents, we used the temporal spectra of the ECMWF 700 mb daily reanalysis dataset for 2006 for T, u, v, h_s, w and z (the same that was analyzed spatially in Chapters 4 and 6). These are shown in Fig. 8.5a, with a blow-up of the high-frequency part in Fig. 8.5b. The overall shape and the transition scales are about the same as for the instrumental series, although the exponents are not necessarily the same; a detailed comparison is given in Table 8.1. Focusing on the high-frequency weather regime in Fig. 8.5a,we see that it is extremely narrow; Fig. 8.5b shows a blow-up of the high-frequency decade. From this we see that regression estimates of the spectral exponent β will depend somewhat on the regime used for the fit; it was therefore decided to use theoretically motivated reference lines rather than regressions – the only exception being the vertical velocity, where a regression over $\omega > (5\text{ days})^{-1}$ was used. The basic motivating theory was for the values 5/3 for u, v and T (as discussed for u, v and as predicted if T is a passive scalar), and for z, using the isobaric value (3.35) as an estimate of the horizontal exponent followed by space-time isotropy. The 5/3 values work very well for u, v, T (the isobaric exponent 2.40 shown in the figure is clearly very poor), but the isobaric value works well for z. In contrast, the exponents $\beta \approx 0.4$ (h_s) and ≈ 1.10 (w) have no clear theoretical explanation, although the value 0.4 for h_s is very close to the isobaric w value. A statistical link between the specific humidity and the vertical wind could be a consequence of both being sensitive to horizontal convergence or temperature fluctuations. In Chapter 10 we discuss the transition and the lower (climate) frequencies in more detail.

All these spectra showed qualitatively similar behaviour, and we have used the ECMWF interim reanalysis to obtain systematic estimates of β_w. What about τ_w, the weather/macroweather transition scale? In this case, it is important to get accurate estimates of the spectra, so we chose the 20CR dataset, which was long enough to yield 280 segments each 180 days long for each $2^\circ \times 2^\circ$ grid point. By averaging the spectra of the daily resolution data over the 280 segments we obtain quite smooth spectra over the range $(180\text{ days})^{-1}$ to $(2\text{ days})^{-1}$. These were good enough to allow for estimates of τ_w, β_w, β_{mw} by using nonlinear regressions on $\log E(\omega)$ with a bilinear function with transition at $\omega_w = \tau_w^{-1}$. For the 700 mb temperature and the surface precipitation rate, we found $\tau_w = 8.5 \pm 1.9$, 4.5 ± 1.5 days respectively. Since there were systematic latitudinal variations, we calculated the mean τ_w and the longitude-to-longitude

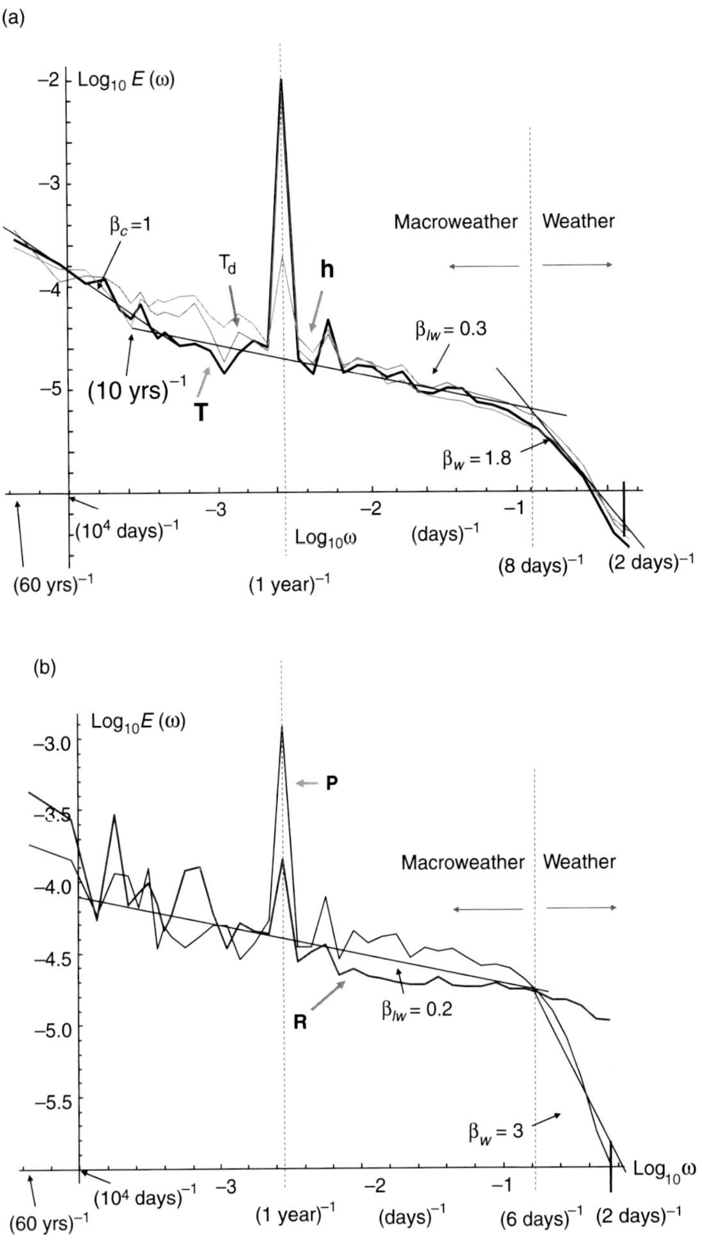

Fig. 8.4 (a) Mean spectrum of daily dew-point temperature (T_d), temperature (T), and relative humidity (h), for 36, 33 and 7 stations respectively (numbers vary due to missing data, only near-complete series were used), from stations with long (60 year; 22 200 days) records, from a 2° grid over the continental USA (the same database as Fig. 8.2c). The low-frequency (climate) and the "plateau" (macroweather) reference lines have slopes −1 and −0.2 respectively. The spectra were averaged over $1dB\omega$ bins (i.e. 10 per order of magnitude in frequency), every 2° from 30° to 50° latitude, from −105° to −71° longitude. The high-frequency reference line has absolute slope $\beta = 1.8$, close to the horizontal β value for humidity and temperature, which are each about 1.8–1.9 (see Table 4.4), and the plateau value is very close to the theory value 0.2. Adapted from Lovejoy and Schertzer (2010). (b) Same as Fig. 8.4a but for variables with slightly shallower "plateau." Mean spectrum of daily pressure (P) and rain amount (R) from 24 stations with long (60 year; 22 200 days) records, from a 2° grid. The reference slopes have absolute slope $\beta = 1$ (low frequency), 0.2 (plateau). At high-frequency, $\beta \approx 3$ close to the horizontal geopotential value (see Table 4.1). Adapted from Lovejoy and Schertzer (2010).

spread every 10°. The result is shown in Fig. 8.5c. It can be seen that there is a minimum at around $\pm 45°$ with a significant increase near the equator and a smaller increase towards the poles. There was also a tendency for the largest τ_w (60, 12 days for T, R respectively) to occur over the tropical Pacific and for the minimum τ_w to occur of the mid-latitude oceans (4, 2 days respectively). Also shown is the theoretically predicted τ_w discussed in the next section.

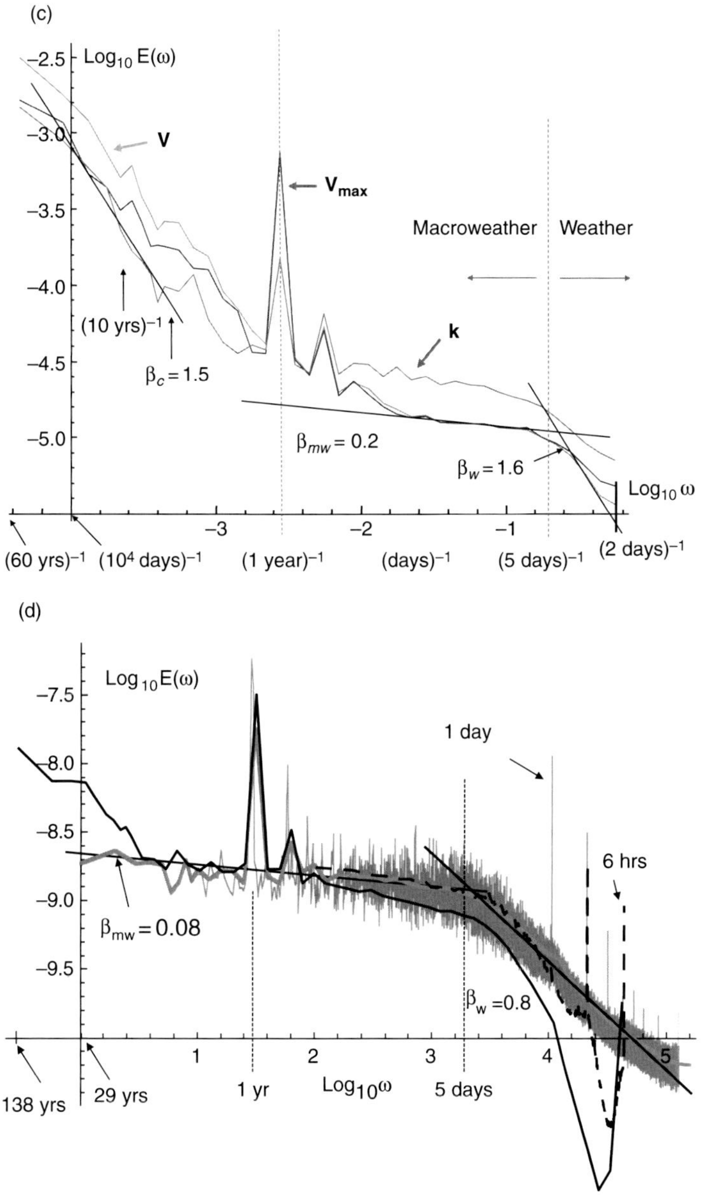

Fig. 8.4 (c) Spectra for the wind speed (v), the maximum wind ("gusts," v_{max}) and the inverse visibility (the effective extinction coefficient, k) for the same database as in Figs. 8.4a, 8.4b. The reference lines have absolute slopes $\beta = 1$, 0.2, 5/3 (left to right; the middle plateau value is theory, the latter is close to the value in Table 4.4). Adapted from Lovejoy and Schertzer (2010). (d) A comparison of the temporal spectra of the CPC data (thin background, noisy curve) and the ECMWF 3-hourly dataset (thick, dashed). The thick grey curve is the CPC spectrum averaged over logarithmically spaced frequency bins (10 per order of magnitude). The long thick curve is from the 20CR at 45° N, from the full 3-hour-resolution data (from 1871–2008). The transition scale from the high-frequency weather regime and low-frequency macroweather regime is indicated by the dashed line at periods of 5 days. The axis is in units such that $\omega = 1$ is (29 years)$^{-1}$; i.e. the full length of the CPC series. There are three reference lines with absolute slopes indicated; the value 0.08 is from the Haar structure function analysis (Fig. 10.14) from 3 months to 29 years. Reproduced from Lovejoy *et al.* (2012).

8.1.3 Estimating the weather: macroweather transition scale τ_w from "first principles"

We have shown evidence that temporal scaling holds from small scales to a transition scale τ_w (w for "weather") of around 5–20 days. Let us now consider the physical origin of this scale. In Fig. 8.1 (the $\omega E(\omega)$ versus logω plot), it appears as a low-frequency spectral "bump" at around 4–5 days; its origin was argued to be due to "migratory pressure systems of synoptic weather-map scale" (Van der Hoven, 1957). The corresponding features at around 4–20 days, notably for temperature and pressure spectra, were termed "synoptic maxima" by Kolesnikov and Monin (1965) and Panofsky (1969), in reference to the similar idea that it was associated

Table 8.1 Temporal scaling exponents from the ECMWF interim reanalysis and from various in-situ estimates. The latter were taken from the literature when no other source was available, and from daily station data that had been selected for a climate study; see the footnotes for details. The bottom rows show the isobaric H values estimated as discussed in the text; they are the zonal values (the only exception being the meridional wind, where the meridional value is given). The C_1, α estimates are from the flux analyses presented earlier, the H are estimated from the spectra (Fig. 8.5b). Note that the ECMWF estimates were made at the (hyper)viscous dissipation scale whereas the in-situ data were estimated in the scaling regime. Recalling the discussion in Chapter 4 (Eqn. 4.14), the difference is a factor ≈ 2.07 for the velocity with possibly similar factors for the other variables. This factor has been applied to increase the in-situ values so as to make them comparable with the reanalyses. Due to the narrow range of frequencies (the Nyquist frequency (2 days)$^{-1}$ to about (5 days)$^{-1}$ before the spectra begin to flatten due to the transition to climate), direct estimates are not accurate and depend on the exact frequency range used. Instead, the spectral exponents were taken from the reference lines in Figs. 8.5a, 8.5b: for T, u, v they are seen to be compatible with the Kolmogorov value 5/3 (ignoring the intermittency corrections of $K(2) \approx 0.16$); for h_s the value $\beta = 0.4$ is close to the data and is the same as the spatial w value. Finally the β value for z was close to the spatial value (see Figs. 8.5a, 8.5b), and was used.

Exponent	Source	h_s	T	u	v	w	z
β	ECMWF	0.40	5/3	5/3	5/3	1.10	3.35
	In-situ	2.20[d]	1.67 ± 0.04[a]	1.68 ± 0.05[b,e]	1.68 ± 0.05[b,e]	—	3.00[c]
C_1	ECMWF	0.10	0.075	0.083	0.086	0.115	0.085
	In-situ	0.09[d]	0.087 ± 0.015[a]	0.088 ± 0.01[b,e]	0.088 ± 0.01[b,e]	—	0.085[c]
α	ECMWF	1.77	1.90	1.85	1.85	1.92	1.90
	In-situ	1.8[d]	1.61[a]	1.5[b,e]	1.5[b,e]		1.7[c]
H (time)	ECMWF	−0.21	1/3	1/3	1/3	0.17	1.26
	In-situ	0.68[d]	0.41 ± 0.03[a]	0.33 ± 0.03[b,e]	0.33 ± 0.03[b,e]	—	1.07[c]
H (space)	ECMWF (space)	0.54	0.77	0.77	0.78	−0.14	1.26
	aircraft	0.51	0.50	1/3[e]	1/3[e]	—	—

[a] The mean and standard deviations were calculated from the three published studies: Finn *et al.* (2001), Schmitt *et al.* (1996), Wang (1995); these are of high-frequency measurements near the surface.

[b] These numbers were calculated from estimates in Schmitt *et al.* (1993; see also Schmitt *et al.*, 1996) and are from high-frequency (hot-wire) data near the surface. We used the estimate $C_{1\varepsilon} \approx 0.25 \pm 0.05$, $\alpha = 1.5 \pm 0.1$. To find C_{1v}, we used the fact that in Schmitt *et al.* (1993), ε was estimated from $\varepsilon = \Delta v^3/l$ and the spatial scale l was estimated from $l = v\Delta t$ where v was the measured speed (i.e. the "Taylor's hypothesis"); therefore we used Eqn. 4.14: $C_{1v} = \eta^{\alpha} C_{1\varepsilon}$ with $\eta = 1/3$, $\alpha = 1.5$. Note that although the C_1 estimate seems robust, the α estimate is lower than that of the spatial estimates, and is therefore probably underestimated; a value $\alpha \approx 1.8$ is probably more realistic.

[c] These are estimates of surface pressure (not geopotential) statistics from 23 daily in-situ series over (near-complete) 60 years in the continental USA (stations were taken every 2 degrees from 30° to 50° NS, −105° to −71° EW; only the longest, near-complete series were used: Fig. 8.4b); the statistics are over the narrow range (2 days)$^{-1}$ (Nyquist frequency) to (5 days)$^{-1}$ (to avoid spectral flattening due to the weather/climate transition).

[d] Same study as c, except only 7 stations were both near complete and 60 years long.

[e] No distinction was made between the zonal and meridional wind components; the same values were used for u, v.

with synoptic scale weather dynamics (see Monin and Yaglom, 1975, for some other early references).

More recently, Vallis (2010) suggested that this scale is the basic lifetime of baroclinic instabilities, which he estimated using $\tau_w = \tau_{Eady}$ where τ_{Eady} is the inverse Eady growth rate: $\tau_{Eady} \approx L_d/U$; U is the typical horizontal wind speed (taken as ≈ 10 m/s) and the deformation rate is $L_d = NH/f_0$, where f_0 is the Coriolis parameter, H is the thickness of the troposphere and N is the mean Brunt–Vaisalla frequency across the troposphere. The Eady growth rate is obtained by linearizing the equations about a hypothetical state with uniform shear and stratification across the troposphere. By taking $H \approx 10^4$ m, $f_0 \approx 10^{-4}$ s^{-1}, and $N \approx 10^{-2}$s^{-1}, one obtains Vallis's estimate $L_d \approx 1000$ km. Using the maximum Eady growth rate then introduces a numerical factor 3.3 so that the actual predicted inverse growth rate is: $3.3\tau_{Eady} \approx 4$ days. Vallis similarly argues that this also applies to the oceans but with $U \approx 10$ cm/s and $L_d \approx 100$ km, yielding $3.3\tau_{Eady} \approx 40$ days. The obvious theoretical problem with using τ_{Eady} to estimate τ_w, is that the former is expected to be valid in quasi-linear systems whereas we have given evidence for the existence of highly heterogeneous vertical and horizontal structures (including strongly nonlinear cascade structures) extending throughout the troposphere to scales substantially larger than L_d. Another difficulty is that although the observed transition scale τ_w is well behaved at the equator (Fig. 8.5c), f_0 vanishes, implying that L_d and τ_{Eady} diverge: using τ_{Eady} as an estimate of τ_w is at best a mid-latitude approximation.

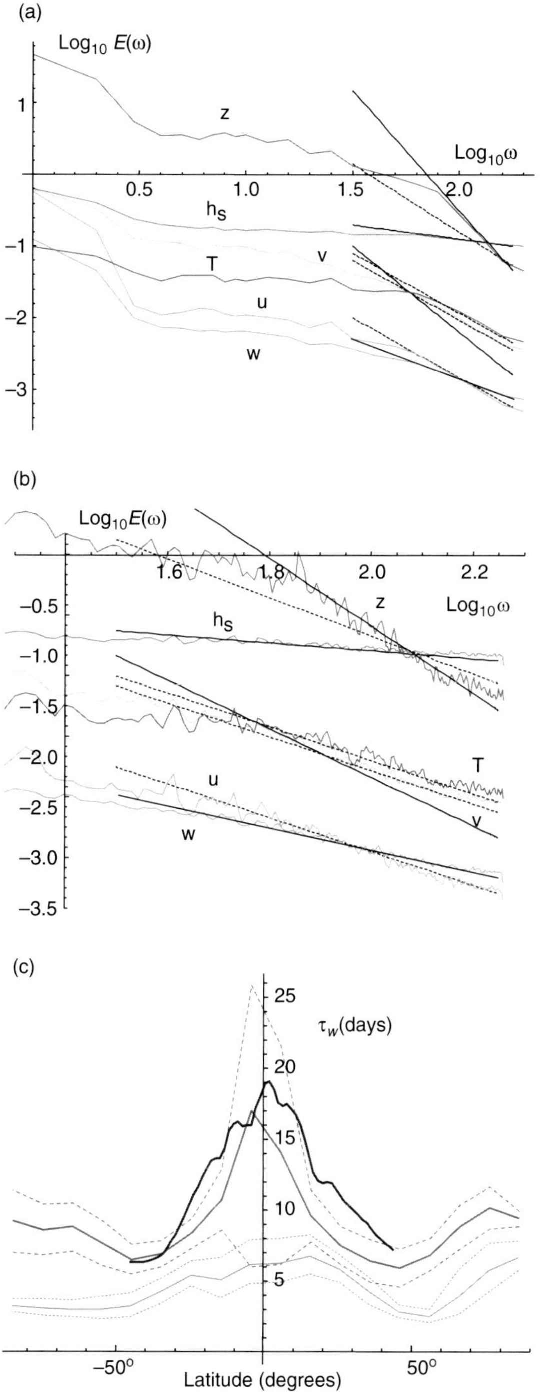

Fig. 8.5 (a) The temporal spectra of the daily 700 mb ECMWF interim reanalysis fields that were analyzed spatially in Chapter 4 (i.e. between ± 45° latitude). The dashed slopes are 5/3, the frequency spectrum is shown estimated using ensemble and spectral averaging (into 10 bins per order of magnitude for all $\omega > 36.5$ cycles/year). The dashed lines have slopes −5/3, the solid lines have slopes −3.35, −0.4, −2.4, −1.1 (top to bottom); they are drawn for $\omega >$ (11 days)$^{-1}$; for a blow-up of the high-frequency (weather) regime, see Fig. 8.5b. These correspond to the spatial z exponent, the spatial w exponent (which accurately fits h_s), the spatial u, $\mathbf{v}$, T exponent and the regression w slope, respectively. The curves top (on left) to bottom are z, h_s (multiplied by 10 from spatial analysis, i.e. the spectra are multiplied by 100), $\mathbf{v}$, T, u and w respectively. Note that the low-frequency rise is due to only two frequencies (1 year)$^{-1}$, (6 months)$^{-1}$, i.e. periodic components at low resolution, not a break in the scaling. Reproduced from Lovejoy and Schertzer (2011).
(b) Same as Fig. 8.5a, showing a blow-up of the high-frequency decade ((20 days)$^{-1}$ to (2 days)$^{-1}$) using ensemble but not spectral averaging, units of ω: cycles/year. As in Fig. 8.5a, the dashed lines have slopes −5/3, the solid lines have slopes −3.35, −0.4, −2.4, −1.1 (top to bottom). Reproduced from Lovejoy and Schertzer (2011).
(c) The variation of τ_w as a function of latitude as estimated from the 138-year-long 20CR reanalyses. The estimates were made by performing bilinear regressions on spectra from 180-day-long series averaged over 280 segments per grid point. The top wide and thick curve shows the mean over all the longitudes, the dashed lines a bit above and below are the corresponding one-standard-deviation spreads. The bottom wide curves are the corresponding results for the surface precipitation rates. Also shown (thick line between ± 45°) is the similarly averaged theoretically predicted eddy turnover times estimated from the tropospherically averaged zonal wind from the year 2006 using the ECMWF reanalysis data (Fig. 8.6a).

Finally, there is no evidence for any special behaviour at length scales near $L_d \approx 1000$ km.

In our space-time scaling framework, it is more natural to divide the whole dynamical range (from millions of years to milliseconds) into various wide scaling ranges (Lovejoy and Schertzer, 1986): see Chapter 10. On log-log plots this synoptic maximum appears instead as a smooth transition, identified in the new framework with the transition between the weather and macroweather, described as the "spectral plateau" (Lovejoy and Schertzer, 1986). Later the term "weather–climate" regime was used but this suggests that there is a transition zone between weather and climate (Lovejoy and Schertzer, 2010). Since we already saw in Fig. 1.9d that it was reproduced by the FIF and GCM control runs (both essentially weather models), the description "low-frequency weather" is more accurate, but the name "macroweather" is less cumbersome (Lovejoy and Schertzer, 2012a) and is used here.

Although the original term "plateau" is somewhat of a misnomer, since the corresponding spectrum isn't perfectly flat, its logarithmic slope is small, and it is essentially independent of C_1, H and is only weakly dependent on α, so that the term is at least qualitatively correct. Surprisingly, it is weakly dependent on the overall range of scale in the plateau (its low-frequency limit). In Appendix 10A we find numerically that for a range of $\approx 10^2$–10^3 (i.e. $\sim$10 days to 3–30 years), $\beta \approx 0.2 - 0.4$; in addition, it can be larger ($\beta \approx 0.6$) for a maritime climate: see Appendix 10D. A similar basic framework involving broad scaling regimes was also adopted by Pelletier (1998), Koscielny-Bunde *et al.* (1998), Talkner and Weber (2000), Ashkenazy *et al.* (2003), Huybers and Curry (2006) and others. We will return to both a transition model and a more detailed analysis of the climate regime in Chapter 10, concentrating below on an explanation for the transition and a justification for terming τ_w a "weather/macroweather" transition.

If there is a statistically well-defined relation between spatial scales and lifetime "eddy turnover times" (Chapter 2), then the lifetime of planetary-scale structures τ_{eddy} is of fundamental importance, i.e. we expect $\tau_w = \tau_{eddy}$. If we evaluate this at planetary scales $L_e = 2 \times 10^7$ m (i.e. we assume that the outer weather scale $L_w \approx L_e$) then we obtain a large-scale velocity $V_w \approx \varepsilon_w^{1/3} L_e^{1/3}$, which is the typical velocity across a structure of size L_e. The corresponding eddy turnover time/lifetime of planetary structures is therefore $\tau_w = \tau_{eddy} = L_e/V_w = \varepsilon_w^{-1/3} L_e^{2/3}$.

If the eddy turnover time is fundamental, this implies that the mean energy flux density ε_w plays a fundamental role in determining the horizontal structures – and hence in characterizing atmospheric complexity. It is therefore interesting to note that one of the prominent strands in complexity theory holds precisely that the "energy rate density" – essentially the same thing as the energy flux density ε – plays a fundamental role in determining the basic level of complexity in the universe from its astrophysical origins to the development of biology and human society (see Chiason, 2010, for a interesting argument for this).

But what determines the globally averaged fundamental flux ε_w? We can estimate the mean ε_w by using the fact that the mean solar flux absorbed by the earth is $\sim$200 W/m^2 (e.g. Monin, 1972; a more modern value is 240 W/m^2). If we distribute this over the troposphere (thickness $\approx 10^4$ m), with mean air density ≈ 0.75 kg/m^3, and we assume a 2% conversion of energy into kinetic energy (Palmén, 1959; Monin, 1972), then we obtain a value $\varepsilon_w \approx 5\times10^{-4}$ m^2/s^3, which is indeed typical of the values measured in small-scale turbulence (Brunt, 1939; Monin, 1972). If we now assume that the horizontal dynamics are indeed dominated by the energy flux, then we can use Kolmogov's formula to extrapolate these first-principle estimates up to planetary scales to estimate the large-scale velocity difference across a hemisphere, and we obtain $V_w \approx 21$ m/s. The corresponding eddy turnover time, lifetime, is therefore $\tau_w = 9.5 \times 10^5$ s ≈ 11 days, i.e. roughly the time associated with synoptic/global-scale phenomena, as discussed in the previous subsection.

Although this "first principles" calculation of the weather velocity V_w and time scales τ_w from the solar energy input is seductive, as far as we can tell it was not proposed until recently (Lovejoy and Schertzer, 2010), presumably because the Kolmogorov law was believed to only hold in its isotropic form so that the relation $\tau \approx \varepsilon^{-1/3} L^{2/3}$ could not possibly apply to such large scales.

To obtain a modern estimate of ε_w we can use the ECMWF interim reanalysis at small scales using east–west, north–south and "isotropic" estimates with the help of the formula $\varepsilon \approx \Delta v^3/\Delta x$, using $\Delta x = 3°$ (i.e. in the scaling range, not too affected by the hyperviscosity). Fig. 8.6a shows the resulting ε_w averaged over the troposphere ($p > 200$ mb) as a function of latitude; since the theory assumes that the fluid density is constant, when averaging over different levels we have weighted the ε_w estimates from the different pressure levels by the corresponding air density. More detailed analyses (Lovejoy

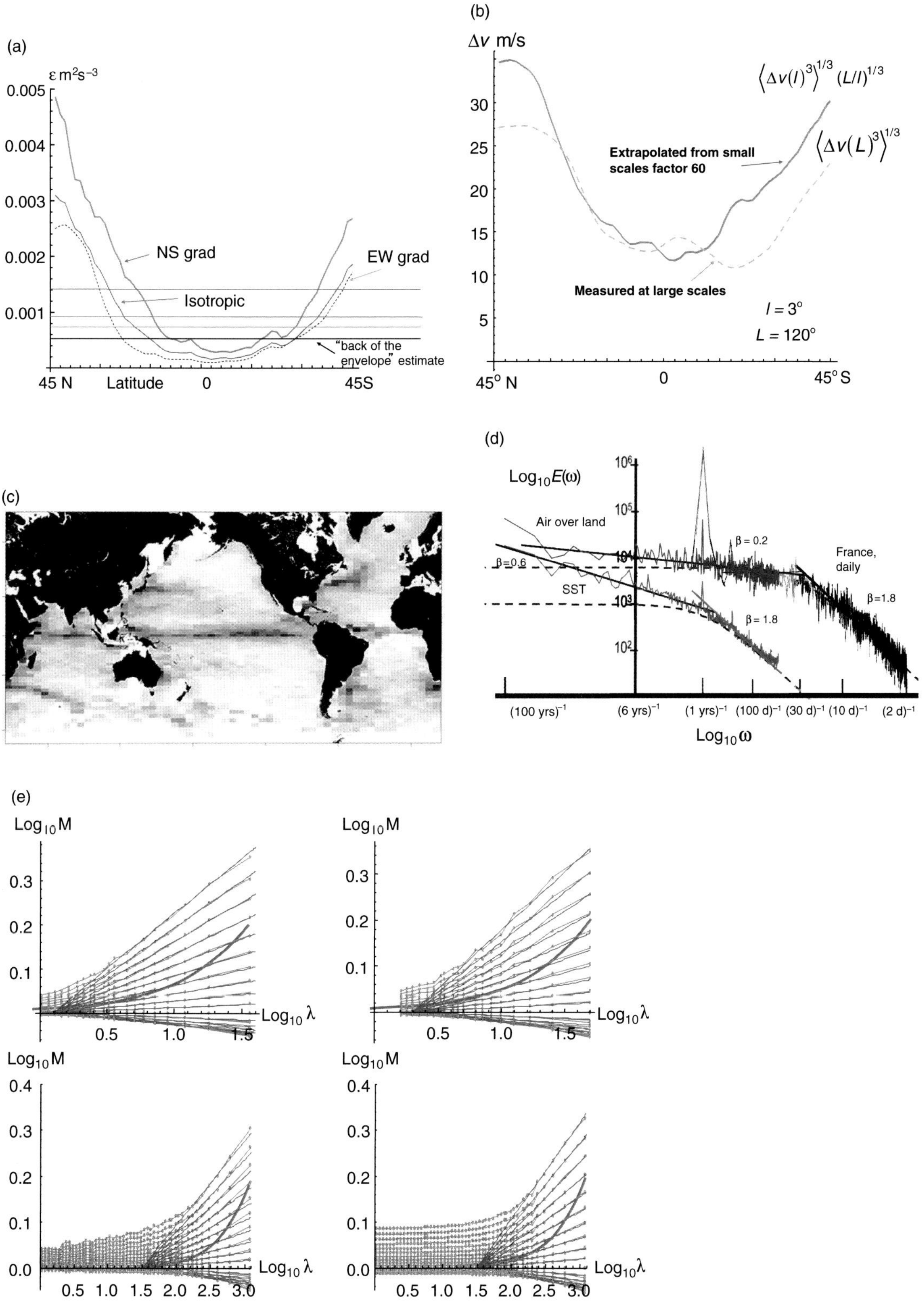
(a)
ε m²s⁻³
0.005
0.004
0.003
0.002
0.001
NS grad
Isotropic
EW grad
"back of the envelope" estimate
45 N
Latitude
0
45S
(b)
Δv m/s
30
25
20
15
10
5
⟨Δv(l)³⟩^1/3 (L/l)^1/3
⟨Δv(L)³⟩^1/3
Extrapolated from small scales factor 60
Measured at large scales
l = 3°
L = 120°
45° N
0
45° S
(c)
(d)
Log10E(ω)
10⁶
10⁵
10⁴
10³
10²
Air over land
β = 0.2
France, daily
β=0.6
SST
β=1.8
β = 1.8
(100 yrs)⁻¹
(6 yrs)⁻¹
(1 yrs)⁻¹
(100 d)⁻¹
(30 d)⁻¹
(10 d)⁻¹
(2 d)⁻¹
Log10ω
(e)
Log10M
0.4
0.3
0.2
0.1
0.0
Log10λ
0.5
1.0
1.5
2.0
2.5
3.0

and Schertzer, 2010) show that the mid-latitudes have mean ε_w up to 10 times those of the equator and that the highest values are mostly in the upper troposphere (near the 300 mb level) where it is about 5–6 times that of the lower atmosphere. Also indicated in Fig. 8.6a is the overall (isotropic, density-weighted) mean 9.25×10^{-4} m^2/s^3 which is within a factor of 2 of the first-principles "back-of-the envelope" estimate above (the comparable figures for the east–west and north–south gradients are 7.50×10^{-4} m^2/s^3 and 1.40×10^{-3} m^2/s^3).

If we compare these estimates with the observed mean "hemispheric antipode" velocity differences (i.e. the opposite side of the earth in the same hemisphere) we find that they follow the same pattern of variation with latitude as the variations of ε (it turns out that this is very close to the true antipode difference). Alternatively we can use the ε_w estimated at 3° (Fig. 8.6a) to infer the 180° difference using the Kolmogorov law. This extrapolated velocity difference is the cube root of the mean cube; it is shown in Fig. 8.6b. We see that the agreement between the predicted hemispheric antipodes difference and the observed difference is overall to within ± 19%. It is especially close between 30° N and 10° S, where the agreement is to within ± 7%. This is remarkable given the simplicity of the assumptions and the theory. We may also compare the overall means to the "first principles" estimate (21 m/s): the mean of the cubic estimate is 17.33 ± 5.7 m/s and the mean "extrapolated" large-scale velocity difference is 20.7 ± 7.4 m/s (the "±" represent the spread in the values for different latitudes). We conclude that the solar energy flux does a very good job of explaining the horizontal wind fluctuations up to planetary scales.

We can now use this estimate of ε_w to determine the lifetime ("eddy turnover time") $\tau_{eddy} = \varepsilon_w^{-1/3} L_e^{2/3} = L_e/V_w$ of the largest (planetary-scale) eddies. Using the isotropic ε_w estimate, we obtain $\tau_{eddy} = 8.7 \times 10^5$s = 10.0 days which is quite close to the "first principles" estimate (11 days) and to those of Radkevitch *et al.* (2008), who used analyses at a scale 6000 km to determine the statistical distribution of eddy turnover times, finding a mean of $\tau_w \approx 9.5 \times 10^5$ s with only a narrow dispersion. We see that τ_{eddy} is indeed a reasonable estimate of the weather/macroweather transition scale τ_w as displayed in the spectra of the previous section; to make this more convincing, we can now use these zonal estimates of ε_w to estimate $\tau_{eddy}(\theta)$ as a function of latitude θ and compare the result with the direct empirical estimates of $\tau_w(\theta)$ in Fig. 8.5c. Over the latitude band covered by the ECMWF interim reanalyses study (± 45°), we see that the agreement between $\tau_{w,T}(\theta)$ (for the 700 mb T field) and $\tau_{eddy}(\theta)$ is to within ± 15%, which is excellent given that the former estimate is for the 20CR 700 mb temperature field from 1871–2008 while the latter is from the zonal wind over all tropospheric pressure levels during 2006 as estimated from the ECMWF

Fig. 8.6 (a) Estimates of ε using gradients of the vector wind at resolution 3° (330 km at equator), all averages over the troposphere (i.e. $p > 200$ mb), as functions of latitude starting at 45° N, the contribution from the different pressure levels have been weighted by the air density. The straight lines are the corresponding latitude averages (isotropic: $\varepsilon = 0.00093$ m^2/s^3. EW: $\varepsilon = 0.00075$ m^2/s^3, NS: $\varepsilon = 0.0014$ m^2/s^3 and the lowest line is the back-of-the-envelope calculation discussed in the text yielding $\varepsilon_w \approx 0.00093$ m^2/s^3). Adapted from Lovejoy and Schertzer (2010). (b) The predicted hemispheric antipode large-scale velocity difference obtained from the isotropic estimate of ε (20.7 ± 7.37 m/s) and Kolmogorov's law (solid line), and the actual hemispheric difference (dashed line: the cube root of the mean cube hemispheric difference, 17.3 ± 5.7 m/s). Adapted from Lovejoy and Schertzer (2010). (c) The distribution of the fluctuation ocean currents at 15 m depth as estimated by drifters at 3° × 2° resolution, 5-day average (remapped here on a 6° × 2° grid), reproduced from Niiler (2001). The mean seems to be about 20 cm/s, although near equator it may be closer to 40cm/s. (d) Ocean and atmospheric plateaus superposed, showing their great similarity. Left: A comparison of the monthly sea surface temperature (SST) spectrum and monthly atmospheric temperatures over land for monthly temperature series from 1911 to 2010 on a 5° × 5° grid (the NOAA NCDC data: see Table 8.2 for details). Only near-complete series (missing less than 20 months out of 1200) were considered: 465 for the SST, 319 for the land series; the missing data were filled using interpolation. The reference slopes correspond to $\beta = 0.2$ (top), 0.6 (bottom left) and 1.8 (bottom right). A transition at 1 year corresponds to a mean ocean $\varepsilon_o \approx 1 \times 10^{-8}$ m^2s^{-3}. The dashed lines are Ornstein–Uhlenbeck processes (of the form $E(\omega) \propto (\omega^2 + \omega_0^2)^{-1}$: see Appendix 10B; ω_0 is a characteristic transition frequency) used as the basis for stochastic linear forcing models. Right: The average of five spectra from 6 year-long sections of a 30-year series from daily temperatures at a station in France (taken from Lovejoy and Schertzer, 1986). The reference line has a slope 1.8 (there is also a faint slope $\beta = 0$ reference line). The relative up/down placement of this daily spectrum with respect to the monthly spectra (corresponding to a constant factor) was determined by aligning the atmospheric spectral plateaus. Reproduced from Lovejoy and Schertzer (2012). (e) Comparison of the cascade analysis of the monthly SST (left column, HadCRUT data) and monthly land series (right column, CRUTEM3 data) for space (zonal), top row ($\lambda = 1$ corresponds to 20000 km), and time, bottom row ($\lambda = 1$ corresponds to 100 years) for the data discussed in Fig. 8.6d. The parameters and sampling details are given in Table 8.2. It is particularly noteworthy that although the land and ocean cascade structures are nearly identical, the corresponding spectra (Fig. 8.6d) are very different, indicating that the intermittency of the land temperatures is controlled by the ocean "weather" variability. The longest "pure" land scale accessible was ≈ 165° at mid-latitudes, hence the smallest accessible scale ratio was $\lambda \approx 10^{0.2}$ in the upper right graph. Since the cascade ranges are not so large, we have superposed (thick lines) the "universal" quasi-Gaussian curve for the envelope of the curves for moments of order $q \le 2$ (Appendix 4A). It may be seen that in all cases the empirical variability is much stronger than would be possible for quasi-Gaussian processes.

interim reanalysis. Note that the agreement is still excellent near the equator where τ_{Eady} diverges. The comparison between $\tau_{eddy}(\theta)$ and $\tau_{w,R}(\theta)$ is not as good, although the latter is nevertheless within a factor of 2 of the former (the mean ratio $\tau_{w,T}(\theta)/\tau_{w,R}(\theta)$ over the latitudes is quite constant: 2.08±0.51). Finally we can note that Fig. 8.8c shows that $\tau_{eddy}(\theta)$ follows very closely the latitudinal dependence of the cascade external scale $\tau_{eff}(\theta)$ for the ECMWF 700 mb temperatures while being about a factor 2 too small.

8.1.4 "Ocean weather," low-frequency ocean "macroweather" and the transition

Although the ocean and the atmosphere have several distinguishing dynamical features, they also have many similarities. In particular, both are large-Reynolds-number turbulent systems and both are highly stratified – albeit due to somewhat different mechanisms. Because of the overarching similarities, it is not surprising that there are debates about the nature of the horizontal ocean velocity (current) spectrum which are analogous to the atmospheric debates discussed at length in Chapter 2. Certainly, starting at the smaller scales with the classical ocean velocity spectra (e.g. Grant *et al.*, 1962; see also Nakajima and Hayakawa, 1982), there is no question that horizontal ocean currents are dominated by the energy flux ε_o and hence follow roughly $E(\omega) \approx \omega^{-5/3}$ (and presumably in the horizontal they follow $E(k) \approx k^{-5/3}$). However, as in the atmosphere, the low wavenumber limits of these Kolmogorov spectra are strongly debated, and the possibility of a quasi-geostrophic (QG) turbulent regime with k^{-3} spectrum has been proposed (see Smith and Vallis, 2001, for an application of QG theory to the ocean). Although surprisingly few current spectra have been published, of late the use of satellite altimeter data to estimate sea surface height (a pressure proxy) has provided relevant empirical evidence and has somewhat revived the debate. At scales where approximate geostrophic equilibrium may pertain, the pressure gradient is proportional to the current so that the surface height (h) and current (v) spectral exponents are related by $\beta_h = 2 + \beta_v$. However, according to Le Traon *et al.* (2008), at least over the scale range accurately covered by the altimeter ($\approx$ 10–300 km), $\beta_h \approx 11/3$, not the QG prediction $\beta_h \approx 5$, implying that $\beta_v \approx 5/3$. However, the debate is far from over, since the value $\beta_v \approx 5/3$ is compatible not only with wide-range horizontal/vertical ocean scaling with $2 < D_{el} < 3$ (i.e. with scaling vertical ocean stratification), but also with a variant of QG called surface quasi-geostrophic turbulence (Blumen, 1978; Held *et al.*, 1995) with $D_{el} = 2$. For our present purposes, and in the absence of a clear understanding of vertical ocean stratification, the main point is that the existing data are compatible with $k^{-5/3}$ horizontal current spectra out to planetary scales, hence with the relevance of the ocean energy flux ε_o.

Although empirically the current spectra (or their proxies) at scales larger than several hundred kilometres are apparently not well known, other spectra – especially those of sea surface temperatures (SST) – are better known and are relevant due to their strong nonlinear coupling with the current. There are many reports of wide-range spatial SST scaling. These include in-situ results such as those of McLeish (1970; Eulerian, $\beta_T \approx 5/3$), Seuront *et al.* (1996; Lagrangian, $\beta_T \approx 2$) and Lovejoy *et al.* (2000; towed instruments, $\beta_T \approx 1.63$, $H = 0.31$, $C_1 = 0.031$, $\alpha = 1.8$). Remote sensing using thermal IR images first from aircraft (Saunders, 1972), and then from satellites (Deschamps *et al.*, 1981, 1984; Park and Chung, 1999) yields respectively $\beta_T \approx 5/3$, ≈ 2.2, ≈ 1.9, ≈ 2, $\approx 1.87 \pm 0.25$ out to distances of $\sim$100 km. At larger scales (out to at least $\sim$500 km), Burgert and Hsieh (1989) found $\beta_T \approx 2.1$ from "cloud-free" satellite data. The satellite data – even if nominally cloud-free – are somewhat smoothed by atmospheric effects, hence their β_T values are probably a little bit too high. Therefore, the monthly averaged in-situ SST data (see Table 8.1 and Fig. 8.6d, discussed below), which yield $\beta_T \approx 1.8$ are evidence that this scaling really does continue up to scales of 5000 km or more, a conclusion that is bolstered by the corresponding spectral and cascade analyses (Figs. 8.6d, 8.6e). This conclusion is all the more plausible since the corresponding H value (≈ 0.50) is almost identical to that of the atmosphere (0.50, 0.41 in space and time; Table 8.1).

Just as for the wind, Kolmogorov spectra for the current may thus be presumed to hold (at least roughly) over wide ranges, indicating the presence of oceanic energy cascades in the horizontal. Just as in the atmosphere, where the stratification is scaling, so it may well also be in the ocean, leading to the likelihood that – as in the atmosphere – the energy flux will dominate the horizontal ocean dynamics. If this is correct, then we can use the same methodology as in the previous subsection – basic turbulence theory combined with the mean ocean energy flux ε_o – to predict that the outer scale τ_o of the ocean regime is equal to the largest ocean eddy lifetime $\tau_{o,eddy}$. Thus, for ocean gyres and eddies of size l, we expect there to be a characteristic eddy turnover time (lifetime) $\tau = \varepsilon_o^{-1/3} l^{2/3}$ with a critical "ocean-weather"/"ocean-macroweather" transition time scale $\tau_o = \tau_{o,eddy} = \varepsilon_o^{-1/3} L_o^{2/3}$ (L_o is the outer spatial scale of the oceans; presumably $L_o \approx L_w \approx L_e = 2 \times 10^7$m). Again, we expect a fundamental difference in the statistics for fluctuations of duration $\Delta t < \tau_o$ – the ocean equivalent of "weather" with a turbulent spectrum with roughly $\beta_o \approx 5/3$ (at least for the current) – and for durations $\tau > \tau_o$, the low-frequency ocean "macroweather" with a corresponding shallow ocean spectral plateau with $\beta \approx <1$.

As a first step in testing the idea, let us attempt to estimate ε_o, the globally averaged ocean current energy flux. As expected, ε_o is highly intermittent, and as far as we know no one has yet attempted to estimate its global average. Early work resulted in statements such as "$\varepsilon \approx 3 \times 10^{-7}$ m^{2}s^{-3} is typical of moderately intense turbulence" (Crawford, 1976). The difficulties with such single-site characterizations were underscored in studies such as Clayson and Kantha (1999), who used turbulence closure models combined with data from the Tropical Heat 1984 and Tropical Heat 1987 experiments to estimate ε_o at depths down to 150 m and at hourly resolutions. When averaged over columns of depth 10–110 m, their 7-day time series fluctuated in the range 3×10^{-7} to 3×10^{-8} m^{2}s^{-3} in 1984, but in 1987, in the same location and over the same duration, it fluctuated over the range 3×10^{-8} to 3×10^{-9} m^{2}s^{-3}, i.e. a difference of factor ≈ 10. Beyond this strong intermittency, this experiment underscores another important aspect of the ocean: the strong depth dependence of ε_o. The Clayson and Kantha (1999) profiles showed ε_o decreasing by a factor of 10–100 over the first ~100 m. Similar findings were made by Moum *et al.* (1995), who determined two ε_o profiles about 11 km apart near the equator and 140° W. Starting at a 15 m depth, and averaging over 5 m sections and over 3.5 days, they found that ε_o decreased from $\approx 5 \times 10^{-6}$ m^{2}s^{-3} to $\approx 10^{-9}$ m^{2}s^{-3} at 175 m and to $\approx 10^{-11}$ m^{2}s^{-3} at 1 km, and that the agreement between measurements in both locations was generally to within a factor of 2. The strong depth dependence of ε_o contrasts with the situation in the atmosphere, which has only relatively small vertical variations in ε_w (a factor of 5–6 from the surface to the maximum at 300 mb) (Lovejoy and Schertzer, 2010). This depth dependency means that the value of ε_o representative of the surface layer dynamics and structures, and the value relevant for deep ocean structures with larger vertical extents, are not trivially the same, a point to which return in Section 8.2.1.

For reference, we could mention that other explicit direct estimates of ε_o have largely been associated with developing new measurement techniques; for example Lien and D'Asaro (2006) compared two different measurement methods, sonar and float accelerometers in Puget sound, finding that ε_o varied over range from 10^{-3} to 10^{-8} m^{2}s^{-3} and was typically about 10^{-7} m^{2}s^{-3} (for 80 s resolution and with a lower measurement bound on $\varepsilon_o \approx 10^{-8}$ m^{2}s^{-3}). A final example includes Matsuno *et al.* (2006), who found ε_o was typically $\approx 10^{-8}$ m^{2}s^{-3} in the China Sea, but highly variable.

Although there is a clear tendency for near-surface experimental values to be in the vicinity of 10^{-8} m^{2}s^{-3}, more systematic estimates are needed for global averages. Fortunately, ocean drifters have been used to map turbulent "eddy kinetic energy" (EKE, the mean square fluctuation velocity) $\approx \Delta v^2$ over much of the globe over boxes 3° × 2° (Fig. 8.6c; Niiler, 2001). Although this is not quite an estimate of ε_o, from these fluctuation data, it is a simple matter to estimate ε_o using $\varepsilon_o = \Delta v^3/l$ and taking the box scale $l \approx 250$ km. From Fig. 8.6c we see that typical root mean square current fluctuations Δv at this scale are 10–20 cm/s, corresponding to $\varepsilon_o \approx 4 \times 10^{-9} - 3 \times 10^{-8}$ m^{2}s^{-3} (although some values exceed 50 cm/s, corresponding to 5×10^{-7} m^{2}s^{-3}). These values are quite consistent with the typical in-situ estimates cited above. More recently, satellite altimeters working in pairs have been used to estimate EKE at a resolution $l \approx 100$ km. Again using $\varepsilon_o = \Delta v^3/l$ and using the mean South China Sea basin estimate of $\Delta v^2 = 360$ cm^2/s^2 (Cheng and Qi, 2010), we obtain $\varepsilon_o \approx 7 \times 10^{-8}$ m^{2}s^{-3} (ranging between a minimum of 3×10^{-9} m^{2}s^{-3} up to a maximum of 4×10^{-7} m^{2}s^{-3}). The mean value is right in the middle of preceding drifter estimates. Using the formula $\tau_o = \tau_{o,eddy} = \varepsilon_o^{-1/3} L_o^{2/3}$ we find that a range of ε_o between 1×10^{-8} and 8×10^{-8} m^{2}s^{-3} corresponds closely to the range of $\tau_o \approx 1$–2 years; as expected, this is somewhat larger than the corresponding value for the atmosphere: in the previous section we found $\varepsilon_w \approx 10^{-3}$ m^{2}s^{-3}, $\tau_w \approx 10$ days.

A way to test the model is to filter out the high-frequency weather variability due to the weather cascade; this is at least partially done by temporal averaging over scales $> \tau_w$; the effect of such averaging is discussed in detail in Section 10.1. A one-month resolution allows us to use many convenient surface temperature datasets; the three we chose are the NOAA NCDC merged land air and sea surface temperature dataset (abbreviated NOAA NCDC, from 1880 on a $5° \times 5°$ grid), the NASA GISS dataset (from 1880 on a $2° \times 2°$ grid) and the HadCRUT3 dataset (from 1850 to 2010 on a $5° \times 5°$ grid). More details on these datasets including references and statistical characterizations of their similarities and differences, and a comparison with the 20CR reanalysis dataset, can be found in Appendix 10C. The NOAA NCDC and NASA GISS series are both heavily based on the Global Historical Climatology Network (Peterson and Vose, 1997), and have many similarities including the use of sophisticated statistical methods to smooth and reduce noise. In contrast, the HadCRUT3 data are less processed, with corresponding advantages and disadvantages. We therefore primarily used the qualitatively more distinct NOAA NCDC and HadCRUT3, although for limited global studies we also used the globally averaged NASA GISS set.

For all datasets and for virtually any given pixel, the series had many missing months, and these were often successive so that interpolation could easily lead to serious biases in the spectra and trace moments. (Indeed, the data were generally only available on sparse fractal sets in time and in space: see Section 3.2.2.) To minimize this problem, we restricted our separate land and SST analyses to the most recent 100 years and selected only those pixels with less than a total of 20 missing months (see Table 8.2 for details). The mean spectra are shown in Fig. 8.6d. While the land spectrum is – as expected – essentially a pure spectral plateau (with $\beta \approx 0.2$, the value cited earlier; the high-frequency air data in the figure are from daily data), we see that the SST spectrum is quite different, displaying a clear transition between two power laws at $\tau_o \approx 1$ year (with $\beta \approx 0.6$, 1.8 for scales $> \tau_o$ and $< \tau_o$ respectively). Note also the rough convergence of the spectra at about a 100-year scale, implying that the land and ocean variability become equal and the hint that there is a low-frequency rise in the land spectrum for periods $> \sim 30$ years. Since above we predicted $\tau_o \approx 1$ year from estimates of ε_o, we see that the break in the empirical spectrum is very close to that predicted, although compared to the land temperature spectral plateau (representing more closely the free atmosphere), the low-frequency ocean plateau β is a little larger, a point to which we return in Chapter 10.

Using two different SST datasets and the DFA analysis method, Monetti *et al.* (2003) obtained very similar results. Translating the DFA exponents to spectral exponents, they found $\beta \approx 1.76 \pm 0.08$ and $\beta \approx 0.06 \pm 0.16$ for high- and low-frequency spectral exponents respectively, with "crossover" at 10 months ($\approx \tau_o$), although they made no attempt to give physical interpretations to any of these values and did not present analyses for scales longer than ~ 35 years. Similar spectra with "crossovers" in the range 2–7 years (even accompanied by a broad peak) are also routinely found in SST and corresponding surface air temperatures averaged over the wide equatorial regions important in the El Niño phenomenon (see e.g. the spectra in AchutaRao and Sperber, 2006). The fact that one commonly has a transition from a high-frequency β near the value 2 to a low-frequency β near the value 0 means that the spectra are not too far from Ornstein–Uhlenbeck (OU) processes of the form $E(\omega) \propto \left(\omega^2 + \omega_0^2\right)^{-1}$, where ω_0 is a characteristic transition frequency. Such processes are essentially the results of integrated Gaussian white noise and they are used as the basis for stochastic linear forcing models discussed in Appendix 10B, and for SST (and other) forecasts over scales of months to about a year. In Fig. 8.6d, we see that the approximation is fairly rough, especially for the SST spectrum. If we consider the spectrum of the first principal compment of the Pacific SST (called the Pacific Decadal Oscillation, PDO, see Figs. 10.8, 10.14, for spectra and structure functions), then the comparison to OU processes is much less favourable. This is also true when compared to the Southern Oscillation Index (SOI), which is a proxy for the El Niño Southern Oscillation phenomena (ENSO; see Fig. 10.8). Also relevant are cascade analyses (Fig. 8.6e), which include for comparison the $q = 2$ envelope of the "universal" quasi-Gaussian trace moment analyses (Appendix 4A), which lies significantly below the data, supporting the cascade hypothesis.

The fact that both oceanic and atmospheric temperatures have high frequency β's with the turbulent value ≈ 1.8 supports this interpretation. It also allows us to directly estimate the ratio of atmospheric to oceanic energy fluxes, $\varepsilon_w/\varepsilon_o$. This can be estimated from the left/right separation of the parallel $\omega^{-1.8}$ lines, which correspond to a factor of ~ 30 in critical time scales, i.e. $\tau_o/\tau_w \approx 30$. If we assume that the spatial outer scale for the atmosphere and oceans is the same ($L_o = L_w \approx L_e$), then we can infer that $\varepsilon_w/\varepsilon_o = (\tau_o/\tau_w)^3 \approx 3 \times 10^4$. Using the

Table 8.2 A comparison of various cascade parameter estimates for land (NOAA NCDC and CRUTEM3, 5° × 5°), ocean (NOAA NCDC and HadSST2, 5° × 5°), whole-planet (NOAA NCDC and HadCRUT3, 5° × 5°), and the 20CR reanalysis (700 mb, averaged over a 8° × 10° grid). All the datasets are monthly averages; the whole-planet entries are also global averages. Due to the large numbers of missing data points (except for the 20CR reanalysis) the data were sampled as indicated in the columns on the right. The β (and derived *H*) estimates are for the whole range except for the ocean in time, where up to ~1 year we find β ≈ 1.8 (Fig. 8.6d), and up to ~5000 km in space. Uncertainties indicated by "±" are estimates based on comparing NOAA NCDC HadSST2 and CRUTEM3 estimates. See Fig. 8.6e for some cascade analyses. Note that all the surface series assumed that the sea surface temperature and the air temperature immediately above it are the same.

		C_1	α	β	H	Outer scale[a]	Data used (dates)	Latitudes	Longitudes
Space	land	0.14 ± 0.01	2.0 ± 0.1	1.75 ± 0.1	0.51 ± 0.03	11 400 km	1911–2010[g] 1722 transects[i]	27.5N to 57.5N	12.5W to 147.5E
	ocean	0.12 ± 0.01	1.9 ± 0.1	1.8 ± 0.1[b]	0.50[b]	16 000 km	1911–2010[h] 5614 transects	52.5E to 82.5W (230°)	12.5S to 12.5N
Time	land	0.12 ± 0.01	1.7 ± 0.1	0.25 ± 0.05	−0.27 ± 0.02	4.1 yr	1911–2010[f] 319 series	Most sampled	Most sampled
	ocean	0.12 ± 0.01	1.8 ± 0.2	1.8[b,e]	0.52[b]	2.1 yr	1911–2010[f] 465 series	Most sampled	Most sampled
	HadCRUT3[j]	0.15	1.9	0.3	−0.21	16 yr	1911–2010[f] 408 series	Most sampled	Most sampled
Whole, planet, (global average, time)[c]	20CR (8×10°)	0.12	1.7	0.64	−0.13	0.8 yr	1880–2008, all	All	All
	NCDC, NASA, HadCRUT3[d]	0.11	1.8	0.64	−0.13	1.6 yr	1880–2008, all	All	All

[a] The outer scales are geometric means of the NOAA NCDC and HadCRUT3 cascade estimates.
[b] These estimates are from the NOAA NCDC data; the spectra (hence β, *H*) of the ocean data were too noisy and were not used.
[c] These are reproduced from Table 10.C.2, which has further global-scale estimates. The C_1, α parameters are based on regressions < 1 year in scale, the β, *H* estimates are from the range of scales of months to ~(10 year)$^{-1}$, after which there is a lower-frequency rise.
[d] These are the statistics obtained from treating the surface in-situ data (NASA GISS, NOAA NCDC, HadCRUT3) as members of an ensemble; these are the resulting ensemble statistics.
[e] This estimate (see Fig. 8.6d) is for the high frequencies (> (1 year)$^{-1}$); at lower frequencies, β ≈ 0.6.
[f] These were all the near-complete series, i.e. those missing fewer than 20 values out of 1200. These missing values were filled by interpolation.
[g] These were the 33-pixel-long transects over land in the geographical area indicated.
[h] These were the complete 46-pixel-long transects over ocean in the geographical area indicated.
[i] The space and time figures here are for the NOAA NCDC data; the corresponding other numbers are: CRUTEM3 transects: 1062 (near complete); HadSST2 transects: 1381 complete; CRUTEM3 series: 161 near complete; HadSST2: 232 series near complete.
[j] These were estimated on the 408 grid points at 5° × 5° spatial resolution with near-complete series. In Table 10.C.2, there is a different set of HadCRUT3 parameters based on the series obtained by first globally averaging the available data.

atmospheric value $\varepsilon_w \approx 10^{-3}$ m^{2}s^{-3}, we find $\varepsilon_o \approx 5 \times 10^{-8}$ m^{2}s^{-3} which is close to the estimates discussed above. If this interpretation is correct, then, since ocean eddies and gyres obey roughly the same turbulent phenomenology as the atmosphere, the time scales $< \tau_o$ indeed correspond to ocean "weather," with the implication that the limits to forecasting the ocean are $\approx \tau_o \approx 1$ year (see Section 9.3).

Although we deliberately put off an in-depth discussion of the climate scales to Chapter 10, a final piece of evidence supporting our picture is worth giving here: the trace moment–cascade analysis. Using the same SST and land datasets used for the spectral analysis above, we can determine the cascade structures (Fig. 8.6e). This not only shows that the spatial temperature scaling continues to near planetary scales (see Table 8.2 for parameter estimates), but also that the temporal cascades are almost identical for land and sea temperatures, with accurate scaling up to about 1 year and with outer cascade scales of about 3.5 years, i.e. fully consistent with our estimate of τ_o above. To put this in perspective, recall that the spectra (Fig. 8.6d) showed qualitatively and quantitatively different SST and air-over-land spectra. Since in

Chapter 10 we shall see that the weather variability for $\tau > \tau_w$ has very low intermittency, the strong intermittency of these monthly averages is presumably almost entirely due to ocean turbulence. This scaling intermittency thus leads – even in the land statistics – to small statistical deviations from perfect power-law scaling for periods less than a year. This was recently noted by Lanfredi *et al.* (2009), who treated the deviations as a (scale-bound) Markov process rather than a scaling one.

8.1.5 The temporal cascade structure

The extrapolation of the small-scale velocity differences to planetary scales was based on the assumption that the energy flux was indeed independent of scale. We can substantiate this by considering the fluxes and their scale-by-scale cascade structures (their statistical moments; see also the analyses in Lovejoy and Schertzer, 2010). Fig. 8.7a shows the first of these: the analysis of the ECMWF fluxes, the analogues of the spatial analyses presented in Fig. 4.1 but in the temporal domain (resolution 1 day up to 1 year). Fig. 8.7b shows the temporal analysis of the 20CR reanalysis corresponding to the spatial analyses in Fig. 4.2c, in this case spanning 6 hours to 138 years. Figs. 8.7c and 8.7d show some results for the temporal analysis of GEM and GFS forecast models spanning 6 hours to 200 days, and 6 hours to 1 year, respectively (see Section 4.2.3 for the corresponding spatial analyses). Fig. 8.7e shows the temporal analyses of the various precipitation products analysed spatially in Figs 4.8a, b (for the TRMM satellite radar at 4 days- 1 year, the ECMWF stratiform rain product, 3 hours to 3 months, and the CPC network, 1 hour to 29 years). Fig. 8.7f shows the temporal cascade structure for the MTSAT at resolutions 1 hour to 2 months and analyzed spatially in Fig. 4.10, and Fig. 8.7g compares the spatial and temporal exponents $K(q)$, showing that they are indeed very close. The fluxes for the ECMWF interim and 20CR reanalysis were taken from fluxes estimated from absolute spatial Laplacians, the others were from second time differences (a detailed discussion and comparison is in Stolle *et al.*, 2012). In all cases, we see the same basic features: a cascade structure which is reasonably well respected up to scales of 5–10 days, with outer cascade scales typically in the range 20–60 days, followed by a flattening at longer time scales. Note that the outer cascade scale τ_{eff} is a bit larger than τ_w, which roughly corresponds to the time scale at which the scaling breaks down, i.e. where the empirical curves diverge from the regression lines.

These cascades use (roughly) daily data that avoid the problems of strong diurnal cycles mentioned in Section 8.1.2. The exception is the hourly MTSAT images, but this is of IR cloud temperatures, which are only moderately affected by the diurnal cycle (in the corresponding spectrum shown in Fig. 8.12a there is only a small diurnal peak). In order to extend the temporal cascade analysis to smaller time scales, we must grapple with the periodic detrending problem indicated in Fig. 8.3c. In fact, to properly demonstrate the cascade structure of these hourly in-situ temperatures, we need to first periodically detrend them, and then estimate the corresponding turbulent flux. The problem is that hourly averaged station measurements are essentially points in space, so there is a large mismatch between the (tiny) spatial and hourly temporal resolutions. In order to estimate fluxes which have a closer match between space and time scales, the simplest method is to use absolute spatial temperature differences between stations. The four stations analyzed in Fig. 8.3c were roughly collinear arranged in the order Lander Wy, Harrison Ne, Whitman Ne, Lincoln Ne. Fig. 8.7h shows the trace moments calculated for the Harrison–Whitman difference ($\approx$ 170 km, left) and the mean of the Lander–Harrison and Whitman–Lincoln differences ($\approx$ 400, 450 km respectively, right). We see that the cascade structure is well respected from scales of 5, 7 days respectively down to inner scales of ~3, 8 hours respectively. These inner scales imply velocities of respectively 170/3 $\approx$ 57 km/h and 425/8 $\approx$ 53 km/h (both $\approx$ 1300 km/day or $\approx$ 15 m/s), which is comparable to other space-time transformation velocities found below: it is comparable to typical wind speeds. At scales shorter than these in time, the spatial resolution is simply too low; hence the small-scale (large λ) deviations from scaling in Fig. 8.7h. These results show that in order to justify taking data at hourly resolutions, the ground network should have stations roughly every 55 km.

In Table 8.1 we have already considered the basic exponents from the ECMWF interim analysis and compared them with estimates from various in-situ data sources. Tables 8.3a, 8.3b, 8.3c and 8.3d compare the exponents and outer scales of the u, T, h fields at 700 mb, with all parameters estimated from the Laplacian fluxes. The main systematic variations are: (a) the α for h is systematically lower than for u, T; (b)

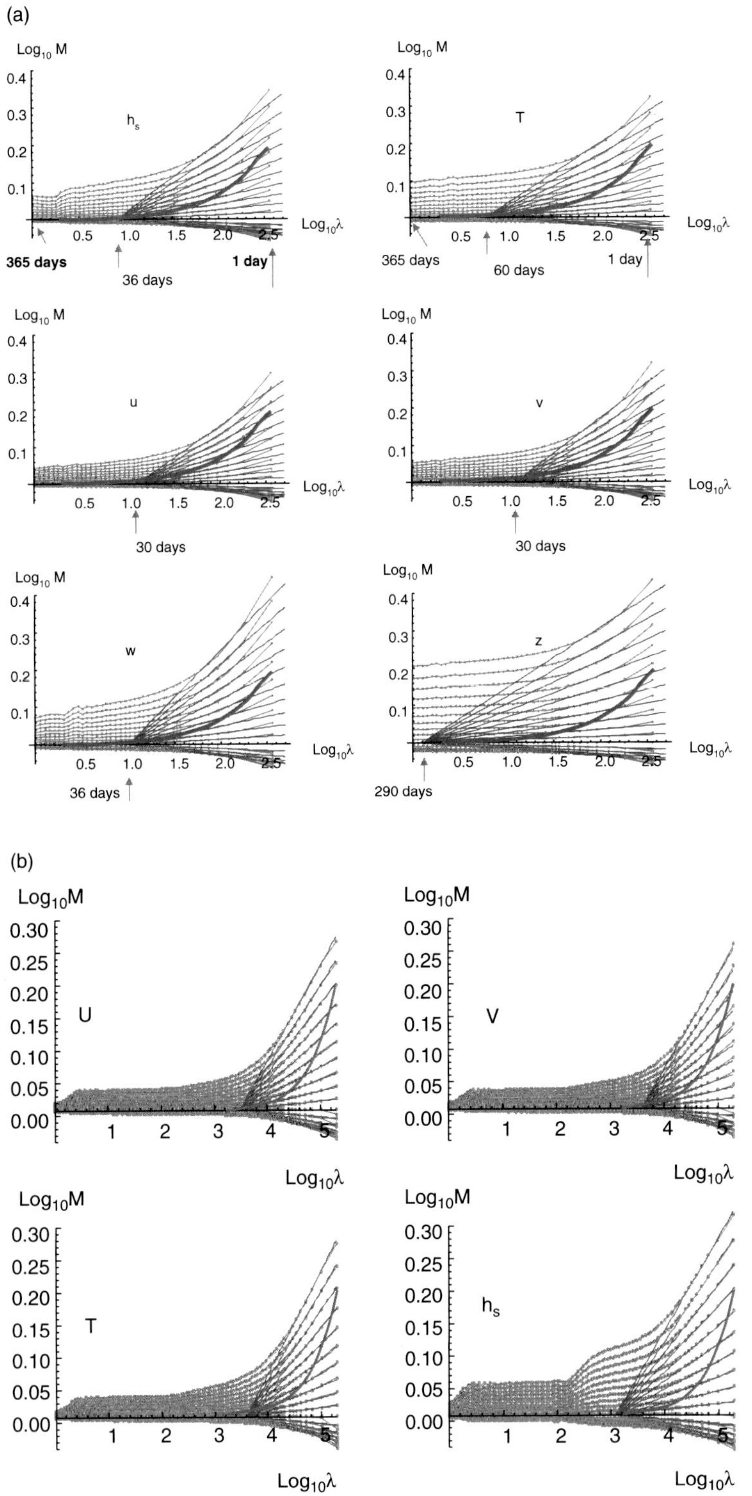

Fig. 8.7 (a) The trace moments of the ECMWF interim reanalyses from daily data for 2006: the same as Figs. 4.1a and 4.1b but for the temporal analyses. $\lambda = 1$ corresponds to 1 year. The effective outer temporal cascade scales (τ_{eff}) are indicated with arrows. Also shown (superposed in a thick line) are the $q = 2$ envelopes for the universal quasi-Gaussian processes; even the u, v fields are significantly more variable. Adapted from Lovejoy and Schertzer (2011).
(b) The trace moments of the spatial Laplacians of Twentieth Century Reanalysis (20CR) products for the band 44–46° N for the zonal wind (upper left), meridional wind (upper right), the temperature (lower left) and specific humidity (lower right) from series at 6-hour resolution. This is the temporal analysis corresponding to the zonal analysis in Fig. 4.2c; the largest scale, $\lambda = 1$, corresponds to 138 years; the parameters of the fits are given in Tables 8.3a, 8.3b, 8.3c, 8.3d. Notice the "bulge" in the h_s moments up to scales of ~1 year, possibly a reflection of the ocean cascade. Also shown (superposed in thick lines) are the $q = 2$ envelopes for the universal quasi-Gaussian processes; all the fields are significantly more variable.

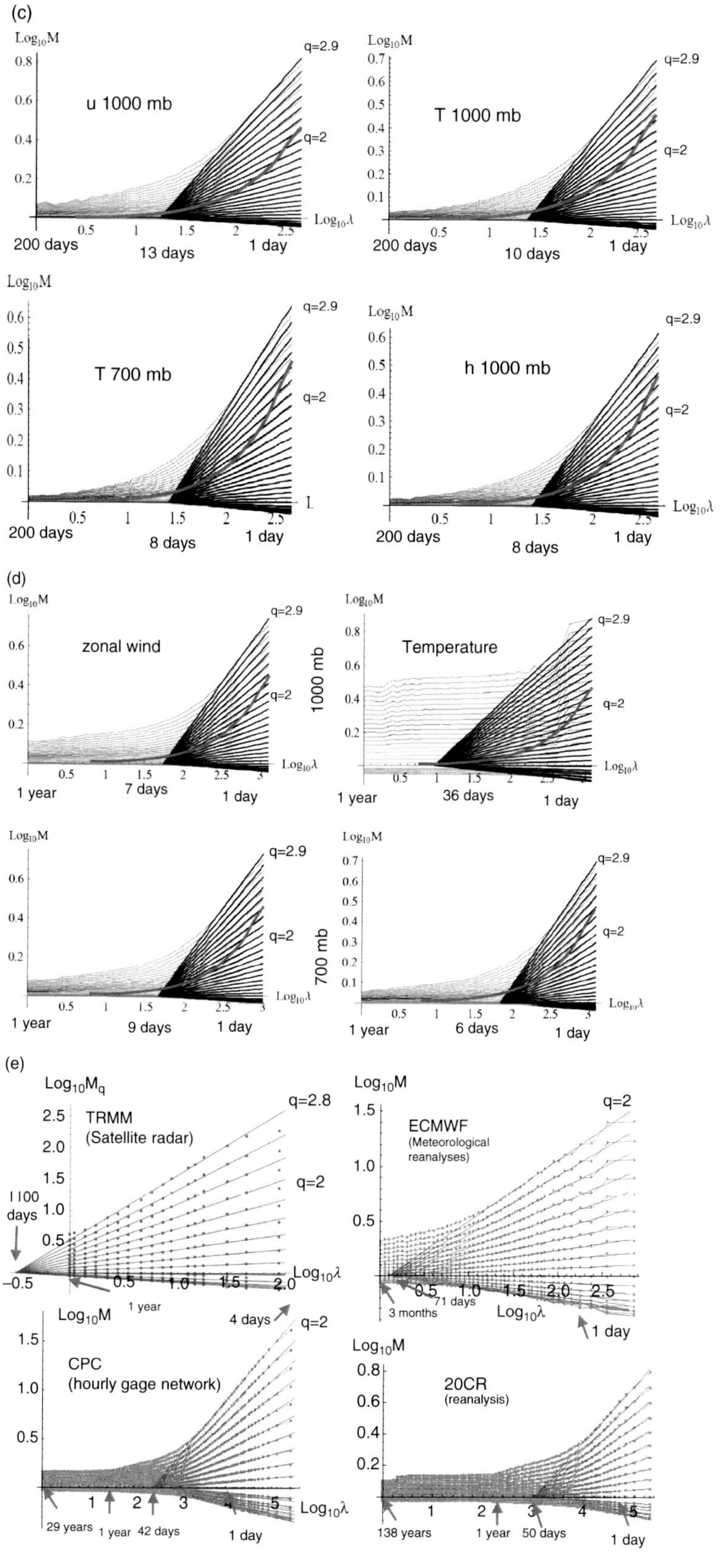

Fig. 8.7 (c) The temporal cascades estimated from the GEM data, every 6 hours, temporal (second derivative) flux estimates. Also shown (superposed in solid lines) are the $q = 2.9$ envelopes for the universal quasi-Gaussian processes; all the fields are significantly more variable. Reproduced from Stolle *et al.* (2012). (d) Analysis of the $t = 0$ GFS meteorological model output: zonal wind at 1000 mb, (top left), and 700 mb (bottom left), and the corresponding plots for the temperature 1000 mb (top right) and 700 mb (bottom right). $\lambda = 1$ corresponds to 1 year. Also shown (superposed in solid lines) are the $q = 2.9$ envelopes for the universal quasi-Gaussian processes; all the fields are significantly more variable. Reproduced from Stolle *et al.* (2012). (e) Temporal analyses of precipitation products. The second time flux for the 100 × 100 km gridded (4-day resolution) TRMM radar satellite rain rate estimates (upper left), for the 3 months of the 3-hourly ECMWF interim stratiform rain product (upper right) and 29 years of NOAA's CPC hourly gridded surface raingauge network (lower left). We have included (lower right) the unique very long 20CR product analyzed at 45° N at 2° resolution in space and 6 hours in time from 1871 to 2008; in it, there is a hint of a second lower intermittency cascade from about 10 days to 1 year. The regressions were performed over the range of scales 8 days to 1 year (TRRM), 6 hours to 10 days (ECMWF), 1 hour to 10 days (CPC) and 6 hours to 4 days (20CR). These are the temporal analyses corresponding to the spatial analyses presented in Figs. 4.8a and 4.8b. Reproduced from Lovejoy *et al.* (2012).

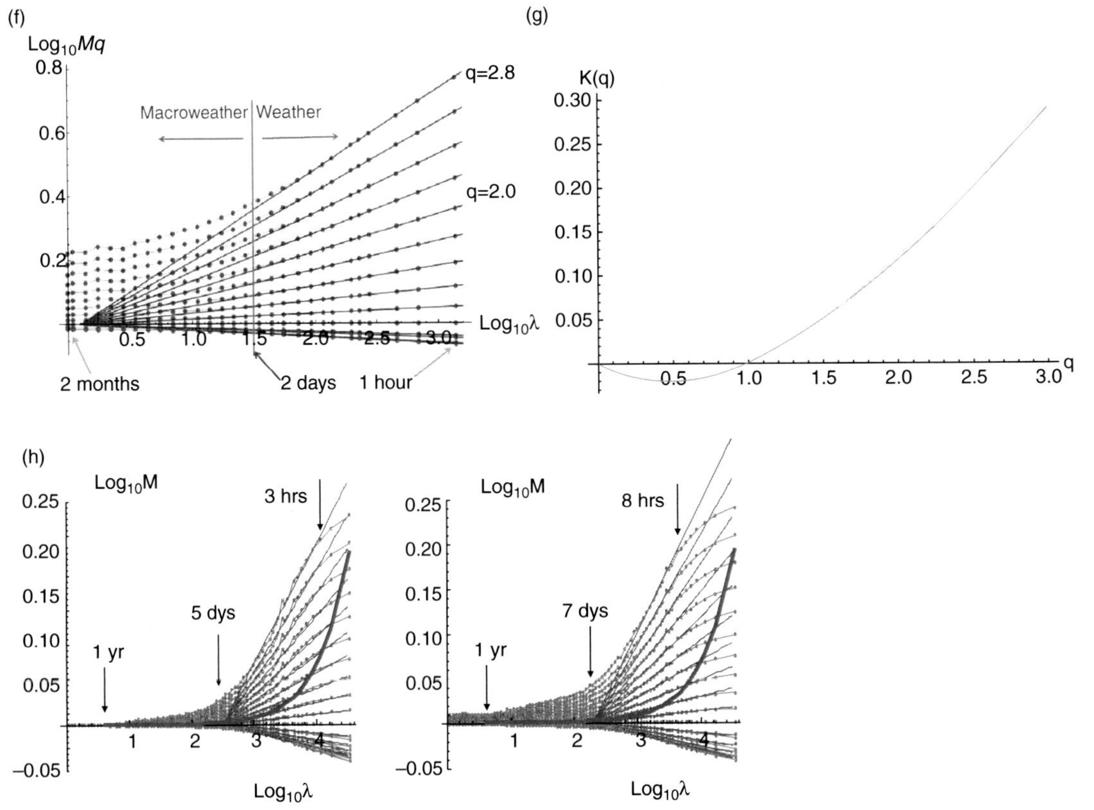

Fig. 8.7 (f) Analysis of MTSAT hourly resolution thermal IR imagery over the Pacific. The temporal analysis of the spatial Laplacian (at 30 km resolution) geostationary MTSAT thermal IR imagery over the Pacific for 2 months. This is the temporal counterpart of Fig. 4.10. The external scale is 48 days. Reproduced from Pinel *et al.* (2012). (g) MTSAT estimates of $K(q)$, showing the near-perfect superposition of horizontal space and time (hence isotropy). Lower line at right is time ($C_1 = 0.073$), upper line is NS/EW ($C_1 = 0.074$). Reproduced from Pinel *et al.* (2012). (h) The fluxes using the diurnally detrended hourly surface temperature series whose spectra are shown in Fig. 8.3c. On the left is the analysis of the absolute Whitman–Harrison temperature difference ($\approx$ 170 km), and on the right is the mean of the two fluxes defined as the Lander–Harrison and Whitman–Lincoln absolute differences ($\approx$ 400, 450 km respectively). The cascade outer scales are indicated as well as the scales where the scaling becomes poor due to the somewhat excessive smoothing introduced by the overly large spatial scales. The regression lines correspond to $C_1 = 0.069$, $\alpha = 1.95$, $C_1 = 0.072$, $\alpha = 2.00$ (left, right). Also shown (superposed in thick lines) are the $q = 2$ envelopes for the universal quasi-Gaussian processes; both fields are significantly more variable.

the basic intermittency parameter (C_1) is about the same (and low) for the ECMWF interim, 20CR and GEM products, and the about the same (and high) for the GFS and ERA40 products; (c) using the same groupings we find that the low C_1 models display high τ_{eff}, and conversely for the high C_1 models; (d) there is a variation in V_{eff} ($= L_{eff}/\tau_{eff}$) corresponding to the variation in τ_{eff}, with ERA40 and GFS about double the 20CR and GEM and about four times the ECMWF interim values. As expected, from one field to another, a given model has fairly constant V_{eff} (the main exception being h for GFS, which is anomalously low). As noted earlier, C_1 determines the rate at which the intermittency builds up from one scale to another, whereas τ_{eff} determines the starting scale of the build up. Thus, a low τ_{eff} partially offsets a high C_1 and vice versa, so that at a given time scale all the model results are actually quite close.

The temporal parameters for the precipitation fields were already shown in Table 4.6. In spite of the fact that the temporal C_1 estimates are quite close to each other (0.33 ± 0.03), the external scales vary considerably, with the TRMM external scale ($\tau_{eff} \approx 3$ years) being about 20 times larger than that of the

Table 8.3a Estimates of the temporal C_1 parameters from (spatial) Laplacian estimates of the fluxes all at 700 mb and between ± 45° (with the exception of the 20CR reanalysis, which was estimated over 44–46° N only). The forecast models GEM, GFS are for the $t = 0$ field. The ERA40, GEM, GFS results are from Stolle *et al.* (2012); the ECMWF interim results are from Lovejoy and Schertzer (2011). Compare these to the corresponding spatial analyses in Table 4.2a.

	ECMWF interim	ERA40	20CR	GEM	GFS
u	0.083	0.14	0.083	0.084	0.11
T	0.075	0.12	0.090	0.084	0.13
h	0.10	0.12	0.083	0.14	0.12

Table 8.3b The corresponding estimates of α.

	ECMWF interim	ERA40	20CR	GEM	GFS
u	1.85	1.7	1.81	1.8	1.8
T	1.90	1.9	1.82	2.0	1.8
h	1.77	1.7	1.74	1.6	1.7

Table 8.3c The corresponding estimates of the outer time scale T_{eff} (in days).

	ECMWF interim	ERA40	20CR	GEM	GFS
u	30	7	16	13	5
T	60	9	13	28	6
h	36	9	40	12	30

Table 8.3d The corresponding estimates of the effective velocity $V_{eff} = L_{eff}/\tau_{eff}$, with V_{eff} given by Table 4.2a and τ_{eff} from Table 8.3c (in km/day).

	ECMWF interim	ERA40	20CR	GEM	GFS
u	420	1700	700	850	1800
T	330	1600	850	300	1400
h	350	1200	870	1000	300

CPC and ECMWF reanalysis products; indeed, it is much larger than any other measured atmospheric external time scale. The obvious explanation is that the TRMM external scale is determined by the ocean variability, and hence $\tau_{eff} \approx \tau_o$ rather than τ_w. This could be a reflection of the fact that the TRMM data are between ± 40° latitude, i.e. predominantly over ocean (Fig. 8.8a) whereas the CRC data are only over land, explaining the fact that it has $\tau_{eff} \approx \tau_w$. This is consistent with $\tau_{eff} \approx 70$ days for the ECMWF interim product, which is more of a land–ocean average with an external scale between the two.

Although there are still anomalies, it seems that cascade exponents in horizontal space and in time are compatible with the hypothesis that they have the same values. Since the α estimates are not too precise (due to the zero rain-rate problem, which tends to seriously bias them towards low α values), we can attempt to substantiate this by comparing spatial and temporal C_1 estimates (see Table 4.6). Some of these are shown in Tables 8.4a and 8.4b. We first consider the data. The most reliable are those measurements in both space and time: in Table 8.4a we therefore show some MTSAT, TRMM satellite

Table 8.4a Comparison of the C_1 estimates of various satellite thermal IR (MTSAT IR, TRMM VIRS5), passive microwave (TMI8), satellite radar rain rates (TRMM, *R*), in-situ rain rates (CPC *R*) and ECMWF interim stratiform rain.

	MTSAT (IR)	TRMM VIRS5 (IR)	TRMM TMI8	TRMM *R*	CPC *R*	ECMWF *R*
time	0.073	0.05	0.06	0.30	0.37	0.34
zonal	0.074	0.05	0.05	0.27	0.49	0.41
meridional	0.074	0.05	0.05	0.32	0.51	0.45

Table 8.4b Comparison of the C_1 estimates of various models, reanalyses and in-situ measurements. The Twentieth-Century Reanalysis (20CR) was at 45° N; all others (except the in-situ estimates) were ± 45° latitude; all were spatial Laplacian flux estimates (except in-situ). The in-situ "zonal data" are in fact from aircraft and are not zonal but simply horizontal. They were multiplied by 2.07 to attempt to correct for the fact that they were scaling, not dissipation range estimates of fluxes (Eqn. 4.15; for details of the in-situ temporal estimates, see the notes to Table 8.1).

		ECMWF interim	ERA40	20CR	GEM	GFS	In-situ
u	**time**	0.083	0.14	0.083	0.084	0.11	0.053
	zonal	0.081	0.096	0.089	0.104	0.082	0.088
T	**time**	0.075	0.12	0.090	0.084	0.13	0.087
	zonal	0.074	0.094	0.088	0.077	0.080	0.107
h	**time**	0.10	0.12	0.083	0.14	0.12	0.09
	zonal	0.095	0.094	0.077	0.100	0.091	0.083

radiance estimates, although for the latter the time resolution was poor (about 2 days for VIRS5, TMI8, 4 days for the radar *R* estimates). These are probably the most reliable simultaneous space and time empirical estimates, and they give C_1 estimates which are sufficiently close that their C_1 values are probably in fact the same. The CPC and ECMWF *R* estimates are less reliable and give less similar estimates. For the state variables, we can consider the models and the reanalyses (Table 8.4b). We see that while the ECMWF interim has virtually identical space and time C_1 estimates (this was already noted in Chapter 4, and this is also true of the 20CR reanalysis) the space and time C_1's for ERA40, GEM and GFS are a little different. These small differences are more likely to reflect limitations of the models rather than true space-time differences.

8.1.6 The latitudinal dependence of the cascade structure from ECMWF interim reanalyses

Up until now, we have taken statistics from ± 45° latitude in order to concentrate on the basic variation with direction (zonal, meridional, temporal). However, a basic aspect of atmospheric dynamics is its latitudinal dependence, due not only to the Coriolis force and strong north–south temperature gradients, but also to varying fractions of ocean and land (Fig. 8.8a). Paradoxically, the fairly limited analysis of latitudinal dependence in Stolle *et al.* (2009) found that latitudinal variations were small; this is presumably because the cascade structure is mostly dependent on the nonlinear interactions whereas the most important north–south effects involve linear terms and boundary conditions. Similarly, we have seen that there are differences in the spectra over land and ocean (Fig. 8.6d) but that the cascade parameters (α, C_1) are essentially unaffected (Fig. 8.6e, Table 8.2). Hence the latitudinal variations in land/ocean distribution will mostly affect H (hence β) rather than α, C_1. For reference, Fig. 8.8a shows how the fraction of land and ocean varies with latitude at a resolution of 5°. Let us now investigate this more systematically here.

In order to study the latitudinal dependence, we broke up the earth into 15° bands and calculated the cascade structure (trace moments) and estimated the corresponding parameters. Fig. 8.8b shows the evolution of the exponents C_1, α, and Fig. 8.8c the external spatial and temporal scales. The main difference visible is a small but systematic change in the outer scales; there is also generally a good degree of north–south symmetry.

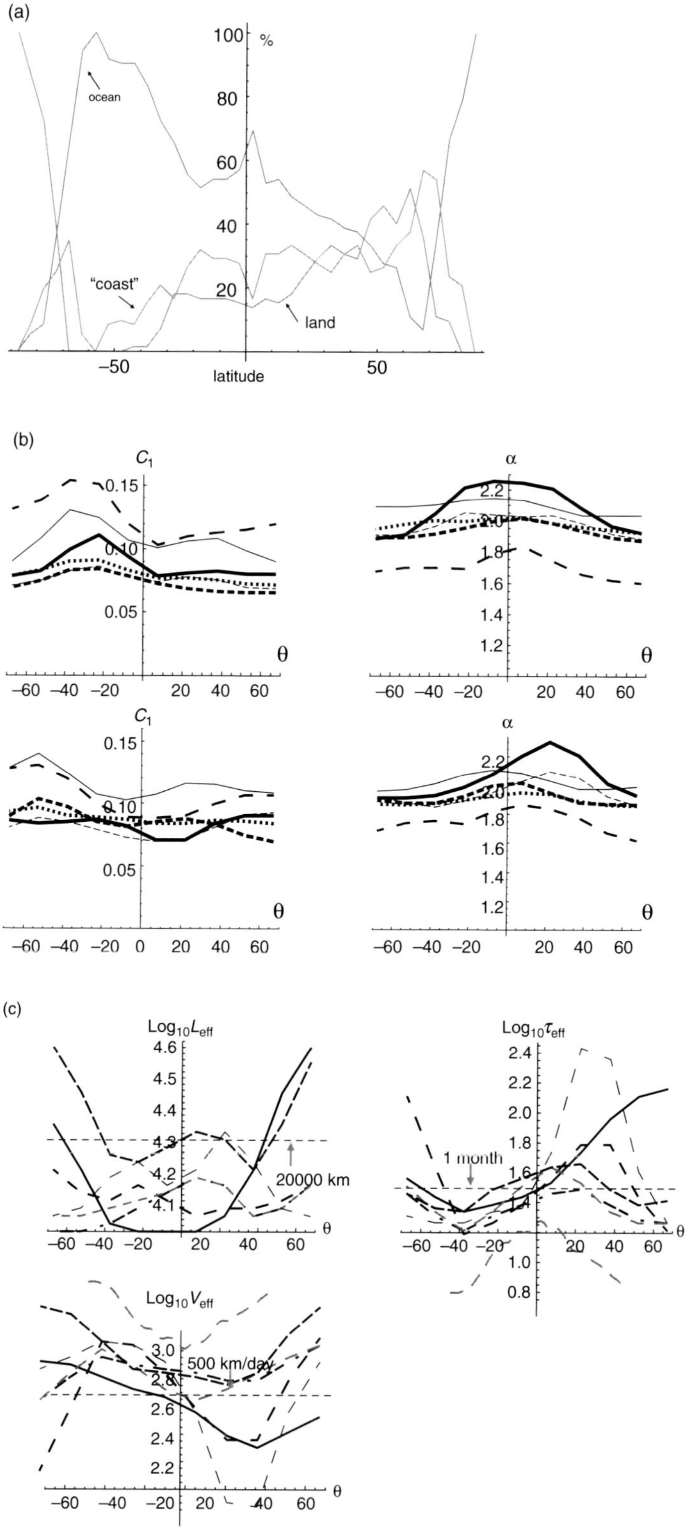

Fig. 8.8 (a) The percentage of ocean, land and "coastal" pixels as functions of latitude at 5° × 5° resolution. The land and ocean were defined as pixels with > 90% land and ocean respectively; the remainder were classified as "coastal." This clearly shows the preponderance of ocean in the southern hemisphere, with land only dominating around 45° N. Since the land is a (fractal) topographic exceedance set, the above will clearly depend on the resolution; the 5° resolution values are given for reference purposes. (b) The cascade exponents C_1, α: the top row is from the spatial (zonal) analysis, the bottom row is from the temporal analysis. From bottom to top in the upper left graph we have the zonal wind (u), the meridional wind (v), the temperature (T), the geopotential height (z), the vertical velocity (w), the specific humidity (h_s) (same lines used throughout the four graphs). The extreme latitude bands (± 75–90°) were not used since the mean map factor is very large and the results were considered unreliable. Reproduced from Lovejoy and Schertzer (2011). (c) The external scales as functions of latitude: upper left is the outer space scale (in units of km) from the zonal cascade analyses; upper right is from the time analysis (in units of days); lower left is their ratio, the effective speed of space-time transformations (in units of km/day). The thin dashed horizontal lines are convenient reference lines; note that 500 km/day = 5.8 m/s. The thick dashed dark and light grey lines are zonal and meridional components of the wind; in the τ_{eff}, V_{eff} plots we also show a short thick dashed line (between ± 45° only) representing the theoretical predictions based on the latitude dependence of the tropospheric averaged ε estimates from Fig. 8.6a (using the EW gradient estimates appropriate to these zonal analyses). As can be seen, the latitudinal variation is nearly exactly reproduced but the values are shifted by a factor ≈ $10^{0.2}$ ≈ 1.6. Reproduced from Lovejoy and Schertzer (2011).

First, for a given field we can compare the spatial and temporal exponents; we find the differences are generally less than 0.02 for C_1 and less than 0.1 for α, which is probably less than the statistically significant level, especially since the sample size of each 15° band is 1/6 of the previously analysed (90°) band, and the range of scaling in the time domain was quite limited (2–8 days; the 1-day value was not used since, as can be see from the plot, it suffers from significant "finite size effects"). In addition, we see that for some of the fields (essentially the geopotential and specific humidity near the equator), the α values are a bit larger than the theoretical maximum ($= 2$) so that the curvature (α) estimates (which were made here using $\alpha = K''(1)/K'(1)$) are not too accurate; their deviations from $\alpha = 2$ are probably not statistically significant. We notice a slight tendency for the intermittency to increase away from the equator, especially in the southern hemisphere; it is sufficiently systematic that it is probably a real change in C_1, probably associated with the greater influence of the ocean, itself intermittent.

Even though the exponents show remarkably little latitudinal variation, that does not imply that the cascade structure is completely independent of latitude. Fig. 8.8c shows the variation of the external space and time scales as well as their ratios: the effective speed V_{eff} needed for space-time transformations. The main noteworthy features are: (a) the latitudinal variations in external scales are relatively small (except perhaps the geopotential height) – in space they are almost all between 10 000 and 20 000 km, in time between 20 and 40 days, in speed between 400 and 1000 km/day; (b) the external scales have significant north–south asymmetry, especially the time scales, with the northern hemisphere having significantly larger τ_{eff}; but this is somewhat mirrored in larger L_{eff} values so that their ratio (V_{eff}) drops in the northern hemisphere. The reason for the asymmetry is presumably due to the differences in data density and in land cover between the two hemispheres (Fig. 8.8a). The figure also shows that the latitudinal variations are well explained by the corresponding ε estimates from Fig. 8.6a.

Box 8.1 World record wind singularities

Until recently, in the gift shop at the summit of Mt. Washington (White Mountains, New Hampshire, USA), one could find on prominent display a brochure describing the legendary summit measurement of the "world record wind" – an average wind of 231 miles per hour. This was recorded on two occasions over a distance of 0.3 miles and translates into = 103 m/s over 4.7 s (Pagliuca, 1934). We are also told that this short gust was embedded in a particularly windy hour whose mean was 69.3 m/s. During this hour, an extreme 5-minute average of 84.0 m/s was obtained as well as extreme 1-minute averages of 85.8 m/s and a 17-second average of 93.9 m/s.

It seems that the 103 m/s record stood until January 22, 2010, when a World Meteorological Organization panel of experts officially announced that on April 10, 1996, on Barrow Island, Australia, Tropical Storm Olivia had "gusts" that surpassed the Mt. Washington record, reaching 113.2 m/s (see www.wmo.int/pages/mediacentre/infonotes/info_58_en.html; the record had been overlooked for over a decade: for the story, see blog.ametsoc.org/uncategorized/mt-washingtons-world-record-wind-toppled). The new record gust occurred within a 5-minute interval whose average wind speed was 48.8 m/s. The high value of this longer average supported the credibility of the higher gusts, and underlines the problem of attempting to define the record speed without reference to resolution. Obviously, the shorter the interval, the easier it is to obtain a large value; from the point of view of multifractals, the true measure of intensity is not the mean at a subjective resolution, but rather the value of the corresponding singularity. It turns out that the term "gust" officially refers to a 3-second average value. So, from a multifractal perspective, what is the true world record?

Aside from its curiosity value, the world record wind gives us an opportunity to test our theory and the parameter estimates for v. Certainly the story of valiant Mt. Washington observers toiling under extreme conditions is far more appealing than that of an anonymous automated recording station whose extreme measurements took over 10 years to be even noticed (on a recent visit to the gift shop, we noted that the 1934 record is still acclaimed as the highest wind measured by *man*!). Let us therefore start with the Mt. Washington record; the following calculations are mostly for illustrative purposes.

Pagliuca (1934) details the heroic efforts that were made to calibrate and record the extreme winds that occurred on April 12, 1934. The 4.7 s average of of $v_{max} = 103$ m/s was recorded using a heated anemometer and "hummer" (note that the corresponding speed of sound at this 2 km altitude at near freezing temperature is about 300 m/s). A gust of 103 m/s lasting 4.7 s seems large; but is it really unexpected? From 1870 (when the

Box 8.1 (*cont.*)

observatory was opened) to 2010 there are 4.4×10^9 seconds, so there have been $\sim 9.5 \times 10^8$ such periods, and the rarest event that would have been detected has a probability of about $p_{obs} \approx 1/(9.5 \times 10^8) \approx 1.1 \times 10^{-9}$. Let us compare this rough detection limit with the theoretical probability of finding such a large gust; we can easily do this using the codimension function and Eqn. (5.16).

The first step is to estimate the overall range of scales; $\lambda = \tau_w/\Delta t$ where τ_w is the outer weather scale. From Fig. 8.6b we see that at 45° N, the mean Δv is 27 m/s; this corresponds to $\tau_w \approx L_e/\Delta v \approx 7.4 \times 10^5$ s, so with $\Delta t = 4.7$ *s* we have $\lambda = 1.57 \times 10^5$. We can use:

$$\Delta v_\lambda / \Delta v_1 = \varphi_{v,\lambda} \lambda^{-H} \tag{8.1}$$

where $\varphi_{v,\lambda}$ is the normalized flux corresponding to the wind (i.e. $\varepsilon^{1/3}$) Eqn. (8.1) yields $\gamma_\varphi = \gamma_v + H$. According to Table 8.1 for the temporal statistics we have $C_{1v} \approx 0.048$, with $\alpha \approx 1.5$–1.9 (see Table 8.1 note *b*) and $H \approx 1/3$. Since the wind was highly "gusting" we can estimate the small-scale 4.7 s record fluctuation as $\Delta v_\lambda \approx 103$ m/s, taking $\Delta v_1 \approx 27$ m/s; we have $\Delta v_\lambda / \Delta v_1 \approx 3.8$ and $\gamma_{v,max} = \text{Log}(3.8)/\text{Log}(1.57 \times 10^5) = 0.111$, so that $\gamma_{\varphi\text{-}max} = 0.445$. In comparison, we have noted the most extreme event for the record has a probability of occurrence of $p_{obs} \approx 1.1 \times 10^{-9}$, so that $c_{max} = c(\gamma_{\varphi,max}) = -\text{Log}(1.1 \times 10^{-9})/\text{Log}(1.57 \times 10^5) = 1.72$.

Proceeding in this way for the maxima at the other resolutions, and estimating the probabilities by assuming that the events were indeed the extremes at the given resolution over the entire period, we obtain the circles in

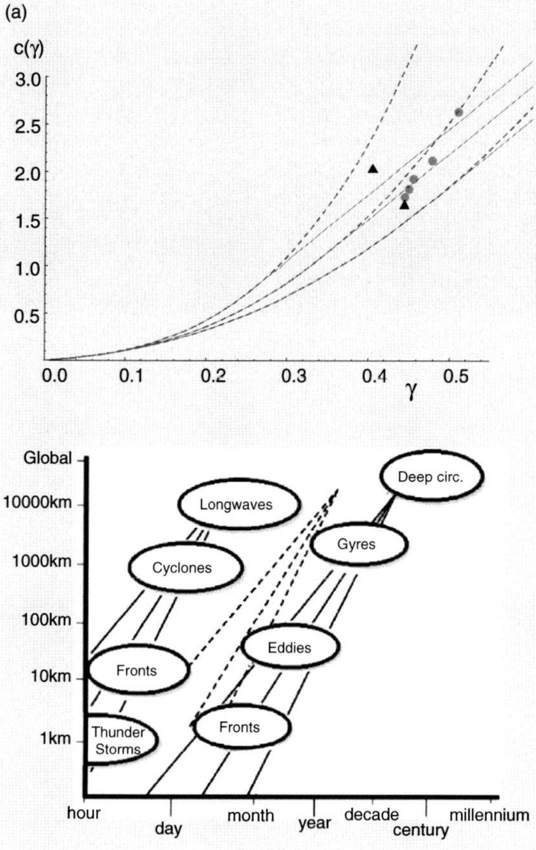

Fig. 8.9 (a) The empirical singularities and codimensions calculated for the Mt. Washington data (circles: 4.7 seconds, 17 seconds, 1 minute, 5 minutes, 1 hour; left to right) and the Tropical Storm Olivia data (triangles, 3 seconds, 5 minutes, left to right). The curves are for $C_1 = 0.048$ and $\alpha = 1.4, 1.6, 1.8$ (left to right). To calculate the singularities, the outer scale τ_w is taken as 7.4×10^5 s, corresponding to $\Delta v = 27$ m/s, $\varepsilon \approx 10^{-3}$ m^2/s^3. To calculate the codimensions, the events are all assumed to be the extremes over the entire period of operation of the stations (from 1870 at Mt. Washington, from 1932 on Barrow Island, to 2010). The straight (solid) asymptotes indicate the dressed codimensions with $q_D = 7$. (b) A (Langrangian, co-moving) Stommel diagram showing the length and time scales associated with typical atmospheric and oceanic dynamics, adapted from Steele (1995). This graph has been slightly modified from the adaptation in Lovejoy *et al.* (2000) by the placement of the global scale at 20 000 km. The centre of each triplet of lines corresponds to the mean behaviour, the other two to various degrees of intermittency as discussed in the text. The left group corresponds to the mean atmospheric value $\varepsilon = 10^{-3}$ m^{2}s^{-3}, ($\tau_w \approx 10$ days) the central (dashed) to the mean ocean surface value $\varepsilon_o = 10^{-8}$ m^{2}s^{-3} ($\tau_o \approx 1$ year), and the right to value 10^{-12} m^{2}s^{-3} ($\tau_o \approx 20$ years) representative of deeper waters. Periods of 100, 1000 years correspond to $\varepsilon \approx 10^{-14}$, 10^{-17} m^{2}s^{-3} and may be appropriate for the very deep structures such as the thermohaline "conveyor belt."

Box 8.1 (*cont.*)

Fig. 8.9a. Interestingly, as measured by the value of the singularity, the 103 m/s result had the lowest, whereas the 1 hour result had the highest order of singularity (0.501 rather than 0.445)! We can now compare this with the theoretical probability $p_{th} \approx \lambda^{-c(\gamma_\varphi)}$ using the theoretical c. The figure shows this for the multifractal parameters indicated above both "bare" and "dressed" codimensions (Eqn. (5.49)) for $\alpha = 1.4, 1.6, 1.8$ with the empirical $q_D \approx 7$ (see Table 5.1). The sensitivity of p_{th} on α is expected since we are interested in singularities far from the mean (values near the mean do however suffice to give accurate estimates of C_1). We can see that if $\alpha \approx 1.6$ then the record values are roughly as expected.

But what about Tropical Storm Olivia? Let us again convert the speeds and resolutions into singularities, noting that the observatory on Barrow Island that measured the new record has only been in operation since 1932. The results for the 5-minute and 3-second gusts are also shown in Fig. 8.9a. We find for the 3-second gust $\gamma = 0.449$, which is very slightly above the Mt. Washington value (0.445), but for the 5-minute Olivia average we find the disappointing value of 0.409, substantially below the Mt. Washington 5-minute value of 0.479. Presumably, going through the Barrow Island record back to 1932 one would find a higher 5-minute average than this. On this basis, although Olivia (barely) succeeded in toppling the Mt. Washington record at high resolution, Mt. Washington still holds the title at 5-minute resolution! Note that we have used the same 45° N mean value $\Delta v = 27$ m/s for both Mt. Washington and Barrow. According to Fig. 8.6b, at 20° S, a value of 12 m/s is more accurate (corresponding to $\tau_w \approx 1.7 \times 10^6$ s), and in this case the new singularity values are 0.503, 0.496 (3 s, 300 s, respectively), so that Olivia also wins the 5-minute title. However, this is equivalent to giving the title on the basis of the varying local climatology, and seems hardly fair. Ultimately we should use the planetary mean ε which is not so well estimated, but which is closer to the value used for Mt. Washington: $\varepsilon \approx 10^{-3}$ m^2/s^3.

We can now turn this around and use the above to obtain a rough estimate of the parameter α (i.e. assuming $q_D \approx 7$, $C_1 = 0.048$). Using the maximum Mt. Washington at 4.7 s, we find $\alpha \approx 1.70$, whereas using the Barrow 3 s maximum, we find $\alpha = 1.76$ (mean: $\alpha = 1.73$). We can then use this to calculate the record winds that would have been recorded at other resolutions; for example at 1 second, we find $\lambda = 7.4 \times 10^5$, $c_{max} = -\mathrm{Log}(4.41 \times 10^9)^{-1}/\mathrm{Log}(7.4 \times 10^5) = 1.643$. With $\alpha = 1.73$, the new $\gamma_{\varphi,max}$ satisfying $c(\gamma_{\varphi,\max}) = 1.643$ is $\gamma_{\varphi,max} \approx 0.4412$, corresponding to $\gamma_{v,max} = 0.1079$, so that $\Delta v \approx 116$ m/s. The analogous calculation near the dissipation scale (taking the resolution = 1 mm) yields $\Delta v \approx 130$ m/s. We therefore see that even these rare extreme events are not so close to the speed of sound; taking into account the Mt. Washington altitude and low temperature, we find the extreme Mach number $Ma \approx 130/300 \approx 0.43$, so that compressibility effects (which depend on Ma^2) are still small.

A particularity of this example is that $\alpha > 1$, so that the singularities are unbounded. In cases where $\alpha < 1$, there is a theoretical maximum order of singularity. For the corresponding theory and applications to temporal record rain rates, see Hubert *et al.* (1993).

8.2 Anisotropic space-time turbulence

8.2.1 Space-time scale functions

We have argued that atmospheric variables – including the wind – have wide-range (anisotropic) scaling statistics and that the spatial scaling of the horizontal wind leads to the temporal scaling of all the fields. Unfortunately, space-time scaling is somewhat more complicated than pure spatial scaling. At meteorological time scales this is because we must take into account the advection of structures and the Galilean invariance of the dynamics. At longer (climate) time scales, this is because we consider the statistics of many lifetimes ("eddy turnover times") of structures. We first consider the shorter time scales; the longer time scales are discussed in Chapter 10. This is a summary of the more detailed discussions in Lovejoy *et al.* (2008), Lovejoy and Schertzer (2010), Pinel *et al.* (2012).

In order to illustrate the formalism, consider the horizontal wind v. In the 23/9D model (Chapter 6), the energy flux ε dominates the horizontal and the buoyancy variance flux φ dominates the vertical so that horizontal wind differences follow:

$$\begin{aligned} \Delta v(\Delta x) &= \varepsilon^{1/3}\Delta x^{H_h}; \quad H_h = 1/3 \quad a \\ \Delta v(\Delta y) &= \varepsilon^{1/3}\Delta y^{H_h}; \quad H_h = 1/3 \quad b \\ \Delta v(\Delta z) &= \phi^{1/5}\Delta z^{H_v}; \quad H_v = 3/5 \quad c \\ \Delta v(\Delta t) &= \varepsilon^{1/2}\Delta t^{H_\tau}; \quad H_\tau = 1/2 \quad d \end{aligned} \tag{8.2}$$

where Δx, Δy, Δz, Δt are the increments in the horizontal, in the vertical and in time respectively.

Eqns. (8.2*a–b*) describe the real-space horizontal Kolmogorov scaling and (8.2*c*) the vertical Bolgiano–Obukhov (BO) scaling for the velocity. As usual, the equality signs should be understood in the sense that each side of the equation has the same scaling properties (recall that the FIF model (Section 5.4.4) interprets this more precisely as a fractional space integration). The anisotropic Corrsin–Obukov law for passive scalar advection is obtained by the replacements $v \rightarrow \rho;\quad \varepsilon \rightarrow \chi^{3/2}\varepsilon^{-1/2}$ where ρ is the passive scalar density, χ is the passive scalar variance flux.

We have included Eqn. (8.2*d*), which is the result for the pure time evolution in the absence of an overall advection velocity; this is the classical Lagrangian version of the Kolmogorov law (Inoue, 1951; Landau and Lifschitz, 1959), it is essentially the result of dimensional analysis using ε and Δt rather than ε and Δx. Although Lagrangian statistics are notoriously difficult to obtain empirically (see however Seuront *et al.*, 1996), they are roughly known from experience and are used as the basis for the space-time or "Stommel" diagrams that adorn introductory meteorology textbooks. Fig. 8.9b gives an example adapted from Steele (1995); another, very similar one with corresponding added reference lines may be found in Schertzer *et al.* (1997a). In the figure, we have added triplets of reference lines, one for the weather, two others for the ocean.

Although the original interpretation was in terms of separate dynamical processes at each scale range (the ellipses), the straight reference lines show that a scaling interpretation works very well using the Lagrangian relation between space and time (Eqn. (8.2*d*); $l = \tau^{3/2}\varepsilon^{-1/3}$). In each triplet of lines the central one (slope $H_t = (1/2)/(1/3) = 3/2$) is a line of constant energy flux showing that the basic Kolmogorov (nonintermittent) scaling model holds remarkably well for both the atmosphere and the ocean. While the atmosphere lines correspond to the measured value $\varepsilon_w = 10^{-3}\ \mathrm{m^2s^{-3}}$, the far-right ocean lines correspond to $\varepsilon = 1 \times 10^{-12}\ \mathrm{m^2s^{-3}}$, which is considerably less than that estimated for the ocean surface in Section 8.1.4 ($\varepsilon_o \approx 1 \times 10^{-8}\ \mathrm{m^2s^{-3}}$: the central dashed lines). The lines converge at 20 000 km and ~10 days (weather), and at 20 000 km, ~1 year (ocean surface), and the third at ~20 years. For the oceans, the value $1 \times 10^{-12}\ \mathrm{m^2s^{-3}}$ is reasonable for depths of ~1 km or more (see Section 8.1.4). The additional lines to the left and right of the central ones give an idea of the fluctuations expected due to multifractal intermittency. Their slopes are $3/(2 + \gamma)$ with singularities γ taken to be $\pm\ C_1$ (C_1 is actually the standard deviation of γ, not ε, when $\alpha = 2$). When $\gamma = +C_1$ (the left-most thin lines, taken here = 0.25 in accord with atmospheric measurements; e.g. Schmitt *et al.*, 1992) the line indicates the effect of the sparse intermittent structures which give the dominant contribution to the mean. When $\gamma = -C_1$, these lines give roughly the space-time relationships for the weaker structures, which at any scale are the most probable (when $\alpha = 2$, i.e. for log-normal multifractals, this statement is exact). As expected, at each scale the weaker structures live longer, and the stronger ones less long.

Following the developments in Chapter 6, we can express the scaling Eqns. (8.2*a–d*) in a single expression valid for any space-time vector displacement $\underline{\Delta R} = (\underline{\Delta r}, \Delta t) = (\Delta x, \Delta y, \Delta z, \Delta t)$ by introducing a scalar function of space-time vectors called the "space-time scale function", denoted $[\![\underline{\Delta R}]\!]$, which satisfies the fundamental (functional) scale equation:

$$[\![\lambda^{-G_{st}}\underline{\Delta R}]\!] = \lambda^{-1}[\![\underline{\Delta R}]\!];\quad G_{st} = \begin{pmatrix} G_s & 0 \\ 0 & H_t \end{pmatrix};$$
$$H_t = (1/3)/(1/2) = 2/3 \tag{8.3}$$

where G_s is the 3 × 3 matrix spatial generator:

$$G_s = \begin{pmatrix} 1 & 0 & 0 \\ 0 & 1 & 0 \\ 0 & 0 & H_z \end{pmatrix} \tag{8.4}$$

(with rows and columns corresponding to (x, y, z); cf. Eqn. (6.16)) and the 4 × 4 matrix G_{st} is the extension to space-time. We have introduced the notation "[[ ]]" for the space-time scale function in order to distinguish from the purely spatial scale function denoted "|| ||". This distinction will be particularly useful in Chapter 9. (Note on notation: we use $H_h/H_v = H_z$ (horizontal/vertical) and $H_h/H_\tau = H_t$ (horizontal/time).)

Using the space-time scale function, we may now write the space-time generalization of the Kolmogorov law (Eqn. (8.2)) as:

$$\Delta v(\underline{\Delta R}) = \varepsilon_{[\![\underline{\Delta R}]\!]}^{1/3}[\![\underline{\Delta R}]\!]^{1/3} \tag{8.5}$$

where the subscripts on the flux indicate the space-time scale over which it is averaged. This anisotropic intermittent (multifractal) generalization of the Kolmogorov law is thus one of the key emergent laws of atmospheric dynamics and serves as a prototype for the emergent laws governing the other fields.

The result analogous to that of Section 6.1.4, the corresponding simple ("canonical") space-time scale function, is:

$$[\![\underline{\Delta R}]\!]_{can} = l_s\left(\left(\frac{\|\underline{\Delta r}\|}{l_s}\right)^2 + \left(\frac{|\Delta t|}{\tau_s}\right)^{2/H_t}\right)^{1/2} \tag{8.6}$$

where $\tau_s = \phi^{-1/2}\varepsilon^{1/2}$ is the "sphero-time" analogous to the sphero-scale $l_s = \phi^{-3/4}\varepsilon^{5/4}$ (see also Marsan *et al.*, 1996). With scale function (Eqn.(8.6)), the fluctuations (Eqn. (8.5)) respect Eqns.(8.2*a*–*d*).

8.2.2 Advection, Galilean invariance and Eulerian statistics

Using the Lagrangian temporal scaling (Eqn. (8.2*d*)) implies $H_\tau \neq 1$, apparently predicting different horizontal and temporal scaling. This is in contradiction with the empirical analyses of the previous section, which showed that horizontal and temporal exponents were very close to each other. However, we are interested in the temporal scaling in the Eulerian frame, and for this we are missing a key ingredient: advection. When studying laboratory turbulence generated by an imposed flow of velocity v and with superposed turbulent fluctuations, Taylor (1938) proposed that the turbulence is "frozen" such that the pattern of turbulence blows past the measuring point sufficiently fast that it does not have time to evolve; i.e. he proposed that the spatial statistics could be obtained from time series by the deterministic transformation $v\Delta t \to \Delta x$ where v is a constant: in the lab v is determined by the fan and by the wind-tunnel geometry. While this transformation has frequently been used in interpreting meteorological series, it can only be properly justified by assuming the existence of a scale separation between small and large scales so that the large scales really do blow the small-scale (nearly "frozen") structures past the observing point. Since we have argued that there is no scale separation in the atmosphere, this is inappropirate.

However, if we are only interested in the statistical relation between time and space, and if the system is scaling, then advection can be taken into account using the Gallilean transformation $\underline{r} \to \underline{r} - \underline{v}t$, $t \to t$, which corresponds to the following matrix A:

$$A = \begin{pmatrix} 1 & 0 & 0 & v_x \\ 0 & 1 & 0 & v_y \\ 0 & 0 & 1 & v_z \\ 0 & 0 & 0 & 1 \end{pmatrix} \tag{8.7}$$

where the mean wind vector has components $\underline{v} = (v_x, v_y, v_z)$ (Schertzer *et al.*, 1997b), and the columns and rows correspond to x, y, z, t. The new "advected" generator is $G_{st,\,advec} = A^{-1}G_{st}A$ and the scale function $[\![\underline{\Delta R}]\!]_{advec}$ which is symmetric with respect to $G_{st,\,advec}$ is: $[\![\underline{\Delta R}]\!]_{advec} = [\![A^{-1}\underline{\Delta R}]\!]$. The canonical advected scale function is therefore:

$$[\![\underline{\Delta R}]\!]_{advec,can} = [\![A^{-1}\underline{\Delta R}]\!]_{can} = l_s\left(\left(\frac{\Delta x - v_x\Delta t}{l_s}\right)^2 + \left(\frac{\Delta y - v_y\Delta t}{l_s}\right)^2 + \left(\frac{\Delta z - v_z\Delta t}{l_s}\right)^{2/H_z} + \left(\frac{\Delta t}{\tau_s}\right)^{2/H_t}\right)^{1/2} \tag{8.8}$$

Note that since $D_{st,\,advec} = TrG_{st,advec} = Tr(A^{-1}G_{st}A) = TrG_{st} = D_{st}$, such constant advection does not affect the elliptical dimension (see however the next section for the "effective" G_{eff}, D_{eff}).

It will be useful to study the statistics in Fourier space. For this purpose we can recall the result from Chapter 6 that the Fourier generator $\widetilde{G} = G^T$ so that:

$$\widetilde{G}_{st,\,advec} = A^T G_{st}^T \left(A^{-1}\right)^T \tag{8.9}$$

The corresponding canonical dimensional Fourier-space scale function is therefore:

$$[\![\underline{K}]\!]_{advec,\,can} = [\![A^T\underline{K}]\!]_{can} = l_s^{-1}\left((k_x l_s)^2 + (k_y l_s)^2 + (k_z l_s)^{2/H_z} + (\tau_s(\omega + \underline{k}\cdot\underline{v}))^{2/H_t}\right)^{1/2} \tag{8.10}$$

In agreement with the fact that the physical space Gallilean transformation $\underline{r} \to \underline{r} - \underline{v}t;\ t \to t$ corresponds to the Fourier-space transformation $\underline{k} \to \underline{k};\ \omega \to \omega + \underline{k}\cdot\underline{v}$.

8.2.3 Advection in the horizontal

Eqn. (8.8) is valid because of the Gallilean invariance of the equations and boundary conditions; it assumes that the advection velocity is essentially constant over the region and independent of scale. We now consider this in more detail. We will only consider horizontal advection (put $w = 0$; the interesting but nontrivial effects of the vertical velocity on the temporal scaling are discussed in Appendix 8A). If we apply the formula over a finite region with relatively well-defined mean horizontal velocity, then it should apply as discussed in Lovejoy *et al.* (2008). But what about applying it to very large-scale, e.g. global-scale, regions where the mean velocity is small (if only because of rough north–south symmetry)? However, even if we consider a flow with zero imposed mean horizontal velocity (as argued by Tennekes, 1975) in a scaling turbulent regime with $\Delta v_l \approx \varepsilon^{1/3} l^{1/3}$, the typical largest eddy, the "weather-scale" $L_w \approx L_e$, will have a mean velocity $V_w \approx \Delta v_w \approx \varepsilon_w{}^{1/3} L_w{}^{1/3}$ and will survive for the corresponding eddy turnover time $\tau_{eddy} = \tau_w = L_w/V_w = \varepsilon_w{}^{-1/3} L_w{}^{2/3}$ estimated as $\approx$ 10 days above. In other words, if there is no break in the scaling then we expect that smaller structures will be advected by the largest structures in the scaling regime.

With this estimate of the horizontal velocities to insert in Eqn. (8.8), let us compare them to the Lagrangian term $(\Delta t/\tau_s)^{1/H_t}$ considering only the temporal variations (i.e. take $\Delta x = \Delta y = \Delta z = 0$) and taking horizontal axes such that the advection term is $V_w \Delta t/l_s$. By definition, the sphero-time τ_s satisfies: $l_s = \varepsilon^{1/2} \tau_s^{1/2}$ and since $\tau_w = V_w^2/\varepsilon_w$ we see that the condition that the pure temporal evolution term is negligible (i.e. that $V_w \Delta t/l_s > (\Delta t/\tau_s)^{3/2}$; using $H_t = 2/3$) is $\Delta t < \tau_w$ so that the term $(\Delta t/\tau_w)^{2/H_t} = (\Delta t/\tau_w)^3$ only becomes important for $\Delta t > \tau_w \approx 10$ days. However, since the physical size of the eddies with lifetime $\Delta t = \tau_w$ is already the size of the planet (L_w), presumably the term ceases to be valid for scales $\Delta t > \tau_w$. Nevertheless, it is possible that it might play a modest role in breaking the scaling for Δt comparable to τ_w, i.e. for the transition from weather to macroweather.

Neglecting this Δt^3 term, we can now use this information to rewrite the horizontal scale function (Eqn. (8.8) with $w = 0$) in terms of L_w, τ_w and V_w instead of l_s, τ_s. We can also allow for some trivial anisotropy corresponding to the scale-independent (east–west)/(north–south) aspect ratio a introduced in Chapter 6. In Table 4.1, this was found to be $\approx 1.6 \pm 0.3$ for the ECMWF interim fluxes and roughly the same for the MTSAT radiances, and about the same for the precipitation fields (Table 4.6) (more general trivial anisotropy – corresponding for example to structures elongated in arbitrary horizontal directions – could easily be introduced if necessary).

First consider overall (nonrandom) advection (v_x, v_y): the square of the nondimensional space-time scale function is:

$$\begin{aligned} [\![\underline{\Delta R}]\!]^2 &= \left(\left(\frac{\Delta x - v_x \Delta t}{L_w} \right)^2 + \left(\frac{\Delta y - v_y \Delta t}{(L_w/a)} \right)^2 \right) \\ &= \left(\frac{\Delta x}{L_w} \right)^2 + \left(\frac{a \Delta y}{L_w} \right)^2 + \left(\frac{v_x^2 + a^2 v_y^2}{L_w{}^2} \right) \Delta t^2 \\ &\quad -2 \left(v_x \frac{\Delta x}{L_w} + a^2 v_y \frac{\Delta y}{L_w} \right) \left(\frac{\Delta t}{L_w} \right) \end{aligned} \qquad (8.11)$$

This is helpful for understanding the effect of averaging over random v_x, v_y.

The statistics of the intensity gradients of real fields are influenced by random turbulent velocity fields and involve powers of such scale functions but with appropriate "average" velocities. Let us now average Eqn. (8.11) over a distribution representing the velocities of various eddies or structures over a given region. In this case we can nondimensionize the variables by the following transformation:

$$\begin{aligned} &\Delta x \to \frac{\Delta x}{L_w}; \quad \Delta y \to \frac{\Delta y}{L_w}; \quad \Delta t \to \frac{\Delta t}{\tau_w}; \\ &\mu_x = \frac{\overline{v_x}}{V_w}; \qquad \mu_y = \frac{\overline{v_y}}{V_w} \end{aligned} \qquad (8.12)$$

The symbols μ_x, μ_y are used for the components of the nondimensional velocity and:

$$V_w = \left(\overline{v_x^2} + a^2 \overline{v_y^2} \right)^{1/2}; \quad \tau_w = \frac{L_w}{V_w} \qquad (8.13)$$

Note that here V_w is a large-scale turbulent velocity whereas $\overline{v_x}$, $\overline{v_y}$ are given by the overall mean advection in the region of interest and $\mu_x < 1$, $\mu_y < 1$ (since $\overline{v^2} > (\overline{v})^2$). The use of the averages (indicated by the overbars) is only totally justified if the second power of the scale function is averaged; presumably, it is some other power that is physically more relevant and there will thus be (presumably small) intermittency corrections (which we ignore). It is now convenient to define:

$$\underline{\mu} = \left(\mu_x, \mu_y \right); \quad |\underline{\mu}|^2 = \mu_x^2 + \mu_y^2 \qquad (8.14)$$

which satisfies $|\underline{\mu}| < 1$. In terms of the nondimensional quantities this yields an "effective" nondimensional scale function:

$$[\![\underline{\Delta R}]\!]_{eff} \approx \left(\underline{\Delta R}^T B \underline{\Delta R}\right)^{1/2};$$
$$B = \begin{pmatrix} 1 & 0 & -\mu_x \\ 0 & a^2 & -a^2\mu_y \\ -\mu_x & -a^2\mu_y & 1 \end{pmatrix};$$
$$\underline{\Delta R} = (\Delta x, \Delta y, \Delta t) \quad (8.15)$$

where the rows and columns correspond to x, y, t (left to right, top to bottom). Using the terminology of Chapter 6, the scale function in Eqn. (8.15) is only "trivially anisotropic" since it is scaling with respect to an "effective" G matrix G_{eff} = the identity; the matrix B simply determines the trivial space-time anisotropy.

8.3 Global space-time scaling in Fourier space

8.3.1 Fourier-space scale functions

In order to test out the correctness of the effective (horizontal-time) global space-time scale function (Eqn. (8.15)) it is convenient to use Fourier techniques. If we assume that the structure function of a field I (e.g. an IR radiance from MTSAT, Chapter 4) is scaling, then this implies the scaling of the spectral density (P_{st} in space-time with dimension $D_{st} = 2 + 1 = 3$), with a (different) Fourier-space (represented by a tilde) scale function (see Section 7.2.1):

$$\left\langle \Delta I(\underline{\Delta R})^2 \right\rangle = [\![\underline{\Delta R}]\!]^{\xi(2)};\ P_{st}(\underline{K}) \propto \left\langle |\tilde{I}(\underline{K})|^2 \right\rangle \approx [\![\underline{K}]\!]^{-s_{st}};\ s_{st} = D_{st} + \xi(2)$$
$$\left\langle \Delta I(\underline{\Delta r})^2 \right\rangle = \|\underline{\Delta r}\|^{\xi(2)};\ P_s(\underline{k}) \propto \left\langle |\tilde{I}(\underline{k})|^2 \right\rangle \approx \|\underline{k}\|^{-s_s};\ s_s = D_s + \xi(2) \quad (8.16)$$

where $\langle \Delta I^2 \rangle$ and 2(1 – P) are Fourier transform pairs (Eqn. (7.65)), $\underline{K} = (\underline{k}, \omega)$, $\underline{k} = (k_x, k_y)$ and $[\![\Delta R]\!]$ satisfies the scaling equation with respect to G_{st}, $\|\underline{\Delta r}\|$ satisfies it with respect to G_s, $[\![\underline{K}]\!]$ with respect to $G^T{}_{st}$ and $\|\underline{k}\|$ with respect to $G^T{}_s$. D_{st}, D_s are the traces of the space-time and space generators G_{st}, G_s. The subscripts *st* are for "space-time," i.e. x, y, t space; the subscript s is for horizontal space, i.e. x, y space; sometimes the subscripts *xyt* and *xy* are used instead. If Eqn. (8.16) is regarded as defining the Fourier scale functions $[\![\underline{K}]\!]$, $\|\underline{k}\|$ then we must have $[\![\underline{K}]\!] \geq 0$, $\|\underline{k}\| \geq 0$ (as for real-space scale functions). This is assured since $P_s(\underline{k})$ and $P_{st}(\underline{K})$ are ≥ 0, in addition, they are related by $P_s(\underline{k}) = \int_{-\infty}^{\infty} P_{st}(\underline{k}, \omega)d\omega$ (see below).

We can similarly define the purely spatial Fourier-space scale function from the spatial ($\underline{k}$) subspace of $\underline{K}$.

Let us use real-space coordinates nondimensionalized as in Eqn. (8.12) and the corresponding nondimensional Fourier-space vector:

$$\underline{K} = (k_x, k_y, \omega);\ K \to (L_w k_x, L_w k_y, \tau_w \omega) \quad (8.17)$$

We can now use the mathematical result (a Tauberian theorem, Box 2.2):

$$|\underline{\Delta R}|^{\xi(2)} \propto \int |\underline{K}|^{-(D+\xi(2))} e^{-i\underline{K}\cdot\underline{\Delta R}} d^D\underline{K} \quad (8.18)$$

Now, using the transformation of variables $\Delta R \to C\Delta R$ (where C is an arbitrary nonsingular real matrix), we obtain the general result:

$$\int (\underline{K}^T B^{-1} \underline{K})^{-(D+\xi(2))/2} e^{-i\underline{K}\cdot\underline{\Delta r}} d^D\underline{K} \propto (\det B)^{-1/2} (\underline{\Delta R}^T B \underline{\Delta R})^{\xi(2)/2};\ B = CC^T \quad (8.19)$$

since $\det B = (\det C)^2$ and C is real, the condition for the validity of the above is $\det B > 0$. In Chapter 9 we shall see that the scale functions for non (space-time) localized (wave-like) behaviour $\det B < 0$. In other words, the real-space and Fourier-space scale function pairs:

$$[\![\underline{\Delta R}]\!] = (\underline{\Delta R}^T B \underline{\Delta R})^{1/2};$$
$$[\![\underline{K}]\!] = (\underline{K}^T B^{-1} K)^{1/2};\ \det B > 0 \quad (8.20)$$

satisfy Eqn. (8.16); Eqn. (8.20) is the relation between their respective "trivial" anisotropies. A relation useful below is that the inverse of the B matrix is:

$$B^{-1} = \frac{1}{1 - \mu_x^2 - a^2\mu_y^2} \begin{pmatrix} 1 - a^2\mu_y^2 & \mu_y\mu_x & \mu_x \\ \mu_y\mu_x & (1 - \mu_x^2)/a^2 & \mu_y \\ \mu_x & \mu_y & 1 \end{pmatrix} \quad (8.21)$$

If we now introduce:

$$\sigma = \sqrt{1 - \left(\mu_x^2 + a^2\mu_y^2\right)} \quad (8.22)$$

we find:

$$[\![\underline{K}]\!]^2 = (\underline{K}^T B^{-1} \underline{K}) = (\omega' + \underline{k}\cdot\underline{\mu})^2 \sigma^{-2} + k_x^2 + a^{-2}k_y^2 \quad (8.23)$$

The corresponding spectral density is:

$$P(k_x, k_y, \omega) = \|\underline{K}\|^{-s} = (\underline{K}^T B^{-1}\underline{K})^{-s/2}$$
$$= [\omega'^2 + \|\underline{k}\|^2]^{-s/2} \quad (8.24)$$

where the primed nondimensional frequency and spatial scale functions are:

$$\omega' = (\omega + \underline{k}\cdot\underline{\mu})/\sigma; \|\underline{k}\| = (k_x^2 + a^{-2}k_x^2)^{1/2} \quad (8.25)$$

The transformation $\omega \to (\omega + \underline{k}\cdot\underline{\mu})/\sigma$ has a simple interpretation: it accounts for both the mean advection and the statistical variability of the latter. $\|\underline{k}\|$ is a scale function that accounts for the squashing by a factor of a in the east-west direction.

This analysis provides some theoretical justification for a series of essentially ad hoc techniques starting with Hubert and Whitney (1971) for measuring "satellite winds," now more accurately called "atmospheric motion vectors" (AMVs). Such techniques using MTSAT and GOES geostationary IR and visible satellite imagery are currently used operationally. Although a number of methods exist, these techniques are mostly based on cross-correlations of sequences of satellite images, i.e. on the real-space counterparts of $P(k_x, k_y, \omega)$; see for example Szantai and Sèze (2008) for a recent overview and comparison. Pinel *et al.* (2012) show that the maximum cross-correlation using Eqn. (8.15) does indeed yield the vector $(\overline{v_x}, \overline{v_y})$, hence providing a theoretical basis for the AMV technique.

8.3.2 One- and two-dimensional subspaces and the analysis of MTSAT thermal IR data

In order to test the above theory it is best to use data spanning as wide a range of scales as possible in space and in the temporal domain below τ_w. Although reanalyses are convenient, they are not ideal due to their somewhat low resolutions. However, a limited space-time analysis of 20CR 700 mb temperature data can be found in Figs. 10.15a, 10.15b, 10.15c (i.e. including spectra in the macroweather and the climate). In the weather regime, a better choice for analysis is the hourly (x, y, t) MTSAT dataset (Fig. 8.10); we can easily calculate the three-dimensional spectral density $P_{st}(k_x, k_y, \omega)$. However, this full 3D function is quite unwieldy; it is best to examine various 1D and 2D subspaces. The easiest way to do this is to recall that the correlation function (R) and the spectral density are Fourier transform pairs:

$$R(\Delta x, \Delta y, \Delta t) = (2\pi)^{-3}\int P_{st}(k_x, k_y, \omega)e^{i(k_x\Delta x + k_y\Delta y + \omega\Delta t)}dk_x dk_y d\omega$$
$$P_{st}(k_x, k_y, \omega) = \int R(\Delta x, \Delta y, \Delta t)e^{-i(k_x\Delta x + k_y\Delta y + \omega\Delta t)}d\Delta x d\Delta y d\Delta t \quad (8.26)$$

(the Wiener–Khinchin theorem, Chapter 2) where $R(\Delta x, \Delta y, \Delta t)$ is the correlation function (this should not to be confused with the space-time coordinate whose vector "lag" $\underline{\Delta R}$ will occasionally be needed). Recall (Eqn. (7.79)) that the correlation function is simply related to

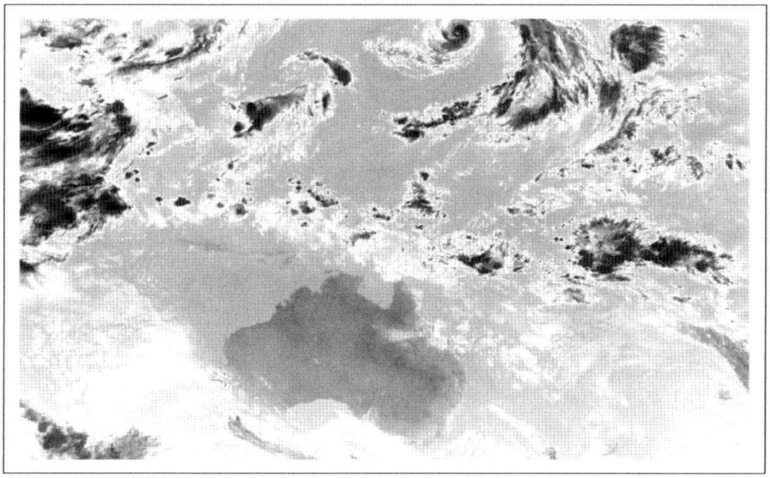

Fig. 8.10 One of the 1440 hourly samples of the MTSAT IR radiances at 5 km resolution from 40° S to 30° N and 80° E to 200° E. The white areas are the coldest (generally clouds), the dark areas are the warmest (generally land). Australia is clearly seen in the lower middle (Pinel, *et al.* 2012).

the usual (second-order, difference) structure function by $S_2(\Delta x, \Delta y, \Delta t) = 2(R(0,0,0) - R(\Delta x, \Delta y, \Delta t))$.

Successively substituting $\Delta t = 0$, $\Delta y = 0$, $\Delta x = 0$ in the above we obtain:

$$R(\Delta x, \Delta y, 0) = (2\pi)^{-2}\int P_{xy}(k_x, k_y)e^{i(k_x\Delta x + k_y\Delta y)}dk_x dk_y;$$
$$P_{xy}(k_x, k_y) = \int P_{xyt}(k_x, k_y, \omega)d\omega$$
$$R(\Delta x, 0, \Delta t) = (2\pi)^{-2}\int P_{xt}(k_x, \omega)e^{i(k_x\Delta x + \omega\Delta t)}dk_x d\omega;$$
$$P_{xt}(k_x, \omega) = \int P_{xyt}(k_x, k_y, \omega)dk_y$$
$$R(0, \Delta y, \Delta t) = (2\pi)^{-2}\int P_{yt}(k_y, \omega)e^{i(k_y\Delta y + \omega\Delta t)}dk_y d\omega;$$
$$P_{yt}(k_y, \omega) = \int P_{xyt}(k_x, k_y, \omega)dk_x \quad (8.27)$$

where $P_{xy}(k_x, k_y) = P_s$, $P_{xt}(k_x, \omega)$, $P_{yt}(k_y, \omega)$ are the spectral densities on the (k_x, k_y), (k_x, ω), (k_y, ω) subspaces respectively. Using results analogous to Eqn. (8.16) we obtain:

$$P_{xy}(k_x, k_y) = \|(k_x, k_y)\|_{xy}^{-(s_{st}-1)};$$
$$P_{xt}(k_x, \omega) = \|(k_x, \omega)\|_{xt}^{-(s_{st}-1)};$$
$$P_{yt}(k_y, \omega) = \|(k_y, \omega)\|_{yt}^{-(s_{st}-1)}$$

where $s_{st} = 3 + \xi(2)$. The corresponding spatial, spectral, nondimensional scale function for the (k_x, k_y) subspace is:

$$\|(k_x, k_y)\|_{xy} = (k_x^2 + a^{-2}k_y^2)^{1/2} \quad (8.28)$$

This equation implies that the purely spatial subspace (k_x, k_y) is anisotropic with a squashing by a factor a in the k_y direction, which is the Fourier correspondence to the real-space squashing by factor a in the x direction. The two wavenumber/frequency subspaces (k_x, ω), (k_y, ω) are obtained by integrating out the third coordinate and have scale functions which correspond to constant mean advection velocities:

$$\|(k_x, \omega)\|_{xt} = (\omega_x^2 + k_x^2)^{1/2};$$
$$\omega_x = (\omega + k_x\mu_x')/\sigma_x;$$
$$\sigma_x = \sqrt{1 - \mu_x^2};$$
$$\mu_x' = \mu_x\sqrt{\frac{1 - \mu_x^2}{\sigma^2 + \mu_y^2}} \quad (8.29)$$

$$\|(k_y, \omega)\|_{yt} = (\omega_y^2 + a^{-2}k_y^2)^{1/2};$$
$$\omega_y = (\omega + k_y\mu_y')/\sigma_y;$$
$$\sigma_y = \sqrt{1 - a^2\mu_y^2};$$
$$\mu_y' = \mu_y\sqrt{\frac{1 - a^2\mu_y^2}{\sigma^2 + \mu_x^2}} \quad (8.30)$$

although with slightly different nondimensional velocities (μ_x', μ_y') than for the corresponding nonintegrated but zero wavenumber subspaces $P_{st}(k_x, 0, \omega)$, $P_{st}(0, k_y, \omega)$ of the full three-dimensional $P_{st}(k_x, k_y, \omega)$ (the variance σ is given by Eqn. (8.22)). In all cases, the anisotropy of the wavenumber/frequency subspaces can be approximated by ellipsoids whose characteristics are determined by the magnitude of the dimensionless horizontal wind, and these ellipsoids have the same shapes at all scales.

We can further reduce the spectra to 1D (denoted E) using:

$$R(\Delta x, 0, 0) = (2\pi)^{-1}\int E(k_x)e^{ik_x\Delta x}dk_x;$$
$$E(k_x) = \int P_{st}(k_x, k_y, \omega)dk_y\, d\omega$$
$$R(0, \Delta y, 0) = (2\pi)^{-1}\int E(k_y)e^{ik_y\Delta y}dk_y;$$
$$E(k_y) = \int P_{st}(k_x, k_y, \omega)dk_x\, d\omega$$
$$R(0, 0, \Delta t) = (2\pi)^{-1}\int E(\omega)e^{i\omega\Delta t}d\omega;$$
$$E(\omega) = \int P_{st}(k_x, k_y, \omega)dk_x\, dk_y \quad (8.31)$$

Using the dimensional (k_x, k_y, ω) this yields:

$$E_x(k_x) \propto |L_w k_x|^{-\beta};\quad E_y(k_y) \propto |L_w k_y/a|^{-\beta};$$
$$E_t(\omega) \propto |\tau_w\omega|^{-\beta};\quad \beta = s - 2 = 1 + \xi(2) \quad (8.32)$$

We can now use these theoretical 1D and 2D formulae to test the full (x, y, t) theory (Eqns. (8.15)–(8.19)). This is done by performing regressions on the various subspaces that can be used to successively estimate the various parameters, which greatly simplifies the parameter estimation. Starting with the 1D analyses (Fig. 8.11a), and following Pinel *et al.* (2012), we can see that the spectra are indeed very close to power laws and remarkably similar to each other if we use the following parameters:

$$s \approx 3.4 \pm 0.1;\quad L_w \approx \tau_w L_{eff,s} \approx 5000\,\text{km};$$
$$\tau_{eff,s} \approx 5\,\text{days};\quad V_w \approx L_{eff,s}/\tau_{eff,s} \approx 11.4 \pm 1.1\,\text{m/s};$$
$$a \approx 1.2 \pm 0.1 \quad (8.33)$$

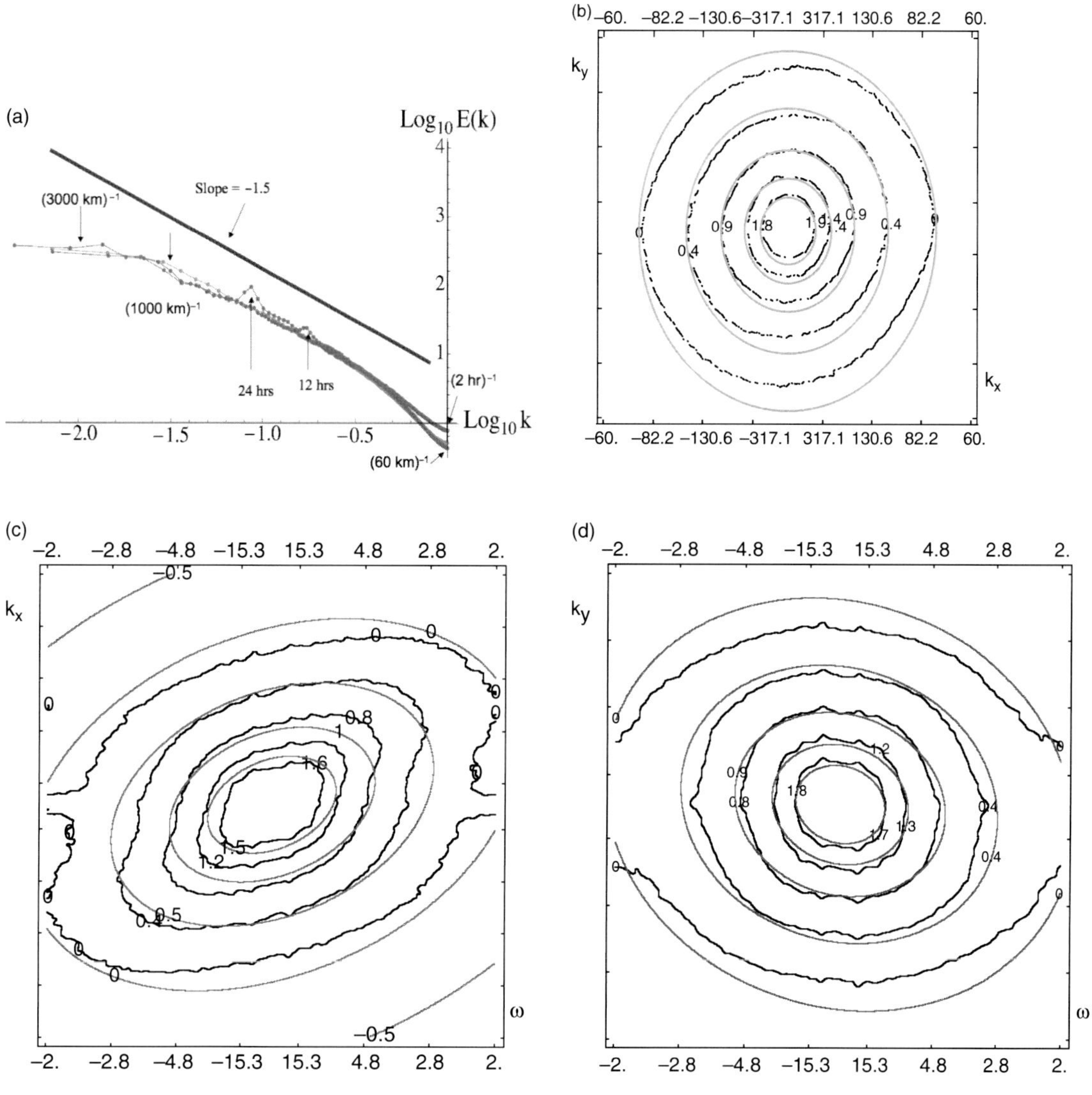

Fig. 8.11 (a) The 1D spectra calculated from the 3D ($P(k_x, k_y, \omega)$) spectral density. The slightly shorter curve with the small but noticeable diurnal peak is $E(\omega)$, the curve just below the arrow $(1000\text{ km})^{-1}$ is $E(k_x)$, the other curve is $E(k_y)$ and the reference line has absolute slope $\beta = 1.5$. (b) Contours of Log$P(k_x, k_y)$: the spatial spectral density. (c) Contours of Log $P(k_x, \omega)$, the zonal wavenumber/frequency subspace. The orientation is a consequence of the mean zonal wind, –3.4 m/s. (d) Contours of Log $P(k_y, \omega)$, the meridional wavenumber/frequency subspace: there is very little if any "tilting" of structures since the mean meridional wind was small: 1.1 m/s. All reproduced from Pinel *et al.* (2012).

the $L_{eff,s}$ is the distance scale where the spectral scaling starts to break down (hence the subscript s) and $\tau_{eff,s} = L_{eff,s}/V_w$ is the corresponding time scale (only V_w is determined directly by the regression and it is comparable to those estimated from the fluxes, Table 4.6). The small deviations from power laws at small and large wavenumbers and frequencies are unimportant "finite size effects" which we demonstrate momentarily. Before proceeding to estimate the mean velocity, we can now use the estimate of s to find H. Using the MTSAT trace moment/flux estimate of $K(2) = 0.12$ (Table 4.6: $C_1 = 0.07$, $\alpha = 1.5$) we obtain: $H = (-3 + s - K(2))/2 = 0.26$, which is near the passive scalar value (1/3) and the H value of many other radiances (see Tables 4.7a, 4.7b).

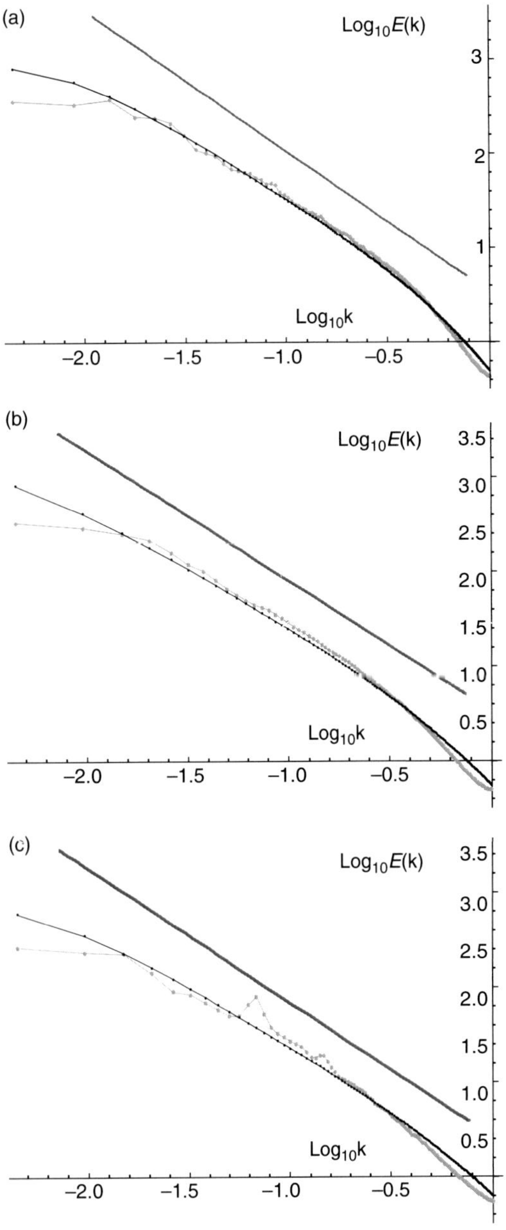

Fig. 8.12 (a) $E(k_x)$ compared with the regression (smooth curve, theory, parameters Eqn. (8.33)) and the line $\beta = 1.5$). The wavenumber range is $(60\text{ km})^{-1}$ to $(13\,000\text{ km})^{-1}$. (b) $E(k_y)$ compared with the regression (smooth curve, theory, same parameters as Fig. 8.12a) and the line $\beta = 1.5$. The wavenumber range is $(60\text{ km})^{-1}$ to $(8000\text{ km})^{-1}$. (c) $E(\omega)$ compared with the regression (smooth curve, theory, same parameters as Fig. 8.12a) and the line $\beta = 1.5$. Note the diurnal spike and its harmonic at $(12\text{ hours})^{-1}$. Frequency range: $(2\text{ h})^{-1}$ to $(276\text{ h})^{-1}$. All reproduced from Pinel *et al.* (2012).

In order to verify the theory more fully, more directly and in order to estimate μ_x, μ_y, we now turn to the various 2D subspaces/sections (Figs. 8.11b, 8.11c, 8.11d). We again see that the comparison is excellent with the additional parameters:

$$\begin{aligned} \mu_x &= -0.3 \pm 0.1; \quad \overline{v_x} = -3.3 \pm 0.1 m/s \\ \mu_y &= 0.1 \pm 0.07; \quad \overline{v_y} = 1.1 \pm 0.08 m/s \end{aligned} \tag{8.34}$$

Hence $\sigma = 0.95 \pm 0.05$.

The relatively small values of the mean winds are consequences of both the near-equator latitudes and also the near-north/south symmetry of the region analysed. We can now return to the issue of low- and high-frequency and wavenumber curvature in the 1D spectrum. Recall that the 1D spectra are obtained by finite sums rather than integrals, and this only over a finite part of Fourier space. There are therefore non-scaling "finite size" effects at both low and high wavenumbers/frequencies. In Figs. 8.12a, 8.12b, 8.12c we successively compare the discretized, numerically integrated 1D spectra (integrated over the part of Fourier space actually observed) based on the theoretical $P(k_x, k_y, \omega)$ with the actual 1D spectra. From these figures we can see that even much of the nonscaling curvature is reproduced as finite size effects at both large and small wavenumbers/frequencies.

However, examination of the 2D spectra shows that there are still residual differences between the empirical and theoretical spectral densities and that it is precisely on these small residual differences between the spectra and the "turbulence background" analysed above that the traditional wave analyses have been made (Wheeler and Kiladis, 1999; Hendon and Wheeler, 2008). Using the same data, we return to the issue of turbulence-generated waves in the next chapter, showing quantitatively how they can help explain the small residuals.

8.4 Space-time relations

8.4.1 Space-time diagrams from the ECMWF interim reanalysis

In Section 8.2.1 we discussed the Lagrangian (comoving)-based relation for the statistics of the lifetimes of structures as functions of their spatial scales, pointing out that they well explained the standard space-time diagrams (Fig. 8.9b). As was mentioned, it is much easier to objectively determine Eulerian (fixed frame) space-time statistics and to produce Eulerian space-

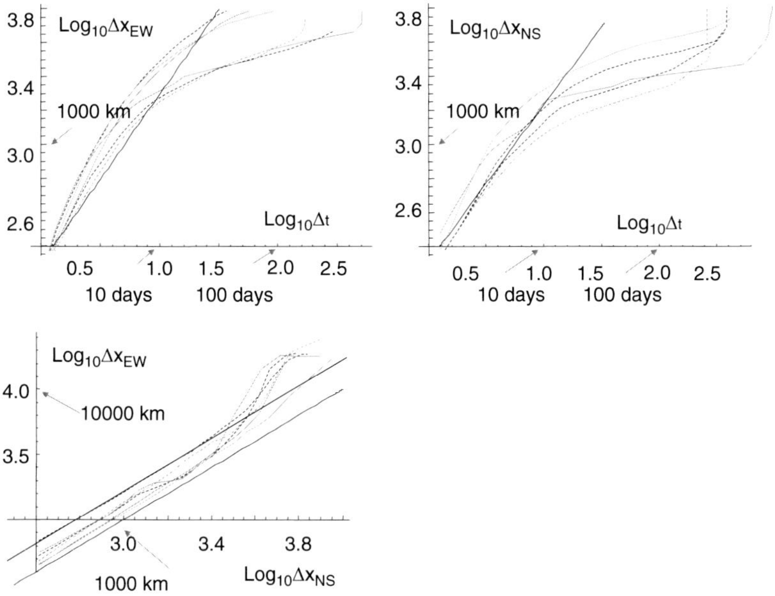

Fig. 8.13 Space-time and space-space plots using the $q = 2$ moments and using $\lambda = \tau_{ref}/\Delta t$ and $\lambda = L_{ref}/\Delta x$ for time and space respectively (east–west and time upper left, north–south and time, upper right, north–south and east–west, lower left). Bottom dashed = h_s, middle dashed = T, top dashed = u, top solid = v, middle solid = w, bottom solid = z. In all cases, the black reference lines have slopes 1; in the space-time diagrams, it corresponds to a speed of ≈225 km/*day*; the spread in the lines indicates a variation over a factor of about 1.6 in speed. In the space-space diagram, the bottom reference line corresponds to isotropy; the top to an aspect ratio of a ≈1.6 difference as discussed in the text. Adapted from Lovejoy and Schertzer (2011).

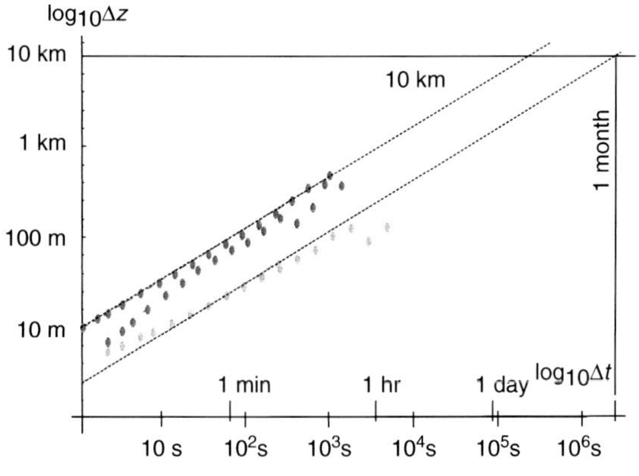

Fig. 8.14 A space-time (vertical/time) diagram obtained from the first-order structure functions of 3 lidar time series at 1 s (top) and 2 s (lower two) resolutions. At the largest scales, the statistics are poor, potentially accounting for the small deviations. We see that the troposphere thickness (which corresponds roughly to planetary sizes in the horizontal) has a time scale of several weeks to a month (see Section 4.1.2). Assuming that $l_s = 1$ m, the top line corresponds to $v = 60$ m/s, the bottom line to 5 m/s (slopes H_z=5/9). If instead l_s = 10 cm, the top line implies 400 m/s, the bottom one to 30 m/s. This is estimated using the formula: $v\Delta t/l_s = (\Delta z/l_s)^{1/H_z}$. Reproduced from Lovejoy and Schertzer (2010b).

time diagrams; indeed, the basic technique was already discussed for estimating space-space diagrams (horizontal/vertical) in Chapter 6. Applying the same technique to space-time, we can use the following implicit relation between length scales L and time scales τ:

$$\left\langle \varphi^{q}_{L_{ref}/L} \right\rangle = \left\langle \varphi^{q}_{\tau_{ref}/\tau} \right\rangle \tag{8.35}$$

Actually, in principle, Eqn. (8.35) gives a different $L - \tau$ relation for each q value. However, in the simplest GSI case where the C_1 and α for the spatial and temporal analyses are the same (as is roughly the case here: see Table 8.1), any q will give the same relationship, although larger values of q will give more statistically accurate results (as long as the moments are not so large as to be spuriously dependent on a few extreme values; recall that for $q = 1$, $<\varphi>$ is independent of scale so that it cannot be used). This was indeed shown to be empirically valid on ERA40 reanalyses (Stolle *et al.*, 2012). Here we chose $q = 2$, which has the advantage that the corresponding $K(2)$ is precisely the intermittency correction necessary for the spectrum; this is needed below.

Fig. 8.13 shows the results for $q = 2$ for the three pairs of directions: NS/EW, EW/time, NS/time. The space-time diagrams show that a linear (constant-velocity) relation between space and time works reasonably well up to 2000–2500 km in space and up to time scales of ~7–10 days in time. After 7–10 days there is a drastic change in the relationship; this is the transition to the macroweather regime. While the space-space diagram shows that structures are typically elongated in the EW direction by factors up to $a = L_{eff,EW}/L_{eff,NS} \approx 1.6$, comparable to the

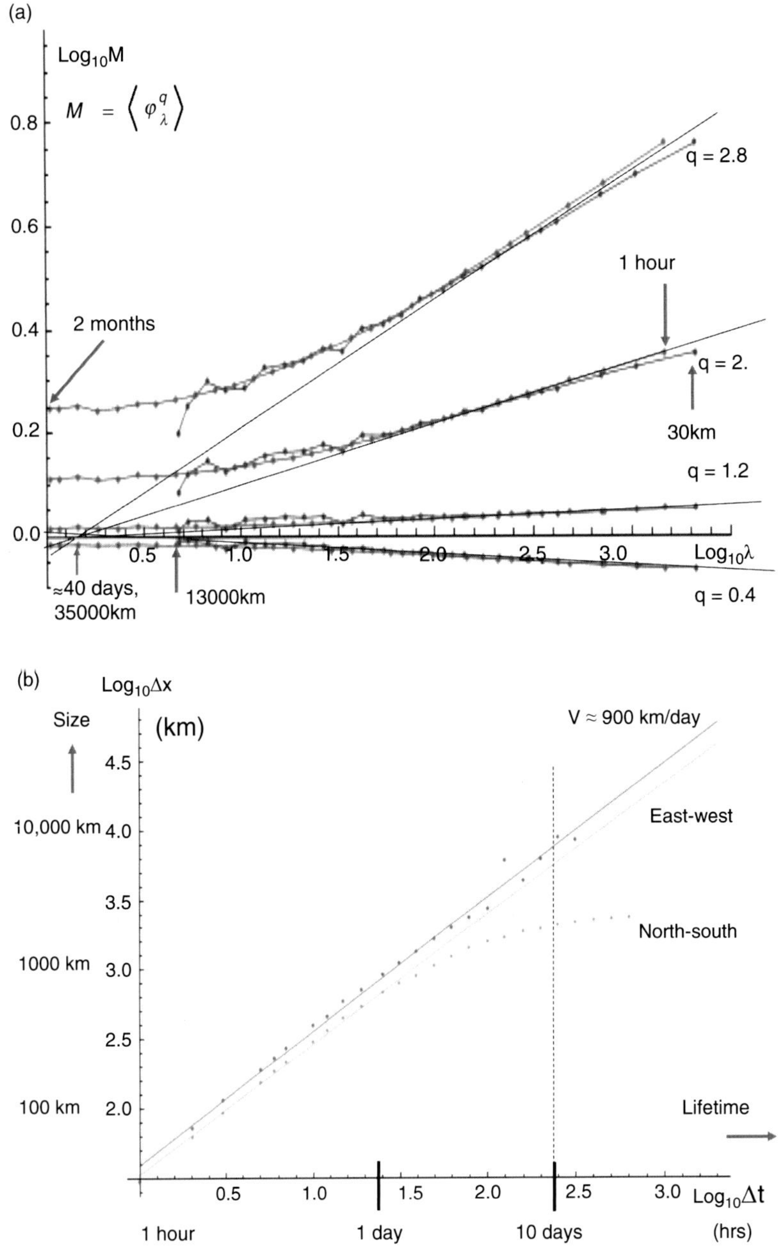

Fig. 8.15 (a) A comparison of $\log_{10}M_q(\Delta t)$ for east–west IR radiance fluxes (shorter and to the left) and time (longer and to the right) for $q = 0.4, 1.2, 2, 2.8$. λ is defined with respect to a time scale of 2 months for the temporal analyses and 20 000 km for the spatial analyses. The spatial $\log_{10}M_q(\Delta x)$ has been shifted so as to superpose as closely as possible on the $\log_{10}M_q(\Delta t)$ curves. The corresponding speed is ~900 km/day (10 m/s) and the outer cascade scale is ~40 days in time, ≈ 35 000 km in space. The deviations from scaling become important at ~5000 km or ~6 days. Compare this with the nearly perfectly scaling Fig. 4.10, which is the geometric mean of the east–west above with the north–south analysis. Reproduced from Pinel *et al.* (2012). (b) The horizontal space-time diagram constructed from Fig. 8.15a (upper curve and straight line) and the corresponding diagram from the north–south M_q (lower). Reproduced from Pinel *et al.* (2012).

value of the (average) ratio discussed above, the space-time diagrams indicate effective space-time transformation speeds $V_{eff} = L_{eff}/\tau_{eff}$ in the range ~200–400 km/day (≈ 2.5–5 m/s).

8.4.2 Space-time diagrams from lidar

In the previous subsection, we used the turbulent fluxes to relate space and time to determine a (Eulerian) space-time diagram for the ECMWF interim reanalyses. This method was particularly pertinent for the reanalyses, since we found in Chapter 6 that the ECMWF fields (as opposed to the fluxes) were more complex, involving possibly spurious horizontal (scaling, nontrivial) anisotropies. However, space-time relations can be established using other statistics: in Section 6.6 we used spectra to relate the east–west and north–south statistics and in Section 6.5.2 we used structure functions to relate the horizontal and vertical for lidar aerosol backscatter. Using this structure function technique and the same lidar data (except for (z, t) rather than (x, z) sections), for a given Δt, let us estimate the corresponding $\Delta z(\Delta t)$ from the solution of the implicit equation using the first-order structure functions $\langle|\Delta B(\Delta z)|\rangle = \langle|\Delta B(\Delta t)|\rangle$ (B is the lidar backscatter ratio, ΔB is a flucutation estimated from differences). For three of the longer (z, t) sections, the results are shown in Fig. 8.14. We see that the data follow reasonably accurately the theoretical curve (assuming horizontal wind dominated temporal statistics and $H_z = 5/9$). In addition, if the sphero-scale is assumed to be 1 m (roughly what was determined for the vertical section data in Fig. 6.21), then we find a horizontal wind in the range 5–60 m/s, which is quite reasonable. We also see that the space-time diagram gives direct evidence that the top of the troposphere ($\approx$ 10 km) corresponds to the outer time scale $\approx$ 2 weeks.

8.4.3 Space-time diagrams from MTSAT thermal IR

We have considered the MTSAT data in some detail already, both the spatial (Chapter 4) and temporal (Section 8.1.5) cascade structures as well as the space-time spectra (Sections 8.3.1, 8.3.2). Fig. 8.15a shows the superposition of the spatial (zonal) trace moments with the temporal trace moments when the fluxes are determined from the spatial Laplacians and there is a left–right displacement corresponding to a horizontal speed of 900 km/day ($\approx$ 10.4 m/s), which is very close to the turbulent velocity $v \approx 9.3$ m/s deduced from the spectra (Section 8.3.2). Interestingly, although the zonal scaling is not so good at the largest scales (the average of the zonal and meridional moments have much better scaling: see Fig. 4.10), the temporal and zonal scaling have nearly identical deviations from pure power-law scaling. The result is that the zonal space-time diagram is nearly perfectly scaling over the entire range (Fig. 8.15b).

8.4.4 Space-time diagrams from TRMM thermal IR

Let us now consider the TRMM data at a 12-hour resolution, averaging over one year ($\approx$ 5300 orbits) of the thermal IR data remapped to 100×100 km grids. The results are shown in Fig. 8.16a. Note that for resolutions below 2 days, the statistics are poor since only a small fraction of the 100 × 100 km "pixels" are visited at such small time intervals. We see that the

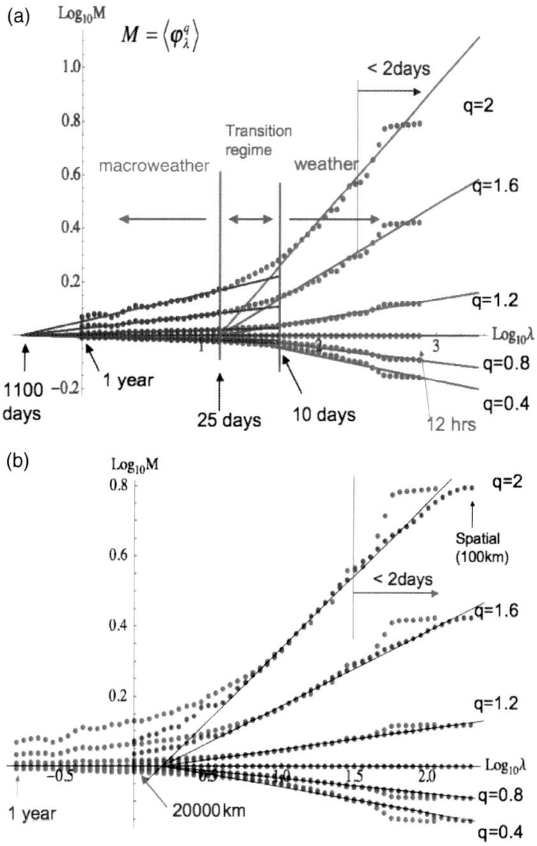

Fig. 8.16 (a) The normalized moments of the TRMM thermal IR data averaged over 100 × 100 km pixels at 12-hour resolution from 5300 orbits (1 year corresponding to $\lambda = 1$). The long time variability has been fitted to a cascade with outer scale at 1100 days, which could be a consequence of the ocean cascade. Adapted from Lovejoy and Schertzer (2010). (b) The same normalized moments of the TRMM thermal IR data as Fig. 8.16a but with temporal and spatial moments superposed, corresponding to a velocity of 400 km/day. The longer series of dots to the left is the temporal analysis (from Fig. 8.16a), the shorter series of dots to the right is the east–west spatial analysis (corresponding to Fig. 4.9b except that the analysis is not along orbit, and is at lower resolution). Reproduced from Lovejoy and Schertzer (2010).

plot can be divided into three regions. Up to about 10 days, the moments are relatively linear, as expected from space-time multiplicative cascade processes. If we extrapolate the lines to larger scales, they cross at about 25 days; the variability at less than 10 days is accurately that which would have resulted from a multiplicative cascade starting at 25 days. To aid the interpretation, Fig. 8.16b shows the superposition of the zonal spatial analysis of the same data. We see that – although the resolution is much lower and hence the scaling region much shorter – up to about 10 000 km and 10–15 days the fit is comparable to that of the MTSAT data (although the velocity is somewhat smaller, ~400 km/day).

8.5 Summary of emergent laws in Chapter 8

To treat time, we first make a straightforward extension from space to space-time, introducing the space-time scale function:

$$[\![T_\lambda \underline{\Delta R}]\!] = \lambda^{-1}[\![\underline{\Delta R}]\!];\ T_\lambda = \lambda^{-G_{st}} \quad (8.36)$$

where:

$$\underline{\Delta R} = (\underline{\Delta r}, \Delta t) \quad (8.37)$$

and G_{st} is the 4 × 4 matrix generator of the space-time scale transformation matrix T_λ and the space-time scale function is $[\![\underline{\Delta R}]\!]$. The space-time fluctuations in the fields are now related to the turbulent fluxes φ by:

$$\Delta f(\underline{\Delta R}) = \varphi_{[\![\underline{\Delta R}]\!]}[\![\underline{\Delta R}]\!]^H \quad (8.38)$$

The FIF model interprets this as a fractional integration of order H, i.e. in Fourier space as a power-law filter, and in real space as a convolution with a power law:

$$\tilde{f}(\underline{K}) = \tilde{\varphi}(\underline{K})[\![\underline{K}]\!]^{-H};\ f = \varphi * [\![\underline{R}]\!]^{-(D_{st}-H)} \quad (8.39)$$

And the structure function and spectral densities are:

$$\left\langle \Delta I(\underline{\Delta R})^2 \right\rangle = [\![\underline{\Delta R}]\!]^{\xi(2)};\ P_{st}(\underline{K}) = \left\langle |\tilde{I}(\underline{K})|^2 \right\rangle \approx [\![\underline{K}]\!]^{-s_{st}};$$
$$s_{st} = D_{st} - \xi(2) \quad (8.40)$$

where D_{st} is the space-time elliptical dimension = $Trace(G_{st})$ and $\xi(2)$ is the structure function exponent for $q = 2$ and $\underline{K} = (k_x, k_y, \omega)$.

Although comoving (Lagrangian) horizontal-time scaling involves an anisotropic (*x,y,t*) space generator G_{xyt}, taking into account horizontal advection and averaging over random advection velocities leads to G_{xyt} as the identity and with trivial anisotropy specified by a 3 × 3 matrix $\underline{B}$ which accounts for advection and trivial zonal/meridional spatial anisotropy. In Fourier space:

$$[\![\underline{\Delta R}]\!] = (\underline{\Delta R}^T B\, \underline{\Delta R})^{1/2};\ [\![\underline{K}]\!] = (\underline{K}^T B^{-1}\, \underline{K})^{1/2} \quad (8.41)$$

where:

$$B = \begin{pmatrix} 1 & 0 & -\mu_x \\ 0 & a^2 & -a^2\mu_y \\ -\mu_x & -a^2\mu_y & 1 \end{pmatrix};$$

$$B^{-1} = \frac{1}{1-\mu_x^2-a^2\mu_y^2}\begin{pmatrix} 1-a^2\mu_y^2 & \mu_y\mu_x & \mu_x \\ \mu_y\mu_x & (1-\mu_x^2)/a^2 & \mu_y \\ \mu_x & \mu_y & 1 \end{pmatrix} \quad (8.42)$$

where μ_x, μ_y are horizontal mean advection velocities nondimensionalized by the RMS ("turbulent") velocity (Eqn. (8.12)) and a is the mean zonal-to-meridional aspect ratio. With these, we obtain the following real-space (horizontal-time) scale functions:

$$[\![\underline{\Delta R}]\!] = (|\underline{\Delta r'}|^2 + \Delta t^2 - 2\underline{\Delta r'}\cdot\underline{\mu}\Delta t)^{1/2};$$
$$[\![\underline{K}]\!] = (\omega'^2 + \|\underline{k}\|^2)^{1/2} \qquad \omega' = (\omega + \underline{k}\cdot\underline{\mu})/\sigma;$$
$$\|\underline{k}\| = (k_x^2 + a^{-2}k_x^2)^{1/2}; \qquad \underline{\Delta r'} = (\Delta x, a\Delta y) \quad (8.43)$$

where the coordinates, wavenumbers and frequencies nondimensionalized by the external scales L_e, τ_w respectively. The (dimensional) structure function and spectral density are:

$$\langle \Delta I(\underline{\Delta R})^2 \rangle \propto (|\underline{\Delta r'}|^2 + \Delta t^2 - 2\underline{\Delta r'}\cdot\underline{\mu}\Delta t)^{\xi(2)/2}$$
$$P_{st}(\underline{K}) \propto (\omega'^2 + \|\underline{k}\|^2)^{-s_{st}/2}; \qquad s_{st} = 3 - \xi(2) \quad (8.44)$$

Appendix 8A: The effect of the vertical wind on the temporal statistics

Dimensional analysis in a Lagrangian frame (Eqn. (8.2*d*)) yields $H = 1/2$, hence a (nonintermittent) spectral exponent $\beta = 1 + 2H = 2$, and there have indeed been several observations of wind spectra roughly of the predicted form $E(\omega) \approx \omega^{-2}$ (see the discussion and references in Radkevitch *et al.*, 2008). However, according to our analysis, such Lagrangian scaling should be observed neither for scales $< \tau_w$ (due to the "sweeping" of small eddies past a fixed observer as they are advected by large eddies) nor for scales $> \tau_w$ (due to the weather/macroweather transition, i.e. the breakdown of the scaling). In this appendix, we give a summary of an explanation for the occasional observation of $\beta \approx 2$ scaling developed in Lovejoy *et al.* (2008), which potentially accounts for this. It is based on the vertical stratification combined with the assumption of $H_w < 0$ scaling of the vertical wind (w). The complications discussed here are important for the smaller temporal scales and in the boundary layer, since for a given time lag Δt their will be a critical height $z_{alt} = w\Delta t$ below which there is a wall-induced change in behaviour.

According to Section 8.2.3, the largest eddies "sweep" (Tennekes, 1975) the smaller ones so that for time scales less than about 2 weeks we can ignore the pure time development term: $(\Delta t/\tau_s)^{2/H_z}$ (see Eqn. (8.8) with $\Delta x = \Delta y = \Delta z = 0$). This just leaves the horizontal and vertical advection terms ($v\Delta t/l_s$ and $(w\Delta t/l_s)^{1/Hz}$). In order to compare them we must take into account the fact that while the mean horizontal wind across a given part of the earth may be relatively large and well defined (and insensitive to the resolution, which mostly affects its fluctuations), the same is not true of the vertical wind. When w is averaged over time scale Δt (denoted $w_{\Delta t}$) since $H_w < 0$ it tends to zero as the region of interest increases in size and with increasing temporal averaging (Δt); in other words, statistically $w_{\Delta t} \approx \Delta t^{H_w}$ where H_w is a small but on average *negative* exponent (see Chapter 10 for more discussion of $H < 0$, and see Table 4.1 for the estimate $H_w \approx -0.14$). Lovejoy *et al.* (2008) argue that statistically the net effect of this is to replace $(\Delta t/\tau_s)^{1/H_t}$ with $(\Delta t/\tau'_s)^{1/H'_t}$ where τ'_s and H'_τ are "effective" parameters, and (putting the "effective vertical velocity" to zero) we may replace G_{st} with an "effective generator" and effective advection matrix:

$$G_{st,\,eff} = \begin{pmatrix} 1 & 0 & 0 & 0 \\ 0 & 1 & 0 & 0 \\ 0 & 0 & H_z & 0 \\ 0 & 0 & 0 & H'_t \end{pmatrix}; \quad A_{eff} = \begin{pmatrix} 1 & 0 & 0 & v_x \\ 0 & 1 & 0 & v_y \\ 0 & 0 & 1 & 0 \\ 0 & 0 & 0 & 1 \end{pmatrix};$$

$$D_{st,\,eff,\,advec} = Tr\left(A_{eff}^{-1} G_{st,\,eff} A_{eff}\right) = TrG_{st,\,eff} = 2 + H_z + H'_t \tag{8.45}$$

with corresponding "effective scale function":

$$[\![\underline{\Delta R}]\!]_{advec,\,eff,\,can} = [\![A^{-1}\underline{\Delta R}]\!]_{eff,\,can} = l_s\left(\left(\frac{\Delta x - v_x\Delta t}{l_s}\right)^2 + \left(\frac{\Delta y - v_y\Delta t}{l_s}\right)^2 + \left(\frac{\Delta z}{l_s}\right)^{2/H_z} + \left(\frac{\Delta t}{\tau'_s}\right)^{2/H'_\tau}\right)^{1/2} \tag{8.46}$$

The exponent H_t' and the value τ_s' depend on the exact (scaling) statistics of the vertical wind, which are not known, although Radkevitch *et al.* (2008) find empirically using lidar (z, t) sections that $H_t' \approx 0.7$ and τ_s' has a large variability but is somewhat larger than the spherotime τ_s (which is also highly variable: see the discussion in Lovejoy *et al.*, 2008). Using meteorological analyses, Radkevitch *et al.* (2008) show that nevertheless the pure temporal development will still dominate at large enough time scales. In any case the global-scale analyses presented earlier were for (x,y,t) radiation fields.

Chapter 9

Causal space-time cascades: the emergent laws of waves, and predictability and forecasting

9.1 Causality

9.1.1 Causal and acausal impulse response functions and fractional derivatives

Up until now, we have treated time as though it were no different from space, and we have deliberately avoided discussion of a crucial difference: causality. Whereas we have typically treated spatial coordinates as though they were left/right symmetric (often a good approximation in the atmosphere), we certainly cannot treat time in this way: the past influences the future, but not the converse! This motivated the introduction of causal space-time cascades (Marsan *et al.*, 1996; Schertzer *et al.*, 1997, 1998).

To see this most simply, consider the Hth-order fractional derivative equation for the impulse response function $g(t)$ (the "Green's function"):

$$\frac{d^H g}{dt^H} = \delta(t) \tag{9.1}$$

where $\delta(t)$ is the usual Dirac delta function. Fourier transforming both sides of the equation, we obtain:

$$\tilde{g}(\omega) = (i\omega)^{-H} \tag{9.2}$$

where we have used the fact that the Fourier transform of the δ function = 1, that the Fourier transform of d/dt is $-i\omega$, and have indicated the Fourier transform by the tilde (~). This g can be used to solve the general inhomogeneous fractional differential equation:

$$\frac{d^H h}{dt^H} = f(t); \quad h = I^H f \tag{9.3}$$

where $h(t)$ is the response to the forcing $f(t)$. We have written the equation both in differential and in the equivalent integral form, where I^H is the Hth-order integral operator, the inverse of d^H/dt^H. The solution of Eqn. (9.3) is thus:

$$\tilde{h} = \tilde{g}\tilde{f} = (i\omega)^{-H}\tilde{f} \overset{F.T.}{\leftrightarrow} h = g * f \tag{9.4}$$

where $*$ indicates convolution (we have used the fact that multiplication in Fourier space corresponds to convolution in real space). Comparing Eqns. (9.3) and (9.4) we see that:

$$h = I^H f = g * f \tag{9.5}$$

where:

$$\tilde{g}(\omega) = (i\omega)^{-H} \overset{F.T.}{\leftrightarrow} g(t) = \frac{\Theta_{Heavi}(t)t^{-(1-H)}}{\Gamma(H)};$$

$$\Theta_{Heavi}(t) = \begin{Bmatrix} 0 & t < 0 \\ 1 & t \geq 0 \end{Bmatrix} \tag{9.6}$$

$\Theta_{Heavi}(t)$ is the Heaviside function. Writing the final solution explicitly, we obtain:

$$h(t) = I_L^H f(t) = \frac{1}{\Gamma(H)} \int_{-\infty}^{t} (t-t')^{H-1} f(t')dt' \tag{9.7}$$

where we have added the subscript L to indicate that the fractional integral is of the "Liouville" type and $\Gamma(H)$ is the usual gamma function (not to be confused with the cascade generator). We can see by inspection that it is causal in the sense that the value of the response $h(t)$ depends only on the forcing for times $t' \leq t$, a consequence of the fact that the Green's function $g = 0$ for $t < 0$. This is indeed a general feature of the time derivative operator.

We can compare this extreme asymmetric fractional derivative (i.e. of only nonzero for $t > 0$) to the symmetric fractional derivative used in the spatial simulation discussed in Chapter 5, which was based on powers of the absolute value: $|\omega|^{-H}$. The latter corresponds to the Riemann–Liouville (RL) fractional integration:

$$I_{RL}^H f(t) = \frac{1}{\Gamma(H)} \int_{-\infty}^{\infty} |t-t'|^{H-1} f(t')dt' \tag{9.8}$$

which is based on the Green's function:

$$\tilde{g}(\omega) = |\omega|^{-H}\sqrt{\frac{2}{\pi}}\sin\frac{\pi}{2}(1-H) \overset{F.T.}{\leftrightarrow} g(t) = \frac{|t|^{H-1}}{\Gamma(H)} \tag{9.9}$$

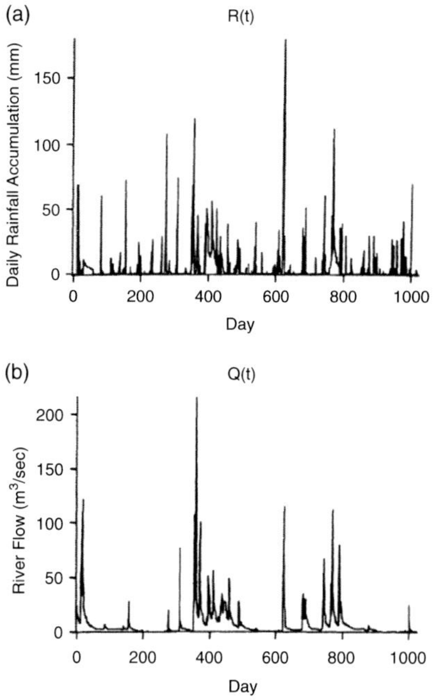

Fig. 9.1 (a) Simultaneous daily rainfall accumulation and river flow for the Le Gardon St. Jean river. Adapted from Tessier *et al.* (1996). (b) The nearby station of Corbes Roc Courbes (France). Adapted from Tessier *et al.* (1996).

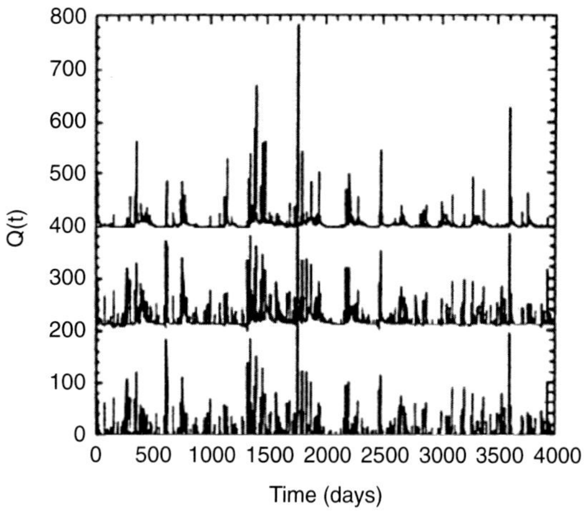

Fig. 9.2 Comparison of rainfall data (bottom), river-flow data (top) and rainfall data fractionally integrated with a causal filter (transfer function) (middle). The fields have been offset for clarity. Reproduced from Tessier *et al.* (1996).

As an example, Tessier *et al.* (1996) proposed that the relation between rainfall $R(t)$ in small river basins could be considered as the forcing of the corresponding river flow $Q(t)$ in the fractional differential equation:

$$\frac{d^H Q}{dt^H} = R(t); \quad Q(t) = \frac{1}{\Gamma(H)} \int_{-\infty}^{t} (t-t')^{H-1} R(t')dt' \tag{9.10}$$

where empirically it was found that $H = 0.3$. In this hydrology context, the Green's function g corresponding to the Liouville fractional integral is called a "transfer function." Physically this convolution corresponds to a specific power-law (scaling) "storage" model for the runoff and groundwater processes, which are thus assumed to be scaling over a wide range. Examples are shown in Figs. 9.1 and 9.2.

9.1.2 Causality in space-time: propagators

If we extend the above discussion to space-time, then the corresponding Green's function/impulse response function is called a "propagator." Let us consider as an example the propagator for the classical wave equation:

$$\left(\nabla^2 - \frac{1}{V^2}\frac{\partial^2}{\partial t^2}\right) g(\underline{r}, t) = \delta(\underline{r}, t) \tag{9.11}$$

where V is the wave speed. Taking the space-time Fourier transform of both sides, we find:

$$\tilde{g}(\underline{k}, \omega) = \left(\omega^2/V^2 - |\underline{k}|^2\right)^{-1} \tag{9.12}$$

Because of the negative sign, the character of this propagator is totally different from those obtained with a positive sign (relevant to the discussion in Chapter 8). Its behaviour is totally dominated by the waves satisfying the relation $\omega^2/V^2 = |\underline{k}|^2$, which makes the propagator singular; this is indeed the significance of this "dispersion" relation. Note that in terms of the scaling symmetries, $\tilde{g}(\underline{k}, \omega)$ (and hence $g(\underline{r}, t)$) are symmetric with respect to $G = 1$ (the identity matrix): $\tilde{g}(\lambda^{-1}(\underline{k}, \omega)) = \lambda^s \tilde{g}(\underline{k}, \omega)$; the solution to this functional equation is simply $\tilde{g}(\underline{k}, \omega) = [[(\underline{k}, \omega)]]^{-s}$ where the space-time scale function $[[(\underline{k}, \omega)]]$ is symmetric with respect to $G = 1$ and in Eqn. 9.12 $s = 2$.

We can now use the propagator to solve the inhomogeneous wave equation:

$$\left(\nabla^2 - \frac{1}{V^2}\frac{\partial^2}{\partial t^2}\right) I(\underline{r}, t) = h(\underline{r}, t) \tag{9.13}$$

for the wave I due to the forcing h:

$$\widetilde{I}(\underline{k},\omega)=\frac{\widetilde{h}(\underline{k},\omega)}{\left(\omega^2/V^2-|\underline{k}|^2\right)} \tag{9.14}$$

The $\widetilde{g}(\underline{k},\omega)$ in Eqn. (9.12) is causal since it was derived directly from a usual (causal) PDE. However, for a given $\widetilde{g}(\underline{k},\omega)$ it is useful to be able to directly determine whether or not it is causal. To see how to do this, take the inverse Fourier transform with respect to ω only. We obtain:

$$\widetilde{g}(\underline{k},t)=\int_{-\infty}^{\infty}\frac{e^{-i\omega t}d\omega}{\left(\omega^2/V^2-|\underline{k}|^2-2i\delta\right)} \tag{9.15}$$

where we have added in an extra $-2i\delta$ term that we will take to zero momentarily. The above integral has two singularities/poles in the lower half complex ω plane at:

$$\omega=-i\delta\pm\sqrt{|k|^2-\delta^2} \tag{9.16}$$

We can evaluate the integral (Eqn. (9.15)) by contour integration by first making a branch cut connecting the two poles by a line (for $\delta>0$, this will be in the lower half complex ω plane) and completing the contour by enclosing the entire upper complex plane (which is analytic). If we now consider $t<0$, then closing the integral along the real ω axis by an (infinite) semicircle in the upper half of the complex ω plane encloses no poles (singularities) so that the $\widetilde{g}(\underline{k},t)=0$ for all $t<0$. However, when $t>0$, to obtain a closed path integral with zero contribution from the infinite semicircular part we must use the lower half ω plane which on the contrary encloses the poles and is hence nonzero. This illustrates the general condition needed for causality: it suffices for the propagator to be analytic in the upper-half ω plane.

The wave equation propagator is unlocalized in space-time since the dominant (divergence/singular) contribution occurs over space-time lines: the dispersion relation $\omega=\pm Vk$. In contrast, the purely spatial Green's function $|k|^{-2}$ obtained by integrating out ω is localized, since it is singular only at the point at the origin so that $|k|=0$. Physically, this corresponds to the fact that wave packets can maintain their spatial coherence (localization) while being delocalized in space-time (the packets can propagate). Let us now consider the propagator corresponding to the turbulent scale functions and spectral densities discussed in Chapter 8. For example, the usual FIF model for the observable I is:

$$I(\underline{r},t)=\varepsilon(\underline{r},t)*g(\underline{r},t);\quad \widetilde{I}(\underline{k},\omega)=\widetilde{\varepsilon}(\underline{k},\omega)\widetilde{g}(\underline{k},\omega) \tag{9.17}$$

where up until now we have taken g as a pure power law of a scale function. We can now ask: what is the (causal) space-time propagator g needed to filter a turbulent flux such that the spectral density of its corresponding observable is filtered by a power law of order H? From Eqn. (9.17) we see that the spectral energy density of the fractionally integrated field I is:

$$P_I(\underline{k},\omega)\propto\langle|\widetilde{I}|^2\rangle=|\widetilde{g}|^2P_\varepsilon(\underline{k},\omega);\quad P_\varepsilon(\underline{k},\omega)\propto\langle|\widetilde{\varepsilon}|^2\rangle \tag{9.18}$$

For space-time localized turbulent fields we therefore have:

$$|\widetilde{g}(\underline{k},\omega)|^2=[[(\underline{k},\omega)]]^{-2H};\quad [[(\underline{k},\omega)]]^2=\omega^2+\|\underline{k}\|^2 \tag{9.19}$$

(i.e. Eqn. (8.24) with mean velocity $\mu=0$, $\sigma=0$ and $a=1$). In order to satisfy both Eqn. (9.19) and respect causality, it suffices to take:

$$\widetilde{g}(\underline{k},\omega)=\left(i\omega+\|\underline{k}\|\right)^{-H} \tag{9.20}$$

(since $\|\underline{k}\|\geq 0$ for all $\underline{k}$). This is clearly localized in space-time since it is only singular at the point $\omega=0$, $\|\underline{k}\|=0$.

Taking the spatially isotropic case $\|\underline{k}\|=|\underline{k}|$ we see that the turbulent propagator Eqn. (9.19) corresponds to the real-space $g(\underline{r},t)$ satisfying:

$$\left(\left(-\nabla^2\right)^{1/2}+\frac{\partial}{\partial t}\right)^H g(\underline{r},t)=\delta(\underline{r},t) \tag{9.21}$$

where the operator $\left(-\nabla^2\right)^{1/2}$ is a (fractional) Laplacian operator, i.e. the real-space differential operator corresponding to multiplication by $|\underline{k}|$ in Fourier space. Note: the fractional operator on the left-hand side is defined by the inverse Fourier transform of $\left(|\underline{k}|+i\omega\right)^H$. This can be compared to the fractional wave equation:

$$\left(\nabla^2-\frac{1}{V^2}\frac{\partial^2}{\partial t^2}\right)^{H/2}g(\underline{r},t)=\delta(\underline{r},t) \tag{9.22}$$

Clearly the fractional wave equation has the same dispersion relation as the usual wave equation (obtained with $H=2$), so we anticipate that many of the properties of the solutions will be the same (i.e. they will be insensitive to the value of H as long as $H>0$ and they will not be localized in space-time; they will propagate).

We can now show that the above propagators (Eqns. (9.21), (9.22)) are indeed causal; indeed, the more general mixed turbulence and fractional wave propagator:

$$\widetilde{g}(\underline{k},\omega)=\left(-i\omega+\|\underline{k}\|\right)^{-H_{tur}}\left(\omega^2/V^2-\|\underline{k}\|^2-2i\delta\right)^{-H_{wav}/2} \quad (9.23)$$

is also causal where $\|\underline{k}\|$ is the spatial scale function (see Chapter 8), and again we will take $\delta \to 0$ after evaluating the integral for $\widetilde{g}(\underline{k}, t)$; the cases $H_{wav} = 0$ corresponds to the localized case just discussed, whereas $H_{tur} = 0$ corresponds to a fractional wave equation (*wav* for "wave" and *tur* for "turbulence"). Applying the same analysis as above, we see that the poles are at:

$$\omega = \frac{-i\|\underline{k}\|V}{-i\delta \pm \sqrt{\|\underline{k}\|^2 - \delta^2}} \quad (9.24)$$

which are all in the lower half ω plane as required for causality. Note that if we include advection, then $\omega \to \omega + \underline{k}.\underline{v}$ which just shifts the singularities parallel to the real axis so that the poles stay in the lower half complex ω plane. We can again make a branch cut between the two poles lying near the real ω axis and another between the third pole on the imaginary axis and either of the other two. The resulting propagator will be single-valued as long as $H_{tur} + H_{wav} < 1$. Again, the entire upper-half ω plane is analytic, so we conclude that the hybrid turbulent wave propagator g in Eqn. (9.23) is causal.

9.1.3 The causal space-time fractionally integrated flux (FIF) model

We have discussed the modifications of the FIF model needed in the propagator which relates the observable I to the flux ε (Eqn. (9.17)); let us denote it with a subscript I: $g_I(\underline{r}, t)$. It only remains to consider the convolution which relates the subgenerator $\gamma_\alpha(\underline{r}, t)$ (which is a noise composed of independent identically distributed extremal Lévy random variables: see Section 5.5) with the generator $\Gamma(\underline{r}, t)$; this must also be causal, and let us denote it by a subscript ε: $g_\varepsilon(\underline{r}, t)$:

$$\Gamma(\underline{r}, t) = \gamma_\alpha(\underline{r}, t) * g_\varepsilon(\underline{r}, t); \quad \widetilde{\Gamma}(\underline{k}, \omega) = \widetilde{\gamma}_\alpha(\underline{k}, \omega)\widetilde{g}_\varepsilon(\underline{k}, \omega) \quad (9.25)$$

The conserved flux ε is then obtained by exponentiation:

$$\varepsilon(\underline{r}, t) = e^{\Gamma(\underline{r}, t)} \quad (9.26)$$

The observable I is then obtained by a final convolution with g_I:

$$I(\underline{r}, t) = \varepsilon(\underline{r}, t) * g_I(\underline{r}, t); \quad \widetilde{I}(\underline{k}, \omega) = \widetilde{\varepsilon}(\underline{k}, \omega)\widetilde{g}_I(\underline{k}, \omega) \quad (9.27)$$

In order to satisfy the scaling symmetries, the propagators $g_\varepsilon(\underline{r}, t), g_I(\underline{r}, t)$ need not be identical; it suffices that both satisfy generalized scale equations:

$$g\left(T_\lambda(\underline{r}, t)\right) = \lambda^{(D_{st}-H)} g(\underline{r}, t); \quad T_\lambda = \lambda^{-G_{st}} \quad (9.28)$$

where for g_ε the order H must be chosen $= D_{st}(1 - 1/\alpha)$ (recall that $D_{st} = \mathrm{Trace}(G_{st})$ is the "elliptical dimension" characterizing the overall stratification of space-time), whereas for g_I any H can be chosen (depending on the field modelled). The relevant solutions of Eqn. (9.28) are powers of scale functions with a Heaviside function (Eqn. (9.6)) $\Theta(t)$ needed to ensure that causality is respected:

$$g(\underline{r}, t) = \Theta_{Heavi}(t)[[(\underline{r}, t)]]^{-(D-H)} \quad (9.29)$$

We note that for ε and for modelling positive fields (such as the passive scalar fields), this is adequate. However, for the velocity field, it may be of interest to have symmetric positive, negative fluctuations; this can be achieved by multiplying g_I by the factor $sign(\Delta x)\, sign(\Delta y)\, sign(\Delta z)$ (a better solution is to use the real part of a complex cascade (Schertzer and Lovejoy, 1995), see Box 5.6).

We have seen from the previous section that the propagators can equally well be specified in Fourier space; according to the discussion in the previous section, for g to be causal, it suffices that the transform $\widetilde{g}(\underline{k}, \omega)$ is purely analytic in the upper-half complex ω plane. Physically, we have argued that the turbulent flux ε should be localized in both space and in space-time in order to reproduce the basic phenomenology that small turbulent "patches" live for a power-law duration of their size. Similarly observables should be spatially localized (otherwise there would be no physical reality behind eddies or wave-packets), but they need not be space-time localized; such a delocalization typically involves a special set of frequencies and wavenumbers which render the propagator singular: the dispersion relation. In this way scaling symmetries can allow for the emergence of waves driven by turbulent fluxes.

For future reference, we can now rewrite the FIF model in terms of (fractional) partial differential equations. For example, with the help of Eqn. (9.21), in the simple case of horizontal spatial isotropy (ignoring the vertical direction), we may write the equation for the FIF generator as:

$$\left((-\nabla^2)^{1/2}+\frac{\partial}{\partial t}\right)^{H_\alpha}\Gamma(\underline{r},t)=\gamma_\alpha(\underline{r},t);$$
$$H_\alpha=D/\alpha'; \quad 1=\frac{1}{\alpha}+\frac{1}{\alpha'} \tag{9.30}$$

where $\gamma_\alpha(\underline{r}, t)$ is an extremal Lévy noise index α, $D=3$ (the dimension of horizontal space-time) and α' is the usual auxiliary variable whose definition is recalled above (for related fractional Lévy equations, see Schertzer *et al.*, 2001). Similarly, the observable I is given by:

$$\left((\ \nabla^2)^{1/2}+\frac{\partial}{\partial t}\right)^{H_I} I(\underline{r},t)=\varepsilon(\underline{r},t); \ \varepsilon=e^\Gamma \tag{9.31}$$

where H_I is the corresponding order of fractional integration for the observable and where for simplicity we have assumed the same space time localized propagators for both ε and I. More generally, in the "mixed" turbulence wave model, we have:

$$\left((-\nabla^2)^{1/2}+\frac{\partial}{\partial t}\right)^{H_{tur}}\left(\nabla^2-\frac{\partial^2}{\partial t^2}\right)^{H_{wav}/2} I(\underline{r},t)=\varepsilon(\underline{r},t);$$
$$\varepsilon=e^\Gamma; \ H_I=H_{tur}+H_{wav} \tag{9.32}$$

Where we have kept the previous (localized) propagator for the turbulent energy flux ε. See Section 9.2.4 for numerical examples.

9.2 The emergent laws of turbulence-generated waves

9.2.1 Classical quasi-linear waves: gravity waves and the Taylor–Goldstein equations

Because of the atmosphere's enormous Reynolds number it is natural to focus on the strongly nonlinear cascade of conserved turbulent fluxes and to consider that these are the drivers for the observables. If both the propagator of the turbulent fluxes themselves (g_ε) and the observables (g_I) were identical and were localized in space-time, the turbulence would not generate unlocalized (wave-like) structures. In this section we show how – again constrained by the scaling symmetries – such turbulence-generated waves may arise as emergent properties. We first review and criticize the standard linearization approach to gravity waves, which are probably the most commonly observed type of atmospheric waves (although, with a few exceptions such as in the boundary layer downstream of topography (lee waves), one typically sees only three or fewer wave oscillations).

Most gravity wave studies are based around the Taylor–Goldstein equations (Goldstein, 1931; Taylor, 1931). One starts with the equations for inviscid, irrotational flow of a vertical atmospheric section in the Boussinesq approximation:

$$\begin{aligned}
&\frac{\partial u}{\partial t}+u\frac{\partial u}{\partial x}+w\frac{\partial u}{\partial z}=-\frac{1}{\rho}\frac{\partial p}{\partial x}\\
&\frac{\partial w}{\partial t}+u\frac{\partial w}{\partial x}+w\frac{\partial w}{\partial z}=-\frac{1}{\rho}\frac{\partial p}{\partial z}-g\\
&\frac{\partial u}{\partial x}+\frac{\partial w}{\partial z}=0\\
&\frac{\partial \rho}{\partial t}+u\frac{\partial \rho}{\partial x}+w\frac{\partial \rho}{\partial z}=0
\end{aligned} \tag{9.33}$$

where only one horizontal coordinate (x) has been retained, ρ is the air density, g is the acceleration of gravity – not a propagator (see e.g. Nappo, 2002, whom we follow below). From top to bottom, these are the equations for the horizontal, vertical momentum, mass conservation, and the last for the thermal energy.

One next introduces a mean and a fluctuating set of variables:

$$q(x,z,t)=q_0(z)+q_1(x,z,t) \tag{9.34}$$

where q represents u, w, ρ, p. Substituting the means and fluctuations into Eqn. (9.33), one then obtains:

$$\begin{aligned}
&\frac{\partial u_1}{\partial t}+u_0\frac{\partial u_1}{\partial x}+w_1\frac{\partial u_0}{\partial z}+u_1\frac{\partial u_1}{\partial x}+w_1\frac{\partial u_1}{\partial z}=-\frac{1}{\rho_0}\frac{\partial p_1}{\partial x}\\
&\frac{\partial w_1}{\partial t}+u_0\frac{\partial w_1}{\partial x}+u_1\frac{\partial w_1}{\partial x}+w_1\frac{\partial w_1}{\partial z}=-\frac{1}{\rho_0}\frac{\partial p_1}{\partial z}-g\frac{\rho_1}{\rho_0}\\
&\frac{\partial u_1}{\partial x}+\frac{\partial w_1}{\partial z}=0\\
&\frac{\partial \rho_1}{\partial t}+u_0\frac{\partial \rho_1}{\partial x}+w_1\frac{\partial \rho_0}{\partial z}+u_1\frac{\partial \rho_1}{\partial x}+w_1\frac{\partial \rho_1}{\partial z}=0
\end{aligned} \tag{9.35}$$

In the first, second and fourth rows, the two terms immediately to the left of the equality sign are of second order in the perturbations and are ignored in

Box 9.1 Numerical simulations of causal processes

In Section 5.5 we discussed simulations of continuous-in-scale but acausal multifractal processes, in Appendix 5B we discussed finite size effects and other practical numerical issues for both acausal and causal simulations, and in Appendix 5C we gave a simple Mathematica code. Let us consider briefly the changes needed to produce causal continuous-in-scale cascade processes.

According to the discussion in Section 9.1, all that needs to be modified in order to obtain a causal simulation is to multiply a noncausal propagator by the appropriate Heaviside function so that $g(\underline{r},t) = 0$ for $t < 0$. Ignoring for the moment the annoying finite size effects at small and large scales, the only other modification is to the normalization constant. This can be easily calculated from the formulae in Appendix 5B.2, which were developed for symmetrical acausal g which effectively satisfy $g(\underline{x},t) = g(\underline{x},-t)$. The causal normalization constant was simply related to the symmetric acausal constant via $N_{Df,c} = N_{D,f}/2$ (Appendix 5B.2). Aside from this straightforward normalization issue, the correction method and constant are unchanged so that there are very few changes in the numerical implementation (see Lovejoy and Schertzer, 2010, for the full details).

In Fig. 9.3, we give an example of a realization of a causal process and – to emphasize that the simulation is of the temporal evolution of a 1D series – we have plotted a sequence of spatial sections. In Appendix 5B.3 we discussed the scaling finding that the causal processes have fewer "finite size" effects, and that the acausal corrections work extremely well. We can extend the simulations to two spatial dimensions and time; Fig. 9.4 shows an example on a 256^3 grid using $H_t = 2/3$; i.e. in a Lagrangian frame. We see that the small structures "live" for far less time than the larger ones. In Section 9.3 we explore how this "memory" can be used to forecast the future state.

In Appendix 5B.3 we evaluated the accuracy of the spectral scaling for 2D causal and acausal processes and concluded that the correction method works well independently of the dimension of the space, and that for the causal extensions the temporal statistics have significantly smaller deviations. This is presumably because of the sharp discontinuity introduced by the Heaviside function, which roughens the simulations along the time axis; this thereby somewhat compensates for its otherwise overly smooth behaviour.

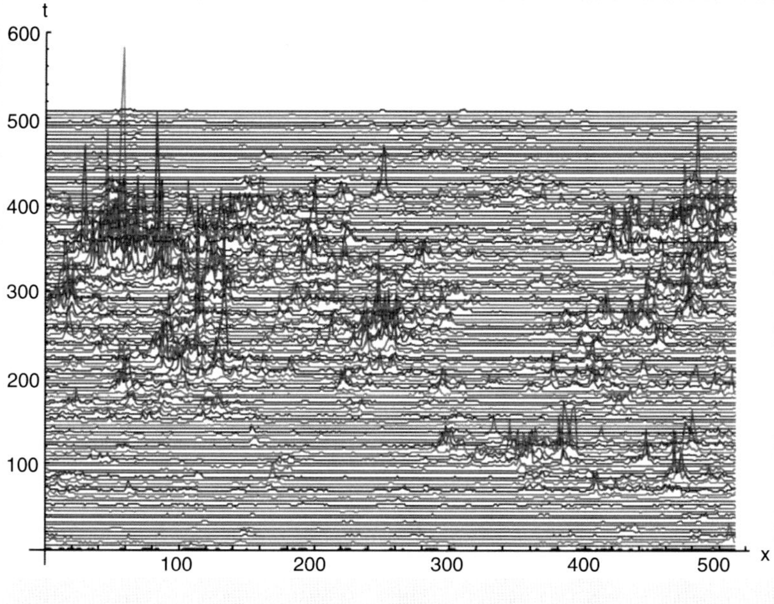

Fig. 9.3 Evolution of (left/right symmetric) 1D spatial sections, series of a causal simulation with $\alpha = 2$, $C_1 = 0.2$. Every second time interval is shown, displaced by two units in the time axis with respect to the previous section. Reproduced from Lovejoy and Schertzer (2010).

Box 9.1 (*cont.*)

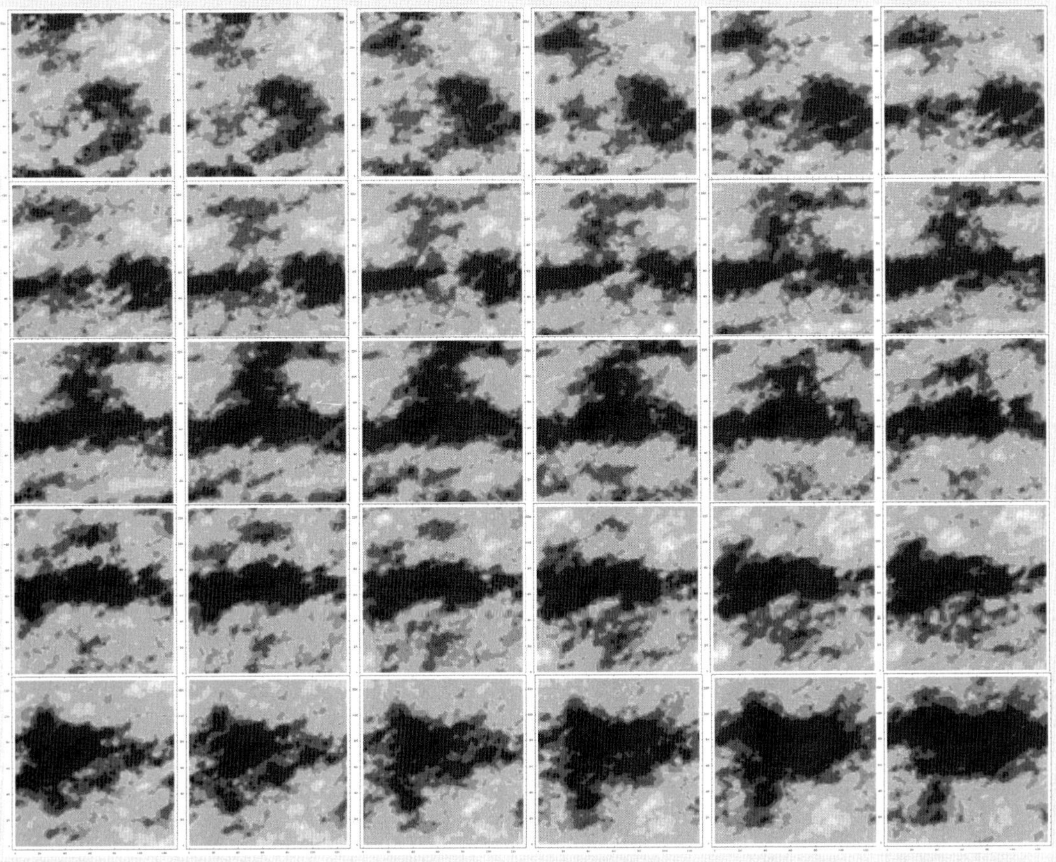

Fig. 9.4 A sequence (from left to right, from top to bottom) simulated on a 256^3 grid showing every fourth time step from step 1 to step 119 (because of the periodicity, the second half, from step 128 on, is artificially correlated with the first half and is not shown). Multifractal parameters: $\alpha = 1.8$, $C_1 = 0.1$, $H = 0.333$, with some horizontal anisotropy ($e = 0.2$, $c = 0.1$) and in a Lagrangian frame ($H_t = 2/3$).

order to yield a linear system. This system is then Fourier transformed with respect to both time and the horizontal coordinate to yield:

$$\frac{d^2\hat{w}}{dz^2} + \left[\frac{k_x^2 N^2}{\omega^2} - k_x^2\right]\hat{w} = 0; \quad \hat{w} = e^{-z/(2H_s)}\widetilde{w};$$

$$N^2 = g\left(\frac{\partial \log\theta_0}{\partial z}\right) \tag{9.36}$$

where N is the Brunt–Vaisaila frequency, θ the potential temperature, $\widetilde{w}(z, k_x, \omega)$ is the horizontal, temporal Fourier transform of the vertical velocity (we have assumed $w_0 = 0$ and so dropped the subscripts: $w = w_1$), and k_x, ω are the corresponding Fourier variables conjugate to x and t. We have used the fact that in the Bousinesq approximation, N^2 is equal to:

$$N^2 = g\left(\frac{\partial \log\rho_0}{\partial z}\right) \tag{9.37}$$

Other assumptions include an exponentially decreasing density (with scale height H_s), the hydrostatic equilibrium of the density fluctuations, and finally the absence of a background horizontal wind field (i.e. $u_0 = 0$: see Nappo, 2002, for more details).

Eqn. (9.36) is the most usual (but still special case) of the Taylor–Goldstein equations for the vertical wind fluctuation. The equations for the other fluctuations can also be derived from this; these are sometimes called "polarization" relations and will not be discussed further. In addition, as in Chapter 8, a mean "background" wind $\underline{u}_0$ can be taken into account by the transformation $\omega \to \omega + \underline{k}\cdot\underline{u}_0$ (cf. Section 8.2.2). Making the final approximation that $N(z) = \text{constant}$, and Fourier transforming with respect to z, we obtain the classic zero background wind ($u_0 = 0$) dispersion relation:

$$\omega = \pm \frac{k_x N}{\left(k_x^2 + k_z^2\right)^{1/2}} \tag{9.38}$$

Note that the low-frequency Rossby wave dispersion relation is nearly of the same form.

Before continuing, let us pause and consider what has been accomplished. One has started with a highly turbulent atmospheric flow whose typical nonlinear terms are roughly 10^{12} times the linear ones (the Reynolds number), yet one has assumed that this is compatible with a *linear* approximation for small perturbations. Although, a priori, the validity of such a linearization seems doubtful at best, the Taylor–Goldstein equations (including the polarization relations) are routinely used as the basis for interpreting various atmospheric datasets. Over time it has spawned a sizeable gravity wave literature. Since the usual response to such theoretical criticism is something along the lines of "well, it seems to work," we will instead attempt to check that the linearization (i.e. dropping the second-order terms in Eqn. (9.35)) is justified, i.e. to pursue an empirically based critique. To this end we can use the dropsonde data discussed in Chapters 4 and 6 to evaluate at least some of the terms in Eqn. (9.35) and to directly empirically check whether or not the neglected (second-order) terms are indeed small compared to the kept terms.

The simplest terms to empirically check are the vertical shear of the horizontal wind, i.e. the assumption that $\partial u_0/\partial z >> \partial u_1/\partial z$ which is used to linearize the horizontal momentum equation (see Eqn. (9.35), top: $w_1\partial u_0/\partial z$ is kept but $w_1\partial u_1/\partial z$ is dropped). To evaluate these terms, we can take advantage of the fact that on many occasions, dropsondes were dropped at 0.3 s intervals, corresponding to ~50 m separation in the horizontal (see Fig. 9.5 for the trajectory of the pair used here). The perturbed and unperturbed winds can be defined as:

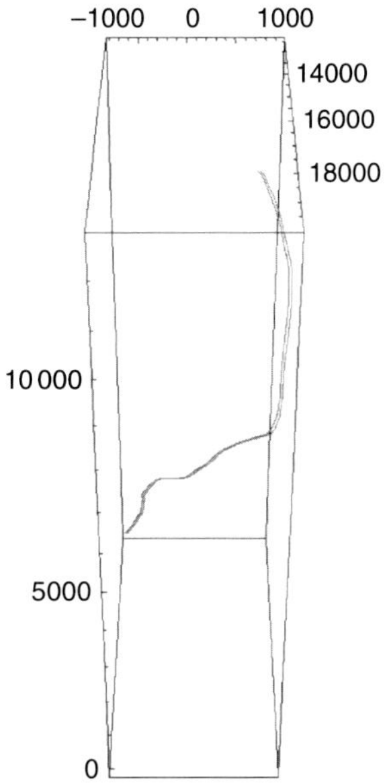

Fig. 9.5 The trajectories of the pair of sondes analysed here, dropped at about 12 km altitude.

$$\begin{aligned} u_0 &= \frac{\left(u_{sonde1} + u_{sonde2}\right)}{2} \\ u_1 &= \frac{\left(u_{sonde1} - u_{sonde2}\right)}{2} \end{aligned} \tag{9.39}$$

In Fig. 9.6 we see the basic means and fluctuations estimated this way at both 80 m and 20 m vertical resolutions. From the figure we see that u_1 can readily be 0.5 m/s, i.e. Δu in the horizontal can readily be ≈ 1 m/s, so that for a 50 m separation this implies that the "fluctuation *Re*" (i.e. the Reynolds number based on Δv rather than on v) is of the order $0.5 \times 10^{-6} \approx 5 \times 10^7$.

In order to see if the linearization can be justified, we may now consider Fig. 9.7, which compares the mean ($\partial u_0/\partial z$) and fluctuation ($\partial u_1/\partial z$) shears for layers at 80 m vertical resolution. We can see that they are typically comparable and much larger than the noise level of the measurements. In order to quantify this further, we can calculate the relative difference (E) at the 80 m resolution:

$$E = \frac{\left|\frac{\partial u_1}{\partial z}\right|_{80} - \left|\frac{\partial u_0}{\partial z}\right|_{80}}{\left|\frac{\partial u_1}{\partial z}\right|_{80} + \left|\frac{\partial u_0}{\partial z}\right|_{80}} \tag{9.40}$$

If $E = 1$, then the mean shear is negligible compared to the perturbation, if $E = -1$, then on the contrary, the

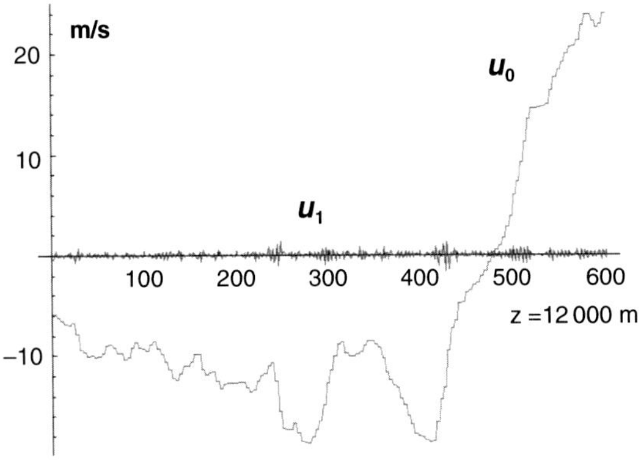

Fig. 9.6 The mean horizontal wind (u_0) and fluctuations (u_1) for the means taken over 80 m thick layers, and for u_1 only for 20 m thick layers. The horizontal axis shows the altitude in units of 20 metres.

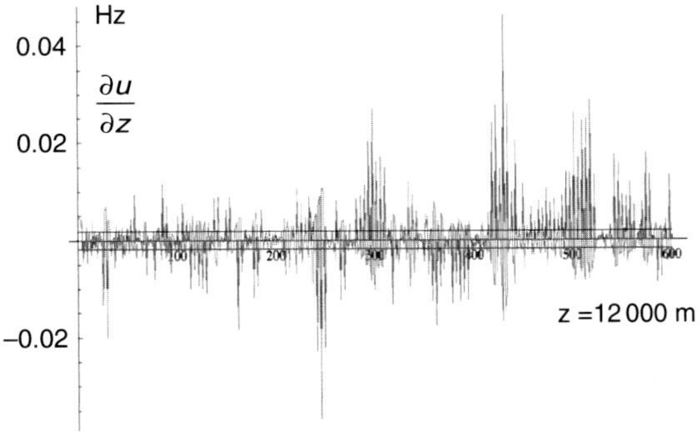

Fig. 9.7 The vertical shears at 80 m vertical resolution; darker is $\partial u_0/\partial z$, lighter the perturbation, $\partial u_1/\partial z$. The solid lines are two-standard-deviation error bars showing that most of the variations are real, not instrumental.

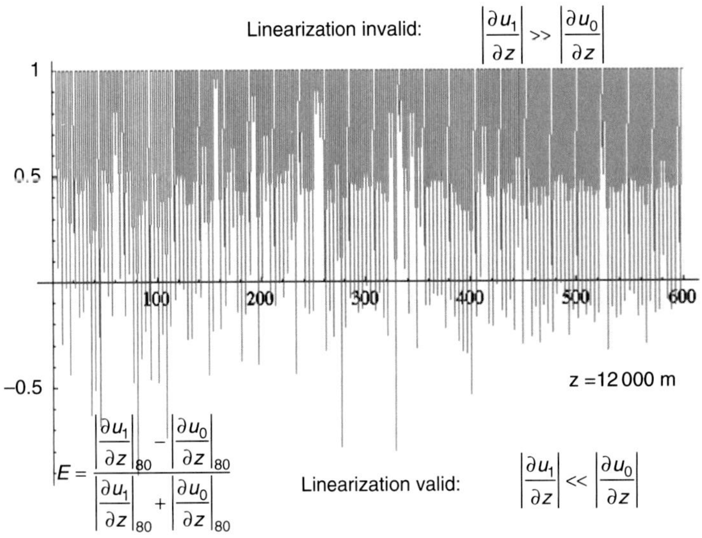

Fig. 9.8 The relative difference of the mean and perturbation shears (E) shown in Fig. 9.7. If $E = 1$, then the mean shear is negligible compared to the perturbation, if $E = -1$, the fluctuation is negligible, justifying the linearization. It can be seen that linearization is generally untenable. The horizontal axis shows the altitude in units of 20 metres.

fluctuation is negligible, justifying the linearization (Fig. 9.8). It can be seen that linearization is generally untenable, and often so by rather large factors. Indeed, it is rather the *mean* shear which should most often be neglected!

So how do we explain the existence of waves, indeed waves whose dispersion relations are presumably not too different from those predicted by the simple theory presented here? The answer is that the dispersion relations, like the other turbulent laws discussed in this book, are emergent laws with their basic characteristics – in this case dispersion laws – determined by the scaling symmetries. In the next section we show quantitatively how this is possible.

9.2.2 The emergence of dispersion laws: an extreme unlocalized (wave) model

Let us now consider a simple FIF model which is only localized in space, but not in space-time. As usual, the propagators must be chosen in Fourier space to respect the appropriate scaling symmetries; for the observable I, let us choose:

$$\widetilde{g}_I(\underline{k}, \omega) = \left(i(\omega - \|\underline{k}\|^{H_t})\right)^{-H/H_t};$$

$$\widetilde{g}_I\left(\lambda^{-G_{st}^T}(\underline{k}, \omega)\right) = \lambda^{H}\widetilde{g}_I(\underline{k}, \omega); \qquad G_{st} = \begin{bmatrix} G_s & 0 \\ 0 & H_t \end{bmatrix} \tag{9.41}$$

Table 9.1 A comparison of flux and wave-like velocity propagators. In both localized and unlocalized cases, the propagators satisfy: $\tilde{g}\left(\lambda^{-G_{st}^T}(\underline{k},\omega)\right) = \lambda^{-H}\tilde{g}(\underline{k},\omega)$. Note that $D_{st} = \mathrm{Tr}(G_{st}) = D_s + H_\tau$, $z = \|\underline{k}\|^{H_t} t$.

Flux-like	Wave-like
(Power law!) Localization in space-time	**Unlocalized**
$\tilde{g}(\underline{k},\omega) = (i\omega + \|\underline{k}\|^{H_t})^{-H/H_t}$	$\tilde{g}(\underline{k},\omega) = \left(i(\omega - \|\underline{k}\|^{H_t})\right)^{-H/H_t}$
$\tilde{g}(\underline{k},t) = \Theta_{Heavi}(t)t^{-1+H/H_t}f(z);\quad f(z) \approx \begin{matrix} z^{H/H_t-1}; & z\approx 0 \\ 1; & z \approx \infty \end{matrix}$	$\tilde{g}(\underline{k},t) = \Theta_{Heavi}(t)t^{-1+H/H_t}f(z);\quad f(z) \approx e^{iz}$
$g(\underline{r},t) = \Theta_{Heavi}(t)[[(\underline{r},t)]]^{-(D_{st}-H)}$	$g(\underline{r},t) \approx \Theta_{Heavi}(t)\dfrac{e^{i(\underline{k}\cdot\underline{r}-\omega(\underline{k})t+\phi(\underline{k}))}}{t^{5/2-H}\left[\det\left(\frac{\partial^2\omega(\underline{k})}{\partial k_i\,\partial k_j}\right)\right]^{1/2}}$
e.g. $[[(\underline{r},t)]]^2 = \|\underline{r}\|^2 + t^{2/H_t}$	$\underline{r} = \underline{v}_g(\underline{k})t;\quad \underline{v}_g(\underline{k}) = \nabla\omega(\underline{k});\quad \omega(\underline{k}) = \|\underline{k}\|^{H_t}$

These are propagators for fractional wave equations (a little different from Eqn. (9.22)). Where G_{st} is the space-time generator of the anisotropy and G_s is the spatial generator satisfying:

$$\|\lambda^{-G_s^T}\underline{k}\| = \lambda^{-1}\|\underline{k}\| \tag{9.42}$$

the form Eqn. (9.41) for the propagator g_I is a little more general than the fractional wave equation propagator, since it allows for $H_t \neq 1$ so that it is useful in a Lagrangian frame where can H_t can be taken as 2/3 (Section 8.2.1). Following our discussion of the propagator for the classical wave equation, we anticipate that the behaviour will be dominated by the ω, $\underline{k}$ which make g_v singular, i.e. those satisfying the dispersion relation:

$$\omega(\underline{k}) = \|\underline{k}\|^{H_t} \tag{9.43}$$

In order to understand the implications of this propagator, it is instructive to take the inverse Fourier transform of $\tilde{g}_I(\underline{k},\omega)$ with respect to ω:

$$\tilde{g}_I(\underline{k},t) = \Theta_{Heavi}(t)t^{-1+H/H_t}e^{i\|\underline{k}\|^{H_t}t} \tag{9.44}$$

(we ignore constant factors). The Heaviside function $\Theta(t)$ and power law shows directly that g_I is a causal temporal fractional integration of order H/H_t of waves; we come to the same conclusion by noting that the upper-half ω plane is analytic (Section 9.1.2).

We can use the standard method of stationary phase (e.g. Bleistein and Handelsman, 1986) to obtain an asymptotic approximation to the space-time convolution for I:

$$I(\underline{\Delta r},t) \approx \tilde{\varepsilon}\left(\underline{k}(\underline{\Delta r}),t\right) *_t \tilde{g}_v\left(\underline{k}(\underline{\Delta r}),t\right) \tag{9.45}$$

where $*_t$ indicates convolution with respect to time only and the propagator is:

$$g_I(\underline{r},t) \approx \tilde{g}_I\left(\underline{k}(\underline{r}),t\right) = \Theta_{Heavi}(t)\frac{e^{i(\underline{k}\cdot\underline{r}-\omega(\underline{k})t+\phi_0(\underline{k}))}}{t^{5/2-H/H_t}\left[\det\left(\frac{\partial^2\omega(\underline{k})}{\partial k_i\,\partial k_j}\right)\right]^{1/2}} \tag{9.46}$$

where ϕ_0 is a phase and:

$$\underline{\Delta r} = \underline{v}_g(\underline{k})t;\quad \underline{v}_g(\underline{k}) = \nabla\omega(\underline{k});\quad \omega(\underline{k}) = \|\underline{k}\|^{H_\tau} \tag{9.47}$$

Eqns. (9.46) and (9.47) should be understood as parametric equations: $\underline{k}$ is the wavevector which satisfies the "ray" equation $\underline{r} = \underline{v}_g(\underline{k})t$, where v_g is the group velocity and $\omega(\underline{k}) = \|\underline{k}\|^{H_t}$ is the dispersion relation (note that we would have obtained $\omega(\underline{k}) = -\|\underline{k}\|^{H_t}$ if we had chosen the equally valid propagator with –ω in the place of ω).

Eqns. (9.46) and (9.47) show that the velocity field is the fractional time integral of wave packets propagating along rays at the group velocity, dispersing and decreasing in amplitude as they travel as t^{-2} (the exponent is $5/2 - H/H_t = 2 > 3/2$; $H_t = 2/3$). The classical time dependence of the attenuation of wave packets is $t^{-3/2}$ so that the waves attenuate a little faster. Also, as usual, the above breaks down when the determinant in the denominator vanishes; these singular curves are the "caustics." Table 9.1 shows the comparison of the flux propagators and the observables.

Although there are two different Green's functions used to obtain v, the overall field is still symmetric with respect to the same generator G_{st}, and the structure function exponent $\xi(q)$ is also unchanged. In addition, the spatial $\|\underline{\Delta r}\|$ – which is the basic physical scale function – can (if necessary) be the same for both g_ε and g_I (it is only the space-time scale function which need be different). Lovejoy *et al.* (2008) give some more information on the statistical properties of the turbulence–wave model, and in the next sections we consider more realistic scaling dispersion relations.

9.2.3* Gravito-turbulence dispersion relations

We have seen that the standard gravity wave model assumes, for a layer of thickness Δz, a uniform stratification characterized by $N^2 = g(\Delta \log\theta/\Delta z)$, and weak nonlinearity leading to the Taylor–Goldstein equations and to the dispersion relation, Eqn. (9.38). In contrast, our turbulence flux-based approach assumes a highly heterogeneous vertical structure whose statistics are determined by the (large-scale averaged) buoyancy variance flux $\phi = g^2[(\Delta\log\theta)^2/\tau_b]_l$ via its effect on l_s (the subscript indicates that the flux is measured at space-time resolution l). The combined ε, ϕ fluxes lead to a physical scale function $\|\underline{\Delta r}\|$, thus to a wave-like propagator (such as Eqn. (9.41)) and hence the dispersion relation Eqn. (9.43).

However, the scale function is fairly general. For example, considering only the vertical (x,z) plane, it is of the form:

$$\|\underline{k}\| = \widetilde{\Theta(\theta)}\|\underline{k}\|_{can};\ \|\underline{k}\|_{can} = l_s^{-1}\left((k_x l_s)^2 + |k_z l_s|^{2/H_z}\right)^{1/2} \tag{9.48}$$

where $\widetilde{\Theta(\theta)}$ is a relatively arbitrary function of direction in the vertical plane; see Section 6.5.1 (here θ is the polar angle in the vertical plane, not the potential temperature, and Θ is not the Heaviside function).

Several of the predictions of gravity wave theory have been at least roughly empirically verified; it is therefore of interest to choose $\widetilde{\Theta(\theta)}$ so that the turbulence/wave theory gives a similar dispersion relation and hence gives similar predictions. Since the classical dispersion relation is symmetric with respect to isotropic scale changes (i.e. with x, z generator $G = \begin{pmatrix} 1 & 0 \\ 0 & 1 \end{pmatrix}$ rather than the anisotropic $G = \begin{pmatrix} 1 & 0 \\ 0 & H_z \end{pmatrix}$), the two dispersion relations cannot be identical. However, they can be chosen to be sufficiently similar so that the new relation can plausibly be compatible with the results of previous atmospheric gravity wave studies.

Using the values $H_t = 2/3$, $H_z = 5/9$ from dimensional analysis, we find that the choice $\widetilde{\Theta(\theta)} = (\cos\theta)^{3/2}$ leads to the following gravity wave-like "gravito-turbulent" dispersion relation:

$$\omega(\underline{k}) = \varepsilon^{1/3}\frac{|k_x|}{\|\underline{k}\|^{1/3}} \tag{9.49}$$

To display the similarity with the classical dispersion relation more clearly, Table 9.2 shows the two special cases which are most commonly tested empirically: near-horizontal and near-vertical propagation.

It can be seen that in both cases, for near-horizontal propagation, the dispersion relation becomes linear in k_x so that the horizontal group velocity is independent of k_x, i.e. it "saturates." This saturation is considered an important empirical confirmation of the Taylor–Goldstein equations and quasi-linear gravity waves; it is the basis of the influential saturated cascade theory (SCT) (Dewan, 1997). In addition, the dependencies on $(\Delta\log\theta)$ are very similar (a 2/5 power instead of a 1/2 power), although it should be recalled that in the turbulence case the potential temperature profile is considered highly variable (turbulent), not linear (smooth). Also, for near-vertical propagation, in both cases ω is independent of k_z. A final physically significant similitude is the fact that in both cases the group velocity has a "restoring" vertical component, i.e. w_g is opposite in sign to ω/k_z, so that for example if the wave front is propagating upward, then the wave energy propagates downwards (in the absence of a mean advection; see Nappo, 2002). The comparison of the group velocities is shown in Fig. 9.9. In Fig. 9.10 we show (x,z) and (t,z) sections of (x,z,t) numerical multifractal simulations, showing the stratified wave-like structures that the model produces, including in the presence of overall advection (these models were actually for a passive scalar, produced by replacing $\varepsilon^{1/3}$ by $\varepsilon^{-1/6}\chi^{1/2}$ where χ is the passive scalar flux). In Fig. 9.11 we show a time sequence, and in Fig. 9.12 we show the effect of changing the sphero-scale and the vertical wind (all these use the gravito-turbulence dispersion/scale function). Finally in Fig. 9.13 we show simulations of horizontal sections with varying scale functions/dispersion relations (by changing the shape of the unit ball B_1 via the function

Table 9.2 A comparison of the standard gravity wave dispersion relations: with a turbulent/wave model with a gravity wave-like choice of $\Theta(\theta)$; a "gravito-turbulence" dispersion relation. To make the comparison more clear, we have expressed the flux ϕ in terms of the potential temperature and g. Recall that l_s is the sphero-scale and τ_b is the time scale of the buoyancy fluctuation.

	Linear theory gravity wave dispersion	"Gravito-turbulent" dispersion
General form	$\omega(\underline{k}) \approx g^{1/2}\left(\frac{\Delta\log\theta}{\Delta z}\right)^{1/2}\frac{\lvert k_x\rvert}{\lvert k\rvert}$ $\lvert k\rvert = (k_x^2 + k_z^2)^{1/2}$	$\omega(\underline{k}) = \varepsilon^{1/3}\frac{\lvert k_x\rvert}{\lVert\underline{k}\rVert^{1/3}}$ $\lVert\underline{k}\rVert = l_s^{-1}\left((l_s k_x)^2 + (l_s k_z)^{18/5}\right)^{1/2}; \quad l_s = \phi^{-3/4}\varepsilon^{5/4}$
Near-horizontal propagation	$\omega(\underline{k}) \approx g^{1/2}\left(\frac{\Delta\log\theta}{\Delta z}\right)^{1/2}\frac{\lvert k_x\rvert}{\lvert k_z\rvert}; \lvert k_x\rvert << \lvert k_z\rvert$	$\omega(\underline{k}) = g^{2/5}\left[\frac{\overline{(\Delta\log\theta)^2}}{\tau_b}\right]^{1/5}_{\lVert\Delta x, \Delta z\rVert}\frac{\lvert k_x\rvert}{\lvert k_z\rvert^{3/5}}; \lvert k_x\rvert << \lvert k_z\rvert^{9/5} l_s^{4/5}$
Near-vertical propagation	$\omega(\underline{k}) \approx g^{1/2}\left(\frac{\Delta\log\theta}{\Delta z}\right)^{1/2}; \lvert k_x\rvert >> \lvert k_z\rvert$	$\omega(\underline{k}) = \varepsilon^{1/3}_{\lVert\Delta x, \Delta z\rVert}\lvert k_x\rvert^{2/3}; \lvert k_x\rvert >> \lvert k_z\rvert^{9/5} l_s^{4/5}$

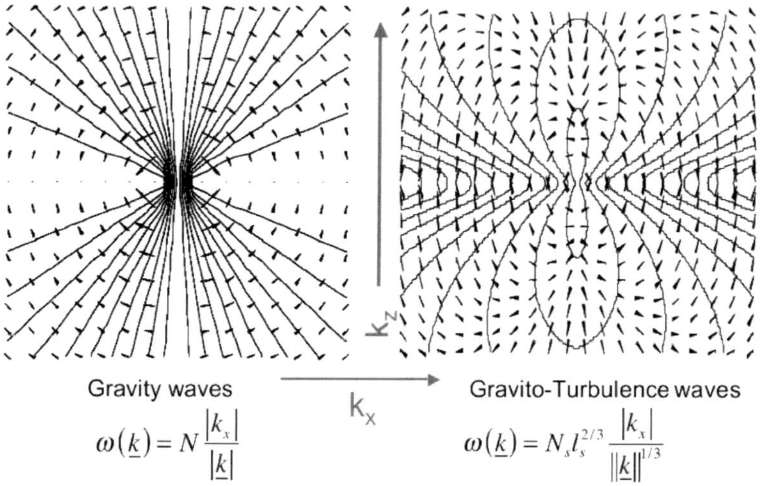

Fig. 9.9 The contour lines of ω, with arrows showing the corresponding gradient (group velocities). The formula for the gravito-tubulence dispersion relation waves is the same as in the text (Table 9.2), with $N_s = \omega_s = (\phi/\varepsilon)^{1/2}$. Note that $N_s l_s^{2/3} = \phi^{1/5}$. Reproduced from Lovejoy *et al.* (2008).

Θ defining the unit vectors), showing that quite realistic morphologies are readily produced.

9.2.4 A mixed turbulence/wave model

We have presented a model in which the space-time propagator corresponds to a fractional integral over waves with a nonlinear turbulent dispersion defined via the physical scale function. The implications for energy transport are very strong: the turbulent energy flux input will generally be transported far from the source via the waves. Since the energy flux is related to the velocity differences via $\varepsilon \approx \Delta v^3/\Delta x$, most of the energy must remain localized for the model to be self-consistent. One way of achieving this energy localization would be if the dispersion relation had a negative imaginary part. However, this would imply a dissipation mechanism which – if too strong – would contradict the picture of a cascade of conservative fluxes upon which the FIF is built. A more satisfying method is to combine the wave with the turbulent scale functions so that the final model has aspects of

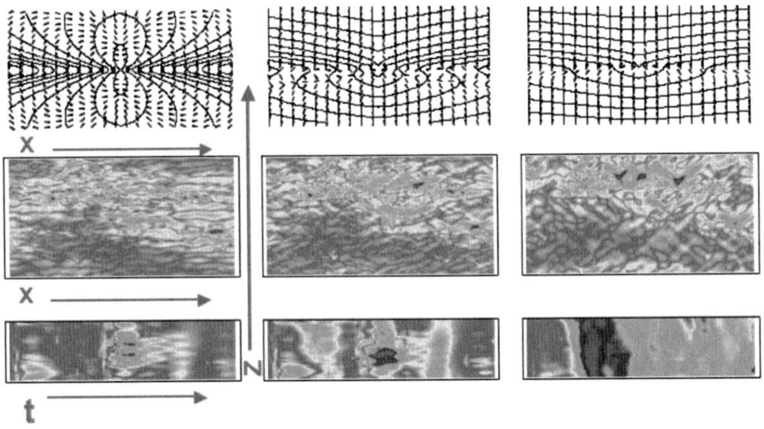
x
x
t

Fig. 9.10 A multifractal simulation of a passive scalar in (x,z,t) space with the observed multifractal parameters ($\alpha = 1.8$, $C_1 = 0.05$) and theoretical values $H_t = 2/3$, $H_z = 5/9$. The simulations show the vertical wind increasing from 0 (left) to 0.25 to 0.5 pixels/time step (only a single time step is shown). The top row shows the dispersion relation (contours), group velocity (arrows), the second is an (x,z) cross-section, and the third row is a (t,z) cross-section. The numerical simulation techniques are based on those described in Chapters 5 and 8. Reproduced from Lovejoy *et al.* (2008). *See colour plate section.*

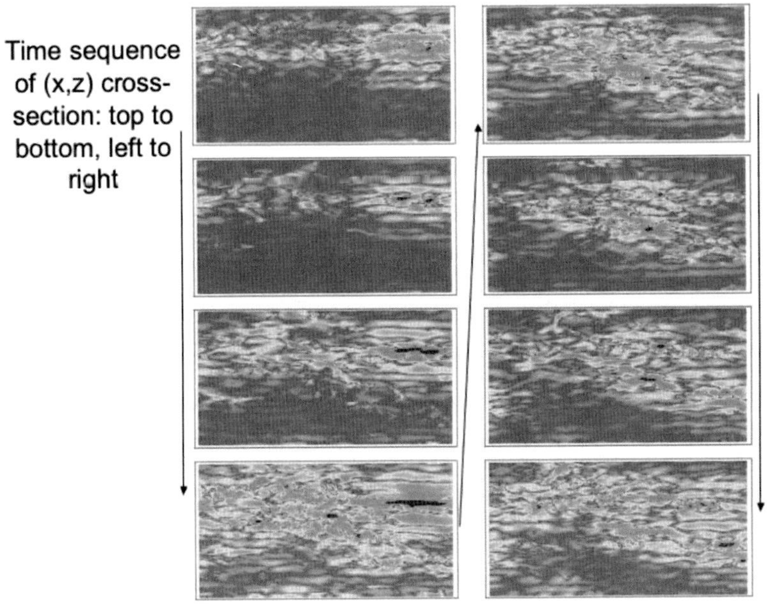
Time sequence of (x,z) cross-section: top to bottom, left to right

Fig. 9.11 Eight time steps in the evolution of the vertical cross-section of a passive scalar component from the zero wind case of Fig. 9.10. The structures are increasingly stratified at larger and larger scales and display wave phenomenology. Reproduced from Lovejoy *et al.* (2008). *See colour plate section.*

both; in this case, the wave energy could considered as "leakage" in analogy with the Lumley–Shur model. A simple way to achieve this with a Lagrangian temporal exponent $H_t = 2/3$ is to use:

$$\tilde{g}_I(\underline{k},\omega) = (-i\omega + \|\underline{k}\|^{H_t})^{-H_{tur}/H_t}[i(\omega - \|\underline{k}\|^{H_t})]^{-H_{wav}/H_t};\quad H_{tur} + H_{wav} = H \tag{9.50}$$

We see that the extreme localized and extreme unlocalized models correspond to $H_{tur} = H$, $H_{wav} = 0$ and $H_{wav} = H$, $H_{tur} = 0$ respectively (recall for the wind, $H = 1/3$, and the subscripts *tur*, *wav* indicate "turbulence" and "waves" respectively) compare this to the Eulerian propagator, Eqn. (9.23). In Fig. 9.14 we show the effect of increasing H_{wav}: one can see how structures become progressively more and more wave-like while retaining the same scaling symmetries, close to observations.

Turning our attention to the Eulerian scale functions discussed in Chapter 8, we saw that with zero mean wind the nondimensional real-space scale function in horizontal space-time (i.e. (x,y,t) space) is isotropic with respect to $G_s = 1$, i.e. it is only

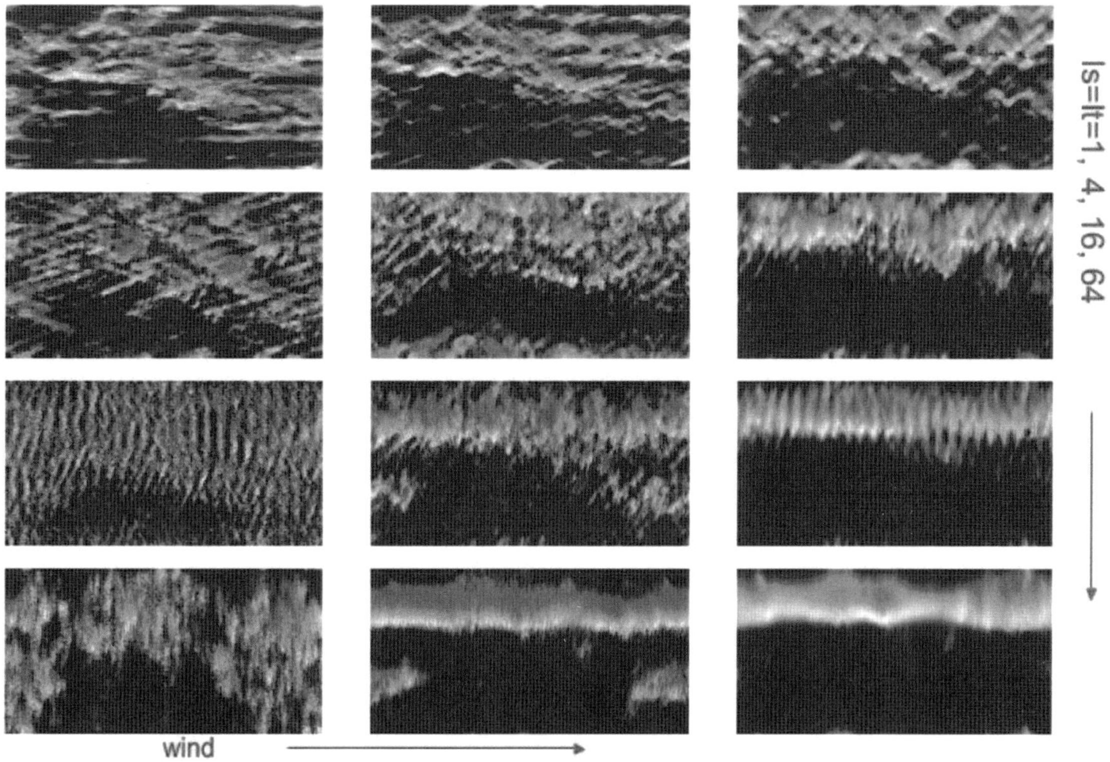

Fig. 9.12 The effect of vertical wind (left to right 0, 0.25, 0.5 pixels/time step), and horizontal and space-time ($l_s = l_t$) sphero-scales increasing top to bottom from 1 pixel to 4, 16, 64 pixels for vertical cross-sections of simulated passive scalar, for the gravito-turbulence dispersion relation described in the text. In order to visualize the time evolution, the successive vertical sections are "stacked" on top of each other; the rendition uses simulated single scattering visible radiation through the stack (it is as though the time dimension is the "depth" of the cloud). Reproduced from Lovejoy *et al.* (2008).

trivially anisotropic so that $H_t = 1$. In this case, rather than use the above model, we could use the propagator of the (anisotropic) "fractional wave equation" for the propagator g($\underline{r}$,t):

$$\left[\frac{\partial^2}{\partial x^2} + a^{-2}\frac{\partial^2}{\partial y^2} - \frac{1}{V^2}\frac{\partial}{\partial t^2}\right]^{H_{wav}/2} g_{wav}(\underline{r}, t) = \delta(\underline{r}, t) \tag{9.51}$$

for some exponent H_{wav} (and wave speed V). By taking Fourier transforms of both sides, it can be seen that:

$$\tilde{g}_{wav}(\underline{k}, \omega) = (\omega^2 - \|\underline{k}\|^2)^{-H_{wav}/2}; \quad \|\underline{k}\|^2 = (k_x^2 + a^{-2}k_y^2) \tag{9.52}$$

so that this wave propagator corresponds to the turbulent propagator $\tilde{g}_{tur}(\underline{k}, \omega) = (i\omega/V + \|\underline{k}\|)^{-H_{tur}}$ with the same spatial part. Notice that the dispersion relation is still essentially the classical $\omega = \pm V\|\underline{k}\|$ although the trivial anisotropy of $\|\underline{k}\|$ makes this somewhat more interesting. The classical dispersion relation is satisfied by Kelvin waves as are inertial gravity (Poincaré) waves in the low Coriolis parameter/ high "effective thickness" limit often invoked at these space-time scales.

This propagator can be "mixed" with the turbulent propagator to yield:

$$\begin{aligned}\tilde{g}_I(\underline{k}, \omega) - \tilde{g}_{tur}(\underline{k}, \omega)\tilde{g}_{wav}(\underline{k}, \omega) \\ = (-i\omega + \|\underline{k}\|)^{-H_t}(\omega^2 V^{-2} - \|\underline{k}\|^2)^{-H_{wav}/2} \\ |\tilde{g}_I|^2 = (\omega^2 + \|\underline{k}\|^2)^{-H_{tur}}(\omega^2 V^{-2} - \|\underline{k}\|^2)^{-H_{wav}}\end{aligned} \tag{9.53}$$

which (as shown in Section 9.1.2) is causal, and which satisfies:

$$\tilde{g}_I\left(\lambda^{-1}(\underline{k}, \omega)\right) = \lambda^H \tilde{g}_I(\underline{k}, \omega); \quad H = H_{tur} + H_{wav} \tag{9.54}$$

In Chapter 8 we allowed for turbulent advection with mean μ and variability accounted for by σ (Eqn. 8.22)

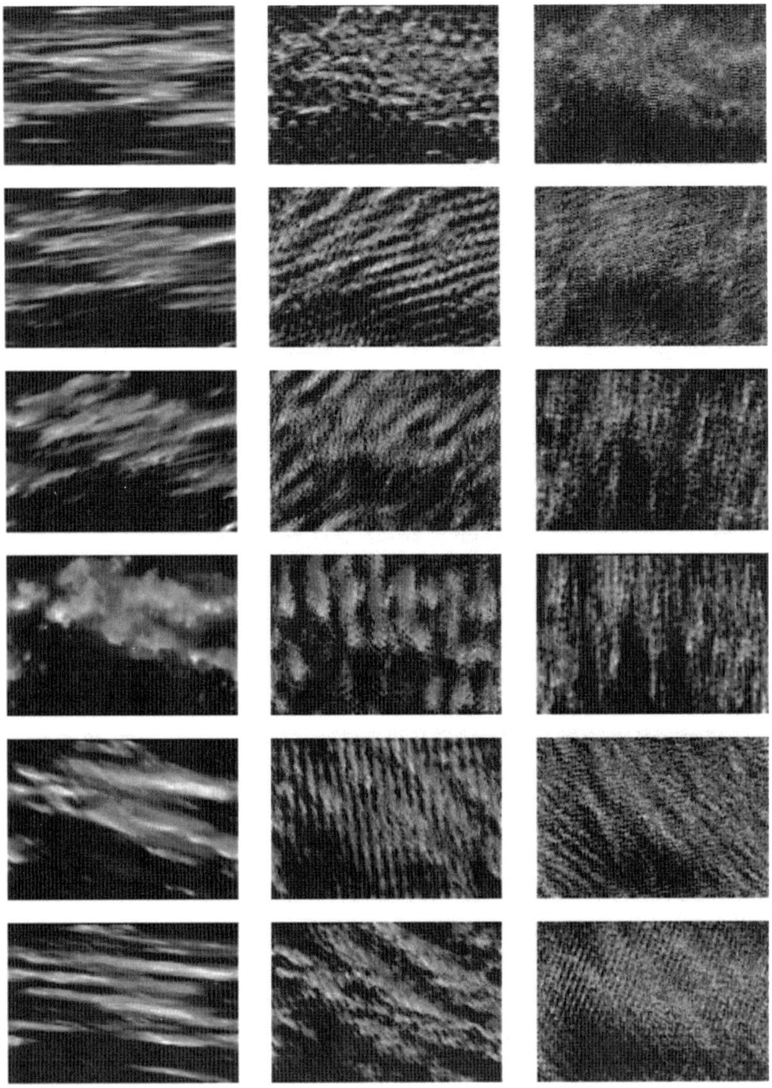

Fig. 9.13 A series of horizontal sections of (x,y,t) passive scalar cloud simulations with horizontal generator $G = \begin{pmatrix} 1.2 & 0.05 \\ -0.05 & 0.8 \end{pmatrix}$ with $H_\tau = 2/3$ as usual. From left to right, the horizontal sphero-scale = 1 pixel, 8, 64 pixels. The horizontal unit ball is characterized by $\Theta(\theta) = 1 + a\mathrm{Cos}(2\theta - 2\theta_0)$ with $a = 0.65$, and from top to bottom, the orientation θ_0 is varied from 0 to $5\pi/6$ in steps of $\pi/6$ (this is the real (x,y) space function). These simulations show how sensitive the morphologies are to the unit balls (i.e. the spatial scale function/dispersion relation). Reproduced from Lovejoy *et al.* (2008).

by replacing ω by $\omega' = (\omega + \underline{k} \cdot \underline{\mu})/\sigma$ (a "background wind" with statistical variability). With this, we obtain:

$$\widetilde{g}_I(\underline{k}, \omega) = \left(-i\omega' + \|\underline{k}\|\right)^{-H_{tur}} \left(\omega'^2/V^2 - \|\underline{k}\|^2\right)^{-H_{wav}/2} \tag{9.55}$$

So that $\widetilde{g}_I(\underline{k}, \omega)$ satisfies $\widetilde{g}_I\left(\lambda^{-1}(\underline{k}, \omega)\right) = \lambda^H \widetilde{g}_I(\underline{k}, \omega)$ with $H = H_{tur} + H_{wav}$.

9.2.5 MTSAT wave behaviour

In Chapter 8 we recalled that the popular Wheeler and Kiladis method of extracting wave behaviour from satellite imagery such as MTSAT is to take the spectral power density, then "average out" the turbulent "background" and examine the deviations to see signs of dispersion type relations notably in (2D) wavenumber/frequency subspaces. Using this as well as cross-spectral analysis between thermal IR and passive microwave fields, Wheeler and Kiladis

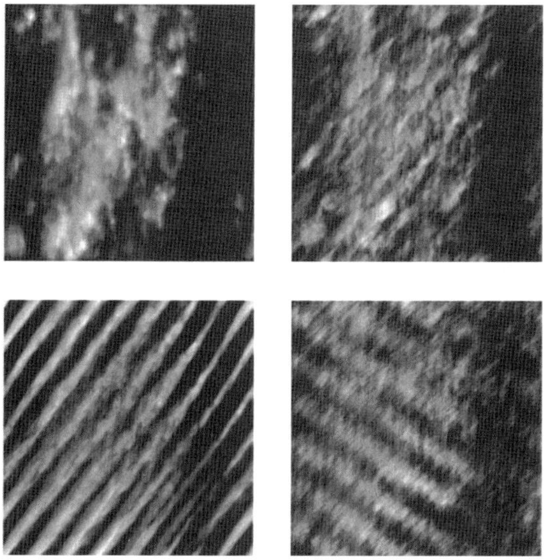

Fig. 9.14 The effect of increasing H_{wav} with $H_{wav} + H_{tur} = H = 0.33$, $H_t = 0.66$; clockwise from the upper left we have $H_{wav} = 0, 0.33, 0.52, 0.38$ (i.e. $H_{tur} = 1/3$, $H_{wav} = 0.33, 0, -0.19, -0.05$), $C_1 = 0.1$, $\alpha = 1.8$. There is a small amount of differential anisotropy characterized by $G = \begin{pmatrix} 0.95 & -0.02 \\ 0.02 & 1.05 \end{pmatrix}$ The horizontal unit ball is characterized by $\Theta(\theta) = 1 + a\mathrm{Cos}(2\theta - 2\theta_0)$ with $a = 0.65$, with $\theta_0 = 0$. The random seed is the same in all cases so that one can see how structures become progressively more and more wave-like while retaining the same scaling symmetries, close to observations. Reproduced from Lovejoy *et al.* (2008). *See colour plate section.*

(1999) claimed evidence for Kelvin waves, equatorial Rossby waves and mixed Rossby–gravity waves. Note that their temporal resolutions differed by about an order of magnitude: their data were daily over a two-year period rather than hourly for two months, so we do not expect to see much evidence for these particular waves.

Let us therefore consider the spectral density corresponding to a purely turbulent ε and a mixed wave/turbulent propagator (Eqn. (9.55)). This model corresponds to the following spectral density:

$$P_I(\underline{k}, \omega) = |\widetilde{g}_I(\underline{k}, \omega)|^2 P_\varepsilon(\underline{k}, \omega) = (\omega'^2 + \|\underline{k}\|^2)^{-H_{tur}} (\omega'^2/V^2 - \|\underline{k}\|^2)^{-H_{wav}} P_\varepsilon(\underline{k}, \omega) \qquad (9.56)$$

According to the model of turbulence-generated waves, for physical reasons the conserved flux ε must be based purely on the turbulent part, so that:

$$P_\varepsilon(\underline{k}, \omega) = (\omega'^2 + \|\underline{k}\|^2)^{-s_\varepsilon/2}; s_\varepsilon = D - K(2) \qquad (9.57)$$

For the MTSAT data, we saw (Figs. 8.7f, 8.7g, 8.11, 8.12) that $K(2) \approx 0.12 \pm 0.01$, $D = 3$, $s_\varepsilon \approx 2.88 \pm 0.01$. With these parameters, combining Eqns. (9.55), (9.56), we now find for the overall spectral density:

$$P_I(\underline{k}, \omega) = (\omega'^2 + \|\underline{k}\|^2)^{-s_\varepsilon - H_{tur}} (\omega'^2/V^2 - \|\underline{k}\|^2)^{-H_{wav}} \qquad (9.58)$$

For the MTSAT data, Pinel and Lovejoy, 2012, found $H_I = (s_I - s_\varepsilon)/2 \approx 0.26 \pm 0.05$ so that even if H_{wav} is of the same order as H_{tur}, the exponent of the turbulent term will be much larger than the wave term and the turbulent term will dominate. This is why to a first approximation (Section 8.3) it was possible to ignore possible wave contributions. However, we can perform regressions to minimize the deviations of log $P_{theory}(k_x,k_y,\omega)$ from the empirical log $P_{data}(k_x,k_y,\omega)$ using the full wave-turbulence model (Eqn. (9.58)) for $P_{theory}(k_x,k_y,\omega)$ and determining the best-fit parameters including H_{wav}. This has been done in Fig. 9.15; in order to show the improvement when using the error minimizing value $H_{wav} = 0.17 \pm 0.04 \approx 1/6$ (hence $H_{tur} = H_I - H_{way} \approx 0.09 \pm 0.06$) we have compared contours on subspaces. With the "optimum" parameters (see Eqn. 8.33, 8.34 and the nondimensional wave speed $V \approx 1.0 \pm 0.8$), the function $P_{data}(k_x,k_y,\omega)$ – which varies over a range of $\sim 10^5$ – is fit to within $\pm 13\%$ (excluding the diurnal spike). Although the improvement in the fits for $H_{wav} = 1/6$ with respect to the pure turbulence model ($H_{wav} = 0$) is not large, the best classical wave model with $H_{wav} = 2$ is clearly quite poor, (even $H_{wav} = 1$ is poor) so that the waves may originate in a fractional wave equation.

9.3 Predictability/forecasting

9.3.1 Predictability limits from dynamical systems theory in low dimensional chaos

In Chapter 1 we mentioned the quiet revolution in weather forecasting implicit in the introduction of ensemble forecasting systems (EFS). With EFS, rather than the deterministic question "what is tomorrow's weather?" forecasters are asking the stochastic question "what are the possible states of tomorrow's weather and what are their probabilities of occurrence?" As currently constituted, EFS are hybrid deterministic/stochastic systems, since both the selection of the initial ensemble and the basic forward integration of the equations are primarily deterministic. We say "primarily" because with the introduction of "stochastic parametrizations" (Buizza

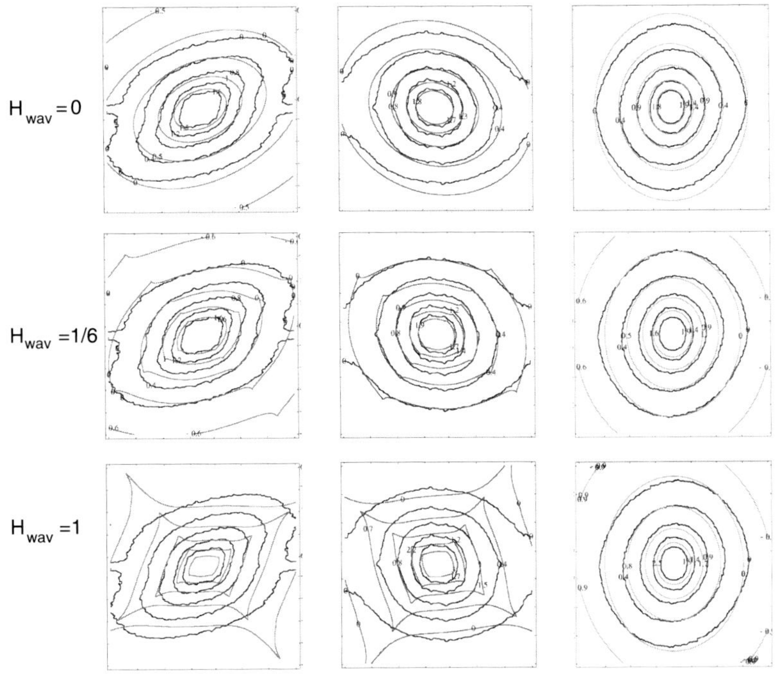

Fig. 9.15 Comparison of the 2D energy densities P_{data} with P_{theory} where the latter is obtained using the wave/turbulence model (Eqn. (9.58)) and the MTSAT data (see fig. 8.11). The three 2D subpaces are considered: from left to right, the subspaces are (ω,k_x), (ω,k_y), (k_x,k_y). The range of ω is $(2\ h)^{-1}$ to $(276\ h)^{-1}$, for k_x it is $(60\ km)^{-1}$ to $(13\,000\ km)^{-1}$ and for k_y it is $(60\ km)^{-1}$ to $(8000\ km)^{-1}$. The parameters are $s_\varepsilon = 3.0$, $H = 0.2$. As we move from the top to bottom row, we increase H_{wav} from zero (top, a pure turbulence model) to $H_{wav} = 1/6$ (middle, the value that minimizes least mean square deviations of the theoretical form from the data) and the bottom row value $H_{wav} = 1$ (the classical integer ordered wave propagator has $H_{wav} = 2$). The parameter values that minimize the regression error are $H_{wav} \approx 0.17 \pm 0.04 \approx 1/6$, $H_{tur} \approx 0.09 \pm 0.06$, $V \approx 1.0 \pm 0.8$ (so that the dimensional wave speed $\approx V_w \approx 11.3 \pm 1.1$ m/s), and the turbulence parameters are the same as in ch. 8: $s_l = 3.4 \pm 0.1$, $s_\varepsilon \approx 2.88 \pm 0.01$, $H \approx 0.26 \pm 0.05$, $\mu_x \approx -0.3 \pm 0.1$ (corresponding to $-3.3 \pm 1.1 m/s$), $\mu_y \approx 0.1 \pm 0.07$ (corresponding to $1.1 \pm 0.8 (m/s)$ hence $\sigma = 0.95 \pm 0.03$ and the turbulence parameters are the same as in Eqn (8.33). Note that the fact that integer H_{wav} gives quite poor fits whereas $H_{wav} = 1/6$ gives quite good fits implies that while the classical dispersion relation may be valid, the classical wave equation does not apply, and the waves are emergent high Reynolds number properties. Reproduced from Pinel and Lovejoy (2012).

et al., 1999; Palmer, 2001) random numbers are introduced in an otherwise deterministic numerical integration scheme, and this at the finest (grid) level. At the moment this is done in a relatively ad hoc manner, but in future it could be based on cascades. However, if cascades are indeed the correct stochastic space-time process, then this also opens another possibility: new, direct techniques of stochastic forecasting. In this section, we review the issue of predictability and forecasting and discuss some particularities of the cascade approach.

The notion of "initial condition sensitivity" became well known due to the work of Lorenz (1963) on his three-component model (corresponding to the first three Fourier components of convection). By the 1980s such exponential error growth became the hallmark of the "deterministic chaos revolution," and it was widely viewed to be a generic property of nonlinear systems.

The idea of exponential error growth emerged from the pioneering work of Lyapunov (1907) and was subsequently generalized into the elegant multiplicative ergodic theorem (MET) (Oseledets, 1968), a cornerstone of chaos theory. A key assumption of the theorem is that temporal averages of a single sample of the process are the same as the average at one time over an ensemble of identical processes, i.e. that the process is "ergodic." From this common geophysical ergodicity assumption Oseledets was able to demonstrate exponential error growth:

$$|\underline{\delta X}(t)| = e^{\mu t} |\underline{\delta X}(0)| \tag{9.59}$$

where $\underline{X}$ is a state vector in phase space and $\delta\underline{X}$ is the vector difference in state between a state and its perturbation. μ is the rate of divergence; the "Lyapunov exponent" (for more details, see the review by Schertzer and Lovejoy, 2004b). Eqn. (9.59) is valid as long as μ is finite, an assumption usually taken for granted. If τ is characteristic time after which predictions are effectively impossible then $\mu = 1/\tau$. This result is generally valid for finite d-dimensional systems. However, if we attempt to generalize this to evolving fields, i.e. to nonlinear partial differential equations (infinite dimensional (functional) spaces), we encounter severe difficulties. In fact, only a few limited extensions have been obtained (Ruelle, 1982). From our discussion in Chapters 2 and 8 we can already anticipate a key difficulty: the characteristic time for an eddy size l is the eddy turnover time, so that $\mu = 1/\tau = l^{-2/3}\varepsilon^{1/3}$ which diverges for small l, thus violating the finiteness assumption of the mathematical derivation of the MET.

9.3.2 Predictability in homogeneous turbulence: the phenomenology of error growth through scales and the MET

The general phenomenology of error growth through scales is rather straightforward: an error or uncertainty initially confined to small scales will progressively "contaminate" large-scale structures through these interactions. This is in sharp contrast to the MET, which does not consider the problem of many nonlinearly interacting spatial scales. The problem of the evolution of spatially extended fields was first theoretically investigated by Thompson (1957). Using initial time-derivatives and various meteorological models, Thompson studied the nonlinear uncertainty growth due to errors in the initial conditions resulting from the limited resolutions of the measurement network and of the models. He estimated the root mean square (RMS) doubling time for small errors to be about two days, whereas Charney (1966), using more elaborate meteorological models, estimated it as five days.

The scale dependency of the predictability times was underlined by Robinson (1971). Indeed, if the notion of characteristic error time τ is still relevant, it should depend on the spatial scale l in a hierarchical manner. For $t > \tau(l)$ two fields initially similar at scale l become quite different (e.g. rather decorrelated) at this scale, but may remain similar at larger scales. This is in agreement with the estimates of the Lyapunov exponent and the characteristic space scale reached by the error at the eddy turnover time τ mentioned above. This shows – contrary to the usual assumption – that unless a break in the scaling occurs, leading to smooth small-scale behaviour, the Lyapunov exponent μ will diverge at small scales.

Let $\underline{u}_1(\underline{r}, t)$ and $\underline{u}_2(\underline{r}, t)$ be two solutions of a nonlinear system (e.g. velocities for Navier–Stokes equations) initially identical, but with a perturbation (error) $\delta\underline{u}(\underline{r}, 0) = \underline{u}_2(\underline{r}, 0) - \underline{u}_1(\underline{r}, 0)$ at $t = 0$, confined to infinitesimally small spatial scales. In terms of the "butterfly effect", the time-evolution of $\delta\underline{u}(\underline{r}, 0)$ corresponds to the effect of butterflies homogeneously distributed in space, rather than the effect of a single butterfly. When the nonlinear interactions preserve the kinetic energy (e.g. Navier–Stokes equations), it is convenient (but not sufficient) to consider both the correlated (kinetic) energy (per unit of mass):

$$e_c(\underline{r}, 0) = \frac{1}{2}\underline{u}_2(\underline{r}, 0)\cdot\underline{u}_1(\underline{r}, 0) \tag{9.60}$$

and the decorrelated energy:

$$e_\Delta(\underline{r}, t) = \frac{1}{2}|\delta\underline{u}(\underline{r}, t)|^2 = \frac{1}{2}|\underline{u}_2(\underline{r}, t) - \underline{u}_1(\underline{r}, t)|^2 \tag{9.61}$$

as well as the total energy e_T and the energy of each solution e_n:

$$e_T(\underline{r}, t) = e_1(\underline{r}, t) + e_2(\underline{r}, t); \quad e_n(\underline{r}, t) = \frac{1}{2}|\underline{u}_n(\underline{r}, t)|^2 \tag{9.62}$$

This implies the relation:

$$e_T(\underline{r}, t) = e_c(\underline{r}, t) + e_\Delta(\underline{r}, t) \tag{9.63}$$

Hence, if the total energy is statistically stationary (conserved on average), there will be a flux of correlated energy $e_c(\underline{r}, t)$ to decorrelated energy $e_\Delta(\underline{r}, t)$. Since the latter corresponds to a linear decomposition of the former with respect to wavenumber k, this also holds for the corresponding energy spectra $E_T(\underline{k}, t) = E_c(\underline{k}, t) + E_\Delta(\underline{k}, t)$. Therefore, the decorrelated energy spectrum $E_\Delta(\underline{k}, t)$ steadily increases in magnitude from large to small wavenumbers, converging to the total energy spectrum $E_T(\underline{k}, t) \approx k^{-5/3}$ (Fig. 9.16). The critical wavenumber $k_e(t)$ of the transition from dominant correlation to dominant decorrelation can be defined by

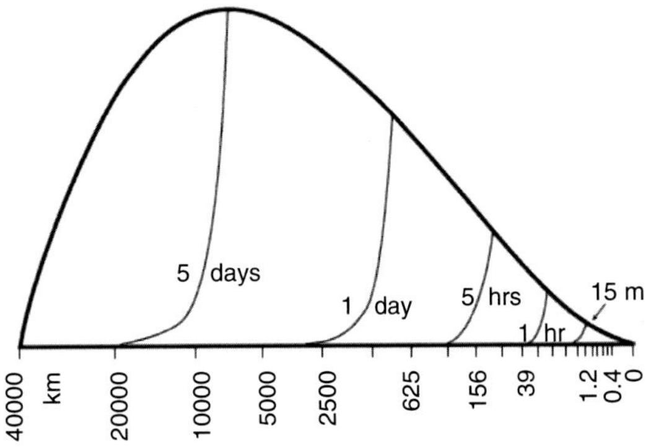

Fig. 9.16 Atmospheric error growth according to a quasi-normal closure simulation (reproduced from Lorenz, 1969). The decorrelated energy spectrum $E_\Delta(k,t)$ initially confined to a few metres (on the right) "pollutes" the larger scales (up to 20 000 km (on the left) after various time intervals (from few minutes to 5 days).

$E_c\left(k_e(t),t\right) = E_\Delta\left(k_e(t),t\right)$, $1/l_e(t)$ and decreases as $k_e(t) \approx t^{-3/2}$.

If the constant of proportionality in the definition of the eddy turnover time, as well as that relating the latter to the error time, is of order unity, then taking "typical values" $\varepsilon \approx 10^{-3}\,\mathrm{m^2 s^{-3}}$ (see Section 8.1.3) and the viscous dissipation scale as $\eta \approx 10^{-3}\,\mathrm{m}$), one obtains $\tau_e(\eta) \approx \varepsilon^{-1/3}\eta^{2/3} = 10^{-1}\,\mathrm{s}$, as well as $\tau_e(l) \approx \tau_e(\eta)(l/\eta)^{2/3}$ and therefore $\tau_e(\eta) = 10$ s; 1/2 h; 28 h; 5.4 days respectively for 1 km, 10^3 km, 10^4 km. These estimates are close to those obtained by Lorenz (1969) (Fig. 9.16), but slightly lower than the numerical (closure, nonintermittent) results obtained by Kraichnan (1970), Leith (1971), Leith and Kraichnan (1972) and Métais and Lesieur (1986).

9.3.3 Divergence of states in multifractal processes

Before discussing the predictability in more theoretical terms, we can already investigate the divergence of multifractal simulations by considering two scalar processes I_1, I_2 which are identical up until $t = t_0$ (they have identical subgenerators $\gamma_1(t) = \gamma_2(t)$ for $t < t_0$) and then diverge, i.e. their subgenerators become statistically independent. Due to the long-range correlations induced by the fractional integrations (for both the generator $\Gamma(t)$ and the second fractional integral on $\varepsilon = e^{\Gamma(t)}$ to produce $I(t)$), there will be strong correlations which will die away as Δt $(= t - t_0)$ increases. Fig. 9.17a gives an example where the processes are identical up until $t_0 = 2^{10} - 2^7$ but

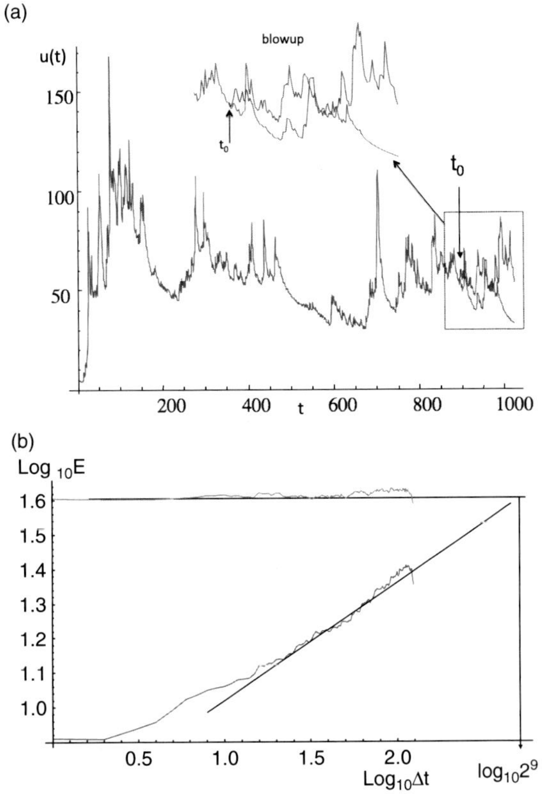

Fig. 9.17 (a) The evolution of a pair of one-dimensional multifractals with $\alpha = 1.8$, $C_1 = 0.1$, $H = 0.333$. The subgenerators are identical up until $t_0 = 2^{10} - 2^7$ (indicated by the arrow) after which they are independent. The insert shows a blow-up. (b) The "error" E defined in Eqn. (9.64) for the average of 1000 realizations with the parameters of Fig. 9.17a indicating a power-law function of the time Δt from divergence t_0. The top curve is for I_1 and I_2 statistically independent, the bottom (power law with reference slope $= H = 0.333$) is for subgenerators identical until $t = t_0$, independent thereafter (as in Fig. 9.17a). As expected, by extrapolation we see that the two are the same at $\Delta t = 2^9$, which is half the length of the series (the simulated series is periodic so that at longer times the dependent series are actually less dependent).

diverge thereafter. In order to quantify this, we can define the "error" E as the absolute difference normalized by the average:

$$E = 2\left\langle \left| \frac{I_2(t_0 + \Delta t) - I_1(t_0 + \Delta t)}{I_2(t_0 + \Delta t) + I_1(t_0 + \Delta t)} \right| \right\rangle \tag{9.64}$$

(using qth powers would simply introduce intermittency corrections depending on $K(q)$). In Fig. 9.17b we see that the error is indeed a power law of the time and diverges with increasing Δt. This is indeed –

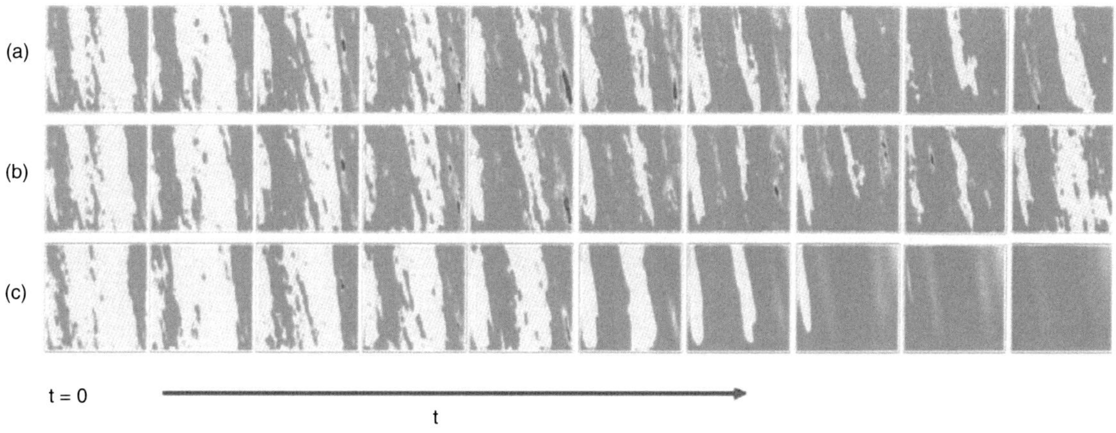

Fig. 9.18 The top two rows (a–b) show successive snapshots ($t = i \times \tau$, $i = 3, 6, 9, \ldots, 27$; τ being the eddy turnover time of the smallest structures) of two simulations (256 × 256 in space) that are identical until time $t = 0$, when their fluxes at small scales become progressively independent step by step due to the sudden independence of the subgenerators at that time. Most of the difference between the two realizations is concentrated in a few "hot spots". The bottom row (c) shows a forecast based on the "memory" of the evolution up to $t = 0$ of (a), i.e. it has the same stochastic subgenerator until time $t = 0$, then it is defined in a deterministic manner to preserve the mean of the flux. Note the more rapid disappearance of small-scale structures. Parameters are $\alpha = 1.5$, $C_1 = 0.2$, $H = 0.1$ (close to those of rain), and the colour scale is logarithmic. The anisotropy of space-time is characterized by $H_t = 2/3$. Reproduced from Schertzer and Lovejoy (2004a). *See colour plate section.*

as expected for multifractals – quite different from the exponential divergence for low-dimensional chaos.

We can extend the simulations to two spatial dimensions and time; Fig. 9.18 shows an example with two simulations (upper rows) and a forecast (bottom row). This visually shows how initially identical structures for $t < 0$ (identical subgenerators $\gamma_1(\underline{r},t) = \gamma_2(\underline{r},t)$ for $t < 0$) have divergent evolutions for $t > 0$ if we allow them to be statistically independent for $t > 0$. We see that for larger and larger t, only the larger structures are common to both simulations since the lifetime of the small structures is comparatively short. In the bottom row, we see a simple forecast which keeps the subgenerator constant for $t > 0$; we see that the forecast deteriorates quickly for the small, less quickly or the large: see the next subsection.

9.3.4 Predictability limits in multifractal cascades

In order to generalize the approach followed in the spectral analysis of predictability to multifractals (Section 9.3.2), we consider the time evolution of a pair of fields of common resolution Λ. They are identical up to the time t_0 when one lets the fluxes become independent at small scales (Schertzer and Lovejoy, 2004a). For simplicity, consider the scalar rain rate $R(x,t)$ illustrated by Fig. 9.17 (time t along the horizontal, location x along the vertical). Figs. 9.19a and 9.19b display a pair of rain-rate fields $R_{1\Lambda}(x,t)$ and $R_{2\Lambda}(x,t)$ and Fig. 9.19d their absolute difference $|\delta R_\Lambda(x,t)|$. One may qualitatively note the role of intermittency: most of the difference $|\delta R_\Lambda(x,t)|$ is due to a small number of extremely large values.

Marsan *et al.* (1996) checked that the spectral analyses of multifractal simulations of a velocity component are in agreement with homogeneous turbulence results. Bursts of violent fluctuations cannot be accounted for using second-order statistical moments, in particular energy spectra; these are evident in Fig. 9.20, which displays an "elementary" decorrelated/error energy spectrum, i.e. not obtained by ensemble averaging, but only over a unique sample. It is no longer as smooth as the ensemble-averaged decorrelated/error energy spectrum $E_\Delta(k,t)$ (e.g. Fig. 9.16), but rather corresponds to a sequence of decorrelation (more generally of independence) bursts at different scales. These bursts result from the fact that although the energetics of the upscale cascade of errors remain basically the same, they do not constrain the largest fluctuations of the errors as much as in the homogeneous turbulence case.

We emphasized that statistics of second-order moments, in particular their correlation, which

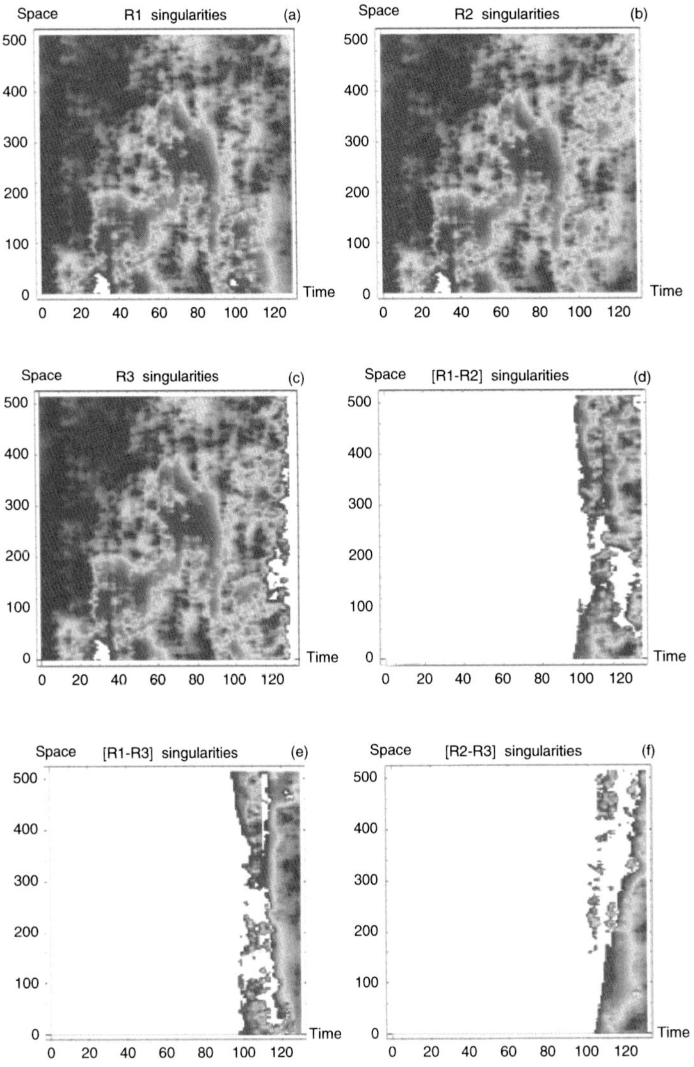

Fig. 9.19 Simulation of multifractal predictability decay for rain field. (a) and (b) are identical up to $t_0 = 64$, after which their fluxes become independent. (c) displays the forecast based on their common past and the deterministic conservation of the flux afterward. Singularities of the fields (i.e. their log divided by the log resolution), as well as of their absolute differences (d–f), are displayed according to the following palette: white for negative singularities; green to yellow for singularities contributing to statistics up to the mean; red for singularities contributing to second- and higher-order moments. Reproduced from Schertzer and Lovejoy (2004b). *See colour plate section.*

corresponds to the correlation energy for a velocity field, are inadequate in accounting for the coevolution of a pair of multifractal fields. Therefore we need to consider a covariance of order q for different values of q. This is rather simple for fluxes, e.g. the respective energy flux densities $\varepsilon_{i\Lambda}$ $(i = 1,2)$ of a pair of velocities $u_i(x,t)$. Up to t_0 the fluxes are identical over the full range of the cascade process (i.e. over the possibly infinite cascade scale ratio Λ). After t_0, they remain rather similar only over a decreasing scale ratio $\lambda(t) \leq \Lambda$, which necessarily follows a power law. More precisely (Schertzer and Lovejoy, 2004b), the latter is defined by the dynamical exponent H_t, which defines the scaling space-time anisotropy (Chapter 8):

$$\begin{aligned} &\lambda(t) = \Lambda; && t \leq t_0 \\ &\lambda(t) = Min\left(\Lambda, \left(T/(t-t_0)\right)^{1/H_t}\right) && t > t_0 \end{aligned} \qquad (9.65)$$

where T is the outer time scale.

As a consequence, one obtains for the (normalized) covariance of order q:

$$C^{(q)}(\varepsilon_{1,\Lambda}, \varepsilon_{2,\Lambda}) = \left\langle \frac{(\varepsilon_{1,\Lambda}, \varepsilon_{2,\Lambda})^q}{\left\langle (\varepsilon_{1,\Lambda})^q \right\rangle \left\langle (\varepsilon_{2,\Lambda})^q \right\rangle} \right\rangle \propto \lambda(t)^{K(q,2)} \qquad (9.66)$$

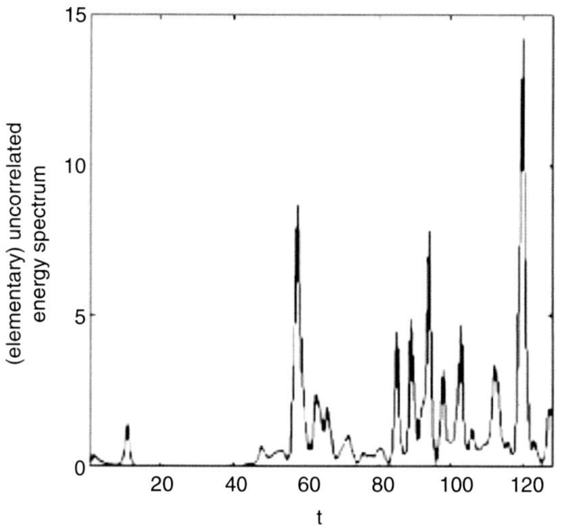

Fig. 9.20 Elementary error energy spectrum displaying decorrelation bursts (reproduced from Schertzer *et al.*, 1997). This elementary error energy spectrum is obtained from a unique realization rather than from an ensemble average. It is no longer as smooth as in Fig. 9.19, but rather corresponds to a sequence of decorrelation bursts at different scales.

with $K(q, 2) \equiv K(2q) - 2K(q)$. The multifractality $K(q,2)$ of the joint field $\varepsilon_{1,\Lambda}\varepsilon_{2,\Lambda}$ is purely defined by that of $\varepsilon_{i,\Lambda}$ (i.e. $K(q)$). The distinctive feature is that instead of being fixed at Λ (as for $\varepsilon_{i,\Lambda} = \varepsilon_{2,\Lambda}$) the range of scale ratios $(1, \lambda(t))$ also has a power-law decay (Eqn. (9.65)). The same occurs for its probability distribution. It is important to appreciate that these power laws are valid for all time scales, not only the large scales. This is in a sharp contrast to the exponential decay predicted for low-dimensional chaos (the MET), which is therefore not relevant for multifractal fields.

We may now explore the question of optimizing forecast procedures so that the decay law of the (normalized) covariance $C^{(q)}(\varepsilon_{F,\Lambda}, \varepsilon_{2,\Lambda})$ of order q of the forecast field $\varepsilon_{F,\Lambda}$ and of the observed field $\varepsilon_{0,\Lambda}$ is as close as possible to the theoretical $C^{(q)}(\varepsilon_{1,\Lambda}, \varepsilon_{2,\Lambda})$ (Eqn. (9.66)). For example, let us point out that the multifractal behaviour of meteorological fields theoretically explains and confirms the recent empirical evidence that stochastic parametrizations do better than deterministic ones (Buizza *et al.*, 1999; Houtekamer *et al.*, 1996), in particular in the EFS framework. It suffices to use the fact that a multifractal field may be defined with the help of a white-noise subgenerator. Indeed, past and future components of a white noise are independent and identically distributed. Therefore, any white noise identically distributed to the past component is obviously a possible future component. The resulting process will in the future keep the same statistical properties, as well as the same scale ratio. However, a future component defined in a deterministic manner cannot have an identical statistical distribution. In particular, its scaling function $K_{det}(q)$ is linear with respect to q, instead of being nonlinear, as is that of the observations $K(q)$. At best, one can only find a (deterministic) procedure to preserve the statistics of a given order q. This is illustrated in Fig. 9.19, obtained by a numerical simulation, where the (deterministic) future component of the noise of the forecast field $R_{3,\Lambda}$ (x,t) (Fig. 9.19c) was defined to preserve the mean ($q = 1$) of the flux. Figs. 9.19e and 9.19f display the drastic loss of all extreme events ($q >> 1$) with respect to the samples $R_{1,\Lambda}$ (x,t) and $R_{2,\Lambda}$ (x,t). More quantitative statements can be readily obtained with the help of the covariance $C^{(q)}(\varepsilon_{F,\Lambda}, \varepsilon_{2,\Lambda})$ of order q. This should encourage the radical EFS evolution to increasingly account for the randomness of meteorological fields at different scales and go beyond ensemble deterministic forecasts and stochastic parametrizations by developing stochastic forecasts.

9.4 Summary of emergent laws in Chapter 9

Causality requires the use of a Heaviside function $\Theta_{Heavi}(t)$ in cascade processes for the turbulent flux $\varphi(\underline{r},t)$:

$$\varphi(\underline{r},t) = N_{D_{st}}^{1/\alpha} C_1^{1/\alpha} \gamma_\alpha(\underline{r},t) * \left[\Theta_{Heavi}(t)\left[\left[(\underline{r},t)\right]\right]_{tur}^{-D_{st}/\alpha}\right];$$
$$\Theta_{Heavi}(t) = \begin{cases} 1; & t > 0 \\ 0; & t \le 0 \end{cases} \qquad (9.67)$$

where γ_α is the subgenerator of extremal Lévy noise index α, C_1 is the codimension of the mean and N_{Dst} is the appropriate normalization constant and $[[(\underline{r},t)]]_{tur}$ is a localized space-time scale for turbulent flux (cf. Eqn. 5.78).

Although, presumably due to its physical nature, the turbulence flux is localized in space-time, the observables I need not have such a strong constraint. While their structures in space continue to be localized, they need not be localized in space-time, so we can also have wave-like propagators. In the FIF model, this can be written:

$$I = \varphi * g_{tur}(\underline{R}) * g_{wav}(\underline{R}); \quad \underline{R} = (\underline{r}, t) \tag{9.68}$$

where:

$$g_{tur}(\underline{r}, t) = \left[\Theta_{Heavi}(t)\left[\left[(\underline{r}, t)\right]\right]_{tur}^{-(D_{st}-H_{tur})}\right] \tag{9.69}$$

and:

$$g_{wav}(\underline{r}, t) = (F.T.)^{-1}[\widetilde{g}_{wav}(\underline{k}, \omega)] \tag{9.70}$$

where both propagators satisfy the scaling equations:

$$\begin{aligned} g_{tur}(T_\lambda \underline{R}) &= \lambda^{s_{tur}} g_{tur}(\underline{R}); \\ g_{wav}(T_\lambda \underline{R}) &= \lambda^{s_{wav}} g_{wav}(\underline{R}) \\ s_{tur} = D_{st} - H_{tur}; &\quad s_{wav} = D_{st} - H_{wav} \end{aligned} \tag{9.71}$$

and:

$$H_{\text{tur}} + H_{\text{wav}} = H; T_\lambda = \lambda^{-G_{st}} \tag{9.72}$$

The unlocalized "wave" space-time propagator g_w is causal if $\widetilde{g_{wav}}(\underline{k}, \omega)$ is analytic in the lower half of the complex w plane. In the fairly general wave case where the waves satisfy a (fractional, anisotropic) wave equation, we have the nondimensional ($V=1$) wave-like (delocalized in space-time) and turbulent-like (localized in space-time) propagators:

$$\widetilde{g}_{wav}(\underline{k}, \omega) = (\omega^2/V^2 - \|\underline{k}\|^2)^{-H_{wav}/2} \tag{9.73}$$

Which implies the dispersion relation:

$$\omega = \pm V\|\underline{k}\| \tag{9.74}$$

where: $\|\underline{k}\|$ is the spatial, Fourier scale function.

A specific example taken from the MTSAT analysis yields a combined turbulent/wave spectral density P_I:

$$\begin{aligned} P_\varphi(\underline{k}, \omega) &= (\omega'^2 + \|\underline{k}\|^2)^{-s_\phi} \\ P_I(\underline{k}, \omega) &= P_\phi(\underline{k}, \omega) \quad (\omega'^2 + \|\underline{k}\|^2)^{-H_{tur}} \\ &(\omega'^2/V^2 - \|\underline{k}\|^2)^{-H_{wav}} \end{aligned} \tag{9.75}$$

(P_φ for the flux) where we have taken into account an overall mean advection $\underline{\mu}$ as well as its statistical variability by the transformation $\omega' = (\omega + \underline{k} \cdot \underline{\mu})/\sigma$ where σ is the standard deviation corrected for spatial anisotropy (Eq. 8.25).

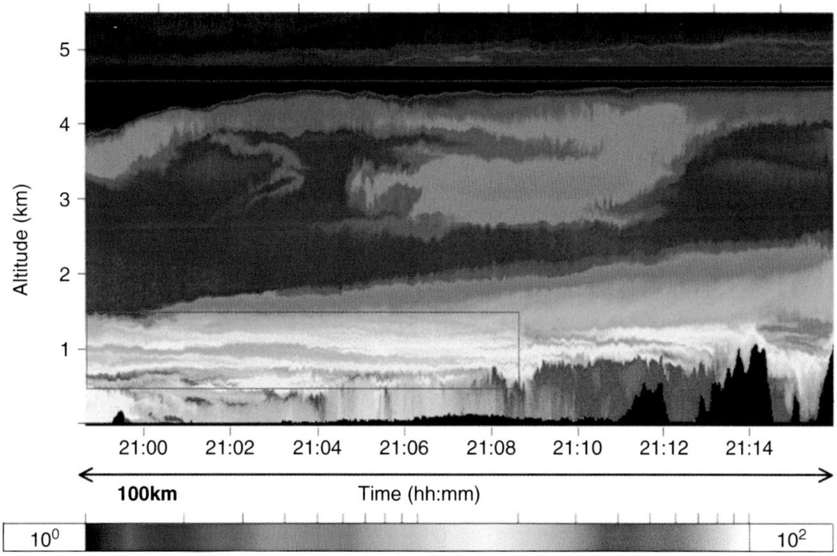

Fig. 1.7 (b) Typical vertical–horizontal lidar cross-section acquired on August 14 2001. The scale (bottom) is logarithmic: darker is for smaller backscatter (aerosol density surrogate), lighter is for larger backscatter. The black shapes along the bottom are mountains in the British Columbia region of Canada. The line at 4.6 km altitude shows the aircraft trajectory. The aspect ratio is 1 : 96. Reproduced from Lilley *et al.* (2004).

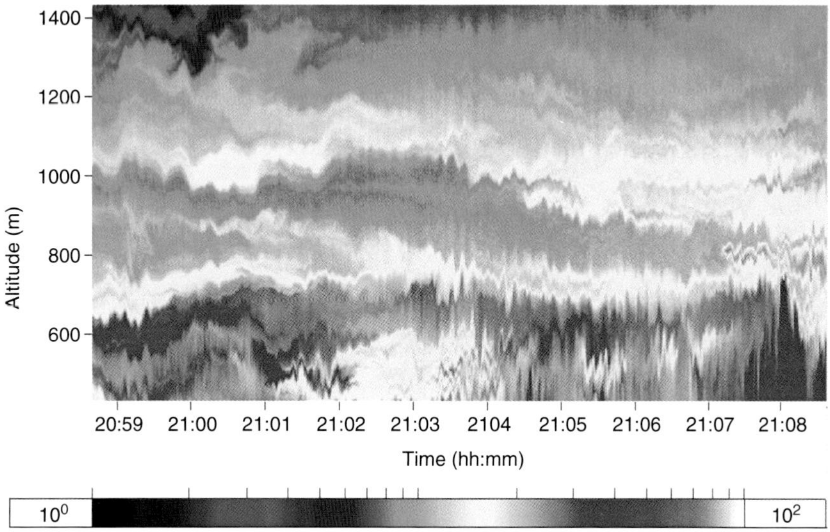

Fig. 1.7 (c) Zoom of Fig. 1.7b showing that at the small scales, structures are beginning to show vertical (rather than horizontal) "stratification" (even though the visual impression is magnified by the 1 : 40 aspect ratio, the change in stratification at smaller and smaller scales is visually obvious). Reproduced from Lilley *et al.* (2004).

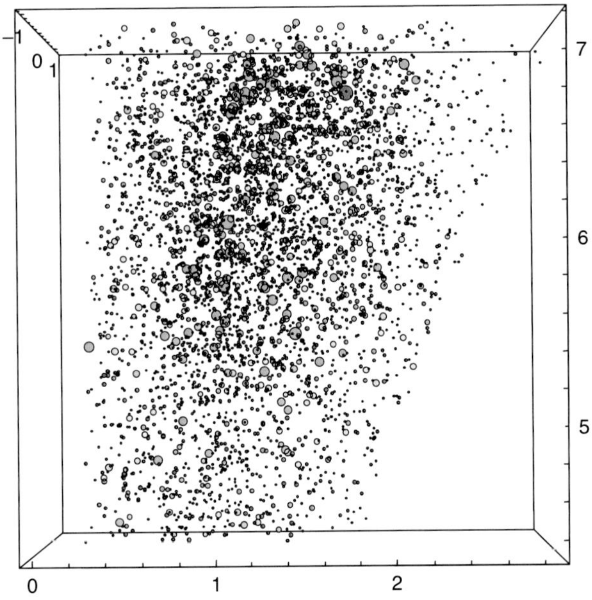

Fig. 1.8 (a) An example of a drop reconstruction. For clarity only the 10% largest drops are shown, only the relative sizes and positions of the drops are correct, the colours code the size of the drops. The boundaries are defined by the flash lamps used for lighting the drops and by the depth of field of the photographs. Adapted from Lovejoy and Schertzer (2008).

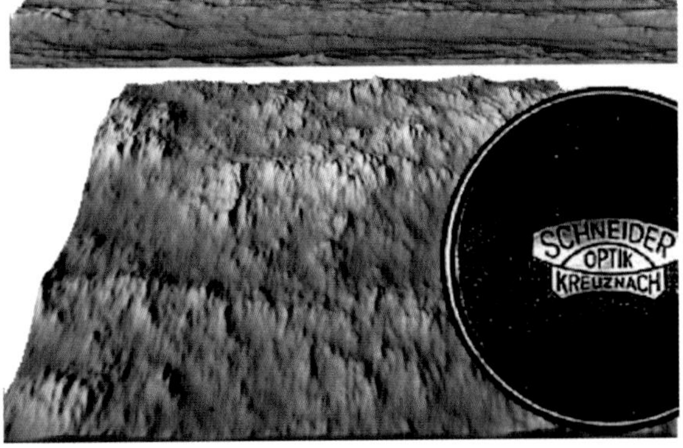

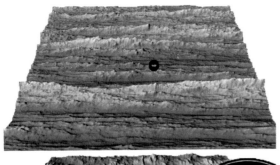

Fig. 1.13 A self-affine simulation illustrating the "phenomenological fallacy": upper and lower images look quite different while having the same generators of the scale-changing operator *G* (see Chapter 6; *G* is diagonal with elements 0.8, 1.2) and the same (anisotropic) statistics at scales differing by a factor of 64 (top and bottom blow-up). The figure shows the proverbial geologist's lens cap at two resolutions differing by a factor of 64. Seen from afar (top), the structures seem to be composed of left-to-right ridges, but closer inspection (bottom) shows that in fact this is not the case at the smaller scales. Reproduced from Lovejoy and Schertzer (2007).

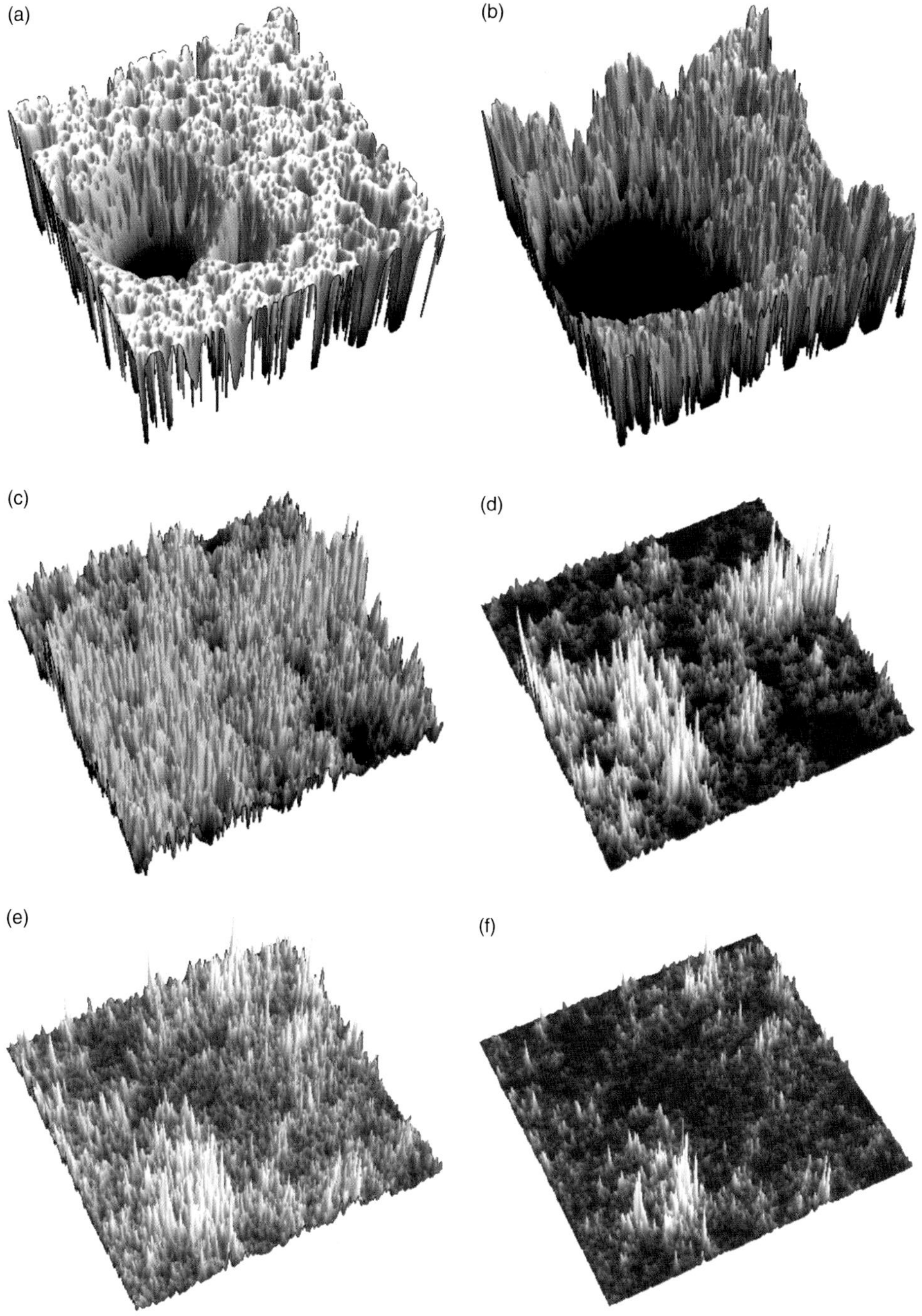

Fig. 5.13 Isotropic realizations in two dimensions with $\alpha = 0.4, 1.2, 2$ (top to bottom) and $C_1 = 0.05, 0.15$ (left to right). The random seed is the same so as to make clear the change in structures as the parameters are changed. The low α simulations are dominated by frequent very low values, the "Lévy holes". The vertical scales are not the same. Reproduced from Lovejoy and Schertzer (2010a).

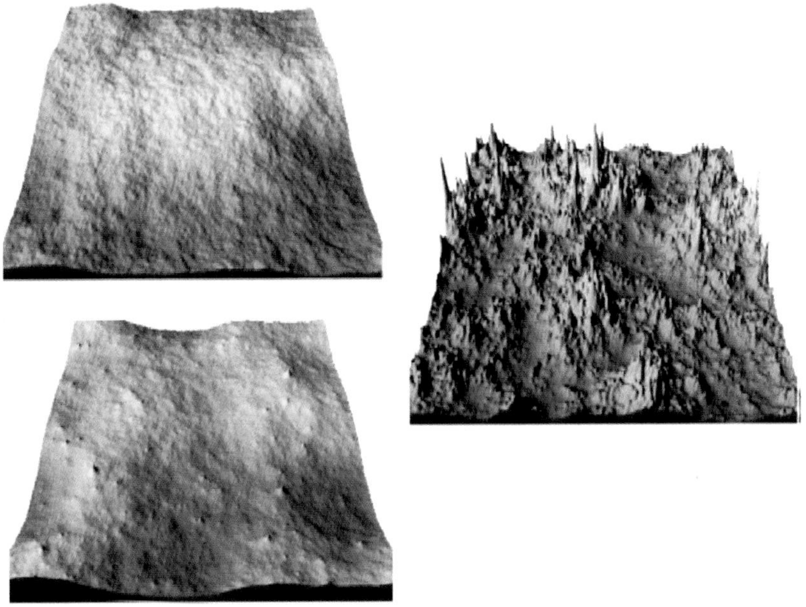

Fig. 5.32 (a) The upper left simulation shows fBm, with $H = 0.7$, lower left fLm with $H = 0.7$, $\alpha = 1.8$, and the right the multifractal FIF with $H = 0.7$, $\alpha = 1.8$, $C_1 = 0.12$ (close to observations for topography, adapted from Gagnon *et al.*, 2006). Note the occasional "spikes" in the fLm which are absent in the fBm; these are due to the extreme power-law tails. (In this fLm positive extremal Lévy variables were used, so there are no corresponding "holes.").

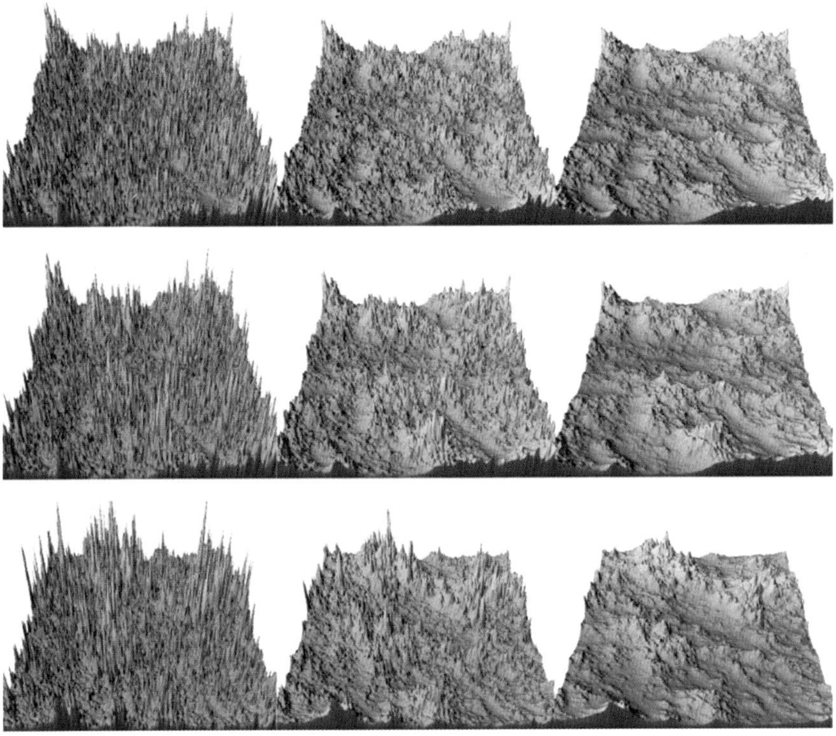

Fig. 5.32 (b) Isotropic (i.e. self-similar) multifractal simulations showing the effect of varying the parameters α and H ($C_1 = 0.1$ in all cases). From left to right, $H = 0.2$, 0.5 and 0.8. From top to bottom, $\alpha = 1.1$, 1.5 and 1.8. As H increases the fields become smoother, and as α decreases one notices more and more prominent "holes" (i.e. low smooth regions). The realistic values for topography ($\alpha = 1.79$, $C_1 = 0.12$, $H = 0.7$) correspond to the two lower right-hand simulations. All the simulations have the same random seed. Reproduced from Gagnon *et al.* (2006).

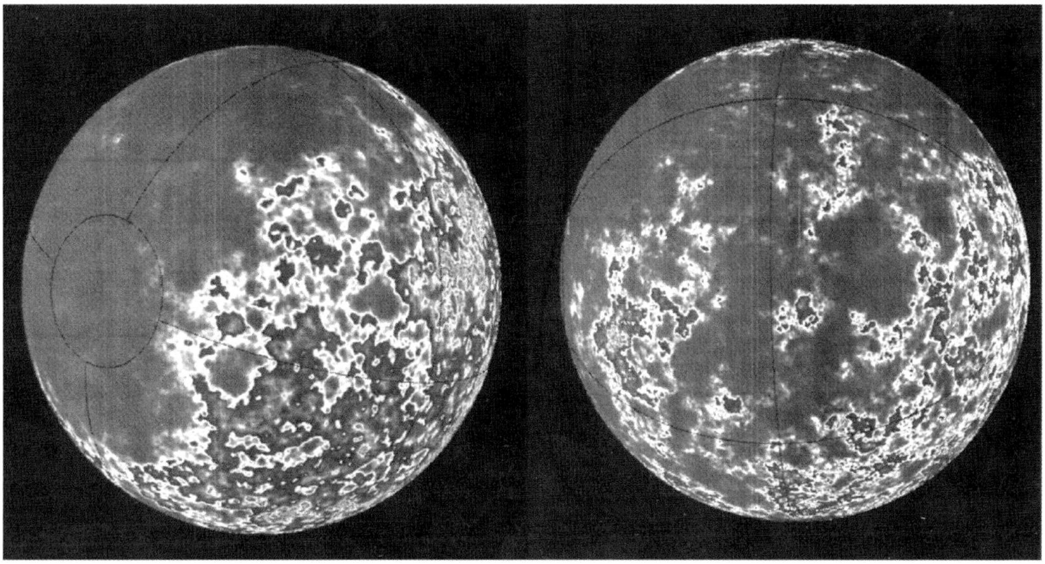

Fig. 5.32 (c) A simulation of an (isotropic) multifractal topography on a sphere using the spherical harmonic method discussed in the appendix (both sides of a single simulation are shown, using false colours). The simulation parameters are close to the measured values: $\alpha = 1.8$, $C_1 = 0.1$, $H = 0.7$ (see Chapter 4). The absence of mountain "chains" and other typical geomorphological features are presumably due to the absence of anisotropy. We thank J. Tan for help with this simulation, adapted from Quattrochi and Goodchild (1997).

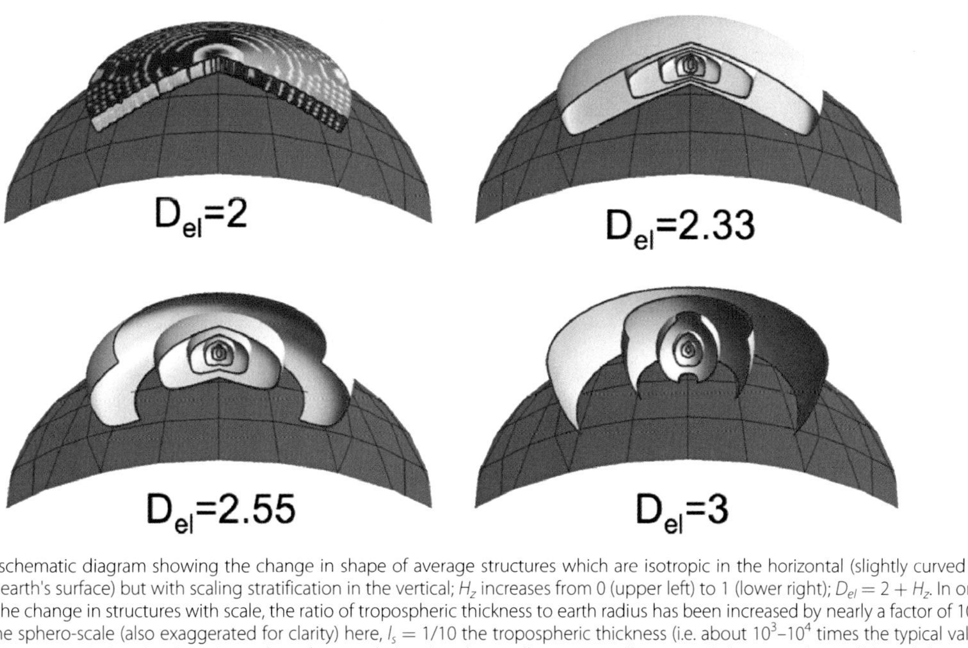

Fig. 6.8 A schematic diagram showing the change in shape of average structures which are isotropic in the horizontal (slightly curved to indicate the earth's surface) but with scaling stratification in the vertical; H_z increases from 0 (upper left) to 1 (lower right); $D_{el} = 2 + H_z$. In order to illustrate the change in structures with scale, the ratio of tropospheric thickness to earth radius has been increased by nearly a factor of 1000. In units of the sphero-scale (also exaggerated for clarity) here, $l_s = 1/10$ the tropospheric thickness (i.e. about 10^3–10^4 times the typical value), the balls shown are ½, 1, 2, 4, 8, 16, 32 times the sphero-scale (so that the smallest is vertically oriented, the second roundish and the rest horizontally stratified). Note that in the $D_{el} = 3$ case, the cross-sections are exactly circles; the small distortion is an effect of perspective due to the mapping of the structures onto the curved surface of the earth. Reproduced from Lovejoy and Schertzer (2010).

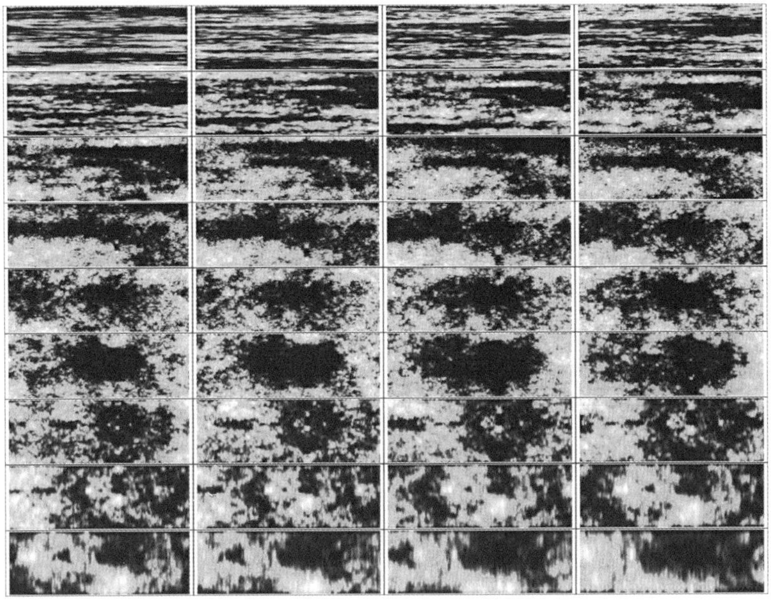

Fig. 6.7 (a) A sequence from a zoom (downscaling, disaggregation) of a stratified universal multifractal cloud model with $\alpha = 1.8$, $C_1 = 0.1$, $H = 1/3$, $H_z = 5/9$. From top left to bottom right each successive cross-section represents a blow-up by a factor 1.31 (total blow-up is a factor $\approx$ 12 000 from beginning to end). If the top left simulation is an atmospheric cross-section 8 km left to right, 4 km thick, then the final (lower right) image is about 60 cm wide by 30 cm high; the sphero-scale is 1 m, as can be roughly visually confirmed since the left–right extent of the simulation second from bottom on the right is 1.02 m, where structures can be seen to be roughly roundish. Reproduced from Lovejoy and Schertzer (2010).

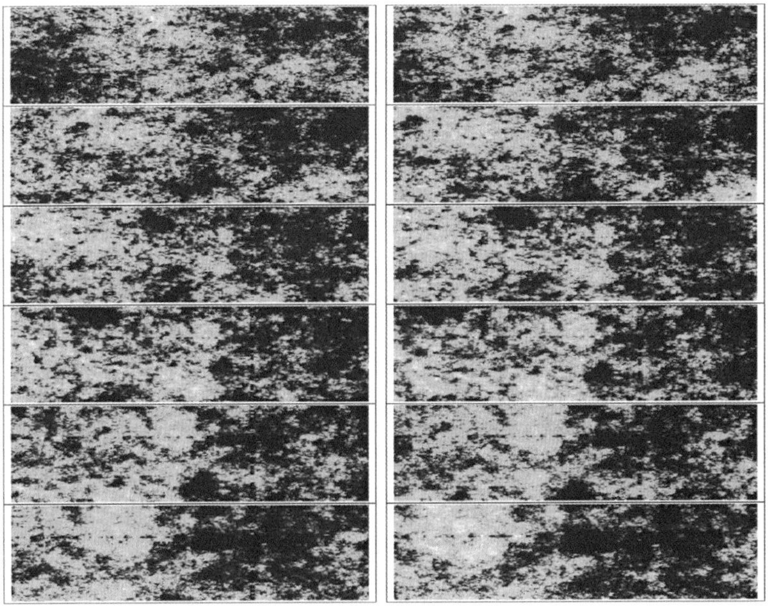

Fig. 6.7 (b) A zoom (downscaling, disaggregation) sequence for an isotropic cloud with the same multifractal parameters as for the anisotropic simulation in Fig. 6.7a, from upper left to lower right. Each image is an enlargement by a factor 1.7 of the previous. As in Fig. 6.7a, the grey shades are "renormalized" separately in each image.

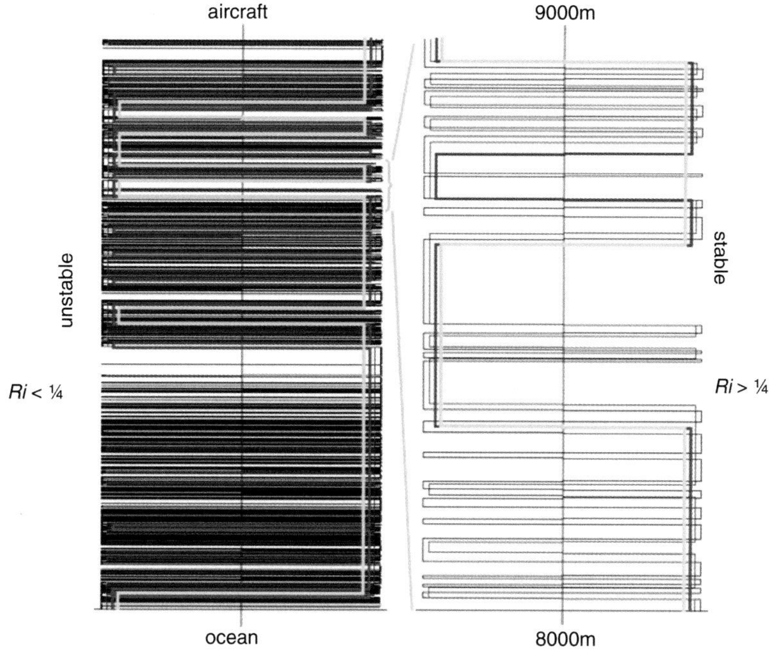

Fig. 6.11 The stability of the atmosphere as determined by a dropsonde using the stability criterion $Ri > 1/4$ where the Richardson number (Ri) is estimated using increasingly thick layers: 5, 20, 80, 320 m thick (black, red, blue, cyan respectively). The figure shows atmospheric columns, the left one from the ocean to 11 520 m (just below the aircraft), while the right is a blow-up from 8000 to 9000 m. The left of each column indicates dynamically unstable conditions ($Ri < 1/4$) whereas the right-hand side indicates dynamically stable conditions ($Ri > 1/4$). The figure reveals a Cantor set-like (fractal) structure of unstable regions. Reproduced from Lovejoy *et al.* (2008).

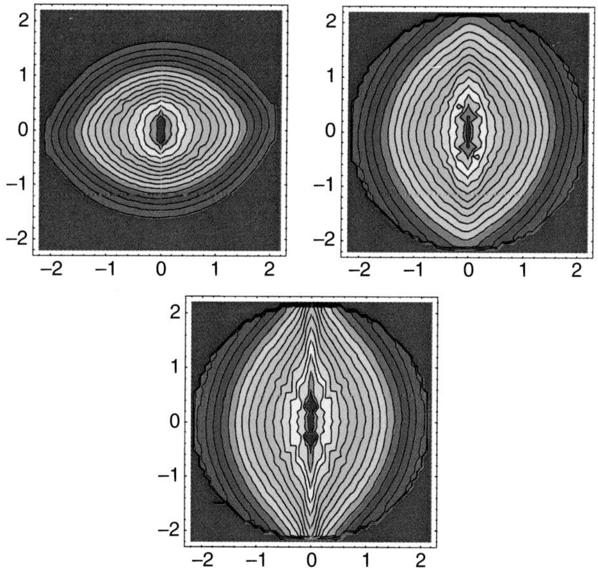

Fig. 6.19 (d) Contour plot of S_1 with different nonlinear transformations in polar coordinates with radius = $\log_{10} r'$ and the polar angle = θ'. Upper left: isotropic turbulence (no transformation, $H_z = 1$); upper right: 23/9D model ($H_z = 5/9$); bottom: gravity waves ($H_z = 1/3$). Note that although the x' (left–right) and z' (up–down) coordinate directions show the distance from the centre of the plot in the corresponding direction, they are not Cartesian coordinates of the plot. Pacific 2001 0815t6, vertical–horizontal cross-section. If H_z is such that Eqn. (6.50) is satisfied, then the spacing between the contours is constant in all directions (close to the upper right, $H_z = 5/9$ case), this criterion does not imply that the contours have shapes independent of scale. Reproduced from Radkevitch *et al.* (2008).

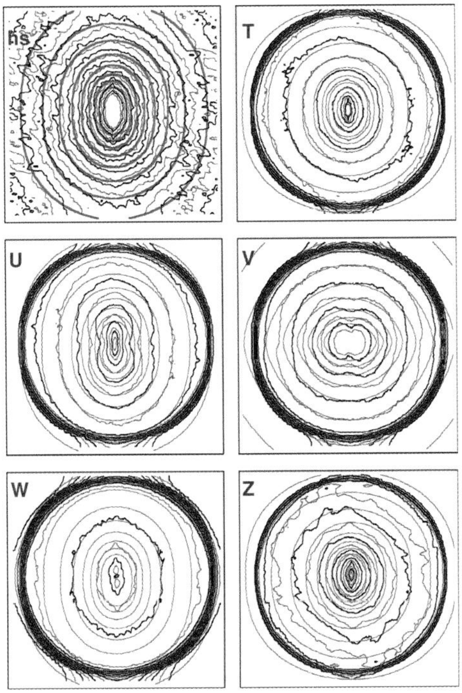

Fig. 6.24 Contour plots of log P. Left–right is k_x, vertical direction is k_y. Upper left is the specific humidity, upper right is the temperature, middle is the zonal (left) and meridional (right) winds. Lower left is the vertical wind, and lower right is the geopotential. Contours of the logarithm of the theoretical canonical scale functions (blue) all have $H_y = 0.8$ (except for v, which has $H_y = 1/0.8$) and the sphero-wavenumbers are $k_s = 60, 30, 60, 30, 60, 30$ respectively for h_s, T, u, v, w, z. Due to the Nyquist frequency, the largest k_y is 30 cycles/90° corresponding to 2 pixels. Due to the 2 : 1 aspect ratio (which compensates for the 2 : 1 change in range of k_x with respect to k_y), a circle the diameter of the square in the figure corresponds to isotropy at a 2-pixel scale. Reproduced from Lovejoy and Schertzer (2011).

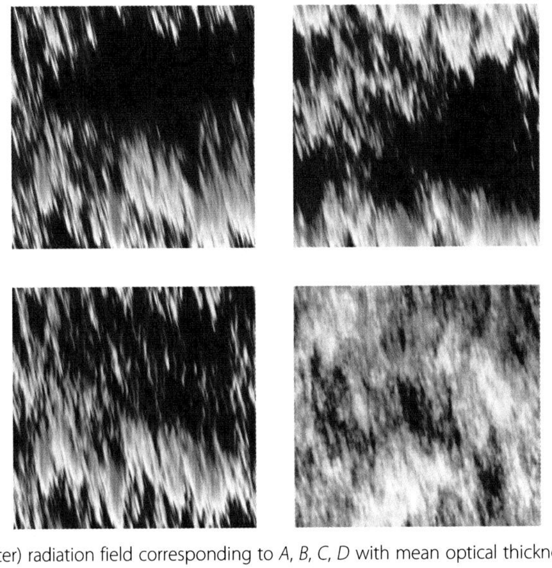

Fig. 7.13 (c) The top (single-scatter) radiation field corresponding to A, B, C, D with mean optical thickness = 10 and the sun incident at 45° from the right.

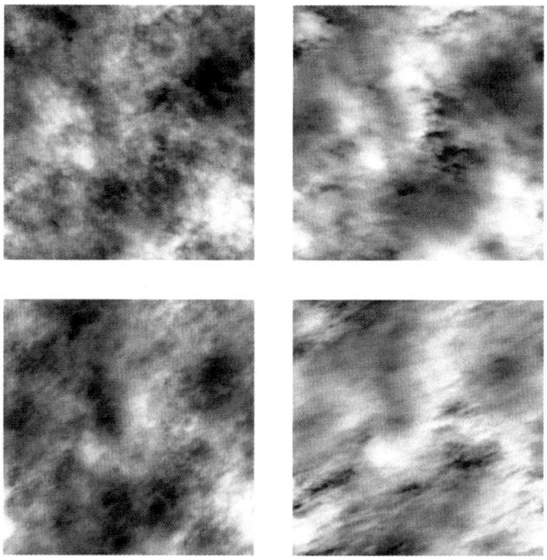

Fig. 7.10 (c) The top view with single scattering radiative transfer; incident solar radiation at 45° from the right, mean vertical optical thickness = 50.

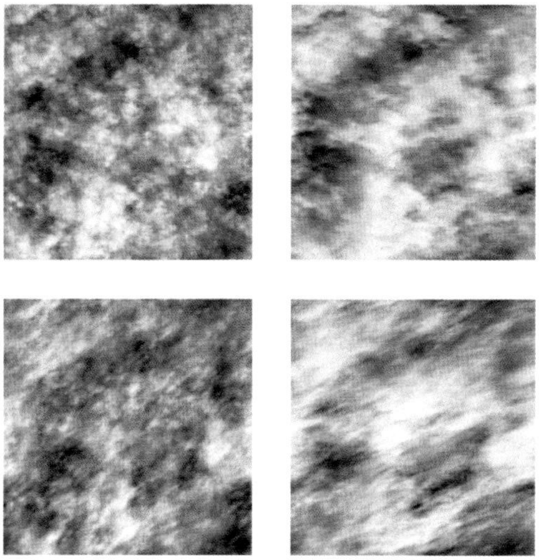

Fig. 7.10 (d) Same as Fig. 7.10c except viewed from the bottom.

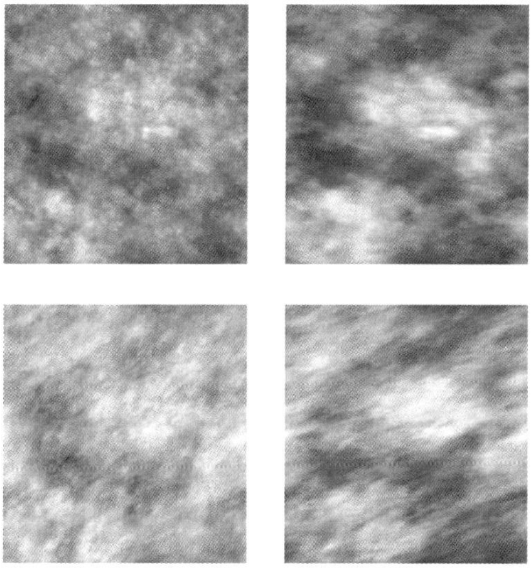

Fig. 7.10 (e) Same as Fig. 7.10d except for a false-colour rendition of a thermal infrared field (assuming a constant extinction coefficient and a linear vertical temperature profile).

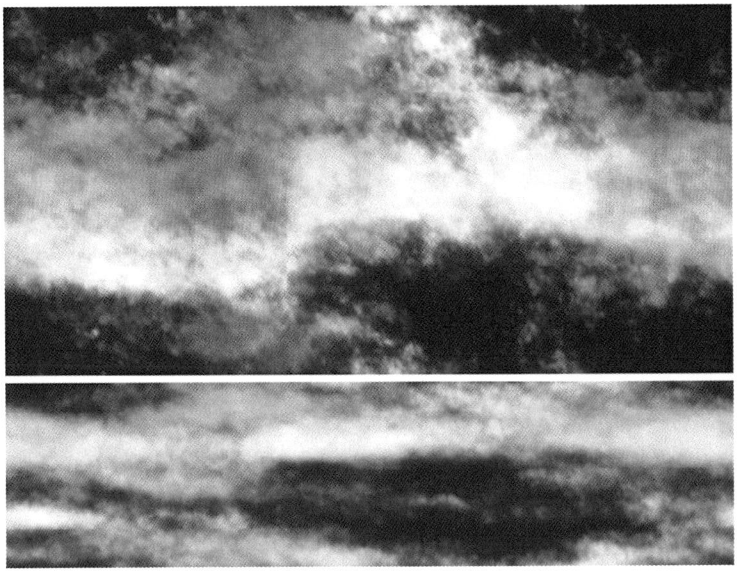

Fig. 7.14 (b) The top is the visible radiation field (corresponding to Fig. 7.14a) looking up (sun at 45° from the right); the bottom is a side radiation fields (one of the 512 × 128 pixel sides), average optical thickness = 10, single scattering only.

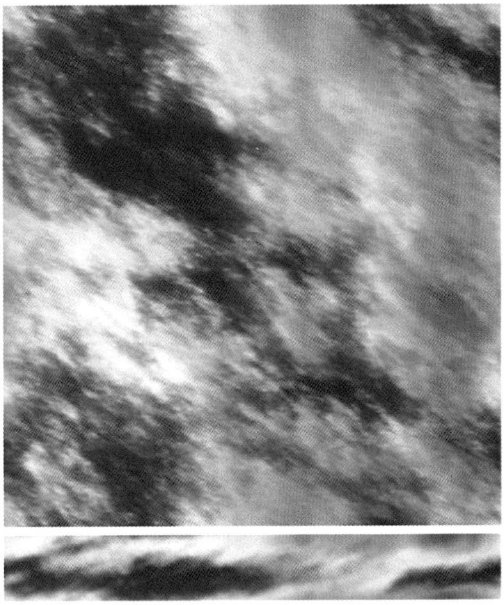

Fig. 7.14 (d) The top is the visible radiation field (corresponding to Fig. 7.14c), looking up (sun at 45° from the right); the bottom is a side radiation fields (one of the 512 × 64 pixel sides), the average optical thickness = 5, single scattering only.

(a)

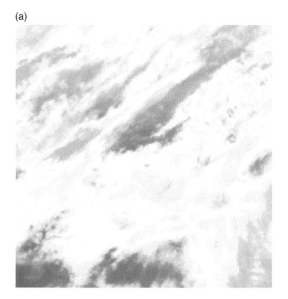

Fig. 7.17 (d) A cloud radiance field in the thermal IR from the NOAA AVHRR satellite off the coast of Florida (512 × 512 pixels, 1.1 km resolution). Various cloud morphologies are visible, although there is a dominant southwest/northeast alignment of large-scale structures. Reproduced from Schertzer and Lovejoy (1996).

(b)

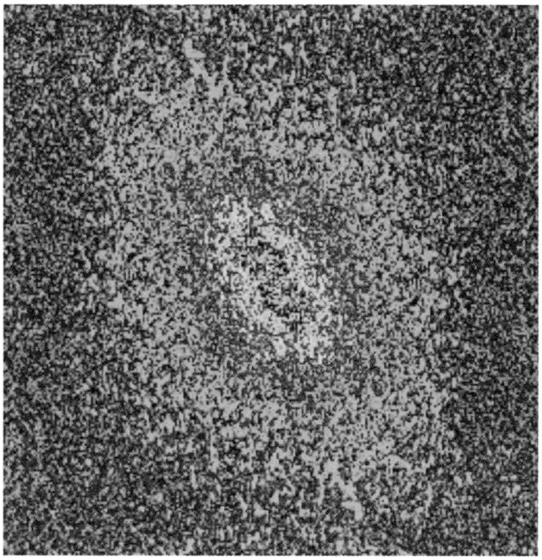

Fig. 7.17 (d) The power spectrum $P(\underline{k})$ of Fig. 7.17a, showing Fourier-space structures roughly perpendicular to the real-space ones in Fig. 7.17a. Reproduced from Schertzer and Lovejoy (1996).

(c)

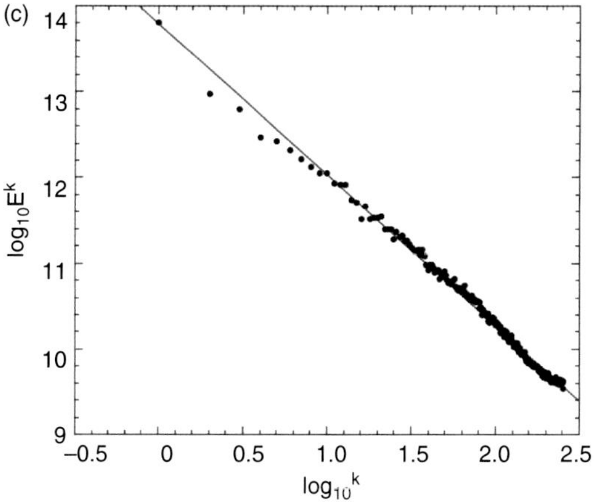

Fig. 7.17 (c) The isotropic power spectrum of the radiances displayed in Fig. 7.17a (the spctral density field, Fig. 7.17b, integrated around circular annuli, cf. Eqn. (6.58)). The spectral slope determined by linear regression is given by $\beta = 1.77$. Reproduced from Schertzer and Lovejoy (1996).

(d)

Fig. 7.17 (c) A multifractal FIF model simulation with $c = -0.1$, $f = -0.15$, $e = 0.15$, $d = 1$. Compare this with Fig. 7.17a and notice that much of the NE/SW texture/morophology is reproduced, but that there are clearly regions of Fig. 7.17a where the morphology is different and where the simulation (which assumes the type of anisotropy is constant across the region) does a poor job. Reproduced from Schertzer and Lovejoy (1996).

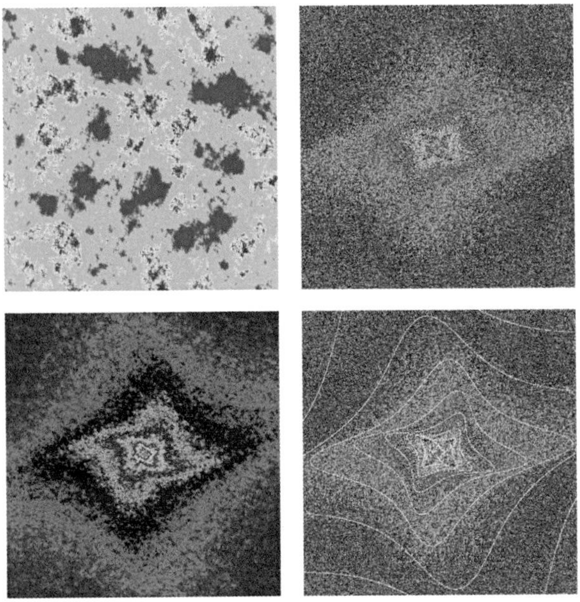

Fig. 7.18 Real-space (top left), spectral energy density (top right), enhanced spectral energy density (bottom left) and spectral energy density with estimated GSI contours (bottom right). This is for a simulation with parameters $s = 2.64$, $c = 0.1$, $f = 0.1$, $e = 0.5$; the measured parameters are $s = 2.63$, $c = 0.05$, $f = 0.08$, $e = 0.51$. Reproduced from Lewis (1993).

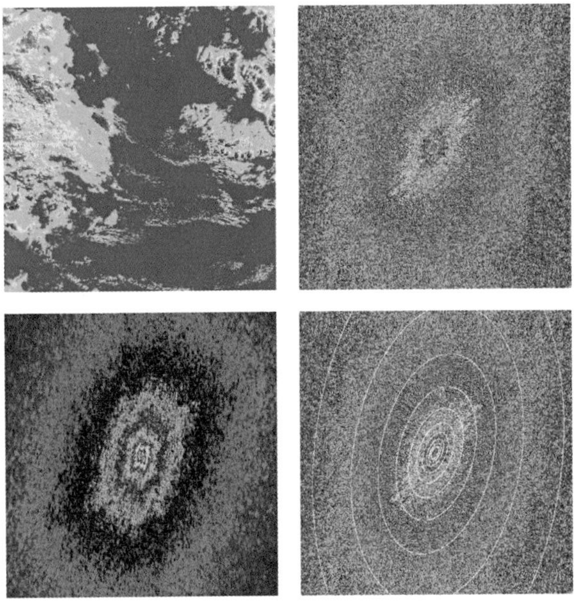

Fig. 7.19 In real space (top left), spectral energy density (top right), enhanced spectral energy density (bottom left) and spectral energy density with estimated GSI contours (bottom right). From AVHRR cloud pictures taken at visible wavelengths; estimated parameters are $s = 2.34$, $c = -0.05$, $f = 0.12$, $e = -0.12$. Reproduced from Lewis (1993).

(a)

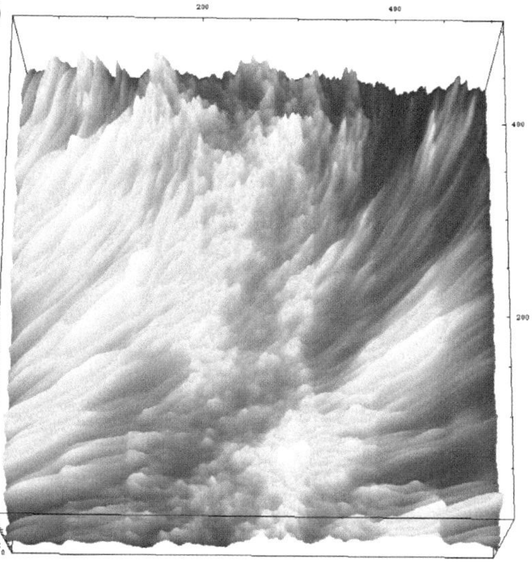

Fig. 7.24 (a) A multifractal simulation of quadratic GSI (with $\underline{g}$ given by the cubic h, Eqn. (7.122)) with $\alpha = 1.8$, $C_1 = 0.1$, $H = 0.33$ and sphero-scale = 256 pixels (the simulation is 512 × 512 pixels). The effect of the varying G is quite subtle.

(b)

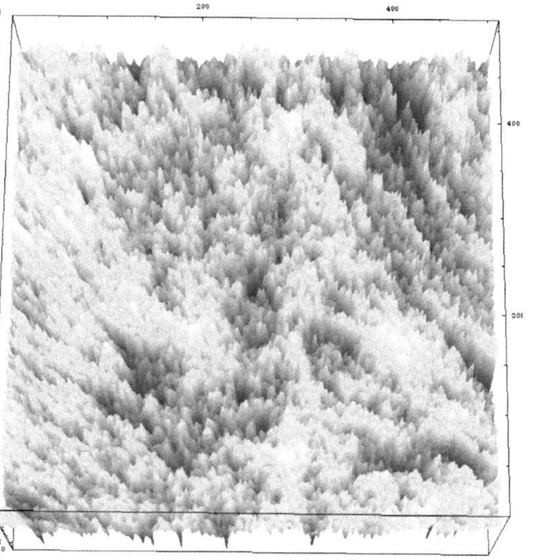

Fig. 7.24 (b) Same as Fig. 7.24a but for $l_s = 1$, $H = 0$.

(c)

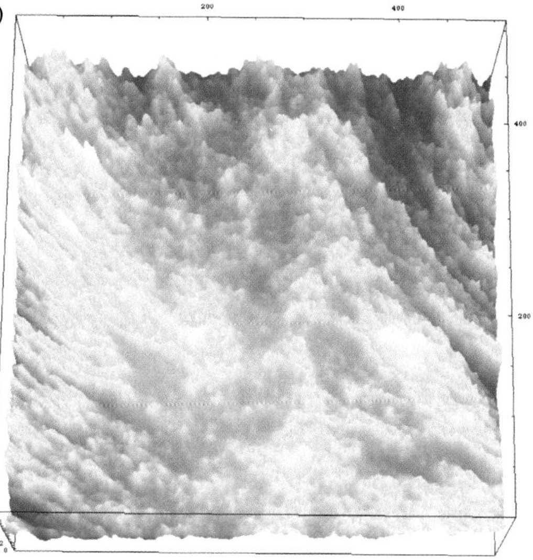

Fig. 7.24 (c) Same as Fig. 7.24a but for $l_s = 1$.

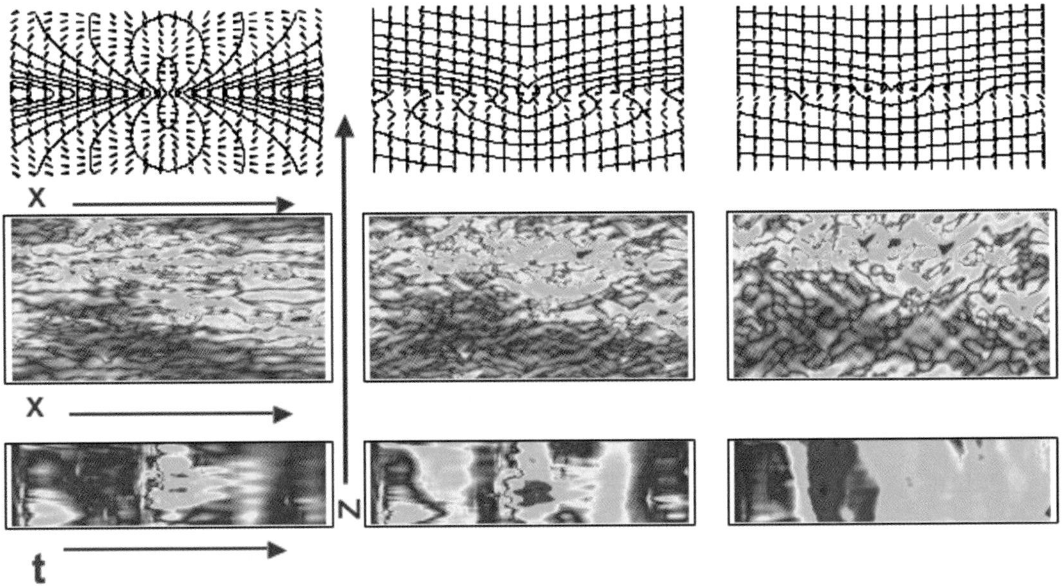
x
x
N
t

Fig. 9.10 A multifractal simulation of a passive scalar in (x,z,t) space with the observed multifractal parameters ($\alpha = 1.8$, $C_1 = 0.05$) and theoretical values $H_t = 2/3$, $H_z = 5/9$. The simulations show the vertical wind increasing from 0 (left) to 0.25 to 0.5 pixels/time step (only a single time step is shown). The top row shows the dispersion relation (contours), group velocity (arrows), the second is an (x,z) cross-section, and the third row is a (t,z) cross-section. The numerical simulation techniques are based on those described in Chapters 5 and 8. Reproduced from Lovejoy *et al.* (2008).

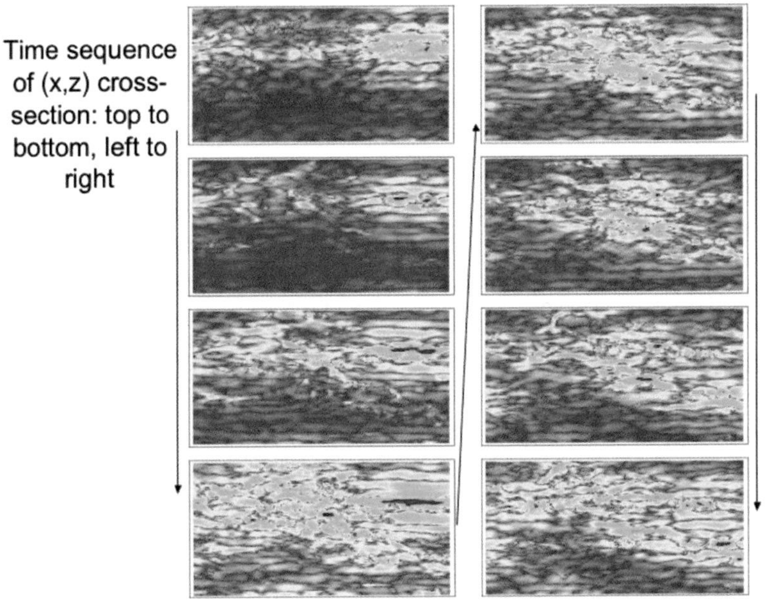
Time sequence of (x,z) cross-section: top to bottom, left to right

Fig. 9.11 Eight time steps in the evolution of the vertical cross-section of a passive scalar component from the zero wind case of Fig. 9.10. The structures are increasingly stratified at larger and larger scales and display wave phenomenology. Reproduced from Lovejoy *et al.* (2008).

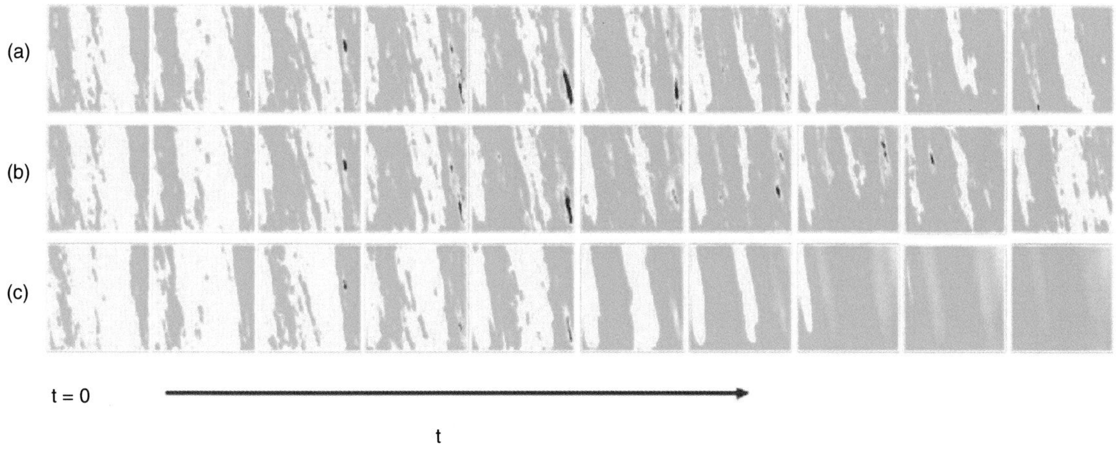

Fig. 9.18 The top two rows (a–b) show successive snapshots ($t = i \times \tau$, $i = 3, 6, 9, \ldots, 27$; τ being the eddy turnover time of the smallest structures) of two simulations (256 × 256 in space) that are identical until time $t = 0$, when their fluxes at small scales become progressively independent step by step due to the sudden independence of the subgenerators at that time. Most of the difference between the two realizations is concentrated in a few "hot spots". The bottom row (c) shows a forecast based on the "memory" of the evolution up to $t = 0$ of (a), i.e. it has the same stochastic subgenerator until time $t = 0$, then defined in a deterministic manner to preserve the mean of the flux. Note the more rapid disappearance of small-scale structures. Parameters are $\alpha = 1.5$, $C_1 = 0.2$, $H = 0.1$ (close to those of rain), and the colour scale is logarithmic. The anisotropy of space-time is characterized by $H_t = 2/3$. Reproduced from Schertzer and Lovejoy (2004a).

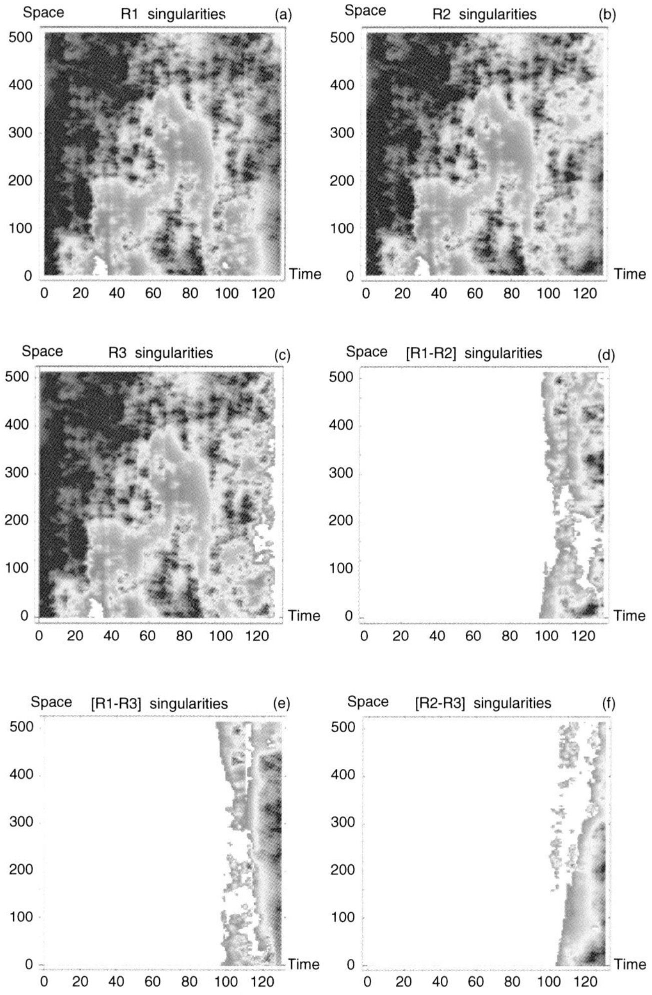

Fig. 9.19 Simulation of multifractal predictability decay for rain field. (a) and (b) are identical up to $t_0 = 64$, after which their fluxes become independent. (c) displays the forecast based on their common past and the deterministic conservation of the flux afterward. Singularities of the fields (i.e. their log divided by the log resolution), as well as of their absolute differences (d–f), are displayed according to the following palette: white for negative singularities; green to yellow for singularities contributing to statistics up to the mean; red for singularities contributing to second- and higher-order moments. Reproduced from Schertzer and Lovejoy (2004b).

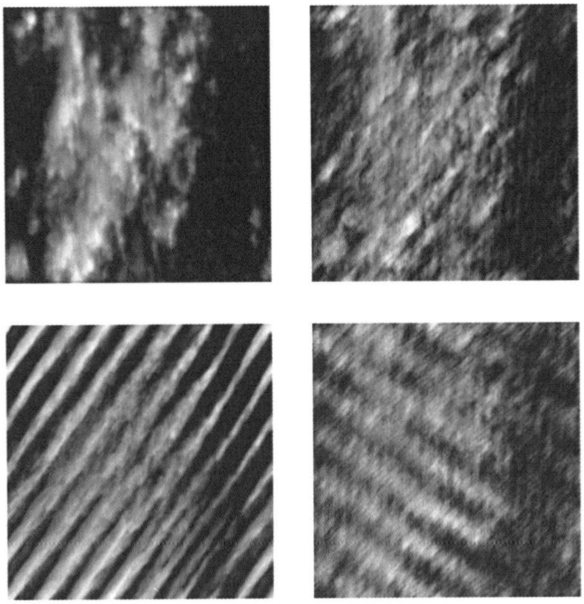

Fig. 9.14 The effect of increasing H_{wav} with $H_{wav} + H_{tur} = H = 0.33$, $H_t = 0.66$; clockwise from the upper left we have $H_{wav} = 0, 0.33, 0.52, 0.38$ (i.e. $H_{tur} = 1/3$, $H_{wav} = 0.33, 0, -0.19, -0.05$), $C_1 = 0.1$, $\alpha = 1.8$. There is a small amount of differential anisotropy characterized by $G = \begin{pmatrix} 0.95 & -0.02 \\ 0.02 & 1.05 \end{pmatrix}$ The horizontal unit ball is characterized by $\Theta(\theta) = 1 + a\mathrm{Cos}(2\theta - 2\theta_0)$ with $a = 0.65$, with $\theta_0 = 0$. The random seed is the same in all cases so that one can see how structures become progressively more and more wave-like while retaining the same scaling symmetries, close to observations. Reproduced from Lovejoy *et al.* (2008).

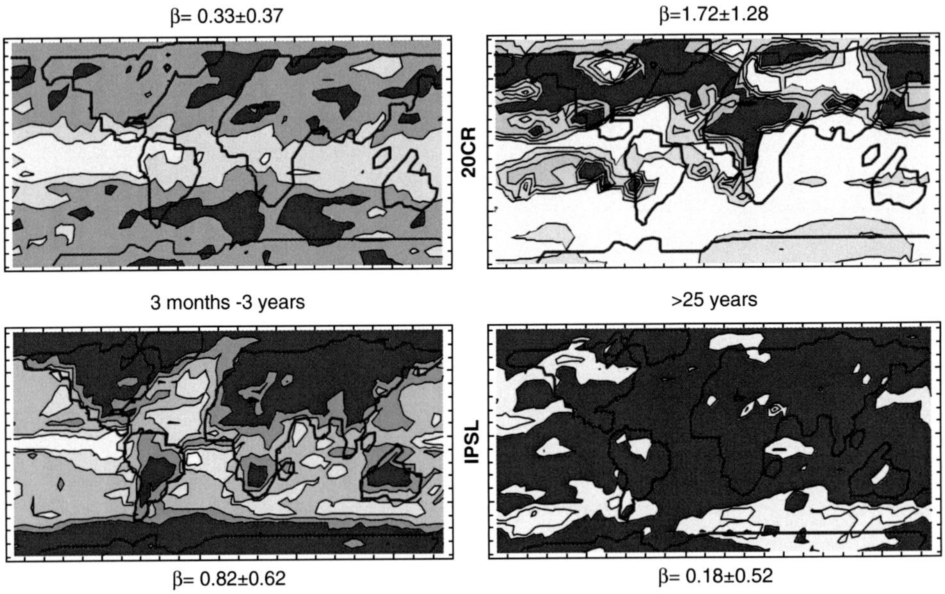

Fig. 11.4 The spatial distribution of the spectral exponents for the reanalyses (top row) and the IPSL control run (bottom row). The left-hand side shows the distribution of the high-frequency exponents, the right-hand side, the low-frequency β's. In all cases the contour lines were at $\beta = 0.5$ (dark blue), 1, 1.5, 2 (white). One can see that Greenland has exceptionally small low-frequency β's (< 0.5) although about 20° to the east the values are quite high (> 1.5: the white area to the right of Greenland). The mean β's and one-standard-deviation spreads are also indicated. The high-frequency 20CR β map is very similar to that of Huybers and Curry (2006) (although they determined β from NCEP reanalyses from 2 months to 30 years).

Chapter

10 The emergent laws of macroweather and the transition to the climate

10.1 What is the climate?

10.1.1 Climate as an emergent scaling process

Meteorological and climatological sciences have become increasingly distinct, yet there is still no generally accepted definition of the climate, or – which is nearly the same thing – what precisely is the distinction between the weather and the climate? And if our notion of the climate is vague, what do we mean by climate *change*?

While atmospheric scientists routinely use the expressions "climate scales" and "meteorological scales," the actual boundary between them is not clear and most improve little upon the dictum:

> The climate is what you expect, the weather is what you get. (the character Lazarus Long in Heinlein, 1973; often attributed to Mark Twain)

This definition is not so different from earlier ideas of climate such as "the synthesis of the weather" (Huschke, 1959) or "the sum total of the weather experienced at a place in the course of a year and over the years" (Lamb, 1972). More recently, E. Lorenz approvingly cited this, saying: "before embarking on a search for an ideal definition, assuming one exists, let me express my conviction that such a definition, when found, must agree in spirit with the statement, 'climate is what you expect'" (Lorenz, 1995). He then proposed several definitions based on dynamical systems theory (see Lovejoy and Schertzer, 2012a).

The Heinlein/Twain dictum is actually quite close to the principal definition given by the US National Academy of Sciences:

> Climate is conventionally defined as the long-term statistics of the weather. (Committee on Radiative Forcing Effects on Climate, 2005)

Which improves on the Almanac only a little by proposing:

> to expand the definition of climate to encompass the oceanic and terrestrial spheres as well as chemical components of the atmosphere.

Another official source attempts to quantify the issue of the exact meaning of "long-term" by tentatively suggesting a month as the basic "inner" climate time scale, but ultimately it seems to yield to a higher authority:

> Climate in a narrow sense is usually defined as the "average weather," or more rigorously, as the statistical description in terms of the mean and variability of relevant quantities over a period of time ranging from months to thousands or millions of years. The classical period is 30 years, as defined by the World Meteorological Organization (WMO). These quantities are most often surface variables such as temperature, precipitation, and wind. Climate in a wider sense is the state, including a statistical description, of the climate system. (Intergovernmental Panel on Climate Change AR4, Appendix I: Glossary, p. 942: Solomon *et al.*, 2007)

At first sight, the last sentence is an interesting addition but is ultimately tautological since it defines the climate as the statistics of the "climate system," which itself is left undefined.

Finally, an attempt at a more comprehensive definition is only a little better:

> Climate encompasses the statistics of temperature, humidity, atmospheric pressure, wind, rainfall, atmospheric particle count and numerous other meteorological elements in a given region over long periods of time. Climate can be contrasted to weather, which is the present condition of these same elements over periods up to two weeks ... Climate ... is commonly defined as the weather averaged over a long period of time. The standard averaging period is 30 years, but other periods may be used depending on the purpose. Climate also includes statistics other than the average, such as the magnitudes of day-to-day or year-to-year variations ... The difference between climate and weather is usefully summarized by the popular phrase "Climate is what you expect, weather is what you get." (Wikipedia, March 2010)

What is new here is the explicit attempt to distinguish weather (periods of less than two weeks) and climate (periods of 30 years or more). However, as with the IPCC definition, these time periods are simply suggestions, with no attempt at physical justification. In any case they leave the intervening factor of 1000 or so in scale (literally!) up in the air.

An obvious problem with these definitions is that they fundamentally depend on subjectively defined averaging scales. This fuzziness is also reflected in numerical climate modelling, since global climate models (GCMs) are fundamentally the same as weather models but at lower resolutions, with a different assortment of subgrid parametrizations, and they are coupled to ocean models and – increasingly – to carbon-cycle, cryosphere and land-use models. Consequently, whether we define the climate as the long-term statistics of the weather, or in terms of the long-term interactions of components of the "climate system," we still need an objective way to distinguish it from the weather. These problems are clearly compounded when we attempt to objectively define climate *change*.

However, there is yet another difficulty with this and allied definitions: they imply that climate dynamics are nothing new; that they are simply weather dynamics at long time scales. This seems naïve, since we know from numerous examples in physics that when processes repeat over wide enough ranges of space or time scale they typically display qualitatively new features, so that over long enough time scales we expect that new climate laws should emerge from the higher-frequency weather laws. These qualitatively new emergent laws could simply be the consequences of long-range statistical correlations in the weather physics in conjunction with qualitatively new climate processes – due to either internal dynamics or to (external) orbital, solar, volcanic or anthropogenic forcings – their nonlinear synergy giving rise to emergent laws of climate dynamics. From the GCM modelling point of view, the weather "boundary conditions" – supposedly quasi-fixed and determining the climate (Bryson, 1997) – turn out instead to be nonlinearly coupled with it (Pielke, 1998): i.e. they constitute new "slow dynamics."

A useful definition of climate should involve a physical basis for the distinction/boundary between weather and climate as well as an identification of each regime with specific mechanisms and a corresponding specific type of variability. Both of these ingredients are provided by the scaling approach. In this framework, weather processes are those whose space-time variability follows the emergent turbulent laws discussed earlier. We have seen in Chapter 8 that the driving solar energy flux, combined with the finite size of the earth, implies a drastic "dimensional transition" in the behaviour at the lifetime of the corresponding atmospheric structures at $\tau_w \approx 10$ days, and for ocean structures at $\tau_o \approx 1$ year. Both time scales not only follow directly from the energy-flux-based theory of the horizontal variability/dynamics (and τ_w, directly from realistic estimates of the solar forcing), but – at least in the case of τ_w – also correspond to the observed sharp change in the scaling of all the atmospheric fields somewhere in the 5–20-day ("synoptic maximum") region, followed by a flatter "plateau" low-frequency macroweather regime. At least for the temperature, over time scales between τ_w and τ_o, the ocean boundary condition has strongly turbulent variability with $C_1 \approx 0.1$, $\alpha \approx 1.8$, $H \approx 0.5$ so that its effect on atmospheric buoyancy and humidity is to increase the intermittency of monthly average temperatures over the land and ocean (see Fig. 8.6e).

Since the atmosphere is a nonlinear dynamical system with interactions and variability occurring over huge ranges of space and time scales, the natural approach is to consider it as a hierarchy of processes each with wide-range scaling, i.e. each with nonlinear mechanisms that repeat scale after scale over potentially wide ranges (see Figs. 1.9c, 1.9d). Following Lovejoy and Schertzer (1986), Schmitt *et al.* (1995), Pelletier (1998), Koscielny-Bunde *et al.* (1998), Talkner and Weber (2000), Blender and Fraedrich (2003), Ashkenazy *et al.* (2003), Huybers and Curry (2006) and Rybski *et al.* (2008), this approach is increasingly superseding earlier approaches that postulated more or less white-noise backgrounds with a large number of spectral "spikes" corresponding to many different quasi-periodic processes. This includes the slightly more sophisticated variant (Mitchell, 1976) which retains the spikes but replaces the white noise with a hierarchy of Ornstein–Uhlenbeck processes (white noises and their integrals: in the spectrum, "spikes" and "shelves"; see Fig. 8.6d). In Appendix 10B we discuss the corresponding stochastic linear forcing approaches. Finally, we could also mention Fraedrich *et al.* (2009), who proposed a "hybrid" composite which includes a single short-range scaling regime.

The resulting trichotomy scaling weather/macroweather/climate model was already empirically illustrated in Fig. 1.9c (see Lovejoy, 2012). The label "weather" for the high-frequency regime seems obvious and requires no further comment. Similarly the lowest frequencies correspond to our usual ideas of multidecadal, multicentennial, multimillennial variability as "climate." But labelling the intermediate region "macroweather" – rather than, say, "high-frequency climate" – needs some justification. The point is perhaps made more clearly with the help of Fig. 1.9d, which shows a blow-up of Fig. 1.9c with both global and locally averaged instrumentally based spectra as well the corresponding spectra from a GCM control run (see the figure caption for details, and Sections 11.3.3 and 11.3.4). While the spectrum of the data (especially those globally averaged) begins to rise for frequencies below $\sim$(10 years)$^{-1}$, the control-run spectrum maintains its relatively flat "plateau"-like behaviour out to at least (500 years)$^{-1}$. Similar conclusions for the control runs of other GCMs at even lower frequencies were found by Fraedrich and Blender (2003), Blender *et al.* (2006), Zhu *et al.* (2006) and Rybski *et al.* (2008) (see Section 11.3.3 for various GCM simulations and a detailed discussion), so that it seems that in the absence of external climate forcing, the GCMs – which are thus essentially macroweather models – reproduce the plateau but not the lower-frequency climate regime with its characteristic strong spectral rise. Similarly, we shall see in the next section that when the FIF cascade models that reproduce the weather-scale statistics are extended to low frequencies they too predict the "plateau," and they too need some new ingredient to yield the lower-frequency climate regime. In this chapter, we therefore focus on low-frequency (macro) weather; the climate regime proper is the subject of Chapter 11.

10.1.2 From the weather to macroweather: a dimensional transition to a new scaling regime characterized by $H < 0$

In Chapter 8 we studied the high-frequency weather regime, finding that the nonconservation/fluctuation exponent H was generally positive, indicating that mean fluctuations in a field f: $\langle\Delta f\rangle = \langle\phi_{\Delta t}\rangle\Delta t^H$ increased with scale Δt (since the mean flux $\langle\phi_{\Delta t}\rangle$ is constant) up to scales of $\tau_w \approx$ 5–20 days, after which the statistics underwent a drastic transition. This was found both in spectra of diverse atmospheric variables and in cascade analyses of the corresponding fluxes (see Figs. 8.2a, 8.2b, 8.2c, 8.3a, 8.3b, 8.3c, 8.4a, 8.4b, 8.4c). Although we noted that at lower frequencies the spectra were fairly flat, we did not consider these low frequencies in much detail except to note the existence of a corresponding ocean spectral plateau at periods longer than $\tau_o \approx 1$ year (Fig. 8.6d). In this section we show that a general characteristic of the new regime is that $H < 0$, implying that fluctuations decrease with scale, so that the corresponding series appear "stable" with fluctuations "dying out" as they are averaged over larger and larger scales. We also show how this behaviour is predicted by generalizing the cascade-based FIF model to long time scales. In order to visually appreciate the typical difference between $H < 0$ and $H > 0$ series, see Figs. 1.9e and 5E1a.

In order to understand the basic features predicted by the model for the weather, the transition and the macroweather regime, we can restrict our attention to a (x, y, t) section of the full (x, y, z, t) model and ignore the complications associated with the intermittency augmented by the ocean which is relevant over the regime $\tau_w < \Delta t < \tau_o$ (see however Appendix 10D). If we rewrite the equation for the cascade generator (see Section 9.1.3, Eqns. (9.25)–(9.29)), nondimensionalizing $\underline{r}$ with $L_w = L_e$ and t with τ_w, then we obtain for the generator $\Gamma(\underline{r}, t) = \log\varepsilon(\underline{r}, t)$:

$$\Gamma(\underline{r}, t) = \int_{\Lambda_w^{-1}}^{1}\int_{B_w}^{B_1} \gamma(\underline{r} - \underline{r}', t - t')g(\underline{r}', t')d\underline{r}'dt' + \int_{1}^{\Lambda_c}\int_{B_w}^{B_1} \gamma(\underline{r} - \underline{r}', t - t')g(\underline{r}', t')d\underline{r}'dt' \quad (10.1)$$

$\Lambda_w = L_w/L_i = \tau_w/\tau_i$ is the total range of meteorological scales (L_i, τ_i are the inner dissipation space and time scales) and $\Lambda_c = \tau_c/\tau_w$ is ratio of the overall outer time scale of the overall high and macroweather process τ_c to the outer time scale of the weather process (τ_c is the outer scale of the macroweather process; it is equal to the inner scale of the climate process; the total range of scales is then $\tau_c/\tau_i = \Lambda_w\Lambda_c$). We ignore the vertical, i.e. $\underline{r} = (x, y)$, so that the spatial domain of integration is an annulus between radii Λ_w^{-1} and 1, i.e. between circles, "balls" B_w and B_1 respectively. Fig. 10.1 shows a schematic with the ranges of integration in Eqn. (10.1). Eqn. (10.1) is a convolution between the subgenerator

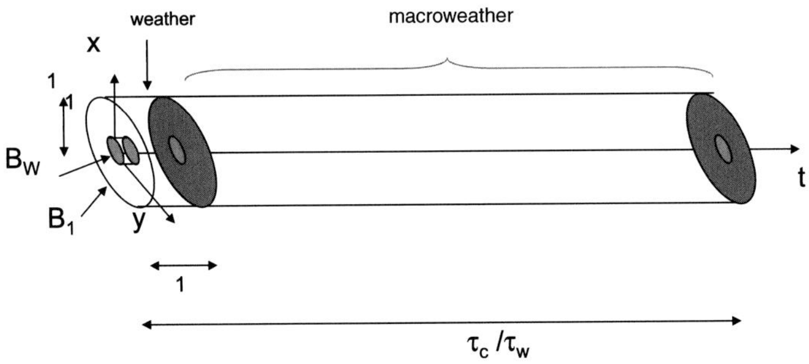

Fig. 10.1 Schematic diagram showing the regions of integration in Eqn. (10.1) and the idea of "dimensional transition" when the region becomes pencil-like (1D, large scales) rather than volume-like (3D, small scales); this is the dimensional transition. The integrations are between the cylinders whose cross-sections are B_W and B_1.

noise γ which represents the "innovations" and the power-law kernel g which represents the interaction strength between scales physically and temporally separated by the space-time interval $(\underline{r}',t')$. For $\Lambda_c = \tau_c / \tau_w >> 1$ we therefore have approximately:

$$\Gamma(\underline{r},t) \approx \Gamma_w(\underline{r},t) + \Gamma_{mw}(t)$$

$$\Gamma_w(\underline{r},t) = \int_{\Lambda_w^{-1}B_w}^{1}\int^{B_1} \gamma(\underline{r}-\underline{r}',t-t')g(\underline{r}',t')d\underline{r}'dt'$$

$$\Gamma_{mw}(t) = \int_1^{\Lambda_c} \overline{\gamma}(t-t')g(0,t')dt' \qquad (10.2)$$

where $\overline{\gamma}(t-t')$ is a spatially integrated Lévy noise. The approximation in Eqn. (10.2) consists in assuming for $t >> |\underline{r}|$ that $g(\underline{r},t) \approx g(0,t)$ so that for long enough time lags the spatial lags are unimportant. $\Gamma_w(\underline{r},t)$ is a 2D (space-time) integral corresponding to the contribution to the variability from the weather regime ($t < 1$, $|r| < 1$), and the second $\Gamma_{mw}(\underline{r},t)$ is a 1D (purely) temporal contribution due to the macroweather regime (*mw* for "macroweather"). This drastic change of behaviour due to the change of space-time dimension over which the basic noise driving the system acts is a kind of "dimensional transition" between weather and macroweather processes. This concept was originally introduced to refer to the hypothetical large scale/small scale (2D/3D) transition required by classical isotropic turbulence approaches in the weather range (Schertzer and Lovejoy, 1984), see section 1.24 and Ch. 6.1. Fig. 10.1 gives a schematic indicating that at small scales the interactions occur over all spatial and temporal intervals (the interaction region is a space-time volume), whereas for long times the interaction region is pencil-like: it is essentially 1D. Physically this is a transition from the high-frequency regime, where both spatial and temporal interactions are important, to a lower-frequency regime where the dynamics are dominated by temporal interactions. In the former case this means between neighbouring structures of all sizes and at their various stages of development, but in the latter case only between very large structures at various stages in their development.

In this simplest model it is this separation into independent additive weather and macroweather generators with correlated noises integrated over spaces of different effective dimensions that is responsible for the statistical difference between weather and the macroweather plateau. At the level of the fluxes it means that the macroweather process multiplicatively modulates the weather process at the larger time scales:

$$\varepsilon_{\Lambda_w,\Lambda_c}(\underline{r},t) \approx e^{\Gamma_w(\underline{r},t)+\Gamma_{mw}(\underline{r},t)} = \varepsilon_{\Lambda_w}(\underline{r},t)\varepsilon_{\Lambda_c}(t) \qquad (10.3)$$

with $\varepsilon_{\Lambda_w}(\underline{r},t)$ having the high-frequency variability, $\varepsilon_{\Lambda_c}(t)$ the low-frequency. The generic result is a "dimensional transition" in the form of a fairly realistic spectral plateau. The notation in Eqn. (10.3) emphasizes the range of scales of the processes, but in what follows it will be more convenient to refer to them by their physics, i.e. weather "*w*", macroweather "*mw*" or the overall process "*w,mw*":

$$\varepsilon_{w,mw}(\underline{r},t) = \varepsilon_w(\underline{r},t)\varepsilon_{mw}(t) \qquad (10.4)$$

To obtain the observables, we again appeal to the FIF model, e.g. $v = g^*\varepsilon$, where we use the Green's function g discussed in Chapter 8. In this case, for $d = 1, 2$ spatial dimensions and time, nondimensionalizing $\underline{r}$ with L_e, and t with τ_w we have:

$$g(\underline{r},t) = \Theta(t)[\![\underline{R}]\!]^{-D/\alpha} = \Theta(t)(|\underline{r}|^2 + t^2)^{-(d+1)/(2\alpha)};$$
$$\underline{R} = (\underline{r},t) \qquad (10.5)$$

with $D = d + 1$ (cf. Eqns. (8.15), (9.29) with $d = 2$ for (x,y,t) space $\mu_x = \mu_y = 0$, $a = 1$). In d spatial dimensions, from Eqn. (10.4), we see that the macroweather generator is:

$$\varepsilon_{mw}(t) = e^{\Gamma_{mw}(t)};\quad \Gamma_{mw} = \left(\frac{C_1}{N_{d+1}}\right)^{1/\alpha} \gamma_\alpha * g_{mw};$$
$$g_{mw} = \Theta(t)t^{-(d+1)/\alpha} \tag{10.6}$$

(N_{d+1} is a normalization constant: see Section 5.4.2) and g_{mw} falls off more rapidly than for a $d = 0$ (pure temporal multifractal) cascade process which requires $g(t) \approx t^{-1/\alpha}$, leading to low intermittency.

The statistical behaviour of this regime is quite complex to analyze and has some surprising properties, which are investigated in detail in Appendix 10A (see also Lovejoy and Schertzer, 2011b). The main characteristics are that (a) although the bare process is still log-Lévy, the weak temporal correlations lead to (slow) central limit convergence (apparently) to Gaussians for the dressed statistics; (b) at large temporal lags Δt the autocorrelations ultimately decay as Δt^{-1}, although very large ranges of scale may be necessary to observe it; (c) since the spectrum is the Fourier transform of the autocorrelation and the transform of a pure Δt^{-1} function has a low (and high) frequency divergence, the actual spectrum of a finite-range macroweather regime depends on the overall range of scales Λ_c (by comparing Fig. 8.5c with Fig. 11.5b, we find empirically for the temperature field that to within a factor of ~2, the mean $\Lambda_c = \tau_c/\tau_w$ over the latitudes ≈ 1100); (d) over surprisingly wide ranges (factors of 100–1000 in frequency for values of Λ_c in the range 2^{10}–2^{16}), one finds "pseudo-scaling" with nearly constant spectral exponents β_{mw} which are typically in the range 0.2–0.4 for realistic values of Λ_c; (e) the statistics are independent of H and C_1 and only weakly dependent on α.

The upshot of this is that we expect a rough scaling with spectral exponent β_{mw} whose value largely depends on the overall range of the plateau regime (Λ_c), pretty much independently of the values of α, C_1 and H. Fig. 10A.6 shows some of the details: for example, using regressions over a range 128 in scale, we obtain β_{mw} = 0.40, 0.33, 0.29, 0.23 with outer scales $\tau_c \approx$ 30, 110, 450, 1800 years (with $\alpha = 1.8$, $\tau_w = 10$ days, i.e. corresponding to $\Lambda_c = 2^{10}, 2^{12}, 2^{14}, 2^{16}$). The (rough) empirical range of $\beta_{mw} \approx$ 0.2–0.4 is thus compatible with $\tau_c \approx > 30$ years. In summary, we therefore find for the overall FIF model:

$$\begin{aligned} &E(k) \approx k^{-\beta_w}; && k > L_w^{-1} \\ &E(\omega) \approx \omega^{-\beta_w}; && \omega > \tau_w^{-1} \\ &E(\omega) \approx \omega^{-\beta_{mw}}; && \tau_c^{-1} < \omega < \tau_w^{-1} \end{aligned} \tag{10.7}$$

where τ_c is the long external scale where the plateau ends and the climate regime begins, and the weather and macroweather spectral exponents are:

$$\begin{aligned} &\beta_w = 1 + 2H - K(2) \\ &0.2 < \beta_{mw} < 0.4 \end{aligned} \tag{10.8}$$

The weather exponent is the usual one (with the usual structure function exponent $\xi(q) = qH - K(q)$), but the macroweather exponent β_{mw} is new. Note in particular that it is independent of H and that $\beta_{mw} > 0$. As we just argued, in the macroweather regime, the intermittency rapidly disappears as we "dress" the process by averaging over scales $> \tau_w$ so that the plateau is roughly quasi-Gaussian and we have an effective macroweather exponent H_{mw}:

$$H_{mw} \approx -(1 - \beta_{mw})/2 \tag{10.9}$$

so that (since $\beta_{mw} < 1$), $H_{mw} < 0$ and the corresponding (generalized) structure function exponent at least approximately satisfies the monofractal linear relation:

$$\xi_{mw}(q) \approx qH_{mw} \tag{10.10}$$

Using $0.2 < \beta_{mw} < 0.4$ corresponding to $-0.4 < H_{mw} < -0.3$, this result already explains the preponderance of spectral plateau β's around the low values already noted in many analyses presented in Section 8.1. However, as we saw in Section 8.1.4 (Fig. 8.3d), the low-frequency ocean (*lo*) plateau has a somewhat higher $\beta_{lo} \approx 0.6$, which implies $H_{lo} \approx -0.2$; presumably one cause is the smaller range τ_c/τ_o compared to τ_c/τ_w ($\tau_o/\tau_w \approx 30$, Fig. 8.6d). In addition, in Appendix 10D we use a simple coupled ocean–atmosphere model to model this as a consequence of double (atmosphere and ocean) dimensional transitions.

As a final comment, we could note that since β_w is often close to the value 2 and β_{mw} close to zero, the combined weather/macroweather regime is not far from the Ornstein–Uhlenbeck (OU) spectrum $E(\omega) \approx \sigma^2/(\omega^2 + a^2)$ where σ is a constant and $a = \tau_w^{-1}$ is the transition frequency. This is the spectrum that results from the solution of linear stochastic systems forced by white noise, it is the basis of the stochastic linear forcing (SLF) approach (e.g. Penland, 1996) to stochastic forecasts in the macroweather regime: see Appendix 10B for more details. However, since $\beta_{mw} > 0$, the Ornstein–Uhlenbeck spectrum has too little low-frequency variability to be fully realistic: see for example Fig. 8.4d, which shows a comparison with empirical air and sea surface temperature spectra, and Fig. 10.14 for the corresponding real-space comparison. Similarly, at high frequencies the fields are intermittent (non-Gaussian) and in addition β_w is not exactly = 2.

10.1.3 Testing the FIF model . . . or how to determine decadal-scale macroweather variability from 1 Hz aircraft data

In order to test the realism of the FIF model in reproducing the macroweather regime, we made a detailed comparison of temperature data and numerical simulations of the FIF weather model. The data were taken at 75° N (taken from the 20CR reanalysis, from 1871–2008). We averaged from six-hourly to daily resolutions, which resulted in a series 50 404 days long for each 2° × 2° longitudinal pixel. The

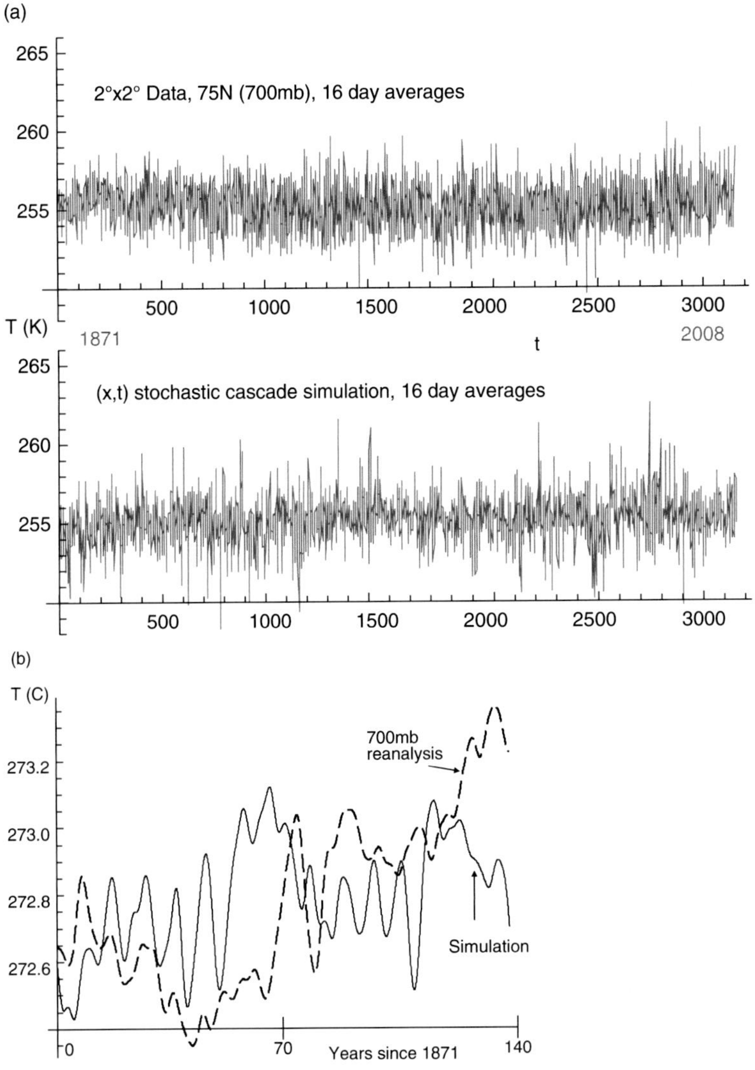

Fig. 10.2 (a) Top: 700 mb temperature data from the 20CR reanalysis (1871–2008) for 2° × 2° pixels at 75° N; the data were annually detrended and averaged over 16 days (1871 at left, 2008 at right); 16 pixels spaced at 10° in longitude were used so as to exactly match the simulation. Bottom: a time series of a single pixel of an (x,t) simulation 16 × 50 404 pixels and then averaged over 16 simulated days; the effective scales of the simulation were 1 day in time and $L_W/16 \approx 1200$ km in space; the parameters were $\alpha = 1.8$, $C_1 = 0.1$, $H_w = 0.5$ (close to the temperature as measured by aircraft, cf. Table 4.4). For graphical purposes, in addition to the standard deviation (= ± 4.05 K), the simulation mean was adjusted to be the same as the data. (b) Comparison of the globally averaged reanalysis temperature with the same simulation realization as in Fig. 10.2a, also averaged over the simulation's (single) spatial dimension (the simulated global temperature). Both were passed through 3-year moving averages to bring out the low-frequency variability. Note the great reduction in the standard deviation (± 0.39 K, cf. ± 4.05 K for the 16-day resolution standard deviation in Fig. 10.2a) of the data. To facilitate the comparison the simulation standard deviation was increased by a factor 1.4 with respect to the calibration of Fig. 10.2a.

simulation was of a simple (x,t) cascade (see Figs. 10.2a, 10.2b) with the parameters $\alpha = 1.8$, $C_1 = 0.1$, $H_w = 0.5$, $\varepsilon_w \approx 10^{-3}\ \mathrm{m}^2/\mathrm{s}^3$ ($\tau_w \approx 10$ days) observed by the NOAA Gulfstream 4 aircraft near 200 mb altitude; see Table 4.8 and see Section 8.1.3 for estimates of ε_w. The simulation was scaled so that its standard deviation coincided with that of the data. The latitude 75° N was chosen both because it is largely dominated by land or ice-covered ocean, so that intermittency effects due to the ocean circulation are not so strong, and because later we discuss and compare it with Greenland paleotemperatures which are nearly at the same latitude (see Section 11.1.2).

Except for the standard deviation of the temperature at 75° N (and for Fig. 10.2a, the mean) there was no attempt whatsoever to "fit" the simulation to the reanalysis data, so we do not expect a perfect data/simulation match. The object was to see how close the above "toy model" of a dimensional transition can account for the atmospheric variability over large ranges of time scales. In order to remove extraneous issues of sample size and series length, a $2^{17} \times 2^4$ simulation was made and the first 50 404 spatial segments were taken (this is a bit less than half the simulation length and so avoids artificial correlations/tendencies due to the periodicity of the simulations). The effective spatial resolution of the simulation was thus $L_w/16 \approx 1200$ km. For comparison, 16 series from pixels spaced at 10° in longitude were also used so that both the data and the simulation were 50 404 × 16 pixels. In Fig. 10.2a we see that the real and simulated series are indeed quite similar in appearance, although the simulation has apparently a few more extremes. It cannot be excluded that these extremes are in fact realistic but are poorly reproduced by the (overly smooth) reanalysis. In Fig. 10.2b we show the same globally averaged reanalysis and "globally averaged" simulation obtained by averaging the simulation over all of its (single) spatial dimension. The main difference between the two is a slight tendency for the reanalysis temperatures to increase during the second half of the series (global warming).

The comparison of the simulation and empirical spectra is shown in Fig. 10.3. For clarity, we have averaged the spectrum over logarithmically spaced bins, 10 per order of magnitude. Over the high-frequency weather regime, the model and data agree quite well. This is not surprising, since the aircraft data that were used to determine the parameters were at these smaller scales (recall that the overall standard deviations are the same; this determines the relative vertical placement of the curves in Fig. 10.3). However, what is not at all trivial is that the low-frequency part of the spectrum – including the mean spectral exponent $\beta_{mw} \approx 0.2$ – is also quite well reproduced; presumably the agreement would be better if the critical external scale was given a small adjustment. Certainly, ensemble averaging over many realizations of the FIF model give a fairly accurate slope $\beta_{mw} = 0.2$ as indicated by the reference line for the theory as well as further data/model comparisons: see Appendix 10A. The high-frequency aircraft parameters thus give a remarkably good model of the temperature

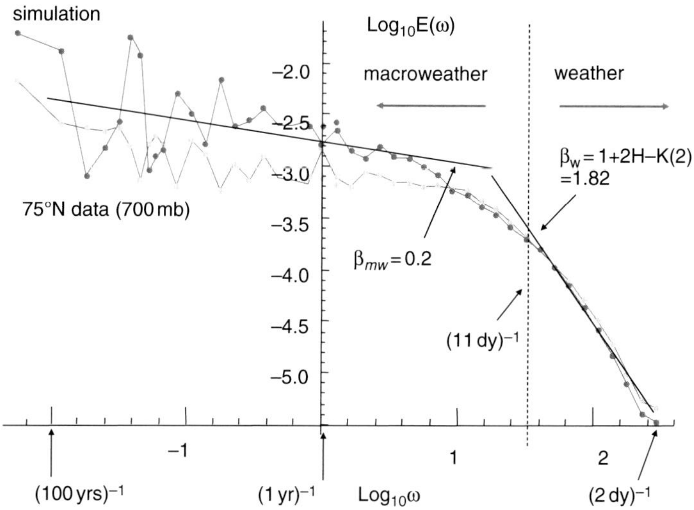

Fig. 10.3 A comparison of the spectrum of the same data as in Fig. 10.2a and 10.2b (bottom), with the simulation (top); both spectra were averaged over logarithmic bins, 10 per order of magnitude. The reference lines have the theoretical slopes β_w, β_{mw}, as indicated in Eqn. (10.7).

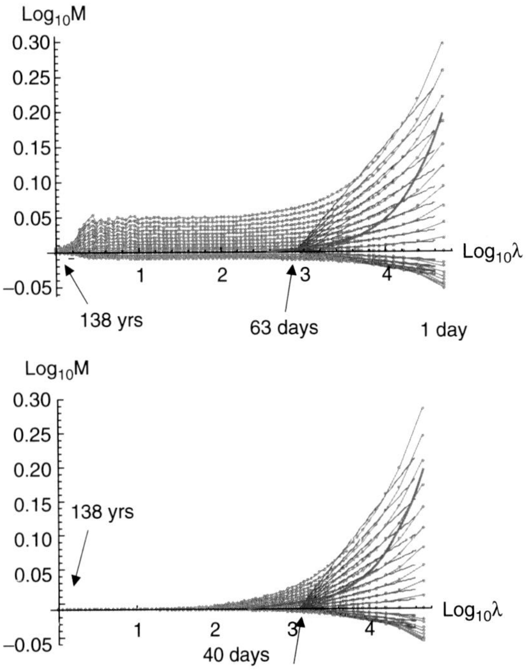

Fig. 10.4 The cascade analysis of the data (top) and simulation (bottom) shown in Figs. 10.1 and 10.2. Notice that the data are significantly more intermittent in the plateau region than the simulation (which has virtually no intermittency (near zero flux moments) for long time periods (to the left, lower figure). The low intermittency of the simulation can be gauged by the comparison with the superposed $q = 2$ envelopes for the universal quasi-Gaussian processes (thick curving lines: see Appendix 4A). The simulation is only a bit more variable, whereas the data are much more variable.

spectra at 75° N out to decadal scales (until the data spectrum starts to rise at around (50 years)$^{-1}$).

In order to check the intermittency characteristics, we refer the reader to Fig. 10.4, where we compare the simulation and empirical trace moments (up to order $q = 2$). We note that at the high frequencies the intermittencies agree quite well (with a slight difference in the effective outer scales that could be improved by adjusting the model outer scale). The main difference is that whereas the model has very low intermittencies at scales larger than τ_w, the data have nearly constant trace moments. While these moments are small (≈ 0.05), there is nevertheless a systematic difference.

The simplest FIF model discussed here thus has fluxes with not enough low-frequency variability. Recall from Fig. 8.6e that one-month averages (with the meteorological variability largely averaged out) display cascades associated with ocean turbulence with $C_1 \approx 0.12$, $\alpha \approx 1.8$ (outer scale ≈ 2 years) so that this higher intermittency in Fig. 10.4 (top) could be at least partially explained by the ocean intermittency; in Section 10.3.2 we discuss a simple "mixed" ocean–atmosphere model that attempts to reproduce this behaviour. If this is true, then Eqns. (10.1)–(10.3) show that even without scaling we would still expect the bare fluxes to have log-Lévy distributions with the same index α; this is presumably true even without including the influence of the ocean intermittency, since Table 8.1 indicates that the ocean α, C_1 are quite close to the weather values. However, as discussed in Appendix 10A, the dressed fluxes do apparently eventually approach quasi-Gaussian limits. This appendix also describes a final simulation/data comparison using real-space fluctuations.

10.1.4 More evidence for the spectral plateau

A fairly general prediction of the atmosphere and ocean dimensional transition models for the plateau is that the macroweather exponents are independent of H and depend only weakly on C_1, α. Since the main state variables all share roughly the parameters $C_1 \approx 0.1$, $\alpha \approx 1.8$, we therefore expect two fundamental plateau spectral exponents: $\beta_{mw} \approx 0.2$–0.4 for those primarily affected by the atmospheric dimensional transition ("continental climates"), and $\beta_{mw} \approx 0.6$ for those affected by both oceanic and atmospheric transitions ("maritime climates"). Therefore a further test of the model is to consider the other fields and compare them at locations where we expect them to be mostly continental or mostly maritime. Figs. 10.5a and 10.5b show the spectra for T, u, v, h_s from two 2°-wide latitude bands, one tropical (centred on 5° N, mostly maritime), the other mid-latitude (centred on 45° N, mostly continental), both 20CR products. We see that for the temperature, wind and humidity, the 45° N spectrum (Fig. 10.5a) follows the continental $\beta_{mw} \approx 0.2$ prediction very well up to about 10 years whereas the 5° N spectrum (Fig. 10.5b) is closer to the maritime value $\beta_{mw} \approx 0.6$ (with the notable exception of the meridional wind, which displays a $\beta_{mw} \approx 0.2$ region as indicated). Overall, we conclude that all the data can be reasonably modelled by one or the other ($\beta_{mw} \approx 0.2$ or $\beta_{mw} \approx 0.6$) regime being dominant. Since both have $H < 0$, the resolution dependence is significant, although not as strong as for the other latitudes

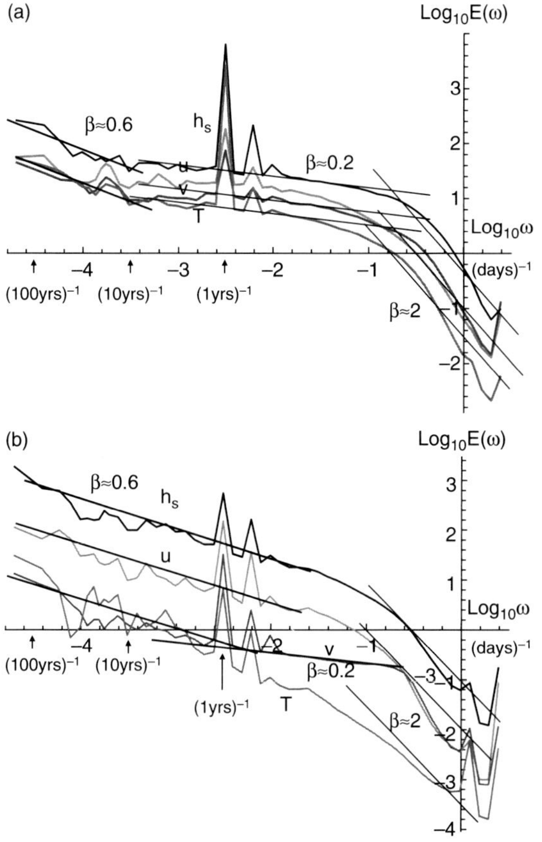

Fig. 10.5 (a) Spectra from 20CR (1871–2008) at 45° N for temperature (*T*), zonal and meridional wind (*u*, *v*) and specific humidity (*hs*). The reference lines correspond to $\beta_{oc} = 0.6$, $\beta_{mw} = 0.2$, $\beta_w = 2$, left to right respectively. (b) Same as Fig. 10.5a but at 5° N. The reference lines correspond to $\beta_{oc} = 0.6$, $\beta_{mw} = 0.2$, $\beta_w = 2$, left to right respectively.

(where $H_{mw} \approx -0.4$). In Appendix 10D we show how to make simple stochastic models of this varying maritime/continental behaviour.

Our basic empirical conclusions are more or less in accord with a growing literature – particularly with respect to the temperature statistics. They are especially close to those of Huybers and Curry (2006), who studied many paleoclimate series as well as the 60-year-long NCEP reanalyses and concluded that for periods of months up to about 50 years, the spectra are scaling with mid-latitude β_{mw} larger than tropical β_{mw} (their values are 0.37 ± 0.05, 0.56 ± 0.08, quite similar to ours). Using observations and climate models at monthly resolutions, Fraedrich and Blender (2003) used the DFA method (see Section 5.5.2; the DFA is very similar to the Haar structure function, Section 10.2.2). They found $\beta_{mw} \approx 0$ over continents but $\beta_{mw} \approx 0.3$ in continental/ocean "transitional regions"; a similar finding was made by Huybers and Curry (2006), who also noted a small mean land–sea difference (in their case of $\Delta\beta_{mw} = 0.2$); see Table 10.1.

Koscielny-Bunde *et al.* (1998), Bunde *et al.* (2004), Eichner *et al.* (2003) and Lennartz and Bunde (2009) gave evidence that over continents β_{mw} was in the range 0.2–0.3 (similar to our estimates) and that the exponents implied the existence of long-range correlations in accord with our continental results. These papers only used lags τ up to one-quarter the length of the series, so the largest τ analyzed was $\approx$ 40 years, which is close to τ_c, a fact that presumably explains why they did not notice the beginning of the climate regime. Broadly similar results were obtained by Blender *et al.* (2006), who argued that at least in some regions such as the North Atlantic (but not Pacific), there is indeed a long-term memory (i.e. for Gaussian processes $\beta \neq 0$) which at these scales they attribute to the thermal inertia of the Atlantic zonally averaged ocean circulation (for evidence of long-range dependencies see also Fraedrich *et al.*, 2009; Franzke 2010, 2012). We could also mention a paper by Lanfredi *et al.* (2009), who considered the small but systematic deviations from perfect DFA scaling over the range 1 month to ~1 year (caused at least partially by the ocean cascade; see also Appendix 10A). Other pertinent analyses are of GCM outputs and historical reconstructions of the northern hemisphere temperatures and tropical ocean averaged temperatures relevant to El Niño (AchutaRao and Sperber, 2006). However, to understand these, we must consider first the scale τ_c at which the plateau ends as well as the geographical distribution of the exponents. We postpone discussion of these until Section 10.2.

To complete this examination of the spectral plateau, let us consider other relevant climate-scale variables. In the model outlined in Chapter 8, it was the balance of incoming and outgoing radiation (modulated by the clouds, abundances of H_2O, aerosols, CO_2, O_3, CH_4, N_2O (Tuck, 2008) and the surface variability, all of which are scaling) that determined the overall energy flux variations that drove the weather. Therefore, we expect ε_w (and by implication ε_o) to be particularly significant for the climate. In Fig. 10.6, we show an estimate from the 20CR reanalysis at 45° N based of the Laplacian of the

Table 10.1 A comparison of various estimates of spectral exponents β and scaling range limits for the air and sea surface temperatures (and their surrogates) and streamflow and precipitation and (where indicated) variables of state. For more data on τ_c, see Box 11.1; for GCM estimates see Royer *et al.* (2008).

	Series length (years)	β_w	τ_w (days)	β_{mw}	τ_c (years)	β_c
Northern hemisphere instrumental (Lovejoy and Schertzer, 1986), 1 month	100	–	–	$\approx 0.2^a$	3	1.8
20th C (6 hours) (700 mb), global, Fig. 1.9c	138	3	10	$\approx 0.6^a$	5	$\approx 1.7^a$
NOAA NCDC, NASA GISS, HadCRUT3 (surface, monthly, Appendix 10.C)	129	–	–	0.2 (over land) 0.6 (SST)	10	1.7
Satellite global, (600 mb), daily, Fig. 10.7	7	3	25	$\approx 0.2^a$	2	$\approx 2^a$
20thC *u*, *v*, *T*, *h*, ε (6 hours) (700 mb), 44 N, Figs. 10.5a, 10.6	138	2^a	5	$\approx 0.2^a$	10	0.6^a
20thC *u*, *v*, *T*, *h* (6 hours) (700 mb), 5 N, Fig. 10.5b	138	2^a	25	$\approx 0.6^a$	–	0.6^a
ECMWF interim (700 mb), year 2006, Fig. 8.5	1	2^a	5	≈ 0.2, 0.6	–	–
Instrumental, daily, USA	60	2^a	7	$\approx 0.2^a$	10	1^a
Precipitation, USA (CPC), Fig. 8.4d	29	0.8^b	5	0.05		
Precipitation, France (Tessier *et al.*, 1996)	4096 days	0.4 ± 0.1	16	0.1 ± 0.1		
River flow (Tessier *et al.*, 1996)	10–30 years	1.3 ± 0.1		0.5 ± 0.1		
River flow (Pandey *et al.*, 1998)	10–80 years	2.4	8	0.72	20	1.7
(Lanfredi *et al.*, 2009); deviations < 1 year considered as scale-bound Markov process	≈ 100			0.26		
Mid-latitude, tropics (Huybers and Curry, 2006)	Composite to 10^6	–	–	0.4 (mid) 0.6 (tropics)	≈ 100	1.6 (mid) 1.3 (tropics)
Fraedrich and Blender, 2003	NCEP reanalyses, 60	–	–	≈ 0 continents, ≈ 0.3 coasts	–	–
Eichner *et al.*, 2003	In-situ daily temperatures analysed up to 40 years			≈ 0.3		
Monetti *et al.*, 2003	SST, 100	1.76 ± 0.08	10 months[c]	0.6 ± 0.16		
AchutaRao and Sperber, 2006	Mean SST over El Niño significant regions, GCM and reanalysis, 30 years	≈ 2	2–7 years	≈ 0	–	–
Rybski *et al.*, 2006	1000 years northern hemisphere temp. (to τ ≈ 200 years)			≈ 0.8 – 1		

Table 10.1 *(cont.)*

	Series length (years)	β_w	τ_w (days)	β_{mw}	τ_c (years)	β_c
Bunde *et al.*, 2004; Koscielny-Bunde *et al.*, 1998; Lennartz and Bunde, 2009	In-situ, 2 weeks to 30 years	_	14	0.2 – 0.3		
Ditlevsen *et al.*, 1996	Ice cores (GRIP)[d]	_	_	≈ 0.2	3	0.8
Lovejoy and Schertzer, 1986	Composite: ice cores, instrumental	1.8	25	≈ 0	3–300[e]	1.8
Pelletier, 1998	Composite: ice core, instrumental	1.5	7	0.5	200	1.5

[a] These values are not from regression lines but from plausible reference lines indicated in the figures; in most cases, the value $\beta_{mw} \approx 0.3$ would work nearly as well.
[b] The exponent depends somewhat on the rain/no rain detection threshold.
[c] This "cross-over" time corresponds to our τ_o, see Chapter 8.
[d] Using monthly resolution GRIP ice-core data for the last 3 kyr.
[e] The 3-year figure used northern hemisphere instrumental series, the 300-year figure was indirectly from Central England temperature series.

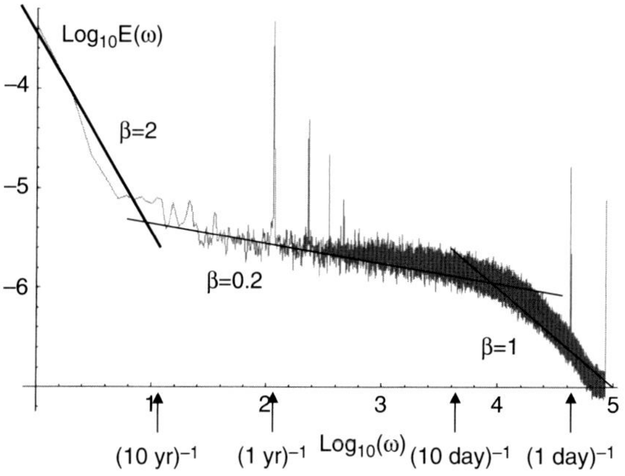

Fig. 10.6 Spectrum from 20CR (1891–2002); the energy flux estimated from the absolute Laplacian of the zonal wind at 700 mb, 42° N. The reference lines have $\beta_c = 2$, $\beta_{mw} = 0.2$, $\beta_w = 1$. Note that the low frequency rise is influenced by the changes in the data quality from the first and second half of the series.

(dominant) zonal wind component. We see that it is similar to the other 20CR fields although the plateau does indeed look very uniform and flat.

Although reanalyses have the great virtue of having excellent statistics, they are potentially subject to unknown biases due to their intrinsic limitations as hybrid data/model "products." Similarly, in-situ networks have other problems (how to fill the "holes" in the fractal networks; Chapter 3). It is therefore of interest to compare our results with those from remotely sensed data. Fig. 10.7 shows an analysis of daily microwave radiance data, a proxy for the 600 mb temperature field, averaged globally from 1979 to 1986. Again, the spectral plateau is clearly visible and is in roughly the same frequency range as for the real temperatures (Table 10.1); interestingly the spectral exponent has the continental value $\beta_{mw} \approx 0.2$ rather than the maritime value ≈ 0.6 (which is the value found for the global 20CR and globally averaged in-situ data: Figs. 10C.2, 10C.3). This discrepancy is potentially important for evaluating global warming, although we must be cautious since the data are only over a seven-year period. Another interesting feature is that the weather regime β_w is close to 3, i.e. close to the global value (Fig. 1.9 c) but significantly higher than the single pixel (local) value for the ECMWF interim and in-situ estimates discussed in Chapter 8, which have $\beta_w \approx 2$.

Finally, we may consider classical climate indices, first the North Atlantic Oscillation (NAO) index. The NAO index is essentially the mean daily pressure difference between the Arctic low and the mid-latitude high, traditionally taken as the pressure difference between Iceland the Azores. The NAO characterizes the strength and direction of westerly winds and storm tracks across the North Atlantic; its statistical properties have been intensively studied, especially with a view to finding characteristic cycles/periodicities (e.g. Lind *et al.*, 2007; Berger, 2008), although Stephenson *et al.* (2000) also considered some scaling alternatives. In Fig. 10.8 we show that the index is also very similar to the other continental fields analysed; again it displays a $\beta_{mw} \approx 0.2$ plateau, showing however some evidence

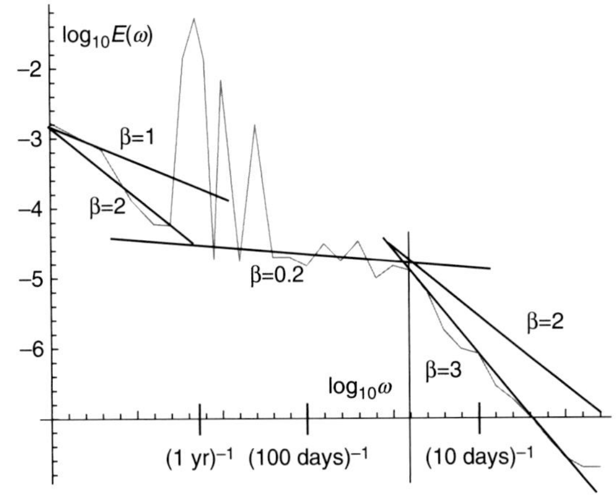

Fig. 10.7 The spectrum of 7 years of $\approx$ 600 mb daily satellite temperature estimates from the Microwave Sounding Unit (MSU) channel 2 global daily average data at 100 km spatial resolution at nadir (from 1979–1985; the other years had too much missing data and were not used). The data were taken from http://discover.itsc.uah.edu/amsutemps/ (page authors R. Spencer and D. Braswell, NSSTC), and the spectrum was averaged in logarithmically spaced frequency bins to reduce the noise. The slopes are for the indicated reference lines and include once again a near $\beta_{mw} \approx 0.2$ spectral plateau. In the very-low-frequency regime and in the plateau the exponents β are close to those of the instrumental temperatures but in the high-frequency regime, they are a bit lower ($\beta_w \approx 3$) than the in-situ local data ($\beta_w \approx 2$). The spikes are at (1 year)$^{-1}$ and subharmonics. While various satellite calibration issues may make it difficult to use such data for inferring global warming or cooling tendencies, such issues are not likely to affect the spectrum, except perhaps at the lowest frequencies.

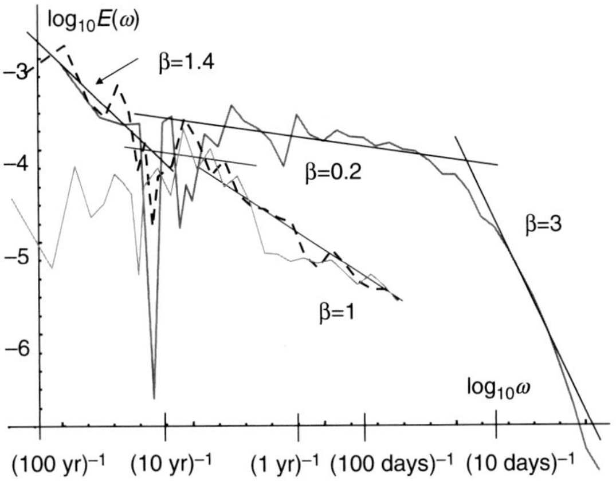

Fig. 10.8 The solid line shows the spectrum of the daily North Atlantic Oscillation (NAO) index (the pressure difference between the Azores high and the Icelandic low), 60 years (1950–2009), from the public NCEP, NOAA database at: ftp://ftp.cpc.ncep.noaa.gov/cwlinks. The dashed line is the spectrum of the Pacific Decadal Oscillation (PDO) at monthly resolution. The thin line is the Southern Oscillation Index (SOI). The reference lines correspond to $\beta = 1.4, 0.2, 3$ (left to right).

for a new low-frequency regime below $\approx$ (10 years)$^{-1}$. The low-frequency low E excursion (for a single frequency value) in Fig. 10.8 is presumably an artefact of the poor statistics near (15 years)$^{-1}$.

In Fig. 10.8, we also show an ocean SST-based index, the Pacific Decadal Oscillation (PDO: we used the monthly series from 1900 to 2010, http://jisao.washington.edu/data_sets/). The PDO is the amplitude of the largest principal component (PC, equivalently, empirical orthogonal function) of the Pacific SST distribution; it characterizes ocean dynamics (Zhang *et al.*, 1997). From about (10 years)$^{-1}$ to lower frequencies it follows the (atmospheric) NAO, but the higher frequencies are closer to the 5° resolution SST spectrum in Fig. 8.6d. The real-space (Haar) structure function of the PDO is compared to those of SST and global temperatures in Fig. 10.14. We also show the spectrum of the Southern Oscillation Index (SOI), which is the difference in pressure between Darwin and Tahiti and which is used as a surrogate for the El Niño–Southern Oscillation (ENSO) phenomenon. We see that at high frequencies the spectrum is very similar to the PDO ($\beta \approx 1$, $H \approx 0$), whereas the two diverge for frequencies $\approx <$ (2 years)$^{-1}$, with the latter having $\beta \approx -0.3$, $H \approx -0.65$ (paleo-ENSO surrogates show that this continues only to about (100 years)$^{-1}$, after which the sign of H becomes positive – the climate regime (J. Emile-Geay, personal communication). The ENSO oscillation phenomenology is thus apparently associated with the transition between “ocean weather” and the ocean macroweather regime at scales $\approx \tau_o$.

10.1.5 Lévy collapse and universality

According to the previous section, if the FIF model is extended beyond $\tau_w \approx 10$ days, we no longer have a scaling weather regime, but rather a new macroweather regime. We have already mentioned that from the analysis of the plateau in Appendix 10A we expect the dressed fluxes to tend to low intermittency variables at scales $\tau >> \tau_w$. However, we also noted in Section 8.1.4 that we see evidence for the influence of the ocean cascade with similar parameters (at least for the temperature) even over land but up to scale $\tau_o \approx 1$ year. This would lead to the continuation of the intermittent multifractal variabilty to scales up to τ_o. In Section 10.1.3 we examined this prediction

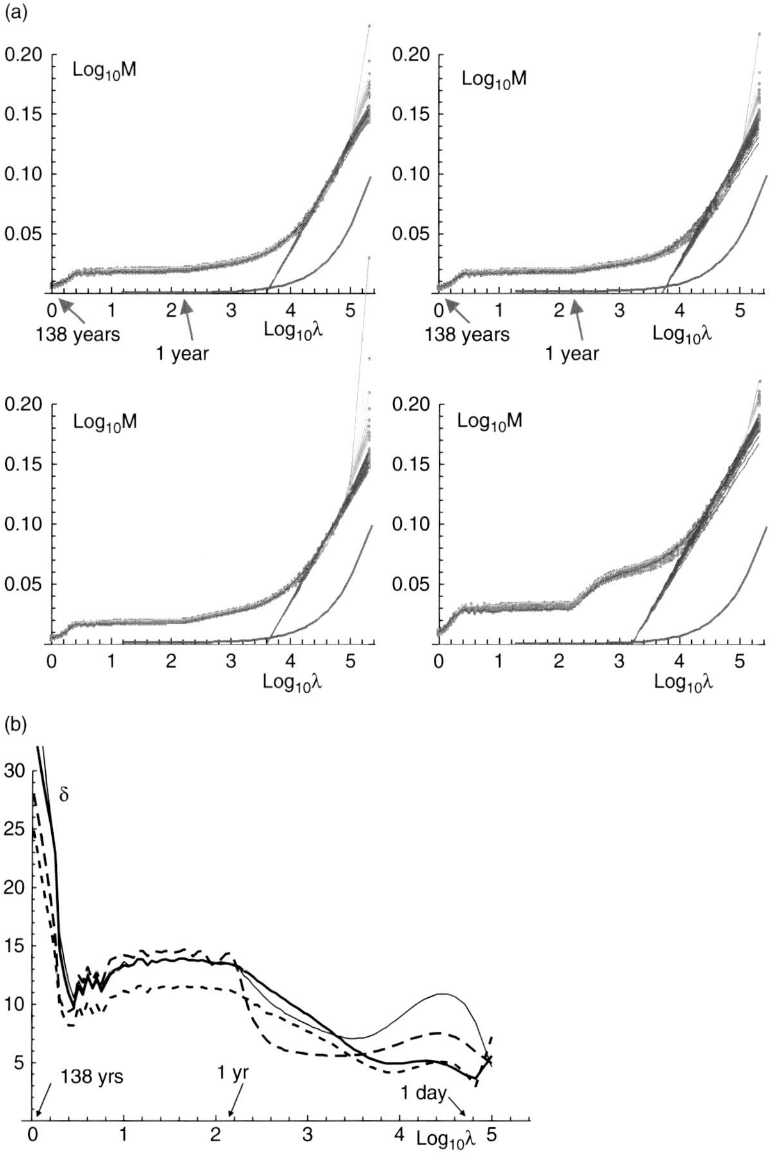

Fig. 10.9 (a) Twentieth-century Lévy collapses (time) of Laplacians of 700 mb, 6-hourly data at 45° N of u, v, T, h_s (upper left, upper right, lower left, lower right, and using $\alpha = 1.9$, 1.9, 1.9, 1.8, respectively). These are the collapses of the cascades shown in Fig. 8.7 b. The C_1 values corresponding to the linear (cascade regime) are: 0.083, 0.082, 0.090, 0.083, respectively. The outer scale is 138 years so that 1 year corresponds to $\log_{10}\lambda \approx 2.15$. The lower curves are the envelopes of the corresponding "collapse" curves for quasi-Gaussian processes (collapsed with $\alpha = 1.8$; see Appendix 4A). (b) The evolution of the percentage spread of the collapse δ (as defined in Eqn. 4.16) for the Twentieth-Century Reanalyses shown in Fig. 10.9a. In top-to-bottom order above the number "1" in "1 day", we have meridional wind (v), specific humidity (h_s), zonal wind (u) and temperature (T).

for the 75° N temperature series. Here, we examine other fields, other latitudes and other products.

Consider the Lévy collapse of the 20CR reanalyses from 1871 to 2008 (Fig. 10.9a): they are quite reasonable. This is quantified by the percentage deviations shown in Fig. 10.9b. We see that up to about 50 years (and ignoring the very small time scale finite size deviations) the spread of the curves for all the logarithms of the moments up to second order are less than about ± 15%. Again, the fluxes are nearly completely independent of scale for periods longer than a year, although for periods longer than about 50 years (corresponding to the poor collapses) the variability starts to decrease. The figure also shows the corresponding curve for quasi-Gaussian processes (such as the Ornstein–Uhlenbeck processes); again they have much lower variability.

For the corresponding collapses for the ECMWF interim reanalyses (corresponding to the spatial collapses in Fig. 4.3), for h, T, u, v, w, z for the 700 mb level in 2006 and for those of daily meteorological station data (34-year-long series) for T, p, h (see the spectra displayed in Fig. 8.2 a, b, c), see Lovejoy and Schertzer (2011a, 2010, respectively).

10.2 Macroweather: its temporal variability, and outer-limit τ_c

10.2.1 Tendency and difference fluctuations and structure functions for $H > 0$ and $H < 0$

In Table 10.1 we gave some limited evidence that τ_c was in the range (10 years)$^{-1}$ to (100 years)$^{-1}$ but its value was poorly discerned because it is near the extreme low-frequency limit of instrumental data. We now attempt to determine τ_c with more precision. Up until now, we have primarily used spectral analysis, since it is a classical, straightforward technique whose limitations are well known, and it was adequate for the purpose of determining the basic scaling regimes in time and in space. We now focus on the low frequencies corresponding to several years to $\approx$ 100 kyr, so it is convenient to study fluctuations in real rather than Fourier space. There are several reasons for this. The first is that we are focusing on the lowest instrumental frequencies, and so spectral analysis provides only a few useful data points – for example on data 150 years long, the time scales longer than 50 years are characterized only by three discrete frequencies $\omega = 1, 2, 3$: Fourier methods are "coarse" at low frequencies. The second is that in order to extend the analysis to lower frequencies it is imperative to use proxies, and these need calibration: the mean absolute amplitudes of fluctuations at a given scale are convenient for statistical calibrations. A third is that the absolute amplitudes are also important for gauging the physical interpretation and hence significance of the fluctuations.

The simplest fluctuation is also the oldest: the difference $(\Delta T(\Delta t))_{diff} = \Delta T(t + \Delta t) - \Delta T(t)$. The corresponding statistical moments $<\Delta T^q>$ are the classical structure functions ("generalized" from second to qth order). According to Eqn. (8.2), the fluctuations follow:

$$\Delta T \stackrel{d}{=} \varphi_{\Delta t} \Delta t^H \qquad (10.11)$$

where $\varphi_{\Delta t}$ is a resolution Δt turbulent flux (in the notation of Chapter 5, $\Delta t = \tau/\lambda$ where τ is the outer scale of the cascade, and λ is a scale ratio and $\stackrel{d}{=}$ means equality in a statistical sense (see Eqn. (5.94)). From this we see that:

$$\langle \Delta T(\Delta t)^q \rangle = \langle \varphi_{\Delta t}^q \rangle \Delta t^{qH} \approx \Delta t^{\xi(q)}; \quad \xi(q) = qH - K(q) \qquad (10.12)$$

where $\xi(q)$ is the structure function with exponent and $K(q)$ is the (multifractal, cascade) intermittency exponent introduced in Chapter 3. Since the turbulent flux has the property that it is independent of scale Δt ($<\phi_{\Delta t}>$ is constant), we have $K(1) = 0$ and $\xi(1) = H$. The physical significance of H is thus that it determines the rate at which mean fluctuations grow ($H > 0$) or decrease ($H < 0$) with scale Δt.

The problem is that the mean difference cannot decrease with increasing Δt, and hence when studying scaling processes with $H < 0$ it is clearly inappropriate: the differences simply converge to a spurious constant depending on the highest frequencies available in the sample. Similarly, when $H > 1$, fluctuations defined as differences grow linearly with Δt, with slope depending on the lowest frequencies present in the sample. In both cases, the exponent $\xi(q)$ is no longer correctly estimated. The problem is that we need a definition of fluctuations such that $<\Delta T(\Delta t)>$ is dominated by frequencies $\approx \Delta t^{-1}$.

As discussed in Section 5.5.1, the need to more flexibly define fluctuations motivated the development of wavelets, and the classical difference fluctuation is only a special case, the "poor man's wavelet." In the weather regime, most geophysical H parameters are indeed in the range 0–1 (Table 4.6), so that fluctuations tend to *increase* with scale and this classical difference structure function is adequate. However, we saw that a prime characteristic of the macroweather regime is precisely that $H < 0$, so that fluctuations *decrease* rather than increase with scale. Hence, for studying this regime, difference fluctuations are inadequate. To change the range of H over which fluctuations are usefully defined one must change the shape of the defining wavelet. In the usual wavelet framework, this is done by modifying the wavelet directly, e.g. by choosing various derivatives of the Gaussian (e.g. the second derivative "Mexican hat") or by choosing them to satisfy some special criterion such as orthogonality. Following this, the fluctuations are calculated as convolutions with fast Fourier (or equivalent) numerical techniques.

A problem with this usual wavelet implementation is that not only are the convolutions numerically cumbersome, but the physical interpretation of the fluctuations is largely lost. In contrast, when $0 < H < 1$, the difference structure function gives direct information on the typical difference ($q = 1$) and typical variations around this difference ($q = 2$) and even typical skewness ($q = 3$) or typical kurtosis ($q = 4$) or, if the probability tail is algebraic, of the divergence of high-order moments of differences. Similarly, when $-1 < H < 0$ one can define the "tendency structure function" (below), which directly quantifies the fluctuation's deviation from zero and whose exponent characterizes the rate at which the deviations decrease when we average to larger and larger scales. These poor man's and tendency fluctuations are also very easy to directly estimate from series with uniformly spaced data and – with straightforward modifications – to irregularly spaced data. While the corresponding wavelets may not be orthogonal, when wavelets are used for statistical characterizations this is generally unimportant.

The study of real-space fluctuation statistics over scale ranges including both the weather and the macroweather regimes therefore requires a definition of fluctuations valid at least over the range $-1 < H < 1$. Before discussing our choice – the Haar wavelet – let us recall the definitions of the difference and tendency fluctuations; the corresponding structure functions are simply the qth moments. The difference/ poor man's fluctuation is thus:

$$\left(\Delta T(\Delta t)\right)_{diff} \equiv |\delta_{\Delta t} T|; \quad \delta_{\Delta t} T = T(t + \Delta t) - T(t) \tag{10.13}$$

where δ is the difference operator. Similarly, the "tendency fluctuation" (Section 5.5.1, Appendix 5E) can be defined using the series with overall mean $(\overline{T(t)})$ removed: $T'(t) = T(t) - \overline{T(t)}$ by:

$$\left(\Delta T(\Delta t)\right)_{tend} = \left| \frac{1}{\Delta t} \sum_{t \le t' \le t+\Delta t} T'(t') \right| \tag{10.14}$$

or, with the help of the summation operator s, equivalently by:

$$\left(\Delta T(\Delta t)\right)_{tend} = \left| \frac{1}{\Delta t} \delta_{\Delta t} ST' \right|; \quad ST' = \sum_{t' \le t} T'(t') \tag{10.15}$$

We can also use the suggestive notation $\left(\Delta T(\Delta t)\right)_{tend} = \overline{\Delta T}$ (see the schematic in Fig. 10.10). When $-1 < H < 0$, $\left(\Delta T(\Delta t)\right)_{tend}$ has a straightforward interpretation in terms of the mean tendency of the data to decrease with averaging, but it is useful only for $-1 < H < 0$. It is also easy to implement: simply remove the *overall* mean and then take the mean over intervals Δt: this is equivalent to taking the mean of the differences of the running sum. Fig. 10.10 shows schematically the difference and tendency structure functions.

Figs. 10.11a, 10.11b show the daily and annual mean structure functions (Fig. 10.11a) and tendency structure functions (Fig. 10.11b) for the same latitudes and for the global average temperatures (after detrending). It can be seen that the same basic regimes can be identified, although the transition points are (unsurprisingly) somewhat different. In particular, considering the annually averaged regular structure functions (Fig. 10.11a), we can begin to see evidence at all latitudes – except perhaps for 75° N – of the beginning of a rise for periods $\Delta t > \approx 5$–10 years. In particular, the evidence is clearer that the global temperature begins a new power law, roughly of the form as indicated (which corresponds to $\beta_c \approx 1.8$), which is very close to the $S(\Delta t) = \langle \Delta T^2 \rangle^{1/2} \approx \Delta t^{\xi(2)/2} \approx 0.092\Delta t^{0.4}$ and $\approx 0.077\Delta t^{0.4}$ laws found in hemispheric and global instrumental temperatures in Lovejoy and Schertzer (1986) corresponding to $\beta_c = 1 + \xi(2) = 1.8$ (this is discussed at length in Chapter 11, and see Table 10C.2). Again – significantly for the interpretation of the Arctic and Antarctic paleotemperatures – the 75° S structure function seems to follow the global climate regime relation whereas the 75° N structure function seems to remain roughly constant, as expected for the spectral plateau. Turning our attention to the tendency structure functions (Fig. 10.11b), we see that even for the 75° N location the predicted $\beta_{mw} \approx 0.2$ regime seems not to hold much beyond a few years (left graph, daily resolution), changing to more of a $\beta_{mw} \approx 0.6$ regime at larger Δt's. Once again, the 75° S curve is somewhat anomalous, being fairly flat (as expected if $H > 0$), and follows the global curve for roughly a year and longer periods. In Section 11.1.2 we will see that this difference between the Arctic and Antarctic locations is reflected in their ice-core paleotemperatures and persists until scales of about 2000 years. From the daily tendencies, we see that the low β_{mw} (≈ 0.2) of the higher latitudes compared to the lower latitudes ($\beta_{mw} \approx 0.6$) implies higher absolute H_{mw}'s

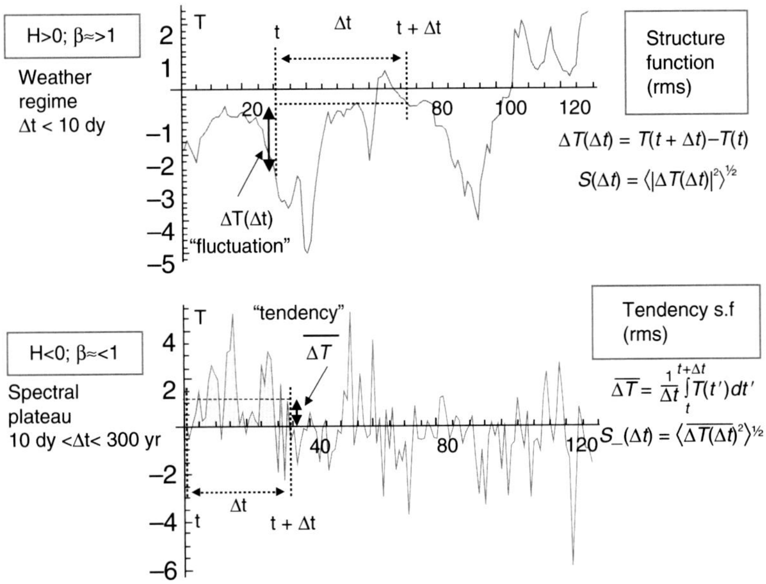

Fig. 10.10 The top shows extracts of the temperature simulation (Fig. 10.2) at full resolution (the weather regime, $H_w = 0.5$), the bottom at 16-day resolution (the macroweather regime, $H_{mw} = -0.4$). One can clearly see their differing characters corresponding to $H > 0$, $H < 0$, respectively. These simulations are used to illustrate the two different types of structure function needed when $H > 0$ (the usual) and $H < 0$, the "tendency" structure function. Whereas the usual structure function yields typical differences in T over an interval Δt, so that the mean is of no consequence, the tendency structure function uses the average of the field with the mean removed. In both cases the $q = 2$ moment was chosen because it is directly related to the spectrum. Taking the square root is useful, since the result is then a direct measure of the "typical" fluctuations in units of K.

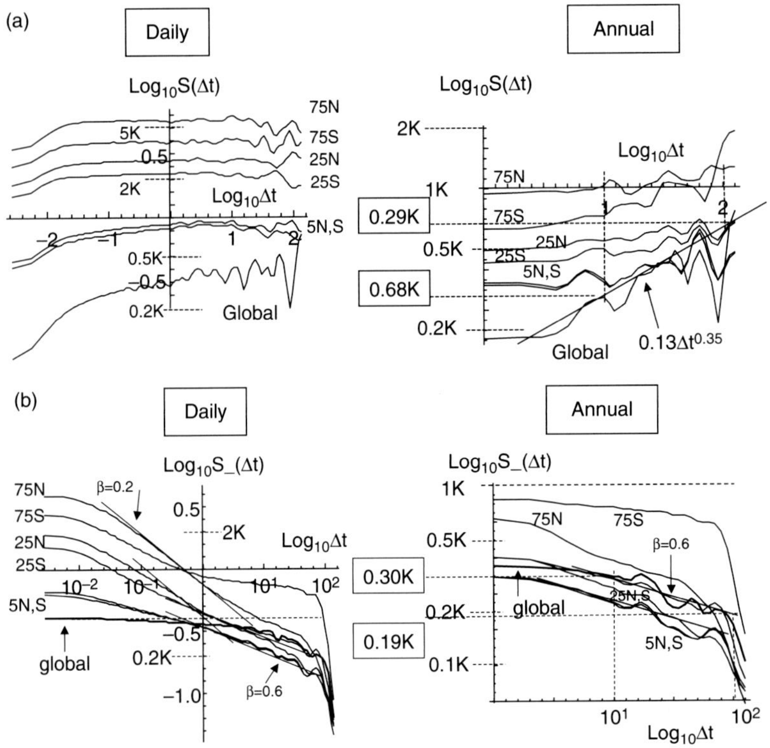

Fig. 10.11 (a) Comparison of the daily (left) and annual (right) RMS difference temperature structure functions for various latitudes indicated just to the right of daily figure (all at 700 mb; from the 20CR reanalysis, 1871–2008). The bottom thick line corresponds to the global average. On the right, we show a power-law approximation to the global structure function (apparently roughly valid for $\Delta t > 5$ years; see below). In addition, the RMS decadal and centennial fluctuations are shown: 0.29 K and 0.68 K, respectively. (b) Same as Fig. 10.11a except for the RMS tendency structure functions $S_{-}(\Delta t)$ (daily, left, and annual, right, with exponents $\xi(2)/2$; the reference lines are labelled with $\beta = 1 + \xi(2)$). The middle thick line corresponds to the global average. On the right, we show the RMS decadal and centennial tendencies: 0.30 K and 0.17 K, respectively.

(−0.4, compared with −0.2) so that the higher-latitude fluctuations (which are comparatively large at daily scales) get smoothed out more rapidly by temporal averaging so that at decadal scales there is not much difference; they all follow the global $\beta_{mw} \approx 0.6$ (except 75° S). Also shown in Fig. 10.11a, 10.11b are decadal and centennial differences and tendencies.

10.2.2 Haar fluctuations and structure functions: a convenient choice for geophysics

The difference and tendency structure functions are easy to implement and to interpret but are only useful over narrow ranges of H. The Haar fluctuation/wavelet is the basis of the orthogonal Haar decomposition (Haar, 1910) in the context of modern wavelet theory; it is a special case of the Daubechies family of orthogonal wavelets (see e.g. Holschneider, 1995, for a mathematical introduction to wavelets, and Torrence and Compo, 1998, for a pedagogical introduction). Haar fluctuations are nearly as simple as the poor man's and tendency fluctuations but are useful over the wider range $-1 < H < 1$ which encompasses all of the fields commonly encountered in geophysics. Although Haar wavelets are frequently mentioned in the wavelet literature, there have been surprisingly few applications (although see several papers in Foufoula-Georgiou and Kumar, 1994, or Ashok *et al.*, 2010, for recent examples; and for a comparison with the DFA technique see Koscielny-Bunde *et al.*, 2006).

An easy way to define the (absolute) Haar fluctuation is by taking the mean of second differences of the running sums:

$$\begin{aligned}\left(\Delta T(\Delta t)\right)_{Haar} &= \left|\frac{2}{\Delta t}\delta^2_{\Delta t/2}s\right| \\ &= \left|\frac{2}{\Delta t}\left(\left(s(t)+s(t+\Delta t)\right)-2s(t+\Delta t/2)\right)\right| \\ &= \left|\frac{2}{\Delta t}\left[\sum_{t+\Delta t/2\le t'\le t+\Delta t} T(t') - \sum_{t\le t'\le t+\Delta t/2} T(t')\right]\right| \end{aligned} \tag{10.16}$$

From this, we see that the Haar fluctuation at resolution Δt is simply the first difference of the series degraded to resolution $\Delta t/2$. In words: the Haar wavelet at scale Δt is simply the difference of the means over the first and second half of the interval. The constant 2 in the definition was chosen so that the Haar structure functions are close in value to the difference and tendency structure functions: see Appendix 5E. From the definition, we can see that for data of resolution Δt, the smallest Haar fluctuation that van be estimated is of resolution $2\Delta t$ (and for the quadratic Haar defined in Appendix 5E, it is $3\Delta t$). Although this is still a valid wavelet (see Fig. 5.35b), it is almost trivial to calculate and (thanks to the summing) the technique is useful for series with $-1 < H < 1$. Note also that, strictly speaking, the Haar wavelet is defined without the division by Δt, which has been added so that the Haar fluctuations have the same scaling exponents as the poor man's and tendency fluctuations.

Haar structure functions can then be defined as the various (qth-order) statistical moments of the Haar fluctuations. The real advantage of these structure functions is for functions with two or more scaling regimes, one with $H > 0$, one with $H < 0$. We shall see that, ignoring intermittency, this criterion is the same as $\beta < 1$ or $\beta > 1$, and hence Haar fluctuations will be useful for the data analyzed which straddle – either at high or low frequencies – the boundaries of the macroweather regime.

Is it possible to "calibrate" the Haar structure function so that the amplitude of typical fluctuations can still be easily interpreted? To answer this, consider the definition of a "hybrid" fluctuation as the maximum of the difference and tendency fluctuations:

$$(\Delta T)_{hybrid} = max\left((\Delta T)_{diff}, (\Delta T)_{tend}\right) \tag{10.17}$$

The "hybrid structure function" is thus the maximum of the corresponding difference and tendency structure functions and therefore has a straightforward interpretation. The hybrid fluctuation is useful if a calibration constant C can be found such that:

$$\left\langle \Delta T(\Delta t)^q_{hybrid}\right\rangle \approx C^q\left\langle \Delta T(\Delta t)^q_{Haar}\right\rangle \tag{10.18}$$

In a pure scaling process with $-1 < H < 1$, this is clearly possible since there is a unique scaling exponent. However, in a case with two or more scaling regimes, this equality cannot be exact, but as we see in the next section, it can still be quite reasonable.

Now that we have defined the Haar fluctuations and corresponding structure function, we can use it to analyse a fundamental climatological series: the global mean surface temperature at monthly resolution. At this resolution, the high-frequency weather variability is largely filtered out and the statistics are dominated first by the macroweather regime ($H < 0$), and then at low enough frequencies by the climate regime ($H > 0$).

Several such series have been constructed. In Appendix 10C we discuss three in detail: NOAA's CDC, NASA's GISP and the Climate Research Unit's HadCRUT3 series. Here we analyze the global averages obtained by averaging over all the available data for the common 129-year period 1880–2008 (taking into account the latitude-dependent map factors). Before analysis, each series was periodically detrended to remove the annual cycle – if this is not done, then the scaling of the structure function near $\Delta t \approx 1$ year will be artificially degraded. This periodic detrending was done by setting the amplitudes of the Fourier components corresponding to annual periods to the "background" spectral values.

Fig. 10.12 shows the comparison of the difference, tendency, hybrid and Haar root mean square (RMS) structure functions $<\Delta T(\Delta t)^2>^{1/2}$, the latter increased by a "calibration" factor $C = 2.2$. Before commenting on the physical implications, let us first make some technical remarks. It can be seen that the "calibrated" Haar and hybrid structure functions are very close; the deviations are $\pm$ 14% over the entire range of nearly a factor 10^3 in Δt. This implies that the indicated amplitude scale of the calibrated Haar structure function in kelvin (K) is quite accurate, and that at least in this case, to a good approximation, the Haar structure function preserves the simple interpretation of the difference and tendency structure functions: in regions where the logarithmic slope is between -1 and 0, it approximates the tendency structure function, whereas in regions where the logarithmic slope is between 0 and 1, the calibrated Haar structure function approximates the difference structure function. For example, from the graph we can see that global scale temperature fluctuations decrease from $\approx$ 0.3 K at monthly scales to $\approx$ 0.2 K at 10 years and then increase to $\approx$ 0.8 K at $\approx$ 100 years. All of the numbers have obvious implications, although note that they indicate the mean overall range of the fluctuations, so that for example the 0.8 K corresponds to $\pm$ 0.4 K etc. Since it was found that $C = 2$ was generally fairly accurate, unless otherwise indicated, the Haar analyses presented in this book were systematically increased by this factor.

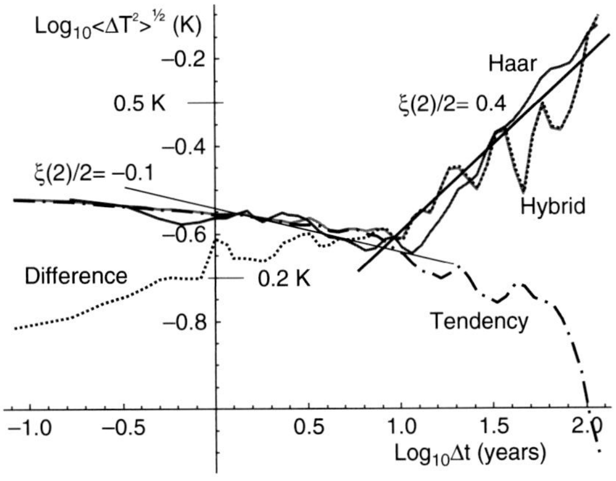

Fig. 10.12 A comparison of the different structure function analyses (root mean square, RMS) applied to the ensemble of three monthly surface series discussed in Appendix 10C (NASA GISS, NOAA CDC, HadCRUT3), each globally and annually averaged, from 1881 to 2008 (1548 dicpoints each). The usual (difference, poor man's) structure function is shown (thin, lower left), the tendency structure function (thin, lower right), the maximum of the two ("Hybrid", thick), and the Haar in dark, medium thickness (as indicated); it has been increased by a factor $C = 10^{0.35} = 2.2$ and the RMS deviation with respect to the hybrid is $\pm$ 14%. Reference slopes with exponents $\xi(2)/2 \approx 0.4, -0.1$ are also shown (corresponding to spectral exponents $\beta = 1+\xi(2) = 1.8, 0.8$, respectively). In terms of difference fluctuations, we can use the global root mean square $\langle \Delta T(\Delta t)^2 \rangle^{1/2}$ annual structure functions (fitted for 129 years $> \Delta t >$ 10 years), obtaining $\langle \Delta T(\Delta t)^2 \rangle^{1/2} \approx 0.08\Delta t^{0.33}$ for the ensemble. In comparison, Lovejoy and Schertzer (1986) found the very similar $\langle \Delta T(\Delta t)^2 \rangle^{1/2} \approx 0.077\Delta t^{0.4}$ using northern hemisphere data (these correspond to $\beta_c = 1.66, 1.8$ respectively). Reproduced from Lovejoy and Schertzer (2012b).

From Fig. 10.12 we also see that the global surface temperatures separate into two regimes at about $\tau_c \approx 10$ years, with negative and positive logarithmic slopes $= \xi(2)/2 \approx -0.1, 0.4$ for $\Delta t < \tau_c$ and $\Delta t > \tau_c$, respectively. Since the spectrum is a second-order moment, theoretically, $\beta = 1 + \xi(2)$, so that $\beta \approx$ 0.8, 1.8. We also analysed the first-order structure function whose exponent $\xi(1) = H$; at these scales the intermittency $K(2) \approx 0.03$ so that $\xi(2) \approx 2H$ so that $H \approx -0.1, 0.4$, confirming that fluctuations decrease with scale in the macroweather regime but increase again at lower frequencies in the climate regime (see Fig. 11.16a,b,c for more intermittency analyses). Note that, ignoring intermittency, the critical value of β discriminating between growing and decreasing fluctuations (i.e. $H < 0$, $H > 0$) is $\beta = 1$.

For more discussion of this example, a comparison with the detrended fluctuation analysis, generalizations to higher-order Haar fluctuations (valid for larger

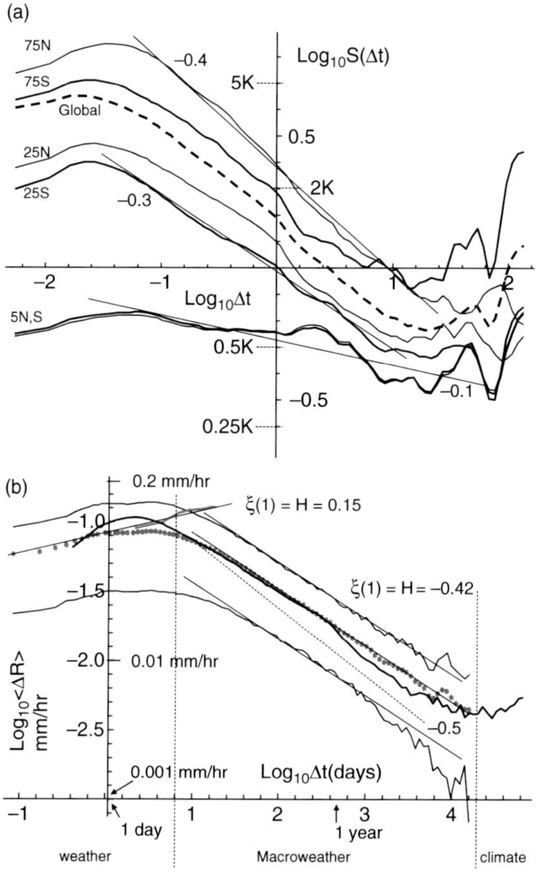

Fig. 10.13 (a) The daily averaged, annually detrended RMS Haar temperature structure functions averaged for various latitudes (indicated at left), northern hemisphere (thin), southern (thick) for the period 1871–2008 (all at resolution 2° × 2°). The reference lines, slope = $\xi(2)/2$, correspond to $\beta = 1 + \xi(2) = 0.2, 0.4, 0.8$ (top to bottom, respectively; the absolute slopes are indicated in the figure itself). The "global" curve is the average over all the pixels, weighted by the map factors. The rise at the left starting at $\Delta t = 2$ days (the smallest lag for daily data for Haar structure functions) is the meteorological regime (the maximum is at $\tau_w \approx 10$ days), the middle is the macroweather regime with minimum at $\tau_c \approx 10$ years, and at the right we see the beginning of the climate regime. This has been "calibrated" by boosting the fluctuations by a factor of 2.2 so that the large Δt part is close to the tendency structure functions shown in Fig. 10.11b; the corresponding temperatures are indicated in degrees *K*. (b) Haar, $q = 1$, 20CR (thick) with 29-year hourly CPC gauges (dots: see Section 4.4.2). For the CPC we also show the corresponding grid point to grid point one-standard-deviation limits (thin) with reference lines slopes $H = -0.42$ (solid) and -0.5 (dashed) corresponding to a Gaussian white noise process. Reproduced from Lovejoy and Schertzer (2012b).

H's) and for a discussion of the robustness of the results with respect to nonscaling perturbations (including linear global warming signals), see Appendix 5E, especially Fig. 5E.6.

10.2.3 Haar structure function analysis of the temporal variability: latitude, land, ocean and principal components

The systematic use of Haar fluctuations allows us to revisit in real space various issues that we have already discussed in the spectral domain. In this section, we consider the temporal scaling of the air temperature as a function of latitude, for air temperature over land, for sea surface temperatures (SST) and for SST principal components. In the next section, we consider the structure of the spatial scaling.

Let us first (re)consider the latitudinal dependence of the grid point scale, daily averaged RMS fluctuations using the RMS Haar structure functions (rather than the difference or tendency structure functions, Figs. 10.11a, 10.11b, or spectra, Figs. 10.5a, 10.5b). Fig. 10.13a shows the result using the same latitude bands as in Fig. 10.11a,10.11b and the same "calibration" as in Fig. 10.12. One can see that the Haar structure functions usefully combine the information from both the difference and tendency structure functions (cf. Figs. 10.11a, 10.11b) but without the limitations of the latter. The same basic three-scaling-regime behaviour is also found in Haar structure function analyses of precipitation (Fig. 10.13b), although for precipitation and temperature the weather regime (extreme left) and climate regime (extreme right) are barely visible since their ranges do not overlap the data range very much. Since there have been claims (e.g. Kantelhardt *et al.*, 2006) that at these scales precipitation is a Gaussian white noise (which theoretically has $H = -0.5$ because the standard deviation increases as $\Delta t^{0.5}$, so that its mean decreases as $\Delta t^{-0.5}$), we have included the corresponding reference line as well as the standard deviation of the structure functions indicating their variation from grid point to grid point. These show fairly convincingly that $H \neq -0.5$: the estimate −0.42 is reasonably accurate over the range of about 1 month to 30 years.

For reference, we have added separate analyses of monthly surface series (Fig. 10.14) separating out air temperature over land from the sea surface; this is the real-space analysis corresponding to the spectra in Fig. 8.6d; the reference lines (slopes −0.4, −0.2, corresponding to those in Fig. 8.6d: $\beta = 0.2, 0.6$ for land and SST, respectively), and the ocean transition scale τ_o is again about a year. We have also included a

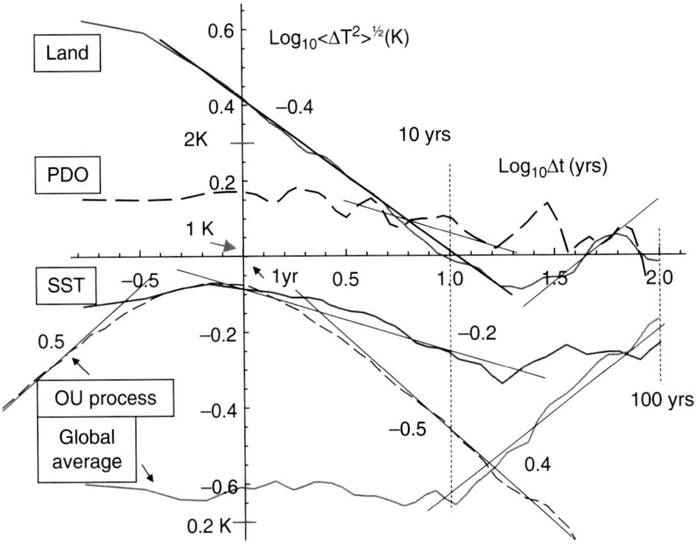

Fig. 10.14 RMS Haar structure function analyses of various monthly temperature series. The land (top), SST (middle) and global average curves are from the NOAA CDC dataset described in Appendix 10C; the land and SST curves are from the 5° resolution, near-complete 100-year data analyzed in Fig. 8.6d. The Pacific Decadal Oscillation (PDO) series is the dominant Pacific SST principal component described in Section 10.1.4 and analyzed spectrally in Fig. 10.8; we can see that its RMS fluctuations are very close to those of the SST but about a factor of 2 larger. For comparison, the results for 50 simulations of Ornstein–Uhlenbeck (OU) processes are also given using simulations with a characteristic time of 36 days. The theoretical asymptotic slopes (0.5, –0.5) are added to show the convergence to theory. For the data, the reference slopes are $\xi(2)/2 = -0.4, -0.2$ or $+0.4$, corresponding to spectral exponents $\beta = 1 + \xi(2) = 0.2, 0.6, 1.8$, respectively (cf. the corresponding spectral reference lines in the figures referred to above; a flat curve here corresponds to $\beta = 1$). We may note that the OU process is a poor approximation – not only to the SST curve but also to the land and PDO curves.

comparison with the Ornstein–Uhlenbeck (OU) process (Appendices 4A, 10B) used in stochastic linear modelling (SLM), comparing it with the SST structure functions which is the closest curve (on this log-log plot, pure OU processes only vary by shifts left–right or up–down; their shapes are the same). We see that the OU structure function is only close to the data over about an octave in scale.

We also added the structure function of the largest principal component of the Pacific SST (the PDO discussed in Section 10.1.4). This is significant because the usual approach to SLM uses a few dozen principal components to form a vector OU process (Appendix 10B). While the individual components of such vector OU processes will not be identical to a scalar OU process (whose shape is given in the figure for reference), since the principal component contributes a large fraction of the overall variance of the process one would nevertheless expect that its structure function would be not too far from the pure OU function shown. It is therefore significant that the PDO and OU structure functions are very different.

Worthy of note in Fig. 10.14 at large Δt is the beginning of a rise in the structure functions corresponding to the climate regime with $\tau_c \approx 10$–25 years. Since the curves are calibrated, their amplitudes are significant. For example, the global average curve (i.e. from the series of globally averaged surface temperatures, not to be confused with the global average over all the $2° \times 2°$ pixel curves shown in Fig. 10.13a), coincides with the ($5° \times 5°$ resolution) SST curve at about 80–100-year scales. Extrapolating to longer times, we may speculate that these will join with the ($5° \times 5°$ resolution) land and PDO curves at 200–300 years, so that at these scales the variation of the entire earth temperature may become dominant with respect to geographical (regional) variability.

10.3 Spatial variability in macroweather and climatic zones

10.3.1 Theoretical considerations in understanding and modelling climatic zones

A fundamental prediction of the weather/macroweather model is a dimensional transition leading to the factorization of the energy (and presumably other) cascading fluxes into a high-frequency space-time weather process $\varepsilon_w(\underline{r},t)$ and a purely temporal macroweather process $\varepsilon_{mw}(t)$: $\varepsilon_{w,mw}(\underline{r},t) = \varepsilon_w(\underline{r},t)\,\varepsilon_{mw}(t)$ (Eqn. (10.4)). While ε_w varies over spatial scales $\Delta x < L_e$, and over time scales $\Delta t < \tau_w$, ε_c is variable only over the range $\tau_w < \Delta t < \tau_c$ (i.e. in spectral terms ε_w, ε_{mw} are "band-limited" over the corresponding wavenumber and frequency intervals). Up until now, we have mainly explored the implications of this in the time domain, showing mainly (theoretically and empirically) that the transition is fundamentally one between an $H > 0$ and an $H < 0$ process at τ_w. It is now time to consider the spatial variability.

If we only consider the model described above, then the basic spatial variability is quite simple. Recall that in the weather domain there is a space-time relation (Section 8.4). This means that the statistics of the temporal average of $\varepsilon_w(\underline{r},t)$ over time scale $\tau < \tau_w$ are the same as the statistics of spatial averages over distance $l = V\tau$, where $V = L_e/\tau_w$. In other words, temporally averaging the weather process flux over scales τ_w is equivalent to spatially averaging it over scale L_e: it essentially smooths out all the spatial variability. Since the macroweather process $\varepsilon_{mw}(t)$ only has temporal variability, this simple model predicts that there would be no significant spatial intermittency in atmospheric fields averaged over time scales much beyond τ_w. Although the fields (with a fractional integration order H) would still be spatially variable, this is not obviously comparable with the existence of climatic regions and zones, i.e. the obvious long-time persistent and strong spatial (geographical) variability.

This suggests that even in the macroweather regime, in order to obtain realistic spatial variability, we must extend the model into the climate domain. The simplest way to do this while keeping the physical idea that the energy flux (modulated by the weather/climate processes) is the dynamical driver, is to include in Eqn. (10.4) a lower-frequency climate energy flux $\varepsilon_c(\underline{r},t)$:

$$\varepsilon_{w,c}(\underline{r},t) = \varepsilon_{w,mw}(\underline{r},t)\varepsilon_c(\underline{r},t)$$
$$\varepsilon_{w,mw}(\underline{r},t) \approx \varepsilon_w(\underline{r},t)\varepsilon_{mw}(t) \tag{10.19}$$

where $\varepsilon_{w,c}(\underline{r},t)$ is the overall weather/climate process, $\varepsilon_{w,mw}(\underline{r},t)$ is the overall weather/macroweather process (Eqn. (10.4)) and $\varepsilon_c(\underline{r},t)$ is the climate process with long time outer scale τ_{lc} where a new "low-frequency climate" process begins. These processes are band-limited so as to only have variability at frequencies $\tau_c^{-1} < \omega < \tau_i^{-1}$, $\tau_{lc}^{-1} < \omega < \tau_c^{-1}$ respectively (τ_i is the inner weather scale). The full model for the observable (in this case the wind v, since ε represents the energy flux) is now given by the usual FIF model but with a modified Green's function:

$$v(\underline{r},t) = g_{w,c}(\underline{r},t) * \varepsilon_{w,c}(\underline{r},t)$$
$$g_{w,c}(\underline{r},t) = \Theta(t)\left(\left[\!\left[(\underline{r},t)\right]\!\right]_{w,mw}^{-(d+1-H_w)} * \left[\!\left[(\underline{r},t)\right]\!\right]_{c}^{-(d+1-H_c)}\right) \tag{10.20}$$

where $\left[\!\left[(\underline{r},t)\right]\!\right]_{w,mw}$ and $\left[\!\left[(\underline{r},t)\right]\!\right]_{c}$ are weather/macroweather and climate-scale functions, band-limited so that $\left[\!\left[(\underline{r},t)\right]\!\right]_{w,mw}$ has little variability at frequencies below τ_c^{-1}, and $\left[\!\left[(\underline{r},t)\right]\!\right]_{c}$ has little variability above τ_c^{-1} (see Eqn. (10.5)), and d is the dimension of space (we'll ignore the vertical direction so that $d = 2$). Note that no explicit treatment of the macroweather regime is needed, since as we saw in Section 10.1, it appeared as a consequence of the dimensional transition at τ_w. Let us (as usual) indicate temporal averaging at scale τ with a subscript:

$$\varepsilon_\tau(\underline{r},t) = \frac{1}{\tau}\int_t^{t+\tau} \varepsilon(\underline{r},t')dt' \tag{10.21}$$

If we now average the full weather/climate process $\varepsilon_{w,c}(\underline{r},t)$ over time scales $\tau_w < \tau < \tau_c$, we obtain:

$$\begin{aligned}\varepsilon_{w,c,\tau}(\underline{r},t) &= \varepsilon_{w,\tau}(\underline{r},t)\varepsilon_{mw,\tau}(t)\varepsilon_{c,\tau}(\underline{r},t)\\ &\approx \varepsilon_{mw,\tau}(t)\varepsilon_{c,\tau_c}(\underline{r}); \quad \tau_w < \tau < \tau_c\\ \varepsilon_{w,c,\tau}(\underline{r},t) &\approx \varepsilon_{c,\tau}(\underline{r},t); \quad \tau > \tau_c\end{aligned} \tag{10.22}$$

where $\varepsilon_{w,c,\tau}(\underline{r},t)$, $\varepsilon_{mw,\tau}(t)$ and $\varepsilon_{c,\tau}(\underline{r},t)$ indicate respectively the overall process and the macroweather and climate processes at resolution τ. The top line in Eqn. (10.22) follows since when $\tau > \tau_w$, we average out the weather process intermittency $\varepsilon_{w,\tau}(\underline{r},t) \approx 1$, and the second line follows since similarly when $\tau > \tau_c$ we average out the macroweather intermittency: $\varepsilon_{mw,\tau}(t) \approx 1$. However, for $\tau < \tau_c$, the climate process is essentially constant in time, so we suppressed the time argument.

Eqn. (10.22) shows that for time scales in the macroweather regime, the fluxes factor into separate spatial and temporal functions. If we assume that the climate flux $\varepsilon_c(\underline{r},t)$ is statistically independent of the macroweather flux $\varepsilon_{mw}(t)$, then we may take qth powers and ensemble averages to obtain:

$$\begin{aligned}\langle \varepsilon(\underline{r},t)^q_{w,c,\tau,l}\rangle &\approx \langle \varepsilon_{mw,\tau}(t)^q\rangle\langle \varepsilon_{c,l}(\underline{r})^q\rangle\\ &\approx \left(\frac{L_e}{l}\right)\lambda^{K_c(q)}; \quad \tau_w < \tau < \tau_c\end{aligned} \tag{10.23}$$

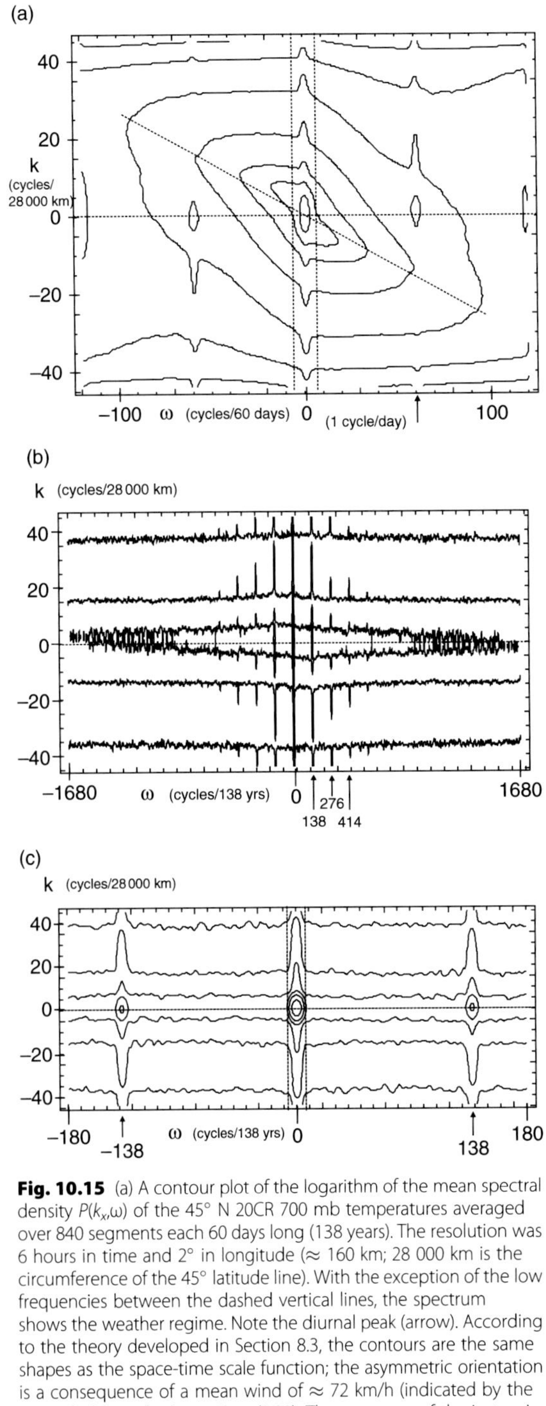

Fig. 10.15 (a) A contour plot of the logarithm of the mean spectral density $P(k_x,\omega)$ of the 45° N 20CR 700 mb temperatures averaged over 840 segments each 60 days long (138 years). The resolution was 6 hours in time and 2° in longitude (≈ 160 km; 28 000 km is the circumference of the 45° latitude line). With the exception of the low frequencies between the dashed vertical lines, the spectrum shows the weather regime. Note the diurnal peak (arrow). According to the theory developed in Section 8.3, the contours are the same shapes as the space-time scale function; the asymmetric orientation is a consequence of a mean wind of ≈ 72 km/h (indicated by the dashed oblique line): see Eqn. (8.29). The exponent of the isotropic spectrum estimated from the 2nd to 30th wavenumber cycles is $\beta_w \approx 1.6$, which is close to that of aircraft and other reanalyses (see Table 8.1). (b) The spectral density of the logarithm $P(k_x,\omega)$ of the same reanalysis temperatures as shown in Fig. 10.15a, only over the macroweather regime (from (138 years)$^{-1}$ to (1 month)$^{-1}$, i.e. 1 and 1680 cycles/138 years, respectively). The central spike is due to cycles 1–7 corresponding to the climate regime (see Fig. 10.15c for a blow-up), and the other spikes are the annual cycle and its harmonics. If the macroweather and climate contribute pure temporal and pure spatial factors which are statically independent, then $P(k_x,\omega) = E(k_x)E(\omega)$ and the contours will be all either horizontal or vertical (parallel to the axes). We see that this is reasonably accurately verified. The main exception is the small slope on the inner contour (affecting the strongest fluctuations), which corresponds to ≈ 100 m/day, i.e. about 10 000 times smaller than the speed in the weather regime (Fig. 10.15a) (this is presumably not an advection speed). (c) A blow-up of the central portion of Fig. 10.15b showing the climate regime in the centre (the region between the dashed vertical lines corresponds to scales > 20 years). Although the temporal scale range is too small, the space-time scaling there appears to be anisotropic, becoming stratified in the k_x direction. This would imply $H_{c,h} > H_{c,\tau}$, in agreement with the estimates $H_{c,h} \approx 1.4$, $H_{c,\tau} \approx 0.4$ (horizontal and temporal H exponents in the climate regime, $H_{c,h} \approx 0.7$ is for the near surface T value). The annual cycle (indicated by the arrows) is noticeably "fat," a consequence of the average number of days per year being noninteger. This frequency spread is the source of difficulty in removing the annual cycle.

where $\varepsilon(\underline{r},t)_{w,c,\tau,l}$ is the flux at time-space resolution τ, l, $K_c(q)$ is the moment scaling exponent for the climate-scale variability, and we have used the fact that macroweather process has low intermittency so that $\langle \varepsilon_{mw,\tau}(t)^q \rangle \approx 1$ (more precisely it is subexponential, i.e. slowly varying in resolution). This is confirmed in Fig. 10.16, where we find that over the entire range 8 days to 44 years it only changes by a factor 2–3. Note that this quasi-Gaussian behaviour implies a linear scaling exponent for the temporal fluctuations (cf. Eqn. (10.9)): we discuss this in the next section. Following our experience in the weather regime, where there are several significant fluxes – not only the energy flux – we expect they will factor in an analogous way.

The model predictions (Eqns. (10.22), (10.23)) can be tested in several ways. A fairly fundamental way is via the spectrum of the flux averaged over the weather regime $\varepsilon_{w,c,\tau_w}(\underline{r},t)$; if Eqn. (10.22) is valid, then this factors as $\varepsilon_{w,c,\tau_w}(\underline{r},t) = \varepsilon_{mw,\tau_w}(t)\varepsilon_{c,\tau_c}(\underline{r})$. If in addition we assume that the factors $\varepsilon_{mw,\tau_w}(t), \varepsilon_{c,\tau_c}(\underline{r})$ are statistically independent, then the spectral density will also factor as $P_{w,c,\tau_w}(\underline{k},\omega) = P_c(\underline{k})E_{mw}(\omega)$, where we have used the notation $E_{mw}(\omega)$ for the 1D low-frequency weather density, P for the wavenumber–frequency spectral density. If the spectral densities of the fluxes factor in this way, then so will the spectral densities of the observables.

To test this, for simplicity we restricted our attention to the zonal direction (k_x) and used the 700 mb temperatures at 45° N from the 20CR reanalysis at

6-hour resolution. The overall range of time scales was a factor of 201 616, so in order to clearly see the behaviour at both high- and low-frequency regimes, the data were analyzed in two ways. For the high-frequency weather regime, we broke the data into 60-day segments and calculated the average spectrum averaged over all the segments (Fig. 10.15a). This confirms that the space-time spectrum is close to that of the theory developed in Chapter 8, including with a mean east–west speed of about 72 km/h (not quite a displacement speed: see the theory). For the macroweather regime, we averaged the data over 15-day segments (roughly τ_w) and then calculated the spectrum (Figs. 10.15b, 10.15c). The contours are indeed nearly parallel to the ω and k_x axes. The only exceptions are the climate regime (see especially the blow-up in Fig. 10.15c) and the central contour of Fig. 10.15b, whose slight diamond shape implies a low speed of $\approx$ 100 m/day - more than 10^4 times smaller than the speed in the weather regime. Overall we see that the approximation $P_{w,c,\tau_w}(k_x,\omega) = E_c(k_x)E_{mw}(\omega)$ is quite well respected except for roughly the first seven periods (corresponding to 138/7 $\approx$ 20 years) where the climate variability begins (Fig. 10.15c). Similar behaviour was found at other latitudes, although near the equator factorization only holds for frequencies below (1 year)$^{-1}$, presumably due to the influence of the ocean (recall $\tau_{lo} \sim$ 1 year).

Figs. 10.15b and 10.15c test both the factorization hypothesis (Eqn. (10.19)) and the statistical independence of the factors. However, Eqn. (10.19) does not require statistical independence, and for $\tau < \tau_c$, it predicts not only factorization but also that the spatial factor is due to the slowly varying stochastic climate process $\varepsilon_c(\underline{r},t)$. This means that over scales $< \tau_c$, $\varepsilon_c(\underline{r},t) \approx \varepsilon_c(\underline{r})$ (roughly independent of time: Eqn. (10.19)), so we expect that the spatial statistics will show signs of converging to the statistics of those of the single realization of the random process $\varepsilon_c(\underline{r})$. Since according to the FIF model, $T(\underline{k}) = \varphi(\underline{k})k^{-H}$, considering the compensated T or (its flux) φ spectra are equivalent; the T spectrum is shown in Fig. 10.16. We can see that as the temporal averaging varies over the range 16 days to 44 years (2^4 to 2^{14} days, a factor 1024) the compensated spectrum only changes by a factor 2–3 (solid lines); indeed, that it seems to converge to a random (but roughly $k^{-2.2}$) spectrum as the averaging increases to τ_c. The spectral exponent $\beta_c \approx 2.2$ implies that $H_c \approx 0.7$ (using $\beta_c = 1 + 2H_c - K_c(2)$). There is even some hint that the 89-year spectrum (dashed

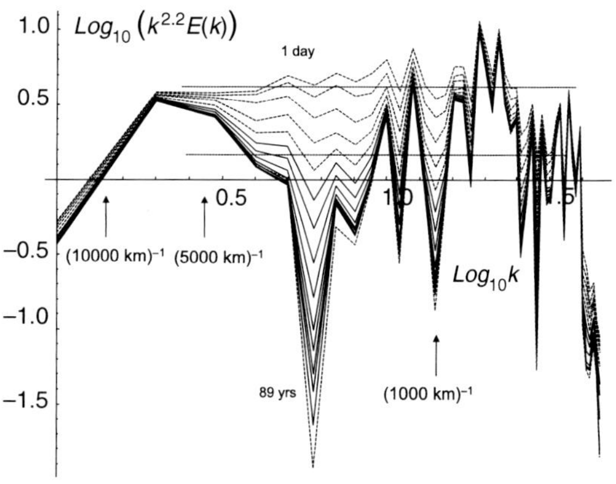

Fig. 10.16 The spatial spectrum of the near-surface time-averaged 20CR temperatures at 45° N compensated by dividing by the theoretical scaling spectrum $k^{-\beta}$ (with $\beta_c = 2.2$). The top dashed lines are in the weather regime (6 hours to 8 days, increasing downwards by factors of 2); the middle solid lines are the spectra of the temperatures averaged over the low frequency weather scales (16 days to 44 years), and the bottom dashed line is the climate scale spectrum at 89 years (2^{15} days). The straight lines are flat reference lines indicating $E(k) \approx k^{-2.2}$ behaviour. The high variability at the extreme right (largest factor $\approx$ 2) in k are presumably finite size effects. The apparently thick line is actually the convergence of the macroweather regime to a well-defined spatial spectrum of a random $\varepsilon_c(\underline{r})$ at $\approx$ 44 years.

bottom curve in the beginning of the climate regime) is becoming variable because of the beginning of the climate regime. We conclude that this initial data analysis gives reasonable support to the model (Eqn. (10.19)); in the next section we give it further support by showing that the temporal and spatial fluctuation statistics are indeed multiplicative and independent.

We have verified to a reasonable degree that the fluxes factor into statistically independent temporal and spatial components (Fig. 10.15b) and that the latter has only low-frequency climate-scale variations (Fig. 10.16). We have already checked that the macroweather process is roughly quasi-Gaussian in the time domain; let us now check that on the contrary it is multifractal, intermittent in the spatial domain (the climate factor): let us test Eqn. (10.23). The easiest way to check this is via the trace moments using various temporal averaging periods on the same 20CR reanalysis data. This is shown in Fig. 10.17, where the data were divided up into segments of duration τ as indicated. The top line shows typical weather-regime trace moments (6 hours, 1 day) whereas the next three (30 days, 1 year, 27 years) are in the macroweather regime, and are somewhat more intermittent, with somewhat larger external scales (this is quantified in Table 10.2). Finally (lower right), we show the remarkably similar climate-regime trace moments at 89-year resolution.

In Fig. 10.17 it is noticeable that as the duration of the averaging increases there is a tendency for the intermittency (as quantified for example by C_1: see Table 10.2) to increase rather than decrease, an effect somewhat augmented by a small but systematic increase in the effective outer scale. The (limited) climate-regime results are very similar but if anything even stronger: see the 89-year graph and parameters. This strong spatial intermittency in both macroweather and the climate regime is a quantitative expression of the fact that at climate scales different "climates" exist in different geographical regions. We return to this in Section 11.1.4. Table 10.2 shows that, as predicted by Eqn. (10.23), there is very little change in the spatial statistics over the entire macroweather regime, indeed into the beginning of the climate regime (which for these data at 45° N starts at about 40 years: see Fig. 11.5b).

Table 10.2 The horizontal parameter estimates the 45° N 700 mb on 20CR temperatures averaged over various temporal resolutions as indicated. C_1, α, L_{eff} are from the trace moment analysis in Fig. 10.16. Over the macroweather regime (between 16 days and 27 years), regression analysis from the trace moment analyses gives the following means and one-standard-deviation spreads in values: $C_1 = 0.115 \pm 0.018$, $\alpha = 1.71 \pm 0.07$; the mean outer scale is 9400 km. The values for the surface (0.995 sigma level) 20CR temperatures are very close with $L_{eff} = 15\,000$ km, $C_1 = 0.098 \pm 0.005$, $\alpha = 1.38 \pm 0.09$.

	C_1	α	L_{eff}
6 hours	0.095	1.65	6400 km
1 day	0.091	1.71	6400 km
30 days	0.093	1.81	10 000 km
1 years	0.118	1.75	10 000 km
27 years	0.136	1.67	10 000 km
89 years	0.130	1.63	13 000 km

10.3.2 Space-time relations in macroweather: joint space-time fluctuation analysis

In the previous section we theoretically showed that in the macroweather regime the spatial variability must be a consequence of a new very-low-frequency

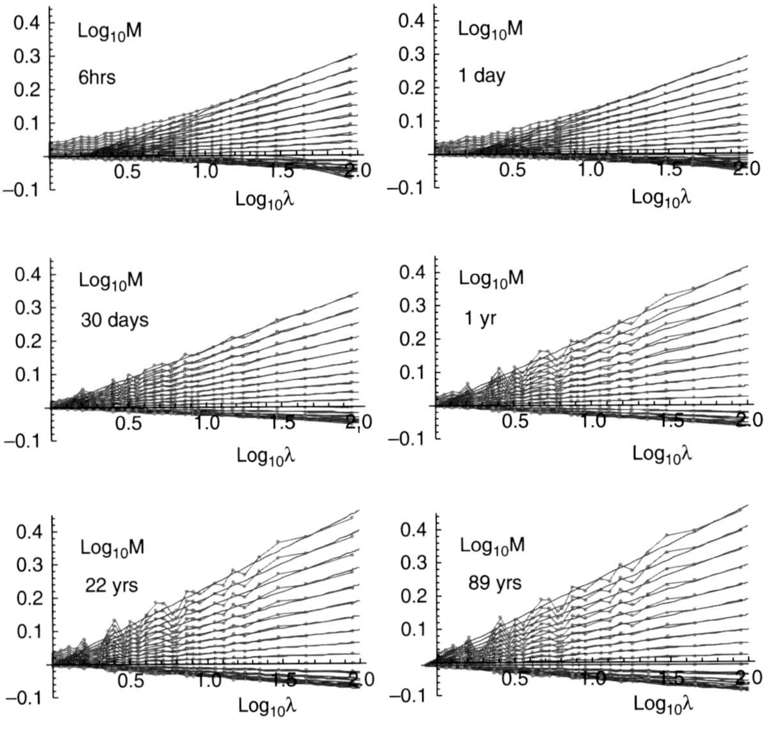

Fig. 10.17 The horizontal trace moment analyses of the 45° N 700 mb on 20CR temperatures averaged over various temporal resolutions as indicated. The scale ratio λ is with respect to a reference scale of 14 000 km (half the earth's circumference at this latitude). It can be seen that the scaling in the macroweather regime (30 days, 1 year, 27 years) and climate regime is excellent out to the largest scales and that the intermittency is fairly constant (see Table 10.2). This used two somewhat overlapping 89-year segments from the beginning and end of the series.

space-time climate process, and we gave evidence that a simple multiplicative model for the turbulent fluxes which included a climate factor was indeed valid. Indeed, if we stick to the range $\tau_w < \tau < \tau_c$, then we found theoretically that the flux at temporal resolution τ factors into independent spatial and temporal factors $\varepsilon_\tau(\underline{r},t) \approx \varepsilon_{mw,\tau}(t)\varepsilon_{c,\tau_c}(\underline{r})$ where $\varepsilon_{mw,\tau}(t)$ has low intermittency and $\varepsilon_{c,\tau_c}(\underline{r})$ on the contrary has fairly high intermittency. By considering the spatial flux moments and spatial spectra we gave support to this multiplicative model. In this section we consider the implications for the fluctuations $\Delta T(\Delta x,\Delta t)$.

Due to the factorization (Eqn. (10.22)) into low-frequency temporal and spatial factors, the behaviour of the fluctuations $\Delta T(\Delta x,\Delta t)$ is quite simple; it is just the product of the corresponding 1D fluctuation statistics, so that we expect structure functions to behave as:

$$S_q(\Delta x,\Delta t)\approx(s_{q,mw})^q\left(\frac{\Delta x}{L_e^*}\right)^{\xi_c(q)}\left(\frac{\Delta t}{\tau_w}\right)^{\xi_{mw}(q)};\ \Delta x<L_e^*;\ \tau_c>\Delta t>\tau_w$$
$$\xi_c(q)\approx qH_c-K_c(q);\qquad H_c\approx 0.7$$
$$\xi_{mw}(q)\approx qH_{mw};\qquad H_{mw}\approx -0.4 \tag{10.24}$$

where S_q is the qth-order (joint) structure function and $s_{q,mw}$ is a proportionality constant with dimensions K. The exponent $H_{mw} \approx -0.4$ was estimated earlier from the temporal Haar structure function (theoretically, Eqn. (10.9), or empirically, Fig. 10.13a, although the corresponding 45° N curve was not shown). $H_c \approx 0.7$ was deduced from the analyses of the previous section, $\beta_c \approx 2.2$ with $K_c(2) \approx 0.2$, (Table 10.2), using the formula $\beta_c = 1 + 2H_c - K_c(2)$. In the above we have used an external scale L_e^* rather than the usual L_e = 20 000 km in anticipation of the fact that we expect there to be latitudinal variations in the outer scale, and also that the actual outer scale is somewhat less than L_e (in these analyses, at 45° N, the half circumference ≈ 14 000 km and we find $L_e^* \approx$ 5000 km).

We can now test Eqn. (10.24) directly by systematically using Haar fluctuations to determine the statistical variation of the fluctuations at joint space-time intervals (Δx, Δt). Using the same 45° N 20CR near-surface temperature dataset, we can first calculate the temporal Haar fluctuation and then analyze it using spatial Haar fluctuations. In order to more clearly bring out the behaviour, we compensated the spatial fluctuations by multiplying them by $\Delta t^{0.4}$ so that the curves for various Δt's collapse on top of each other whenever they satisfy the theoretical temporal scaling (Eqn. (10.24)). When this is done for every factor of 2 in time scales from days to 138 years, we obtain Fig. 10.18. The compensated macroweather curves (16 days to 44 years, solid lines) are indeed tightly bunched over this macroweather range, but are noticeably dispersed for the smaller meteorological scales and the longer climatological scales. In addition to the theoretically predicted bunching, the fluctuations are roughly power law in space up to ~5000 km with the predicted spatial scaling exponents $\xi_c(2)/2 \approx H_c - K_c(2)/2 \approx 0.6$ (see the reference lines). After $L_e^* \approx$ 5000 km the fluctuations apparently "saturate" but nevertheless remain tightly bunched (so that the temporal scaling still approximately holds even at these larger spatial scales). The calibrations used in the figures and in the estimates of the constants $s_{2,mw}$, $s_{1,mw}$ were determined such that at the scale of the maximum (L_e^*) the Haar fluctuations were equal to the corresponding differences.

Also shown in Fig. 10.18 are the structure functions corresponding to the weather regime (2 days to 8 days, bottom) and the climate regime (44 years to 138 years, top): these are indicated by the dashing; they are dispersed because they both have $H \neq -0.4$. The weather-regime structure functions have clear maxima: this is because of the space-time relation already observed in Fig. 10.15a. On the large Δx side of the maximum, the weather curves approximately follow the form of Eqn. (10.24) but with $H_{w,h} \approx -0.5$, $H_{w,\tau} \approx 0.6$ (weather regime, horizontal and temporal exponents).

Finally, for the climate curves (Fig. 10.18, top), we see that the compensated structure functions for 89 and 138 years clearly separate from the macroweather functions, and it is significant that, as predicted by Eqns. (10.22) and (10.23), over both macroweather and climate scales the basic shape of the spatial structure function is unchanged. Indeed, we expect that the main change in the climate regime is that the temporal $H_{mw} < 0$ is replaced by a temporal $H_{c,\tau} > 0$, i.e. with the same power-law spatial prefactor (in Chapter 11 we give evidence that for the temperature in time, $H_{c,\tau} \approx 0.4$).

Rather than viewing structure functions of Δx for fixed Δt, we can instead view the structure functions for fixed Δx and varying Δt (Fig. 10.19). Here the dashed lines have slopes −0.4 and are spaced in the vertical corresponding to $\xi(2)/2$ = 0.6. The main

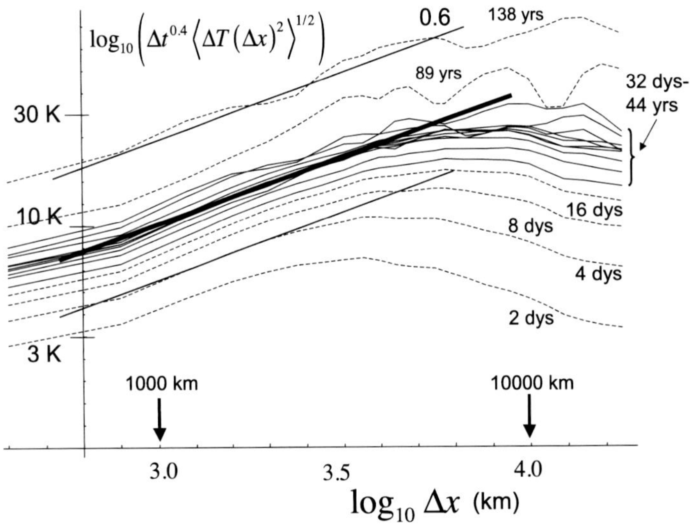

Fig. 10.18 The compensated spatial RMS Haar structure functions using the 20CR near-surface temperature fluctuation fields (0.995 sigma) defined by usual (linear) Haar wavelets at temporal resolutions varying from 12 hours to 138 years at 45° N. The reference lines have slopes $\xi_c(2)/2 = 0.6$ (a bit smaller than H_c since $C_1 \approx 0.1$). The original data had 2° longitudinal and 1-day temporal resolutions so that the highest-resolution time-space fluctuations were $2 \times 1 = 2$ days and $2 \times 2° = 4° = 470$ km. Each curve shows the analysis for temporal Haar fluctuations with resolutions varying from 2 to 2^{15} days ($\approx$ 89 years) indicated on the graph (the 138-year curve is only $2^{15.6}$ days). Each curve has been compensated by dividing by the theoretical behaviour $\Delta t^{-\xi_\tau(2)/2} = \Delta t^{-0.4}$ (the units of Δt here are the number of 2-day periods). The thick curves are from 32 days to 44 years and correspond to the macroweather regime, and we can see that with the exception of the single lowest factor of two in scale (finite size effects) these are all tightly bunched together near the thick reference line (slope $\xi_h(2)/2 = 0.6$), indicating that they follow closely the behaviour in Eqn. (10.22) up to about 5000 km. The weather-regime (2 days to 16 days, bottom) and climate-regime (89 years to 138 years, top) curves are all indicated by dashing and separate out quite clearly from the macroweather curves since they both have $H \neq -0.4$.

deviation from the theoretical behaviour is for the finest (4° longitude) spatial resolution at annual scales, this is an artefact of the imperfect detrending of the strong annual cycle. This can be compared with Fig. 10.13a.

Combining the information from Figs. 10.18 and 10.19, we conclude that Eqn. (10.24) is roughly valid for $\Delta x < L_e^* = 5000$ km and for $\tau_w < \Delta t < 44$ years with $H_{mw} = -0.4$, $H_c = 0.7$, $C_1 \approx 0.11$, $\alpha \approx 1.75$, $L_{eff} \approx 10\,000$ km. Since $\Delta t > \tau_e$ and $H_{mw} < 0$ and $\Delta x < L_e$ and $H_c > 0$, we see that the maximum possible value of $S_q(\Delta x,\Delta t) = S_q(L_e^*,\tau_e) = (s_{q,mw})^q$, and hence $s_{q,mw}$ is the maximum (mean) fluctuation. Considering the $q = 2$ moments, we find empirically that $s_{2,mw} \approx 6\ K$, $s_{1,mw} \approx 5\ K$ (for the RMS and $q = 1$ moments respectively).

Eqn. (10.24) is thus a fundamental law of the macroweather regime; it describes how temperature fluctuations decrease with temporal scale and increase with spatial scale. As was done in the weather regime, the structure functions can again be used to define space-time relations: for a given temperature fluctuation S_q, what are the corresponding possible Δx, Δt? Here, according to Eqn. (10.24), the relation $S_q(\Delta x,\Delta t) = constant$ implies (generalized) hyperbolic contours: considering the $q = 1$ case, this yields:

$$\Delta x = L_e^*\left(\frac{S_1}{s_{1,mw}}\right)^{1/H_c}\left(\frac{\Delta t}{\tau_w}\right)^{-H_{mw}/H_c}\quad ;\quad \tau_c > \Delta t > \tau_w;\ S_1 < s_{1,mw} \tag{10.25}$$

where since $H_{mw}/H_c \approx -0.6$, we obtain $\Delta x \approx \Delta t^{0.6}$. Note that this is quite unlike the (generally ellipsoidal) structure functions that we found in the meteorological regime (see the contours in Fig. 10.15a and Eqns. (8.15), (8.16)).

By linking spatial and temporal variability, Eqns. (10.24), (10.25) suggest several applications. For example, if our objective is to model temperature fluctuations to an accuracy of $S_1 = 0.1$ K at time scales at the beginning of the macroweather regime ($\Delta t = \tau_w$), then, using the above estimates for the parameters, Eqn. (10.25) tells us that we must have a spatial resolution of roughly 20 km. However, if we seek

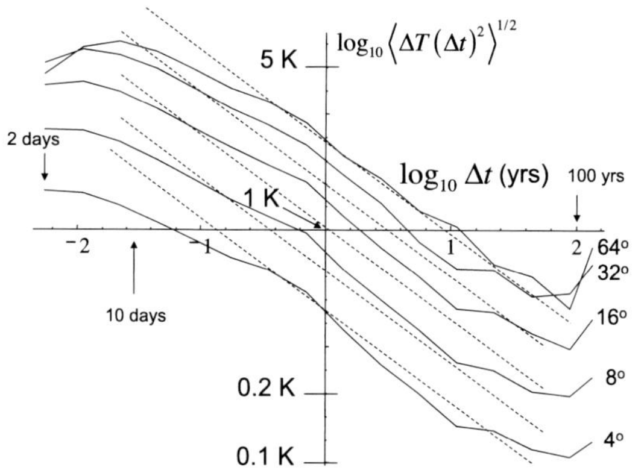

Fig. 10.19 The same structure function analysis of joint space-time fluctuations as in Fig. 10.18, but here each curve corresponds to a different spatial fluctuation scale, with the thick curves roughly in the spatial scaling part of Fig. 10.18 (32° corresponds to ≈ 3500 km). The equally spaced straight dashed lines have slopes $H_{mw} = -0.4$ and vertical interline spacings corresponding $\xi_c(2)/2 = 0.6$: the same exponents as in Fig. 10.15 and Eqn. (10.24). Although the data were detrended (annually), there is a residual variation which perturbs the temporal scaling, strongly for the smallest spatial scales (see the bump in the 4°, 8° curves near 1 year).

this accuracy only at scales $\Delta t = 1$ year, and taking $\Delta t/\tau_w = 36$, we find that the much coarser resolution ≈ 150 km is adequate. Alternatively, when analyzing regional data (e.g. $\Delta x = 5000$ km) at annual scales ($\Delta t = 1$ year) for signs of climate change (anthropogenic or otherwise), Eqn. (10.24) tells us that one expects natural fluctuations of ≈ ± 1 K so that only larger fluctuations would be statistically significant. To make proper statistical tests in order to determine confidence limits and the like, we could simply exploit the fact that the moments in Eqn. (10.24) implicitly define the probabilities (quasi-Gaussian in time, multifractal in space). It would be interesting to verify this directly with numerical climate models. (Note that these values are for the 20CR at the surface; the numbers for 700 mb are somewhat different.)

10.4 Summary of emergent laws in Chapter 10

In this chapter we investigated the consequences of the FIF model for $\Delta t >> \tau_w$, where τ_w is the weather time scale identified with the lifetime (eddy turnover time) of planetary-sized structures (≈ 10 days). If we assume that the outer time scale of the cascade is $\tau_c >> \tau_w$, then due to the finite size of the earth there will be a "dimensional transition" at τ_w. Nondimensionalizing space by L_e, and time by τ_w, so that $\Lambda_c = \tau_c/\tau_w$ and $\Lambda_w = L_e/L_i$ ("*i*" for "inner scale"), we may the write the cascade generator Γ as a sum of a weather and macroweather component:

$$\Gamma_{w,mw} = \Gamma_w + \Gamma_{mw}$$
$$\varepsilon_{w,mw} = \varepsilon_w \varepsilon_{mw};\ \varepsilon_{w,mw} = e^{\Gamma_{w,mw}};\ \varepsilon_w = e^{\Gamma_w};\ \varepsilon_{mw} = e^{\Gamma_{mw}} \tag{10.26}$$

with:

$$\Gamma_w(\underline{r},t) = \left(C_1 N_{d+1}^{-1}\right)^{1/\alpha} \int_{\Lambda_w^{-1}}^{1}\int_{B_w}^{B_1} \Theta(t-t')\frac{\gamma_\alpha(\underline{r}',t')}{\left(|\underline{r}-\underline{r}'|^2+(t-t')^2\right)^{(d+1)/(2\alpha)}}d\underline{r}'dt'$$

$$\Gamma_{mw}(t) \approx \left(C_1 N_{d+1}\right)^{1/\alpha}\int_1^{\Lambda_c}\Theta(t-t')\frac{\gamma_\alpha(t')dt'}{(t-t')^{(d+1)/\alpha}} \tag{10.27}$$

where B_w is the dissipation-scale "ball" (i.e. a circle radius Λ_w^{-1} in $d = 2$ spatial dimensions), B_1 is the corresponding (planetary)-scale ball (recall that the nondimensional $x = 1$ corresponds to planetary scales L_e) and N_{d+1} is a normalization constant (= π in one spatial dimension plus time, 2π in two spatial dimensions plus time). The corresponding second characteristic functions of the bare processes at a scale ratio λ from the outer scale of the weather process (= 1) and at the same scale ratio from the outer scale of the macroweather process (which is also the inner scale of the climate process = Λ_c) are:

$$K_w(q) = Log\langle\varepsilon_{w,\lambda}^q\rangle = \frac{C_1 q^\alpha}{\alpha-1}N_{d+1}^{-1}\int_{\lambda^{-1}}^{1}\int_{B_\lambda}^{B_1}\frac{d\underline{r}dt}{\left(|\underline{r}|^2+t^2\right)^{(d+1)/2}}$$

$$K_{mw}(q) = Log\langle\varepsilon_{mw,\lambda}^q\rangle \approx \frac{C_1 q^\alpha}{\alpha-1}N_{d+1}^{-1}\int_{\Lambda_c/\lambda}^{\Lambda_c}\frac{dt}{t^{(d+1)}} \tag{10.28}$$

These second characteristic functions are to base e, not base λ, so that the exponent in the integrand in the top (weather) equation gives a logλ behaviour (as required for multifractal weather processes). Whenever $d > 0$, the kernel for K_{mw} falls off too quickly: there will instead be a λ^{-d} fall-off. The consequences of this are that (a) the above bare and dressed statistics are quite different from each other, with the latter approaching the usual Gaussian limit although with (b) long-range correlations: the autocorrelation function has an asymptotic (large Δt) behaviour $R(\Delta t) \approx \Delta t^{-1}$, but the convergence requires very wide ranges of scale; (c) the spectrum is the Fourier transform of R, so that it has a low-frequency divergence leading to "pseudo-scaling," i.e. near power-law scaling but with exponents that depend on the outer scale Λ_c (not on C_1 or H and only weakly on α).

With Λ_c in the empirically relevant range (e.g. 30 years $< \tau_c <$ 2000 years, corresponding to $2^{10} < \Lambda_c < 2^{16}$), we find that the macroweather spectral exponent β_{mw} is in the range $0.4 < \beta_{mw} < 0.2$, which is roughly the same as the empirically measured exponents. These values of β_{mw} correspond to exponents following the usual macroweather exponent H_{mw}:

$$H_{mw} \approx -(1 - \beta_{mw})/2; \quad -0.4 < H_{mw} < -0.3 \tag{10.29}$$

(the dressed process is nearly nonintermittent, so that unlike the weather scales, which generally have $H > 0$, the macroweather regime generally has $H < 0$). A coupled atmosphere–ocean model leads to $\beta_{mw} \approx 0.6$ for frequencies in the low-frequency ocean regime $\omega < \omega_{lo} \approx (1\ \text{year})^{-1}$.

The pure weather/macroweather model does a very good job at predicting the temporal statistics up to a scale $\approx \tau_c$, where a new climate regime begins. However, we pointed out that its predictions for the spatial statistics of temporal averages of time scales $\tau > \tau_w$ are completely unrealistic; it predicts only very mild variability, which is contradicted by the strong spatial heterogeneity associated with climatic zones. We were therefore led – if only in order to understand the macroweather regime – to introduce a lower-frequency climate factor, the (multifractal) process $\varepsilon_c(\underline{r}, t)$.

$$\varepsilon_{w,c}(\underline{r}, t) = \varepsilon_{w,lq}(\underline{r}, t)\varepsilon_c(\underline{r}, t) \approx \varepsilon_w(\underline{r}, t)\varepsilon_{mw}(t)\varepsilon_c(\underline{r}, t) \tag{10.30}$$

Using space-time spectra, structure functions and spatial trace moments, we showed empirically that this was reasonably compatible with 20CR temperature reanalyses, including the hypothesis that $\varepsilon_c(\underline{r}, t)$ is statistically independent of the other factors. Using the fact that $\varepsilon_{mw}(t)$ is quasi-Gaussian and assuming that $\varepsilon_c(\underline{r}, t)$ is multifractal, this predicts the form of the joint space-time fluctuations. Combining this with the simplified form for the weather-scale fluctuations, we have the following structure functions:

$$S_{q,w}(\underline{\Delta r},\Delta t)=(s_{q,w})^q[\![(\underline{\Delta r},\Delta t)]\!]_w^{qH_w-K_w(q)}; \quad \tau_i<\Delta t<\tau_w$$

$$[\![(\Delta x,0,0,\Delta t)]\!]_{w,can}=\left(\left(\frac{\Delta x}{L_e^*}\right)^2+\left(\frac{\Delta t}{\tau_w}\right)^2\right)^{1/2}$$

$$S_{q,mw}(\underline{\Delta r},\Delta t)=(s_{q,mw})^q\|\underline{\Delta r}\|_c^{qH_c-K_c(q)}\left(\frac{\Delta t}{\tau_w}\right)^{qH_{mw}}; \quad \tau_c>\Delta t>\tau_w$$

$$\|(\Delta x,0,0)\|_{c,can}=\left|\frac{\Delta x}{L_e^*}\right| \tag{10.31}$$

all valid for $\Delta x < L_e^*$, where L_e^* is the effective outer scale estimated at 45° N to be about 5000 km (compared to the half-circumference of ~14 000 km), although the small spatial scale limit of the climate factor may be several hundred kilometres: see Section 11.2.4. The s_q's are proportionality constants with dimensions K discussed below. In Eqn. (10.31), we have first given the expected general form of the structure functions $S_q(\underline{\Delta r},\Delta t)=\langle\Delta T(\underline{\Delta r},\Delta t)^q\rangle$ for general space-time vectors $(\Delta r,\Delta t)$, in terms of space-time scale and space functions ($[\![(\underline{\Delta r},\Delta t)]\!]$,$\|\underline{\Delta r}\|$ respectively). For illustration purposes, just below S_q, we have given the example of the simple (canonical) scale function of a single spatial lag Δx. For simplicity, the weather-regime equation (subscript w) canonical-scale function ignores overall advection; for temperature, we found $H_w \approx 0.5$, $C_{1w} = 0.087$, $\alpha_w = 1.61$ (we used Eqn. (8.16) with a zero mean advection velocity in the matrix B. For the exponent estimates, see Chapter 8, Table 8.1. The macroweather structure function (subscript mw, defined by Haar wavelets) is discussed above; we estimated $H_c \approx 0.7$, $H_{mw} \approx -0.4$ and $C_1 \approx 0.11$, $\alpha \approx 1.75$ (Table 10.2). In the Chapter 11 we extend this to the climate regime.

The constants $s_{q,w}$, $s_{q,mw}$ are constrained so that the structure functions are continuous at the boundaries between the weather and macroweather regimes. Taking $\Delta x = L_e^*$ (or almost equivalently considering the structure functions of the global averages), and for simplicity considering only $q = 1$, we have:

$$\langle\Delta T(L_e^*, \tau_w)\rangle_w = \langle\Delta T(L_e^*, \tau_w)\rangle_{mw}; \quad s_{1,w} \approx s_{1,mw} \tag{10.32}$$

Using the 45° N data, we find that these equations work remarkably well. For example we verify

$s_{1,w} \approx s_{1mw} \approx 5\ K$ (e.g. Fig. 10.19 for slightly larger s_2 values). For the spectra corresponding to Eqn. (10.31), see Section 11.1.5.

The macroweather result implies that although the temporal intermittency is small, nevertheless the spatial intermittency is moderately large through the macroweather regime right into the climate regime, a subject to which we return in the next chapter. For a somewhat different approach, using the temporal evolution of the universal multifractal parameters α and C_1, as well as the resulting evolution of the extremes, (see Royer *et al.*, 2008, 2009, Schertzer *et al.*, 2012). Indeed, if the high-quality space-time data existed at those scales, we could test the prediction that the main change in the climate regime is that the temporal macroweather exponent $H_{mw} < 0$ is replaced by a temporal climate exponent $H_{t,c} > 0$. By fixing the mean fluctuation (say S_1), Eqn. (10.28) can be used to define space-time relations in the macroweather regime. These may help for example to determine the necessary spatial resolution of models once the size (S_1) and temporal resolutions (Δt) are specified, or to construct statistical tests to detect regional climate change.

Appendix 10A: The dimensional transition asymptotic scaling of cascades in the macroweather regime

10A.1 The basic bare and dressed statistical behaviour in the macroweather regime

In Chapter 8 we saw empirically, and in Chapter 10 we saw theoretically, that beyond the weather scale τ_w there is a drastic "dimensional transition"; for time scales $<\tau_w$ we obtain the usual turbulent statistics, while for scales $\tau_w>$ the spectrum becomes fairly flat, often with $\beta \approx 0.2$–0.4, and this for all the fields irrespective of their β at scales $<\tau_w$ (e.g. Fig. 10.5a, 10.5b). In Section 10.2 we saw that if we assume that the temporal outer scale $\tau_c >> \tau_w$ then we can extrapolate the turbulent model well beyond weather scales, indeed, perhaps to $\tau \approx 10$–100 years (see Sections 8.1.2 and 10.1.2). In this appendix, we discuss a surprising consequence: that although the statistical properties are indeed asymptotically scaling (power laws), the actual "effective" exponent can depend on the cascade scale range – and yet nevertheless be fairly accurately scaling over 2–3 orders of magnitude, i.e. physically, from 10 days to 10 years or more.

We have already seen in Section 10.1.2 that the generator for the flux separates into the sum of a weather and macroweather contribution corresponding to the multiplicative modulation of the weather flux by a macroweather flux. If we nondimensionalize time by the weather scale τ_w we find that for nondimensional times $t = \lambda >> 1$, we can ignore the spatial degrees of freedom and the second characteristic function of the bare flux climate ε_λ is given by:

$$\varepsilon_{mw} = e^{\Gamma_{mw}}; \quad \Gamma_{mw} = \left(\frac{C_1}{\alpha - 1} N_{d+1}^{-1}\right)^{1/\alpha} \gamma_\alpha * g$$

$$K_{mw}(q) = \log\langle \varepsilon_{mw,\lambda}^q \rangle = \frac{C_1 q^\alpha}{\alpha - 1} N_{d+1}^{-1} \int_{\Lambda_c/\lambda}^{\Lambda_c} \frac{dt}{t^{(1+d)}}$$

$$= \frac{C_1 q^\alpha}{N_{d+1} d(\alpha - 1)} \Lambda_c^{-d} (\lambda^d - 1) \qquad (10.33)$$

where we have used the propagator $g(\underline{r}, t) = \Theta(t)(|\underline{r}|^2 + t^2)^{-(1+d)/(2\alpha)}$ (see Eqn. (10.5)), Θ is the Heaviside function needed to maintain causality, d is the number of spatial dimensions. Ignoring the vertical, this gives $g(0,t)^\alpha = \Theta(t)t^{-2}$ ($d=1$ spatial dimension), $g(0,0,t)^\alpha = \Theta(t)t^{-3}$ ($d=2$ spatial dimensions). N_{d+1} is the causal normalization constant in $d+1$ dimensions: $= 2\pi/2$ in $d=1$, $4\pi/2$ in $d=2$ (see Appendix 5B; the factor 2 is due to the Heaviside function). Note that the logarithms are taken to the base e rather than the base λ as in the usual cascade regime (which is recovered if we take $d=0$ in the above and replace the power laws by logarithms on the right). The above bare nonnormalized $K(q)$ functions can be normalized in the usual way by taking $K(q) \to K(q) - qK(1q)$ (see Eqn. (5.87)), i.e. we replace q^α by $q^\alpha - q$ in Eqn. (10.33). Since $C_1 < d+1$, the exponent $K_{mw}(q)$ is typically quite small so that the exponentiation needed to obtain $\langle \varepsilon_{mw,\lambda}^q \rangle$ yields:

$$\langle \varepsilon_{mw,\lambda}^q \rangle \approx 1 + K_{mw}(q) = 1 + \frac{C_1(q^\alpha - q)}{N_{d+1} d(\alpha - 1)} \Lambda_c^{-d} (\lambda^d - 1) \qquad (10.34)$$

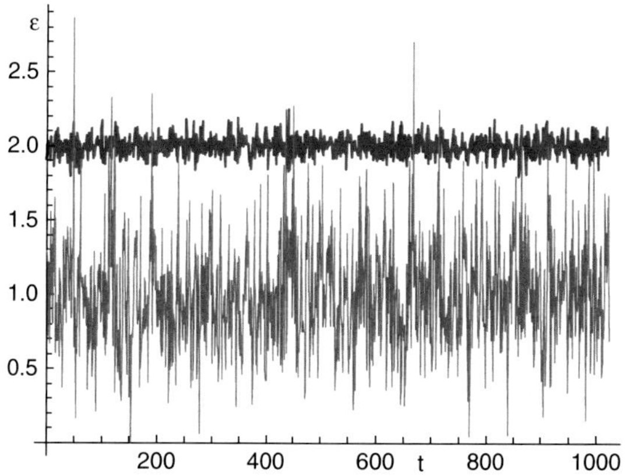

Fig. 10A.1 A realization of a bare climate cascade with $\lambda = 2^{10}$ (thin line) and the corresponding dressed process (thick line, offset by one unit for clarity) at the same resolution obtained by continuing the cascade to $\Lambda_c = 2^{16}$ and then averaging over a scale ratio of 2^6.

We thus see that the moments depend only weakly on C_1, q, λ. In any case, as usual, it is the dressed properties that are most important, and because of the weak correlations (studied explicitly below) we in fact (presumably) obtain central limit theorem convergence; this can be seen visually in the example in Fig. 10A.1, where we compare a bare and dressed realization of such a process with $\Lambda_c = 2^{16}$ which has been averaged over a scale ratio of 2^6.

10A.2 Autocorrelations: asymptotic Δt^{-1} scaling

In order to better understand the scale dependence of the macroweather-regime cascades, let us determine the autocorrelation function, from which the spectrum can be found by Fourier transformation (the Wiener–Khinchin theorem, Eqn. (3.31)). In this case, we can follow the development of Appendix 5B, where we calculated the generator of the autocorrelation function. In particular (see Eqn. (5.139)), considering just the climate regime, in nondimensional coordinates, after integrating out the small-scale spatial degrees of freedom we obtain:

$$S(\Delta t) \approx \int_1^{\Lambda_c} \left(g(t) + g(t - \Delta t) \right)^{\alpha} dt; \quad \begin{matrix} g(t) = t^{-(d+1)/\alpha}; & t \geq 1 \\ g(t) = 0; & t < 1 \end{matrix} \tag{10.35}$$

S is the temporal part of the second characteristic function of the autocorrelation (not to be confused with the structure function). Eqn. (10.35) takes into account causality and is only the climate contribution (it ignores the weather scales $0 < t < 1$). Recall that the normalized autocorrelation function $R_n(\Delta t)$ is:

$$R_n(\Delta t) = R(\Delta t)/R(\Lambda_c); \quad R(\Delta t) = e^{\frac{C_1}{\alpha - 1} N_d^{-1} S(\Delta t)} \tag{10.36}$$

Considering just the $d = 1$ case for the moment, this yields:

$$S(\Delta t) \approx \int_1^{\Delta t - 1} \frac{dt}{t^2} + \int_{\Delta t + 1}^{\Lambda_c} \left(\frac{1}{t^{2/\alpha}} + \frac{1}{(t - \Delta t)^{2/\alpha}} \right)^{\alpha} dt \tag{10.37}$$

The first term yields 1 - Δt^{-1}, while the second yields:

$$S(\Delta t) \approx 1 - \Delta t^{-1} + \Delta t^{-1} \int_{(1+\Delta t)^{-1}}^{(1+\Delta t/\Lambda_c)^{-1}} \frac{dr}{r^2} (1 + r^{2/\alpha})^{\alpha} dr \tag{10.38}$$

where we have used the transformation of variables:

$$r = \frac{t - \Delta t}{t}; \quad \frac{dt}{t^2} = \frac{dr}{\Delta t} \tag{10.39}$$

Since there is no singularity at $r = 1$, we can take Λ_c large so that to high accuracy, we can take the upper bound $r = 1$ and rewrite the integral as:

$$S(\Delta t) \approx 1 - \Delta t^{-1} + \Delta t^{-1} \left[\begin{matrix} \int_0^1 \frac{dr}{r^2} \left((1 + r^{2/\alpha})^{\alpha} - 1 \right) dr \\ - \int_0^{\Delta t^{-1}} \frac{dr}{r^2} \left((1 + r^{2/\alpha})^{\alpha} - 1 \right) dr + \int_{\Delta t^{-1}}^1 \frac{dr}{r^2} \end{matrix} \right] \tag{10.40}$$

For $\alpha < 2$, the first term in the brackets is a simple α-dependent constant:

$$A = \int_0^1 \frac{dr}{r^2} \left((1 + r^{2/\alpha})^{\alpha} - 1 \right) dr \tag{10.41}$$

(e.g. $A = 16.379$ for $\alpha = 1.8$), while the third term in brackets is simply Δt - 1. The second term in the brackets can be evaluated by using the binomial expansion:

$$\int_0^{\Delta t^{-1}} \frac{dr}{r^2} \left((1 + r^{2/\alpha})^{\alpha} - 1 \right) dr = \frac{\alpha}{2/\alpha - 1} \Delta t^{1 - 2/\alpha} + O(\Delta t^{1 - 4/\alpha}) \tag{10.42}$$

Therefore, we finally obtain:

$$S(\Delta t) \approx 2 + (A-2)\Delta t^{-1} - \frac{\alpha}{2/\alpha - 1}\Delta t^{-2/\alpha} - O(\Delta t^{-4/\alpha}) \tag{10.43}$$

Just the first two Δt-dependent terms yield an excellent approximation.

Turning to the more realistic case $d = 2$ (two horizontal dimensions), the same technique yields:

$$S(\Delta t) \approx \frac{1}{2}(1 - \Delta t^{-2}) + \Delta t^{-2}\int_{(1+\Delta t)^{-1}}^{(1+\Delta t/\Lambda_c)^{-1}} \frac{dr}{r^3}(1-r)(1+r^{3/\alpha})^{\alpha} dr \tag{10.44}$$

Although the full analysis is more complex, we can see from Eqn. (10.38) that the main contribution for large Δt will come from the $1/r^2$ singularity, which once again yields a dominant Δt^{-1} term so that $S(\Delta t)$ asymptotes to the value 1/2. Note that again the upper limit of integration in Eqn. (10.38) can with high accuracy be replaced by 1 (i.e. $\Lambda_c \to \infty$). Once again, since $S << 1$, we may use the approximation:

$$R(\Delta t) = e^{\frac{C_1}{\alpha-1}N_{d+1}^{-1}S(\Delta t)} \approx 1 + \frac{C_1}{\alpha - 1}N_{d+1}^{-1}S(\Delta t) \tag{10.45}$$

We may thus numerically determine the range of Δt^{-1} behaviour by using the autocorrelation function normalized by the limiting value as $\Delta t \to \infty$, i.e. using the relation: $R(\Delta t)/R(\infty) - 1 \approx \frac{C_1}{\alpha-1}N_{d+1}^{-1}\left(S(\Delta t) - S(\infty)\right) \propto \Delta t^{-1}$. This relation (compensated by multiplying it by Δt) is confirmed in Fig. 10A.2 (thick lines) using direct numerical integration of Eqn. (10.38). It can be seen that convergence to the theoretical asymptotic behaviour is very slow, especially for the larger α values. However, the situation is actually worse than this slow convergence indicates, since for finite Λ_c it is not the autocorrelation function normalized by $R(\infty)$ that is important, but rather (as in Eqn. (10.36)) by $R(\Lambda_c)$, the largest in the cascade regime, and due to the extremely slow fall-off implied by the Δt^{-1} behaviour, this significantly modifies the large Δt behaviour as shown by the thin lines in Fig. 10A.2. Because of this additional large Δt effect, the pure Δt^{-1} scaling only occurs at very large Δt (recall that even being generous, empirically $\Lambda_c \approx < 2^{12}$ ($\approx$ 100 years/10 days)).

10A.3 Spectra, pseudo-scaling

We have seen that the asymptotic autocorrelation function has rather slow convergence to a Δt^{-1} form; from the point of view of the spectrum (the Fourier transform), this is particularly problematic since the spectrum of a pure Δt^{-1} function diverges at both low wavenumbers (also in fact at high wavenumbers, but this is not relevant here). We may therefore anticipate that the actual spectrum will be quite sensitive to the outer scale limit Λ_c. To investigate this, we studied both Monte Carlo simulations (40 realizations of a multifractal in one spatial dimension and time), and the corresponding ensemble spectrum $E(\omega)$ obtained numerically for a simulation length $\Lambda_c = 2^{12}$ (Fig. 10A.3). From the figure, there are three key points to note. First, note that the simulation and theoretical ensemble average spectrum calculated numerically from Eqn. (10.38) agree extremely well. Second, that over a remarkably wide range – over a factor $\approx$ 100–1000 – the spectrum is nearly linear on a log-log plot, i.e. it is "nearly" a power law; it is easy to see that this behaviour could easily be mistaken for a real scaling regime. Third, that this $d = 1$ exponent is low (here, $\beta \approx 0.4$) and close to the values found empirically ($\beta \approx 0.2$–0.4, see Sections 8.1.2 and 10.1.5, Table 10.1).

Returning to two spatial dimensions and calculating the ensemble averaged spectra directly from Eqn. (10.38) for various Λ_c, we obtain Fig. 10A.4. We can see that for the example given with $\alpha = 1.8$, the apparently scaling low-frequency regime is again of length varying by a factor of 100–1000 depending

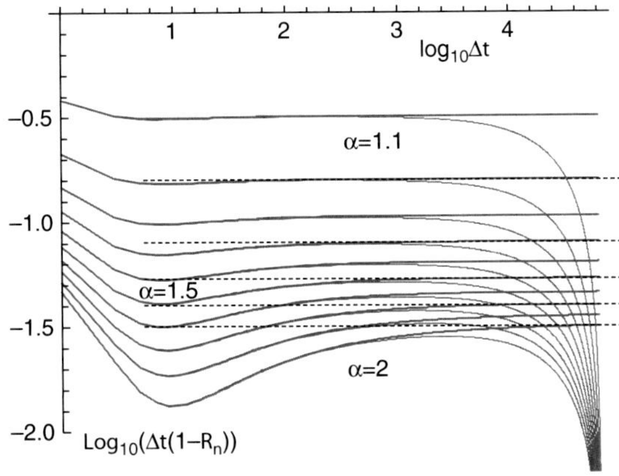

Fig.10A.2 The log of the compensated by Δt^{-1}, normalized autocorrelation function in two spatial dimensions for $C_1 = 0.4$. The thick lines are when R_n is estimated using $R_n(\Delta t) = R(\Delta t)/R(\infty)$, the thin lines are for $R_n(\Delta t) = R(\Delta t)/R(\Lambda_c)$ (important for the power spectrum over a finite range) $\Lambda_c = 2^{16}$, showing the extreme sensitivity of the large scales to the length of the series (Λ_c). The dashed black lines are flat reference lines corresponding to the asymptotic Δt^{-1} behaviour. Each curve is for a different α value, increasing from 2 (bottom) to 1.1 (top) at intervals of 0.1.

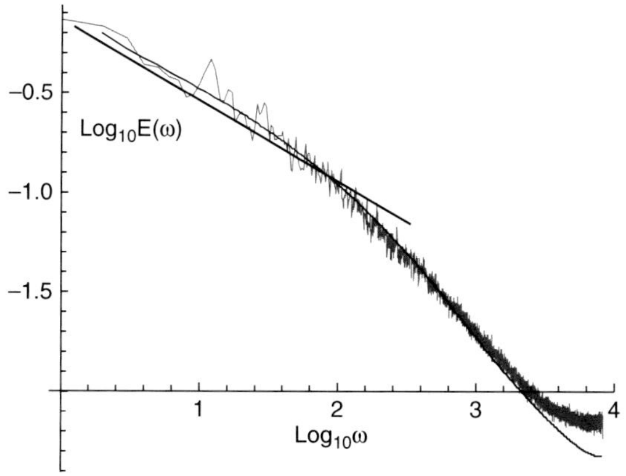

Fig 10A.3 Comparison of Monte Carlo results for 40 realizations $\alpha = 1.8$, $C_1 = 0.1$, $2^{14} \times 2^6$ (time, space) with the corresponding theory (solid line) for one spatial dimension. For $\omega > 2^{14}/2^6 = 2^8$ (i.e. the transition is expected at $\log_{10}\omega \approx 2.4$), there is some disagreement since the spatial degrees of freedom (neglected for the solid line) begin to be important at high frequencies. The solid reference line has a slope −0.4 which holds well for a factor of 100 or so in scale.

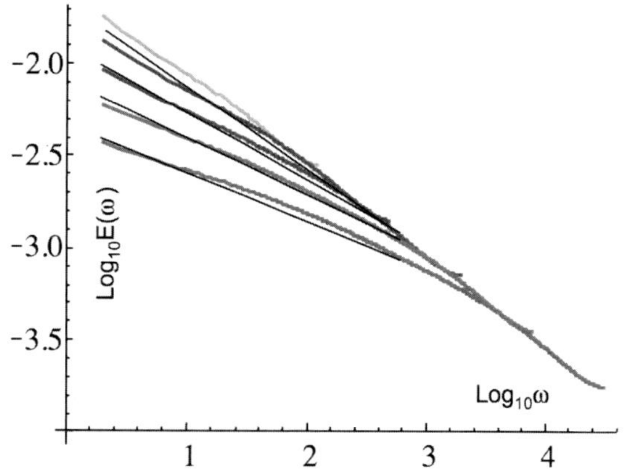

Fig 10A.4 The theoretical plateau power spectrum for $\alpha = 1.8$, two spatial dimensions ($d = 2$) for outer scales of $2^8, 2^{10}, 2^{12}, 2^{14}, 2^{16}$ (top to bottom). The lines have root mean square slopes fit up for $\omega \leq 512$, with values −0.45, −0.38, −0.32, −0.27 (the 2^8 line is not given since the corresponding simulation isn't long enough; see the next graph for a quantification).

on how stringent we are about the linearity of the log-log spectrum (since S is small, virtually identical spectra are found for other values of C_1: see Eqn. (10.45)). Since the scaling regime is only approximate, it could be called "pseudo-scaling."

In the figure we also see that the pseudo-scaling of the first factor 100–1000 depends to some extent on Λ_c as well as the maximum frequency (ω_{max}) used for the regression. In order to quantify this, we show in Fig. 10A.5 the regression β estimates as functions of ω_{max} for various Λ_c, and for $\alpha = 1.5, 1.8$ (the main empirical range). It can be seen that for ω_{max} between 4 and 512, the slopes are typically constant to $\pm$ 0.05 and the standard deviations of the residuals of the fits are all < 0.019. At the same time, β is a weak function of Λ_c; β changes by roughly 0.1 for every factor of 16 in Λ_c. We also note that taking $\Lambda_c \approx$ 100–300 years/10 days $\approx 2^{12}$–2^{14}, so that β is in the range 0.2–0.4.

The final piece in the puzzle is to consider the plateau for the observables, the fractionally integrated fields. It turns out that the climatological contribution to the fractional integration (i.e. the contribution for $t > 1$) is totally unimportant; Fig. 10A.6 shows that the low-frequency spectra of the flux and the observables are the same, and hence the results above on the spectral plateau apply. The reason is that fractional integration order H for the observable I_{mw}: $I_{mw}(t) = \Theta(t)\varepsilon_{mw}(t) * t^{-(d+1-H)}$ has essentially no effect on the spectrum since when $d - H > 0$, the

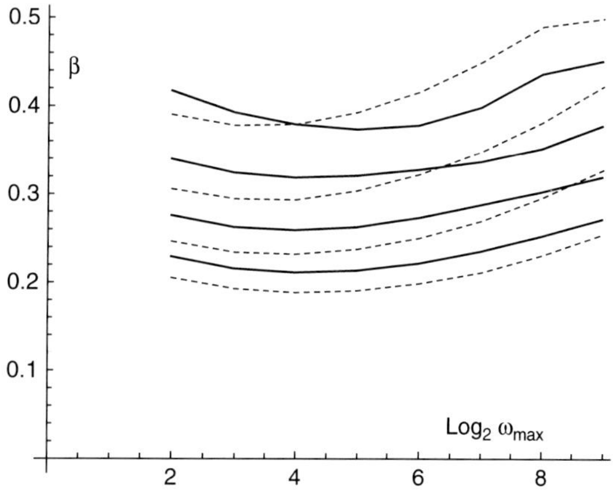

Fig. 10A.5 The regression estimates of the spectral exponent β (the absolute logarithmic slope) with fits from $\omega = 4$ to $\omega = \omega_{max}$, the latter increasing from left to right in two spatial dimensions ($d = 2$). The lines from top to bottom are for outer scales Λ_c increasing from $2^{10}, 2^{12}, 2^{14}, 2^{16}$. The solid lines are for $\alpha = 1.8$, the dashed lines for $\alpha = 1.5$, all for two spatial dimensions. Over all the regression ranges, the standard deviations of the residuals are between $\pm$ 0.008 and $\pm$ 0.014 ($\alpha = 1.8$) and $\pm$ 0.011 to $\pm$ 0.019 ($\alpha = 1.5$).

one-dimensional Fourier transform of $t^{-(d+1-H)}$ does not converge at small t so that $t^{-(d+1-H)}$ truncated at $t = 1$ is completely dominated by the truncation scale details.

In summary, we obtain spectra which are very nearly scaling over 2–3 orders of magnitude in scale whose exponents are independent of C_1, H, only weakly dependent on α and weakly dependent on the overall range of scales of the regime (Λ_c).

10A.4 Further comparison with the data using structure functions

In Section 10.1.3 we gave some spectral and (x,t) cascade comparison between the low-frequency regime of the FIF model and 20CR 700 mb temperatures at 75° N. We briefly return to this now using both usual (difference) and tendency structure functions. In order to apply either of these structure functions to the data, it is first necessary to remove ("detrend") the strong periodicities at 1 year and 6 months. There are various ways of doing this; here we simply removed the corresponding individual Fourier modes and decreased them to the average level of the background spectrum. Fig. 10A.7 shows the resulting two structure functions. At small time scales, we see that the usual structure function increases with the theoretically expected exponent but that it levels off as expected in the spectral plateau regime where $H < 0$. By contrast, the tendency structure function (right-hand side) is constant over the small Δt weather regime where $H > 0$, but decreases systematically at longer and longer time scales, accurately following the theoretical climate regime exponent (Eqn. (10.8)). As can be seen, the agreement is excellent, especially considering that no parameter adjustments were made beyond the calibration by the overall standard deviation of the reanalysis temperatures at 75° N; the model is completely based on 1 Hz aircraft data. The tendency structure function directly shows the strong effect of temporal averaging/resolution on smoothing out the fluctuations.

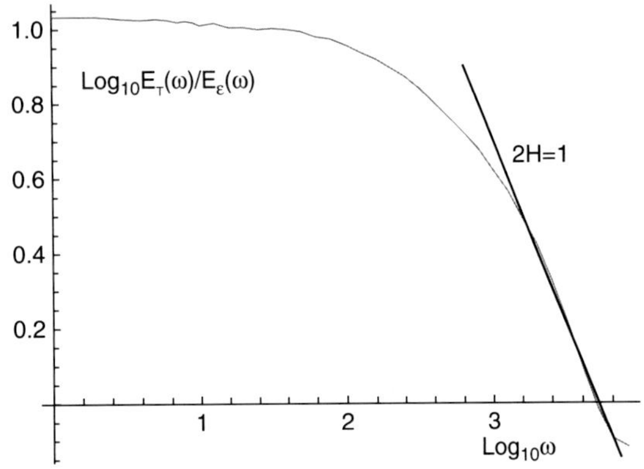

Fig. 10A.6 Using the same MonteCarlo realizations as in Fig. 10A.3, we show the ratio of the spectra of the fractionally integrated temperature field (*ET*) (which was obtained by fractionally integrating, order $H = 1/2$) the spectrum $E\varepsilon$ of the corresponding flux ε over 40 realization. Since the (single) spatial dimension was 2^6 (compared to 2^{14} for time), the transition scale is $\omega_w \approx 2^8$ (note, $\log_{10}2^8 = 2.4$). We see that for the low frequencies, the fractional integration is totally ineffective; the spectra are the same to within a constant factor.

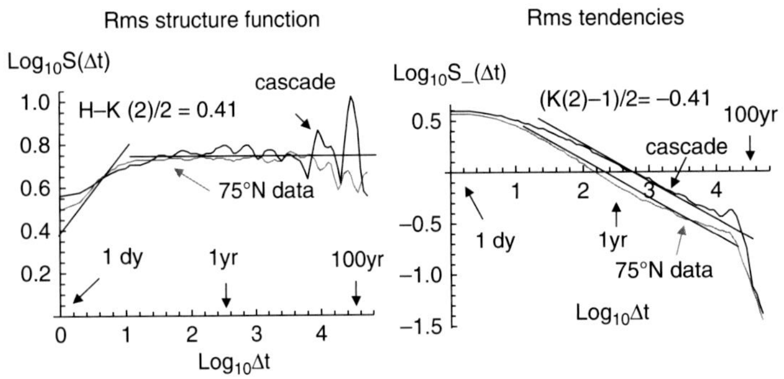

Fig. 10A.7 Comparison of the simulation (top) and data (bottom) for both the root mean square of the usual structure function, *S* (left) and of the tendency structure function, S_{-} (right); the analysis uses the daily data. The reference lines have the theoretical slopes indicated.

Appendix 10B: Stochastic linear forcing paradigm versus the fractionally integrated flux model

The stochastic scaling approach presented here turns out to be similar in many ways to another stochastic approach called the "stochastic linear forcing" (SLF) paradigm developed by Hasselmann (1976), Penland (1996), Newman *et al.* (2003), Sardeshmukh and Sura (2009) and others. The idea of the SLF approach is to use the weather/macroweather scale separation at τ_w to define rapidly varying fluctuations with respect to lower-frequency (e.g. weekly, monthly) averages and to exploit the short-range correlations of the resulting fluctuations, and the macroweather spectral "plateau" region with small β_{mw} (Hasselmann, 1976; Penland, 1996). SLF models, when interpreted as providing the most probable forecast, assume the plateau statistics to be nearly Gaussian white noise; however, Gaussianity is not required for most of SLF's diagnostic products, as long as the dynamical description is well approximated as linear. Today, for many purposes – including SST anomalies, diabatic heating rates and El Niño – SLF techniques are among the best available for forecasting (Penland and Sardeshmuhk, 1995; Newman *et al.*, 2003; Sardeshmuhk *et al.*, 2000). At present, the impact on SLF techniques of possible long-range correlations in the spectral plateau region, or deviations from Gaussian distributions, are not clear.

In Section 9.2 we recalled that inhomogeneous space-time partial differential equations can be solved by finding a propagator and convolving it with the forcing, and showed how to rewrite the FIF model as a solution to an inhomogeneous partial differential equation. Although the result was essentially formal (not directly useful in applications), this rewriting of the model is helpful in understanding the similarities and differences between the FIF and SLF approaches.

There are various ways to introduce the SLF approach. Perhaps the simplest is to first consider a scalar process $f(t)$ evolving in time according to the simple linear equation:

$$\frac{df}{dt} = -af + \sigma\eta \tag{10.46}$$

where η is a unit white "δ" correlated Gaussian noise:

$$\langle \eta(t)\eta(t')\rangle = \delta(t - t') \tag{10.47}$$

and σ is the amplitude. Taking a as a constant, f is then an Ornstein–Uhlenbeck process. The discrete time analogues are the popular AutoRegressive (AR) processes. Fourier transforming, we see that:

$$\widetilde{f}(\omega) = \frac{\sigma}{(i\omega + a)}\widetilde{\eta}(\omega) \tag{10.48}$$

The spectrum is obtained by multiplying the above by its complex conjugate and then ensemble averaging:

$$\langle \widetilde{f}(\omega)\widetilde{f}(\omega')\rangle = \delta(\omega + \omega')E_f(\omega); \quad E_f(\omega) = \frac{\sigma^2}{(\omega^2 + a^2)} \tag{10.49}$$

where for the forcing we have used $E_\eta(\omega) = 1$. From Eqn. (10.49) we see that if $a = 1/\tau_w$ is the characteristic frequency associated with the weather/macroweather transition, then the Ornstein–Uhlenbeck process has a spectrum fairly close to that of the atmospheric variables (see the fit to SST and air temperature spectra in Fig. 8.6d with $\beta_{lo} \approx 0.6$, $\beta_{mw} \approx 0.2$). In particular, for $\omega << 1/\ \tau_w$ the spectrum is close to the roughly constant spectral plateau: $E_f(\omega) \approx \frac{\sigma^2}{a^2}$, i.e. $\beta_{mw} = 0$, whereas for $\omega >> 1/\ \tau_w$ it is a power law $E_f(\omega) \approx \sigma^2\omega^{-\beta}$, i.e. with $\beta_w = 2$, which is indeed close to that observed for T, h spectral exponents (see Table 8.1). Although it is not quite right for the wind ($\beta_w \approx 5/3$) and other variables, this can be easily remedied by considering the fractional extension of the original equation:

$$\left(a+\frac{d}{dt}\right)^{\beta/2} f=\sigma\eta;\quad E_f(\omega)=\frac{\sigma^2}{(\omega^2+a^2)^{\beta/2}} \tag{10.50}$$

(we define the fractional differential operator via the inverse Fourier transform of $(a+i\omega)^{\beta/2}$. This is still linear in f and retains the flat plateau for $\omega << a$ but which modifies the exponent of the high-frequency regime so that β can be essentially arbitrary. Note that a less flattering comparison with data is obtained by considering the principal component of the SST whose spectrum for the Pacific Ocean (the Pacific Decadal Oscillation, PDO) is shown in Fig. 10.8 and which is quite far from this form (this is confirmed in the real-space analysis: Fig. 10.14). Perhaps even more difficult to account for in the SLF framework is the spectrum of the SOI index (the El Niño surrogate: Fig. 10.8), which makes a transition from $\beta \approx 1$ to $\beta \approx -0.3$ at around 1–2 years (i.e. at the critical ocean scale τ_o, not the weather scale τ_w).

The key feature of the linear stochastic differential equations (10.46), (10.50) is the separation of scales at $a = 1/\tau_w$. SLF approaches exploit the spectral plateau in order to model El Niño and other low-frequency phenomena. They generalize the scalar Ornstein–Uhlenbeck process by considering an N component state vector $\underline{I}(t)$ which is the solution of the vector extension of Eqn. (10.38):

$$\left(A+\frac{d}{dt}\right)^{H_{SLM}} \underline{I}=\underline{B}\eta;\quad H_{SLM}=1 \tag{10.51}$$

where the characteristic frequency a has been replaced by a characteristic matrix $A(t)$. In the simplest case, $\underline{B}$ is an N component vector and η is a scalar white noise. If needed, $\underline{B}$ could be an $N \times M$ component matrix and η an M component vector of white noises and an external forcing vector could also be added. We have written the equation in a slightly more general fractional form in order to compare it with the FIF approach. The state vector $\underline{I}$ is quite general; in practical applications, it is most efficient to take its components as the principal empirical orthogonal functions (EOFs) of sea-surface temperature, diabatic heating and possibly other fields (see e.g. Penland and Sardeshmuhk, 1995).

From the partial differential equation form of the FIF model (Eqns. (9.30), (9.31)) let us now write the FIF equations for a single Fourier component $\underline{k}$ of the generator:

$$\left(|\underline{k}|+\frac{d}{dt}\right)^{H_\alpha} \Gamma(\underline{k},t)=\gamma_\alpha(\underline{k},t) \tag{10.52}$$

and of the observable:

$$\left(|\underline{k}|+\frac{d}{dt}\right)^{H_I} I(\underline{k},t)=\varepsilon(\underline{k},t) \tag{10.53}$$

with $\varepsilon = e^{\Gamma}$ as usual.

If we keep only a finite number of the wavevector components we can write $I(\underline{k},t)$ in vector form: $(\underline{I}(t))_{\underline{k}} = I(\underline{k},t)$, we can write the flux as: $(\underline{\varepsilon}(t))_{\underline{k}} = \varepsilon(\underline{k},t)$ and we can write the (diagonal) matrix $\underline{A}$: $(\underline{A}(t))_{\underline{kk}} = |\underline{k}|$ so that:

$$\left(\underline{A}+\frac{d}{dt}\right)^{H_I} \underline{I}(t)=\underline{\varepsilon}(t) \tag{10.54}$$

The form of the FIF model (Eqn. (9.31)) and Eqn. (10.53) is the same as the fractional extension of the SLF model (Eqn. (10.51)) with the exception that the driving force is a highly intermittent (multifractal) turbulent flux ε rather than a (calm) Gaussian η.

In the weather regime, due to the scaling, fluctuations at time scale τ are dominated by spatial structures of scale $l = \tau^{3/2}\varepsilon^{1/2}$ (i.e. τ is the lifetime of size l structures). However, in forecasting there will be an inverse cascade of error growth (Section 9.3) so that an essential feature is that without a scale break, forecasts involve a wide range of spatial scales. If the whole continuum of wavenumbers is important then there is little to be gained by the SLF approach, but at low frequencies we can hope to exploit the weather/macroweather transition where the scaling is broken thanks to the finite size of the earth.

Specifically, we see that for frequencies $< 1/\tau_w$ we expect that only wavenumbers $|\underline{k}| \approx 1/L_w$ will be important, so that if the FIF is valid, then the matrix A will be composed of elements of order $1/L_w$ (depending on the exact choice of state vector components).

The SLF model presented in Eqn. (10.50) is based on correlated additive noise. Extensions to correlated additive and multiplicative (CAM) noise are straightforward and notably yield power-law probabilities for the components $\underline{I}$ (even though the forcing is still Gaussian; Sardeshmukh and Sura, 2009). For example if $a = a(t)$ and $\sigma = \sigma(t)$ then the simple CAM model for scalar f is:

$$\frac{df}{dt} = af + \sigma\eta f \tag{10.55}$$

where again η is a white noise. This is equivalent to the additive equation for the CAM generator:

$$\frac{d\Gamma_f}{dt} = a + \sigma\eta; \quad f = e^{\Gamma_f} \tag{10.56}$$

Which is very similar to Eqn. (10.51) for the FIF generator (if the latter has $H_\alpha = 1$). Due to the similarities of the FIF and SLF approaches, and the proven ability of the latter to make useful climate forecasts, one may reasonably hope that FIF-based models could be used advantageously for the same purpose.

Appendix 10C: A comparison of monthly surface temperature series

We have extensively used the Twentieth Century Reanalysis (20CR) products at 700 mb in order to characterize various atmospheric fields; the 700 mb level was chosen so as to be fairly representative of the free atmosphere, i.e. without the sensitivity of the surface fields to topography or coastal discontinuities. The 20CR dataset has several interesting characteristics including its coverage of a range of scales from 6 hours to 138 years (a factor $> 2 \times 10^5$), with no missing data and characteristics that are are relatively homogeneous over time. However, this advantage comes at a price: it is based purely on surface pressure and monthly sea surface temperature data, thus raising the question of how representative is it of the real atmosphere. Compo *et al.* (2011) give a partial answer by making specific comparisons with the National Centers for Environmental Prediction (NCEP) reanalysis products, but systematic statistical studies have not yet been made (see, however, Ferguson and Villarini, 2012). In order to further answer this question, in this appendix we will compare it to three surface temperature datasets, each with their own advantages and disadvantages.

The three we have chosen are the NOAA NCDC (National Climatic Data Center) merged land, air and sea surface temperature dataset (abbreviated NOAA NCDC below), from 1880 on a $5° \times 5°$ grid (see Smith *et al.*, 2008, for details), the NASA GISS (Goddard Institute for Space Studies) dataset (from 1880 on a $2° \times 2°$ (Hansen *et al.*, 2010) and the HadCRUT3 dataset (from 1850 to 2010 on a $5° \times 5°$ grid). HadCRUT3 is a merged product created out of the HadSST2 (Rayner *et al.*, 2006) Sea Surface Temperature (SST) dataset and its companion dataset of atmospheric temperatures over land, CRUTEM3 (Brohan *et al.*, 2006). Both the NOAA NCDC and the NASA GISS data were taken from http://www.esrl.noaa.gov/psd/; the others from http://www.cru.uea.ac.uk/cru/data/temperature/. The NOAA NCDC and NASA GISS are both heavily based on the Global Historical Climatology Network (Peterson and Vose, 1997), and have many similarities including the use of sophisticated statistical methods to smooth and reduce noise. In contrast, the HadCRUT3 data are less processed, with corresponding advantages and disadvantages. (Note added in proof: Haar analysis of the space-time densities of the HadCRUT3 measurements shows that they are sparse in both space and time with fractal codimensions of ≈ 0.25, 0.2 respectively.)

To avoid the problem of missing data (each pixel in each dataset suffered from this problem), in this appendix we only consider the globally averaged series obtained by averaging over all the available data over the common 129-year period 1880–2008 (taking into account the latitude-dependent map factors; see Section 8.1.4 for further characterizations, especially for the land–SST differences). The resulting series are shown in Fig. 10C.1a; we can see that – as expected – the series are quite similar, and the 20CR series is not much different from the others. Before looking at the similarities and differences scale by scale, we can consider the "bulk" statistical characterizations of the series (with their means removed) (Table 10C.1). Along the diagonal we see that the series have quite similar characteristics: the amplitudes of the variation around a zero mean are $\approx \pm\ 0.3$ K with the 20CR a little higher – this is not surprising: the atmosphere at 700 mb is a bit more variable than at the surface. Along the off-diagonals we see the typical inter-series differences; for the surface series, these are all around $\pm\ 0.1$ K, noticeably smaller than for the surface–20CR differences, which are around $\pm\ 0.27$ K. Since the 20CR series are not at the surface the larger difference is not surprising; however, the differences between the surface series is surprisingly large given that there has

Table 10C.1 Some of the bulk statistical characteristics of the series after their means have been removed. The off-diagonal elements are the standard deviations of differences and the diagonal elements are the standard deviations of the corresponding series. These are for monthly, global values; units = K.

	NOAA NCDC	NASA GISS	HadCRUT3	20CR
NOAA NCDC	0.328	0.107	0.083	0.275
NASA GISS	0.107	0.284	0.099	0.262
HadCRUT3	0.083	0.099	0.304	0.275
20CR	0.275	0.262	0.275	0.382

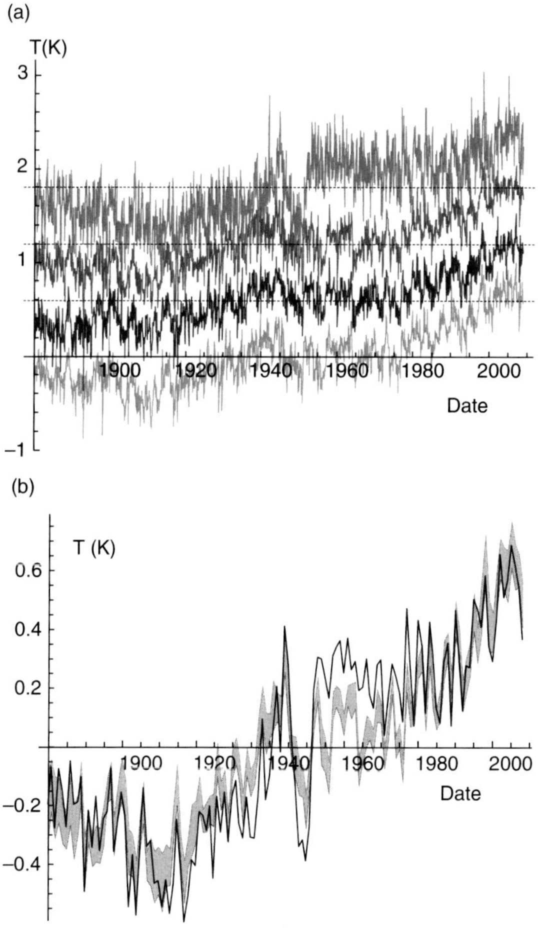

Fig. 10C.1 (a) The monthly series over the common part of their domains: 1880–2008 (129 years), bottom to top: NOAA NCDC, NASA GISS, HadCRUT3, 20CR. Each series had its mean removed and then was displaced by 0.6 K for clarity; the dashed lines are the displaced axes. (b) Annual averages from Fig 10C.1a. The grey line of variable thickness indicates the mean of the monthly resolution one-standard-deviation spreads of the three surface series; the thin dark line is the 20CR series.

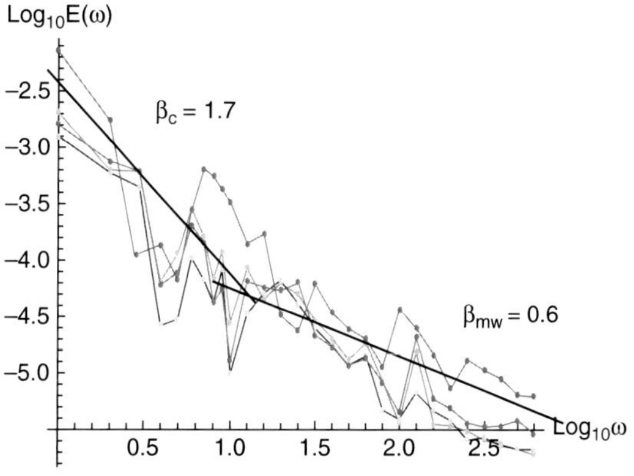

Fig. 10C.2 The spectra (averaged over logarithmically spaced bins, 10 per order of magnitude). The units are such that $\omega = 1$ corresponds to (129 yrs)$^{-1}$; note the annual spike, (1 year)$^{-1}$ is at 2.11 on the $\log_{10}\omega$ axis).

been a huge amount of spatial and temporal averaging. A slightly different comparison is given in Fig. 10C.1b, which shows the annual averages of the previous. Here, the three surface series have been used to estimate the standard deviation at 1-month resolutions; the shaded area shows the annual mean one-standard-deviation limits. We can see that the 20CR series is mostly within or close to the bounds, with the notable exception of the period ~1920–1945 (where it is mostly below) and 1945–1970 (where it is above).

Let us now consider the scale by scale properties; Fig. 10C.2 shows the spectra. These are quite similar, showing roughly $\beta_{mw} \approx 0.6$ up to ≈ 10 years, and for $\omega <$ (10 years)$^{-1}$, roughly $\beta_c = 1.7$, although clearly there is too much scatter at these low frequencies for precise exponent estimates. It is now interesting to consider both the signed and absolute difference spectra; for both the surface series (Fig. 10C.3a) and between the 20CR and the surface series

Table 10C.2 Comparison of the parameters for the monthly global averages (fits up to 1 year), along with the global annual structure functions (fitted for $\Delta t > 10$ years).

	C_1	α	H	β	Outer scale (years)	Annual RMS structure ($\Delta t > 10$ years)
NOAA NCDC	0.116	1.72	−0.12	0.69	1.6	$0.10\Delta t^{0.29}$
NASA GISS	0.123	1.66	−0.13	0.62	1.0	$0.05\Delta t^{0.47}$
HadCRUT3	0.124	1.71	−0.13	0.61	1.6	$0.12\Delta t^{0.21}$
20CR	0.116	1.58	−0.07	0.66	0.8	$0.09\Delta t^{0.38}$
Ensemble surface	0.114	1.71	−0.13	0.64	1.6	$0.08\Delta t^{0.33}$
Ensemble surface differences	0.108	1.66	−0.16	0.50	1.3	

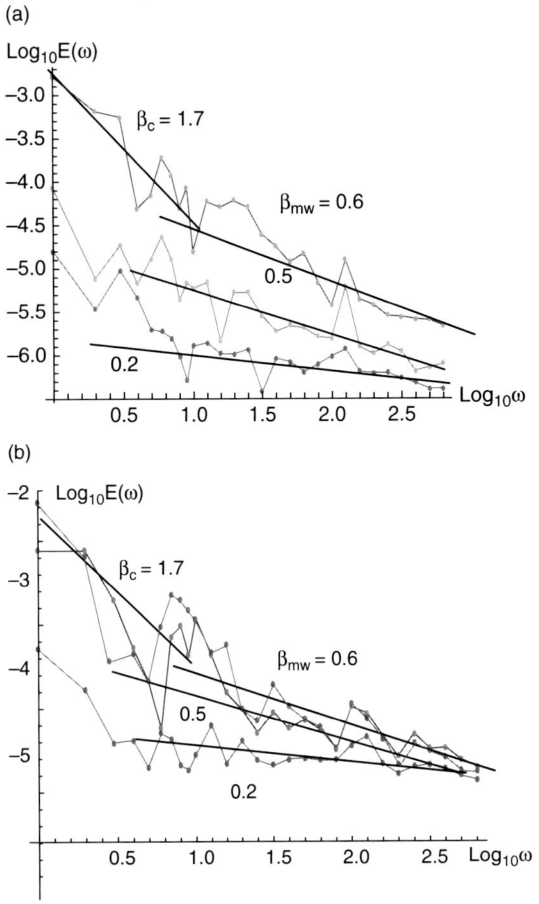

Fig. 10C.3 (a) Ensemble spectra for the three surface series (top); the ensemble spectra of the same series of the three (signed) differences (middle) and the ensemble spectra of the same series using the three absolute differences (bottom). Notice that the differences are not white noise (flat), but are themselves scaling with exponents only a little lower than for the series themselves. (b) Analogous to Fig. 10C.3a, but for the 20CR spectrum (top) and the spectrum of the ensemble of differences with respect to the three surface spectra (middle), and absolute differences (bottom). The basic behaviours are the same as for the three surface series although the differences are larger.

(Fig. 10C.3b). In both cases, the differences are *not* flat white noises, they are themselves scaling with nearly the same exponents as for the original series themselves: see Table 10C.2 for exponents and outer scale of the differences. The absolute differences which characterize the "flux of differences" are a little flatter (lower β), but they are still not white noises. This indicates that there are long-range dependencies in differences between the series; it is hard to avoid the conclusion that there must therefore be long-range dependencies in the *biases* between the series (both the surface and 20CR series). This implies spurious tendencies at all scales.

Finally, we can consider the trace moment analysis of the series (Fig. 10C.4). Since the meteorological variability has been filtered out by the monthly and global averaging, as discussed in Section 8.1.4, up to 1–2 years this reflects the ocean variability with critical transition time $\tau_o \approx 1$ year. Again, the results are very similar from one series to another, with the 20CR series being typical although with a slightly smaller outer scale (as expected, since it is less influenced by the SST cascade). Quantitative comparisons can be found in Table 10C.2. We can see that the multifractal parameters are remarkably stable – with for example C_1 varying by less than 0.01 for all four

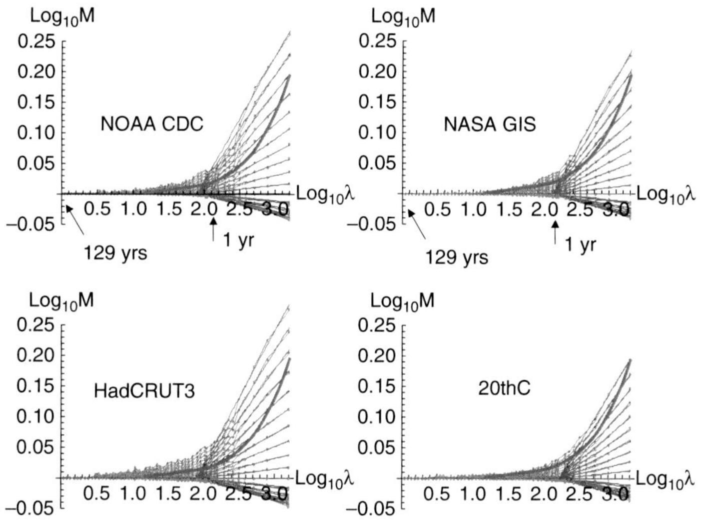

Fig. 10C.4 Trace moments based on the absolute second differences of the globally averaged monthly series. The corresponding parameters are in Table 10C.2. The thick curves indicate the $q = 2$ envelopes of quasi-Gaussian processes (Appendix 4A). We can see that the 20CR data are close to quasi-Gaussian (although the scaling seems quite good at high λ, presumably due to the ocean cascade), the HadCRUT3 data are reasonably more variable, and the other datasets are in between.

products – and H varying only over a range of 0.06. Interestingly, the differences also display nearly identical cascade structures as the series themselves; the bottom line in Table 10C.2 indicates the parameters. Also shown in the table are the root mean square structure functions (i.e. the square root of the $q = 2$ structure functions); these were estimated for periods $\Delta t > 10$ years, i.e. the climate regime (see Section 10.2.3); again, although there is a lot of scatter, the 20CR regression is quite close to the others, supporting its use in understanding the climate-scale variability.

Appendix 10D: Coupled ocean–atmosphere modelling

10D.1 A coupled ocean–atmosphere interface model and the ocean plateau

From the analysis of sea surface and land temperatures at 5° × 5° and 1-month resolution (Fig. 8.3d, Section 8.1.4, Fig. 10.14) we have seen that the ocean is very similar to the atmosphere (including for T roughly the same parameters: $\alpha \approx 1.8$, $C_1 \approx 0.1$, $H \approx 0.5$), the major difference being that the ocean velocities are about 30 times lower, corresponding to energy fluxes about 10^5 times lower and an outer time scale τ_o about 30 times larger. Although the ocean and atmosphere both show spectral plateaus at the expected time scales, the low-frequency ocean exponent β_{lo} was 0.6, i.e. a bit steeper than the corresponding atmospheric β_{mw}, which was typically in the range 0.2–0.4. Following the analysis in Appendix 10A we should perhaps not be surprised, since β_{mw} was not a true asymptotic scaling exponent but only a "pseudo-scaling" exponent depending on the overall scale range and perhaps other low-frequency details. Since the range of scales between τ_c and τ_o is about 30 times smaller than between τ_c and τ_w, we expect that β_{lo} is somewhat higher than β_{mw}; according to Fig. 10A.6, it could easily be 0.5–0.6. However, there is another effect which tends to increase the ocean surface variability and hence β_{lo} – the fact that the latter is coupled to the atmosphere. However, it turns out to be fairly simple to make a model that approximates to this coupling and resultant variability. In this section we outline a simple coupled ocean–atmosphere model that reproduces this somewhat steeper spectral ocean plateau as a consequence of the double ocean and atmosphere dimensional transitions.

For motivation, recall that the currents and winds are physically coupled, and hence it is natural to suppose that the flux controlling the sea surface temperature depends on both; the simplest model is the product of an uncoupled atmospheric flux φ_a with an uncoupled (i.e. statistically independent) oceanic flux φ_o. The idea is that these primarily represent the result of internal atmospheric and oceanic dynamics, but that the surface flux relevant for the temperature at the interface is the product of the two:

$$\varphi_s = \varphi_a \varphi_o \tag{10.57}$$

Using the product flux φ_s, the surface temperature T_s is given by the usual FIF model:

$$T_s = \varphi_s * [\![\Delta \underline{R}]\!]^{-(d-H)} \tag{10.58}$$

where $\underline{R} = (x,y,t)$, $d = 3$, $H \approx 0.5$. In order to roughly reproduce the empirical trace moments and spectra, we used the values $\alpha = 1.8$, $C_1 = 0.1$, $H = 0.5$, in accord with the analyses in Chapter 8 for both φ_a and φ_o (see Figs. 8.6d, 8.6e and Table 8.2). In this simple model, we assume that the oceanic and atmospheric circulations and hence φ_a, φ_o are statistically independent; however, both modulate the flux relevant to the surface temperatures so that the latter is nevertheless "coupled" to both.

If the cascades controlling each of the fluxes acted over identical scale ranges λ, then the statistical moments would yield $\langle \varphi_s^q \rangle = \langle \varphi_a^q \rangle \langle \varphi_o^q \rangle$, which would imply $\lambda^{K_s(q)} = \lambda^{K_a(q)} \lambda^{K_o(q)}$ and hence $K_s(q) = K_a(q) + K_o(q)$. In the example considered, with $\alpha_a = \alpha_o = \alpha$, $C_{1a} = C_{1o} = C_1$, we would obtain $C_{1s} = 2\, C_1$, $\alpha_s = \alpha$, which would imply that the intermittency over the ocean had double the C_1 of that over land – whereas according to Table 8.2 it is almost unchanged. More generally, if the ocean and atmosphere high-frequency ("weather") regimes operated over scale ranges that overlap significantly, then the corresponding intermittency would be too high to be compatible with the observations. However, this is only true if the overlap occurs over scale ranges which are unaffected by the dimensional transitions, and because of the large difference in planetary scale velocities $\lambda_{wo} = V_w / V_o \approx 30$ (see Section 8.1.4) this must be carefully considered. In particular, the observed difference in velocities implies that at the weather scale τ_w – corresponding to atmospheric structures of planetary extent (L_e = 20 000 km) – the ocean cascade has developed over a range $\tau_o/\tau_w = \lambda_w \approx 30$, i.e. down to structures/eddies of size $L_e/\lambda_w = 670$ km.

Although the cascade presumably starts at $\tau_c >> \tau_o > \tau_w$, the conclusion of Section 10.1.2 on the dimensional transition and the spectral plateau is that the cascade only fully develops its intermittency over ranges where both temporal and spatial degrees of freedom contribute to the generator Γ (Eqns. (10.1), (10.2)). Therefore, we need only avoid overlapping cascade ranges that give these full contributions to the intermittency. A simple model that ensures that the overlap is only over these weakly intermittent ranges is to simulate φ_o, φ_a from time scales τ_c down to τ_w and over spatial scales from L_e down to L_e/λ_w and then to couple φ_o and φ_a at that scale; T_s is then obtained by fractional integration as in Eqn. (10.58). The physical significance of coupling the fluxes in this way is that the ocean surface temperature is effectively modulated by the mean atmospheric flux, averaged both globally and over the lifetime of the planetary-scale atmospheric structures.

Fig. 10D.1 shows the spectrum of the resulting coupled model; we see that it fits the observations very well. At the high frequencies, it follows the theoretical spectrum $\beta = 1 + 2H - K(2) \approx 1.82$ (i.e. with $K(2) = 0.18$; an overlap of the cascade range would increase $K(2)$ to 0.37 and hence $\beta \approx 1.63$), whereas at the lower frequencies it follows $\beta_{lo} \approx 0.6$ reasonably well. Presumably, it is a variant on the pseudo-scaling discussed in Appendix 10 A, although in this case the consequence of the weak intermittency associated with both the ocean and atmospheric dimensional transitions, both acting at frequencies $< \omega_o \approx (1\ \text{year})^{-1}$. We might also note that a trace moment analysis yields $C_1 \approx 0.13$, $\alpha \approx 1.95$, $\tau_{eff} \approx 5$ years, which is quite close to the data (Table 8.2) but with C_1 and α a little larger than the theoretical values. The outer scale is also somewhat too large (recall $\tau_o = 1$ year), but we have already noted a tendency for the spectral break scale to be a somewhat shorter period than the estimated outer cascade scales. As for the pure FIF dimensional transition model in Section 10.1.3, no attempt has been made to obtain optimum parameter estimates or to introduce ad hoc changes to improve the fit with the data. The key point is that the "double dimensional transition," i.e. both ocean and atmosphere, can potentially explain the value $\beta_{lo} \approx 0.6$.

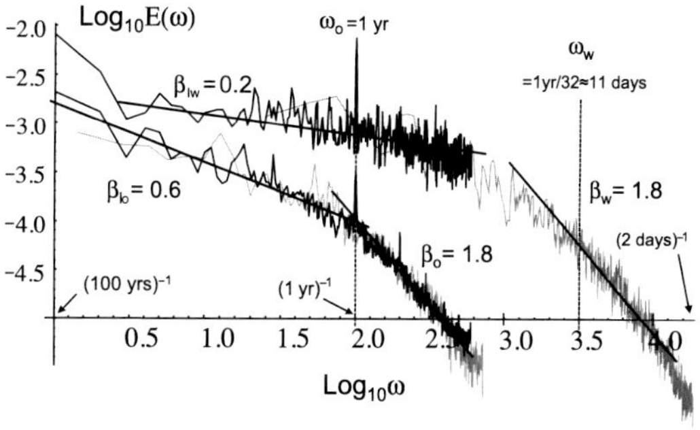

Fig. 10D.1 A comparison of SST and land surface temperatures (from the monthly NOAA NCDC data discussed in Section 8.1.4) with multifractal simulations in time and in one horizontal spatial dimension. The far right spectrum (grey) is the result of 10 simulations of an atmospheric model with $\tau_w = 11$ days, $\alpha = 1.8$, $C_1 = 0.1$, $H = 0.5$ simulated over the range 4 years/$2^{11} \approx 1$ day and $L_w/2^4 \approx 1000$ km (more precisely, this is the $a = 5/6$ land model described in the text; i.e. the top curve from Fig. 10D.2a). The thick spectrum (lower left) is from 10 realizations of the coupled ocean–atmosphere model described in the text with the same exponents, and with $\tau_o = 1$ year, simulated over grids with temporal resolution 64 years/$2^{13} \approx 3$ days and spatial resolution $L_w/2^2 \approx 5000$ km. The high-frequency reference slopes are roughly the theory value:
$\beta = 1 + 2H - K(2) = 1.82$.

10D.2 A simple model of maritime and continental temperature regimes

When averaged over scales of a month, the sea surface temperature model discussed in the preceding section yielded both a realistic spectrum (Fig. 10D.1) and a realistic cascade structure with outer scale corresponding to τ_o, i.e. to the ocean cascade. However, all meteorological (e.g. daily) resolution data show evidence for the meteorological scale cascades, i.e. with outer scales of the order of at most a few times τ_w (see Sections 8.1.5, 8.1.6), and this for all latitudes (although with some variation: see Fig. 8.8c). The presence of this high-frequency cascade intermittency – even over latitudes strongly dominated by oceans – can be understood if there is always a purely atmospheric contribution to the temperature variability. Similarly, the statistics of monthly averaged temperatures – even over land – show cascades with outer scales of the order of τ_o (Fig. 8.6 e, Table 8.2), indicating the ubiquitous influence of oceanic-induced intermittency. In this section, we show that a simple model can yield the observed behaviours – even when only fairly small oceanic or land components are present (i.e. maritime or continental regimes).

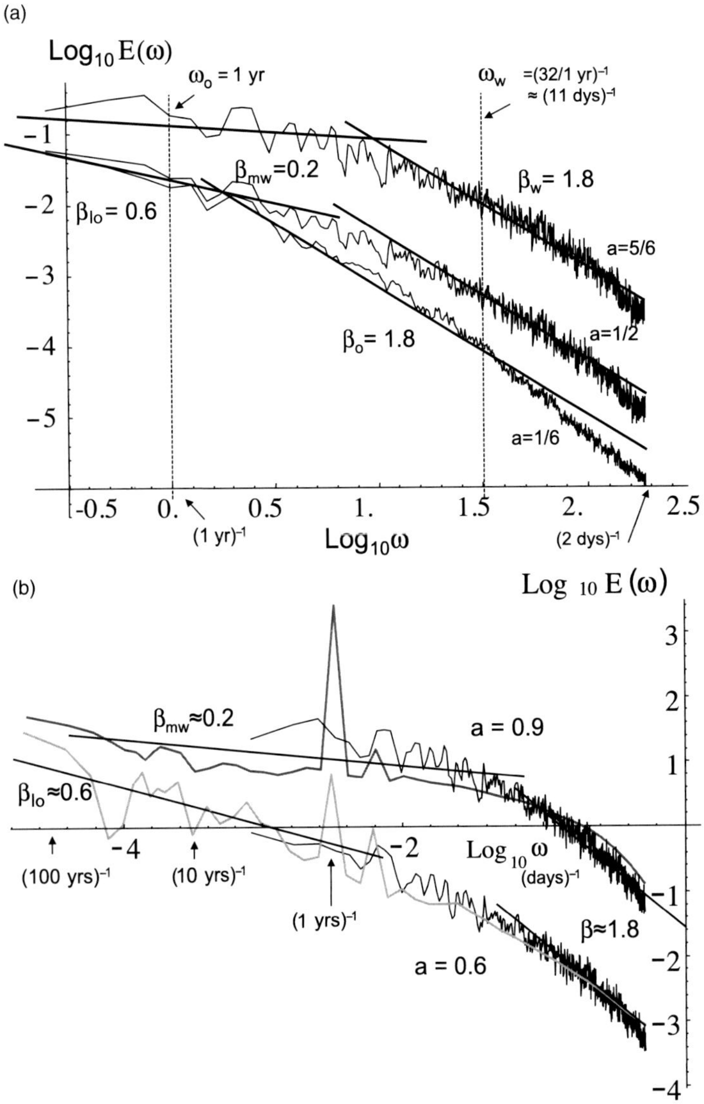

Fig. 10D.2 (a) Spectra from 10 realizations of the linear combination of ocean and atmosphere models, with the latter weighted by the coefficient a indicated next to the spectra. $a \approx 5/6$ is close to the land, SST observations in Fig. 8.6 d. (b) Comparison of the 20CR spectra of the 700 mb temperature field (daily resolution to 138 years) at 5° N (thick grey, bottom) and at 45°N (thick grey, top), both averaged over 10 bins per order of magnitude in frequency with the mixed maritime–continental model discussed in the text, with weights $a = 0.6$ (thin, bottom right) and $a = 0.9$, (thin, top right).

Consider a linear combination of a nondimensional temperature T_s from the ocean cascade model of the previous section and T_a from a classical pure atmospheric cascade (FIF model); each can be normalized to have the same variability at the highest resolution: $\langle \Delta T_a^2 \rangle = \langle \Delta T_s^2 \rangle$. We now combine them with a weight a:

$$T(x, y, t) = aT_a(x, y, t) + (1 - a)T_s(x, y, t) \qquad (10.59)$$

When a is zero, the spectrum is of the maritime shape, e.g. as in the simulation in Fig. 10D.1. When $a = 1$, it is the pure atmosphere model, and Fig. 10D.2a shows the result. Fig. 10D.2b shows a comparison of the model with data when the 5° N 20CR spectrum is used as a proxy for pure maritime spectra (see Fig. 8.8a; about 70% ocean, 15% land, 15% "coastal") and when the 45° N spectrum is used as a proxy for continental spectrum (40% land, 30% ocean, 30% "coast"). In order to compare the statistical properties of this mixed maritime–continent model with the empirical spectra, we can multiply the nondimensional T by an empirical amplitude

Table 10D.1 Parameters for the linear combination of (*x*, *t*) space ocean and atmosphere models from the analyses in Fig. 10D.3. Both have $C_1 = 0.1$, $\alpha = 1.8$, $H = 0.5$. The ocean model had 2^{11} (time) × 2^9 (space) pixels with outer simulation scale $= 4\tau_o = 4$ years, inner simulation time scale 4 years/2^{11} ≈ 0.7 days; in space the outer scale is $L_e = 20000$ km, the minimum scale is $L_e/2^9 \approx 40$ km. The atmospheric model was 2^{11} (time) × 2^4 (space) pixels; the duration was the same (= 4 years) with $\tau_w = \tau_o/32 \approx 11$ days, the spatial resolution was $L_e/2^4 \approx$ 1250 km. The spatial resolution of the ocean model was then degraded by factor 2^5 so that the two models would have the same resolutions. The parameter *a* is the weight of the atmosphere model, i.e. the temperature is $T = aT_a + (1 - a)T_s$. As the table shows, as long as the parameter *a* is not too small, the basic parameters of the system at full resolution and degraded by 32 resolution are roughly correct, although the outer scales are a little too big (they should be 11 days and 1 year for τ_w, τ_o respectively).

a	C_1	α	Outer time scales
1/6 Daily	0.11	1.92	$\tau_w = 58$ days
Monthly	0.11	2.01	$\tau_o = 5$ years
1/2 Daily	0.10	1.93	$\tau_w = 46$ days
Monthly	0.085	1.98	$\tau_o = 1.6$ years
5/6 Daily	0.12	1.80	$\tau_w = 36$ days
Monthly	0.09	1.94	$\tau_o = 1$ year

(e.g. the typical empirical standard deviation of fluctuations at the smallest data resolution); this is equivalent to moving the spectra up and down on a log-log plot. We see that the model does a reasonable job of reproducing the different spectra shapes over the entire range of 1 day to the external model scale (however, from the previous sections,we know that at lower frequencies the model will continue to follow the $\beta_{mw} \approx 0.2$ and $\beta_{lo} \approx 0.6$ curves; there was no need for larger-scale simulations). In the figure, the coefficient *a* was estimated by hand; the aim was simply to demonstrate that the shape of the observed spectra falls in between the pure coupled ocean– atmosphere model (Section 10.3.1) and a pure atmosphere model (Section 10.1.3) and can be reasonably modelled by a combination of the two.

The final test of this mixed model is to show that we at least roughly recover the cascade structure and parameters of both the atmosphere and ocean using daily and monthly averaged data, i.e. that at daily resolution the oceanic intermittency is negligible whereas at monthly resolution atmospheric intermittency is sufficiently averaged out so that the ocean cascade structure can be recovered. Fig. 10D.3 illustrates this on the simulations and Table 10D.1 shows

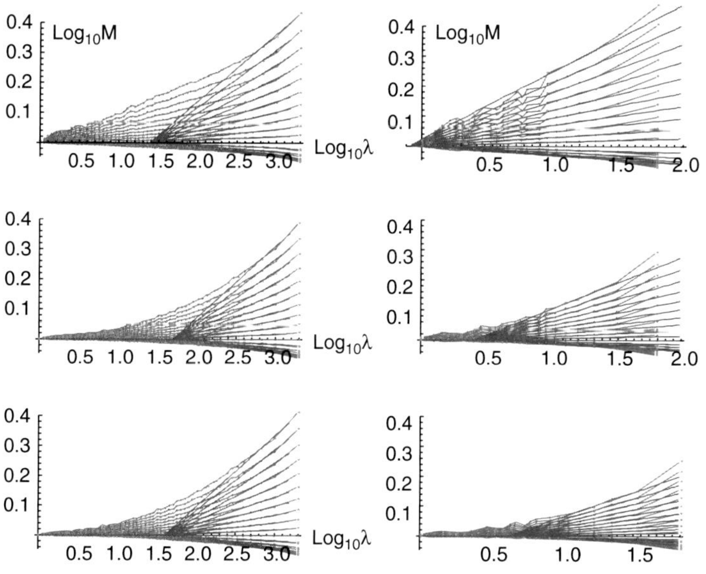

Fig. 10D.3 Trace moment analyses of the linear ocean–atmosphere models with coefficient *a* increasing from *a* = 1/6 (top) to *a* = 1/2 (middle) to *a* = 5/6 (bottom). The left column shows the analysis at the full (~ daily) resolution, the right at the equivalent of monthly resolution (degraded by a factor of 32 in time). The parameters for the regressions are given in Table 10D.1, where it can be seen that the intermittency parameters and outer scales of the ocean and atmosphere can be reasonably well estimated from the analyses at the different resolutions.

the parameter estimates. Consider first the daily resolution data (left column): in the top graph ($a = 1/6$) we see that the range of atmospheric cascades is somewhat limited due to the strong oceanic intermittency, but it becomes more dominant as a increases to 5/6 (bottom). These graphs should be compared to those at monthly scales (Fig. 8.6e, bottom row), and at daily scales (e.g. Figs. 8.7a, 8.7b). At 1-month resolution (Fig. 10D.3, right column), we see the converse: that the cascade with roughly yearly outer scale is diminished in strength as a increases from 1/6 to 5/6. These qualitative changes are confirmed quantitatively in Table 10D.1, which shows that the basic exponents C_1, α only suffer from a small bias, although the outer scales τ_w, τ_o are more sensitive to a. Presumably, this mixed maritime–continent model is only an approximation to the real atmosphere; however, it does show that simple mixtures of this type can go a long way not only towards explaining the variability over wide ranges of time scales, but also in explaining geographical variations in parameters (see e.g. Section 8.1.6).

Chapter 11 The climate

11.1 Multidecadal to multimillennial scaling: instruments and multiproxies

11.1.1 A diversity of approaches to climate dynamics

Over the last 50 years, climate science has evolved from a largely descriptive branch of geography into a vibrant and quantitative atmospheric science. At first, this evolution was driven by new paleo- and satellite data and by the development of climate models of various types. The data from ice and ocean cores in particular showed unequivocally that over the last 2–3 million years – the Pleistocene – there had been numerous quasi-periodic glacial/interglacial transitions – the ice ages – while satellite data from other planets underlined climate diversity and complexity. While the original "General Circulation Models" (GCMs) were too crude to be very realistic, even early on they were used to test the notion of anthropogenic global warming (Manabe and Wetherald, 1975).

More important for this chapter were the first long ice cores (Dansgaard *et al.*, 1969; Hays *et al.*, 1976). These demonstrated that climate variations were anything but smooth and regular. As the resolutions of the paleodata improved, it became increasingly clear that the climate involved fluctuations over huge ranges of scale and included drastic and sudden ("abrupt") transitions (see Figs. 11.11, 11.12 and Dansgaard, 2004, for a historical account). It is therefore not surprising that in the early 1980s, in the midst of the nonlinear revolution, highly simplified, low-dimensional chaotic energy balance and other climate models were developed (including the famous "Daisyworld": Watson and Lovelock, 1983) that graphically demonstrated how the evolution of life itself can alter the climate through negative feedback). This was the beginning of the (initially) deterministic Dynamical Systems approach to climate modelling (e.g. Ghil, 1981). Although this approach continues, it has recently begun to mutate into a rather different (stochastic) Random Dynamical Systems approach (e.g. Chekroun *et al.*, 2010). As discussed in Chapter 2, the early 1980s also saw the development of scaling, fractal and multifractal approaches to the atmosphere, and this included an early attempt to put weather and climate variability into a scaling framework (Lovejoy and Schertzer, 1984; 1986).

But in the last 30 years the real driver of climate science has undoubtedly been social and political: the spectre of global warming and the implications of climate change – whether anthropogenic or otherwise. In order to understand and to assess the extent of the warming, and to predict future trends, massive resources were committed to the development of ever more complex GCMs: a term which over time has come to mean "global climate models." In order to be able to feed the models with the requisite large quantities of data, increasingly sophisticated statistical treatments were required, but these were typically single-scale techniques – such as 3D or 4D variational data assimilation techniques or principal component analyses. They were rarely accompanied by even traditional spectral techniques that would have provided knowledge of the wide dynamically significant range of scales and the corresponding statistics. In any case, even in the relatively rare instances when scaling analyses have been used, they often had quite limited objectives – such as demonstrating the efficacy of novel analysis techniques – and they consequently generally led to isolated results. This was well illustrated in Chapter 10 where – with one or two partial exceptions – the application of scaling analyses failed to even lead to a clear recognition of the fundamental distinction between macroweather and the climate. Below, this same tendency is apparent in the analysis of ice-core data, where even the existence of several scaling analyses has been insufficient to generate a coherent picture of the corresponding climate scaling.

For the reasons outlined in Chapters 1, 2 and 10, here we explore the scaling approach to the climate regime while recognizing at the outset the frustrating paucity of our knowledge. Yet today this poverty is not so much a consequence of lack of data or of computer or even human resources, but rather a consequence of the nearly exclusive focus on traditional GCM models. Whereas massive resources have been devoted to GCMs, scaling approaches to climate – including those described below – have in general been virtually unfunded. Their current underdeveloped status is therefore hardly surprising, yet, as we show, their implications are quite exciting and promise to put climate science on stronger empirical and theoretical basis and to provide new – more rigorous since physically and more directly empirically based – statistical means for assessing anthropogenic effects (e.g. by using knowledge of the statistical structure to construct classical statistical tests).

Whereas in weather – and even macroweather – there was a theoretical connection (however imperfect) with the usual dynamical equations, in the climate regime – by (our) definition – new forcings and/or new slow dynamics are important and this connection is weaker. However, it is still reasonable to assume that certain basic aspects of the weather/macroweather dynamics carry over to the climate: for example, that climate dynamics involve large (not small) numbers of degrees of freedom – hence justifying stochastic approaches – but also that whatever the exact dynamics are, they are strongly nonlinear, acting over a wide range of scales, and hence presumably scaling. In addition, they are quite possibly most efficient when acting between structures of neighbouring scales – and hence have cascade phenomenologies with multifractal statistics. Indeed, in Section 10.3 we showed how the spatial structure of the macroweather regime could be understood and modelled by generalizing the multiplicative scaling flux approach into the climate regime. Therefore, in exploring the climate regime we still have a plausible theoretical framework, even if it is less solid.

11.1.2 The relation between macroweather and the climate

In Chapter 10, we argued that it is misleading to consider that there is a direct transition from weather to climate at $\tau_w \approx 10$ days, that the real transition is to an intermediate qualitatively different lower-frequency "macroweather" regime and that what should properly be called the "climate" only begins at a critical scale of $\tau_c \approx 10$–30 years. Whereas for scales either below τ_w or beyond τ_c fluctuations tend to grow with scale ($H > 0$), in macroweather, for scales between τ_w and τ_c, they decrease with scale ($H < 0$). Based on our discussion of the qualitatively comparable ($H > 0$) weather regime, the natural explanation is the existence of a new mechanism of internal variability involving nonlinearly coupled spatial degreees of freedom, each with a lifetime that grows with scale. Possible candidate mechanisms which are not currently part of GCM systems or are poorly modeled are land ice and deep ocean currents, whose spatial extents fluctuate over wide ranges of spatial scales, and perhaps over the necessary range of time scales.

If the basic cascade model which reproduces the weather-regime statistics is assumed to hold up to the climate scale τ_c, then the basic consequence – which we verified in Chapter 10 – is that there is a drastic weather/macroweather change ("dimensional transition") in the temporal statistics at τ_w: as a consequence, the weather cascade fluxes approximately factorize into separate and statistically independent weather and macroweather processes: $\varepsilon_{w,mw}(\underline{r}, t) = \varepsilon_w(r, t)\varepsilon_{mw}(t)$ (Eqn. (10.4)). In Chapter 10, the predictions of this equation in the temporal domain were given copious support, yet consideration of the spatial statistical structure led us rapidly to the conclusion that this equation cannot be the full story. This is because it predicts that if we temporally average $\varepsilon_{w,mw}(\underline{r}, t)$ (and hence the observables) over scales $> \tau_w$ this would completely smooth out the *spatial* variability as well – and this contradicts the strong spatial multifractal variability observed as climate zones.

A model with realistic spatial statistics was, however, easy to achieve. All we needed was to include a further lower-frequency space-time factor $\varepsilon_c(\underline{r}, t)$ so that the full weather/climate model becomes $\varepsilon_{w,c}(\underline{r}, t) = \varepsilon_w(\underline{r}, t)\varepsilon_{mw}(t)\varepsilon_c(\underline{r}, t)$. The implication of this augmented model for the macroweather regime is that the latter will inherit the strong (multifractal) climate-scale spatial variability of $\varepsilon_c(\underline{r}, t)$ – and this in spite of the fact that temporal intermittency of macroweather is small. Using this simple model of space-time atmospheric variability we were therefore able to predict the statistics of fluctuations as well as the power spectra (Section 10.3). Finally, we indicated how the model could be used to define a climate state as the average of meteorological variables over scales up to τ_c (climate "normals"). "Climate change" then has a precise meaning as the variation of this minimally variable state at still longer time scales. The model predicts that this state varies strongly in both space and at (longer) times.

In this chapter, we focus on the "climate" defined as the variability at scales greater than τ_c up to a critical "low-frequency climate" scale $\tau_{lc} \approx 100$ kyr, where we find that the pattern repeats with H again apparently becoming negative at longer times. Since instrumental records only cover the beginning of this climate regime, we are forced to rely heavily on various surrogates. While these are problematic, they are detailed enough to give us a fairly coherent (if incomplete) picture of the temporal statistics of the climate regime (the spatial statistics require large numbers of geographically distributed paleotemperatures, and this is challenging). However, if we attempt to go further into the low-frequency climate regime at scales of several hundred thousand years, the quantity and quality of the relevant surrogates is very small: the low-frequency climate regime is therefore out of our scope. (Note added in proof: the 800 kyr EPICA Antarctic core yields $H_{lc} \approx -0.8$.)

The chapter is organized as follows. We first consider the relevant instrumental data, even though they do not allow us to go much beyond multidecadal variability. To understand the multicentennial variability, we require "multiproxy" surrogates, and these are discussed in Section 11.1.5, where the fluctuation statistics are analyzed primarily using (Haar) structure functions. Multimillennial and longer-period variability requires ice and ocean core paleotemperatures; these combined with all the other analyses (including over the weather scales) allow us to construct a convincing three-scaling-regime composite picture of atmospheric variability spanning the (turbulent) weather regime out to $\approx$ 100 kyr and a more general model for the corresponding space-time variability (Section 11.2). In Section 11.3, we discuss possible causes of the climate variability, and examine the variability of external forcings (solar, volcanic and orbital), focusing on their (neglected) scale dependence and implied (stochastic) scaling climate sensitivities. This allows us to consider the variability of unforced and forced GCMs and to attempt to answer the question as to whether current GCMs predict the climate or rather macroweather. In Section 11.4 we conclude with a summary and a statement of the fundamental space-time atmospheric statistics covering the weather, macroweather and climate regimes.

11.1.3 Epoch-to-epoch climate variability and the Holocene exception

Study of the climate regime requires indirect longer-time-scale sources of data so as to clearly discern the multidecadal, multicentennial and multimillennial variability. The problem is complex for several reasons. First, because beyond the strong spatial multifractal geographical variability associated with climate zones (Section 10.3), the climate state and the critical transition scale τ_c also have significant epoch-to-epoch variability; second, because proxy data sources have nontrivial problems of interpretation; and third, because during the last century or so the natural variability has been perturbed by anthropogenic effects, and this can bias our results. In the next two subsections we consider the epoch-to-epoch variations and then revisit the geographic variations. The problems of interpretation and anthropogenic effects are discussed in Section 11.1.5.

Consider the long stretch of relatively mild and stable conditions since the retreat of the last ice sheets about 11.5 kyr ago, the "Holocene." This epoch has been claimed to be exceptionally "stable"; it has even been suggested that such stability is a precondition for the invention of farming and thus for civilization itself (Petit *et al.*, 1999). It is therefore possible that the statistics from paleodata such as ice cores sampled over 100 kyr or longer periods may not be as pertinent as we would like for understanding the current epoch. Although we return in more detail to the ice-core data in Section 11.2, let us briefly consider the high-resolution GRIP core, which shows striking differences between the Holocene and previous epochs in central Greenland (Fig. 11.1). Even a cursory visual inspection of the figure confirms the relative absence of low-frequency variability in the current 10 kyr section as compared to previous 10 kyr sections.

To quantify this epoch-to-epoch variability we can turn to Fig. 11.2, which compares the RMS Haar structure functions for both GRIP (Arctic) and Vostok (Antarctic) cores for both the Holocene 10 kyr section for the mean and spread of the eight earlier 10 kyr sections. The GRIP Holocene curve is clearly exceptional, with the fluctuations decreasing with scale out to $\tau_c \approx 2$ kyr in scale and with $\xi(2)/2 \approx -0.3$. This implies a spectral exponent near the macroweather value $\beta \approx 0.4$, although it seems that for large Δt we have $\xi(2)/2 \approx 0.4$ ($\beta \approx 1.8$). It is therefore plausible that the main difference is that τ_c is much larger than

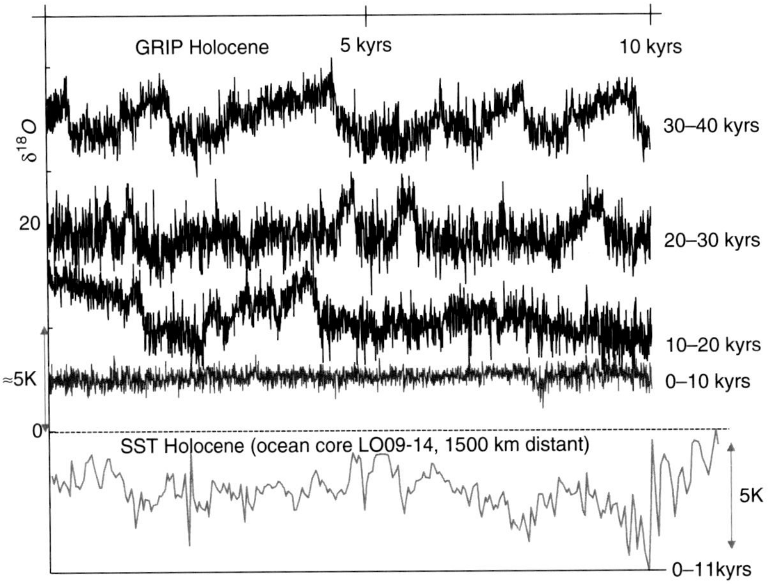

Fig. 11.1 The top part shows four successive 10 kyr sections of the 5.2-year resolution GRIP data, the most recent to the oldest from bottom to top. Each series is separated by 10 mils in the vertical for clarity (vertical units: mils – i.e. parts per thousand of isotope excess). For reference, a 5 K corresponding temperature spread is also shown using a calibration constant of 0.5 K/mil. We see that the bottom Holocene GRIP series is indeed relatively devoid of low-frequency variability compared to the previous 10 kyr sections, a fact confirmed by statistical analysis discussed in the text and shown in Fig. 11.2. In contrast, the bottom curve shows the (much lower resolution but on the same scale) paleo-SST curve from ocean core LO09–14 (Berner *et al.*, 2008), taken from a location only 1500 km distant and displaying far larger variability: see Fig. 11.2. Adapted from Lovejoy and Schertzer (2012a).

for the other series (see Table 11.2 for quantitative comparisons). The Greenland Holocene exceptionalism is quantified by noting that the corresponding RMS fluctuation function ($S(\Delta t)$) is several standard deviations below the average of the previous eight 10 kyr sections. In comparison (to the right in the figure) the Holocene period of the Vostok core is also somewhat exceptional, although less so: up to $\tau_c \approx 1$ kyr it has $\xi(2)/2 \approx -0.3$ ($\beta \approx 0.4$) and it is more or less within one standard deviation limits of its mean, although τ_c is still large. Beyond scales of ≈ 1 kyr its fluctuations start to increase; Table 11.1 quantifies the differences. We corroborated this conclusion by an analysis of the 2 kyr (yearly resolution) series from other (nearby) Greenland cores (as described in Vinther *et al.*, 2008), where Blender *et al.* (2006) also obtained $\beta \approx 0.2$–0.4 and similar low β estimates for the Greenland GRIP, GISP2 cores over the last 3 kyr (see Table 10D.1).

We can also compare the intermittency of the Greenland Holocene with that of the other 10 kyr periods and with that of Antarctica. Fig. 11.3 shows the cascade analyses, which show remarkably similar cascade structures: it would seem that – contrary to the spectrum and structure functions which depend on H – the temporal intermittency in the Holocene is relatively "typical." This conclusion is supported by the detailed consideration of the corresponding exponents for the Holocene: we find $C_1 = 0.078$, 0.060 (GRIP, Vostok), compared to the mean of the preceding 10 kyr periods (0.081 ± 0.008, 0.12 ± 0.02) and the outer scales (for the Holocene: $\tau_{eff} = 250$, 300 years and for the other 10 kyr sections 380 ± 140 and 2500 ± 1000 years, GRIP and Vostok respectively). The α's do not vary much; for this and the other parameters, see Table 11.2, where we also compare the exponents with those of other paleo series. Note that Fig. 11.3 also shows the comparison with quasi-Gaussian processes indicating that the paleotemperatures are not much more variable. In themselves these cascade analyses are not therefore so conclusive, but the results – at least for C_1 – are roughly supported by the structure function analyses of intermittency in Section 11.2.3.

The fluctuation analyses (Fig. 11.2) convincingly demonstrate that the Holocene was exceptionally stable at the GRIP site in Summit Greenland. Nevertheless, the significance of this for our understanding of the natural variations of northern hemisphere temperatures is doubtful. Indeed, on the basis of paleo-SST reconstructions just 1500 km southeast of Greenland (Andersen *et al.*, 2004; see also Berner *et al.*, 2008) it was concluded that the latter was on the contrary "highly unstable." Using several ocean cores as proxies, a Holocene SST reconstruction was produced (see Fig. 11.1, bottom) which includes a

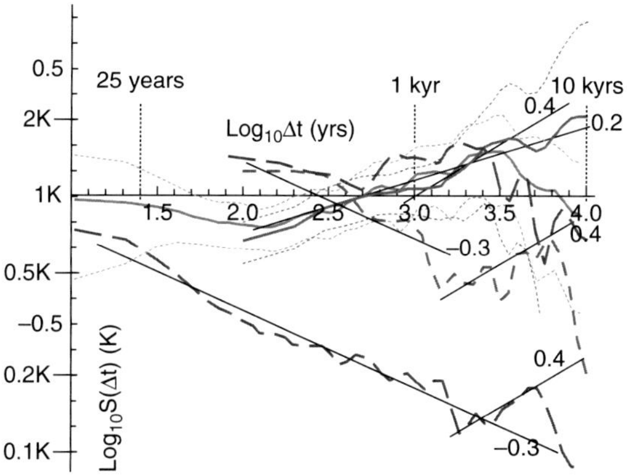

Fig. 11.2 A comparison of the RMS Haar structure function ($S(\Delta t)$) for both Vostok and GRIP high-resolution cores (resolutions 5.2 and 50 years respectively over the last 90 kyr). The Haar fluctuations were calibrated and are accurate to ≈ ± 20%. For Vostok we used the Petit *et al.* (1999) calibration, for GRIP, 0.5 K/mil. These series were broken into 10 kyr sections. The thick dashed lines show the most recent of these (roughly the Holocene). The top thick dashed line is the (Berner *et al.*, 2008) paleo-SST series; the middle thick dashed line is from Vostok; the bottom thick dashed line is from GRIP. The thick continuous lines are the $S(\Delta t)$ of the ensemble of eight 10–90 kyr GRIP (longest) and Vostok (shorter). The one-standard-deviation variations about the mean are indicated by dotted lines. Also shown are reference lines with slopes $\xi(2)/2 = -0.3, 0.2, 0.4$, corresponding to $\beta = 0.4, 1.4, 1.8$ respectively. Although the Holocene is exceptional for the GRIP and Vostok series, for GRIP it is exceptional by many standard deviations. However, the paleo-SST curve (from only ≈ 1500 km away) is quite different and is very close to the pre-Holocene GRIP results, presumably a consequence of the strong spatial intermittency (see Table 11.4 for a comparison). For the Holocene we can see that $\tau_c \approx 1$ kyr for Vostok, and ≈ 2 kyr for GRIP, although for the previous 80 kyr we find $\tau_c \approx 100$ years for both. Adapted from Lovejoy and Schertzer (2012a).

difference between maximum and minimum of roughly 6 K and "typical variations" of 1–3 K. Indeed, although at lower resolution the variability of this Holocene SST series visually resembles the pre-Holocene GRIP series, in Fig. 11.2 we show this quantitatively: the mean SST structure function is very close to that of the mean GRIP function over the previous eight 10 kyr sections, including the amplitude of the millennium scale fluctuations of ≈ 1–2 K, which is about the amount needed to explain the glacial/interglacial temperature swings. In comparison, the mean Holocene RMS temperature fluctuations deduced from the GRIP core in the last 10 kyr are roughly ≈ 0.2 *K*.

These apparently conflicting results are presumably consequences of the geographical/spatial climate intermittency discussed in Section 10.3. The juxtaposition of the radically different neighbouring SST and ice-core Holocene experiences thus lends support to statistical conclusions based on the variability over longer periods (e.g. the spectra and structure functions averaged over the whole cores). In the next section, we further examine the question of geographical variability which confirms (among other things) that the region a little to the east of Greenland is indeed much more unstable ($\beta>1$, $H>0$).

11.1.4 Geographic climate variability, spatial intermittency, τ_c, β_{mw}, β_c

The high spatial climate intermittency implied in Section 10.3 and the contrast between the Greenland and the near-Greenland sea surface paleotemperatures highlights the need to better understand the

Table 11.1 Comparison of various paleo exponents estimated using the Haar structure function over successive 10 kyr periods: Vostok at 50-year resolution, GRIP at 5.2 year resolution, all regressions over the scale ranges indicated. The Holocene is the most recent period (0–10 kyr). Note that while the Holocene exponents are estimates from individual series, the 10–90 kyr exponents are the means of the estimates from each 10 kyr section and (to the right), the exponent of the ensemble mean of the latter. Note that the mean of the exponents is a bit below the exponent of the mean, indicating that a few highly variable 10 kyr sections can strongly affect the ensemble averages. For the Holocene, the separate ranges < 2 kyr and Δt > 2 kyr were chosen because according to Fig. 11.2, $\tau_c \approx$1- 2 kyr. For comparison, the (Berner *et al.*, 2008) paleo-SST data have $\tau_c \approx 500$ years and $H \approx 0.2$, $\beta \approx 1.4$.

	H				**β**			
	Holocene		**10–90 kyr**		**Holocene**		**10–90 kyr**	
Range of regressions	100 yr – kyr	2 kyr – 10 kyr	100 yr – 10 kyr	100 yr – 10 kyr ensemble	100 yr – 2 kyr	2 kyr – 10 kyr	100 yr – 10 kyr	100 yr – 10 kyr ensemble
GRIP	−0.25	0.21	0.14 ± 0.18	0.17	0.43	1.33	1.14 ± 0.33	1.20
Vostok	−0.40	0.38	0.19 ± 0.28	0.31	0.18	1.76	1.29 ± 0.51	1.49

geographical variability of the climate regime. In Section 10.1.4 and in Table 10.1 we considered various data analyses primarily in the macroweather regime. A limitation of these analyses is that they made little attempt to distinguish macroweather from the lower-frequency climate regime which even in local temperatures starts to be apparent at $\omega_c \approx (10\ \text{years})^{-1} - (30\ \text{years})^{-1}$ (see Figs. 10.12, 10.13). In order to clarify this, let us consider the geographical distribution of the scaling exponents in more detail, using the 20CR reanalysis and comparing exponents estimated in low- ($\omega<(25\ \text{years})^{-1}$) and high- ($(3\ \text{years})^{-1} < \omega < (3\ \text{months})^{-1}$) frequency regions (chosen to avoid the anticipated transition at around $\omega_c \approx (10\ \text{years})^{-1}$). The 138-year-long monthly averaged 20CR 700 mb temperature data were again used to estimate both low- and high-frequency exponents: $2° \times 2°$ data were sampled every 8° in latitude and every 10° in longitude. The geographical distributions are shown in Fig. 11.4 (top row). Whereas at the high frequencies the main pattern visible is the latitudinal variation (with the mean in the expected range of $0.2 < \beta < 0.6$ for land and ocean (i.e. $H \approx (\beta - 1)/2$ in the range −0.4 to −0.2), at the low frequencies we see a rather different picture, with both a strong north–south exponent gradient and hints that the low β values are concentrated over land and the high ones over oceans. In particular, we can see that Greenland is in a particularly low β regime, consistent with its anomalous Holocene paleotemperature behaviour. The latitudinal variation of the β's is shown in Fig. 11.5a (top row). The high-frequency curve is nearly symmetric about the equator and is almost identical to that of Huybers and Curry (2006), who used regressions from $(2\ \text{months})^{-1}$ to $(30\ \text{years})^{-1}$, but the low-frequency curve shows an

Table 11.2 The parameter estimates for the four datasets shown in Fig. 11.18 interpolated or averaged to 50-year resolutions, all over the last 90 kyr only. For reference we have given the mean resolutions from which the 50-year series were generated; the values are over the period 0–90 kyr BP.

Location	Resolution (mean) (years)	C_1	α	τ_{eff} (years)
GRIP high resolution, Greenland	5.2	0.108	1.68	900
GRIP 55 cm, Greenland	18	0.122	1.50	2260
NGRIP	50	0.101	1.70	1800
Vostok, Antartica	72	0.108	1.68	5600
Ensemble	50	0.12	1.62	2800

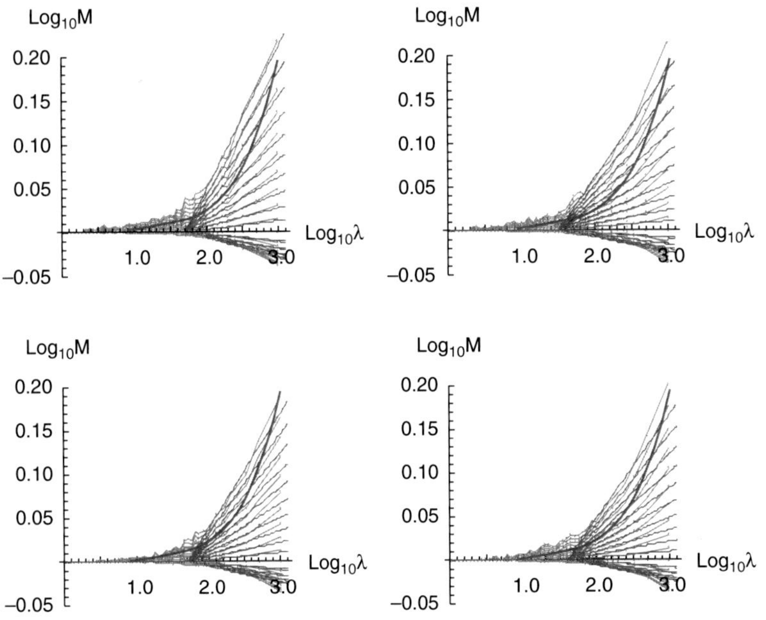

Fig. 11.3 GRIP high resolution. The first four sections of 10 kyr high resolution (same axes); the upper left is the most recent (Holocene) 10 kyr section. Moving left to right, top to bottom we display the moments $q = 0$ to $q = 2$ (intervals 0.2) for the next 10 kyr periods, up to 30–40 kyr (bottom right). The reference scale (corresponding to $\lambda = 1$ in the figure) is 10 kyr, the mean outerscale $\tau_{eff} = 380 \pm 140$ years. The superposed curves are the envelopes of the trace moments of the quasi-Gaussian processes as discussed in Appendix 4A. It can be seen that the data are not far from quasi-Gaussian, so the cascade parameter estimates are not too reliable, although the C_1 estimates are supported by the fluctuation analyses in Section 11.2.3.

Box 11.1 Using paleoclimate data to statistically constrain models

The traditional way of comparing paleoclimate data and climate models is to take the most realistic models and then drive them with the most realistic "reconstructed" climate forcings (e.g. the Last Millennium project). The model outputs are then compared with paleo series. Ignoring for simplicity the spatial coordinate, consider a paleo reconstruction $T_p(t)$ and a model series $T_m(t)$. We are used to considering $T_p(t)$ as a constraint on $T_m(t)$ in a direct manner: perfect models and data satisfy $T_m(t) = T_p(t)$. A consequence is that the fluctuations ΔT of perfect models and data satisfy:

$$\Delta T_m(\Delta t) = \Delta T_p(\Delta t) \tag{11.1}$$

where the fluctuations can be defined via absolute differences: $\Delta T(\Delta t) = |T(t+\Delta t) - T(\Delta t)|$, or – as we discussed in Chapter 10 – often more advantageously by Haar wavelets.

There are several reasons why exact data–model comparisons are overly ambitious. Some of these are: (a) the different space-time resolutions of the data and models, (b) nontrivial paleo "calibration" and measurement noise issues, (c) because of sensitive dependence on initial conditions (the "butterfly effect"), the model reconstructions are not really expected to be realistic in a deterministic sense. As a consequence, the best we can expect is $T_m(t) \approx T_p(t)$ and often instead, comparisons are often achieved with the help of low-pass filtered series. Alternatively, the trends in the models and data are compared at various significant times t over durations Δt; these can be defined either by linear regressions of $T(t)$ over intervals Δt or more simply by fluctuation gradients $\Delta T(\Delta t)/\Delta t$.

Using data to directly constrain the models in this deterministic way is often too demanding. One simple solution is to replace the deterministic Eqn. (11.1) by its statistical version:

$$\Delta T_m(\Delta t) \stackrel{d}{=} \Delta T_p(\Delta t) \tag{11.2}$$

where the $\stackrel{d}{=}$ sign indicates equality in probability distribution, i.e. $a \stackrel{d}{=} b$ if and only if $\Pr(a > s) = \Pr(b > s)$, where s is an arbitrary threshold and Pr indicates probability. Although Eqn. (11.2) is a consequence of Eqn. (11.1) it is much weaker, since two series T_m, T_p whose fluctuations respect Eqn. (11.2) may not even be statistically correlated with each other.

A useful way to exploit Eqn. (11.2) is to calculate statistical averages over various powers. The qth-order "structure function" S_q is particularly convenient:

$$S_{q,m}(\Delta t) = S_{q,p}(\Delta t); \quad S_q(\Delta t) = \langle \Delta T(\Delta t)^q \rangle \tag{11.3}$$

Since the power spectrum $E(\omega)$ is essentially the Fourier transform of $S_2(\Delta t)$, taking $q = 2$ implies that the latter are also equal: $E_p(\omega) = E_m(\omega)$ (Box 2.2). If the fluctuations are quasi-Gaussian then this fully specifies the fluctuation statistics; however, they are generally far from Gaussian. When this is important, the full statistics can be studied by considering higher and higher values of q which allows us to characterize the intermittency and the extremes. It is this flexibility and ease of interpretation that makes the study of fluctuations superior to the use of power spectra.

Eqns. (11.2) and (11.3) are necessary consequences of Eqn. (11.1), yet they may be true even if ΔT_m and ΔT_p are not even statistically correlated with each other; they provide *necessary* but not *sufficient* constraints on the models. Another way to look at this is that since Eqns. (11.2) and Eqns. (11.2) are functions of the time lag Δt, the way the fluctuations vary with Δt characterizes the type of corresponding dynamics (and this can conveniently be quantified by using exponents, i.e. the type of scaling).

Using Eqns. (11.2) and (11.3) in place of Eqn. (11.1) has several advantages. For example, (statistically independent) measurement noises would simply add their noise variance σ^2 to the paleodata variance $\langle \Delta T_p(\Delta t)^2 \rangle$, which would thus mostly affect the small lags, and different linear calibration constants correspond to vertical shifts in log-log plots of $\langle \Delta T_p(\Delta t)^2 \rangle^{1/2}$ against Δt. Since the type of statistical variability (i.e. the exponents, the type of scaling) is expected to be robust, the use of statistical constraints has the further advantage that exact geographical co-locations of T_p and T_m are not required, nor do we need perfect paleochronologies (although they should be fairly linear).

anomaly north of $\approx 30°$ N. Also shown (bottom row) is a comparison with the (quite different) output of a 500-year GCM control run (discussed in Secton 11.3.2).

Rather than divide the frequency domain into predefined regimes and estimate the corresponding exponents, we can do the opposite: find the statistically optimum transition scale between two different

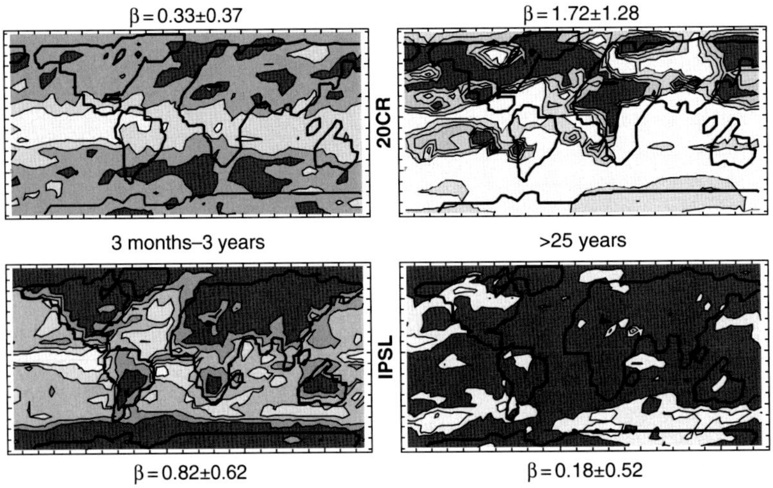

Fig. 11.4 The spatial distribution of the spectral exponents for the reanalyses (top row) and the IPSL control run (bottom row). The left-hand side shows the distribution of the high-frequency exponents, the right-hand side, the low-frequency β's. In all cases the contour lines were at $\beta = 0.5$ (dark blue), 1, 1.5, 2 (white). One can see that Greenland has exceptionally small low-frequency β's (< 0.5) although about 20° to the east the values are quite high (> 1.5: the white area to the right of Greenland). The mean β's and one-standard-deviation spreads are also indicated. The high-frequency 20CR β map is very similar to that of Huybers and Curry (2006) (although they determined β from NCEP reanalyses from 2 months to 30 years). *See colour plate section.*

regimes, each defined with different fixed exponents. Taking the low-frequency regime to have $\xi(1) = H = -0.4$ ($\beta \approx 0.2$) and the high-frequency regime to have $\xi(1) = H = 0.4$ ($\beta \approx 1.8$), regression analysis on the 20CR data yields an estimate of the optimum transition scale τ_c (Fig. 11.5b). First-order Haar structure functions (rather than spectra) were used since they had better scale resolution at the corresponding low frequencies. We could note that the τ_c estimates were quite insensitive to the exact choices of H for the high- and low-frequency regimes. These analyses are thus consistent with the idea that the main geographical variation is the macroweather climate transition scale τ_c and that the apparent variations in low-frequency β (estimated for time scales longer than 25 years: i.e. below τ_c for much of the northern hemisphere) may be biased by using regressions which straddle the transition scale. Also shown are the analogous estimates for the 20CR precipitation field. We see that, except at very high latitudes, the precipitation τ_c are somewhat higher than the temperature τ_c and are more north–south symmetric. However, they are within roughly factors of two of each other over the whole range.

Although clearly the low-frequency β estimates are "noisy," being based on only five discrete frequencies (a scale range of factor 138/25), the resulting mean behaviour turns out to be quite close to the variability required to explain the interglacials (see Fig. 1.9c, 11.10 and Section 11.2.2), so that the results are quite reasonable. If this analysis is correct then it implies that the regions where the most intensive empirical analyses have been made – Greenland and the northern hemisphere – are precisely regions where there are spectral plateaus which are anomalously long (large τ_c's) with consequently anomalously low β's. Since the surface series show that the spectra of the mean northern and mean southern hemisphere temperatures are not too different (see the error bars in Fig. 11.5b), the main difference is in the longitude-to-longitude variability (which is higher in the northern hemisphere; the mean β is significantly lower than the β of the mean).

Further information about the macroweather/climate transition is given in Fig. 11.6, which compares the 20CR estimates from Fig. 11.5a with those of various longer "multiproxy" series discussed in the next section as well as those of the GRIP and Vostok cores already discussed in Fig. 11.1, 11.2. We can see that generally the more recent multiproxy series have higher β's which are in accord with the conclusion that the low variability of the Greenland proxies is exceptional and that a value of $\beta > 1$ (probably in the range 1.4–2) is more realistic. This is discussed in the next section.

11.1.5 Multiproxy temperature data, centennial-scale variability and twentieth-century warming

Although there is consensus that the interglacial Holocene period is warmer than the preceding glacial period, we have noted the conflicting claims of

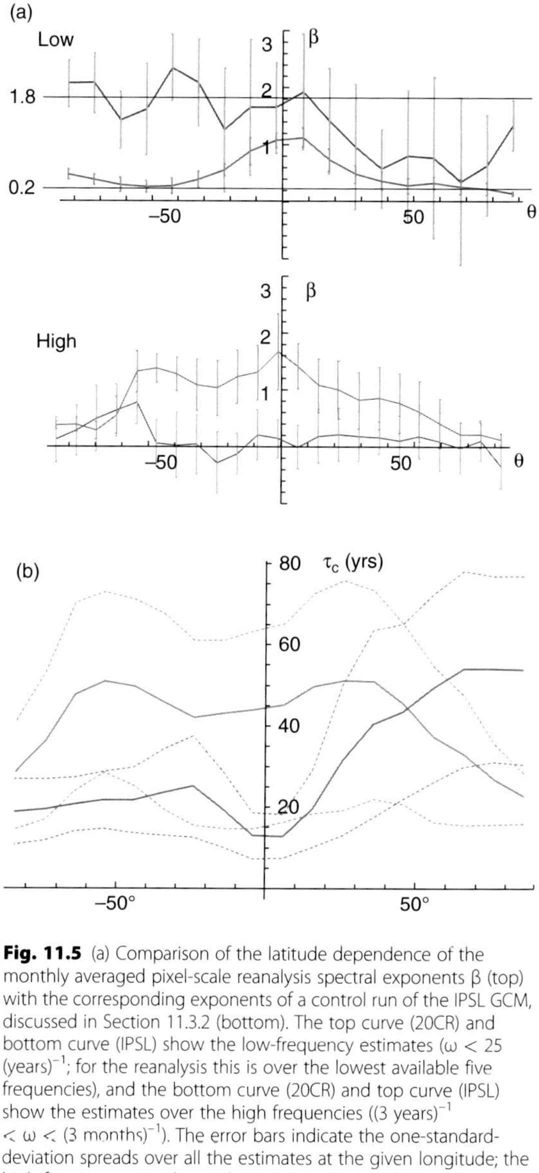

Fig. 11.5 (a) Comparison of the latitude dependence of the monthly averaged pixel-scale reanalysis spectral exponents β (top) with the corresponding exponents of a control run of the IPSL GCM, discussed in Section 11.3.2 (bottom). The top curve (20CR) and bottom curve (IPSL) show the low-frequency estimates ($\omega < 25$ (years)$^{-1}$; for the reanalysis this is over the lowest available five frequencies), and the bottom curve (20CR) and top curve (IPSL) show the estimates over the high frequencies ((3 years)$^{-1}$ $< \omega <$ (3 months)$^{-1}$). The error bars indicate the one-standard-deviation spreads over all the estimates at the given longitude; the high-frequency spreads are about ± 0.3 whereas the low frequency spreads are about ± 0.6, ± 0.2 (20CR, IPSL, respectively, reflecting the greater length of the IPSL estimates, 500 years). A key point to note is that the relative positions of the of high- and low-frequency curves are inverted: whereas the 20CR low-frequency β's are much larger than the corresponding IPSL β's, the opposite is true for the high-frequency β's. For reference, we have also shown on the graph as lines the values of the exponents of the globally averaged temperatures from Fig. 10.12. The Greenland paleotemperatures are at roughly 75° N, where we note particularly low β values. (b) Latitudinal dependence (mean thick, one standard deviation thin dashed) of the critical climate transition time τ_c for 20CR 700 mb temperature (bottom) and surface precipitation (top). This is estimated from bilinear regression of slopes $\xi(1) = H = -0.4, +0.4$ on the $q = 1$ Haar structure functions. One can see that the precipitation τ_c's are generally somewhat larger and roughly north–south symmetric whereas the temperature τ_c's are somewhat asymmetric, qualitatively the same as the variation in low-frequency β's (top row, Fig. 11.5a). This figure can be compared with Fig. 8.5c, which shows the corresponding 20CR estimates of τ_w.

Holocene stability or instability based on Greenland and near Greenland data: this has hindered the development of a clear idea of the type of variability (i.e. whether $H > 0$ or $H < 0$) during the last 10 kyr. Similarly, the recent instrumental period – especially with the help of the 20CR reanalysis (Figs. 11.5a, 11.5b, 11.6) – has shown that it is dangerous to geographically generalize from the Greenland Holocene.

The key to linking the long but geographically limited ice-core series with the short but global-scale instrumental series is the intermediate category of "multiproxy temperature reconstructions." These series, pioneered by Mann *et al.* (1998, 1999), have the potential of capturing "multicentennial" variability over at least the (data-rich) northern hemisphere. Multiproxies are typically at annual resolutions and combine a variety of different data types ranging from tree rings, ice cores, lake varves, boreholes, ice melt stratigraphy, pollen, Mg/Ca variation in shells, ^{18}O in foraminifera, diatoms, stalagmites (in caves), biota and historical records. In what follows, we analyze eight of the longest of these; see Fig. 11.7a for the series and Table 11.3 for some of statistical characteristics and descriptions.

Before reviewing the results, let us discuss some of the technical issues that continue to drive the development of new series. Consideration of the original series (Mann *et al.*, 1998; extended back to AD 1000 in Mann *et al.*, 1999 and updated in Mann *et al.*, 2008) illustrates both the technique and its attendant problems. A basic difficulty is in getting long series that are both temporally uniform and spatially representative. For example, the original six-century-long multiproxy series presented in Mann *et al.* (1998) had 112 indicators going back to 1820, 74 to 1700, 57 to 1600 and only 22 to 1400. Since only a small number of the series go back more than two or three centuries, the series' "multicentennial" variability depends critically on how one takes into account the loss of data at longer and longer time intervals. When it first appeared, the Mann *et al.* series created a sensation by depicting a "hockey-stick"-shaped graph of temperature, with the fairly flat "handle"

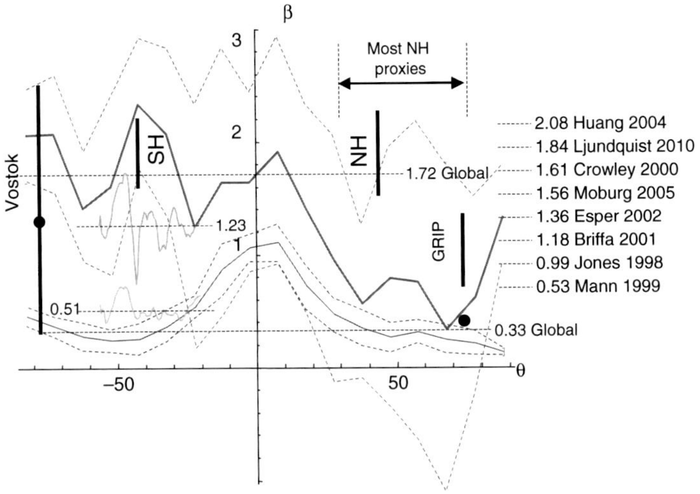

Fig. 11.6 The spectral exponent as a function of latitude for the 138-year monthly averaged 20CR temperature at 700 mb and various other proxy estimates discussed in Section 11.1.4. The thick and thin continuous lines are the mean exponents for frequencies $\omega < (25\ \text{years})^{-1}$, and for $(3\ \text{months})^{-1} < \omega < (3\ \text{years})^{-1}$ respectively (from Fig. 11.5a). The dashed lines indicate the one-standard-deviation longitude-to-longitude variations (corresponding to the map in Fig. 11.4). The vertical bars marked "NH, SH" are the exponents of the mean northern and mean southern hemisphere. Each of the three surface series (1880–2008) was used; the centre of the bar is the mean exponent for $\omega < (25\ \text{years})^{-1}$ and the length of the bar indicates the series-to-series variation. The dashed line indicating "global" is the mean over all the latitudes of the pixel-by-pixel exponents (spread ± 1.28, not shown). On the right are indicated the regression estimates for the eight annual reconstructions discussed in the text and Table 11.3. The regressions for the reconstructions are for $(480\ \text{years})^{-1} < \omega < (25\ \text{years})^{-1}$ (the period 1500–1979). The rough range of latitudes where almost all the proxies are situated is indicated. Also indicated by circles are the Holocene (last 10 kyr) GRIP (at 5.2-year resolution, 72.57° N) and Vostok (at 300-year resolution, 78.45° S) β's; the vertical bars are the one-standard-deviation variations of β for the 8 × 10 kyr periods 10–90 kyr (GRIP) and the previous 41 × 10 kyr period 10 – 420 kyr (Vostok). Finally, the multiproxy estimates of South American annual temperatures (at 0.5° resolution; Neukom *et al.*, 2010) are shown, both high and low frequency (respectively the short bottom and top curves between 20° and 60° S), as well as the overall average values (horizontal dashed lines). The high-frequency 20CR β distribution with latitude is almost the same as that of Huybers and Curry (2006), who fit NCEP reanalysis β's from 2 months to 30 years. The near convergence of the high- and low-frequency β's over part of the northern hemisphere is probably mostly due to a larger and highly variable transition scale τ_c.

continuing from AD 1000 until a rapid twentieth-century increase. This led to the famous conclusion – echoed in the IPPC AR3 (Houghton *et al.*, 2001) – that the twentieth century was the warmest century of the millennium, that the 1990s was the warmest decade, and that 1998 was the warmest year. This success encouraged the development of new series using larger quantities of more geographically representative proxies (Jones *et al.*, 1998), the introduction of new types of data (Crowley and Lowery, 2000), in some cases the more intensive use of pure dendrochronology (Briffa *et al.*, 2001), or the latter combined with an improved methodology (Esper *et al.*, 2002).

However, the interest generated by reconstructions also attracted criticism. In particular, McIntyre and McKitrick (2003) claimed there were several flaws in the Mann *et al.* (1998) data collection and in the application of the principal component analysis technique which it had borrowed from intelligence testing and econometrics. After reprocessing, the same proxies were claimed to yield series with significantly larger low-frequency variability, including the reappearance of the famous "medieval warming" period at around AD 1400, which had been attenuated in the original. Later, an additional comment underlined the sensitivity of the methodology to low-frequency red noise variability present in the calibration data (McIntyre and McKitrick, 2005). This was modelled with Markov (exponentially decorrelating) processes, so its low frequency effect is presumably smaller than that which would have been found using the more appropriate scaling noises. Other work in this period, notably by von Storch *et al.* (2004) using "pseudo-proxies" (i.e. the simulation of the whole calibration process with the help of GCMs), similarly underlined the nontrivial issues involved in extrapolating multiproxy calibrations into the past (see also Hegerl *et al.*, 2007). After many exchanges in the literature, it now seems that the original Mann

Table 11.3 Comparison of parameters estimated from the multiproxy data from 1500–1979 (480 years). The Ljundqvist high-frequency numbers are not given, since the series has decadal resolution. Note that the β for several of these series was estimated in Rybski *et al.* (2006) but no distinction was made between macroweather and climate, so the entire series were used, resulting in generally lower β's. Also shown (bottom line) is a South American estimate.

	β (high freq (4–10 years)$^{-1}$)	β (lower freq than (25 years)$^{-1}$)	H_{high} (4–10 years)	H_{low} (>25 years)	C_1	α	τ_{eff} (years)
Jones *et al.*, 1998	0.52	0.99	−0.27	0.063	0.104	1.64	15
Mann *et al.*, 1998, 1999	0.57	0.53	−0.22	−0.13	0.100	1.67	30
Crowley and Lowery, 2000	2.28	1.61	0.72	0.31	0.105	1.65	15
Briffa *et al.*, 2001	1.19	1.18	0.15	0.13	0.092	1.64	20
Esper *et al.*, 2002	0.88	1.36	0.01	0.22	0.092	1.72	15
Huang, 2004	0.94	2.08	0.02	0.61	0.090	1.70	20
Moberg *et al.*, 2005	1.15	1.56	0.09	0.32	0.094	1.69	15
Ljundqvist, 2010	–	1.84	–	0.53	0.098	1.67	95
Neukom *et al.*, 2010	0.34 ± 0.37	1.23 ± 0.18	−0.24	0.12	0.096	1.85	22

et al. (1998) results are reasonably robust: see Wahl and Ammann (2007).

Beyond the potential social and political implications of the debate, the scientific upshot was that increasing attention had to be paid to the preservation of the low frequencies. One way to do this is to use borehole data which, when combined with the use of the equation of heat diffusion, have essentially no calibration issues whatsoever. Huang (2004) used 696 boreholes (only back to AD 1500, roughly the limit of this approach) to augment the original (Mann *et al.*, 1998) proxies so as to obtain more realistic low-frequency variability. Similarly, in order to give proper weight to proxies with decadal and lower resolutions (especially lake and ocean sediments), Moberg *et al.* (2005) used wavelets to separately calibrate the low- and high-frequency proxies. Once again the result was a series with increased low-frequency variability. Finally, Ljundqvist (2010) used a more up-to-date and more diverse collection of proxies to produce a decadal-resolution series going back to AD 1. The low-frequency variability of the new series was sufficiently large that it even included a third-century "Roman warm period" as the warmest century on record and permitted the conclusion that "the controversial question whether Medieval Warm Period peak temperatures exceeded present temperatures remains unanswered" (Ljundqvist, 2010).

In this context, let us quantitatively analyse the eight series cited above using the Haar structure function. We concentrate here on the period 1500–1979 for several reasons: (a) because it is common to all eight reconstructions; (b) being relatively recent, it is more reliable (it has lower uncertainties); and (c) it avoids the medieval warm "anomaly" and thus the possibility that the low-frequency variability is artificially augmented by the possibly unusual warming in earlier centuries. The resulting series are shown in Fig. 11.7a and the Haar structure functions in Fig. 11.7b, where we have grouped the structure functions into the five pre-2003 and three post-2003 reconstructions. Up to about 200 years the basic shapes of the curves are quite similar to each other – and indeed to the surface temperature $S(\Delta t)$ curves back to 1881 (Fig. 10.12: the ensemble of the CDC, GISS and HadCRUT3 series discussed earlier). However, quite noticeable for the pre-2003 constructions is the systematic drop in RMS fluctuations for $\Delta t > \sim 200$ years, which contrasts with their continued rise in the post-2003 reconstructions. This difference is also clearly visible in the low-pass filtered series shown in Fig. 11.7a. This confirms the above analysis to the effect that the post-2003 analyses were apparently more careful in their treatments of multicentennial variability. To both quantify this and put it into perspective, we may return to Fig. 11.6, which indicates the low-frequency ($\omega < (25\ \text{years})^{-1}$) regression exponents on a graph of β as a function of latitude. It can be seen that the post-2003 β's are roughly the same as the 20CR exponents, especially when it is recalled that almost all the series contributing to the

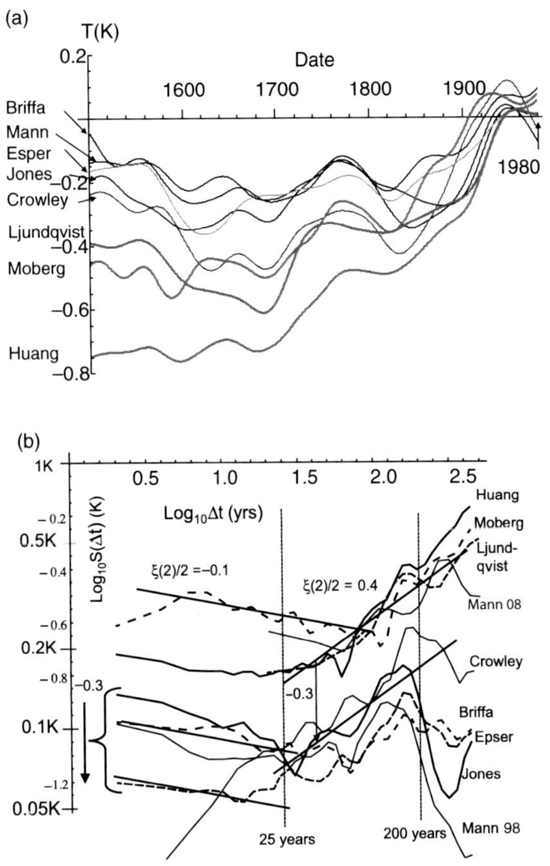

Fig. 11.7 (a) Northern hemisphere multiproxy temperature anomaly series, analysed here from 1500 to 1980, with a 30-year ($\approx \tau_c$) running average filter to bring out the climate-scale variability; the anomalies series were adjusted so that the 1960–1980 averages coincide. The thick curves are the post-2003 series, the thin curves, the pre-2003 series, each identified by the lead author indicated on the left. Note that the running average is not accurate at the extreme (beginning and end) 30-year parts of the graph. (b) RMS temperature fluctuations ($S(\Delta t)$ with exponent $\xi(2)/2$ as estimated using Haar wavelets for the eight multiproxy temperature series for the northern hemisphere discussed in Table 11.3. The ninth (Mann) was added in proof but is not discussed further. The fluctuation temperature scale has been calibrated using usual (difference), "poor man's" wavelets for $\xi(2) > 0$, and tendency structure functions for $\xi(2) < 0$ and is accurate to within $\approx \pm$ 20%. The structure functions have been divided into two groups corresponding to pre-2003 and post-2003 analyses (note that the Ljundqvist (2010) series had decadal resolution and is not shown for $\Delta t < 10$ years). For clarity the pre-2003 group was displaced downwards by 0.3, corresponding to a factor of 2 in fluctuations. Reference lines corresponding to $\beta = 1.8$ ($\xi(2) = 0.4$) and $\beta = 0.8$ ($\xi(2) = -0.1$) have been added. With the exception of the Crowley and Lowery (2000) series – whose high frequencies are not reliable – the shape exponents are very similar between 2 years (the shortest lag) and nearly 200 years, diverging for longer lags with the variability of the older series decreasing rapidly in contrast to the newer series, whose variability continues to increase roughly with the same power law, exponent $\xi(2)/2 = 0.4$.

multiproxies come from data between latitudes 30° and 80° N. (Note added in proof: we recently analysed the Mann *et al.* (2008) reconstruction but only managed to include it in Fig. 11.7b, where it is in rough accord with the other post-2003 reconstructions.)

As usual, a more complete characterization of the series requires going beyond the spectral β or conservation exponent H, to also characterize the intermittency. Fig. 11.8 shows the corresponding cascade structures and Table 11.3 gives the corresponding parameter estimates. The intermittency is generally low for $\Delta t > \sim 30$ years (although Mann *et al.* (1998) and Huang (2004) are partial exceptions), and we also note that the C_1, α parameters are very similar to each other and also to those of the surface temperatures (see Table 10C.2, Fig. 10C.4). The comparison with the envelope of the quasi-Gaussian trace moments shows that cascade conculsions are not strong. This may partly explain why the outer scales have $\tau_{eff} \approx$ 15–30 years, which is significantly larger than for the surface series (cf. $\tau_{eff} \approx 3$ years, Table 10C.2), the main exception is the low-resolution series (Ljundqvist, 2010). In this case, the Haar structure function is perhaps better suited to studying the intermittency (see Section 11.2.3).

In order to improve the statistics, we can make ensemble averages of the pre- and post-2003 structure functions and compare them with the ensemble average of the instrumental global surface series (Fig. 11.9). This figure confirms that the basic behaviour: small Δt scaling with $\beta \approx 0.8$ followed by large Δt scaling with $\beta \approx 1.8$ is displayed by all the data, and the pre-2003 structure function drops off precipitously for $\Delta t > \sim 200$ years (see also Fig. 11.7b). Notable are: (a) the transition scale in the global instrumental temperature, at $\tau_c \approx 10$ years, is somewhat smaller than that found in the reconstructions ($\tau_c \approx$ 40–100 years); and (b) the amplitudes of the reconstruction RMS fluctuations are about a factor of two lower than for the global instrumental series. The reason for the amplitude difference is not at all clear, since the monthly and annually averaged Haar structure functions of the instrumental series are virtually identical (the temporal resolution is not an issue), and similarly the difference between the northern and the southern hemisphere instrumental $S(\Delta t)$ functions is much smaller than this (only about 15%).

Before continuing, let us attempt to address the other exception discussed earlier: that of the twentieth

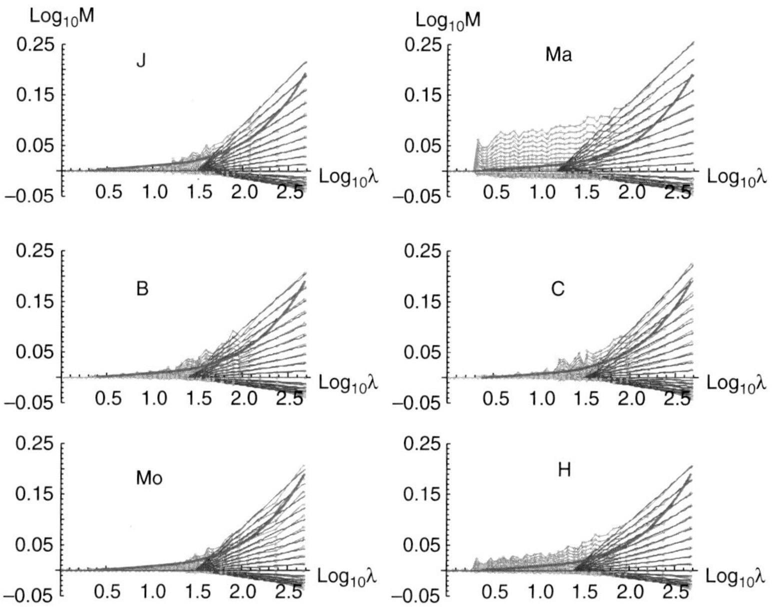

Fig. 11.8 Trace moment analysis of the annually resolved series (with the exception of the Esper series, omitted for reasons of space): J = Jones, Ma = Mann 1998, B = Briffa, C = Crowley, Mo = Moberg, H = Huang. The parameters are quite similar to each other and are given in Table 11.3. Note that the intermittency ends near the cascade outer scales ($\approx$ 30 years) except for the Mann and to a lesser extent the Huang series, where it is significant even at the largest scales. The superposed curves are the envelopes of the trace moments of the quasi-Gaussian processes as discussed in Appendix 4A. It can be seen that with the probable exception of the Mann proxy (upper right), that the data are not far from quasi-Gaussian, so the cascade parameter estimates are not reliable.

century, which – especially since 1970 – is somewhat warmer than the nineteenth. It has been recognized that this warming causes problems for the calibration of the proxies (e.g. Ljundqvist, 2010), and it will clearly contribute to the RMS multicennial variability in Fig. 11.7b. In order to demonstrate that the basic type of statistical viability is not an artefact of the inclusion of exceptional twentieth-century temperatures in Fig. 11.9, we also show the corresponding Haar structure functions for the period 1500–1900. Truncating the instrumental series at 1900 would result in a series only 20 years long, so the closest equivalent for the surface series was to remove overall linear trends, and then redo the analysis. As expected, the figure shows that all the large Δt fluctuations are reduced, but that the basic scaling behaviours are apparently not affected. We conclude that the type of variability as characterized by the scaling exponents is a robust – if difficult to accurately determine – statistic.

Another assessment of the low-frequency variability can be made by comparing typical instrumental, multiproxy and ice-core paleotemperature data (Fig. 11.10). In this figure, we have superposed the calibrated Vostok deuterium-based RMS temperature fluctuations with RMS multiproxy and RMS surface series fluctuations. We see that extrapolating the latter out to 30–50 kyr is quite compatible with the Vostok

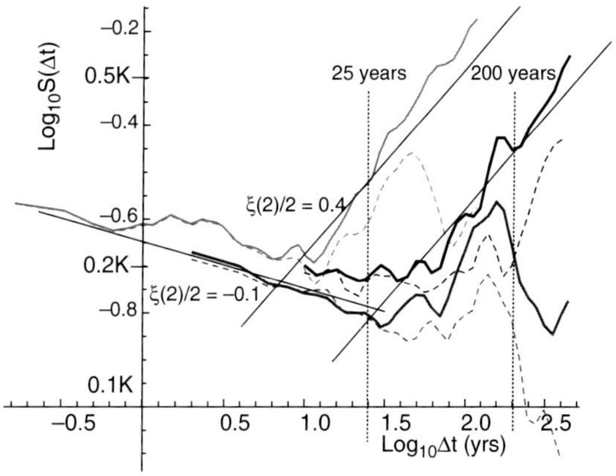

Fig. 11.9 The RMS Haar fluctuation for the mean of the pre- and post-2003 series from 1500 to 1979 (bottom and middle solid lines respectively and excluding the Crowley series because of its poor resolution), along with the mean of the globally averaged monthly resolution surface series (NOAA CDC, NASA GISS, HadCRUT3) (solid, top). In order to assess the effect of the twentieth-century warming, the structure functions for the multiproxy data were recalculated from 1500–1900 only (the dashed lines that join the solid lines at small lags) and for the instrumental surface series with their linear trends from 1880–2008 removed (the data from 1880–1899 are too short to yield a meaningful $S(\Delta t)$ estimate for the lower frequencies of interest). Although in all cases the large Δt variability is reduced, the basic power-law trend seems to remain, although the transition scale τ_c increases (especially for the post-2003 reconstructions). Note that the decrease in $S(\Delta t)$ for the linearly detrended surface series over the last factor of 2 or so in lag Δt is a pure artefact of the detrending. Reference lines corresponding to $\beta = 0.8$ and 1.8 have been added. Reproduced from Lovejoy and Schertzer (2012a).

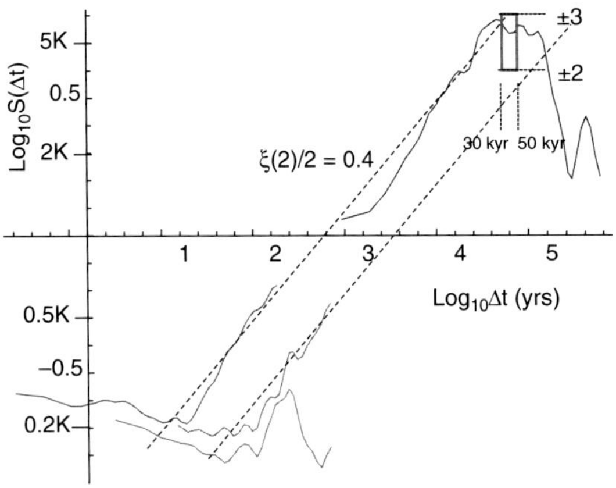

Fig. 11.10 RMS Haar fluctuations for the mean monthly global surface series (left), the mean pre-2003 and mean post-2003 proxies (bottom and middle left, respectively) as well as the mean Vostok $S(\Delta t)$ function over the last 420 kyr interpolated to 300-year resolution and using the Petit *et al.* (1999) calibration (upper right). Also shown is the "interglacial window," the probable typical range of fluctuations and quasi periods of the glacial/interglacials. Reproduced from Lovejoy and Schertzer (2012a).

core and with the "interglacial window." This "window" frames the rough amplitude and quasi period of the glacial/interglacial transition that one expects on the basis of a wide variety of proxy indicators (Lovejoy and Schertzer, 1986). One way to estimate this is to consider the change in temperature since the Last Glacial Maximum (LGM), 19–23 kyr BP. For changes in the global mean, Schmittner *et al.* (2011) find 3 K, and J. Annan (personal communication) finds 4.5 K. Since the variations are about 1.5–2 times larger at high latitudes, these new results are compatible with the indicated window. Although the Vostok $S(\Delta t)$ curve is from the entire 420 kyr record (not just the Holocene), this certainly makes it plausible that the low-frequency variability displayed in the post-2003 reconstructions is indeed more realistic than the contrasting relative lack of variability in the pre-2003 reconstructions.

11.2 Scaling up to 100 kyr: a composite overall scaling picture of atmospheric variability

11.2.1 Ice-core paleotemperatures

Beyond scales of one or two centuries, there are very few direct instrumental series; the longest is the Central England series from 1659 onwards (Manley, 1974), whose spectrum follows reasonably well the "maritime" macroweather exponent $\beta_{mw} \approx 0.5$–0.6 up to ≈ 100 years (see Lovejoy and Schertzer, 1986; Pelletier, 1998, for scaling analyses of the monthly series). As a consequence, virtually all our multicentennial and longer-time-scale information comes from various surrogates. The most reliable are the paleotemperatures, especially those from isotope records obtained from ice cores (ocean core records are also pertinent but their resolutions are lower and they are less directly related to the temperature).

Ice cores can be used as proxies because heavy water – especially molecules containing ^{18}O or deuterium (D) – evaporate with more difficulty than the lighter "normal" H_2O molecules (containing only ^{16}O and ^{1}H), and this differential evaporation rate is itself temperature-dependent. Accordingly, the deficit in parts per thousand, or "mils" of ^{18}O or D in snow, and hence in ice cores, is calibrated by linear regressions of the latter on modern-day temperature records in the regions where the cores are taken. Today, thanks to several ambitious international projects, many cores exist, particularly in the Greenland and Antarctic ice caps. The most famous are probably the GRIP (Greenland Ice Core Project) and Vostok (Antarctica) cores (which we briefly considered in Section 11.1.2), each of which is over 3 km long (limited by the underlying bedrock) and goes back 240 and 420 kyr, respectively. Near the top of the cores, individual annual cycles can be discerned (in some cases going back over 10 000 years); below that the shearing of ice layers and diffusion between the increasingly thin annual layers makes such direct dating impossible, and models of the ice flow and compression are required. Various "markers" (such as dust layers from volcanic eruptions) are also used to help fix the core chronologies. The result is a highly variable relation between depth and chronology: for a detailed discussion, see Box 11.3; and see Fig. 5.21 for probability distributions of fluctuations and its implications for abrupt climate changes including the "Dansgaard–Oeschger" events, some of which are shown in Fig. 11.11.

Consider the GRIP (summit) location $\delta^{18}O$ relatively high-resolution (5.2-year) data in Fig. 11.11. The dataset is 17 551 points long and spans the period from the present to 91 kyr BP (BP = "before present"). We can see that the present interglacial warming (the "Holocene") corresponds to the particularly high $\delta^{18}O$ values around -36 to -34 mils.

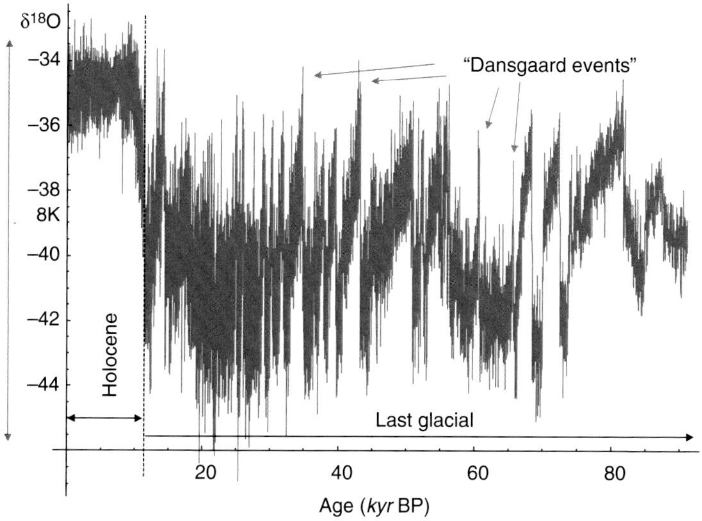

Fig. 11.11 GRIP (summit) high-resolution $\delta^{18}O$ series (in parts per thousand) for the last 91 kyr (time axis = kyr before present, which is the origin). The sharp spikes are "Dansgaard events" (often called "Dansgaard–Oeschger events"): although of decadal scale, they can be comparable in magnitude to the glacial/interglacial variation in temperature. See Fig. 5.21 for the probability distribution of changes: these events roughly correspond to the hyperbolic tail of the distribution (see Box 11.3) A temperature scale is indicated using a 0.5 K/mil calibration. Data courtesy of P. Ditlevsen, University of Copenhagen.

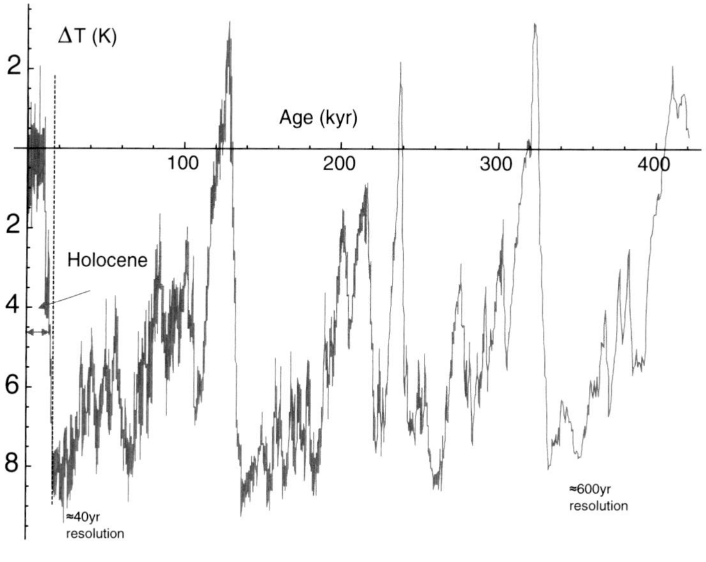

Fig. 11.12 Vostok paleotemperature reconstruction from δD measurements every 1 m in depth for the last 420 kyr (time axis = kyr before present). See Fig. 5.21 for probabilities.

At around 10–12 kyr BP, the signal does drop, but hardly in a step-function-like way. Indeed, the preceding glacial period (further to the right in the graph) is far from being a constant low value; there are several 50–100-year intervals (Dansgaard–Oeschger events) where the signal varies by 3–4 mil, i.e. by as much as the glacial/interglacial transition itself (cf. Steffensen *et al.*, 2008); some changes are claimed to be as much as 2–4 K in one year). Due to their hyperbolic probability tails, these abrupt changes are expected in scaling processes (see Fig. 5.21). To consider longer periods we can turn to Fig. 11.12 from the Antarctic Vostok core. From this, we see that the signal is pseudo-periodic, i.e. with a broad spectral maximum with a period of about 100 kyr.

With these caveats concerning the variable-resolution chronologies, we can move on to studying the spectra themselves. Fig. 11.13a shows the GRIP high-resolution spectrum over the last 91 kyr, "compensated" by dividing by $\omega^{-1.4}$ so that a $E(\omega) \approx \omega^{-1.4}$

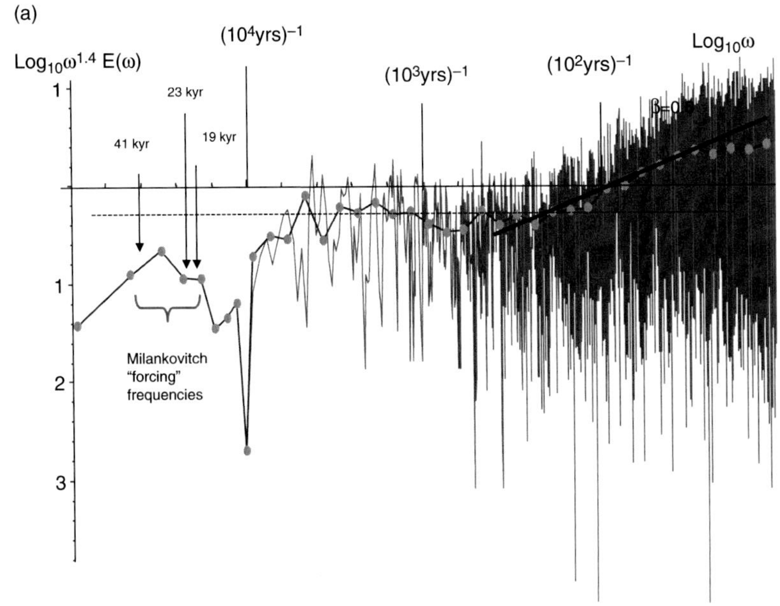

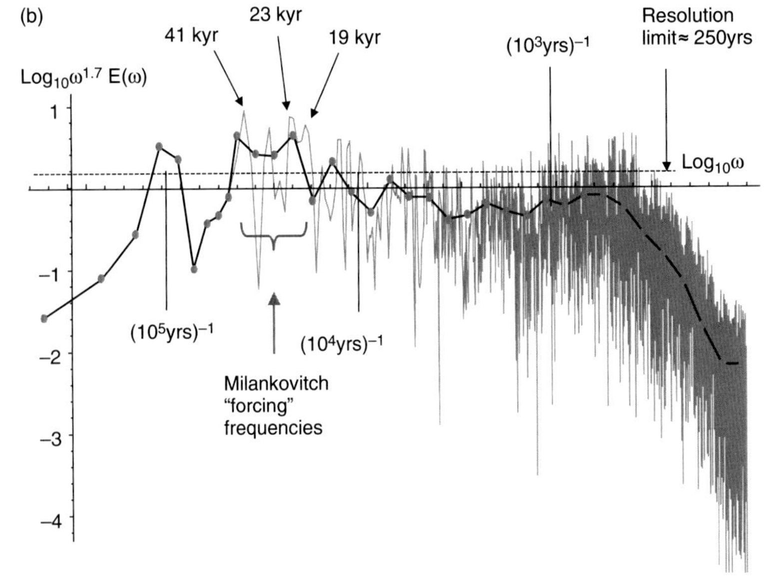

Fig. 11.13 (a) GRIP high-resolution power spectrum compensated by $\omega^{1.4}$. The slope of the black line is +1.2, corresponding to $\beta_{mw} = 0.2$; the arrows show the principal astronomical (Milankovitch) forcing frequencies (19, 23 kyr are the centres of the two precessional bands, 41 kyr is the centre of the obliquity band). Note that these cycles are a bit below the background (indicated by the dashed line). (b) The power spectrum of the Vostok paleotemperature series interpolated to 50 years and compensated by $\omega^{1.7}$, with arrows showing the principal astronomical (Milankovitch) forcing frequencies. In this case they are a little above the background (dashed) line.

form would appear as a horizontal line. Starting at the highest frequencies (about (10 years)$^{-1}$ in this case), we see that down to $\omega \approx$ (300 years)$^{-1}$, the spectrum rises above the line corresponding roughly to a $\omega^{-0.2}$ scaling form. From (300 years)$^{-1}$ to about (10 kyr)$^{-1}$ the spectrum follows roughly the $\omega^{-1.4}$ form (as pointed out by Schmitt *et al.*, 1995, for a lower-resolution GRIP spectrum using a somewhat earlier chronology; see Table 11.4 for similar spectral estimates and Fig. 1.9a for an uncompensated graph). Interestingly, for the slightly longer periods corresponding to the three main astronomical (Milankovitch, 1941) forcings at 19, 23 kyr (precessional) and 41 kyr (obliquity), we see that the signal retreats *below* the

Table 11.4 A comparison of various estimates of the spectral exponents β_c of the climate regime and series lengths and resolutions. The last four rows are for the (anomalous) Holocene only (see Section 11.1.2, Fig. 11.2, and Section 11.3 for GCMs).

Series	Authors	Series length (kyr)	Resolution (yr)	β_c
Composite ice cores, instrumental	Lovejoy and Schertzer, 1986	Composite: minutes to 10^6 years	1000	1.8
$\delta\ ^{18}O$ from GRIP, Greenland	Schmitt *et al.*, 1995	123	200	1.4
$\delta^{18}O$ from GRIP, Greenland	Ditlevsen *et al.*, 1996	91	5	1.6
Composite, Vostok, Antarctica (ice core, instrumental)	Pelletier, 1998	10^{-5} to 1000	0.1 to 500	2
$\delta^{18}O$ from GRIP, Greenland	Wunsch, 2003	100	100	1.8
Planktonic $d^{18}O$ ODP677, Panama basin	Wunsch, 2003	1000	300	2.3
CO_2, Vostok, Antarctica	Wunsch, 2003	420	300	1.5
$\delta^{18}O$ from GISP, Greenland	Ashkenazy *et al.*, 2003	110	100	1.3
$\delta^{18}O$ from GRIP, Greenland	Ashkenazy *et al.*, 2003	225	100	1.4
$\delta^{18}O$ from Taylor, Antarctica	Ashkenazy *et al.*, 2003	103	100	1.8
δD from Vostok	Ashkenazy *et al.*, 2003	420	100	2.1
Composite, mid-latitude	Huybers and Curry, 2006	10^{-4} to 1000	0.1 to 10^3	1.6
Composite tropics	Huybers and Curry, 2006	10^{-4} to 1000	0.1 to 10^3	1.3
$\delta^{18}O$ from GRIP, Greenland	This book	91	5	1.4
δD from Vostok, Antarctica	This book	420	300	1.7
$\delta^{18}O$ from GRIP, Greenland	Blender *et al.*, 2006	3	3	0.4
$\delta^{18}O$ from GISP2, Greenland	Blender *et al.*, 2006	3	3	0.7
$\delta^{18}O$ from GRIP, Greenland (last 10 kyr only)	This book	10	5	0.2
Paleo-SST near Greenland (last 11 kyr only)	Berner *et al.*, 2008, analysed in this book	10.6	40	1.4

background. These are the main periodicities in the top of the atmosphere solar forcing. This 91 kyr series is clearly too short to obtain good estimates of the spectrum at theses frequencies; we can therefore turn to Fig. 11.13b from the Vostok core. Here, the compensation is by $\omega^{-1.7}$, which does a fairly good job at flattening the spectrum. The main exceptions are: (a) at the high frequencies, which are polluted by the resolution issues discussed above (the mean Vostok resolution over the entire length is $\approx$ 127 years, but the oldest sections have resolutions of about 400 years: see Fig. 11.14); and (b) at the low frequencies, which have a genuine fall-off below about $(100\ \text{kyr})^{-1}$ as confirmed for example by longer ocean-core paleo series, which can go back more than 1000 kyr. The difference in exponents ($\beta \approx$ 1.7 rather than 1.4) may be a consequence of the greatly changing altitude of the GRIP ice sheet surface as the glacials/interglacials came and went.

In the compensated Vostok spectrum, we see that the astronomical cycles are indeed visible – but only barely – above the scaling "background." We also see evidence for the much weaker earth orbit eccentricity fluctuations with characteristic frequency of about $(100\ \text{kyr})^{-1}$, but again this is barely visible above the background; see Section 11.3.1 for a discussion of these anomalies in the Milankovitch theory. The predominance of this background was noted from the 1980s onwards. The case for their relative dynamical insignificance was particularly strongly made by Wunsch (2003), who performed scaling analyses of various paleo spectra, and who specifically attempted to determine the portion of the variance which could be accounted for by spectral bands near the three main Milankovitch frequencies. He concluded that even with a liberal account of these bands, they represent no more than 10% of the total. Wunsch went on to propose various (nonscaling) Markov processes as models of the temperature. Certainly, the compensated spectra support the view that while traces of astronomical forcings are indeed discernible, they are barely so, and in any case it is not obvious how they could explain the key features, which are (a) the variance-containing scaling "backgrounds" and (b) the fairly broad, not narrow $(100\ \text{kyr})^{-1}$ maximum. The basic picture has now been supported by several empirical studies (see Table 11.4 for a summary). Athough it is natural to suppose that the background is associated with nonlinear internal climate variability, the more usual assumption is that it is on the contrary due to external climate forcings (see Section 11.3.1 for further discussion).

Of course, in principle, even very narrow band forcing could lead to a broad turbulent-like spectrum: this is indeed the standard paradigm for isotropic turbulence that we discussed in Chapter 2. However, just as in the atmosphere, where the actual solar forcing is modulated by cloud cover and other scaling processes resulting in a forcing with wide-range scaling, so we expect that narrow-band forcing is also an unrealistic idealization; in Section 11.3.1, this is confirmed for solar and volcanic forcings. In any event, no matter what the forcing, the fact that the background is scaling and dominates any narrow-band contributions as described above underlines the importance of this presumably internal scaling variability.

Finally, we could mention another scaling analysis technique which has been applied to various climate and paleoclimate series: the rescaled range technique (Mandelbrot and Wallis, 1969). While this method may have certain advantages over traditional second-order structure function (variance) analysis, for multifractals its relationship to the usual exponents ($\xi(q)$, $K(q)$) is still unclear. Therefore, while the analyses of Bodri (1994) and King (2005) confirm the scaling of a variety of series over various scale ranges, their exponents could not be directly compared with those discussed here.

11.2.2 The multiple scaling regime model: an empirical composite of temporal temperature variability from seconds to 100 kyr

To clarify our ideas about the variability, it is useful to combine data over huge ranges of scale into a single composite analysis (such as the spectra shown in Fig. 1.9c). In this section we discuss the variability only in the time domain; in Section 11.2.4 we extend this to space-time.

The best-known – and apparently the first – attempt to produce a wide-scale-range picture of atmospheric variability was that by Mitchell (1976). He produced what is still the most ambitious single composite spectrum of atmospheric variability to date: it ranged from hours to the age of the earth ($\approx 10^{-4}$ to 10^{10} years). Given the rudimentary quality of the data at that time, he admitted that his composite was mostly an "educated guess." In order to accommodate the wide range of scales it was "necessary to resort to logarithmic coordinates"; however, there was no implication that any of the underlying physical processes might be scaling. In Mitchell's classical framework, in addition to several periodic processes (many of which have since fallen from favour), over a succession of fairly narrow scale ranges, the climate was dominated by numerous distinct physical processes. These were each considered to be stochastic with fixed time constants. He explained his idea as follows:

> As we scan the spectrum from the short-wave end toward the longer wave regions, at each point where we pass through a region of the spectrum corresponding to the time constant of a process that adds variance to the climate, the amplitude of the spectrum increases by a constant increment across all substantially longer wavelengths. In other words, each stochastic process adds a shelf to the spectrum at an appropriate wavelength. (Mitchell, 1976)

The mathematical model that Mitchell presumably had in mind was thus a hierarchy of Ornstein–Uhlenbeck processes yielding "shelves," not scaling processes (see Appendix 10C).

The first attempt at constructing a composite model based on scaling symmetries was Lovejoy and

Box 11.2 Emergent laws and global climate change: natural variability, climate forecasting, anthropogenic warming and model structural uncertainty

Thanks to its social, economic and political implications, the term "climate change" has increasingly become synonymous with anthropogenic causation, with legislators regularly passing laws to "prevent climate change." Even in scientific discourse, there is a regrettable tendency to forget that the climate has been changing for billions of years, that irrespective of the fate of our species, it will continue to change.

The idea that the natural state of the climate is unchanging is an unfortunate conceptual side effect of the dominant deterministic atmospheric paradigm. We have seen that out to millennia – and presumably out to their long time limits – GCM outputs remain nontrivially variable (scaling). However, at least their control run variability is of the weak macroweather type (Fig. 1.9d, Section 11.4) so that fluctuations tend to cancel out, diminishing with scale ($H < 0$), yielding the appearance of unchanging stability. As we move to scales beyond a few years (the outer limit of the $H > 0$ "ocean weather" regime associated notably with the El Niño phenomenon) and notwithstanding possible quasi-periodic multidecadal, centennial and even millennial oscillations (e.g. Mann *et al.* 1995; Isono *et al.*, 2009), changes are increasingly attributed to external "climate forcings" rather than to internal nonlinear variability. This is because, for GCMs, the climate is essentially a boundary value problem (Bryson, 1997) whereas the boundaries are in fact part of a coupled slow dynamics (Pielke, 1998).

Yet the need to properly understand the natural variability is becoming urgent. This is because in principle predictions of anthropogenic warming must be verified via rigorous statistical testing of the observations against the null hypothesis. This means that we must make a specific hypothesis about the probability that the atmosphere would naturally behave in the way that is observed. Only if the probability is low enough should we reject the hypothesis that the observed changes are natural in origin. In addition, the systematic comparison of model and natural variability in the preindustrial era is the best way to fully address the issue of "model structural uncertainty," to be sure that the models are not missing important slow processes (see Section 11.4).

The current lack of such statistical testing is the nugget of legitimate criticism often inarticulately expressed by many of the "climate sceptics" when they refer to the need for "empirically based approaches" or hold up events such as "medieval warming" as evidence that the current warming is of natural origin. Of course, the main response to this scepticism has been to argue (a) that the warming is so strong and evident that it could not be natural (i.e. that statistical hypothesis testing is superfluous), and (b) that the model warming predictions are robust (i.e. "certain" enough to be trusted – even though the quantification of the degree of certainty is itself a model product). While we agree with the first point, this does not obviate the need for testing, and even without GCMs, the second can be at least be plausibly argued on physical grounds using simplified greenhouse arguments going back to Arrhenius (1896).

A consequence is a malaise in the community that was embarrassingly exposed in the "climategate" burglary. Even if we agree that there is warming and that we are its cause, there is the uneasy feeling that the scientific issues are not completely settled. Both within the scientific community and in scientists' public pronouncements, the failure to satisfactorily address the issue of natural variability has constant repercussions. How to explain that the earth can sometimes have prolonged periods of cooling in the midst of anthropogenic warming? Was this winter's record mild temperature evidence for anthropogenic influence – or was it simply a natural occurrence? What should we tell the press about yesterday's catastrophic flood?

As should by now be clear, the theme of this book – emergent stochastic laws – could provide precisely the tools needed to address this issue. In the absence of anthropogenic influences the natural variability is identified with the dynamics, it is directly statistically quantified, climate is not reduced to a boundary value problem. The variability is characterized either by probability distributions or by statistical moments, and these are specified as functions of the space-time scale. In principle, this can be used to perform stochastic climate forecasts. Compared with the relatively massive (albeit inadequate) societal investment in deterministic approaches, the research effort into natural variability has been small, yet it could settle the issue of climate change in a scientifically more satisfactory manner. And we would know with confidence whether anthropogenic effects are worse – or better – than they are currently believed to be.

Schertzer (1984, 1986), whose analyses clarified the following points: (a) the distinction between the variability of regional- and global-scale temperatures, with the latter having particularly long scaling regimes; (b) that there was a scaling range for global averages between scales of about 3 years (τ_c in the notation here) and 40–50 kyr with an exponent $\beta_c \approx$ 1.8; (c) that a scaling regime with this exponent could

Box 11.3 The variable resolution of paleotemperatures and its implications

In Fig. 11.12 we showed the Vostok deuterium proxy at a fixed depth resolution of 1 m. Moving from left to right one can clearly see the effect of the reduced temporal resolution: the curve becomes progressively smoother. This is an artefact of the compression of the ice column; in this case, the publicly available data shown were taken at 1 m intervals and the age of each was dated by the nontrivial techniques mentioned above. Interestingly, analysis of the $\delta^{18}O$ or D anomaly as a function of depth (rather than time) gives a series with statistical characteristics (i.e. H, C_1, α) which are more or less independent of depth.

In order to see how important this resolution effect is, we can turn to Fig. 11.14, which shows the temporal resolution Δt (as determined by the official chronology) as a function of the age of the sample. We have included the publicly available GRIP data at constant 55 cm depth resolution (from which the GRIP high-resolution data were obtained by subsampling). From the figure we see that the resolutions vary over several orders of magnitude yet in a highly intermittent manner. This type of variability is reminiscent of that of the dropsondes (Chapter 6). In fact, spectral analysis shows that the resolution function $\Delta t(z) = t(z+\Delta z) - t(z)$ (where z is the depth, Δz is the depth increment) is itself roughly scaling (with spectral exponents $\beta \approx 1$ (GRIP55) and $\beta \approx 2$ (Vostok)). In addition – as with the dropsondes – the resolution is highly intermittent, being roughly multifractal with $C_1 \approx 0.07$, 0.03 respectively (Fig. 11.15). A significant difference from the dropsonde case is that the resolution decreases relatively monotonically with depth.

In order to study the effect of variable resolutions, we performed multifractal simulations of the effect of this intermittent resolution variability; we conclude that the main consequence is that the intermittency of the series is increased: C_1 by about 0.02–0.04. As concerns the spectrum, if the series are interpolated, then there is not much bias unless we go to scales directly affected by the poor resolution (although it is important to use linear, not higher-order interpolation techniques). Our conclusion is that overall, for scales smaller than roughly twice the mean resolution, we get artificially low spectral densities corresponding to the overly smoothed series. There have been several other attempts to take the variable paleoclimate resolutions into account. For example, Witt and Schumann (2005) used wavelets, Davidsen and Griffin (2010) used (monofractal) fractional Brownian Motion as a model, and Karimova *et al.* (2007) used (mono)fractal interpolation. However, if we consider the extremes, then the effect of the interpolation is important. The problem is the treatment of the abrupt transitions (e.g. the Dansgaard–Oeschger events in Fig. 11.11). In the uninterpolated series (isotope ratios as functions of depth), these are associated with "fat" power-law probability tails (Fig. 5.21). However, the interpolation and uniform resampling in time may smooth out these abrupt changes enough so as to replace the fat tails by "long tails," i.e. roughly lognormal, log-Lévy tails, but still extreme compared to classical (exponential) "thin tails." From our viewpoint, the abrupt transitions in the climate record are prima facie evidence for the prevalence of fat-tailed fluctuation distributions in climate records.

Other proxy data can also be used to get a better idea of the climate variability: see Table 11.5 for more examples.

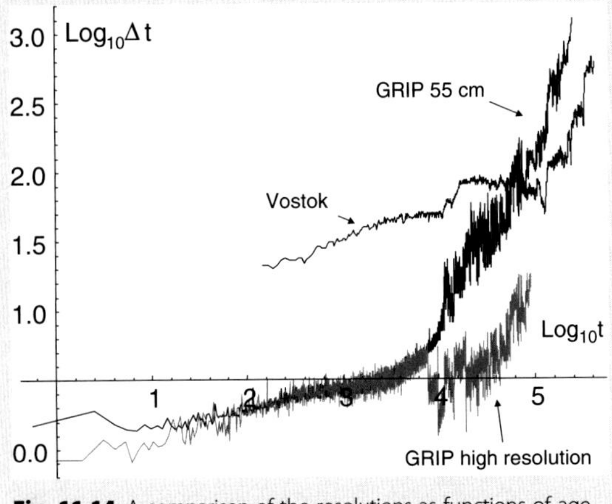

Fig. 11.14 A comparison of the resolutions as functions of age using the published age models. The oldest parts of the cores are: GRIP55: 247 618 yr BP, Vostok: 420 888 yr BP, GRIPhigh = 91 139 yr BP. $\Delta t = t(z+\Delta z) - t(z)$, where z is the depth at intervals of $\Delta z = 0.55$ m for GRIP55 and 1 m for Vostok variable for GRIPhigh.

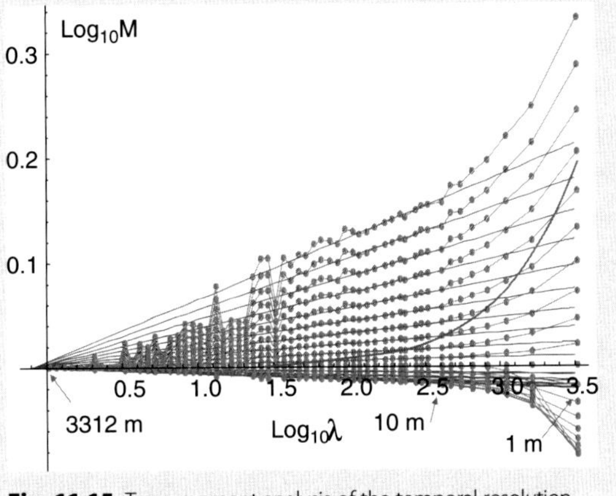

Fig. 11.15 Trace moment analysis of the temporal resolution dependence of the Vostok 1 m resolution core (length 3312 m). The variable analysed is the sequence of inter-layer time intervals Δt. The parameters are: $C_1 = 0.026$, $\alpha = 2$, the outer scale = 4170 m. Note that the scaling is reasonable for scales above ≈ 10 m. Also shown (thick curve) is the envelope of the $q = 2$ quasi-Gaussian processes (Appendix 4A); the data are quite far from this.

Table 11.5 A comparison of scaling exponents and scales for various paleoclimate datasets. Unless otherwise stated, β is for the low frequencies and α, C_1 for the high-frequency regimes.

Location	Type of data	Period (kyr BP)	No. of points	Resolution (years)	β_c^a	H^b	C_1^c	α^c	τ_{eff} (years)[c]
GRIP high resolution, Greenland[d]	$\delta^{18}O$	0–91	17551	5	1.38 ± 0.22	0.20 ± 0.12	0.079 ± 0.01	1.85 ± 0.16	380 ± 180
GRIP 55cm, Greenland[e]	$\delta^{18}O$	0–247	5426	45	1.46 ± 0.33	0.24 ± 0.17	0.13 ± 0.06	1.81 ± 0.04	2200 ± 1500
NGRIP[f]	$\delta^{18}O$	0–123	4918	50	1.4	0.21	0.095	1.85	2000
NGRIP, gas bubbles[g]	Using $\delta^{15}N$, CH_4 in bubbles for reconstruction	30–80		1	1.5	–	0.06	1.2	200
Renland, Greenland[h]	$\delta^{18}O$	≈ 0–59	946	20	1.5	0.26	0.13	1.5	930
Vostok, Antartica[i]	Paleotemperature (δD)	0–420	3312	125	1.71 ± 0.25	0.37 ± 0.13	0.075 ± 0.024	1.82 ± 0.15	3200 ± 800
Dome C, Antarctica[j]	100 kHz conductivity	≈0- 45	38419	1	0.6	-0.18	0.12	1.94	50

[a] The β's are estimated over the range (250 years)$^{-1}$ and lower, C_1's, α's and τ_{eff} over (400 years)$^{-1}$ and higher. The only data with reliable β's over this range are from the GRIP high-resolution set, which yields $\beta_{mw} = 0.55 \pm 0.32$ from ≈ 5 years to 400 years (using only last 50 kyr for reliable resolutions).

[b] The H's are estimated from the low-frequency β's (the previous column) using $H = (\beta - 1 + K(2))/2$, and $K(2)$ is estimated from the C_1 and α, also at low frequencies; these were all in range 0.01–0.02 and 1.9–2 respectively are not given in the table (the C_1, α, T_{eff} values are for the high-frequency range only). Using the high-frequency β estimate (see note *a*), with C_1, α over the same range we can estimate $H = -0.15 \pm 0.13$.

[c] These are estimated from the high-frequency trace moment (cascade) analyses for time scales < 400 years.

[d] The Greenland Ice Core Project (GRIP) ice core at Summit. The dating (transforming a depth in the ice to an age) is done by annual layer counting in the top and ice-flow modelling further down. The way the ice was processed is to cut it in 55 cm pieces = "1 bag." Then for the measurements it was cut into 1, 2, 4 or 8 pieces (more further down) in order to get an approximately even temporal resolution down through the core. This record dates back 91 kyr, which is not all the way to the bottom, but the part below this point is now known to be corrupted. We thank P. Ditlevsen, University of Copenhagen, for kindly supplying this dataset; this is the same data he analyzed in Ditlevsen *et al.*, 1996 (see also Ditlevsen, 2004). The β, H exponent estimates are from the eight older 10 kyr sections since the most recent had anomalously low values (Fig. 11.2). The uncertainties are the spreads (one standard deviation).

[e] From the Centre for Ice and Climate, University of Copenhagen: http://www.gfy.ku.dk/~www-glac/data/grip18o.txt. This dataset contains 55 cm average $\delta^{18}O$ values from the GRIP core on the ss09 time scale. The exponent estimates are from 15 sections of 10 kyr; the uncertainties are the spreads (one standard deviation). Data provided by NCAR/EOL under sponsorship of the National Science Foundation from http://data.eol.ucar.edu/cgi-bin/codiac/fgr_form/id=106.ARCSS008 (see Dansgaard *et al.*, 1989, 1993; Greenland Ice Core Project, 1993; Grootes *et al.*, 1993; Johnsen *et al.* 1997). Deuterium measurements have been performed on three adjacent cores, 3G, 4G and 5G (see Jouzel *et al.*, 1997). Temperature differences with respect to the mean recent time value (i.e. corresponding departure from the mean deuterium value) were used.

[f] These data were available interpolated to 50-year resolution from the IGBP PAGES/World Data Center for Paleoclimatology Data Contribution Series # 2004–059. NOAA/NGDC Paleoclimatology Program, Boulder, CO, USA. Located at 75.10° N and 42.32° W, elevation of 2917 m, ice thickness of 3085 m. The NGRIP drilling started in 1996, and bedrock was reached in July 2003 (see North Greenland Ice Core Project, 2004).

[g] Based on trapped gas reconstruction of temperatures: www.ncdc.noaa.gov/paleo/icecore/greenland/ngrip/ngrip-data.htm (Huber *et al.*, 2006). The North Greenland Ice Core Project (NGRIP) site is located at 75.10° N and 42.32° W, elevation of 2917 m, ice thickness of 3085 m. Data from the Centre for Ice and Climate, University of Copenhagen: www.gfy.ku.dk/~www-glac/data/grip18o.txt. The resolution is nominally 1 year but the spectrum has a strong scale break at about (200 years)$^{-1}$ where β increases from 1.5 to about 4, which is much too small. On the other hand, the cascade outer scale is about the same scale so the β and cascade parameters (C_1, α) were estimated over different ranges; hence no H estimate is given.

[h] Data from the University of Copenhagen: www.gfy.ku.dk/~www-glac/data/grip18o.txt (see Vinther *et al.*, 2008; Johnsen *et al.*, 1992).

[i] Vostok: data available from the National Climatic Data Center: hurricane.ncdc.noaa.gov/pls/paleo/ftpsearch.icecore. The exponent estimates are from 15 sections of 10 kyr, the uncertainties are the spreads (one standard deviation) (see Petit *et al.*, 1999).

[j] The Dome C cores are from Antarctica and cover the period 45–0 kyr BP. The conductivity data are from 100 kHz dielectric profiling measurements from the EDC96 EPICA Dome C core at 2 cm resolution. From NOAA Paleoclimatology Program and World Data Center for Paleoclimatology, Boulder, CO: hurricane.ncdc.noaa.gov/pls/paleo/ftpsearch.icecore (see Wolff *et al.*, 1999).

Table 11.6 Comparison of macroweather (β_{mw}) and climate (β_c) exponents and transition scales from various instrumental/paleo composite statistical analyses. The large τ_c values in the top two rows are from data north of 30° N and are probably anomalously large (see below).

	β_{mw}	β_c	Local τ_c	global τ_c
Lovejoy and Schertzer, 1986	< 1 (central England)	1.8 (poles)	≈ 400 years	≈ 5 years
Pelletier, 1998	0.5 (continental North America)	1.7 (Antarctica)	≈ 300 years	_
Huybers and Curry, 2006 (tropical sea surface)	0.56 ± 0.08 (NCEP reanalysis)	1.29 ± 0.13 (several different paleotemperatures)	≈100 years	_
Huybers and Curry, 2006 (high-latitude continental)	0.37 ± 0.05 (NCEP reanalysis)	1.64 ± 0.04 (several different paleotemperatures)	≈ 100 years	_

potentially quantitatively explain the magnitudes of the temperature swings between interglacials: the "interglacial window."

Similar scaling composites but in Fourier space were proposed by Pelletier (1998) and, more recently, by Huybers and Curry (2006), who made a more data-intensive study of the scaling of many different types of paleotemperatures collectively spanning the range of about 1 month to nearly 10^6 years (Table 11.6). The results are qualitatively very similar, including the positions of the scale breaks; the main innovations are (a) the increased precision on the β estimates and (b) the basic distinction made between continental and oceanic spectra including their exponents. We could also mention the composite of Fraedrich *et al.* (2009), which is a modest adaptation of that of Mitchell (1976), innovating by introducing a single scaling regime spanning the two orders of magnitude from ≈ 3 to ≈ 100 years (with $\beta \approx 0.3$), although surprisingly exhibiting a *decrease* (rather than an increase) in variability at frequencies lower than this.

Fig. 11.16a shows an updated composite where we have combined the 20CR reanalysis spectra (both local, single grid point and global) with the GRIP 55 cm and GRIP high-resolution spectra (both for the last 10 kyr and averaged over the last 90 kyr), the three surface global temperature series and the mean post-2003 multiproxies. For reference, we have also included the 500-year control run of the Institut Pierre Simon Laplace (IPSL) GCM used in the IPCC AR4 (Solomon *et al.*, 2007) (using fixed external forcing at preindustrial greenhouse gas levels), which is discussed in Section 11.3.2. In Fig. 11.16a we used difference structure functions so that the interpretation is particularly simple, although a consequence is that all the logarithmic slopes are > 0. We therefore also give the corresponding Haar structure function analysis, which gives correct exponent estimates over the whole range (Fig. 11.16b and below).

Key points to note in Fig. 11.16a are (a) the use of annually averaged instrumental data, and (b) the distinction made between globally and locally averaged quantities. Also shown is the "glacial/interglacial window," which is a rough delineation of the time scales and amplitudes of the interglacials (Δt is the half quasi-period, and for a white noise, S is double the amplitude). The calibration of the paleotemperatures is thus constrained so that it goes through the window at large Δt but joins up to the *local* instrumental $S(\Delta t)$ at small Δt (see the discussion around Fig. 11.10). Interestingly, we see that the local month-to-month temperature variations are of about the same order as the mean temperature variations between glacials and interglacials.

In addition, as discussed in Section 11.1.2, since the last 10 kyr GRIP fluctuations are anomalously low (in Fig. 11.16a see the nearly flat Holocene curve, compared with the full 91 kyr curve), the calibration must be based on this flatter $S(\Delta t)$. Starting at $\tau_c \approx$ 10 years, one can plausibly extrapolate the global $S(\Delta t)$'s using $H = 0.4$ ($\beta \approx 1.8$), all the way to the interglacial window (with nearly an identical S as in Lovejoy and Schertzer, 1986), although the northern hemisphere multiproxy series do not extrapolate quite as well, possibly because of their higher intermittency (see Section 11.2.3). The local temperatures extrapolate (starting at $\tau_c \approx 20$ years) with a lower exponent corresponding to $\beta \approx 1.4$, which is

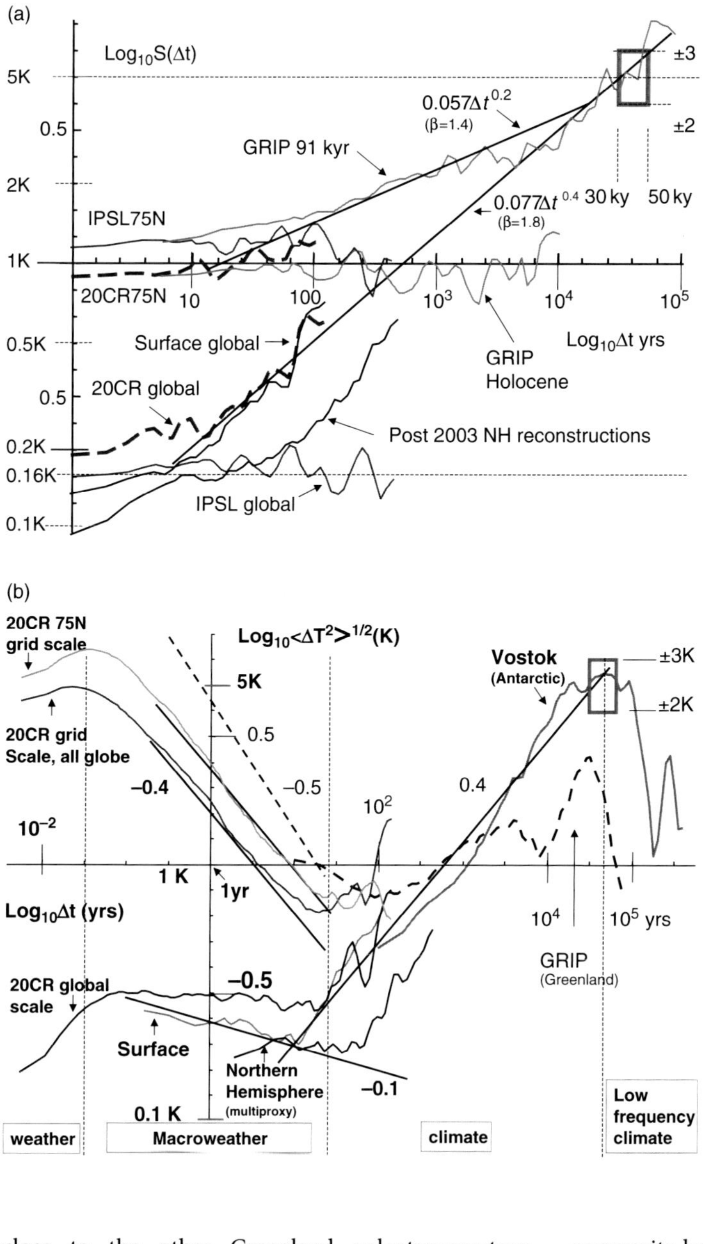

Fig. 11.16 (a) Comparison of the RMS structure function $S(\Delta t)$ of the high-resolution (5.2-year) GRIP, IPSL, 20CR mean surface series, mean of the three post-2003 Northern hemisphere reconstructions for globally averaged temperatures (bottom left set) and the mean at Greenland latitudes (upper set), all using fluctuations defined as differences (poor man's wavelet) so that the vertical scale directly indicates typical changes in temperature. In addition, the GRIP data are divided into two groups: the Holocene (taken as the last 10 kyr, lower) and the entire 91 kyr of the high-resolution GRIP series (upper). The GRIP $\delta^{18}O$ data have been calibrated by lining up the Holocene structure function with the mean 75° N 20CR reanalysis structure function (corresponding to $\approx$ 0.65 K/mil). When this is done, the 20CR and surface mean global structure functions can be extrapolated with exponent $H \approx 0.4$ (see the corresponding line) to the "interglacial window" (box at top right) corresponding to half pseudo-periods between 30 and 50 kyr with variations ($= \pm S/2$) between $\pm$ 2 and $\pm$ 3 K. This line corresponds to spectral exponents $\beta = 1.8$ (this is exactly the same line as proposed in Lovejoy and Schertzer, 1986). Finally, we show a line with slope $\xi(2)/2 = 0.2$ corresponding to the GRIP $\beta = 1.4$; we can see that extrapolating it to 50 kyr explains the local temperature spectra quite well. (b) The equivalent of Fig. 11.16a except for the RMS Haar structure function rather than the RMS difference structure function and including daily-resolution 20CR data (see Fig. 11.17a for the first-order Haar structure function). At the left top we show grid-point-scale (2° × 2°) daily-scale fluctuations for both 75° N and globally averaged along with reference slope $\xi(2)/2 = -0.4 \approx H$ (20CR, 700 mb). On the lower left, we see, at daily resolution, the corresponding globally averaged structure function. Also shown are the average of the three in-situ surface series (Fig. 10.12) as well as the post-2003 multiproxy structure function (Fig. 11.9). At the right we show both the GRIP (55 cm resolution, with calibration constant 0.5 K/mil) and the Vostok paleotemperature series. Also shown is the interglacial "window." All reproduced from Lovejoy and Schertzer (2012a).

close to the other Greenland paleotemperature exponents (Table 11.4), presumably reflecting the fact that the Antarctic temperatures are better surrogates for global rather than local temperatures; these exponents are all averages over spectra of series of $\approx$ 100 kyr or more in length.

Although the Haar structure function composite (Fig. 11.16b) tells essentially the same story, we have extended the overall scale range of the composite by including daily-resolution 20CR data. In addition, because of its ability to capture the statistics of fluctuations over a wider range of H exponents, we can clearly see the transition scales and can even glimpse the beginning of a low-frequency climate regime with $H_{lc} < 0$ at scales beyond $\tau_c \approx 100$ kyr. (Note added in proof: analysis of the 800 kyr EPICA Antarctic series shows $H_{lc} \sim -0.8$.)

11.2.3 Revisiting temporal multifractal climate intermittency

Up until now we have focused on either the spectrum or its real-space equivalent, the RMS fluctuations, although occasionally we have considered the (predicted) power-law tails of probability distributions of extreme fluctuations (e.g. Fig. 5.21), although these may be artificially biased by interpolation or other data-processing artefacts. Of course, if the climate process was quasi-Gaussian, then $K(q) = 0$, $\xi(q) = qH$ and the exponent $\xi(2) = 2H = \beta - 1$ would be sufficient for a complete characterization of the statistics. However, after the introduction of each new

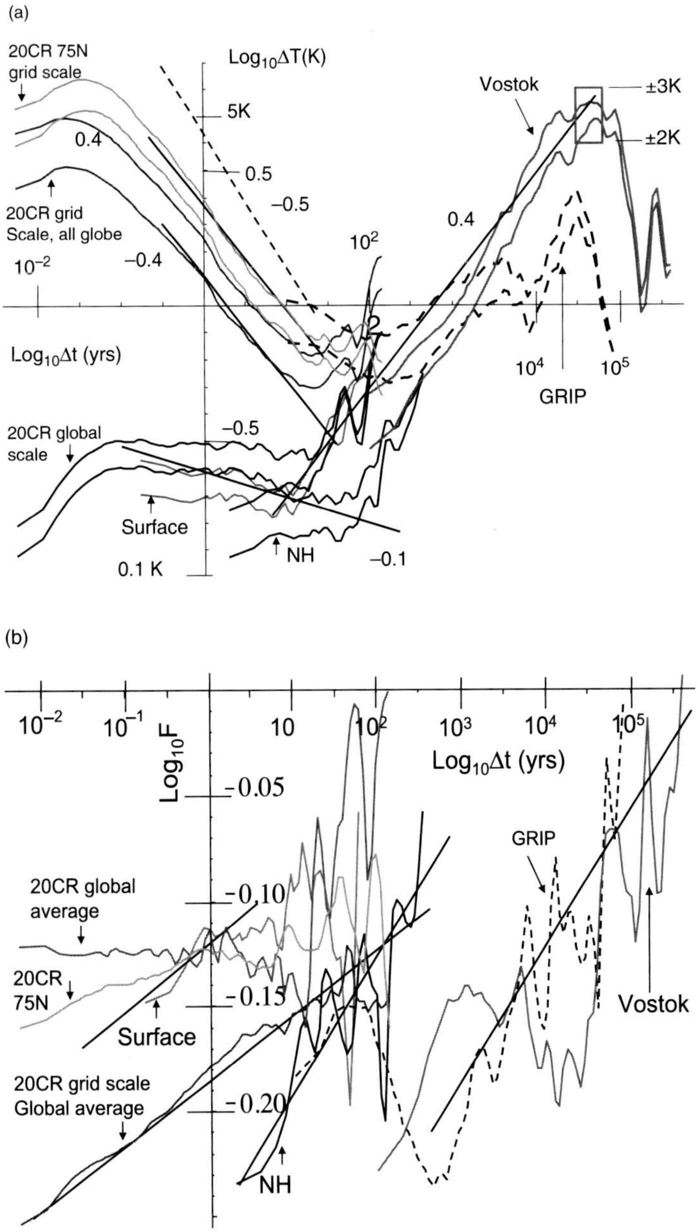

Fig. 11.17 (a) A composite showing (calibrated) Haar structure functions for both the mean fluctuation (logarithmic slope $\xi(1)$, bottom of each pair) and the RMS fluctuation (logarithmic slope $\xi(2)/2$, top of each pair). On the left we use the 20CR daily data at one-grid-point resolution averaged over the longitudes at 75° N (top left), averaged over all longitudes and over the globe (bottom left). In the middle, the mean of the three surface series from Fig. 10.12 and Appendix 10C; at the bottom, the post-2003 northern hemisphere (NH) reconstructions, and right, the GRIP 5.2-year resolution series (dashed) back to 91 kyr and the Vostok series, interpolated to 50 years, back to 420 kyr. Solid reference slopes −0.4, −0.1, 0.4 correspond (ignoring intermittency) to $\beta = 0.2, 0.8, 1.8$ respectively. Also shown is a dashed reference slope −0.5 corresponding to Gaussian white noise. The tendency for the mean and RMS curves to converge is due to intermittency, the rate of convergence has exponent $= K(2)/2 \approx C_1$; see below. The box indicates the "glacial/interglacial window" discussed earlier. (b) The same data as in Fig. 11.17a: this figure attempts to isolate the intermittency near the mean by estimating the function $F = \langle \Delta T \rangle \left(\langle \Delta T^{1-\Delta q} \rangle / \langle \Delta T^{1+\Delta q} \rangle \right)^{1/(2\Delta q)}$ with $\Delta q = 0.1$: in the small Δq limit, $F \approx \Delta t^{C1}$. This exploits the relation $K'(1) = C_1 = \xi(1) - \xi'(1)$, see Eqn. (10.11). Reference lines with slopes 0.03 (left pair) and 0.065 (right pair) are shown. Note the relatively low intermittency (roughly flat) lines associated with the surface global average temperatures and the 75° N 20CR series.

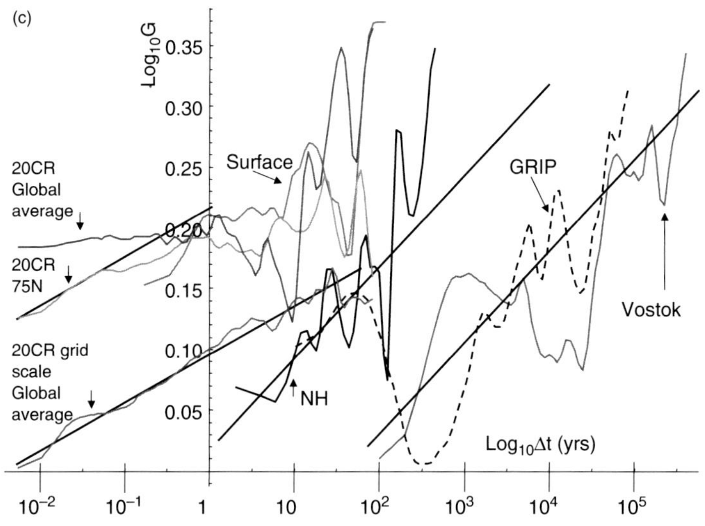

Fig. 11.17 (c) Same as Figs. 11.17a, 11.17b but for a function G whose exponent is the second derivative $-\xi''(1) = K''(1)$: $G = (\langle\Delta T^{1+2\Delta q}\rangle\langle\Delta T^{1-2\Delta q}\rangle/\langle\Delta T\rangle^2)^{-1/(2\Delta q)^2}$, in the small Δq limit, $G \approx \Delta t^{K''(1)}$ so that the logarithmic slope of G is an estimate of $K''(1)$; here we took $\Delta q = 0.1$. Using Eqn. (3.28), we can use this to obtain $\alpha = K''(1)/C_1$, using the C_1 from the exponents estimated in Fig. 11.17b. Reference lines with slopes 0.045 (left pair) and 0.08 (right pair) are shown, corresponding to $\alpha = 1.25, 1.5$ respectively. Note the relatively low intermittency (roughly flat) lines associated with the surface global average temperatures and the 75° N 20CR series.

dataset we have systematically presented flux-based analyses of the intermittency and we have seen that generally $K(q) > 0$ for $q>1$ for a conservative flux so that the choice of $q = 2$ is somewhat subjective.

A drawback of these flux analyses for characterizing intermittency is that it is hard to get an overall wide-scale-range picture: each dataset is used to estimate fluxes at its smallest scale, so that in each scaling regime, the fluxes are physically distinct; one cannot just superpose trace moment analyses from different datasets at widely different resolutions. Another consequence of using fluxes (and hence the cascade parameters and outer scale) is that they are estimated at the smallest scales and can be unduly influenced by noise in the data. This is particularly problematic for proxy data.

It is therefore of interest to develop complementary ways to quantify the intermittency by directly using the fluctuations – e.g. via the Haar structure functions. A straightforward way to quantitatively gauge the importance of the intermittency is to compare the RMS and $q = 1$ moments: $\langle\Delta T^2\rangle^{1/2}$ compared to $\langle\Delta T\rangle$. In Fig. 11.17a we present the results for many of the series discussed earlier. Since $(\langle\Delta T^2\rangle^{1/2}/\langle\Delta T\rangle) \approx \Delta t^{\xi(2)/2-\xi(1)} = \Delta t^{-K(2)/2}$ and $K(2) > 0$, we see that for each pair of curves there is indeed a tendency for them to converge as Δt increases.

In order to quantify this in a theoretically more satisfactory way, we can directly construct a function $F \approx \Delta t^{C1}$ where $C_1 = K'(1)$ is the exponent characterizing the multifractality/intermittency near the mean ($q = 1$):

$$F = \langle\Delta T\rangle(\langle\Delta T^{1-\Delta q}\rangle/\langle\Delta T^{1+\Delta q}\rangle)^{1/(2\Delta q)} \tag{11.4}$$

In the small Δq limit and assuming $k(1)=0$, it is easy to see that $F \approx \Delta t^{C_1}$ (here we numerically use $\Delta q = 0.1$); see Fig. 11.7b. Note that if $\alpha = 2$, then we have $K(2)/2 = C_1$ (see Eqn. (3.46)); here, with a somewhat smaller α, this is still approximately true. Since C_1 is small – roughly in the range 0 to 0.065 – we see that the curves are very "noisy": we need very good statistics to clearly discern the relatively small variations in F. From Fig. 11.17b several points can be made. First, the behaviours tend to be qualitatively different in the three different regimes, although there are exceptions – the 20CR one-grid-point resolution globally averaged data seem to have a constant intermittency exponent straddling the weather/macroweather boundary. Second, at a 2° × 2° scale, the fluctuations averaged over 75° N are larger in amplitude (Fig. 11.17a) but are less intermittent than the corresponding globally averaged fluctuations. Third, the post-2003 northern hemisphere reconstructions are much more intermittent than the global surface series. Fourth, the paleotemperatures are quite intermittent ($C_1 \approx 0.065$; close to the value obtained by Schmitt *et al.*, 1995, for the low resolution GRIP core). Fifth, the 20CR 75° N and surface series have very low slopes (C_1) over most of their ranges, giving some justification for quasi-Gaussian treatments.

For a more complete characterization of the intermittency, we can estimate the second derivative

Box 11.4 Paleocascades

If the paleodata are scaling, then we may anticipate that they will display corresponding cascade structures; we have already shown this for part of the high-resolution GRIP core in Fig. 11.3, although the intermittency was not much stronger than that of quasi-Gaussian processes. In Fig. 11.18 we compare the trace moments from four series using a common length (90 kyr) and common resolution (50 years). Along with the GRIP high-resolution and Vostok data discussed above, we also show the analysis of the publicly available GRIP 55 cm data and an analysis of the publicly available series from the North GRIP core which was finished in 2003 and which has a more stable chronology (North Greenland Ice Core Project, 2004). From the figure we see that there is evidence for cascade structures with outer scales $\approx$ 1–2 kyr; the scaling is reasonably good for periods smaller than about 300 years, which is roughly the scale corresponding to the change in the spectrum (Fig. 11.13a) from $\beta_c \approx 1.4$ to $\beta_{mw} \approx 0.2$. The detailed cascade parameter estimates are given in Table 11.2. It can be seen that the key parameter estimates (C_1, α) are fairly stable (with $C_1 \approx 0.11$, $\alpha \approx 1.7$), the main variation being in the outer scales, which are seen to be sensitive to the resolution (cf. the difference between GRIP high and GRIP 55 cm).

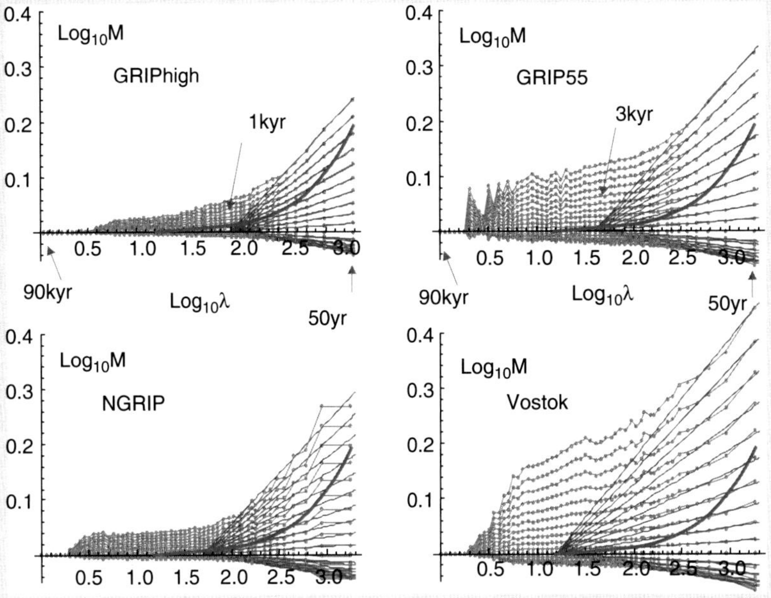

Fig. 11.18 The four paleotemperature datasets transformed by interpolation (or, for the GRIP high-resolution series, by averaging) to 50-year resolutions: the most recent 90 kyr only. The superposed thick curves are the envelopes of the trace moments of the quasi-Gaussian processes as discussed in Appendix 4A. It can be seen that with the possible exception of the GRIP high-resolution (5.2-year) data, the data are not close to being quasi-Gaussian.

$\xi''(1) = -K''(1)$ (Eqn. (10.12)), which can be used to determine $\alpha = K''(1)/C_1$ (from Eqn. (3.28)). By analogy with the estimate of C_1, we can determine the function G (Fig. 11.17c), whose exponent is $K''(1)$. From Fig. 11.17c we obtain a very similar graph to Fig. 11.17b – this is exactly as expected for universal multifractals: the ratio of the absolute slopes is simply α. Doing this, we obtain $\alpha = 1.5, 1.25$ (respectively for the macroweather and climate regimes). The latter may be compared with the estimate $\alpha \approx 1.6$ for GRIP (low resolution) in Schmitt *et al.* (1995).

11.2.4 Space-time temperature variability in the climate regime and a theoretical composite

In Chapter 10, we saw that the pure weather/macroweather model does a very good job at predicting the temporal statistics up to scale $\approx \tau_c$. However, we pointed out that its macroweather prediction of spatial homogeneity is completely unrealistic, being contradicted by the strong spatial heterogeneity associated with climatic zones. We were therefore led – if only in order to understand the macroweather regime – to introduce a lower-frequency space-time climate process, the (multifractal) $\varepsilon_c(\underline{r}, t)$. We hypothesized that this process had strong spatial variability and was multiplicatively linked with the weather/macroweather ("w,mw") process $\varepsilon_{w,mw}(\underline{r}, t)$ so that the entire weather/climate ("w,c") process could be written:

$$\varepsilon_{w,c}(\underline{r}, t) = \varepsilon_{w,mw}(\underline{r}, t)\varepsilon_c(\underline{r}, t) \approx \varepsilon_w(\underline{r}, t)\varepsilon_{mw}(t)\varepsilon_c(\underline{r}, t) \tag{11.5}$$

where we have used the decomposition (Eqn. (10.4)) on the far right (the various factors in Eqn. (11.5) are appropriately band-limited in the frequency domain). Using space-time spectra and spatial trace moments, we showed empirically that this equation was reasonably compatible with 20CR temperature reanalyses, including the additional hypothesis that $\varepsilon_c(\underline{r}, t)$ is statistically independent of the other factors. Using the fact that $\varepsilon_{mw}(t)$ is quasi-Gaussian and assuming that $\varepsilon_c(\underline{r}, t)$ is multifractal, we estimated the horizontal climate exponents $H_c \approx 0.7$, $C_{1,c} \approx 0.11$, $\alpha_c \approx 1.75$. This successfully predicted the form of the joint space-time fluctuations in the macroweather regime (Section 10.3).

Thanks to the temporal analysis of the climate spectra and structure functions in Section 11.2.2, we now have an estimate of the temporal climate fluctuation exponent $H_{c,\tau} \approx 0.4$ (Fig. 11.16b), and thanks to the intermittency analysis in Section 11.2.3 we have an estimate of the temporal intermittency exponents: $C_{1c,\tau} \approx 0.065$, $\alpha_{c,\tau} \approx 1.5$ (Figs. 11.17a, 11.17b). We thus have enough data to give a rough but full (horizontal) space-time statistical description of the atmosphere including the climate regime. Since we have evidence that the climate process is multifractal in time and also in space, the main remaining assumption concerns its nature. Is it a true space-time scaling process analogous to the weather process, or is it more like the macroweather process with statistically independent spatial and temporal factors? On the basis of admittedly little theoretical or empirical information, we hypothesize that the former is the correct model, i.e. that the climate is a genuine space-time process with corresponding space-time scale function. With this assumption, we obtain the bottom line of the following set of equations governing atmospheric variability:

$$\begin{aligned}
&S_{q,w}(\underline{\Delta r}, \Delta t) = (s_{q,w})^q [[(\underline{\Delta r}, \Delta t)]]_w^{qH_w - K_w(q)}; \quad \tau_i < \Delta t < \tau_w \\
&[[(\Delta x, 0, 0, \Delta t)]]_{w,can} = \left(\left(\frac{\Delta x}{L_e^*}\right)^2 + \left(\frac{\Delta t}{\tau_w}\right)^2\right)^{1/2} \\
&S_{q,mw}(\underline{\Delta r}, \Delta t) = (s_{q,mw})^q \|\underline{\Delta r}\|_c^{qH_s - K_c(q)} \left(\frac{\Delta t}{\tau_w}\right)^{qH_{mw}}; \tau_w < \Delta t < \tau_c \\
&\|(\Delta x, 0, 0)\|_{c,can} = \left|\frac{\Delta x}{L_e^*}\right| \\
&S_{q,c}(\underline{\Delta r}, \Delta t) = (s_{q,c})^q [[(\underline{\Delta r}, \Delta t)]]_c^{qH_c - K_c(q)}; \tau_c < \Delta t < \tau_{lc} \\
&[[(\Delta x, 0, 0, \Delta t)]]_{c,can} = \left(\left(\frac{\Delta x}{L_e^*}\right)^2 + \left(\frac{\Delta t}{\tau_{lc}}\right)^{2/H_{c,t}}\right)^{1/2}
\end{aligned} \tag{11.6}$$

In Eqn. (11.6) we have expressed the structure function $S_q(\underline{\Delta r}, \Delta t) = \langle \Delta T(\underline{\Delta r}, \Delta t)^q \rangle$ first for general space-time vectors ($\underline{\Delta r}$, Δt) and in terms of general space-time and spatial scale functions ($[[(\underline{\Delta r}, \Delta t)]]$, $\|\underline{\Delta r}\|$ respectively) and then given the example of the simplest canonical scale function in one spatial dimension. These formulae are valid for $\Delta x < L_e{}^*$, where $L_e{}^*$ is the effective outer scale – estimated at 45° N to be about 5000 km (compared to the half-circumference of $\approx$ 14 000 km). The new climate regime formula is the bottom row: for completeness we have included the weather and macroweather formulae as well (top two rows). $H_{c,t} = H_c/H_{c,\tau}$ is the climate space-time anisotropy exponent analogous to $H_z = H_h/H_v$ which characterizes the vertical scaling in the weather regime. The empirical parameters are summarized in Table 11.7. These are the simplest equations using canonical scale functions, i.e. we have ignored mean winds and the vertical and meridional directions as well as some latitudinal, geographical variations (all of which can in principle be handled with generalized scale invariance). With this caveat, Eqn. (11.6) is compatible with all we know both theoretically and empirically about atmospheric fluctuations.

The s_q's are proportionality constants with dimensions K discussed below. The (top) weather regime equation uses the canonical scale function for the temperature. For simplicity we used Eqn. (8.16) with a zero mean advection velocity in the matrix B. For the weather exponents, we found $H_w \approx 0.51$,

Table 11.7 A comparison of spatial and temporal exponents and temporal outer scales for the temperature fluctuations in Eqn. (11.6). The spatial outer scales are of the order of the planetary scale and vary with latitude; at 45° N, they are about $L_e^* \approx 5000$ km. For the amplitude of the fluctuations in Eqn. (11.6), we have theoretically $s_{1,w} \approx s_{1,mw}$ with value ≈ 5 *K*, and $s_{1,c} \approx s_{1,w}(\tau_c/\tau_w)^{H_{mw}} \approx 0.06$ *K*. For the weather parameters, see Chapter 8, Table 8.1, Fig. 8.5c; for the macroweather, see Section 10.3.2 and Table 10.2; for the climate parameters, see Table 11.4 and Figs. 11.5b, 11.16a, 11.17a, 11.17b. The macroweather–climate spatial exponent ≈ 0.7 is the estimate from the 20CR 0.995 sigma (near-surface) field, which may be biased due to the reanalysis issues discussed in Chapter 4 (see Table 4.3); the value $H \approx 0.5$ (Table 8.2) may be more appropriate.

	H		C_1		α		Outer time scale
Regime	**Space**	**Time**	**Space**	**Time**	**Space**	**Time**	
Weather	0.7	0.51	0.1	0.087	1.4	1.61	5–20 days
Macroweather	0.7	–0.4	0.1	0	1.4	–	20–40 years
Climate	0.7	0.4	0.1	0.065	1.4	1.5	30–50 kyr

$C_{1w} = 0.087$, $\alpha_w = 1.61$ (see Chapter 8, Table 8.1). The (middle row) macroweather structure function (defined by Haar wavelets, 20CR surface) is discussed in Section 10.3.2, where we estimated $H_c \approx 0.7$, $H_{mw} \approx -0.4$ and $C_{1,mw,\tau} = C_{1,c,\tau} \approx 0.11$, $\alpha_{mw\tau} = \alpha_{,\tau} \approx 1.75$ (Table 10.2 the 700 mb value $H_c \approx 1.4$ is probably spuriously high). The (bottom) climate-regime formula uses the canonical scale function for self-affine anisotropic scaling processes; since the horizontal H_c is much bigger than the temporal $H_{c,\tau}$ (Eqn. (8.6)), empirically we have $H_{c,t} \approx 0.7/0.4 = 1.75$. We should note that strictly speaking, Eqn. (11.6) implies that the ratio of the spatial to temporal climate C_1's is also equal to $H_{c,t} \approx 1.75$, indeed the actual C_1 values are respectively ≈ 0.1, ≈ 0.065, ratio 1.55, close to $H_{c,t}$ (Fig. 11.17b).

Let us now consider the constants $s_{q,w}, s_{q,mw}, s_{q,c}$. These are constrained so that the structure functions are continuous at the boundaries between the regimes. Taking $\Delta x = L_e^*$ (or almost equivalently considering the structure functions of the global averages), and for simplicity considering only $q = 1$, we have at $\Delta t = \tau_w, \tau_c$:

$$\langle \Delta T(L_e^*, \tau_w) \rangle_w = \langle \Delta T(L_e^*, \tau_w) \rangle_{mw}; \quad s_{1,w} \approx s_{1,mw}$$
$$\langle \Delta T(L_e^*, \tau_c) \rangle_{mw} = \langle \Delta T(L_e^*, \tau_c) \rangle_c; \quad s_{1,mw} \left(\frac{\tau_c}{\tau_w}\right)^{H_{mw}} \approx s_{1,c} \tag{11.7}$$

Using the 45° N data, 20CR, surface, we find that these equations work remarkably well. For example, we verify $s_{1,w} \approx s_{1,mw} \approx 5$ K (e.g. Fig. 10.19 for slightly larger s_2 values) and with $\tau_c/\tau_{mw} \approx 10^3$, $H_{mw} \approx -0.4$ we obtain the prediction $s_{1,c} \approx 0.06$ *K*, which is indeed close to the observed value (Figs. 11.16a, 11.16b; see also Table 10C.2 and Lovejoy and Schertzer, 1986), nearly the same values for the slightly larger $s_{2,c}$. Finally, using $H_c \approx 0.4$, and $\tau_{lc}/\tau_c \approx 30$ kyr/30 years $\approx 10^3$, we find the prediction $\langle \Delta T(\tau_{lc}) \rangle \approx 5$ *K*, which is roughly the value of the glacial/interglacial "window" (Figs. 11.16a, 11.16b).

The above relations do not depend on the type of the assumed space-time climate relation. However, if the scale function is of the form hypothesized in Eqn. (11.6), it has interesting implications because it predicts the following space-time relations in the climate regime:

$$\Delta x = L_e^* \left(\frac{\Delta t}{\tau_{lc}}\right)^{1/H_{c,t}}; \quad \tau_c < \Delta t < \tau_{lc} \tag{11.8}$$

Taking the smallest time scale in the climate regime, $\Delta t = \tau_c$, yields a characteristic inner spatial climate scale:

$$\Delta x_c = L_e^* \left(\frac{\tau_c}{\tau_{lc}}\right)^{1/H_{c,t}} \tag{11.9}$$

Using $\tau_c/\tau_{lc} \approx 10^{-3}$, $L_e^* \approx 5000$ km, $H_{c,t} \approx 0.7/0.4 = 1.75$, we find $\Delta x_c \approx 100$ km. Below Δx_c, the climate process is expected to be smooth, homogeneous; this represents the characteristic size of the smallest climatic zone. Although this conclusion is perhaps surprising, it appears to be compatible with the spatial analyses in Fig. 10.18.

Perhaps the best way to think about the relation in Eqn. (11.8) is that for a climate zone (i.e. a multifractal singularity) of a given spatial extent Δx, Eqn. (11.8) gives us its "lifetime" or typical length of time before its identity is lost: $\Delta t = \tau_{lc}(\Delta x/L_e^*)^{H_{c,t}}$. This would mean that well-defined climate structures of size 100 km typically maintain their identity for about $\tau_c \approx 30$ years whereas those 5000 km in size maintain them for about $\tau_{lc} \approx$ 30–50 kyr. This corresponds to climatic zones with boundaries shifting at rates of $\approx$ 3 km/year and $\approx$ 100 m/year respectively.

If this analysis is correct, then this could lead to simplifications in climate modelling since there would be no need to make simulations at much higher resolutions. Similarly, when studying temporal variations at scales Δt, there would be no need to collect data at spatial resolutions higher than $\Delta x = L_e^* (\Delta t/\tau_{lc})^{1/H_{c,t}}$.

The space-time climate structure function proposed in Eqn. (11.6) – and its extensions to include meridional variations – is perhaps *the* fundamental law of climate variability. Although it is apparently compatible with all the information we have on the climate, its exact form must be considered tentative, being mostly based either on spatial data at resolutions higher than τ_c, (Section 10.3) or on temporal paleodata averaged over hemispheric scales (Section 11.1.5) or at single points in space (Section 11.2.2). Clearly a convincing test would require data with both high spatial resolution and a significant range of scales in the climate regime – and, given the large climate databases which are now available, such verification may be possible in the near future.

From the theory developed in Chapters 6 and 8 it is easy to write down the spectra corresponding to the fluctuation formulae, and we obtain the following basic spectral densities:

$$\begin{aligned} P_w(k,\omega) &\approx \left((kL_e^*)^2 + (\omega\tau_w)^2\right)^{-s_w/2}; \quad \tau_i^{-1} > \omega > \tau_w^{-1} \\ P_{mw}(k,\omega) &\approx (kL_e^*)^{-\beta_c}(\omega\tau_w)^{-\beta_{mw}}; \quad \tau_c^{-1} < \omega < \tau_w^{-1} \\ P_c(k,\omega) &\approx \left((kL_e^*)^2 + (\omega\tau_{lc})^{2/H_{c,t}}\right)^{-s_c/2}; \quad \tau_{lc}^{-1} < \omega < \tau_c^{-1} \end{aligned} \tag{11.10}$$

(all are valid for horizontal wavenumber $k > L_e^{*-1}$, although P_{mw}, P_c presumably with the restriction $k < \Delta x_c^{-1}$: see Eqn. (11.9)). For the generalization to include a mean advection velocity, see Eqns. (8.29), (8.30): essentially $\omega \to (\omega + \mu_x k_x)/\sigma_x$, where μ_x σ_x are determined by advection, and the outer scale of the climate regime where a new low-frequency climate regime begins is τ_{cl}. The space-time spectral density exponents are $\beta_c = 1 + \xi_c(2)$, $s_w = 1 + \beta_w = 2 + \xi_w(2)$, $s_c = \beta_c + H_{c,t}$ (recall $H_{c,t} = H_{c,\tau}/H_c$); see Eqns. (6.60)–(6.66) for this (apparently) self-affine space-time case. The joint spectral density for the weather regime is given in the simplified form (with no mean advection); in the climate regime – as with the corresponding structure function – it is largely a guess (and ignores complications such as possible "climate advection"). However, the 1D spectra (which can be obtained by integrating $P_c(k_x,\omega)$: see Eqn. (6.66)) can be given with more confidence: $E_c(k) \approx k^{-\beta_c}$, $E_c(\omega) \approx \omega^{-\beta_{c,\tau}}$ with $\beta_{c,\tau} = 1 + H_{c,\tau}\,(\beta_c - 1)/H_c$ – the spatial behaviour due to the theory and limited empirical evidence discussed above, and the temporal from the empirical evidence discussed earlier in this chapter. For the temperature, the empirical exponents are roughly $\beta_w \approx 2$, $\beta_c \approx 2.2$, $\beta_{mw} \approx 0.2$, $\beta_{c,\tau} \approx 1.8$ (the latter is the temporal exponent in the climate regime: Table 11.4).

11.3 Climate forcings and global climate models

11.3.1 The scaling of CO_2 concentrations and solar, volcanic and orbital forcings

Starting at τ_c, climate variability continues up to scales of the age of the earth: a factor $\approx 10^8$. Although in this chapter we have restricted our attention to scales up to the glacial/interglacial transition scale τ_{lc} where a new "low-frequency climate" ("*lc*") regime begins, τ_{lc}/τ_c is nevertheless a large ratio $\approx 10^3$–10^4 ($\approx$ 50–100 kyr/ 10–30 years). In general – and in keeping with the approach used throughout the earlier chapters – we expect that any basic relevant physical processes over this range will be scaling. Why then at τ_c do the stable macroweather fluctuations with $H < 0$ give way to unstable, lower-frequency climate fluctuations with $H > 0$? Let us now examine this question.

Mechanisms governing the climate are traditionally classified as either internal or external, although the usual supposition is that there is a nonlinear combination of both, e.g. the amplification of external causation by nonlinear internal "feedback" mechanisms. In the literature, the discussion of these issues has been strongly tinted by the development of GCMs and their response to various "climate forcings," which are introduced as changing boundary conditions, and the focus has been very much on possible external mechanisms. However, if the supposed amplification factors are large – as we show they must be – then it will be hard to distinguish nominally external paradigms from purely internal ones.

Let us begin by considering the various possible external drivers as functions of scale. These forcings may be classified according to whether they are

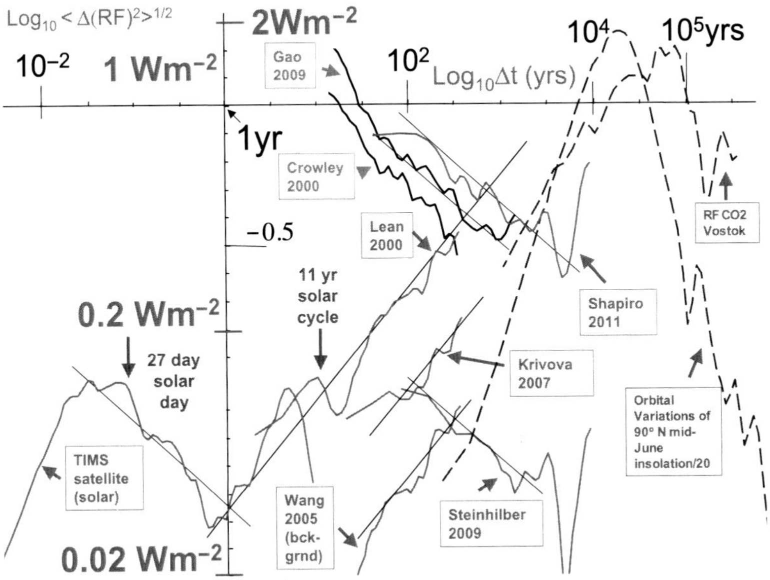

Fig. 11.19 Comparison of RMS Haar fluctuations for various solar, volcanic, orbital and CO_2 data in units of radiative forcing (R_F, units W/m^2). For the solar radiances, the values of estimated total solar irradiance (TSI) were converted into R_F using albedo = 0.7 and a geometric factor 1/4 (yielding an overall reduction factor of 0.175). The TIMS satellite data is for 8.7 years from 2003 to the present at a 6-hour resolution; the data are from http://eobadmin.gsfc.nasa.gov/Features/SORCE/sorce_07.php. Note that the Lean (2000) reconstruction includes the 11-year solar cycle whereas the Wang (2005) curve is only for the background. The Krivova 2007 curve has a 10-year resolution. The Shapiro (2011) curve (the last 8963 years) was degraded to 20-year resolution to average out the solar cycle; the Steinhilber (2009) curve was at a 40-year resolution over the last 9300 years. The volcanic series were from reconstructions of stratospheric sulfates using ice-core proxies. The Vostok paleo-CO_2 series were converted to R_F using 3.7 W/m^2 per CO_2 doubling (IPCC AR4: Solomon *et al.*, 2007), the solar insolation at the north pole on June 15th was divided by 20 but is not a true R_F. The orbital variation curve was interpolated to 100-yesar resolution and the low- and high-frequency fall-offs have logarithmic slopes −1, 1, i.e. they are the minimum and maximum possible for the usual (linear) Haar fluctuations. All the structure functions have been increased by a factor of 2 (i.e. without changing their relative amplitudes) so that the temperature fluctuations are roughly "calibrated" with the difference and tendency fluctuations as discussed in Section 10.2.2. Reproduced from Lovejoy and Schertzer (2012b).

scaling or nonscaling This is useful because nonscaling climate forcings – i.e. at well-defined frequencies – would leave strong signatures in the form of breaks in the temperature (and other) scalings. However, we have seen that over the range of time scales between $\tau_c \approx$ 10–30 years and $\tau_{lc} \approx$ 50–100 kyr, the temperature is at least roughly scaling; similarly, in Fig. 1.9d, we saw that the unforced GCM "control runs" are scaling for all scales greater than $\tau_w \approx$ 10–30 days, i.e. they do not have any characteristic time scales (see Section 11.3.3). Since in the previous sections we saw no evidence for any such strong break, we conclude that the relevant forcings are themselves likely to be scaling.

There is of course an important nonscaling exception: the narrow-band orbital forcings at scales somewhat shorter but close enough to the upper time scale τ_{lc} so that the break they induce is plausibly compatible with the observations. We are of course referring to the orbital "Milankovitch" forcings, and we saw that there is wide consensus that these are responsible for the observed break at τ_{lc}, although even this is not at all trivial. For example, in Section 11.2.1 we noted that the strongest of these – the precessional and obliquity forcings – have only weak signals in the paleotemperature records (Figs. 11.13a, 11.13b). However, if the internal variability is strong enough – as in the case of fully developed turbulence discussed in Chapter 2 – it could, at least in principle, be responsible for a (wide-range) scaling temperature response at higher frequencies.

Interestingly, the main signal in the temperature is nearer 100 kyr, corresponding to the weaker orbital eccentricity variations, yet even this poses a further problem, since signs of the much stronger eccentricity variation at about 400 kyr are virtually absent in the paleoclimate record (these are respectively the "100 kyr" and "400 kyr" problems: see Ganopolski and Calov, 2011, and references therein). Note that while

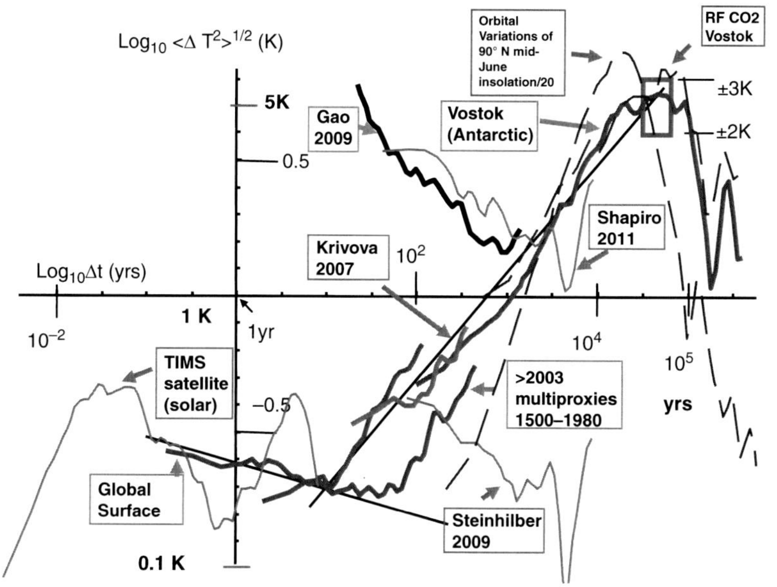

Fig. 11.20 The RMS structure functions of the main forcings from Fig. 11.19 were converted into RMS temperature structure functions using a unique (and scale-independent) climate sensitivity $\lambda = 4.5$ K/(Wm^{-2}). The reference lines have slopes of −0.1 and +0.4. It can be seen that the main orbital insolation fluctuations occur at time scales roughly 3–4 times smaller than the main temperature fluctuations. The CO_2 curve is probably not a true forcing since cross-spectral analysis shows that the CO_2 fluctuations lag the temperature fluctuations over most of the range. Reproduced from Lovejoy and Schertzer (2012b).

this is true for these high-latitude signals, the precessional signal is seen quite strongly in various tropical climate records such as speleothems (see especially Wang *et al.*, 2007). Since the overall (global) change in solar insolation is very small, the exact mechanism that would link the orbital variations to temperature fluctuations is still actively researched (see e.g. Berger *et al.*, 2005). To see the scale dependence, Fig. 11.19 shows Haar structure function analysis of the solar irradiance variations at the north pole (every June 15th) determined from astronomical calculations (Berger and Loutre, 1991). Unlike the other structure functions shown, this does not correspond to a true radiative forcing; the structure function of this deterministic forcing is shown in order to indicate its dominant time scales. One sees that the variability is confined to a fairly narrow range of scales, and in Fig. 11.20 we see that this range is about 3–4 times smaller than that of the peak in the paleotemperature variability. This is the 100 kyr problem: a strong nonlinear feedback mechanism is therefore needed just to explain the lower-frequency variability out to (100 kyr)$^{-1}$.

Even if we accept that the orbital forcing is indeed the "pacemaker of the ice ages" (Hays *et al.*, 1976), there remains the higher-frequency issue as to the origin of the strong continuous (near-scaling) background spectrum; much higher than the highest orbital frequency ($\approx$ (19 kyr)$^{-1}$: Figs. 11.13a, 11.13b, Wunsch, 2003; Section 11.2.1). Although proposals have been made to explain how various higher-frequency harmonics and related "combination tones" might arise (e.g. Ghil, 1994), these mechanisms at best account for discrete frequencies – only a small fraction of the continuous spectrum – and hence a small fraction of the overall variance. This still leaves us with the need to find a scaling mechanism – or at the very least one that would yield the observed broad spectral response.

At first sight, an attractive possibility is to invoke greenhouse gas forcings. CO_2 has indeed varied significantly over precisely the same scale range as the temperature. For example, from the Vostok paleo-CO_2 series we can estimate the implied "radiative forcing." The radiative forcing is defined as the effective change in the balance of incoming and outcoming energy flux (in W/m^2), in this case due to the (logarithmic) change in CO_2 concentrations. Using the value 3.7 W/m^2 for a CO_2 doubling (recommended in the IPCC AR4: Solomon *et al.*, 2007), the time series of paleo-CO_2 concentrations can be converted into an effective radiative forcing series. Fig. 11.19 shows the Haar structure function analysis of this, indicating typical RMS values of $\approx$ 1 W/m^2 at

100 kyr scales. To within a constant factor (see Fig. 11.20) this is very nearly the same as the corresponding Vostok temperature structure function, and cross-spectral analysis of the Vostok paleotemperature and paleo-CO_2 concentrations shows that over the whole range up to frequencies of at least $\approx$ (6 kyr)$^{-1}$, the temperature and CO_2 have coherences $\approx$ 0.8–0.9, i.e. near the maximum (= 1), well above the statistically significant level (using four consecutive 105 kyr sections yields a significance level = 0.5; see Appendix 6B for definitions). At the same time, however, for frequencies > (50 kyr)$^{-1}$, we find that the phase of the CO_2 fluctuations lags by a fairly constant 70 ± 20° with respect to the temperature fluctuations (the same basic conclusion was reached by Mudelsee, 2001). We can therefore conclude that – contrary to contemporary anthropogenic CO_2 – the paleo-CO_2 is essentially a "follower" not a "driver" In any case, it is sufficient to explain the temperature variations if the climate sensitivity (see below) is high enough ($\approx$ 4.5 K/W/m^2, i.e. about 4–5 times larger than the usual estimates). While this statement is true for the Vostok series, things are not so simple for the global temperature over the last 20 kyr (Shakun *et al.*, 2012). In addition, Ganopolski and Calov (2011) claim that it is needed for explaining the dominance of the (100 kyr)$^{-1}$ frequency over both higher and lower orbital forcing frequencies (i.e. for solving the 100, 400 kyr "problems"). We may also note that the near constancy of the phase lag with frequency is compatible with a scaling mechanism for the temperature–CO_2 coupling.

The main external forcings that have been proposed over the frequency range $\approx$ (10 years)$^{-1}$ to $\approx$ (10 kyr)$^{-1}$ are thus solar and volcanic (we neglect here the possible impacts of changing land use and land cover). Ever since the Chinese discovered sunspots in 384 BC, the sun has been known to be variable, but quantifying this is extremely difficult. Since 1980, a series of satellites have estimated the total solar irradiance (TSI), yet the relative calibrations are not known with sufficient accuracy to establish the decadal and longer scale variability. Amazingly, as recently as 2011, thanks to the Total Irradiance Monitor Satellite (TIMS) from the Solar Radiation and Climate Experiment (SORCE), the absolute solar constant was found to be nearly 5 W/m^2 smaller than previously believed (0.36% less; see Kopp and Lean, 2011). This new estimate was possible thanks to a more stable (satellite-borne) measurement technique that yielded a relative precision as high as one part in 10^5 per year. In Fig. 11.19, we show the Haar RMS structure function for this. Although the data do not yet exist for a full 11-year solar cycle, we see clearly a maximum near that scale as well as another near the 27 (earth)-day-long solar "day."

To go beyond the TIMS data to lower frequencies requires proxy-based "reconstructions." To date, the principal ones are based on either sunspot numbers or on ^{10}Be from ice cores; Fig. 11.19 shows structure functions from several of these using both techniques (surrogates based on ^{14}C from tree rings also exist). The earliest (Lean, 2000) used a two-component model, one of which had an 11-year cycle based on the recorded sunspots back to 1610, while the other was a "background" based on observations of sun-like stars. Combining the two results leads to an annual series featuring an overall 0.21% variation in the background since the seventeenth century "Maunder minimum"; this series was used as the forcing in various simulations in the IPCC AR3 (Houghton *et al.*, 2001). As can be seen from the structure functions in Fig. 11.19, this reconstruction actually meshes quite nicely with the TIMS data and indicates that the fluctuations are roughly scaling out to the limiting scale ($\approx$ 390 years), with exponent $\xi(2)/2 \approx 0.4$, i.e. close to the temperature exponent (Fig. 11.20; all the irradiance series had small intermittencies so that $\xi(2)/2 \approx \xi(1) \approx H$). This reconstruction is therefore compatible with a roughly linear relation between irradiance and temperature fluctuations, i.e. with a scale-independent amplification mechanism (see Section 11.3.2).

Unfortunately, the key background irradiance component of the Lean (2000) series was based on only a small number of stars. It was therefore revised (Wang *et al.*, 2005), yielding typical fluctuations about 4–5 times lower (Fig. 11.19: only the background was used in this analysis; see Table 11.8 for a comparison of various parameters). Also shown in the figure is the somewhat later intermediate (but still sunspot-based) estimate by Krivova *et al.* (2007): this series (at 10-year resolution) yielded a variation of 0.1% since the Maunder minimum with typical fluctuations essentially in between the previous two; again it has roughly scaling behaviour, (again) with $\xi(2)/2 \approx 0.4$ (Fig. 11.19).

With the new intermediate Krivova series, a consensus began to form that the 0.1% Maunder minimum

Table 11.8 A comparison of various climate radiative forcings (R_F) shown in Fig. 11.19 converted into W/m^2. The exponents were estimated to the nearest 0.1 and the prefactors A are for the formula $\langle(\Delta R_F)^2\rangle^{1/2} = A\Delta t^{\xi(2)/2}$. When Δt expressed is in years, A is in W/m^2.

Series type	Physical basis	Reference	Series length (years)	Series resolution (years)	Scale range analysed (years)	Prefactor A (W/m^2)	ξ (2)/2 $\approx H$*
Solar	Sunspot-based	Lean, 2000	$\approx$ 400	1	10–400	0.035	0.4
Solar		Wang *et al.*, 2005		1	10–400	0.0074	0.4
		Krivova *et al.*, 2007		10	20–400	0.015	0.4
	TIMS satellite		8.7	6 hours	1–8	0.04	0.4
	^{10}Be	Steinhilber *et al.*, 2009	9300	5 years smoothed to 40 years	80–9300	0.4	−0.3
		Shapiro *et al.*, 2011	9000	1 year, smoothed to 20 years	40 – 9000	3.5	−0.3
Volcanic	Volcanic Indices, ice cores, radiance models	Crowley, 2000	1000	1 year, smoothed to 30 years	60–1000	2.0	−0.3
	Ice-core sulfates, radiance models	Gao *et al.*, 2008	1500	1 year, smoothed 30 years	60–1000	2.5	−0.3

*The solar series all have low intermittencies ($C_1 < \sim 0.03$) so that $\xi(2)/2 \approx H$, whereas the Crowley and Goa *et al.* volcanic series have high intermittencies: $C_1 = 0.16, 0.17$, respectively so that $H \approx \xi(2)/2 + C_1 \approx -0.2$ (with $\alpha = 1.4, 1.6$, $\tau_{eff} = 200, 750$ years).

variation was realistic, and it was recommended in the IPCC AR4 (Solomon *et al.*, 2007). However, soon after this, the situation changed dramatically with the publication of two long reconstructions (Steinhilber *et al.*, 2009; Shapiro *et al.*, 2011) based on ^{10}Be concentrations rather than sunspots. Both series were $\approx$ 9 kyr long and used ice-core ^{10}Be concentrations to estimate the flux of cosmic rays, itself a proxy for the state of the solar magnetic field and hence of solar activity. Both were calibrated so as to be compatible with the satellite observations since 1980 but each involved rather different assumptions, notably about a hypothetical unperturbed "quiescent" solar state. The structure function analyses of these reconstructions (Fig. 11.19) are remarkable for two reasons. First, they differ from each other by a large factor ($\approx$ 8–9: see Table 11.8); second, their variabilities as functions of scale are quite the opposite to the sunspot-based estimates discussed earlier: rather than $\xi(2)/2 \approx H \approx 0.4$, they have $\xi(2)/2 \approx H \approx -0.3$! While the large factor between them attracted much attention, the change in the sign of H was not noticed, even though it is probably more important. Just as an exponent $H \approx 0.4$ could potentially (if appropriate amplification mechanisms were found: see below) explain the multicentennial and multimillennial temperature variability, only amplification mechanisms which increase quite strongly with scale could link increasing temperature fluctuations with decreasing forcings. We could also mention the hybrid (partially sunspot, partially ^{10}Be) reconstructions (Vieira *et al.*, 2011), but the statistics are those discussed above for the periods since AD 1610 and before AD 1610 respectively.

The IPCC AR4s recommended reduction of the amplitude of the hypothetical solar forcing from 0.2% to 0.1% of the Maunder minimum (i.e. from Lean, 2000, to Krivova *et al.*, 2007) had the effect of promoting explosive volcanism to the status of the most promising nonathropogenic driver over the last two millennia. Volcanoes mainly influence the climate through the emission of sulfates that reflect incoming solar radiation. When these are lofted into the stratosphere they can persist for months or years after an eruption. Two main volcanic reconstructions exist (Crowley, 2000; Gao *et al.*, 2008; Crowley *et al.*, 2008), and both use proxies based on ice-core particulate concentrations. The first is 1000 years long at annual resolution and uses volcanic indices (nonlinearly) scaled with the historical Krakatao and Tambora eruptions. First sulfate concentrations are estimated from the indices and then, with the help of models, the corresponding global radiative forcings are estimated; the resulting structure functions are shown in Fig. 11.19. As can be seen, the variability is remarkably similar to that of the ^{10}Be solar variabilities with $\xi(2)/2 \approx -0.3$, i.e. the fluctuations decrease with scale and nearly coincide with the Shapiro *et al.* (2011) solar forcing structure function. Similarly, we show the analysis of the slightly longer (1500 years) series (Gao *et al.*, 2008). This was originally given in terms of the total mass of stratospheric sulfates (in units of Tg); but for illustration purposes we have converted this to an equivalent radiative forcing by scaling it to the Crowley (2000) series so that they both have the same means (the series have essentially 30-year resolutions since a "background" was removed using a 31-year running average). The RMS Haar structure function results for the two series are very similar. Once again, since the volcanic forcing decreases rapidly with time scale, any mechanism responsible for temperature variations must on the contrary involve an amplification which strongly increases with scale. Note that here the intermittency is fairly strong (Table 11.8).

11.3.2 Stochastic and scaling climate sensitivities

To make these analyses even more quantitative, we can transform the radiative forcings into temperature variations by using the "climate sensitivity" λ, which in this section is not a ratio but rather the proportionality constant between the radiative forcing (R_F, in Wm^{-2}) and the climate response:

$$\Delta T = \lambda \Delta R_F \tag{11.11}$$

Here we have only considered temperature sensitivities, but the concept is more general. To avoid confusion, it is useful at this point to mention that the climate modelling community often uses an alternative definition of climate sensitivity in order to compare the performance of individual climate models. In this more restrictive sense, the climate sensitivity is the global mean temperature increase in response to a doubling of the atmospheric CO_2 concentration from a preindustrial value.

The definition (Eqn. (11.11)) is usually interpreted deterministically to refer to a change ΔT caused by a specific radiative forcing ΔR_F. While this definition is convenient in the case of deterministic GCM modelling, it is quite problematic for empirical studies because at best we have statistical correlations between observed (typically paleo) ΔT and (inferred) ΔR_F records (e.g. Chylek and Lohmann, 2008). To interpret our forcing and temperature statistics it is therefore convenient to introduce a stochastic definition of climate sensitivity:

$$\Delta T \overset{d}{=} \lambda \Delta R_F \tag{11.12}$$

where, as usual, $\overset{d}{=}$ means equality in probablity distribution (i.e. the random variables a, b satisfy $a \overset{d}{=} b$ if and only if $\Pr(a > s) = \Pr(b > s)$ for all s). Notice that while both deterministic and stochastic definitions (Eqns. (11.11), (11.12)) predict that the statistical moments are related by the equation $\langle \Delta T^q \rangle = \lambda^q \langle (\Delta R_F)^q \rangle$, the stochastic definition does not actually require that R_F and T are even correlated. A convenient interpretation is therefore that the stochastic (Eqn. (11.12)) be regarded as an upper bound on the deterministic λ with equality in case of full correlation.

The advantage of adopting a stochastic λ is that by fixing its value we may convert Fig. 11.19 into equivalent temperature fluctuations: Fig. 11.20 shows the resulting superpositions when a value $\lambda = 4.5$ K/(Wm^{-2}) is used throughout. To put this value in perspective, let us estimate λ_0 determined from the simplest energy balance model involving a homogeneous atmosphere and radiative equilibria (i.e. without feedbacks). We have:

$$\sigma T^4 = (1 - \alpha) Q \tag{11.13}$$

where σ is the Stephan–Boltzmann constant, $\alpha \approx 0.3$ is the albedo, T the earth absolute temperature and Q the solar radiative forcing. These values imply $\lambda_0 \approx 0.3$ K/(Wm^{-2}) so that a large "feedback factor" (amplification) of factor $f = \lambda/\lambda_0 = 4.5/0.3 \approx 15$ is necessary to justify the overlaps shown in the figure.

From Eqn. (11.12), and for simplicity only considering the mean ($q = 1$) behaviour, we see that if $\langle \Delta T(\Delta t) \rangle \propto \Delta t^{H_T}$ and $\langle \Delta R_F(\Delta t) \rangle \propto \Delta t^{H_{RF}}$, then the sensitivity is also a power-law function of time scale:

$$\lambda(\Delta t) = \langle \Delta T(\Delta t) \rangle / \langle \Delta R_F(\Delta t) \rangle \propto \Delta t^{H_\lambda};$$
$$H_\lambda = H_T - H_{RF} \qquad (11.14)$$

If we take $H_{RF} \approx -0.3$ (volcanic and ^{10}Be solar estimates), $H_{RF} \approx 0.4$ (sunspot-based solar) and $H_T \approx 0.4$, then we find $H_\lambda \approx 0.7$ and ≈ 0 respectively. From Fig. 11.19 we see that while the volcanic and (Shapiro *et al.*, 2011) solar forcings require a feedback factor $f \approx 0.3$ at 30-year scales, this would increase 60-fold to roughly a factor of 20 at 10 kyr. If we consider instead the scale-independent amplification factors ($H_\lambda \approx 0$), i.e. the Krivova and Wang reconstructions, we find the large factors $f \approx 15$, 30 respectively. However, for this to apply at multimillennial scales, it assumes the continued growth of solar variability, which would reach the order of several Wm^{-2} at 10 kyr scales (Fig. 11.19).

Our method of estimating the climate sensitivity requires a stochastic definition of λ (Eqn. (11.12)) rather than the usual deterministic one which postulates sustained (at least slowly varying, deterministic) forcing. This difference should be borne in mind when comparing estimates of λ with the more classical estimates based on the latter assumptions (see e.g. Ganopolski and Schneider von Deimling, 2008, and references therein, for discussion and debate on this). We should note that a theoretical consequence of power-law sensitivities is that the corresponding time-dependent energy balance models must be either strongly nonlinear or of fractional order (rather than linear and integer order, as is usually assumed).

With the help of our scaling framework, we can also consider the recent proposal that changes in cloud cover amplified by the Pacific Decadal Oscillation (PDO) might explain a large part of the climate variability at least in the twentieth century, i.e. up to centennial scales (Spencer and Braswell, 2008). According to this hypothesis, the key low-frequency driver is the PDO; however, analysis of the PDO (Zhang *et al.*, 1997; Figs. 10.8, 10.14) shows that at climate scales its variability is very close to that of the classical (atmospheric) North Atlantic Oscillation index (without even much of a peak at decadal scales!). Indeed, Fig. 10.14 shows that the PDO tracks the SST fluctuations quite closely (although it is about a factor of 2 larger). However, the most important point is probably that the PDO and SST only track the global average temperature up to $\approx$ 10 years. At longer time scales, they diverge, and only reconverge at $\approx$ 80 years (SST) and perhaps (by extrapolating Fig. 10.14) for the PDO at $\approx$ 300 years. Viewed from the point of view of its decadal to centennial scale-by-scale variability, the PDO is thus unable to explain the low-frequency variability – at least not without again invoking a strong scale-dependent amplification mechanism.

While we are on the topic of scaling solar forcing and scaling atmospheric temperature responses, we could mention the papers of Scaffeta and West (e.g. 2005, 2007, 2008). After estimating scaling exponents of various solar and temperature anomaly series and finding that they are similar to each other, they argue on the basis of a speculative principle of "complexity matching" that the series are related, concluding for example that much of the 1980–2002 global warming is due to solar variability. While this and other specific empirical claims have already been criticized (e.g. Lean, 2006), the scaling aspects of their work have not been addressed. The basic problem is that rather than simply statistically analyzing the data using standard multifractal analysis they use a very indirect data analysis technique. Their motivation (Scafetta and Grigolini, 2002) is the surprising claim that all existing analysis methods assume a finite variance (i.e. they ignore multifractals and the corresponding analysis methods). They then adopt a purely additive scaling framework to analyze random walks constructed from the original time series (rather than the time series themselves). Deviations of these random-walk exponents from quasi-Gaussian ones are then taken as evidence of the existence of underlying Lévy distributions. Although they thus make the strong claim that the series have $q_D < 2$ (hence infinite variance), it is significant that they do not actually estimate any empirical probability distributions (see e.g. Scafetta and West, 2003), so their anomalous exponents presumably are simply symptoms of multifractality; indeed, we saw in Fig. 5.21 that temperature series may have long tails ($q_D \approx 5$), but that their variances are apparently finite.

11.3.3 The unforced low-frequency variability of GCM control runs

We have argued that the variability of the atmosphere out to τ_c is dominated by weather dynamics, that there are neither significant new internal mechanisms of variability nor important new sources of external forcing. This enables us to objectively define the "climate state" of a given atmospheric field as its average over the whole range of weather ($H > 0$) and the stable ($H < 0$) macroweather scales up to τ_c. Whereas macroweather corresponds to our idea of "long-term weather," the even lower-frequency changes of such relatively stable "climate normal" states correspond to our idea of "climate change." The transition between the macroweather regime and the climate is thus a transition from a stable regime where fluctuations decrease with scale to an unstable regime where on the contrary they increase with scale. In this section we discuss to what extent GCMs are consistent with these two scaling regimes.

We have already mentioned that the expression "GCM" originally denoted "general circulation model," i.e. a model of atmospheric dynamics designed to reproduce meteorological variability. Later, as ambitions grew, the usage gradually changed to mean "global climate model." However, for predicting climate scales, at a minimum, atmosphere models must be coupled to ocean circulation models; hence the term "atmosphere–ocean general circulation model" (AOGCM). However, today's models are typically further coupled with carbon cycle models as well as with ice and land-use models, so that the term "climate forecasting system" is increasingly used even though such systems in fact constitute "global climate models." In the following, we will refer to the latter with the traditional expression GCM, and Table 11.9a gives some of the technical details of the GCMs discussed in this section.

In order to model climate and climate change, GCMs must therefore adequately reproduce the rising low-frequency variability at time scales $> \tau_c$; otherwise, their claim to model the climate is unjustified. In Fig. 1.9d we have shown that the IPSL GCM "control run" – with essentially constant "forcing," i.e. with constant orbital and solar parameters, no volcanism, constant greenhouse gases and fixed land use – generates a macroweather spectrum all the way to its low-frequency limit, $(500\ \mathrm{yr})^{-1}$. Fig. 11.21a shows RMS Haar structure functions from this 500-year IPSL control run as well as from a 3000-year control run of the Ensemble Forecasting System (EFS: Table 11.9a; Jungclaus *et al.*, 2010), along with the low-frequency extension of the stochastic FIF cascade model discussed in Section 10.1. These structure functions are compared to the corresponding empirical functions for surface temperatures, reanalyses and multiproxies (all at global resolutions). We can clearly see a strong divergence between the empirical and model $S(\Delta t)$ for $\Delta t > \sim 10$–30 years. With the exception of a spurious 2–4-year scale "bump" in the EFS $S(\Delta t)$, the models do a reasonable job over the macroweather regime (i.e. from about 1 month up to $\tau_c \approx 10$–30 years). Beyond that, however, their RMS fluctuations continue to decline whereas the empirical fluctuations start to rise. The grid-point-scale analyses of the control runs (see Fig. 11.21c, discussed below) lead us to exactly the same conclusion; indeed, the low-frequency exponents are all near the same value, corresponding to $\beta \approx 0.2$.

Table 11.9a Some of the details of the various climate simulations systems whose outputs are analyzed here.

Model system	Model components and references	GCM characteristics	Experiment	Series length
ECHO-G (von Storch *et al.*, 2004)	ECHAM4 (Roeckner *et al.*, 1996); HOPE-G, (Wolff *et al.*, 1997)	19 vertical levels, T30, (3.75° resolution)	"Erik the Red," AD 1000 to present, ≈ 0.25% solar forcing	1000 years
Ensemble Forecasting System (EFS) (Jungclaus *et al.*, 2010)	ECHAM5 GCM MPIOM ocean model, (Jungclaus *et al.*, 2006); carbon cycle module HAMOCC5 (Wetzel *et al.*, 2006); land surface scheme JSBACH (Raddatz *et al.*, 2007)	19 levels, T31 (3.75° resolution)	Millennium, solar forcing 0.1%, 0.25%, AD 1000 to present	1000 years with full forcing, 3000-year control run
IPSL climate system model: IPSL-CM4	LMDZ GCM (Hourdin *et al.*, 2006); ORCA2 ocean model (Madec *et al.*, 1998); LIM sea-ice model (Fichefet and Morales Maqueda, 1997); ORCHIDEE land-surface model (Krinner *et al.*, 2005)	19 level 2.5° x3.75° grid	Control run, 1910–2410, for IPCC AR4	500 years

Table 11.9b A summary of scaling studies of temperatures in GCMs. All the estimates were made using the DFA method; the spectral exponent β was determined from $\beta = 2\alpha - 1$ where α is the conventional DFA exponent (this expression ignores intermittency corrections).

	Model	Model characteristics	Series length	Range of scales in analysis	β_{mw}
Fraedrich and Blender, 2003 (with IPCC scenario IS92a greenhouse gas emissions)	ECHAM4/ OPYC	19 levels, T42 OPYC ocean model includes sea ice with rheology	1000 years	240 years	≈0 continents, ≈0.3 coasts, ≈ 1 for oceans
	HadCM3	19 levels, 2.5° x3.75°	1000 years	240 years	Same to within ≈0.2
Zhu *et al.*, 2006 (preindustrial control runs)	GFDL	31 levels, T63	500 years	500 years	≈1
	ECHAM5/MPIOM	24 levels, 2°x2° (land), 1°x1° (ocean)	500 years	500 years	≈1 midAtlantic overturning
Blender *et al.*, 2006; Fraedrich *et al.*, 2009	CSIRO atmosphere-ocean model under present-day conditions	9 levels, R 21 horizontal resolution	10 000-year simulation	3 kyr	0.2 – 0.8 depending on location
Vyushin *et al.*, 2004; Rybski *et al.*, 2008 (one control simulation, one with historical drivers)	ECHO-G=ECHAM4/ HOPE-G	19 vertical levels, T30	1000-year simulated temperature records	≈ 200 years	Land 0.2–0.4, ocean 0.4–0.7

The strong divergence of the GCM control runs and empirical (surface data, reanalysis, multiproxy) $S(\Delta t)$'s raises the question as to whether or not current GCMs are at all capable of reproducing the climate regime. Are they missing either a critical new source of (slow) internal variability or a new coupling with the external forcings? In the latter case the corresponding feedbacks must be large and quite likely increasing with scale. Since we are searching for scaling mechanisms that generate fluctuations over a wide range of scales, it is almost certain that the mechanism would involve spatial interactions over a wide range such that the lifetimes of structures grow with scale. As mentioned in Section 11.1.1, a possible candidate (not included in standard coupled GCM–cryosphere models) is land ice.

Before describing our own analyses further, let us review the relevant literature on long-term GCM statistics (Table 11.9b). There have been several studies of the low-frequency behaviour of GCMs, including some on "ultra-long" 10 kyr runs. The basic conclusions have been pretty uniform: the low-frequency behaviour was scaling, predominantly with $0 < \beta < 0.6$ (roughly $-0.5 < H < -0.3$, i.e. in the same range as our control runs) and with ocean values a little higher than for land. The exponents were also found to be robust: for example, with a fixed scenario they were insensitive to the use of different models, in the same model, to the addition of greenhouse gases (Fraedrich and Blender, 2003), or in the last 1000 years in the northern hemisphere, to constant or to historically changing drivers (Rybski *et al.*, 2008). Finally, models with sophisticated sea-ice rheology also had similar scaling (Fraedrich and Blender, 2003). Although the authors did not pose this question, in no cases and at no geographical location was there evidence of an end to the low-frequency macroweather regime. Apparently, the global-scale IPSL and EFS control-run analyses in Figs. 11.21a and 11.21c are typical.

The above is a summary of the overall (global) situation, but – as expected – due to the strong spatial intermittency (Sections 10.3, 11.1.4) there is quite a lot of regional variability. Before considering the forced runs, let us therefore consider the distribution of low- and high-frequency β's in the IPSL GCM to

Table 11.10 Local β estimates: low is $\omega < (25\ \text{years})^{-1}$, high is $(3\ \text{months})^{-1} > \omega > (3\ \text{years})^{-1}$. Change in low-frequency exponents due to detrending: IPSL: −0.50 ± 0.54; 20CR: 0.08 ± 1.17. All were monthly series, periodically detrended for annual and harmonics down to 3 months. The trend removal indicated here is a linear trend for the entire length of series, different for each pixel from first to last element in each series. The IPSL is more sensitive to trend removal precisely because of its lower β value.

	IPSL	20CR
Low frequency	0.18 ± 0.52	1.72 ± 1.28
Low frequency (trend removed)	0.68 ± 0.56	1.80 ± 1.27
High frequency	0.82 ± 0.62	0.33 ± 0.37
High frequency (trend removed)	1.03 ± 0.65	0.44 ± 0.32

see how closely the geographical variations of the simulations and the reanalyses agree. Figs. 11.4 and 11.5a show the results for the IPSL model with β's defined for low ($\omega < (25\ \text{years})^{-1}$) and high ($(3\ \text{years})^{-1} < \omega < (3\ \text{months})^{-1}$) frequencies (bottom rows); they are in many ways the opposite of that of the 20CR reanalysis (top rows). Whereas the distribution of high-frequency β's corresponds reasonably well to the continents, it is rather the low-frequency 20CR β's that have (limited) land/ocean correlations. Whereas the 20CR β's show a marked increase as we pass from high to low frequencies, the IPSL β's do exactly the opposite, tending instead to decrease to very low values at low frequencies. This inversion is also evident in Fig. 11.5a, which shows the latitudinal variations. A final point is that unlike the low-frequency β 20CR reanalysis which is fairly north–south asymmetric, the IPSL β's (both low- and high-frequency) are roughly symmetrical with respect to maxima near the equator. Table 11.10 summarizes the exponent estimates, including the fairly small changes introduced when linear trends (series by series) were removed (this was done in an attempt to quantify the sensitivity of the analyses to twentieth-century warming trends). As expected, this trend removal only affects the low β results, but hardly at all the 20CR results, which have higher β. For these local β estimates, the IPSL is very close to the other GCMs; the β's for the globally averaged series can be estimated from Fig. 11.21a: $\beta \approx 1.8$, 0.2 for low- and high-frequency empirical series, and $\beta \approx 0.2$ for $\omega < (2\ \text{years})^{-1}$ for IPSL. From this clear global model/data comparison we again conclude that without special forcing, the current GCMs appear to model macroweather rather than the climate.

11.3.4 Do (forced) GCMs predict the climate . . . or macroweather?

If control runs do not correctly reproduce the low-frequency variability, what about forced runs with more realistic low-frequencies? To answer this question, we consider simulations from the Last Millennium project whose aim was to reproduce the climate since AD 1000. We focus on two different models, the ECHO-G "Erik the Red" simulation (von Storch *et al.*, 2004) and the results of two experiments on the more recent Earth Forecasting System model (EFS: Jungclaus *et al.*, 2010; see Table 11.9b).

Since the earth's orbital parameters have changed little in the last 1000 years, if we exclude the twentieth century, the key forcings are volcanic and solar variability. Both EFS and ECHO-G simulations used similar volcanic forcings. In comparison, as discussed above, the correct solar forcing is much less certain, and in these two models it is deduced indirectly since 1627 from correlations between the solar constant and the observed sunspot numbers. Before this, the total solar irradiance (TSI) was estimated from the abundance of ^{14}C isotopes found in tree rings (in a manner somewhat similar to that of ^{10}Be, discussed in Section 11.3.1). The amplitudes of the solar forcings are thus described in terms of percentages of variation of the solar constant since the seventeenth-century "Maunder minimum": this value fixes the calibration constant. Values of 0.1% and 0.25% are considered respectively low and high solar forcing values (see Section 11.3.1, and Krivova and Solanki, 2008, for a recent review). In these terms, the ECHO-G simulations were "high," close to 0.25%, whereas the EFS simulations were run at both 0.1% and 0.25% levels.

We considered the full simulation (to the present, Figs. 11.21b, 11.21c) as well as only the pre-1900 variability so as to focus on the natural variability without strong anthropogenic effects (Figs. 11.21d, 11.21e), considering both globally averaged temperatures (Figs. 11.21b, 11.21d) and grid-scale averages (Figs. 11.21c, 11.21e). The key conclusions are:

(a) The EFS global-scale full series analyses have significant spurious high-frequency variability in the 2–4-year range, but the effect disappears in the pre-1900 series analysis. Similarly, the EFS 0.25% (but not 0.1%) multidecadal, multicentennial scale variability (global and grid scales) is too high, but this disappears completely in the pre-1900 analyses. In fact, the latter have overall variability very close to the corresponding control run (Figs. 11.21a, 11.21c), which is much too weak. The strong variability of the 0.25% EFS run is thus presumably an artefact of the strong twentieth-century warming. The fact that the twentieth century can give a strong contribution to the variability was underscored by the analysis of

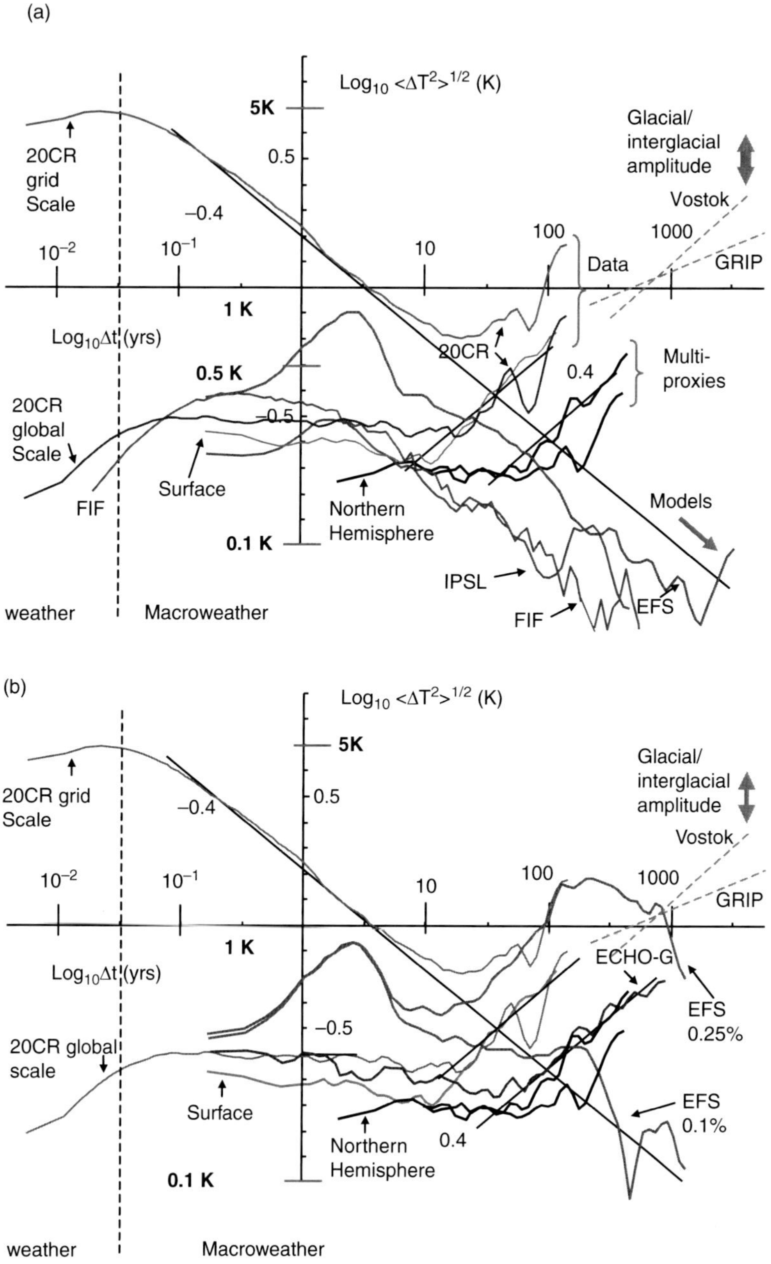

Fig. 11.21 (a) Comparison of the RMS Haar structure functions for temperatures from instrumental ("data", daily 20CR, monthly surface series), multiproxies (post 2003, yearly resolution) GCM control runs (thick, monthly) and the FIF stochastic model (thin). The data are averaged over hemispheric or global scales (except for the 20CR 2° × 2° grid-scale curve added for reference). The surface curve is the mean of three surface series (NASA GISS, NOAA CDC and HadCRUT3, all 1881–2008); the 20CR curves are from the 700 mb level (1871–2008). The IPSL is a 500-year control run, the EFS is from a 3000-year control run; the "bump" at 2–4 years is a broad model quasi-periodic artefact. The multiproxies are from the three post-2003 reconstructions: two curves are shown, the top from 1500–1980, the bottom from 1500–1900, showing the effect of the twentieth-century data. The reference lines have slopes $\xi(2)/2$ so that $\beta = 1 + \xi(2) = 0.2, 0.4, 1.8$. The amplitude of the Haar structure functions has been calibrated using standard and tendency structure functions and is accurate to within ± 25%. At the upper right we have sketched the Vostok and GRIP paleocurves (see Fig. 11.16b) and have indicated the likely glacial/interglacial mean temperature contrast (difference) by the arrows.
(b) Comparison of the RMS Haar structure functions of global-scale ECHO-G and EFS GCM simulations (the latter with both 0.1% and 0.25% solar forcing levels), with the models analyzed over their entire ranges including the twentieth century. Also shown are the 20CR, surface and multiproxy curves as in Fig. 11.21a. Although the 0.25% solar forcing curve has very large multidecadal, multicentennial variability, this is due to high simulated twentieth-century temperatures (see Figs. 11.21d, 11.21e, the analyses before 1900). It can be seen that the ECHO-G simulation is quite close to the northern hemisphere multiproxy reconstructions. Again the reference lines show slopes $\xi(2)/2 = \pm 0.4$.

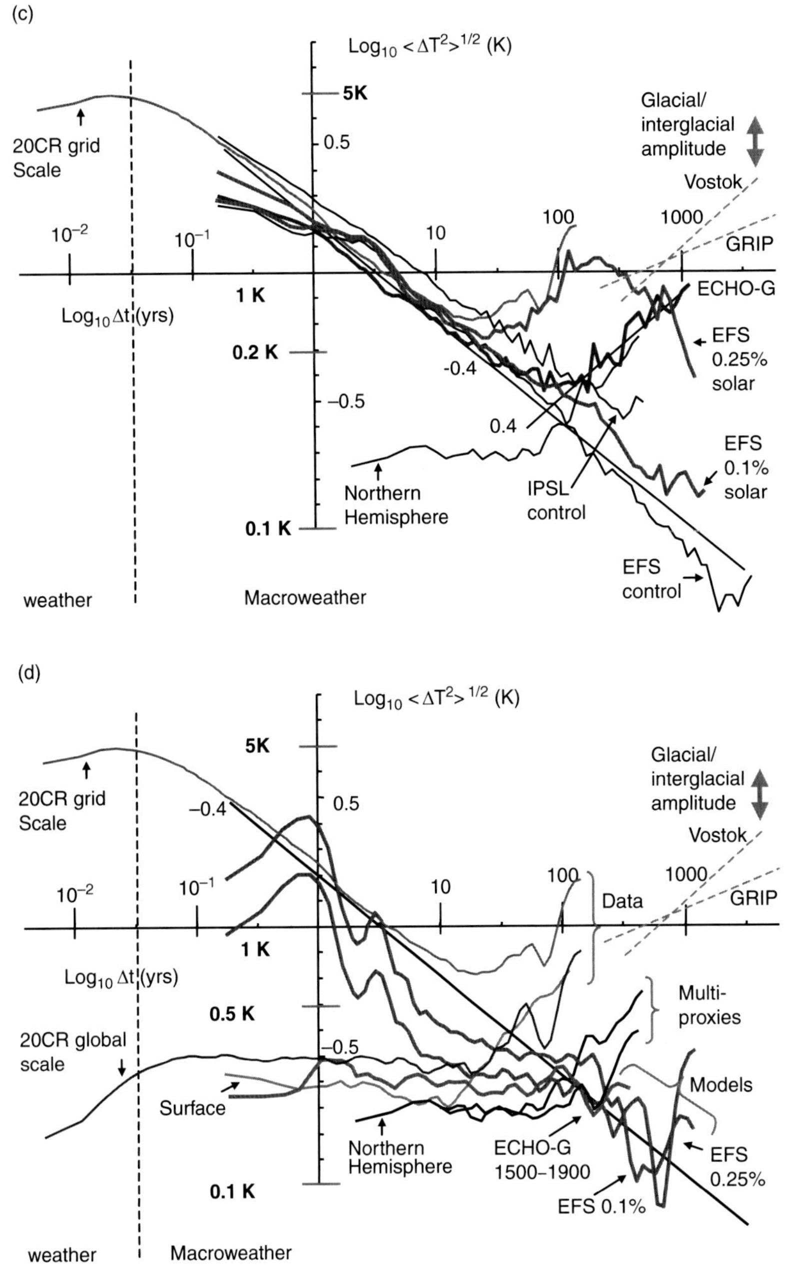

Fig. 11.21 (c) As Fig. 11.21b except for the grid-scale (2° × 2°) variability. The main difference is the wide range of scaling of all the models with $\xi(2)/2 = -0.4$ (see the reference line). Also shown are the $\xi(2)/2 = \pm 0.4$ reference lines; again ECHO-G is quite close to the reconstructions. (d) As Fig. 11.21b (i.e. global averages) except that only pre-1900 model data are analyzed. Note in particular that the strong variability of the 0.25% EFS simulation in Fig. 11.21b has disappeared completely; before 1900, the different levels of solar forcing are not very consequential, the overall variability of the EFS models is too low. Note also that the ECHO-G simulation (shown only from 1500–1900 to avoid both early "spin-up" and later twentieth-century effects) is more realistic than the EFS but has lost most of its low-frequency variability.

IPSL simulations of various future greenhouse gas emission scenarios, with the corresponding $S(\Delta t)$ showing strong responses starting at about $\Delta t > \sim 10$ years with exponents H equal to the theoretical maxima (i.e. $H = 1$ for the Haar, $H = 2$ for the quadratic Haar) structure functions – i.e. they dominate the natural variability.

(b) The full-length ECHO-G global- and pixel-scale simulations have $S(\Delta t)$'s that are close to the empirical (either 20CR or multiproxy) curves. However, whereas the empirical $S(\Delta t)$'s seem to

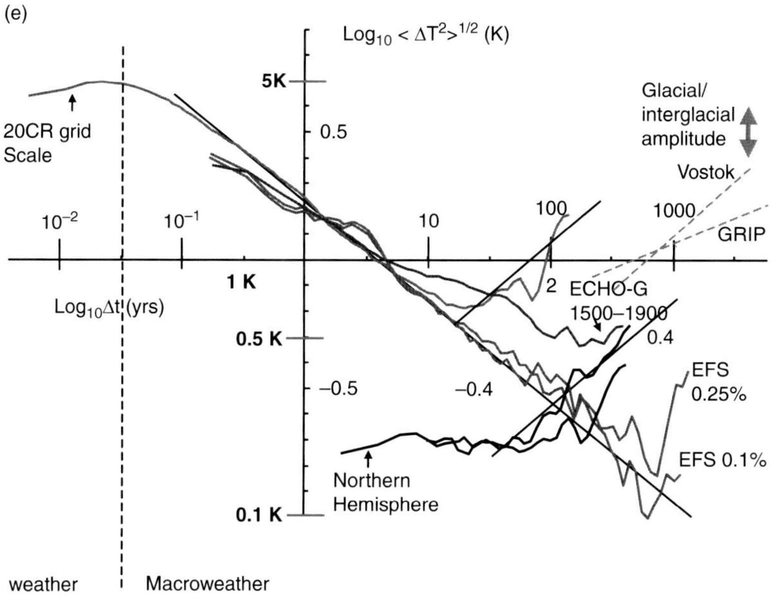

Fig. 11.21 (e) As Fig. 11.21d except for grid-scale analyses. Again, the EFS model has low frequencies that are too weak, but even ECHO-G has weak variability and the low-frequency tendency is not clear (i.e. is it starting to rise at $\Delta t \approx 500$ years?).

be robust (i.e. they continue to displayed strong variability in the pre-1900 period with only a small reduction), the ECHO-G 1500–1900 global series has overly weak variability. This is also true of the grid-scale analyses (Fig. 11.21e): indeed, the large Δt value of $S(\Delta t)$ of the grid-scale ECHO-G series is about the same as that of the northern hemisphere multiproxy $S(\Delta t)$ (which is from data smoothed over a much larger area); it is much lower than the 20CR grid-scale variability.

(c) The pre-1900 grid-scale low-frequency variability of all the GCMs (with the possible exception of the ECHO-G model) decreases with increasing Δt and the EFS $S(\Delta t) \approx \Delta t^{-0.4}$, i.e. the predicted low-frequency scaling behaviour ($\beta \approx 0.2$).

We have noted in Figs. 11.21b and 11.21c that the full-length ECHO-G simulation has relatively realistic multicentennial variability (close to the post-2003 multiproxies). The simulation has roughly the same τ_c and H as the data and the northern hemisphere multiproxies. So what is special about the ECHO-G simulation? In the IPCC AR4 (Solomon *et al.*, 2007), 12 different Millennium simulations are compared and it is noted that indeed the ECHO-G has significantly stronger low-frequency variability than any of the others (however, of these, only two were full-blown GCMs). This outlier status prompted Osborn *et al.* (2006) to perform a special study using a simplified energy-balance-based climate model in order to determine the reasons. These authors concluded that there were two problems with the ECHO-G simulations: (a) they had questionable initialization, so that the first two centuries ($\approx$ AD 1000–1200) had spuriously high temperatures; and (b) they did not include anthropogenic sulfates (which tend to lower the temperatures) so that their twentieth-century climate was also spuriously warm (this is why in Figs. 11.21d and 11.21e we only show the ECHO-G for the period 1500–1900, which is free of both of these problems). (Note added in proof: recent GISS-E simulations substantiate these conclusions (G. A. Schmidt *et al.*, in preparation, Lovejoy *et al.*, 2012.)

If, as we have argued, the high-variability scenario is more likely, how is it that several studies successfully empirically validated – even ultra-long – GCM scaling using unique macroweather scaling regimes? Part of the answer is a consequence of a limitation of the detrended fluctuation analysis (DFA) technique which they all used. Unlike the poor man's (and to a lesser extent) Haar wavelets, whose fluctuations are easy to interpret, DFA fluctuations are defined with respect to RMS deviations from polynomial regressions with respect to integrals of T. In addition, there is no indication that the authors realized the strong implications of the $\beta < 1$ result: that the fluctuations in T tend to *decrease* rather than *increase* with scale. Indeed, these papers were mostly intent on establishing the existence of long-range statistical dependency, which for Gaussian processes implies $\beta \neq 0$. Finally, they apparently did not *expect* to find a new low-frequency climate regime.

A more concrete reason that the $\tau > \tau_c$ regime was missed in the empirical analyses was perhaps that the typical scale range of the data analyses was restricted to only one-quarter of the available time series length and was thus too short. For example, Eichner *et al.* (2003) used station series up to 155 years long, yet the DFA analyses were limited to lags only up to $\tau < 40$ years, which were too close to τ_c to allow the new regime to be detected. The failure to detect the climate regime was also a consequence of the fact that most of the long series were from the anomalous region north of 30° N. This northern anomaly – probably caused by the large fraction of land at these latitudes – made the low-frequency regime more difficult to detect than with global data (see Fig. 11.5a, which shows that τ_c is quite a bit larger north of 30° N). Finally, when paleodata were analysed, they were from Greenland sites, which as we saw already in Section 11.2.1 were exceptional in that the $\beta \approx 0.2$ behaviour continues up to ≈ 2 kyr, which is apparently not at all representative of the hemisphere.

To conclude, it would seem then that there are two opposing hypotheses about the variability in the last millennium: a low- and a high-variability hypothesis. The low-variability hypothesis relies on the low-variability pre-2003 proxies supported by the GCM models and standard internal couplings and forcings. Since the low variability is postulated to continue to at least many thousands of years, the glacial/interglacial transitions are explained purely by orbital forcings whose effects are implicitly only over relatively narrow ranges of scales, especially near $(100\ \text{kyr})^{-1}$. This hypothesis thus has difficulty explaining not only the broad-spectrum variability of the paleotemperatures but also the observed natural variability on centennial and millennial scales.

In contrast, the high-variability hypothesis relies on a scaling framework for the climate variability – including the relative constancy of the exponents $\beta_c \approx 1.8$, $H_c \approx 0.4$ as measured from several different sources. This includes the multidecadal, instrumental data, the post-2003 multicentennial multiproxies but also the multimillennial paleotemperatures (with the caveats about the Holocene and geographic variability of the transition scale τ_c). This high-variability hypothesis can also cautiously appeal to the existence of at least one GCM reconstruction (ECHO-G) with apparently comparable low-frequency variability, indicating that GCMs may be compatible with this (although the 1500–1900 part of the ECHO-G simulation is still too weakly variable). Interestingly, this high-variability hypothesis implies that anthropogenic effects at scales $\Delta t > \sim 10$ years may be much stronger ... or much weaker than otherwise expected. This is because, depending on the relative sign of the strong natural variability, either it could mask an otherwise very strong anthropogenic effect, or on the contrary it could artificially enhance a much weaker one.

11.4 The atmosphere in a nutshell: a summary of emergent laws in Chapter 11

In Chapter 10 we defined the climate as the regime with time scales beyond macroweather where new dynamics and/or forcings became dominant. In this chapter, we covered some of the issues this raises:

(a) The (inner) transition scale τ_c: what is it, and how does it vary (in space and from epoch to epoch)?
(b) The climate outer scale τ_{lc} where a new low-frequency climate zone begins.
(c) The nature of the temporal variability (the scaling and scaling exponents).
(d) The constraint on the exponents and the scaling imposed by the glacial/interglacial window (i.e. the amplitude of the temperature variations at the end of the climate regime at τ_{lc}).
(e) The proposal of a fundamental space-time scaling law governing the climate variability (in real and Fourier space).
(f) The scaling of the various possible forcings (including solar and volcanic).
(g) A stochastic definition of climate sensitivity, its interpretation and its empirical evaluation, including its scaling exponents.
(h) The (macroweather) scaling of unforced GCMs.
(i) We attempted to answer the question: do forced GCMs predict the climate or macroweather?

From the point of view of emergent laws, the most important contribution from this chapter is the proposal of a precise law for the space-time variability (Section 11.2.4):

$$S_{q,c}(\underline{\Delta r}, \Delta t) = \langle \Delta T(\Delta x, \Delta t)^q \rangle_c$$
$$= (s_{q,c})^q [[(\underline{\Delta r}, \Delta t)]]_c^{qH_c - K_c(q)}$$
$$[[(\Delta x, 0, 0, \Delta t)]]_{c,\,can} = \left(\left(\frac{\Delta x}{L_e^*}\right)^2 + \left(\frac{\Delta t}{\tau_{lc}}\right)^{2/H_{c,t}} \right)^{1/2};$$
$$\tau_{lc} > \Delta t > \tau_c$$
$$H_{c,t} = H_c / H_{c,\tau} \qquad (11.15)$$

(we have given the general formula in terms of space-time scale functions, and then the 1D canonical scale function). For the temperature, the empirically estimated parameters are: $L_e^* \approx 5000$ km, $s_{1,c} \approx 0.06\ K$, $\tau_{lc} \approx 30$–50 kyr, $H_c \approx 0.7$, $H_{c,\tau} \approx 0.4$, $H_{c,t} \approx 1.75$, $C_{1,c} \approx 0.11$, $\alpha_c \approx 1.75$. From the earlier chapters of the book, it is clear that although Eqn. (11.15) displays only the simplest canonical space-time scale function $[[(x,t)]] = \left((\Delta x/L_e^*)^2 + (\Delta t/\tau_{lc})^{2/H_{c,t}}\right)^{1/2}$ with a single spatial coordinate, it is not restricted to this form and – if they are needed – generalizations which maintain the observed scalings are straightforward. We deliberately chose to write this simplest space-time law explicitly so that the basic structure is as accessible as possible. Note that the spectral density corresponding to Eqn. (11.15) is easy to determine: see Eqn. (8.16).

Although, due to its very definition, the theoretical framework for understanding the climate-regime variability is more tenuous than for the higher-frequency weather and macroweather regimes, we nevertheless justified Eqn. (11.15) on the basis that the climate shared certain key features with the weather regime, in particular:

(a) A large number of degrees of freedom.
(b) That scaling holds over wide range ($\tau_{lc}/\tau_c \approx 10^3$–$10^4$).
(c) That the energy flux (modulated by the weather/climate processes) is still the dynamical driver, that there other coupled processes (which are needed to account for the observables other than the wind) which have analogous mathematical structure.
(d) Basing ourselves on empirical evidence of multifractal cascade statistics in space and in time, we argued that the climate-regime dynamics were again governed by multiplicative cascades and that the full weather/climate process was the product of a weather/macroweather process (out to τ_c) and a climate process (from scale τ_c to τ_{lc}): $\varepsilon_{w,c}(\underline{r},t) = \varepsilon_{w,mw}(\underline{r},t)\varepsilon_c(\underline{r},t)$(recall from Chapter 10 that the weather/low-frequency weather flux itself factors: $\varepsilon_{w,mw}(\underline{r},t) = \varepsilon_w(\underline{r},t)\varepsilon_{mw}(t)$).
(e) The assumption that the climate process is a true space-time process like the weather, that space and time are not disconnected as they are in the macroweather regime, hence that a space-time scale function exists. Eqn. (11.15) displays only the simplest (canonical) version of this.

We have come full circle. We have systematically investigated the space-time variability of the atmosphere over the full range of scales from seconds to hundreds of thousands of years, from centimetres to the size of the planet. To complete the theoretical description, we display the weather (top) and macroweather laws (bottom) corresponding to Eqn. (11.15):

$$S_{q,w}(\underline{\Delta r}, \Delta t) = (s_{q,w})^q [[(\underline{\Delta r}, \Delta t)]]_w^{qH_w - K_w(q)}; \quad \tau_i < \Delta t < \tau_w$$
$$[[(\Delta x, 0, 0, \Delta t)]]_{w,\,can} = \left(\left(\frac{\Delta x}{L_e^*}\right)^2 + \left(\frac{\Delta t}{\tau_w}\right)^2 \right)^{1/2}$$
$$S_{q,mw}(\underline{\Delta r}, \Delta t) = (s_{q,mw})^q \|\underline{\Delta r}\|_c^{qH_c - K_c(q)} \left(\frac{\Delta t}{\tau_w}\right)^{qH_{mw}};$$
$$\tau_w < \Delta t < \tau_c$$
$$\|(\Delta x, 0, 0)\|_{c,\,can} = \left|\frac{\Delta x}{L_e^*}\right| \qquad (11.16)$$

The parameters (again for the temperature) are: $H_w \approx 0.5$, $C_{1w} = 0.087$, $\alpha_w = 1.61$, $H_{mw} \approx -0.4$, $s_{1,w} \approx s_{1,mw} \approx 5\ K$, $\tau_c \approx 40$ years (at 45° N), and $\tau_i < 1\ s$. Although these are somewhat simplified (e.g. see Chapter 6 for the vertical, Chapter 8 for the effects of advection) Eqns. (11.15) and (11.16) nevertheless capture in a nutshell the basic space-time variability of the atmosphere over huge ranges of space-time scales.

Box 11.5 Summary of the main results of this book

We are living in a golden age of atmospheric data, models and theory; this manna has finally made it possible to understand the basic space-time variability of atmospheric fields over wide ranges of scales. The key development is the finding that the vertical stratification is scaling over much of its range, but with exponents quite different from those in the horizontal. By (theoretically) eliminating the "meso-scale gap" predicted by isotropic turbulence theory, it removes the last obstacle to recognizing the wide-range horizontal scaling of the atmosphere. A model which well reproduces the empirical statistics involves dynamics which are governed by energy flux cascades in the horizontal and buoyancy variance flux cascades in the vertical. This paradigm is a generalization of the classical higher-level "emergent" turbulence laws of Kolmogorov, Obukhov, Corsin and Bolgiano; it implies that the solar energy flux (modulated by clouds, water vapour and gases) drives the atmosphere up to planetary scales, and it predicts that the lifetime of the largest structures is about 10 days. According to extensive statistical analysis of atmospheric reanalyses (including the ECMWF, ERA40 and interim products and forecast models including GFS and GEM) this and conventional modelling share many features including wide-range scaling and cascade structures with very similar parameters.

When this model is extended beyond $\approx$ 10 days (τ_w from the weather into the macroweather regime), it predicts a "dimensional transition" resulting in a spectral plateau: a wide range with a roughly flat frequency spectrum (although with long-range statistical dependencies – not a white noise!). Macroweather is thus a stable intermediate regime (with fluctuations decreasing with scale, $H < 0$) whereas both weather and climate regimes are unstable (with $H > 0$). This basic prediction has been statistically verified on very large-scale reanalyses (every 6 hours from 1871 to 2008) and on high-resolution paleotemperature series from ice cores; indeed, even the simplest model with only three fundamental parameters reproduces the statistical variability of the temperature quite well up to time scales of the order of decades (τ_c).

This scaling paradigm thus directly unifies both weather and macroweather dynamics. In addition, the physical principle upon which it is based – scale invariance – can be used to extend the picture to include the even lower-frequency climate regime: the three-regime model extending to about 100 kyr. The overall weather/climate model is the product of a weather/macroweather process (flux) and a climate process, and the observables such as the wind and temperature are obtained as fractional integrals of this flux. We have thus obtained fundamental high-level emergent laws for the statistics of all orders, of the space-time fluctuations in all three regimes: the "atmosphere in a nutshell." With the exception of the weak intermittency temporal macroweather statistics, all the variability was multifractal with parameters we estimated using a wide variety of data, reanalyses and paleodata.

References

Preface

Manneville, P. (2010). *Instabilities, Chaos And Turbulence*, 2nd edn. London: Imperial College Press.

Chapter 1

Bak, P. (1996). *How Nature Works*. New York, NY: Copernicus Press.

Bak, P., Tang, C., and Weiessenfeld, K. (1987). Self-organized criticality: an explanation of 1/f noise. *Physical Review Letters* **59**, 381–4.

Balmino, G. (1993). The spectra of the topography of the Earth, Venus and Mars. *Geophysical Research Letters* **20**, 1063–6.

Balmino, G., Lambeck, K., and Kaula, W. M. (1973). A spherical harmonic analysis of the Earth's topography. *Journal of Geophysical Research* **78**, 478–81.

Batchelor, G. K. (1953). *The Theory of Homogeneous Turbulence*. Cambridge: Cambridge University Press.

Bell, T. H. (1975). Statistical features of sea floor topography. *Deep Sea Research* **22**, 883–91.

Berkson, J. M., and Matthews, J. E. (1983). Statistical properties of seafloor roughness. In N. G. Pace, ed., *Acoustics and the Sea-Bed*. Bath: Bath University Press, pp. 215–23.

Berrisford, P., Dee, D., Fielding, K., *et al.* (2009). *The ERA-Interim Archive*. ERA Report Series. Reading: European Centre for Medium Range Weather Forecasts.

Berry, M. V., and Hannay, J. H. (1978). Topography of random surfaces. *Nature* **273**, 573.

Bjerknes, V. (1904). Das Problem der Wettervorhersage, betrachtet vom Standpunkte der Mechanik und der Physik. *Meteorologische Zeitschrift* **21**, 1–7. Translation by Y. Mintz: The problem of weather forecasting as a problem in mechanics and physics. Los Angeles, 1954. Reprinted in Shapiro & Grønås, 1999, pp. 1–4.

Buizza, R., Miller, M., and Palmer, T. N. (1999). Stochastic representation of model uncertainties in the ECMWF Ensemble Prediction System. *Quarterly Journal of the Royal Meteorological Society* **125**, 2887–908.

Charney, J. G. (1971). Geostrophic turbulence. *Journal of the Atmospheric Sciences* **28**, 1087–95.

Chigirinskaya, Y., Schertzer, D., Lovejoy, S., Lazarev, A., and Ordanovich, A. (1994). Unified multifractal atmospheric dynamics tested in the tropics. Part 1: horizontal scaling and self organized criticality. *Nonlinear Processes in Geophysics* **1**, 105–14.

Feigenbaum, M. J. (1978). Quantitative universality for a class of nonlinear transformations. *Journal of Statistical Physics* **19**, 25–52.

Fjortoft, R. (1953). On the changes in the spectral distribution of kinetic energy in two dimensional, nondivergent flow. *Tellus* 7, 168–76.

Fox, C. G., and Hayes, D. E. (1985). Quantitative methods for analyzing the roughness of the seafloor. *Reviews of Geophysics* **23**, 1–48.

Gagnon, J.-S., Lovejoy, S., and Schertzer, D. (2006). Multifractal earth topography. *Nonlinear Processes in Geophysics* **13**, 541–70.

Gilbert, L. E. (1989). Are topographic data sets fractal? *Pure and Applied Geophysics* **131**, 241–54.

Grassberger, P., and Procaccia, I. (1983a). Characterization of strange attractors. *Physical Review Letters* **50**, 346–9.

Grassberger, P., and Procaccia, I. (1983b). Measuring the strangeness of strange atractors. *Physica D* **9**, 189–208.

Greenland Ice Core Project (1993). Climate instability during the last interglacial period recorded in the GRIP ice core. *Nature* **364**, 203–7.

Grossman, S., and Thomae, S. (1977). Invariant distributions and stationary correlation functions of one-dimensional discrete processes. *Zeitschrift für Naturforschung A* **32**, 1353–63.

Heisenberg, W. (1948). On the theory of statistical and isotropic turbulence. *Proceedings of the Royal Society A* **195**, 402–6.

Kalnay, E. (2003). *Atmospheric Modelling, Data Assimilation and Predictability*. Cambridge: Cambridge University Press.

Kolmogorov, A. N. (1933). *Grundbegriffe der Wahrscheinlichkeitsrechnung*. Berlin: Julius Springer.

Kolmogorov, A. N. (1941). Local structure of turbulence in an incompressible liquid for very large Reynolds numbers. *Proceedings of the USSR Academy of Sciences* **30**, 299–303. [English translation: *Proceedings of the Royal Society A* 1991, **434**, 9–17.]

Kraichnan, R. H. (1967). Inertial ranges in two-dimensional turbulence. *Physics of Fluids* **10**, 1417–23.

Landau, L. (1944). On the problem of turbulence. *Comptes Rendus (Doklady) de l'Académie des Sciences de l'URSS* **44**, 311–14.

Lazarev, A., Schertzer, D., Lovejoy, S., and Chigirinskaya, Y. (1994). Unified multifractal atmospheric dynamics tested in the tropics: part II, vertical scaling and Generalized Scale Invariance. *Nonlinear Processes in Geophysics* **1**, 115–23.

Lesieur, M. (1987). *Turbulence in Fluids*. Dordrecht: Martinus Nijhoff.

Lilley, M., Lovejoy, S., Strawbridge, K., and Schertzer, D. (2004). 23/9 dimensional anisotropic scaling of passive admixtures using lidar aerosol data. *Physical Review E* **70**, 036307.

Lorenz, E. N. (1963). Deterministic nonperiodic flow. *Journal of the Atmospheric Sciences* **20**, 130–41.

Lorenz, E. N. (1975). Climate predictability. In: *The Physical Basis of Climate and Climate Modelling*. WMO GARP Publication Series 16. Geneva: World Meteorological Organisation, pp. 132–6.

Lovejoy, S., and Schertzer, D. (1986). Scale invariance in climatological temperatures and the spectral plateau. *Annales Geophysicae* **4B**, 401–10.

Lovejoy, S., and Schertzer, D. (1998). Stochastic chaos and multifractal geophysics. In F. M. Guindani and G. Salvadori, eds., *Chaos, Fractals and Models 96*. Genova: Italian University Press, pp. 38–52.

Lovejoy, S., and Schertzer, D. (2007). Scale, scaling and multifractals in geophysics: twenty years on. In A. A. Tsonis, J. B. Elsner, eds., *Nonlinear Dynamics in Geosciences*. New York, NY: Springer, pp. 311–37.

Lovejoy, S., and Schertzer, D. (2008). Turbulence, rain drops and the l 1/2 number density law. *New Journal of Physics* **10**, 075017–32. doi:075010.071088/071367–072630/075010/075017/075017.

Lovejoy, S., and Schertzer, D. (2011). Space-time cascades and the scaling of ECMWF reanalyses: fluxes and fields. *Journal of Geophysical Research* **116**, D14117. doi:10.1029/2011JD015654.

Lovejoy, S., and Schertzer, D. (2012). Low frequency weather and the emergence of the climate. In A. S. Sharma, A. Bunde, D. Baker, and V. P. Dimri. eds., *Complexity and Extreme Events in Geosciences*. AGU monographs.

Lovejoy, S., Currie, W. J. S., Tessier, Y., *et al.* (2000). Universal multfractals and ocean patchiness: phytoplankton, physical fields and coastal heterogeneity. *Journal of Plankton Research* **23**, 117–41.

Lovejoy, S., Schertzer, D., and Tessier, Y. (2001). Multifractals and resolution independent remote sensing algorithms: the example of ocean colour. *International Journal of Remote Sensing* **22**, 1191–234.

Lovejoy, S., Tarquis, A., Gaonac'h, H., and Schertzer, D. (2007). Single- and multiscale remote sensing techniques, multifractals and MODIS derived vegetation and soil moisture. *Vadose Zone Journal* **7**, 533–46.

Lovejoy, S., Schertzer, D., Lilley, M., Strawbridge, K. B., and Radkevitch, A. (2008). Scaling turbulent atmospheric stratification. I: Turbulence and waves. *Quarterly Journal of the Royal Meteorological Society* **134**: 277–300. doi:10.1002/qj.201.

Lovejoy, S., Watson, B. P., Grosdidier, Y., and Schertzer, D. (2009a). Scattering in thick multifractal clouds. Part II: multiple scattering. *Physica A* **388**, 3711–27.

Lovejoy, S., Tuck, A. F., Schertzer, D., and Hovde, S. J. (2009b). Reinterpreting aircraft measurements in anisotropic scaling turbulence. *Atmospheric Chemistry and Physics* **9**, 5007–25.

Lovejoy, S., Tuck, A. F., and Schertzer, D. (2010). The horizontal cascade structure of atmospheric fields determined from aircraft data. *Journal of Geophysical Research* **115**, D13105.

Lynch, P. (2006). *The Emergence of Numerical Weather Prediction: Richardson's Dream*. Cambridge: Cambridge University Press.

Mandelbrot, B. B. (1967). How long is the coastline of Britain? Statistical self-similarity and fractional dimension. *Science* **155**, 636–8.

Mandelbrot, B. B. (1977). *Fractals, Form, Chance and Dimension*. San Francisco, CA: Freeman.

Mandelbrot, B. B. (1983). *The Fractal Geometry of Nature*. San Francisco, CA: Freeman.

Maxwell, J. C. (1890). Molecules. In W. D. Niven, ed., *The Scientific Papers of James Clerk Maxwell*. Cambridge: Cambridge University Press, p. 373.

Monin, A. S. (1972). *Weather Forecasting as a Problem in Physics*. Boston, MA: MIT Press.

Monin, A. S., and Yaglom, A. M. (1975). *Statistical Fluid Mechanics*. Boston, MA: MIT Press.

Obukhov, A. M. (1941a). On the distribution of energy in the spectrum of turbulent flow. *Doklady Akad. Nauk SSSR* **32**, 22–4.

Obukhov, A. M. (1941b). Spectral energy distribution in a turbulent flow. *Akad. Nauk SSSR Ser. Geog. Geofiz.* **5**, 453–66.

Onsager, L. (1945). The distribution of energy in turbulence [abstract only]. *Physical Review* **68**, 286.

Palmer, T. (2001). Predicting uncertainty in numerical weather forecasts. In R. P. Pierce, ed., *Meteorology at the Millennium*. London: Royal Meteorological Society, pp. 3–13.

Palmer, T. N. (2012). Towards the probabilistic Earth-system simulator: a vision for the future of climate and weather prediction. *Quarterly Journal of the Royal Meteorological Society* **138**, 841–61. doi:10.1002/qj.1923.

Palmer, T. N., and Williams, P., eds. (2010). *Stochastic Physics and Climate Models*. Cambridge: Cambridge University Press.

Perrin, I. (1913). *Les Atomes*. Paris: NRF-Gallimard.

Pinel, J. (2012). The anisotropic space-time scaling of the atmosphere: turbulence and waves. McGill University, Montreal (under review).

Poincaré, H. (1892). *Les Méthodes Nouvelles de la Méchanique Céleste, vol. 1*. Paris: Gautier-Villars.

Richardson, L. F. (1922). *Weather Prediction by Numerical Process*. Cambridge: Cambridge University Press. Republished by Dover, 1965.

Richardson, L. F. (1926). Atmospheric diffusion shown on a distance-neighbour graph. *Proceedings of the Royal Society A* **110**, 709–37.

Richardson, L. F. (1961). The problem of contiguity: an appendix of statistics of deadly quarrels. *General Systems Yearbook* **6**, 139–87.

Richardson, L. F., and Stommel, H. (1948). Note on eddy diffusivity in the sea. *Journal of Meteorology* **5**, 238–40.
Rodgers, C. D. (1976). Retrieval of atmospheric temperature and composition from remote measurements of thermal radiation. *Reviews of Geophysics* **14**, 609–24.
Rodriguez-Iturbe, I., and Rinaldo, A. (1997). *Fractal River Basins*. Cambridge: Cambridge University Press.
Sachs, D., Lovejoy, S., and Schertzer, D. (2002). The multifractal scaling of cloud radiances from 1m to 1km. *Fractals* **10**, 253–65.
Sayles, R. S., and Thomas, T. R. (1978). Surface topography as a nonstationary random process. *Nature* **271**, 431–4.
Schertzer, D., and Lovejoy, S. (1985a). Generalised scale invariance in turbulent phenomena. *Physico-Chemical Hydrodynamics Journal* **6**, 623–35.
Schertzer, D., and Lovejoy, S. (1985b). The dimension and intermittency of atmospheric dynamics. In B. Launder, ed., *Turbulent Shear Flow 4*. New York, NY: Springer, pp. 7–33.
Schertzer, D., and Lovejoy, S. (1987). Physical modeling and analysis of rain and clouds by anisotropic scaling of multiplicative processes. *Journal of Geophysical Research* **92**, 9693–714.
Schertzer, D., Tchiguirinskaia, I., Lovejoy, S., *et al.* (2002). Which chaos in the rainfall–runoff process? *Hydrological Sciences Journal* **47**, 139–47.
Schertzer, D. and Lovejoy, S. (2003). Chaos et turbulence en Météorologie et Hydrologie. *C.R. Acad. Agri. Fr.* **89**(2).
Schertzer, D. and Lovejoy, S. (2006). Multifractals en Turbulence et Géophysique. *Irruption des Géométries Fractales dans la Science*. G. Belaubreet al. Eds. Paris, Editions de l'Académie Europééene Interdisciplinaire des Sciences (AEIS): 189–209.
Steinhaus, H. (1954). Length, shape and area. *Colloquium Mathematicum* **III**, 1–13.
Taylor, G. I. (1935). Statistical theory of turbulence, parts I–IV. *Proceedings of the Royal Society A* **151**, 421–78.
Tchiguirinskaia, I. (2002). Scale invariance and stratification: the unified multifractal model of hydraulic conductivity. *Fractals* **10**, 329–34.
Tchiguirinskaia, I. Lu, S., Molz, F. J., Williams, T. M., and Lavallee, D. (2002). Multifractal versus monofractal analysis of wetland topography. *Stochastic Environmental Research and Risk Assessment* **14**, 8–32.
Venig-Meinesz, F. A. (1951). A remarkable feature of the Earth's topography. *Proc. K. Ned. Akad. Wet. Ser. B Phys. Sci.* **54**, 212–28.
von Weizacker, C. F. (1948). Das Spektrum der Turbulenz bei grossen Reynoldsschen Zahlen. *Zeitschrift für Physik* **124**, 614.
Watson, B. P., Lovejoy, S., Grosdidier, Y., and Schertzer, D. (2009). Scattering in thick multifractal clouds. Part I: overview and single scattering. *Physica A* **388**, 3695–710.
Whitney, H. (1936). Differentiable manifolds. *Annals of Mathematics* **37**, 645–80.

Chapter 2

Bacmeister, J. T., Eckermann, S. D., Newman, P. A., *et al.* (1996). Stratospheric horizontal wavenumber spectra of winds, potnetial temperature, and atmospheric tracers observed by high-altitude aircraft. *Journal of Geophysical Research* **101**, 9441–70.
Boer, G. J., and Shepherd, T. G. (1983). Large scale two-dimensional turbulence in the atmosphere. *Journal of the Atmospheric Sciences* **40**, 164–84.
Charney, J. G. (1971). Geostrophic turbulence. *Journal of the Atmospheric Sciences* **28**, 1087–95.
Cho, J. Y. N., and Lindborg, E. (2001). Horizontal velocity structure functions in the upper troposphere and lower stratosphere. I: Observations. *Journal of Geophysical Research* **106**, 10223–32.
Corrsin, S. (1951). On the spectrum of isotropic temperature fluctuations in an isotropic turbulence. *Journal of Applied Physics* **22**, 469–73.
Cvitanovic, P. (1984). Universality in chaos. In: P. Cvitanovic, ed., *Universality in Chaos*. Bristol: Adam Hilger, pp. 3–34.
Davidson, P. A. (2004). *Turbulence: an Introduction for Scientists and Engineers*. Oxford: Oxford University Press.
Dewdney, A. K. (1985). A computer microscope zooms in for a close look at the most complicated object in mathematics. *Scientific American* **253**, 16–24.
Feller, W. (1972). *An Introduction to Probability Theory and its Applications, Vol. 2*. New York, NY: Wiley.
Frehlich, R. G., and Sharman, R. D. (2010). Equivalence of velocity statistics at constant pressure or constant altitude. *Geophysical Research Letters* **37**, L08801. doi: 10.01029/02010GL042912.
Freilich, M. H., and Chelton, D. B. (1986). Wavenumber spectra of Pacific winds measured by the Seasat scatterometer. *Journal of Physical Oceanography* **16**, 741–57.
Frisch, U., and Morf, R. (1981). Intermittency in nonlinear dynamics and singularities at complex times. *Physical Review A*, **23**, 2673–705.
Gage, K. S., and Nastrom, G. D. (1986). Theoretical interpretation of atmospheric wavenumber spectra of wind and temperature observed by commercial aircraft during GASP. *Journal of the Atmospheric Sciences* **43**, 729–40.
Gao, X., and Meriwether, J. W. (1998). Mesoscale spectral analysis of in situ horizontal and vertical wind measurements at 6 km. *Journal of Geophysical Research* **103**, 6397–404.
Goldreich, O., and Wigderson, A. (2008). Computational complexity. In T. Gowers, ed., *The Princeton Companion to Mathematics*. Princeton, NJ: Princeton University Press, pp. 575–604.
Guttenberg, N., and Goldenfeld, N. (2009). Friction factor of two-dimensional rough-boundary turbulent soap film flows. *Physical Review E* **79**, 065306.

Hamilton, K., Takahashi, Y. O., and Ohfuchi, W. (2008). Mesoscale spectrum of atmopsheric motions investigated in a very fine resolution global general ciruculation model. *Journal of Geophysical Research* **113**, D18110. doi: 18110.11029/12008JD009785.

Högström, U., Smedman, A. N., and Bergström, H. (1999). A case study of two-dimensional stratified turbulence. *Journal of the Atmospheric Sciences* **56**, 959–76.

Julian, P. R., Washington, W. M., Hembree, L., and Ridley, C. (1970). On the spectral distribution of large-scale atmospheric energy. *Journal of the Atmospheric Sciences* **27**, 376–87.

Kerr, R. M., and King, G. P. (2009). Evidence for a midlatitude meso-scale downscale energy cascade from the marine boundary layer. www.eng.warwick.ac.uk/staff/rmk/kerr/KK_20may09.pdf.

Kevlahan, N. K. -R., and Farge, M. (1997). Vorticity filaments in two-dimensional turbulence: creation, stability and effect. *Journal of Fluid Mechanics* **346**, 49–76.

King, G. P., and Kerr, R. M. (2010). Second-order and third-order structure functions calculated from 10 years of QuikSCAT winds over the Pacific Ocean. *European Geosciences Union General Assembly 2010*, EGU2010, poster 15601.

Kolmogorov, A. N. (1941). Local structure of turbulence in an incompressible liquid for very large Reynolds numbers. *Proceedings of the USSR Academy of Sciences* **30**, 299–303. [English translation: *Proceedings of the Royal Society A* 1991, **434**, 9–17.]

Kraichnan, R. H. (1967). Inertial ranges in two-dimensional turbulence. *Physics of Fluids* **10**, 1417–23.

Kraichnan, R. H. (1971). An almost-Markovian Galilean-invariant turbulence model. *Journal of Fluid Mechanics* **47**, 513–24.

Lacorta, G., Aurell, E., Legras, B., and Vulpiani, A. (2004). Evidence for a $k^{-5/3}$ spectrum from the EOLE Lagrangian balloons in the lower stratosphere. *Journal of the Atmospheric Sciences* **61**, 2936–42.

Laplace, P. S. (1886). *Théorie Analytique des Probabilités*. Paris: Gauthier-Villars. (Originally published 1812.)

Lesieur, M. (1987). *Turbulence in Fluids*. Dordrecht: Martinus Nijhoff.

Lesieur, M., and Schertzer, D. (1978). Amortissement auto-similaire d'une turbulence à grand nombre de Reynolds. *Journal de Mécanique* **17**, 609–46.

Lévy, P. (1925). *Calcul des Probabilités*. Paris: Gautier-Villars.

Lilly, D. K. (1983). Stratified turbulence and the mesoscale variability of the atmopshere. *Journal of the Atmospheric Sciences* **40**, 749–61.

Lilly, D. K. (1989). Two-dimensional turbulence generated by energy sources at two scales. *Journal of the Atmospheric Sciences* **46**, 2026–30.

Lindborg, E. (1999). Can the atmospheric kinetic energy spectrum be explained by two-dimensional turbulence? *Journal of Fluid Mechanics* **388**, 259–88.

Lindborg, E., and Cho, J. (2001). Horizontal velocity structure functions in the upper troposphere and lower stratosphere, II. Theoretical considerations. *Journal of Geophysical Research* **106**, 10233–41.

Lovejoy, S., and Schertzer, D. (2010). Towards a new synthesis for atmospheric dynamics: space-time cascades. *Atmospheric Research* **96**, 1–52.

Lovejoy, S., Schertzer, D., and Tuck, A. F. (2004). Fractal aircraft trajectories and nonclassical turbulent exponents. *Physical Review E* **70**, 036306.

Lovejoy, S., Tuck, A. F., Hovde, S. J., and Schertzer, D. (2007). Is isotropic turbulence relevant in the atmosphere? *Geophysical Research Letters* **34**, L15802. doi:10.1029/2007GL029359.

Lovejoy, S., Schertzer, D., Lilley, M., Strawbridge, K. B., and Radkevitch, A. (2008). Scaling turbulent atmospheric stratification. I: Turbulence and waves. *Quarterly Journal of the Royal Meteorological Society* **134**: 277–300. doi:10.1002/qj.201.

Lovejoy, S., Tuck, A. F., Schertzer, D., and Hovde, S. J. (2009). Reinterpreting aircraft measurements in anisotropic scaling turbulence. *Atmospheric Chemistry and Physics* **9**, 5007–25.

Lovejoy, S., Tuck, A. F., and Schertzer, D. (2010). The horizontal cascade structure of atmospheric fields determined from aircraft data. *Journal of Geophysical Research* **115**, D13105.

Lynch, P. (2006). *The Emergence of Numerical Weather Prediction: Richardson's Dream*. Cambridge: Cambridge University Press.

Mamrosh, R. D., Daniels, T. S., and Moninger, W. R. (2006). Aviation applications of TAMDAR aircraft data reports. *12th Conference on Aviation Range and Aerospace Meteorology*. Atlanta, GA: American Meteorological Society.

Millionshtchikov, M. (1941). On the theory of homogeneous isotropic turbulence. *Doklady Akad. Nauk SSSR* **32**, 615–18.

Moninger, W. R., Mamrosh, R. D., and Pauley, P. M. (2003). Automated meteorological reports from commercial aircraft. *Bulletin of the American Meteorological Society* **84**, 203–16.

Morel, P., and Larchevêque, M. (1974). Relative dispersion of constant level balloons in the 200 mb general circulation. *Journal of the Atmospheric Sciences* **31**, 2189–96.

Ngan, K., Straub, D. N., and Bartello, P. (2004). Three-dimensionalization of freely-decaying two-dimensional flows. *Physics of Fluids* **16**, 2918–32. http://dx.doi.org/10.1063/1.1763191.

Obukhov, A. (1949). Structure of the temperature field in a turbulent flow. *Izv. Akad. Nauk SSSR. Ser. Geogr. I Geofiz* **13**, 55–69.

Obukhov, A. M. (1941). On the distribution of energy in the spectrum of turbulent flow. *Doklady Akad. Nauk SSSR* **32**, 22–4.

Palmer, T. N. (2001). A nonlinear dynamical perspective on model error: a proposal for non-local stochastic-dynamic

parameterisation in weather and climate prediction models. *Quarterly Journal of the Royal Meteorological Society* **127**, 279–304.
Palmer, T. N., and Williams, P., eds. (2010). *Stochastic Physics and Climate Models*. Cambridge: Cambridge University Press.
Patoux, J., and Brown, R. A. (2001). Spectral analysis of QuikScat surface winds and two dimensional turbulence. *Journal of Geophysical Research* **106**, 23995–4005.
Peitgen, H. O., and Richter, P. H. (1986). *The Beauty of Fractals*. New York, NY: Springer.
Pinel, J., Lovejoy, S., Schertzer, D., and Tuck, A. F. (2012). Joint horizontal–vertical anisotropic scaling, isobaric and isoheight wind statistics from aircraft data. *Geophysical Research Letters* **39**, L11803. doi:10.1029/2012GL051698.
Richardson, L. F. (1926). Atmospheric diffusion shown on a distance-neighbour graph. *Proceedings of the Royal Society A* **110**, 709–37.
Rose, H. A., and Sulem, P. L. (1978). Fully developed turbulence and statistical mechanics. *Journal de Physique* **39**, 441–73.
Schertzer, D. (2009). Interactive comment on "Comment on "Reinterpreting aircraft measurements in anisotropic scaling turbulence" by Lovejoy *et al.* (2009)" by E. Lindborg *et al. Atmospheric Chemistry and Physics Discussions* **9**, C8605–10.
Schertzer, D., and Lovejoy, S. (1991). Nonlinear geodynamical variability: multiple singularities, universality and observables. In D. Schertzer and S. Lovejoy, eds., *Non-Linear Variability in Geophysics: Scaling and Fractals*. Dordrecht: Kluwer, pp. 41–82.
Schertzer, D., Larchevêque, M., and Lovejoy, S. (1998). Beyond multifractal phenomenology of intermittency: nonlinear dynamics and multifractal renormalization. In F. M. Guindani and G. Salvadori, eds., *Chaos, Fractals and Models 96*. Genova: Italian University Press, pp. 53–64.
Schertzer, D. and Lovejoy, S. (1993). *Lecture Notes: Nonlinear Variability in Geophysics 3: Scaling and Multifractal Processes in Geophysics*. Cargèse, France, Institut d'Etudes Scientifique de Cargèse.
Schertzer, D., Tchiguirinskaia, I., Lovejoy, S., and Hubert, P. (2010). No monsters, no miracles: in nonlinear sciences hydrology is not an outlier! *Hydrological Sciences Journal* **55**, 965–79.
Schertzer, D. and Lovejoy, S. (2011). Multifractals, Generalized Scale Invariance and Complexity in Geophysics. *International Journal of Bifurcation and Chaos* **21**(12): 3417–3456.
Schertzer, D., Tchiguirinskaia, I., Lovejoy, S., and Tuck, A. F. (2011). Quasi-geostrophic turbulence and generalized scale invariance: a theoretical reply to Lindborg. *Atmospheric Chemistry and Physics Discussions* **11**, 3301–20.
Schertzer, D., Tchiguirinskaia, I., Lovejoy, S., and Tuck, A. F. (2012). Quasi-geostrophic turbulence and generalized scale invariance: a theoretical reply. *Atmospheric Chemistry and Physics* **12**, 327–36.
Skamarock, W. C. (2004). Evaluating mesoscale NWP models using kinetic energy spectra. *Monthly Weather Review* **132**, 3020.
Smith, K. S. (2004). Comment on: "The k^{-3} and $k^{-5/3}$ energy spectrum of atmospheric turbulence: quasigeostrophic two-level model simulation". *Journal of the Atmospheric Sciences* **61**, 937–41.
Strauss, D. M., and Ditlevsen, P. (1999). Two-dimensional turbulence properties of the ECMWF reanalyses. *Tellus* **51A**, 749–72.
Takayashi, Y. O., Hamilton, K., and Ohfuchi, W. (2006). Explicit global simulation of the meso-scale spectrum of atmospheric motions. *Geophysical Research Letters* **33**, L12812. doi: 10.1029/2006GL026429.
Taylor, G. I. (1935). Statistical theory of turbulence, parts I–IV. *Proceedings of the Royal Society A* **151**, 421–78.
Tung, K. K., and Orlando, W. W. (2003). The k^{-3} and $k^{-5/3}$ energy spectrum of atmospheric turbulence: quasigeostrophic two-level model simulation. *Journal of the Atmospheric Sciences* **60**, 824–35.
Xu, Y., Fu, L., Tulloch, R. (2011). The global characteristics of the wavenumber spectrum of ocean surface wind. *Journal of Physical Oceanography* **41**, 1576–82.

Chapter 3

Aitchison, J., and Brown, J. A. C. (1957). *The Lognormal Distribution, With Special Reference to its Uses in Economics*. Cambridge: Cambridge University Press.
Batchelor, G. K., and Townsend, A. A. (1949). The Nature of turbulent motion at large wave-numbers. *Proceedings of the Royal Society of London* A **199**, 238–55.
Brax, P., and Peschanski, R. (1991). Levy stable law description on intermittent behaviour and quark-gluon phase transitions. *Physics Letters B* **253**, 225–30.
Cahalan, R. (1994). Bounded cascade clouds: albedo and effective thickness. *Nonlinear Processes in Geophysics* **1**, 156–67.
Chigirinskaya, Y., and Schertzer, D. (1996). Cascade of scaling gyroscopes: Lie structure, universal multifractals and self-organized criticality in turbulence. In W. Woyczynski and S. Molchansov, eds., *Stochastic Models in Geosystems*. New York, NY: Springer-Verlag, pp. 57–82.
Chigirinskaya, Y., Schertzer, D., and Lovejoy, S. (1997). Scaling gyroscopes cascade: universal multifractal features of 2D and 3D turbulence. Paper presented at Fractals and Chaos in Chemical Engineering, CFIC 96, Rome.
de Wijs, H. J. (1953). Statistics of ore distribution: (2) Theory of binomial distribution applied to sampling and engineering problems. *Geologie en Mijnbouw* **15**, 12–24.
Doswell, C. A., and Lasher-Trapp, S. (1997). On measuring the degree of irregularity in an observing network. *Journal of Atmospheric and Oceanic Technology* **14**, 120–32.

Falconer, K. (1990). *Fractal Geometry: Mathematical Foundations and Applications*. New York, NY: Wiley.
Frisch, U., Sulem, P. L., and Nelkin, M. (1978). A simple dynamical model of intermittency in fully develop turbulence. *Journal of Fluid Mechanics* **87**, 719–24.
Gabriel, P., Lovejoy, S., Schertzer, D., and Austin, G. L. (1988). Multifractal Analysis of resolution dependence in satellite imagery. *Geophysical Research Letters* **15**, 1373–6.
Giordano, F., Ortosecco, I., and Tramontana, L. (2006). Fractal structure of marine measuring networks. *Il Nuovo Cimento C* **18**, 177–82. doi: 10.1007/BF02512018.
Gledzer, E. B., Dolzhansky, E. V., and Obukhov, A. M. (1981). *Systems of Fluid Mechanical Type and their application*. Nauka, Moscow (in Russian).
Grassberger, P. (1983). Generalized dimensions of strange attractors. *Physics Letters A* **97**, 227–30.
Gupta, V. K., and Waymire, E. (1993). A statistical analysis of mesoscale rainfall as a random cascade. *Journal of Applied Meteorology* **32**, 251–67.
Hentschel, H. G. E., and Procaccia, I. (1983). The infinite number of generalized dimensions of fractals and strange attractors. *Physica D* **8**, 435–44.
Kahane, J. P. (1985). Sur le chaos multiplicatif. *Annales des Sciences Mathématique du Québec* **9**, 435.
Kapteyn, J. C. (1903). *Skew Frequency Curves in Biology and Statistics*. Astronomical Laboratory, Noordhoff., Groningen.
Kida, S. (1991). Log stable distribution and intermittency of turbulence. *Journal of the Physical Society of Japan* **60**, 5–8.
Kolmogorov, A. N. (1941a). Logarithmically normal distribution of fragmentary particle sizes. *Doklady Akad. Nauk SSSR* **31**, 99–101.
Kolmogorov, A. N. (1941b). *Doklady Akad. Nauk SSSR* **30**, 301–5.
Kolmogorov, A. N. (1941c). *Doklady Akad. Nauk SSSR* **31**, 538–40.
Kolmogorov, A. N. (1941d). Local structure of turbulence in an incompressible liquid for very large Reynolds numbers. *Proceedings of the USSR Academy of Sciences* **30**, 299–303. [English translation: *Proceedings of the Royal Society A* 1991, **434**, 9–17.]
Kolmogorov, A. N. (1962). A refinement of previous hypotheses concerning the local structure of turbulence in viscous incompressible fluid at high Reynolds number. *Journal of Fluid Mechanics* **13**, 82–5.
Korvin, G., Boyd, D. M., and O'Dowd, R. (1990). Fractal characterization of the South Australian gravity station network. *Geophysical Journal International* **100**, 535–9.
Lamb, H. (1963). *Hydrodynamics*. Cambridge: Cambridge University Press.
Lévy, P. (1925). *Calcul des Probabilités*. Paris: Gautier-Villars.
Lopez, R. E. (1979). The lognormal distribution and cumulus cloud populations. *Monthly Weather Review* **105**, 865–72.
Lovejoy, S. (2010). Why the bare/dressed cascade distinction matters: interactive comment on "Reconstruction of sub-daily rainfall sequences using multinomial multiplicative cascades by L. Wang *et al.*" *Hydrology and Earth System Sciences Discussions* 7, C1–4.
Lovejoy, S., and Schertzer, D. (1990). Our multifractal atmosphere: a unique laboratory for nonlinear dynamics. *Physics in Canada* **46**, 62–71.
Lovejoy, S., and Schertzer, D. (2006). Multifractals, cloud radiances and rain. *Journal of Hydrology* **322**, 59–88.
Lovejoy, S., Schertzer, D., and Ladoy, P. (1986). Fractal characterisation of inhomogeneous measuring networks. *Nature* **319**, 43–4.
Lovejoy, S., Schertzer, D., and Tsonis, A. A. (1987). Functional box-counting and multiple elliptical dimensions in rain. *Science* **235**, 1036–8.
Mandelbrot, B. B. (1967). How long is the coastline of Britain? Statistical self-similarity and fractional dimension. *Science* **155**, 636–8.
Mandelbrot, B. B. (1974). Intermittent turbulence in self-similar cascades: divergence of high moments and dimension of the carrier. *Journal of Fluid Mechanics* **62**, 331–50.
Mandelbrot, B. B. (1975). Stochastic models for the earth's relief, the shape and the fractal dimension of the coastlines, and the number-area rule for islands. *Proceedings of the National Academy of Sciences of the USA* **72**, 3825–8.
Mandelbrot, B. B. (1977). *Fractals, Form, Chance and Dimension*. San Francisco, CA: Freeman.
Mandelbrot, B. B. (1983). *The Fractal Geometry of Nature*. San Francisco, CA: Freeman.
Mandelbrot, B. B. (1989). Fractal geometry: what is it and what does it do?. In D. J. T. M. Fleischman and R. C. Ball, eds., *Fractals in the Natural Sciences*. Princeton, NJ: Princeton University Press, pp. 3–16.
Matheron, G. (1970). Random functions and their applications in geology. In D. F. Merriam, ed., *Geostatistics*. New York, NY: Plenum Press, pp. 79–87.
Mazzarella, A., and Tranfaglia, G. (2000). Fractal characterisation of geophysical measuring networks and its implication for an optimal location of additional stations: an application to a rain-gauge network. *Theoretical and Applied Climatology* **65**, 157–63.
McAlister, D. (1879). The law of the geometric mean. *Proceedings of the Royal Society* **29**, 367–76.
Meneveau, C., and Sreenivasan, K. R. (1987). Simple multifractal cascade model for fully developed turbulence. *Physical Review Letters* **59**, 1424–7.
Monin, A. S., and Yaglom, A. M. (1975). *Statistical Fluid Mechanics*. Boston, MA: MIT Press.
Nicolis, C. (1993). Optimizing the global observational network: a dynamical approach. *Journal of Applied Meteorology* **32**, 1751–9.
Novikov, E. (1994). Infinitely divisible distributions in turbulence. *Physical Review E* **50**, R3303–5.

Novikov, E. A., and Stewart, R. (1964). Intermittency of turbulence and spectrum of fluctuations in energy-disspation. *Izv. Akad. Nauk. SSSR. Ser. Geofiz.* **3**, 408–12.

Obukhov, A. M. (1941a). On the distribution of energy in the spectrum of turbulent flow. *Doklady Akad. Nauk SSSR* **32**, 22–4.

Obukhov, A. M. (1941b). Spectral energy distribution in a turbulent flow. *Akad. Nauk SSSR Ser. Geog. Geofiz.* **5**, 453–66.

Obukhov, A. M. (1962). Some specific features of atmospheric turbulence. *Journal of Geophysical Research* **67**, 3011–14.

Onsager, L. (1945). The distribution of energy in turbulence [abstract only]. *Physical Review* **68**, 286.

Orszag, S. A. (1970). Indeterminacy of the moment problem for intermittent turbulence. *Physics of Fluids* **13**, 2211–12.

Samorodniitsky, G., and Taqqu, M. S. (1994). *Stable Non-Gaussian Random Processes: Stochastic Models with Infinite Variance.* New York, NY: Chapman and Hall.

Schertzer, D., and Lovejoy, S. (1983). Elliptical turbulence in the atmosphere. *Fourth Symposium on Turbulent Shear Flows.* Karlshule, West Germany.

Schertzer, D., and Lovejoy, S. (1984). On the dimension of atmospheric motions. In T. Tatsumi, ed., *Turbulence and Chaotic Phenomena in Fluids.* Amsterdam: Elsevier, pp. 505–12.

Schertzer, D., and Lovejoy, S. (1985). The dimension and intermittency of atmospheric dynamics. In B. Launder, ed., *Turbulent Shear Flow 4.* New York, NY: Springer, pp. 7–33.

Schertzer, D., and Lovejoy, S. (1987). Physical modeling and analysis of rain and clouds by anisotropic scaling of multiplicative processes. *Journal of Geophysical Research* **92**, 9693–714.

Schertzer, D., and Lovejoy, S. (1989). Nonlinear variability in geophysics: multifractal analysis and simulation. In L. Pietronero, ed., *Fractals: Physical Origin and Consequences.* New York, NY: Plenum, pp. 49–79.

Schertzer, D., and Lovejoy, S. (1991a). *Non-Linear Variability in Geophysics: Scaling and Fractals.* Dordrecht: Kluwer.

Schertzer, D., and Lovejoy, S. (1991b). Nonlinear geodynamical variability: multiple singularities, universality and observables. In D. Schertzer and S. Lovejoy, eds., *Non-Linear Variability in Geophysics: Scaling and Fractals.* Dordrecht: Kluwer, pp. 41–82.

Schertzer, D., and Lovejoy, S. (1996). Lecture notes: resolution dependence and multifractals in remote sensing and geographical information systems. McGill University, Montreal, June **10**, 1996.

Schertzer, D., and Lovejoy, S. (1997). Universal multifractals do exist! *Journal of Applied Meteorology* **36**, 1296–303.

Schertzer, D., Lovejoy, S., Visvanathan, R., Lavallée, D., and Wilson, J. (1988). Multifractal analysis techniques and rain and cloud fields. In D. A. Weitz, L. M. Sander and B. B. Mandelbrot, eds., *Fractal Aspects of Materials: Disordered Systems.* Materials Research Society, pp. 267–70.

Schertzer, D., Lovejoy, S., Lavallée, D., and Schmitt, F. (1991). Universal hard multifractal turbulence: theory and observation. In R. Z. Sagdeev, U. Frisch, A. S. Moiseev and A. Erokhin, eds., *Nonlinear Dynamics of Structures.* Singapore: World Scientific, pp. 213–35.

Schertzer, D., Lovejoy, S., Schmitt, F., Chigirinskaya, Y., and Marsan, D. (1997). Multifractal cascade dynamics and turbulent intermittency. *Fractals* **5**, 427–71.

Schertzer, D., Tchiguirinskaia, I., Lovejoy, S., and Hubert, P. (2010). No monsters, no miracles: in nonlinear sciences hydrology is not an outlier! *Hydrological Sciences Journal* **55**, 965–79.

Schmitt, F., Schertzer, D., Lovejoy, S., and Brunet, Y. (1993). Estimation of universal multifractal indices for atmospheric turbulent velocity fields. *Fractals* **1**, 568–75.

She, Z. S., and Leveque, E. (1994). Universal scaling laws in fully developed turbulence. *Physical Review Letters* **72**, 336–9.

Steinhaus, H. (1960). *Mathematical Snapshots.* New York, NY: Oxford University Press.

Tessier, Y., Lovejoy, S., and Schertzer, D. (1994). The multifractal global raingauge network: analysis and simulation. *Journal of Applied Meteorology* **32**, 1572–86.

Venugopal, V., Roux, S. G., Foufoula-Georgiou, E., and Arneodo, A. (2006). Revisiting multifractality of high-resolution temporal rainfall using a wavelet-based formalism, *Water Resources Research* **42**, W06D14. doi: 10.1029/2005WR004489.

Welander, P. (1955). Studies on the general development of motion in a two-dimensional, ideal fluid. *Tellus* 7, 141–56.

Wielicki, B. A., and Parker, L. (1992). On the determination of cloud cover from satellite sensors: the effect of sensor spatial resolution. *Journal of Geophysical Research* **97** (D12), 12799–823.

Wood, R., and Field, P. R. (2011). The distribution of cloud horizontal sizes. *Journal of Climate* **24**, 4800–16.

Yaglom, A. M. (1966). The influence on the fluctuation in energy dissipation on the shape of turbulent characteristics in the inertial interval. *Sov. Phys. Dokl.* **2**, 26–30.

Chapter 4

Berrisford, P., Dee, D., Fielding, K., *et al.* (2009). *The ERA-Interim Archive.* ERA Report Series. Reading: European Centre for Medium Range Weather Forecasts.

Boer, G. J., and Shepherd, T. G. (1983). Large scale two-dimensional turbulence in the atmosphere. *Journal of the Atmospheric Sciences* **40**, 164–84.

Borde, R., and Isaka, H. (1996). Radiative transfer in multifractal clouds. *Journal of Geophysical Research* **101**, 29461–78. doi:10.1029/96JD02200.

Carsteanu, A., and Foufoula-Georgiou, E. (1996). Assessing dependence among the weights in a mulitplicative cascade model of temporal rainfall. *Journal of Geophysical Research* **101**, 26363–70.

Davis, A., Lovejoy, S., Gabriel, P., Schertzer, D., and Austin, G. L. (1990). Discrete angle radiative transfer. Part III: numerical results on homogeneous and fractal clouds. *Journal of Geophysical Research* **95**, 11729–42.

Davis, A., Lovejoy, S., and Schertzer, D. (1991). Discrete-angle radiative transfer in a multifractal medium. *SPIE Proceedings* **1558**, 37–59.

de Lima, I. (1998). *Multifractals and the Temporal Structure of Rainfall*. Wageningen.

de Lima, I., and Grasman, J. (1999). Multifractal analysis of 15-min and daily rainfall from a semi-arid region in Portugal. *Journal of Hydrology* **220**, 1–11.

de Montera, L., Barthès, L., Mallet, C., and Golé, P. (2009). Rain universal multifractal parameters revisited with dual-beam spectropluviometer measurements. *Journal of Hydrometeorology* **10**, 493–506.

de Montera, L., Verrier, S., Mallet, C., and Barthès, L. (2010). A passive scalar-like model for rain applicable up to storm scale. *Atmospheric Research* **98**, 140–7.

Deidda, R., Benzi, R., and Siccardi, F. (1999). Multifractal modeling of anomalous scaling laws in rainfall. *Water Resoures Research* **35**, 1853–67.

Douglas, E. M., and Barros, A. P. (2003). Probable maximum precipitation estimation using multifractals: application in the eastern United States. *Journal of Hydrometeorology* **4**, 1012–24.

Gabriel, P., Lovejoy, S., Davis, A., Schertzer, D., and Austin, G. L. (1990). Discrete angle radiative transfer. Part II: renormalization approach to scaling clouds. *Journal of Geophysical Research* **95**, 11717–28.

Gagnon, J. -S., Lovejoy, S., and Schertzer, D. (2006). Multifractal earth topography. *Nonlinear Processes in Geophysics* **13**, 541–70.

Gaonac'h, H., Lovejoy, S., and Schertzer, D. (2003). Resolution dependence of infrared imagery of active thermal features at Kilauea volcano. *International Journal of Remote Sensing* **24**, 2323–44.

Garcia-Marin, A. P., Jimenez-Hornero, F. J., and Ayuso-Munoz, J. L. (2008). Universal multifractal description of an hourly rainfall time series from a location in southern Spain. *Atmosfera* **21**, 347–55.

Gires, A., Tchiguirinskaia, I., Schertzer, D., and Lovejoy, S. (2012). Influence of the zero-rainfall on the assessment of the multifractal parameters. *Advances in Water Resources*, **45**, 13–25

Güntner, A., Olsson, J., Calver, A., and Gannon, B. (2001). Cascade-based disaggregation of continuous rainfall time series: the influence of climate. *Hydrology and Earth System Sciences* **5**, 145–64.

Gupta, V. K., and Waymire, E. (1993). A statistical analysis of mesoscale rainfall as a random cascade. *Journal of Applied Meteorology* **32**, 251–67.

Harris, D., Menabde, M., Seed, A., and Austin, G. (1996). Multifractal characterization of rain fields with a strong orographics influence. *Journal of Geophysical Research* **101**, 26405–14.

Harvey, D. A., Gaonac'h, H., Lovejoy, S., and Schertzer, D. (2002). Multifractal characterization of remotely sensed volcanic features: a case study from Kilauea volcano, Hawaii. *Fractals* **10**, 265–74.

Hoang, C. T., Tchiguirinskaia, I., Schertzer, D., *et al.* Assessing the high frequency quality of long rainfall series. *Journal of Hydrology*, **438–439**, 39–51.

Hubert, P., Tessier, Y., Lovejoy, S., *et al.* (1993). Multifractals and extreme rainfall events. *Geophysical Research Letters* **20**, 931–4.

Hubert, P., Biaou, A., and Schertzer, D. (2002). De la meso-echelle à la micro-echelle: desagregation/agregation multifractale et spatio-temporelle des precipitations. Report, Armines-EdF.

Hurst, H. E. (1951). Long-term storage capacity of reservoirs. *Transactions of the American Society of Civil Engineers* **116**, 770–808.

Kiely, G., and Ivanova, K. (1999). Multifractal analysis of hourly precipitation. *Physics and Chemistry of the Earth Part B* **24**, 781–6.

Ladoy, P., Schmitt, F., Schertzer, D., and Lovejoy, S. (1993). Variabilité temporelle des observations pluviométriques à Nimes. *Comptes Rendues de l'Académie des Sciences* **317, II**, 775–82.

Larnder, C. (1995). Observer problems in multifractals: the example of rain. Unpublished MS thesis, McGill University.

Lilley, M., Lovejoy, S., Desaulniers-Soucy, N., and D. Schertzer, D. (2006). Multifractal large number of drops limit in rain. *Journal of Hydrology* **328**, 20–37. http://dx.doi.org/10.1016/j.jhydrol.2005.11.063.

Lovejoy, S. (1982). Area perimeter relations for rain and cloud areas. *Science* **187**, 1035–7.

Lovejoy, S., and Schertzer, D. (1995). Multifractals and rain. In Z. W. Kunzewicz, ed., *New Uncertainty Conceps in Hydrology and Water Ressources*. Cambridge: Cambridge University Press, pp. 62–103.

Lovejoy, S., and Schertzer, D. (2006). Multifractals, cloud radiances and rain. *Journal of Hydrology* **322**, 59–88.

Lovejoy, S., and Schertzer, D. (2008). Turbulence, rain drops and the l 1/2 number density law. *New Journal of Physics* **10**, 075017–32. doi:075010.071088/071367–072630/075010/075017/075017.

Lovejoy, S., and Schertzer, D. (2011). Space-time cascades and the scaling of ECMWF reanalyses: fluxes and fields. *Journal of Geophysical Research* **116**, D14117. doi:10.1029/2011JD015654.

Lovejoy, S., Gabriel, P., Davis, A., Schertzer, D., and Austin, G. L. (1990). Discrete angle radiative transfer. Part I: scaling and similarity, universality and diffusion. *Journal of Geophysical Research* **95**, 11699–715.

Lovejoy, S., Watson, B., Schertzer, D., and Brosamlen, G. (1995). Scattering in multifractal media. In L. Briggs, ed., *Particle Transport in Stochastic Media*. Portland, OR: American Nuclear Society, pp. 750–60.

Lovejoy, S., Schertzer, D., and Silas, P. (1998). Diffusion in one-dimensional multifractal porous media. *Water Resources Research* **34**, 3283–91.

Lovejoy, S., Schertzer, D., and Stanway, J. D. (2001). Direct evidence of planetary scale atmospheric cascade dynamics. *Physical Review Letters* **86**, 5200–3.

Lovejoy, S., Tarquis, A., Gaonac'h, H., and Schertzer, D. (2007). Single- and multiscale remote sensing techniques, multifractals and MODIS derived vegetation and soil moisture. *Vadose Zone Journal* 7, 533–46.

Lovejoy, S., Schertzer, D., and Allaire, V. (2008). The remarkable wide range scaling of TRMM precipitation. *Atmospheric Research* **90**, 10–32.

Lovejoy, S., Schertzer, D., Allaire, V., *et al.* (2009a). Atmospheric complexity or scale by scale simplicity? *Geophysical Research Letters* **36**, L01801. doi:01810.01029/02008GL035863.

Lovejoy, S., Tuck, A. F., Hovde, S. J., and Schertzer, D. (2009b). The vertical cascade structure of the atmosphere and multifractal drop sonde outages. *Journal of Geophysical Research* **114**, D07111. doi:07110.01029/02008JD010651.

Lovejoy, S., Watson, B. P., Grosdidier, Y., and Schertzer, D. (2009c). Scattering in thick multifractal clouds. Part II: multiple scattering. *Physica A* **388**, 3711–27.

Lovejoy, S., Tuck, A. F., and Schertzer, D. (2010). The horizontal cascade structure of atmospheric fields determined from aircraft data. *Journal of Geophysical Research* **115**, D13105.

Lovejoy, S., Pinel, J., and Schertzer, D. (2012). The global space-time cascade structure of precipitation: satellites, gridded gauges and reanalyses. *Advances in Water Resources* **45**, 37–50. http://dx.doi.org/10.1016/j.advwatres.2012.03.024.

Mandelbrot, B. B., and Wallis, J. R. (1968). Noah, Joseph and operational hydrology. *Water Resources Research* **4**, 909–18.

Marguerite, C., Schertzer, D., Schmitt, F., and Lovejoy, S. (1998). Copepod diffusion within multifractal phytoplankton fields. *Journal of Marine Systems* **16**, 69–83.

Meakin, P. (1987). Random walks on multifractal lattices. *Journal of Physics A* **20**, L771–7.

Menabde, M., Harris, D., Seed, A., Austin, G., and Stow, D. (1997). Multiscaling properties of rainfall and bounded random cascades. *Water Resources Research* **33**, 2823–30.

Naud, C., Schertzer, D., and Lovejoy, S. (1997). Fractional Integration and radiative transfer in multifractal atmospheres. In W. Woyczynski and S. Molchansov, eds., *Stochastic Models in Geosystems*. New York, NY: Springer-Verlag, pp. 239–67.

Olsson, J. (1998). Evaluation of a scaling cascade model for temporal rain-fall disaggregation. *Hydrology and Earth System Sciences* **2**, 19–30.

Olsson, J., and Berndtsson, R. (1998). Temporal rainfall disaggregation based on scaling properties. *Water Science Technology* **37**, 73–9.

Olsson, J., and Niemczynowicz, J. (1996). Multifractal analysis of daily spatial rainfall distributions. *Journal of Hydrology* **187**, 29–43.

Onof, C., and Arnbjerg-Nielsen, K. (2009). Quantification of anticipated future changes in high resolutionesign rainfall for urban areas. *Atmospheric Research* **92**, 350–63.

Over, T. M., and Gupta, V. K. (1994). Statistical analysis of Mesoscale rainfall: dependence of a random cascade generator on large-scale forcing. *Journal of Applied Meteorology* **33**, 1526–42.

Over, T. M., and Gupta, V. K. (1996). A space-time theory of mesoscale rainfall using random cascades. *Journal of Geophysical Research* **101**, 26319–31.

Pathirana, A., and Herath, S. (2002). Multifractal modelling and simulation of rain fields exhibiting spatial heterogeneity. *Hydrology and Earth System Sciences* **6**, 695–708.

Pathirana, A., Herath, S., and Yamada, T. (2003). Estimating rainfall distributions at high temporal resolutions using a multifractal model. *Hydrology and Earth System Sciences* 7, 668–79.

Paulson, K. S., and Baxter, P. D. (2007). Downscaling of rain gauge time series by multiplicative beta cascade. *Journal of Geophysical Research* **112**, D09105.

Penland, C. (1996). A stochastic model of IndoPacific sea surface temperature anomalies. *Physica D* **98**, 534–58.

Pinel, J., S. Lovejoy, D. Schertzer, 2014: The horizontal space-time scaling and cascade structure of the atmosphere and satellite radiances, **Atmos. Resear.**, **140–141,** 95–114, doi.org/10.1016/j.atmosres.2013.11.022.

Rupp, D. E., Keim, R. F., Ossiander, M., Brugnach, M., and Selker, J. S. (2009). Time scale and intensity dependency in multiplicative cascades for temporal rainfall disaggregation. *Water Resources Research* **45**, W07409.

Sachs, D., Lovejoy, S., and Schertzer, D. (2002). The multifractal scaling of cloud radiances from 1m to 1km. *Fractals* **10**, 253–65.

Sardeshmukh, P., Compo, G. P., and Penland, C. (2000). Changes in probability assoicated with El Niño. *Journal of Climate* **13**, 4268–86.

Schertzer, D., and Lovejoy, S. (1987). Physical modeling and analysis of rain and clouds by anisotropic scaling of multiplicative processes. *Journal of Geophysical Research* **92**, 9693–714.

Schertzer, D., and Lovejoy, S. (1997). Universal multifractals do exist! *Journal of Applied Meteorology* **36**, 1296–303.

Schertzer, D., Bernardara, P., Biaou, A., Tchiguirinskaia, I., Lang, M., Sauquet, E., Bendjoudi, H., Hubert, P., Lovejoy, S., and Veysseire, J. M. (2006). Extremes and multifractals in hydrology: results, validation and prospects. *Houille Blanche* **5**: 112–119.

Schertzer, D., Veyssere, J. M., Hallegate, S., *et al.* (unpublished). Hydrological extremes and multifractals: from GEV to MEV? Available from the authors.

Schmitt, F., Schertzer, D., Lovejoy, S., and Brunet, Y. (1996). Universal multifractal structure of atmospheric temperature and velocity fields. *Europhysics Letters* **34**, 195–200.

Serinaldi, F. (2010). Multifractality, imperfect scaling and hydrological properties of rainfall time series simulated by continuous universal multifractal and discrete random cascade models. *Nonlinear Processes in Geophysics* **17**, 697–714.

Stolle, J., Lovejoy, S., and Schertzer, D. (2009). The stochastic cascade structure of deterministic numerical models of the atmosphere. *Nonlinear Processes in Geophysics* **16**, 1–15.

Stolle, J., Lovejoy, S., and Schertzer, D. (2012). The temporal cascade structure and space-time relations for reanalyses and global circulation models. *Quarterly Journal of the Royal Meteorological Society*, **138**, 1895–1913.

Strauss, D. M., and Ditlevsen, P. (1999). Two-dimensional turbulence properties of the ECMWF reanalyses. *Tellus* **51A**, 749–72.

Sun, X., and Barros, A. P. (2010). An evaluation of the statistics of rainfall extremes in rain gauge observations and satellite-based and reanalysis products using universal multifractals. *Journal of Hydrometeorology* **11**, 388–404.

Tessier, Y., Lovejoy, S., and Schertzer, D. (1993). Universal multifractals: theory and observations for rain and clouds. *Journal of Applied Meteorology* **32**, 223–50.

Tessier, Y., Lovejoy, S., Hubert, P., Schertzer, D., and Pecknold, S. (1996). Multifractal analysis and modeling of rainfall and river flows and scaling, causal transfer functions. *Journal of Geophysical Research* **101**, 26427–40. doi:10.1029/96JD01799.

Veneziano, D., Furcolo, P., and Iacobellis, V. (2006). Imperfect scaling oftime and spacetime rainfall. *Journal of Hydrology* **322**, 105–19.

Verrier, S., de Montera, L., Barthès, L., and Mallet, C. (2010). Multifractal analysis of African monsoon rain fields, taking into account the zero rain-rate problem. *Journal of Hydrology* **389**, 111–20.

Verrier, S., Mallet, C., and Barthès, L. (2011). Multiscaling properties of rain in the time domain, taking into account rain support biases. *Journal of Geophysical Research* **116**, D20119.

Wang, L., Onof, C., and Maksimovic, C. (2010). Reconstruction of sub-daily rainfall sequences using multinomial multiplicative cascades. *Hydrology and Earth System Sciences Discussions* **7**, 5267–97.

Watson, B. P., Lovejoy, S., Grosdidier, Y., and Schertzer, D. (2009). Scattering in thick multifractal clouds. Part I: overview and single scattering. *Physica A* **388**, 3695–710.

Weissman, H., and Havlin, S. (1988). Dynamics in multiplicative processes. *Physical Review B* **37**, 5994–6.

Chapter 5

Arneodo, A., Decoster, N., and Roux, S. G. (1999). Intermittency, log-normal statistics, and multifractal cascade process in high-resolution satellite images of cloud structure. *Physical Review Letters* **83**, 1255–8.

Ashok, V., Balakumaran, T., Gowrishankar, C., Vennila, I. L. A., and Nirmalkumar, A. (2010). The fast Haar wavelet transform for signal and image processing. *International Journal of Computer Science and Information Security* 7, 126–30.

Austin, L. B., Austin, G. L., Schuepp, P. H., and Saucier, A. (1991). Atmospheric boundary layer variability and its implications for CO2 flux measurements. In S. Lovejoy and D. Schertzer, eds., *Nonlinear Variability in Geophysics*. Dordrecht: Kluwer, pp. 157–65.

Bacry, A., Arneodo, A., Frisch, U., Gagne, Y., and Hopfinger, E. (1989). Wavelet analysis of fully developed turbulence data and measurement of scaling exponents. In M. Lessieur and O. Métais, eds., *Turbulence and Coherent Structures*. Dordrecht: Kluwer, pp. 703–18.

Bak, P., Tang, C., and Weiessenfeld, K. (1987). Self-organized criticality: an explanation of 1/f noise. *Physical Review Letters* **59**, 381–4.

Basu, S., Foufoula-Georgiou, E., and Porté-Agel, F. (2004). Synthetic turbulence, fractal interpolation and large-eddy simulations. *Physical Review E* **70**, 026310.

Bendjoudi, H., Hubert, P., Schertzer, D., and Lovejoy, S. (1997). Interprétation multifractale des courbes intensité–durée–fréquence des précipitations, Multifractal point of view on rainfall intensity–duration–frequency curves. *C. R. A. S. (Sciences de la terre et des planetes/Earth and Planetary Sciences)* **325**, 323–6.

Bernardara, P., Lang, M., Sauqet, E., Schertzer, D., and Tchiguirinskaia, I. (2007). *Analyse Mulifractale en Hydrologie, Applications aux Series Temporelles*. Versailles, France, Editions Quae.

Bernardara, P., Schertzer, D., Sauquet, E., Tchiguirinskaia, I., and Lang, M., (2008). The flood probability distribution tail: how heavy is it? *Stoch. Environ. Resear. and Risk Analysis* **22**(1): 95–106.

Bleistein, N., and Handelsman, R. A. (1986). *Asymptotic Expansions of Integrals*. Mineola, NY: Dover.

Bunde, A., Kropp, J., and Schellnhuber, H. J., eds. (2002). *The Science of Disasters: Climate Disruptions, Heart Attacks and Market Crashes*. Berlin: Springer.

Bunde, A., Eichner, J. F., Kantelhardt, J. W., and Havlin, S. (2005). Long-term memory: a natural mechanism for the clustering of extreme events and anomalous residual times in climate records. *Physical Review Letters* **94**, 048701.

Calvet, L. E., and Fisher, A. J. (2001). Forecasting multifractal volatility. *Journal of Econometrics* **105**, 27–58.

Calvet, L. E., and Fisher, A. J. (2008). *Multifractal Volatility: Theory, Forecasting and Pricing*. New York, NY: Academic Press.

Chigirinskaya, Y., and Schertzer, D. (1997). Cascade of scaling gyroscopes: Lie structure, universal multifractals and self-organized criticality in turbulence. In W. Woyczynski and S. Molchansov, eds., *Stochastic Models in Geosystems*. New York, NY: Springer-Verlag, pp. 57–82.

Chigirinskaya, Y., Schertzer, D., Lovejoy, S., Lazarev, A., and Ordanovich, A. (1994).Unified multifractal atmospheric dynamics tested in the tropics. Part 1: horizontal scaling and self organized criticality. *Nonlinear Processes in Geophysics* **1**, 105–14.

Chigirinskaya, Y., Schertzer, D., Salvadori, G., Ratti, S., and Lovejoy, S. (1998). Chernobyl ^{137}Cs cumulative soil deposition in Europe: is it multifractal? In F. M. Guindani and G. Salvadori, eds., *Chaos, Fractals and Models 96*. Genova: Italian University Press, pp. 65–72.

Clauset, A., Shalizi, C. R., and Newman, M. E. J. (2009). Power law distributions in empirical data. *SIAM Review* **51**, 661–703.

Davis, A., Marshak, A., Wiscombe, W., and Cahalan, R. (1996). Multifractal characterization of intermittency in nonstationary geophysical signals and fields. In J. Harding, B. Douglas, and E. Andreas, eds., *Current Topics in Nonstationary Analysis*. Singapore: World Scientific, pp. 97–158.

De Lima, I. (1998). *Multifractals and the Temporal Structure of Rainfall*. Wageningen.

Deidda, R. (2000). Rainfall downscaling in a space-time multifractal framework. *Water Resources Research* **36**, 1779–94.

Dubrulle, B. (1994). Intermittency in fully developed turbulence: log-Poisson statistics and generalized scale covariance. *Physical Review Letters* **73**, 959–62.

Feller, W. (1972). *An Introduction to Probability Theory and its Applications, Vol. 2*. New York, NY: Wiley.

Finn, D., Lamb, B., Leclerc, M. Y., *et al.* (2001). Multifractal analysis of plume concentration fluctuations in surface layer flows. *Journal of Applied Meteorology* **40**, 229–45.

Foufoula-Georgiou, E., and Kumar, P., eds. (1994). *Wavelets in Geophysics*. San Diego, CA: Academic Press.

Gagnon, J.-S., Lovejoy, S., and Schertzer, D. (2006). Multifractal earth topography. *Nonlinear Processes in Geophysics* **13**, 541–70.

Garcia-Marin, A. P., Ayuso-Munoz, J. L., Jiménez-Hornero, F. J., and Estévez, J. (2012). Selecting the best IDF model using the multifractal approach. *Hydrological Processes* doi: 10.1002/hyp.9272.

Georgakakos, K. P., Carsteanu, A. A., Starvedent, P. L., and Cramer, J. A. (1994). Observation and analysis of midwestern rain rates. *Journal of Applied Meteorology* **33**, 1433–44.

Grassberger, P. (1983). Generalized dimensions of strange attractors. *Physics Letters A* **97**, 227–30.

Grauer, R., Krug, J., and Marliani, C. (1994). Scaling of high-order structure functions in magnetohydrodynamic turbulence. *Physics Letters A* **195**, 335–8.

Halsey, T. C., Jensen, M. H., Kadanoff, L. P., Procaccia, I., and Shraiman, B. (1986). Fractal measures and their singularities: the characterization of strange sets. *Physical Review A* **33**, 1141–51.

Harris, D., Menabde, M., Seed, A., and Austin, G. (1996). Multifractal characterization of rain fields with a strong orographics influence. *Journal of Geophysical Research* **101**, 26405–14.

Hentschel, H. G. E., and Procaccia, I. (1983). The infinite number of generalized dimensions of fractals and strange attractors. *Physica D* **8**, 435–44.

Holschneider, M. (1995). *Wavelets: an Analysis Tool*. Oxford: Clarendon Press.

Huang, Y. X., Schmitt, F. G., Hermand, J.-P. ***et al.***(2011). Arbitrary-order Hilbert spectral analysis for time series possessing scaling statistics: comparison study with detrended fluctuation analysis and wavelet leaders. *Physical Review E* **84**, 016208. doi:10.1103/PhysRevE.84.016208.

Hubert, P., Bendjoudi, H., Schertzer, D., and Lovejoy, S. (2001). Multifractal taming of extreme hydrometeorological events. Proceedings of a symposium on extraordinary floods held at Reyjavik, Iceland, July 2000.

Kantelhardt, J. W., Koscielny-Bunde, E., Rego, H. H. A., Havlin, S., and Bunde, S. (2001). Detecting long range correlations with detrended flucutation analysis. *Physica A* **295**, 441–54.

Kantelhardt, J. W., Zschiegner, S. A., Koscielny-Bunde, E., *et al.* (2002). Multifractal detrended fluctuation analysis of nonstationary time series. *Physica A* **316**, 87–114.

Kiely, G., and Ivanova, K. (1999). Multifractal analysis of hourly precipitation. *Physics and Chemistry of the Earth Part B* **24**, 781–6.

Koscielny-Bunde, E., Bunde, A., Havlin, S., *et al.* (1998). Indication of a universal persistence law governing atmospheric variability. *Physical Review Letters* **81**, 729–32.

Koutsoyiannis, D., and Montanari, A. (2007). Statistical analysis of hydroclimatic time series: Uncertainty and insights. *Water Resources Res.* **43**, W05429. doi: 10.1029/2006wr 005592.

Ladoy, P., Lovejoy, S., and Schertzer, D. (1991). Extreme variability of climatological data: scaling and intermittency. In S. Lovejoy and D. Schertzer, eds., *Nonlinear Variability in Geophysics*. Dordrecht: Kluwer, pp. 241–50.

Ladoy, P., Schmitt, F., Schertzer, D., and Lovejoy, S. (1993). Variabilité temporelle des observations pluviométriques à Nimes. Comptes Rendues de l'Académie des Sciences **317, II**, 775–82.

Lavallée, D., Lovejoy, S., and Schertzer, D. (1991). On the determination of the codimension function. In D. Schertzer and S. Lovejoy, eds., *Non-Linear Variability in Geophysics: Scaling and Fractals*. Dordrecht: Kluwer, pp. 99–110.

Lazarev, A., Schertzer, D., Lovejoy, S., and Chigirinskaya, Y. (1994). Unified multifractal atmospheric dynamics tested in the tropics: part II, vertical scaling and Generalized Scale Invariance. *Nonlinear Processes in Geophysics* **1**, 115–23.

Lennartz, S., and Bunde, A. (2009). Trend evaluation in records with long term memory: Application to global warming. *Geophysical Research Letters* **36**, L16706.

Lesieur, M., and Schertzer, D. (1978). Amortissement auto-similaire d'une turbulence à grand nombre de Reynolds. *Journal de Mécanique* **17**, 609–46.

Lévy-Véhel, J., Daoudi, K., and Lutton, E. (1994). Fractal modeling of speech signals. *Fractals* **2**, 379–82.

Lovejoy, S., and Schertzer, D. (1986). Scale invariance in climatological temperatures and the spectral plateau. *Annales Geophysicae* **4B**, 401–10.

Lovejoy, S., and Schertzer, D. (2006a). Multifractals, cloud radiances and rain. *Journal of Hydrology* **322**, 59–88.

Lovejoy, S., and Schertzer, D. (2006b). Stereophotography of rain drops and compound poisson -cascade processes. Cloud conference, American Meteorological Society, Madison, WI., pp. 14.14.11–14.14.19.

Lovejoy, S., and Schertzer, D. (2007). Scale, scaling and multifractals in geophysics: twenty years on. In A. A. Tsonis, J. B. Elsner, eds., *Nonlinear Dynamics in Geosciences*. New York, NY: Springer, pp. 311–37.

Lovejoy, S., and Schertzer, D. (2008). Turbulence, rain drops and the l 1/2 number density law. *New Journal of Physics* **10**, 075017–32. doi:075010.071088/071367–072630/075010/075017/075017.

Lovejoy, S., and Schertzer, D. (2010a). On the simulation of continuous in scale universal multifractals, part I: spatially continuous processes. *Computers and Geosciences* **36**, 1393–403.

Lovejoy, S., and Schertzer, D. (2010b). On the simulation of continuous in scale universal multifractals, part II: space-time processes and finite size corrections. *Computers and Geosciences* **36**, 1404–13.

Lovejoy, S., and Schertzer, D. (2012a). Haar wavelets, fluctuations and structure functions: convenient choices for geophysics. *Nonlinear Processes in Geophysics* **19**, 1–14.

Lovejoy, S., and Schertzer, D. (2012b). Low frequency weather and the emergence of the climate. In A. S. Sharma, A. Bunde, D. Baker, and V. P. Dimri, eds., *Complexity and Extreme Events in Geosciences*. AGU monographs.

Lovejoy, S., Schertzer, D., and Tessier, Y. (2001). Multifractals and resolution independent remote sensing algorithms: the example of ocean colour. *International Journal of Remote Sensing* **22**, 1191–234.

Lovejoy, S., Tuck, A. F., and Schertzer, D. (2010). The horizontal cascade structure of atmospheric fields determined from aircraft data. *Journal of Geophysical Research* **115**, D13105.

Lovejoy, S., Pinel, J., and Schertzer, D. (2012). The global space-time cascade structure of precipitation: satellites, gridded gauges and reanalyses. *Advances in Water Resources*. http://dx.doi.org/10.1016/j.advwatres.2012.03.024.

Mallat, S., and Hwang, W. (1992). Singularity detection and processing with wavelets. *IEEE Transactions on Information Theory* **38**, 617–43.

Marsan, D., Schertzer, D., and Lovejoy, S. (1996). Causal space-time multifractal processes: predictability and forecasting of rain fields. *Journal of Geophysical Research* **101**, 26333–46.

Mydlarski, L., and Warhaft, Z. (1998). Passive scalar statistics in high-Peclet-number grid turbulence. *Journal of Fluid Mechanics* **358**, 135–75.

Olsson, J. (1995). Limits and characteristics of the multifractal behavior of a high-resolution rainfall time series. *Nonlinear Processes in Geophysics* **2**, 23–9.

Onof, C., and Arnbjerg-Nielsen, K. (2009). Quantification of anticipated future changes in high resolutionesign rainfall for urban areas. *Atmospheric Research* **92**, 350–63.

Oswiecimka, P., Kwapien, J., and Drozdz, S. (2006). Wavelet versus detrended fluctuation analysis of multifractal structures. *Physical Review E* **74**, 16103. doi:10.1103/PhysRevE.74.016103.

Pandey, G., Lovejoy, S., and Schertzer, D. (1998). Multifractal analysis including extremes of daily river flow series for basis five to two million square kilometres, one day to 75 years. *Journal of Hydrology* **208**, 62–81.

Parisi, G., and Frisch, U. (1985). A multifractal model of intermittency. In M. Ghil, R. Benzi and G. Parisi, eds., *Turbulence and Predictability in Geophysical Fluid Dynamics and Climate Dynamics*. Amsterdam: North Holland, pp. 84–8.

Pecknold, S., Lovejoy, S., Schertzer, D., Hooge, C., and Malouin, J. F. (1993). The simulation of universal multifractals. In J. M. Perdang and A. Lejeune, eds., *Cellular Automata: Prospects in Astronomy and Astrophysics*. Singapore: World Scientific (1993), pp. 228–67.

Pecknold, S., Lovejoy, S., and Schertzer, D. (1997a). The morphology and texture of anisotropic multifractals using generalized scale invariance. In W. Woyczynski and S. Molchansov, eds., *Stochastic Models in Geosystems*. New York, NY: Springer-Verlag, pp. 269–312.

Pecknold, S., Lovejoy, S., Schertzer, D., and Hooge, C. (1997b). Multifractals and the resolution dependence of remotely sensed data: generalized scale invariance and geographical information systems. In D. A. Quattrochi and M. F. Goodchild, eds., *Scale in Remote Sensing and GIS*. Boca Raton, FL: Lewis, pp. 361–94.

Peng, C.-K., Buldyrev, S. V., Havlin, S., *et al.* (1994). Mosaic organisation of DNA nucleotides. *Physical Review E* **49**, 1685–9.

Quattrochi, D. A., and Goodchild, M. F., eds. (1997). *Scale in Remote Sensing and GIS*. Boca Raton, FL: Lewis.

Radulescu, M. I., Mydlarski, L. B., Lovejoy, S., and Schertzer, D. (2002). Evidence for algebraic tails of probability distributions in laboratory-scale turbulence. In I. P. Castro, P. E. Hancock, and T. G. Thomas, eds., *Advances in Turbulence IX*. Proceedings of the 9th European Turbulence Conference, Southampton, UK, July 2–5 2002. Barcelona: CIMNE, p. 891.

Rényi, A. (1970). *Probability Theory*. New York: North-Holland.

Salvadori, G., Ratti, S., Belli, G., Lovejoy, S., and Schertzer, D. (1993). Multifractal and Fourier analysis of Seveso pollution. *Journal of Toxicological and Environmental Chemistry* **43**, 63–76.

Sardeshmukh, P. D., and Sura, P. (2009). Reconciling non-gaussian climate statistics with linear dynamics. *Journal of Climate* **22**, 1193–207.

Schertzer, D., and Lovejoy, S. (1985). The dimension and intermittency of atmospheric dynamics. In B. Launder, ed., *Turbulent Shear Flow 4*. New York, NY: Springer, pp. 7–33.

Schertzer, D., and Lovejoy, S. (1987). Physical modeling and analysis of rain and clouds by anisotropic scaling of multiplicative processes. *Journal of Geophysical Research* **92**, 9693–714.

Schertzer, D., and Lovejoy, S. (1989). Nonlinear variability in geophysics: multifractal analysis and simulation. In L. Pietronero, ed., *Fractals: Physical Origin and Consequences*. New York, NY: Plenum, pp. 49–79.

Schertzer, D., and Lovejoy, S. (1991). Nonlinear geodynamical variability: multiple singularities, universality and observables. In D. Schertzer and S. Lovejoy, eds., *Non-Linear Variability in Geophysics: Scaling and Fractals*. Dordrecht: Kluwer, pp. 41–82.

Schertzer, D., and Lovejoy, S. (1992). Hard and soft multifractal processes. *Physica A* **185**, 187–94.

Schertzer, D., and Lovejoy, S. (1993). Lecture notes: nonlinear variability in geophysics 3: scaling and mulitfractal processes in geophysics. Institut d'Etudes Scientifique de Cargèse, Cargèse, France.

Schertzer, D., and Lovejoy, S. (1995). From scalar cascades to Lie cascades: joint multifractal analysis of rain and cloud processes. In R. A. Feddes, ed., *Space/Time Variability and Interdependence for Various Hydrological Processes*. New York, NY: Cambridge University Press, pp. 153–73.

Schertzer, D., and Lovejoy, S. (1996). Lecture notes: resolution dependence and multifractals in remote sensing and geographical information systems. McGill University, Montreal, June 10, 1996.

Schertzer, D., Lovejoy, S., Lavallée, D., and Schmitt, F. (1991). Universal hard multifractal turbulence: theory and observation. In R. Z. Sagdeev, U. Frisch, A. S. Moiseev and A. Erokhin, eds., *Nonlinear Dynamics of Structures*. Singapore: World Scientific, pp. 213–35.

Schertzer, D., Lovejoy, S., and Lavallée, D. (1993). Generic multifractal phase transitions and self-organized criticality. In J. M. Perdang and A. Lejeune, eds., *Cellular Automata: Prospects in Astronomy and Astrophysics*. Singapore: World Scientific (1993), pp. 216–27.

Schertzer, D., Lovejoy, S., and Schmitt, F. (1995). Structures in turbulence and multifractal universality. In M. Meneguzzi, A. Pouquet, and P. L. Sulem, eds., *Small-Scale Structures in 3D and MHD Turbulence*. New York, NY: Springer-Verlag, pp. 137–44.

Schertzer, D., Lovejoy, S., Schmitt, F., Chigirinskaya, Y., and Marsan, D. (1997). Multifractal cascade dynamics and turbulent intermittency. *Fractals* **5**, 427–71.

Schertzer, D., Larchevêque, M., and Lovejoy, S. (1998). Beyond multifractal phenomenology of intermittency: nonlinear dynamics and multifractal renormalization. In F. M. Guindani and G. Salvadori, eds., *Chaos, Fractals and Models 96*. Genova: Italian University Press, pp. 53–64 .

Schertzer, D., Larchevêque, M., Duan, J., Yanovsky, V. V., and Lovejoy, S. (2001). Fractional Fokker–Planck equation for nonlinear stochastic differential equation driven by non-Gaussian Lévy stable noises. *Journal of Mathematical Physics* **42**, 200–12.

Schertzer, D., Bernardara, P., Biaou, A., *et al.* (2006). Extrêmes et multifractals en hydrologie: résultats, validations et perspectives. *Houille Blanche* **5**, 112–19.

Schertzer, D., Tchiguirinskaia, I., Lovejoy, S., and Hubert, P. (2010). No monsters, no miracles: in nonlinear sciences hydrology is not an outlier! *Hydrological Sciences Journal* **55**, 965–79.

Schmitt, F., Schertzer, D., Lovejoy, S., and Brunet, Y. (1993). Estimation of universal multifractal indices for atmospheric turbulent velocity fields. *Fractals* **1**, 568–75.

Schmitt, F., Schertzer, D., Lovejoy, S., and Brunet, Y. (1994). Empirical study of multifractal phase transitions in atmospheric turbulence. *Nonlinear Processes in Geophysics* **1**, 95–104.

Schulz, M. (2002). On the 1470-year pacing of Dansgaard–Oeschger warm events. *Paleoceanography* **17**, 10.1029/2000PA000571.

Schuster, H. G. (1988). *Deterministic Chaos*, 2nd edn. New York, NY: VCH.

Serrano, E., and Figliola, A. (2009). Wavelet leaders: a new method to estimate the multifractal singularity spectra. *Physica A* **388**, 2793–805.

She, Z. S., and Leveque, E. (1994). Universal scaling laws in fully developed turbulence. *Physical Review Letters* **72**, 336–9.

Szépfalusy, P., Tél, T., Csordas, A., and Kovas, Z. (1987). Phase transitions associated with dynamical properties of chaotic systems. *Physical Review A* **36**, 3525–8.

Taqqu, M., Tevcrovsky, V., and Willinger, W. (1995). Estimators for long-range dependence: an empirical study. *Fractals* **3**, 785–98.

Tchiguirinskaia, I., Schertzer, D., Lovejoy, S., and Veysseire, J. M. (2006). Wind extremes and scales: multifractal insights and empirical evidence. In J. Peinke, P. Schaumann, and S. Barth, eds., *Wind Energy: Proceedings of the Euromech Colloquium*. Berlin: Springer-Verlag.

Tél, T. (1988). Fractals, multifractals, and thermodynamics. *Zeitschrift für Naturforschung* **43a**, 1154–74.

Tessier, Y. (1995). Multifractal objective analysis of rain and clouds. Physics, McGill University.

Tessier, Y., Lovejoy, S., and Schertzer, D. (1993). Universal multifractals: theory and observations for rain and clouds. *Journal of Applied Meteorology* **32**, 223–50.

Tessier, Y., Lovejoy, S., and Schertzer, D. (1994). The multifractal global raingauge network: analysis and simulation. *Journal of Applied Meteorology* **32**, 1572–86.

Tessier, Y., Lovejoy, S., Hubert, P., Schertzer, D., and Pecknold, S. (1996). Multifractal analysis and modeling of rainfall and river flows and scaling, causal transfer functions. *Journal of Geophysical Research* **101**, 26427–40. doi:10.1029/96JD01799.

Torrence, T., and Compo, G. P. (1998). A practical guide to wavelet analysis. *Bulletin of the American Meteorological Society* **79**, 61–78.

Tuck, A. F. (2008). *Atmospheric Turbulence: a Molecular Dynamics Perspective*. Oxford: Oxford University Press.

Tuck, A. F. (2010). From molecules to meteorology via turbulent scale invariance. *Quarterly Journal of the Royal Meteorological Society* **136**, 1125–44.

Tuck, A. F., Hovde, S. J., and Bui, T. P. (2004). Scale invariance in jet streams: ER-2 data around the lower-stratospheric polar night vortex. *Quarterly Journal of the Royal Meteorological Society* **130**, 2423–44.

Veneziano, D., and Furcolo, P. (2003). Marginal distribution of stationary multifractal measures and their Haar wavelet coefficients. *Fractals* **11**, 253–70. doi:10.1142/S0218348X03002051.

Verrier, S. (2011). Modélisation de la variabilité Spatiale et temporelle des précipitations a la sub-mésoéchelle pare une approche multifractale. LATMOS, Université de Versailles St-Quentin-en-Yvelines.

Wilson, J., Schertzer, D., and Lovejoy, S. (1991). Physically based modelling by multiplicative cascade processes. In D. Schertzer and S. Lovejoy, eds., *Non-Linear Variability in Geophysics: Scaling and Fractals*. Dordrecht: Kluwer, pp. 185–208.

Chapter 6

Adelfang, S. I. (1971). On the relation between wind shears over various intervals. *Journal of Applied Meteorology* **10**, 156–9.

Allen, S J., and Vincent, R. A. (1995). Gravity wave activity in the lower atmopshere: seasonal and latituidanl variations. *Journal of Geophysical Research* **100**, 1327–50.

Arad, I., Dhruva, B., Kurien, S., *et al.* (1998). Extraction of anisotropic contributions in turbulent flows. *Physical Review Letters* **81**, 5330–3.

Arad, I., L'vov, V. S., and Procaccia, I. (1999). Correlation functions in isotropic and anisotropic turbulence: the role of the symmetry group. *Physical Review E* **59**, 6753–65.

Ashkenazi, S., and Steinberg, V. (1999). Spectra and statistics of velocity and temperature fluctuations in turbulent convection. *Physical Review Letters* **83**, 4760–3.

Bolgiano, R. (1959). Turbulent spectra in a stably stratified atmosphere. *Journal of Geophysical Research* **64**, 2226–9.

Brunt, D. (1927). The period of vertical osciallations in the atmosphere. *Quarterly Journal of the Royal Meteorological Society* **53**, 30–2.

Charney, J. G. (1971). Geostrophic turbulence. *Journal of the Atmospheric Sciences* **28**, 1087–95.

Cho, J. Y. N., and Lindborg, E. (2001). Horizontal velocity structure functions in the upper troposphere and lower stratosphere. I: Observations. *Journal of Geophysical Research* **106**, 10223–32.

Dalaudier, F., Sidi, C., Crochet, M., and Vernin, J. (1994). Direct evidence of "sheets" in the atmospheric tmepreature field. *Journal of the Atmospheric Sciences* **51**, 237–48.

Deidda, R. (2000). Rainfall downscaling in a space-time multifractal framework. *Water Resources Research* **36**, 1779–94.

Dewan, E. (1997). Saturated-cascade similtude theory of gravity wave sepctra. *Journal of Geophysical Research* **102**, 29799–817.

Dewan, E., and Good, R. (1986). Saturation and the "universal" spectrum vertical profiles of horizontal scalar winds in the stratosphere. *Journal of Geophysical Research* **91**, 2742–8.

Endlich, R. M., Singleton, R. C., and Kaufman, J. W. (1969). Spectral analyses of detailed vertical wind profiles. *Journal of the Atmospheric Sciences* **26**, 1030–41.

Fjortoft, R. (1953). On the changes in the spectral distribution of kinetic energy in two dimensional, nondivergent flow. *Tellus* **7**, 168–76.

Fritts, D., Tsuda, T., Sato, T., Fukao, S., and Kato, S. (1988). Observational evidence of a saturated gravity wave spectrum in the troposphere and lower stratosphere. *Journal of the Atmospheric Sciences* **45**, 1741–59.

Gage, K. S., and Nastrom, G. D. (1986). Theoretical interpretation of atmospheric wavenumber spectra of wind and temperature observed by commercial aircraft during GASP. *Journal of the Atmospheric Sciences* **43**, 729–40.

Gardner, C. S. (1994). Diffusive filtering theory of gravity wave spectra in the atmosphere. *Journal of Geophysical Research* **99**, 20601–22.

Gardner, C. S., Hostetler, C. A., and Franke, S. J. (1993). Gravity wave models for the horizontal wave number spectra of atmospheric velocity and density flucutations. *Journal of Geophysical Research* **98**, 1035–49.

Garratt, J. R. (1992). *The Atmospheric Boundary Layer*. Cambridge: Cambridge University Press.

Garrett, C., and Munk, W. (1972). Space-time scales of internal waves. *Geophysical Fluid Dynamics* **2**, 225–64.

Gires, A., Schertzer, D., Tchiguirinskaia, I., *et al.* (2011). Impact of small scale rainfall uncertainty on urban discharge forecasts. In *Weather Radar and Hydrology*. Proceedings of a symposium held in April, Exeter, UK. International Association of Hydrological Sciences.

Gregg, M. (1991). The study of mixing in the ocean: a brief history. *Oceanography* **4**, 39–45.

Harrison, R. G., and Hogan, R. J. (2006). In-situ atmospheric turbulence measurement using the terrestrial magnetic field: a compass for a radiosonde. *Journal of Atmospheric and Oceanic Technology* **23**, 517–23.

Hock, T. F., and Franklin, J. L. (1999). The NCAR GPS dropsonde. *Bulletin of the American Meteorological Society* **80**, 407–20.

Hoskins, B. J., James, I. N., and White, G. H. (1983). The shape, propagation and mean-flow interaction of large-scale weather systems. *Journal of the Atmospheric Sciences* **40**, 1595–612.

Hoskins, B. J., McIntyre. M. E., and Robertson, A. W. (1985). On the use and signficiance of isentropic potential vorticity maps. *Quarterly Journal of the Royal Meteorological Society* **111**, 877–946.

Hostetler, C. A., and Gardner, C. S. (1994). Observations of horizontal and vertical wave number spectra of gravity wave motions in the stratosphere and mesosphere over the mid-Pacific. *Journal of Geophysical Research* **99**, 1283–302.

Hovde, S. J., Tuck, A. F., Lovejoy, S., and Schertzer, D. (2011). Vertical scaling of temperature, wind and humidity fluctuations: dropsondes from 13 km to the surface of the Pacific Ocean. *International Journal of Remote Sensing* **32**, 5891–918.

Kolmogorov, A. N. (1941). Local structure of turbulence in an incompressible liquid for very large Reynolds numbers. *Proceedings of the USSR Academy of Sciences* **30**, 299–303. [English translation: *Proceedings of the Royal Society A* 1991, **434**, 9–17.]

Kraichnan, R. H. (1967). Inertial ranges in two-dimensional turbulence. *Physics of Fluids* **10**, 1417–23.

Kurien, S., L'vov, V. S., Procaccia, I., and Sreenivasan, K. R. (2000). Scaling structure of the velocity statistics in atmospheric boundary layers. *Physical Review E* **61**, 407–21.

Lacorta, G., Aurell, E., Legras, B., and Vulpiani, A. (2004). Evidence for a $k^{-5/3}$ spectrum from the EOLE Lagrangian balloons in the lower stratosphere. *Journal of the Atmospheric Sciences* **61**, 2936–42.

Lamperti, J. (1962). Semi-stable stochastic processes. *Transactions of the American Mathematical Society* **104**, 62–78.

Landahl, M. T., and Mollo-Christensen, E. (1986). *Turbulence and random processes in fluid mechanics.* Cambridge: Cambridge University Press.

Lazarev, A., Schertzer, D., Lovejoy, S., and Chigirinskaya, Y. (1994). Unified multifractal atmospheric dynamics tested in the tropics: part II, vertical scaling and Generalized Scale Invariance. *Nonlinear Processes in Geophysics* **1**, 115–23.

Lesieur, M. (1987). *Turbulence in Fluids.* Dordrecht: Martinus Nijhoff.

Lilley, M., Lovejoy, S., Schertzer, D., Strawbridge, K. B., and Radkevitch, A. (2008). Scaling turbulent atmospheric stratification. II: Spatial stratification and intermittency from lidar data. *Quarterly Journal of the Royal Meteorological Society* **134**, 301–15. doi:10.1002/qj.1202.

Lilly, D. K. (1986a). The structure, energetics and propagation of rotating convective storms. Part I: energy exchange with the mean flow. *Journal of the Atmospheric Sciences* **43**, 113–25.

Lilly, D. K. (1986b). The structure, energetics and propagation of rotating convective storms. Part II: helicity and storm stabilization. *Journal of the Atmospheric Sciences* **43**, 126–40.

Lovejoy, S., and Schertzer, D. (2007). Scaling and multifractal fields in the solid earth and topography. *Nonlinear Processes in Geophysics* **14**, 1–38.

Lovejoy, S., and Schertzer, D. (2010). Towards a new synthesis for atmospheric dynamics: space-time cascades. *Atmospheric Research* **96**, 1–52.

Lovejoy, S., and Schertzer, D. (2011). Space-time cascades and the scaling of ECMWF reanalyses: fluxes and fields. *Journal of Geophysical Research* **116**, D14117. doi:10.1029/2011JD015654.

Lovejoy, S., Schertzer, D., and Tuck, A. F. (2004). Fractal aircraft trajectories and nonclassical turbulent exponents. *Physical Review E* **70**, 036306.

Lovejoy, S., Tuck, A. F., Hovde, S. J., and Schertzer, D. (2007). Is isotropic turbulence relevant in the atmosphere? *Geophysical Research Letters* **34**, L15802. doi:10.1029/2007GL029359.

Lovejoy, S., Tuck, A. F., Hovde, S. J., and Schertzer, D. (2008). Do stable atmospheric layers exist? *Geophysical Research Letters* **35**, L01802.

Lovejoy, S., Tuck, A. F., Hovde, S. J., and Schertzer, D. (2009a). The vertical cascade structure of the atmosphere and multifractal drop sonde outages. *Journal of Geophysical Research* **114**, D07111. doi:07110.01029/02008JD010651.

Lovejoy, S., Tuck, A. F., Schertzer, D., and Hovde, S. J. (2009b). Reinterpreting aircraft measurements in anisotropic scaling turbulence. *Atmospheric Chemistry and Physics* **9**, 5007–25.

Lovejoy, S., Watson, B. P., Grosdidier, Y., and Schertzer, D. (2009c). Scattering in thick multifractal clouds. Part II: multiple scattering. *Physica A* **388**, 3711–27.

Lovejoy, S., Schertzer, D., and Tuck, A. F. (2010a). Why anisotropic turbulence matters: another reply to E. Lindborg. *Atmospheric Chemistry and Physics Discussions* **10**, 7495–506.

Lovejoy, S., Tuck, A. F., and Schertzer, D. (2010b). The horizontal cascade structure of atmospheric fields determined from aircraft data. *Journal of Geophysical Research* **115**, D13105.

Lumley, J. L. (1964). The spectrum of nearly inertial turbulence in a stably stratified fluid. *Journal of the Atmospheric Sciences* **21**, 99–102.

Moeng, C., and Wyngaard, J. C. (1986). An analysis of closures for pressure-scalar i covariances in the convenctive boundary layer. *Journal of the Atmospheric Sciences* **43**, 2499–513.

Monin, A. S., and Yaglom, A. M. (1975). *Statistical Fluid Mechanics.* Boston, MA: MIT Press.

Morel, P., and Larchevêque, M. (1974). Relative dispersion of constant level balloons in the 200 mb general circulation. *Journal of the Atmospheric Sciences* **31**, 2189–96.

Muschinski, A., and Wode, C. (1998). First in situ evidence for coexisting submeter temperature and humidity sheets in the lower free troposphere. *Journal of the Atmospheric Sciences* **55**, 2893–906.

Nappo, C. J. (2002). *An Introduction to Gravity Waves.* Amsterdam: Academic Press.

Nastrom, G. D., and Gage, K. S. (1983). A first look at wave number spectra from GASP data. *Tellus* **35**, 383–8.

Nastrom, G. D., and Gage, K. S. (1985). A climatology of atmospheric wavenumber spectra of wind and temperature by commercial aircraft. *Journal of the Atmospheric Sciences* **42**, 950–60.

Nogueira, M., Barros, A. P., and Miranda, P. M. (2012). Stochastic downscaling of numerically simulated spatial rain and cloud fields using a transient multifractal approach. *Geophysical Research Abstracts* **14**, EGU2012-5772-2011.

Obukhov, A. (1959). Effect of Archimedean forces on the structure of the temperature field in a turbulent flow. *Dokl. Akad. Nauk SSSR* **125**, 1246.

Osborne, T. (1998). Finestructure, microstructure and thin layers. *Oceanography* **11**, 36–43.

Paulson, K. S., and Baxter, P. D. (2007). Downscaling of rain gauge time series by multiplicative beta cascade. *Journal of Geophysical Research* **112**, D09105.

Radkevitch, A., Lovejoy, S., Strawbridge, K. B., and Schertzer, D. (2007). The elliptical dimension of space-time atmospheric stratification of passive admixtures using lidar data. *Physica A* **382**, 597–615.

Radkevitch, A., Lovejoy, S., Strawbridge, K. B., Schertzer, D., and Lilley, M. (2008). Scaling turbulent atmospheric stratification. III: Space–time stratification of passive scalars from lidar data. *Quarterly Journal of the Royal Meteorological Society* **134**, 317–35. doi:10.1002/qj.203.

Richardson, L. F. (1920). The supply of energy from and to atmospheric eddies. *Proceedings of the Royal Society A* **97**, 354–73.

Riehl, H., and Malkus, J. S. (1958). On the heat balance in the equatorial trough zone. *Geophysica (Helsinki)* **6**, 503–38.

Rotta, J. C. (1951). Statistische Theorie nichthomogener Turbulenz. *Zeitschrift für Physik* **129**, 547–72.

Schertzer, D., and Lovejoy, S. (1983). Elliptical turbulence in the atmosphere. *Fourth Symposium on Turbulent Shear Flows*. Karlshule, West Germany.

Schertzer, D. & Lovejoy, S. (1984). On the dimension of atmospheric motions, *Turbulence and Chaotic Phenomena in Fluids*, ed. Tatsumi, T. Elsevier Science Publishers B. V., Amsterdam, pp. 505–512.

Schertzer, D. & Lovejoy, S. (1985) Generalised scale invariance in turbulent phenomena, *Physico-Chem. Hydrodyn.* J. **6**, 623–635.

Schertzer, D., M. Larcheveque, J. Duan and S. Lovejoy (1999). Genaralized Stable Multivariate Distribution and Anisotropic Dilations. Minneapolis, *IMA Preprint #1666*. http://www.ima.umn.edu/preprints/dec99/1666.pdf, U. of Minnesota.

Schertzer, D., and Lovejoy, S. (1985). The dimension and intermittency of atmospheric dynamics. In B. Launder, ed., *Turbulent Shear Flow 4*. New York, NY: Springer, pp. 7–33.

Schertzer, D., and Lovejoy, S. (1987). Physical modeling and analysis of rain and clouds by anisotropic scaling of multiplicative processes. *Journal of Geophysical Research* **92**, 9693–714.

Schertzer, D., and Lovejoy, S. (2011). Multifractals, generalized scale invariance and complexity in geophysics. *International Journal of Bifurcation and Chaos* **21**, 3417–56.

Shang, X. D., and Xia, K. Q. (2001). Scaling of the velocity power spectra in turbulent thermal convection. *Physical Review E* **64**, 065301.

Shur, G. (1962). Eksperimental'nyye issledovaniya energeticheskogo spektra atmosfernoy turbulentnosti. *Tsentral'naya Aerologicheskaya Observatoriya Trudy* **43**, 79.

Sreenivasan, K. R. (1991). On local isotropy of passive scalars in turbulent shear flows. In J. C. R. Hunt, O. M. Phillips, and D. Willliams, eds., *Turbulence and Stochastic Processes: Kolmogorov's Ideas 50 Years On*. London: Royal Society, pp. 165–82.

Takayashi, Y. O., Hamilton, K., and Ohfuchi, W. (2006). Explicit global simulation of the meso-scale spectrum of atmospheric motions. *Geophysical Research Letters* **33**, L12812. doi: 10.1029/2006GL026429.

Taylor, G. I. (1935). Statistical theory of turbulence, parts I–IV. *Proceedings of the Royal Society A* **151**, 421–78.

Tsuda, T., Inoue, T., Kato, S., *et al.* (1989). MST radar observations of a saturated gravity wave spectrum. *Journal of the Atmospheric Sciences* **46**, 2440–7.

Väisälä, V. (1925). Über die Wirkung der Windschwankungen auf die Pilotbeobachtungen. *Soc. Sci. Fenn. Comment. Phys.-Math.* **2**, 19–37.

Van Atta, C. (1991). Local isotropy of the smallest scales of turbulent scalar and velocity fields. In J. C. R. Hunt, O. M. Phillips, and D. Willliams, eds., *Turbulence and Stochastic Processes: Kolmogorov's Ideas 50 Years On*. London: Royal Society, pp. 139–147.

Van Zandt, T. E. (1982). A universal spectrum of buoyancy waves in the atmosphere. *Geophysical Research Letters* **9**, 575–8.

Yano, J. (2009). Interactive comment on "Reinterpreting aircraft measurements in anisotropic scaling turbulence" by S. Lovejoy *et al. Atmospheric Chemistry and Physics Discussions* **9**, S162–6.

Chapter 7

Barker, H. W., and Davies, J. A. (1992). Cumulus cloud radiative properties and the characteristics of satellite radiance wavenumber spectra. *Remote Sensing of Environment* **42**, 51–64.

Beaulieu, A., Gaonac'h, G., and Lovejoy, S. (2007). Anisotropic scaling of remotely sensed drainage basins: the differential anisotropy scaling method. *Nonlinear Processes in Geophysics* **14**, 337–50.

Bouchaud, J. P., and Georges, A. (1990). Anomalous diffusion in disordered media: statistical mechanisms, models and physics applications. *Physics Reports* **195**, 127–293.

Busygin, V. P., Yevstatov, N. A., and Feigelson, E. M. (1973). Optical propeorties of cumulus clouds and radiant fluxes for cumulus cloud cover. *Izv. Acad. Sci. USSR Atmos. Oceanic Phys* **9**, 1142–51.

Cahalan, R. (1994). Bounded cascade clouds: albedo and effective thickness. *Nonlinear Processes in Geophysics* **1**, 156–67.

Cahalan, R. F. (1989). Overview of fractal clouds. In A. Deepak, H. Flemming, and J. Theon, eds., *Advances in Remote Sensing Retrieval Methods*. Hampton, VA: A. Deepak.

Cahalan, R. F., Ridgeway, W., Wiscoombe, W. J., Bell, T. L., and Snider, J. B. (1994). The albedo of fractal stratocumulus clouds. *Journal of the Atmospheric Sciences* **51**, 2434–55.

Cahalan, R. F., Oreopoulos, L., Marshak, A., *et al.* (2005). The I3RC: bringing together the most advanced radiative transfer tools for cloudy atmospheres. *Bulletin of the American Meteorological Society* **86**, 1275–93.

Chandrasekhar, S. (1950). *Radiative Transfer*. Oxford: Clarendon Press.

Davis, A., Lovejoy, S., Gabriel, P., Schertzer, D., and Austin, G. L. (1990). Discrete angle radiative transfer. Part III: numerical results on homogeneous and fractal clouds. *Journal of Geophysical Research* **95**, 11729–42.

Ferlay, N., and Isaka, H. (2006). Multiresolution analysis of radiative transfer through inhomogeneous media. Part I: Theoretical development. *Journal of the Atmospheric Sciences* **63**, 1200–12.

Gabriel, P., Lovejoy, S., Schertzer, D., and Austin, G. L. (1986). Radiative transfer in extremely variable fractal clouds. *Paper presented at 6th Conference on Atmospheric Radiation*, American Meteorological Society, Williamsburg, VA.

Gabriel, P., Lovejoy, S., Davis, A., Schertzer, D., and Austin, G. L. (1990). Discrete angle radiative transfer. Part II: renormalization approach to scaling clouds. *Journal of Geophysical Research* **95**, 11717–28.

Gabriel, P. M., Tsay, S. -C., and Stephens, G. L. (1993). A Fourier-Riccati approach to radiative transfer. Part I: Foundations. *Journal of the Atmospheric Sciences* **50**, 3125–47.

Gagnon, J.-S., Lovejoy, S., and Schertzer, D. (2006). Multifractal earth topography. *Nonlinear Processes in Geophysics* **13**, 541–70.

Havlin, S., and Ben-Avraham, D. (1987). Diffusion in disordered media. *Advances in Physics* **36**, 695–798.

Lewis, G. (1993). *The Scale Invariant Generator Technique and Scaling Anisotropy in Geophysics*. Montreal: McGill University.

Lewis, G., Lovejoy, S., Schertzer, D., and Pecknold, S. (1999). The scale invariant generator technique for parameter estimates in generalized scale invariance. *Computers and Geosciences* **25**, 963–78.

Lovejoy, S., and Schertzer, D. (1985). Generalized scale invariance and fractal models of rain. *Water Resources Research* **21**, 1233–50.

Lovejoy, S., and Schertzer, D. (1989). Fractal clouds with discrete angle radiative transfer. In J. Lenoble and J. F. Geleyn, eds., *IRS '88: Current Problems in Atmospheric Radiation: Proceedings of the International Radiation Symposium, Lille, France, 18–24 August 1988*. Hampton, VA: Deepak, pp. 99–102.

Lovejoy, S., and Schertzer, D. (2007). Scale, scaling and multifractals in geophysics: twenty years on. In A. A. Tsonis, J. B. Elsner, eds., *Nonlinear Dynamics in Geosciences*. New York, NY: Springer, pp. 311–37.

Lovejoy, S., Gabriel, P., Davis, A., Schertzer, D., and Austin, G. L. (1990). Discrete angle radiative transfer.. Part I: scaling and similarity, universality and diffusion. *Journal of Geophysical Research* **95**, 11699–715.

Lovejoy, S., Schertzer, D., and Watson, B. (1993). Radiative transfer and multifractal clouds: theory and applications. In S. Keevallik and O. Karner, eds., *IRS '92: Current Problems in Atmospheric Radiation: Proceedings of the International Radiation Symposium, Tallinn, Estonia, 3–8 August 1992*. Hampton, VA: Deepak, pp. 108–11.

Lovejoy, S., Watson, B., Schertzer, D., and Brosamlen, G. (1995). Scattering in multifractal media. In L. Briggs, ed., *Particle Transport in Stochastic Media*. Portland, OR: American Nuclear Society, pp. 750–60.

Lovejoy, S., Schertzer, D., and Silas, P. (1998). Diffusion in one-dimensional multifractal porous media. *Water Resources Research* **34**, 3283–91.

Lovejoy, S., Watson, B. P., Grosdidier, Y., and Schertzer, D. (2009). Scattering in thick multifractal clouds. Part II: multiple scattering. *Physica A* **388**, 3711–27.

Marguerit, C., Schertzer, D., Schmitt, F., and Lovejoy, S. (1998). Copepod diffusion within multifractal phytoplankton fields. *Journal of Marine Systems* **16**, 69–83.

Marshak, A., and Davis. A. B., eds. (2005). *3D Radiative Transfer in Cloudy Atmospheres*. Berlin: Springer.

McKee, T., and Cox, S. K. (1976). Simulated radiance patterns for finite cubic clouds. *Journal of the Atmospheric Sciences* **33**, 2014–20.

Meakin, P. (1987). Random walks on multifractal lattices. *Journal of Physics A* **20**, L771–7.

Mechem, D. B., Kogan, Y. L., Ovtchinnikov, M., *et al.* (2002). Large-eddy simulation of PBL stratocumulus: comparison of multi-dimensional and IPA longwave radiative forcing. *Paper presented at 12th ARM Science Team Meeting*, St. Petersburg, FL.

Naud, C., Schertzer, D., and Lovejoy, S. (1997). Fractional integration and radiative transfer in multifractal atmospheres. In W. Woyczynski and S. Molchansov, eds., *Stochastic Models in Geosystems*. New York, NY: Springer-Verlag, pp. 239–67.

Pecknold, S., Lovejoy, S., and Schertzer, D. (1996). Universal multifractal landscape topography: data analysis and simulations. *Annales Geophysicae* **14**, C633.

Pecknold, S., Lovejoy, S., and Schertzer, D. (1997). The morphology and texture of anisotropic multifractals using generalized scale invariance. In W. Woyczynski and S. Molchansov, eds., *Stochastic Models in Geosystems*. New York, NY: Springer-Verlag, pp. 269–312.

Pflug, K., Lovejoy, S., and Schertzer, D. (1993). Generalized scale invariance, differential rotation and cloud texture. *Journal of Atmospheric Sciences* **50**, 538–53.

Preisendorfer, R. W., and Stephens, G. I. (1984). Multimode radiative transfer in finite optical media, I fundamentals. *Journal of the Atmospheric Sciences* **41**, 709–24.

Schertzer, D., and Lovejoy, S. (1985). Generalised scale invariance in turbulent phenomena. *Physico-Chemical Hydrodynamics Journal* **6**, 623–35.

Schertzer, D., and Lovejoy, S. (1991). Nonlinear geodynamical variability: multiple singularities, universality and observables. In D. Schertzer and S. Lovejoy, eds., *Non-Linear Variability in Geophysics: Scaling and Fractals.* Dordrecht: Kluwer, pp. 41–82.

Schertzer, D., and Lovejoy, S. (1996). Lecture notes: resolution dependence and multifractals in remote sensing and geographical information systems. *McGill University, Montreal, June* **10**, 1996.

Schertzer, D., and Lovejoy, S. (2011). Multifractals, generalized scale invariance and complexity in geophysics. *International Journal of Bifurcation and Chaos* **21**, 3417–56.

Schertzer, D., Schmitt, F., Naud, C., ***et al.*** (1997). New developments and old questions in multifractal cloud modeling, satellite retrievals and anomalous absorption. Paper presented at 7th ARM Science Team Meeting, San Antonio, TX.

Schertzer, D., Larchevêque, M., Duan, J., and Lovejoy, S. (1999). Generalized stable multivariate distributions and anisotropic dilations. IMA Preprint #1666. http://www.ima.umn.edu/preprints/dec99/1666.pdf.

Schertzer, D., Lovejoy, S., and Hubert, P. (2002). An introduction to stochastic multifractal fields. In A. Ern and Liu Weiping, eds., *Mathematical Problems in Environmental Science and Engineering.* Series in Contemporary Applied Mathematics, vol. 4. Beijing: Higher Education Press, pp. 106–79.

Tessier, Y., Lovejoy, S., and Schertzer, D. (1993). Universal multifractals: theory and observations for rain and clouds. *Journal of Applied Meteorology* **32**, 223–50.

Watson, B. P., Lovejoy, S., Grosdidier, Y., and Schertzer, D. (2009). Scattering in thick multifractal clouds. Part I: overview and single scattering. *Physica A* **388**, 3695–710.

Weinman, J. A., and P. N. Swartzrauber (1968). Albedo of a striated medium of isotrpically scattering particles. *Journal of the Atmospheric Sciences* **25**, 497–501.

Weissman, H., and Havlin, S. (1988). Dynamics in multiplicative processes. *Physical Review B* **37**, 5994–6.

Welch, R. M., Cox, S. K., and Davis, J. M. (1980). *Solar Radiation and Clouds.* Boston, MA: American Meteorological Society.

Chapter 8

AchutaRao, K., and Sperber, K. R. (2006). ENSO simulation in coupled ocean–atmosphere models: are the current models better? *Climate Dynamics* **27**, 1–15.

Ashkenazy, Y., Baker, D. R., Gildor, H., and Havlin, S. (2003). Nonlinearity and multifractality of climate change in the past 420,000 years. *Geophysical Research Letters* **30**, 2146.

Blumen, W. (1978). Uniform potential vorticity flow: Part I. Theory of wave interactions and two-dimensional turbulence. *Journal of the Atmospheric Sciences* **35**, 774–83.

Brunt, D. (1939). *Physical and Dynamical Meteorology.* New York, NY: Cambridge University Press.

Burgert, R. and Hsieh, W. W. (1989). Spectral analysis of the AVHRR sea surface temperature variability off the west coast of Vancouver Island. *Atmosphere-Ocean* **27**, 577–87.

Cheng, X., and Qi, Y. (2010). Variations of eddy kinetic energy in the South China Sea. *Journal of Oceanography* **66**, 85–94.

Chiason, E. J. (2010). Energy rate density as a complexity metric and evolutionary driver. *Complexity* **16**, 27–40.

Clayson, C. A., and Kantha, L. H. (1999). Turbulent kinetic energy and its dissipation rate in the equatorial mixed layer. *Journal of Physical Oceanography* **29**, 2146–66.

Crawford, W. R. (1976). Turbulent energy dissipation in the Atlantic equatorial undercurrent. *PhD thesis*, University of British Columbia, Vancouver, BC).

Deschamps, P. Y., Frouin, R., and Wald, L. (1981). Satellite determination of the mesoscale variability of the sea surface temperature. *Journal of Physical Oceanography* **11**, 864–70.

Deschamps, P. Y., Frouin, R., and Crepon, M. (1984). Sea surface temperature of the coastal zones of France observed by the HCMM-Satellite. *Journal of Geophysical Research* **89**, 8123–49.

Finn, D., Lamb, B., Leclerc, M. Y., *et al.* (2001). Multifractal analysis of plume concentration fluctuations in surface layer flows. *Journal of Applied Meteorology* **40**, 229–45.

Goldman, J. L. (1968). *The Power Spectrum in the Atmosphere Below Macroscale.* Houston, TX: Institue of Desert Research, University of St. Thomas.

Grant, H. L., Steward, R. W., and Moillet, A. (1962). Turbulence spectra from a tidal channel. *Journal of Fluid Mechanics* **2**, 263–72.

Held, I. M., Pierrehumbert, R. T., Garner, S. T., and Swanson, K. L. (1995). Surface quasi-geostrophic dynamics. *Journal of Fluid Mechanics* **282**, 1–20.

Hendon, H. H., and Wheeler, M. (2008). Some space-time spectral analyses of tropical convection and planetary waves. *Journal of the Atmospheric Sciences* **65**, 2936–48.

Hubert, L. F., and Whitney, L. F. (1971). Wind estimation from geostationary-satellite pictures. *Monthly Weather Review* **99**, 665–72.

Hubert, P., Tessier, Y., Lovejoy, S., *et al.* (1993). Multifractals and extreme rainfall events. *Geophysical Research Letters* **20**, 931–4.

Huybers, P., and Curry, W. (2006). Links between annual, Milankovitch, and continuum temperature variability. *Nature* **441**, 329–32.

Hwang, H. J. (1970). Power density spectrum of surface wind speed on Palmyra island. *Monthly Weather Review* **98**, 70–4.

Inoue, E. (1951). On the turbulent diffusion in the atmosphere. *Journal of the Meteorological Society of Japan* **29**, 32.

Kolesnikov, V. N., and Monin, A. S. (1965). Spectra of meteorological field fluctuations. *Izvestiya, Atmospheric and Oceanic Physics* **1**, 653–69.

Koscielny-Bunde, E., Bunde, A., Havlin, S., *et al.* (1998). Indication of a universal persistence law governing atmospheric variability. *Physical Review Letters* **81**, 729–32.

Landau, L. D., and Lifschitz, E. M. (1959). *Fluid Mechanics*. Oxford: Pergamon.

Lanfredi, M., Simoniello, T., Cuomo, V., and Macchiato, M. (2009). Discriminating low frequency components from long range persistent fluctuations in daily atmospheric temperature variability. *Atmospheric Chemistry and Physics* **9**, 4537–44.

Le Traon, P. Y., Klein, P., Hua, B. L., and Dibarboure, G. (2008). Do altimeter wavenumber spectra agree with the interior or surface quasigeostrophic theory? *Journal of Physical Oceanography* **38**, 1137–42.

Lien, R.-C., and D'Asaro, E. A. (2006). Measurement of turbulent kinetic energy dissipation rate with a Lagrangian float. *Journal of Atmospheric and Oceanic Technology* **23**, 964–76.

Lovejoy, S., and Schertzer, D. (1986). Scale invariance in climatological temperatures and the spectral plateau. *Annales Geophysicae* **4B**, 401–10.

Lovejoy, S., and Schertzer, D. (2010). Towards a new synthesis for atmospheric dynamics: space-time cascades. *Atmospheric Research* **96**, 1–52.

Lovejoy, S., and Schertzer, D. (2011). Space-time cascades and the scaling of ECMWF reanalyses: fluxes and fields. *Journal of Geophysical Research* **116**, D14117. doi:10.1029/2011JD015654.

Lovejoy, S., and Schertzer, D. (2012a). The climate is not what you expect. *Bulletin of the American Meteorological Society*, in press.

Lovejoy, S., and Schertzer, D. (2012b). Low frequency weather and the emergence of the climate. In A. S. Sharma, A. Bunde, D. Baker, and V. P. Dimri, eds., *Complexity and Extreme Events in Geosciences*. AGU monographs.

Lovejoy, S., Currie, W. J. S., Tessier, Y. *et al.* (2000). Universal multfractals and ocean patchiness: phytoplankton, physical fields and coastal heterogeneity. *Journal of Plankton Research* **23**, 117–41.

Lovejoy, S., Schertzer, D., Lilley, M., Strawbridge, K. B., and Radkevitch, A. (2008). Scaling turbulent atmospheric stratification. I: Turbulence and waves. *Quarterly Journal of the Royal Meteorological Society* **134**: 277–300. doi:10.1002/qj.201.

Lovejoy, S., Pinel, J., and Schertzer, D. (2012). The global space-time cascade structure of precipitation: satellites, gridded gauges and reanalyses. *Advances in Water Resources*. http://dx.doi.org/10.1016/j.advwatres.2012.03.024.

Marsan, D., Schertzer, D., and Lovejoy, S. (1996). Causal space-time multifractal processes: predictability and forecasting of rain fields. *Journal of Geophysical Research* **101**, 26333–46.

Matsuno, T., Lee, J.-S., Shimizu, M., Kim, S.-H., and Pang, I.-C. (2006). Measurements of the turbulent energy dissipation rate and an evaluation of the dispersion process of the Changjiang Diluted Water in the East China Sea. *Journal of Geophysical Research* **111**, C11S09.

McLeish, W. (1970). Spatial spectra of ocean surface temperature. *Journal of Geophysical Research* **75**, 6872–7.

Monetti, R. A., Havlin, S., and Bunde, A. (2003). Long-term persistance in the sea surface temperature fluctuations. *Physica A* **320**, 581–9.

Monin, A. S. (1972). *Weather Forecasting as a Problem in Physics*. Boston, MA: MIT Press.

Monin, A. S., and Yaglom, A. M. (1975). *Statistical Fluid Mechanics*. Boston, MA: MIT Press.

Moum, J. N., Gregg, M. C., Lien, R. C., and Carr, M. E. (1995). Comparison of turbulent Kinetic energy dissipation rate estimates from two ocean microstructure profiler. *Journal of Atmospheric and Oceanic Technology* **12**, 346–66.

Nakajima, H., and Hayakawa, N. (1982). A cross-correlation analysis of tidal current, water temperature and salinity records. *Journal of the Oceanographical Society of Japan* **38**, 52–6.

Niiler, P. (2001). The world ocean surface circulation. In G. Siedler, J. Church, and J. Gould, eds., *Ocean Circulation and Climate: Observing and Modelling Global Oceans*. San Diego, CA: Academic Press, pp. 193–204.

Pagliuca, S. (1934). The great wind of April 11–12, 1934, on Mount Washington, N.H., and its measurement. Part I: winds of superhurricane force, and a heated anemometer for their measurement during ice forming conditions. *Monthly Weather Review* **62**, 186–9.

Palmén, E. (1959). On the maintenance of kinetic energy in the atmosphere. In B. Bolen, ed., *The Atmosphere and the Sea in Motion*. Oxford: Oxford University Press, pp. 212–24.

Panofsky, H. A. (1969). The spectrum of temperature. *Radio Science* **4**, 1101–9.

Panofsky, H. A., and Van der Hoven, I. (1955). Spectra and cross-spectra of velocity components in the mesometeorlogical range. *Quarterly Journal of the Royal Meteorological Society* **81**, 603–6.

Park, K.-A., and Chung, J. Y. (1999). Spatial and temporal scale variations of sea surface temperature in the East Sea using NOAA/AVHRR data. *Journal of Oceanography* **55**, 271–88.

Pelletier, J. D. (1998). The power spectral density of atmospheric temperature from scales of 10^{-2} to 10^{6} yr. *Earth and Planetary Science Letters* **158**, 157–64.

Pelletier, J. D., and Turcotte, D. L. (1997). Long-range persistence in climatological and hydrological time series: analysis, modeling and application to drought hazard assessment. *Journal of Hydrology* **203**, 198–208.

Peterson, T. C., and Vose, R. S. (1997). An overview of the Global Historical Climatology Network temperature database. *Bulletin of the American Meteorological Society* **78**, 2837–49.

Pinel, J., S. Lovejoy, D. Schertzer, 2014: The horizontal space-time scaling and cascade structure of the atmosphere and satellite radiances, **Atmos. Resear.**, **140–141**, 95–114, doi.org/10.1016/j.atmosres.2013.11.022.

Pinus, N. Z. (1968). The energy of atmospheric macro-turbulence. *Izvestiya, Atmospheric and Oceanic Physics* **4**, 461.

Radkevitch, A., Lovejoy, S., Strawbridge, K. B., Schertzer, D., and Lilley, M. (2008). Scaling turbulent atmospheric stratification. III: Space–time stratification of passive scalars from lidar data. *Quarterly Journal of the Royal Meteorological Society* **134**, 317–35. doi:10.1002/qj.203.

Robinson, G. D. (1971). The predictability of a dissipative flow. *Quarterly Journal of the Royal Meteorological Society* **97**, 300–12.

Saunders, P. M. (1972). Space and time variability of temperature in the upper ocean. *Deep-Sea Research* **19**, 467–80.

Schertzer, D., Lovejoy, S., Schmitt, F., Chigirinskaya, Y., and Marsan, D. (1997a). Multifractal cascade dynamics and turbulent intermittency. *Fractals* **5**, 427–71.

Schertzer, D., Schmitt, F., Naud, C., *et al.* (1997b). New developments and old questions in multifractal cloud modeling, *satellite retrievals and anomalous absorption. Paper presented at 7th ARM Science Team Meeting*, San Antonio, TX.

Schmitt, F., Lovejoy, S., Schertzer, D., Lavallée, D., and Hooge, C. (1992). *Les premières estimations des indices de multifractalité dans le champ de vent et de temperature. Comptes Rendues de l'Académie des Sciences 314*, **II**, 749–54.

Schmitt, F., Schertzer, D., Lovejoy, S., and Brunet, Y. (1993). Estimation of universal multifractal indices for atmospheric turbulent velocity fields. *Fractals* **1**, 568–75.

Schmitt, F., Schertzer, D., Lovejoy, S., and Brunet, Y. (1996). Universal multifractal structure of atmospheric temperature and velocity fields. *Europhysics Letters* **34**, 195–200.

Seuront, L., Schmitt, F., Schertzer, D., Lagadeuc, Y., and Lovejoy, S. (1996). Multifractal analysis of eulerian and lagrangian variability of physical and biological fields in the ocean. *Nonlinear Processes in Geophysics* **3**, 236–46.

Smith, K. S., and Vallis, G. K. (2001). The scales and equilibration of mid-ocean eddies: Freely evolving flow, *Journal of Physical Oceanography* **31**, 554–71.

Steele, J. H. (1995). Can ecological concepts span the land and ocean domains? In T. M. Powell and J. H. Steele, eds., *Ecological Time Series*. New York, NY: Chapman and Hall, pp. 5–19.

Stolle, J., Lovejoy, S., and Schertzer, D. (2009). The stochastic cascade structure of deterministic numerical models of the atmosphere. *Nonlinear Processes in Geophysics* **16**, 1–15.

Stolle, J., Lovejoy, S., and Schertzer, D. (2012). The temporal cascade structure and space-time relations for reanalyses and global circulation models. *Quarterly Journal of the Royal Meteorological Society*, **138**, 1895–1913.

Szantai, A., and Sèze, G. (2008). *Improved extraction of low-level atmospheric motion vectors over west-Africa from MSG images. 9th International Winds Workshop, 14–18 April 2008*, Annapolis, MD, USA, pp. 1–8.

Talkner, P., and Weber, R. O. (2000). Power spectrum and detrended fluctuation analysis: application to daily temperatures. *Physical Review E* **62**, 150–60.

Taylor, G. I. (1938). The spectrum of turbulence. *Proceedings of the Royal Society A* **164**, 476–90.

Tchiguirinskaia, I., Schertzer, D., Lovejoy, S., and Veysseire, J. M. (2006). Wind extremes and scales: multifractal insights and empirical evidence. In J. Peinke, P. Schaumann, and S. Barth, eds., *Wind Energy: Proceedings of the Euromech Colloquium*. Berlin: Springer-Verlag.

Tennekes, H. (1975). Eulerian and Lagrangian time microscales in isotropic turbulence. *Journal of Fluid Mechanics* **67**, pp. 561–7.

Vallis, G. (2010). Mechanisms of climate variaiblity from years to decades. In T. N. Palmer and P. Williams, eds., *Stochastic Physics and Climate Modelling*. Cambridge: Cambridge University Press, pp. 1–34.

Van der Hoven, I. (1957). Power spectrum of horizontal wind speed in the frequency range from.0007 to 900 cycles per hour. *Journal of Meteorology* **14**, 160–4.

Vinnichenko, N. K. (1969). The kinetic energy spectrum in the free atmosphere for 1 second to 5 years. *Tellus* **22**, 158–66.

Vinnichenko, N. K. and Dutton, J. A. (1969). Empirical studies of atmospheric structure and spectra in the free atmosphere. *Radio Science* **4**, 1115–26.

Wang, Y. (1995). Measurements and multifractal analysis of turbulent temperature and velocity near the ground. *Atmospheric and Oceanic Sciences*, McGill University, Montreal.

Wheeler, M., and Kiladis, G. N. (1999). Convectively coupled equatorial waves: analysis of clouds and temperature in the wavenumber-frequency domain. *Journal of the Atmospheric Sciences* **56**, 374–99.

Chapter 9

Bleistein, N., and Handelsman, R. A. (1986). *Asymptotic Expansions of Integrals*. Mineola, NY: Dover.

Buizza, R., Miller, M., and Palmer, T. N. (1999). Stochastic representation of model uncertainties in the ECMWF Ensemble Prediction System. *Quarterly Journal of the Royal Meteorological Society* **125**, 2887–908.

Charney, J. G. (1966). The feasibility of a global observation and analysis experiment. *Bulletin of the American Meteorological Society* **47**, 200–20.

Dawson, A., Palmer, T. N., and Corti, S. (2012). Simulating regime structures in weather and climate prediction models, *Geophys. Resear. Lett.*, **39**, L21805.

Dewan, E. (1997). Saturated-cascade similtude theory of gravity wave sepctra. *Journal of Geophysical Research* **102**, 29799–817.

Goldstein, S. (1931). On the stability of superposed streams of fluids of different densities. *Proceedings of the Royal Society A* **132**, 524–48.

Houtekamer, P., Lefaivre, L., Derome, J., and Ritchie, H. (1996). A system simulation approach to ensemble prediction. *Monthly Weather Review* **124**, 1225–42.

Kraichnan, R. H. (1970). Instability in fully developed turbulence. *Physics of Fluids* **13**, 569–75.

Leith, C. E. (1971). Atmospheric predictability and two-dimensional turbulence. *Journal of the Atmospheric Sciences* **28**, 145–61.

Leith, C. E., and Kraichnan, R. H. (1972). Predictability of turbulent flows. *Journal of the Atmospheric Sciences* **29**, 1041–58.

Lorenz, E. N. (1963). Deterministic nonperiodic flow. *Journal of the Atmospheric Sciences* **20**, 130–41.

Lorenz, E. N. (1969). The predictability of a flow which possesses many scales of motion. *Tellus* **21**, 289–307.

Lovejoy, S., and Schertzer, D. (2010). On the simulation of continuous in scale universal multifractals, part II: space-time processes and finite size corrections. *Computers and Geosciences* **36**, 1404–13.

Lovejoy, S., Schertzer, D., Lilley, M., Strawbridge, K. B., and Radkevitch, A. (2008). Scaling turbulent atmospheric stratification. I: Turbulence and waves. *Quarterly Journal of the Royal Meteorological Society* **134**: 277–300. doi:10.1002/qj.201.

Lyapunov, M. A. (1907). Problème général de la stabilité du mouvement. *Annales de la Faculté des Sciences de Toulouse* **9**, 203–474.

Marsan, D., Schertzer, D., and Lovejoy, S. (1996). Causal space-time multifractal processes: predictability and forecasting of rain fields. *Journal of Geophysical Research* **101**, 26333–46.

Métais, O., and Lesieur, M. (1986). Statistical predictability of decaying turbulence. *Journal of the Atmospheric Sciences* **43**, 857–70.

Nappo, C. J. (2002). *An Introduction to Gravity Waves.* Amsterdam: Academic Press.

Oseledets, V. I. (1968). A multiplicative ergodic theorem: Lyapunov characteristic numbers for dynamical systems. *Transactions of the Moscow Mathematical Society* **19**, 197–231.

Palmer, T. N. (2001). A nonlinear dynamical perspective on model error: a proposal for non-local stochastic-dynamic parameterisaton in weather and climate prediction models. *Quarterly Journal of the Royal Meteorological Society* **127**, 279–304.

Pinel, J. 2013, The space-time structure of the atmosphere, phD thesis, McGill University.

Pinel, J., S. Lovejoy, 2014: Atmospheric waves as scaling, turbulent phenomena **Atmos. Chem. Phys. (in press)**.

Robinson, G. D. (1971). The predictability of a dissipative flow. *Quarterly Journal of the Royal Meteorological Society* **97**, 300–12.

Ruelle, D. (1982). Characteristic exponents and invariant manifolds in Hilbert space. *Annals of Mathematics* **115**, 243–90.

Schertzer, D. and S. Lovejoy (1995). From scalar cascades to Lie cascades: joint multifractal analysis of rain and cloud processes. *Space/time Variability and Interdependance for various hydrological processes.* R. A. Feddes. New-York, Cambridge University Press: 153–17.

Schertzer, D., S. Lovejoy, F. Schmitt, I. Tchiguirinskaia and D. Marsan (1997). Multifractal cascade dynamics and turbulent intermittency. *Fractals* 5(3): 427–471.

Schertzer, D., M. Larchevêque and S. Lovejoy (1998). Beyond Multifractal Phenomenology of Intermittency: Nonlinear Dynamics and Multifractal Renormalization. *Chaos, Fractals and Models 96.* G. Iuculano, Italian University Press: 53–64.

Schertzer, D., and Lovejoy, S. (2004a). Space-time complexity and multifractal predictability. *Physica A* **338**, 173–86.

Schertzer, D., and S. Lovejoy, S. (2004b). Uncertainty and predictability in geophysics: chaos and multifractal insights. In R. S. J. Sparks and C. J. Hawkesworth, eds., *State of the Planet: Frontiers and Challenges in Geophysics.* Washington, DC: American Geophysical Union, pp. 317–34.

Schertzer, D., Lovejoy, S., Schmitt, F., Chigirinskaya, Y., and Marsan, D. (1997). Multifractal cascade dynamics and turbulent intermittency. *Fractals* **5**, 427–71.

Schertzer, D., Larchevêque, M., Duan, J., Yanovsky, V. V., and Lovejoy, S. (2001). Fractional Fokker–Planck equation for nonlinear stochastic differential equation driven by non-Gaussian Lévy stable noises. *Journal of Mathematical Physics* **42**, 200–12.

Taylor, G. I. (1931). Effect of variation in density on the stability of superposed streams of fluid. *Proceedings of the Royal Society A* **201**, 499–523.

Tessier, Y., Lovejoy, S., Hubert, P., Schertzer, D., and Pecknold, S. (1996). Multifractal analysis and modeling of rainfall and river flows and scaling, causal transfer functions. *Journal of Geophysical Research* **101**, 26427–40. doi:10.1029/96JD01799.

Thompson, P. D. (1957). Uncertainty of intial state as a factor in the predictability of large scale atmospheric flow patterns. *Tellus* **9**, 275–95.

Wheeler, M., and Kiladis, G. N. (1999). Convectively coupled equatorial waves: analysis of clouds and temperature in the wavenumber-frequency domain. *Journal of the Atmospheric Sciences* **56**, 374–99.

Chapter 10

AchutaRao, K., and Sperber, K. R. (2006). ENSO simulation in coupled ocean–atmosphere models: are the current models better? *Climate Dynamics* **27**, 1–15.

Ashkenazy, Y., Baker, D. R., Gildor, H., and Havlin, S. (2003). Nonlinearity and multifractality of climate change in the past 420,000 years. *Geophysical Research Letters* **30**, 2146.

Ashok, V., Balakumaran, T., Gowrishankar, C., Vennila, I. L. A., and Nirmalkumar, A. (2010). The fast Haar wavelet transform for signal and image processing. *International Journal of Computer Science and Information Security* 7, 126–30.

Berger, W. H. (2008). Solar modulation of North Atlantic Oscillation assisted by the tides? *Quaternary International* **188**, 24–30.

Blender, R., and Fraedrich, K. (2003). Long time memory in global warming simulations. *Geophysical Research Letters* **30**, 1769.

Blender, R., Fraedrich, K., and Hunt, B. (2006). Millennial climate variability: GCM-simulation and Greenland ice cores. *Geophysical Research Letters*, **33**, L04710.

Brohan, P., Kennedy, J. J. Harris, I., Tett, S. F. B., and Jones, P. D. (2006). Uncertainty estimates in regional and global observed temperature changes: a new dataset from 1850. *Journal of Geophysical Research* **111**, D12106.

Bryson, R. A. (1997). The paradigm of climatology: an essay. *Bulletin of the American Meteorological Society* **78**, 450–6.

Bunde, A., Eichner, J. F., Havlin, S., *et al.* (2004). Comment on "Scaling of Atmosphere and Ocean Temperature Correlations in observations and Climate Models". *Physical Review Letters* **92**, 039801.

Committee on Radiative Forcing Effects on Climate (2005). *Radiative Forcing of Climate Change: Expanding the Concept and Addressing Uncertainties.* Washington, DC: National Academies Press.

Compo, G. P., Whitaker, J. S., Sardeshmukh, P. D., *et al.* (2011). The Twentieth Century Reanalysis Project. *Quarterly Journal of the Royal Meteorological Society* **137**, 1–28.

Ditlevsen, P. D., Svensmark, H. and Johnsen, S. (1996). Contrasting atmospheric and climate dynamics of the last-glacial and Holocene periods. *Nature* **379**, 810–12.

Eichner, J. F., Koscielny-Bunde, E., Bunde, A., Havlin, S., and Schellnhuber, H. -J. (2003). Power-law persistance and trends in the atmosphere: A detailed study of long temperature records. *Physical Review E* **68**, 046133.

Ferguson, C. R., and Villarini, G. (2012). Detecting inhomogeneities in the twentieth century reanalysis over the central United States. *Journal of Geophysical Research* **117**, D05123 doi:10.1029/2011JD016988.

Foufoula-Georgiou, E., and Kumar, P., eds. (1994). *Wavelets in Geophysics.* San Diego, CA: Academic Press.

Fraedrich, K., and Blender, K. (2003). Scaling of atmosphere and ocean temperature correlations in observations and climate models. *Physical Review Letters* **90**, 108501–4.

Fraedrich, K., Blender, R., and Zhu, X. (2009). Continuum climate variability: long-term memory, scaling, and 1/f-noise. *International Journal of Modern Physics B* **23**, 5403–16.

Franzke, C. (2010). Long-range dependence and climate noise characteristics of Antarctica temperature data. *Journal of Climate* **23**, 6074–81.

Franzke, C. (2012). Nonlinear trends, long-range dependence and climate noise properties of temperature. *Journal of Climate* doi: dx.doi.org/10.1175/JCLI-D-11-00293.1.

Haar, A. (1910). Zur Theorie des orthogonalen Funktionsysteme. *Mathematische Annalen* **69**, 331–71.

Hansen, J., Ruedy, R., Sato, M., and Lo, K. (2010). Global surface temperature change. *Revies of Geophysics* **48**, RG4004.

Hasselmann, K. (1976). Stochastic climate models. Part I: theory. *Tellus* **28**, 474–85.

Heinlein, R. A. (1973). *Time Enough for Love.* New York, NY: G. P. Putnam's Sons.

Holschneider, M. (1995). *Wavelets: an Analysis Tool.* Oxford: Clarendon Press,

Huschke, R. E., ed. (1959). *Glossary of Meteorology.* Boston, MA: American Meteorological Society.

Huybers, P., and Curry, W. (2006). Links between annual, Milankovitch, and continuum temperature variability. *Nature* **441**, 329–32.

Kantelhardt, J. W., Koscielny-Bunde, E., Rybski, D., *et al.* (2006). Long-term persistence and multifractality of precipitation and river runoff record. *Journal of Geophysical Research* **111**, D01106.

Koscielny-Bunde, E., Bunde, A., Havlin, S., *et al.* (1998). Indication of a universal persistence law governing atmospheric variability. *Physical Review Letters* **81**, 729–32.

Koscielny-Bunde, E., Kantelhardt, J. W., Braun, P., Bunde, A., and Havlin, S. (2006). Long-term persistence and multifractality of river runoff records: detrended fluctuation studies. *Journal of Hydrology* **322**, 120–37.

Lamb, H. H. (1972). *Climate: Past, Present, and Future. Vol. 1, Fundamentals and Climate Now.* London: Methuen.

Lanfredi, M., Simoniello, T., Cuomo, V., and Macchiato, M. (2009). Discriminating low frequency components from long range persistent fluctuations in daily atmospheric temperature variability. *Atmospheric Chemistry and Physics* **9**, 4537–44.

Lennartz, S., and Bunde, A. (2009). Trend evaluation in records with long term memory: Application to global warming. *Geophysical Research Letters* **36**, L16706.

Lind, P. G., Mora, A., Haase, M., and Gallas, J. A. C. (2007). Minimizing stochasticity in the NAO index. *International Journal of Bifurcation and Chaos* **17**, 3461–6.

Lorenz, E. N. (1995). Climate is what you expect. http://eaps4.mit.edu/research/Lorenz/Climate_expect.pdf.

Lovejoy, S. (2012). What is the Climate? EOS (in press)

Lovejoy, S., and Schertzer, D. (1986). Scale invariance in climatological temperatures and the spectral plateau. *Annales Geophysicae* **4B**, 401–10.

Lovejoy, S., and Schertzer, D. (2010). Towards a new synthesis for atmospheric dynamics: space-time cascades. *Atmospheric Research* **96**, 1–52.

Lovejoy, S., and Schertzer, D. (2011a). Space-time cascades and the scaling of ECMWF reanalyses: fluxes and fields. *Journal of Geophysical Research* **116**, D14117. doi:10.1029/2011JD015654.

Lovejoy, S., and Schertzer, D. (2011b). Precipitation and the dimensional transition from weather to low frequency weather. *Paper presented at 58th World Statistics Congress of the International Statistical Institute*, Dublin, Ireland.

Lovejoy, S., and Schertzer, D. (2012a). The climate is not what you expect. *Bulletin of the American Meteorological Society*, in press.

Lovejoy, S., and Schertzer, D. (2012b). Low frequency weather and the emergence of the climate. In A. S. Sharma, A. Bunde, D. Baker, and V. P. Dimri, eds., *Complexity and Extreme Events in Geosciences*. AGU monographs.

Mitchell, J. M. (1976). An overview of climatic variability and its causal mechanisms. *Quaternary Research* **6**, 481–93.

Monetti, R. A., Havlin, S., and Bunde, A. (2003). Long-term persistance in the sea surface temperature fluctuations. *Physica A* **320**, 581–9.

Newman, M., Sardeshmukh, P. D., Winkler, C. R., and Whitaker, J. S. (2003). A study of subseasonal predictability. *Monthly Weather Review* **131**, 1715–32.

Pandey, G., Lovejoy, S., and Schertzer, D. (1998). Multifractal analysis including extremes of daily river flow series for basis five to two million square kilometres, one day to 75 years. *Journal of Hydrology* **208**, 62–81.

Pelletier, J. D. (1998). The power spectral density of atmospheric temperature from scales of 10^{-2} to 10^{6} yr. *Earth and Planetary Science Letters* **158**, 157–64.

Penland, C. (1996). A stochastic model of IndoPacific sea surface temperature anomalies. *Physica D* **98**, 534–58.

Penland, C., and Sardeshmuhk, P. D. (1995). The optimal growth of tropical sea surface temperature anomalies. *Journal of Climate* **8**, 1999–2024.

Peterson, T. C., and Vose, R. S. (1997). An overview of the Global Historical Climatology Network temperature database. *Bulletin of the American Meteorological Society* **78**, 2837–49.

Pielke, R. (1998). Climate prediction as an initial value problem. *Bulletin of the American Meteorological Society* **79**, 2743–6.

Rayner, N. A., Brohan, P., Parker, D. E., *et al.* (2006). Improved analyses of changes and uncertainties in marine temperature measured in situ since the mid-nineteenth century: the HadSST2 dataset. *Journal of Climate* **19**, 446–69.

Royer, J. F., Biaou, A., Chauvin, F., Schertzer, D., and Lovejoy, S. (2008). Multifractal analysis of the evolution of simulated precipitation over France in a climate scenario. *Comptes Rendus Géoscience* **340**, 431–40.

Royer, J.-F., F. Chauvin, S. Lovejoy, D. Schertzer and I. Tchiguirinskaia (2009). Multiscale analysis of the impact of climate change on rainfall over France. *Road Map Towards a Flood Resilient Urban Environment.* E. Pasche et al. Eds. Hamburg, Hamburger Wasserbauschriftien, pp. 62–70.

Rybski, D., Bunde, A., Havlin, S., and von Storch, H. (2006). Long-term persistance in climate and the detection problem. *Geophysical Research Letters* **33**, L06718.

Rybski, D., Bunde, A., and von Storch, H. (2008). Long-term memory in 1000-year simulated temperature records. *Journal of Geophysical Research* **113**, D02106.

Sardeshmukh, P., Compo, G. P., and Penland, C. (2000). Changes in probability assoicated with El Niño. *Journal of Climate* **13**, 4268–86.

Sardeshmukh, P. D., and Sura, P. (2009). Reconciling non-gaussian climate statistics with linear dynamics. *Journal of Climate* **22**, 1193–207.

Schertzer, D. & Lovejoy, S. (1984). On the dimension of atmospheric motions, *Turbulence and Chaotic Phenomena in Fluids*, ed. Tatsumi, T. Elsevier Science Publishers B. V., Amsterdam, pp. 505–512.

Schertzer, D. & Lovejoy, S. (1985 a). Generalised scale invariance in turbulent phenomena. *Physico-Chem. Hydrodyn. J.* **6**, 623–635.

Schertzer, D. & Lovejoy, S. (1985 b). The dimension and intermittency of atmospheric dynamics. *Turbulent Shear Flow 4*, ed. Launder, B. (Springer-Verlag), pp. 7–33.

Schertzer, D, J. F. Royer, I. Tchiguirinskaia and S. Lovejoy (2012). Multifractal downscaling of climate scenarios: a physically-based and self-consistent stochastic downscaling methodology. *10th International Conference on Hydroinformatics*. Hamburg, Germany, July 14–18, 2012.

Schmitt, F., Lovejoy, S., and Schertzer, D. (1995). Multifractal analysis of the Greenland Ice Core Project climate data. *Geophysical Research Letters* **22**, 1689–92.

Smith, T. M., Reynolds, R. W., Peterson, T. C., and Lawrimore, J. (2008). Improvements to NOAA's historical merged land-ocean surface temperature analysis (1880–2006). *Journal of Climate* **21**, 2283–93.

Solomon, S., Qin, D., Manning, M., *et al.*, eds. (2007). *Climate Change 2007: The Physical Science Basis. Working Group I Contribution to the Fourth Assessment Report of the Intergovernmental Panel on Climate Change*. Cambridge: Cambridge University Press.

Stephenson, D. B., Pavan, V., and Bojariu, R. (2000). Is the North Atlantic Oscillation a random walk? *International Journal of Climatology* **20**, 1–18.

Talkner, P., and Weber, R. O. (2000). Power spectrum and detrended fluctuation analysis: application to daily temperatures. *Physical Review E* **62**, 150–60.

Tessier, Y., Lovejoy, S., Hubert, P., Schertzer, D., and Pecknold, S. (1996). Multifractal analysis and modeling of rainfall and river flows and scaling, causal transfer functions. *Journal of Geophysical Research* **101**, 26427–40. doi:10.1029/96JD01799.

Torrence, T., and Compo, G. P. (1998). A practical guide to wavelet analysis. *Bulletin of the American Meteorological Society* **79**, 61–78.

Tuck, A. F. (2008). *Atmospheric Turbulence: a Molecular Dynamics Perspective*. Oxford: Oxford University Press.

Zhang, Y., Wallace, J. M., and Battisti, D. S. (1997). ENSO-like interdecadal variability 1900–93. *Journal of Climate* **10**, 1004–20.

Zhu, X., Fraedrich, K., and Blender, R. (2006). Variability regimes of simulated Atlantic MOC. *Geophysical Research Letters* **33**, L21603.

Chapter 11

Andersen, C., Koç, N., and Moros, M. (2004). A highly unstable Holocene climate in the subpolar North Atlantic: evidence from diatoms. *Quaternary Science Reviews* **23**, 2155–66.

Arrhenius, S. (1896). On the influence of carbonic acid in the air upon the temperature on the ground. *Philosophical Magazine* **41**, 237–76.

Ashkenazy, Y., Baker, D. R., Gildor, H., and Havlin, S. (2003). Nonlinearity and multifractality of climate change in the past 420,000 years. *Geophysical Research Letters* **30**, 2146.

Berger, A., and Loutre, M. F. (1991). Insolation values for the climate of the last 10 million years. *Quaternary Sciences Reviews* **10**, 297–317.

Berger, A., Mélice, J. L., and Loutre, M. F. (2005). On the origin of the 100-kyr cycles in the astronomical forcing. *Paleoceanography* **20**, PA4019.

Berner, K., Koç, N., Divine, D., Godtliebsen, F., and Moros, M. (2008). A decadal-scale Holocene sea surface temperature record from the subpolar North Atlantic constructed using diatoms and statistics and its relation to other climate parameters. *Paleoceanography* **23**, PA2210.

Blender, R., Fraedrich, K., and Hunt, B. (2006). Millennial climate variability: GCM-simulation and Greenland ice cores. *Geophysical Research Letters*, **33**, L04710.

Bodri, L. (1994). Fractal analysis of climatic data: mean annual temperature records in Hungary. *Theoretical and Applied Climatology* **49**, 53–7.

Briffa, K. R., Osborn, T. J., Schweingruber, F. H., *et al.* (2001). Low-frequency temperature variations from a northern tree ring density network. *Journal of Geophysical Research* **106**, 2929–41.

Bryson, R. A. (1997). The paradigm of climatology: an essay. *Bulletin of the American Meteorological Society* **78**, 450–6.

Chekroun, M. D., Simonnet, E., and Ghil, M. (2010). Stochastic climate dynamics: random attractors and time-dependent invariant measures. *Physica D* **240**, 1685–700.

Chylek, P., and Lohmann, U. (2008). Aerosol radiative forcing and climate sensitivity deduced from the last glacial maximum to Holocene transition. *Geophysical Research Letters* **35**, L04804.

Crowley, T. J. (2000). Causes of climate change over the past 1000 years. *Science* **289**, 270–7.

Crowley, T. J., and Lowery, T. S. (2000). How warm was the Medieval Warm period? *Ambio* **29**, 51–4.

Crowley, T. J., Zielinski, G., Vinther, B., ***et al.*** (2008). Volcanism and the Little Ice Age. *PAGES Newsletter* **16**, 22–3.

Dansgaard, W. (2004). *Frozen Annals: Greenland Ice Sheet Research*. Odder: Narayana Press.

Dansgaard, W., Johnsen, S. J., Møller, J., and Langway, C. C. (1969). One thousand centuries of climate record from Camp Century on the Greenland ice sheet. *Science*, **166**, 377–80.

Dansgaard, W., White, J. W. C., and Johnsen, S. J. (1989). The abrupt termination of the Younger Dryas climate event. *Nature* **339**, 532–4.

Dansgaard, W., Johnsen, S. J., Clausen, H. B., *et al.* (1993). Evidence for general instability of past climate from a 250 kyr ice-core record. *Nature* **264**, 218–20.

Davidsen, J., and Griffin, J. (2010). Volatility of unevenly sampled fractional Brownian motion: an application to ice core records. *Physical Review E* **81**, 016107.

Ditlevsen, P. D. (2004). *Turbulence and Climate Dynamics*. Copenhagen: Frydenberg.

Ditlevsen, P. D., Svensmark, H., and Johnsen, S. (1996). Contrasting atmospheric and climate dynamics of the last-glacial and Holocene periods. *Nature* **379**, 810–12.

Eichner, J. F., Koscielny-Bunde, E., Bunde, A., Havlin, S., and Schellnhuber, H. -J. (2003). Power-law persistance and trends in the atmosphere: A detailed study of long temperature records. *Physical Review E* **68**, 046133.

Esper, J., Cook, E. R., and Schweingruber, F. H. (2002). Low-frequency signals in long tree-ring chronologies for reconstructing past temperature variability. *Science* **295**, 2250–3.

Fichefet, T., and Morales Maqueda, M. A. (1997). Sensitivity of a global sea ice model to the treatment of ice thermodynamics and dynamics. *Journal of Geophysical Research* **102**, 12609–46.

Fraedrich, K., and Blender, K. (2003). Scaling of atmosphere and ocean temperature correlations in observations and climate models. *Physical Review Letters* **90**, 108501–4.

Fraedrich, K., Blender, R., and Zhu, X. (2009). Continuum climate variability: long-term memory, scaling, and 1/f-noise. *International Journal of Modern Physics B* **23**, 5403–16.

Ganopolski, A., and Calov, R. (2011). The role of orbital forcing, carbon dioxide and regolith in 100 kyr glacial cycles. *Climate of the Past* 7, 1415–25.

Ganopolski, A., and Schneider von Deimling, T. (2008). Comment on "Aerosol radiative forcing and climate sensitivity deduced from the Last Glacial Maximum to Holocene transition" by Petr Chylek and Ulrike Lohmann. *Geophysical Research Letters* **35**, L23703.

Gao, C. G., Robock, A., and Ammann, C. (2008). Volcanic forcing of climate over the past 1500 years: and improved ice core-based index for climate models. *Journal of Geophysical Research* **113**, D23111.

Ghil, M. (1981). Internal climatic mechanisms participating in glaciation cycles. In A. Berger, ed., *Climatic Variations and Variability: Facts and Theories*. Dordrecht: D. Reidel, pp. 539–57.

Ghil, M. (1994). Cryothermodynamics: the chaotic dynamics of Paleoclimate. *Physica D* 77, 130–59.

Greenland Ice Core Project (1993). Climate instability during the last interglacial period recorded in the GRIP ice core. *Nature* **364**, 203–7.

Grootes, P. M., Stuiver, M., White, J. W. C., Johnsen, S., and Jouzel, J. (1993). Comparison of oxygen isotope records from the GISP2 and GRIP Greenland ice cores. *Nature* **366**, 552–4.

Hays, J. D., Imbrie, J., and Shackleton, N. J. (1976). Variations in the earth's orbit: pacemaker of the ice ages. *Science* **194**, 1121–32.

Hegerl, G. C., Crowley, T. J., Allen, M., *et al.* (2007). Detection of human influence on a new, validated 1500-year temperature reconstruction. *Journal of Climate* **20**, 650–66.

Houghton, J. T., Ding, Y., Griggs, D. J., *et al.*, eds. (2001). *Climate Change 2001: the Scientific Basis, Contribution of Working Group I to the Third Assessment Report of the Intergovernmental Panel on Climate Change*. Cambridge: Cambridge University Press.

Hourdin, F., Musat, I., Bony, S., *et al.* (2006). The LMDZ4 general circulation model: climate performance and sensitivity to parametrized physics with emphasis on tropical convection. *Climate Dynamics* **27**, 787–813.

Huang, S. (2004). Merging information from different resources for new insights into climate change in the past and future. *Geophysical Research Letters* **31**, L13205.

Huber, C., Leuenberger, M., Spahni, R., *et al.* (2006). Isotope calibrated Greenland temperature record over Marine Isotope Stage 3 and its relation to CH4. *Earth and Planetary Science Letters* **243**, 504–19.

Huybers, P., and Curry, W. (2006). Links between annual, Milankovitch and continuum temperature variability. *Nature* **441**, 329–32.

Isono, D., Yamamoto, M. Irino, T. ***et al.*** (2009). The 1500-year climate oscillation in the midlatitude North Pacific during the Holocene. *Geology* **37**, 591–4.

Johnsen, S. J., Clausen, H. B., Dansgaard, W., *et al.* (1992). A "deep" ice core from east Greenland. *Meddelelser om Grønland* **29**, 3–29.

Johnsen, S. J., Clausen, H. B., Dansgaard, W., *et al.* (1997). The d18O record along the Greenland Ice Core Pproject deep ice core and the problem of possible Eemian climatic instability. *Journal of Geophysical Research* **102**, 26397–26410.

Jones, P. D., Briffa, K. R., Barnett, T. P., and Tett, S. F. B. (1998). High-resolution paleoclimatic records for the last millennium: interpretation, integration and comparison with General Circulation Model control-run temperatures. *The Holocene* **8**, 455–71.

Jouzel, J., Alley, R. B., Cuffey, K. M., *et al.* (1997). Validity of the temperature reconstruction from water isotopes in ice cores. *Journal of Geophysical Research* **102**, 26471–87.

Jungclaus, J. H., Keenlyside, N., Botzet, M., *et al.* (2006). Ocean Circulation and tropical variability in the coupled model ECHAM5/MPIOM. *Journal of Climate* **19**, 3952–72.

Jungclaus, J. H., Lorenz, S. J., Timmreck, C., *et al.* (2010). Climate and carbon-cycle variability over the last millennium. *Climate of the Past* **6**, 723–37.

Karimova, L., Kuandykov, Y., Makarenko, N., Novak, M. M., and Helama, S. (2007). Fractal and topological dynamics for the analysis of paleoclimatic records. *Physica A* **373**, 737–46.

King, M. R. (2005). Fractal analysis of eight glacial cycles from an Antarctic ice core. *Chaos, Solitons and Fractals* **25**, 5–10.

Kopp, G., and Lean, J. L. (2011). A new, lower value of total solar irradiance: evidence and climate significance. *Geophysical Research Letters* **38**, L01706.

Krinner, G., Viovy, N., de Noblet-Ducoudré, N., *et al.* (2005). A dynamic global vegetation model for studies of the coupled atmosphere-biosphere system. *Global Biogeochemical Cycles* **19**, GB1015.

Krivova, N. A., and Solanki, S. K. (2008). Models of solar irradiance variations: current status. *Journal of Astrophysics and Astronomy* **29**, 151–8.

Krivova, N. A., Balmaceda, L., and Solanki, S. K. (2007). Reconstruction of solar total irradiance since 1700 from the surface magnetic field flux. *Astronomy and Astrophysics* **467**, 335–46.

Lean, J. L. (2000). Evolution of the Sun's spectral irradiance since the Maunder Minimum. *Geophysical Research Letters* **27**, 2425–8.

Lean, J. L. (2006). Comment on "Estimated solar contribution to the global surface warming using the ACRIM TSI satellite composite" by N. Scafetta and B. J. West. *Geophysical Research Letters* **33**, L15701.

Ljundqvist, F. C. (2010). A new reconstruction of temperature variability in the extra-tropical northern hemisphere during the last two millennia. *Geografiska Annaler: Physical Geography* **92**A, 339–51.

Lovejoy, S., and Schertzer, D. (1984). 40 000 years of scaling in climatological temperatures. *Meteorological Science and Technology* **1**, 51–4.

Lovejoy, S., and Schertzer, D. (1986). Scale invariance in climatological temperatures and the spectral plateau. *Annales Geophysicae* 4B, 401–10.

Lovejoy, S., and Schertzer, D. (2012a). Low frequency weather and the emergence of the climate. In A. S. Sharma, A. Bunde, D. Baker, and V. P. Dimri, eds., *Complexity and Extreme Events in Geosciences*. AGU monographs.

Lovejoy, S., and Schertzer, D. (2012b). Stochastic and scaling climate sensitivities: solar, volcanic and orbital forcings,

Geophysical Research Letters, **39**, L11702. doi:10.1029/2012GL051871.

Lovejoy, S., Schertzer, D., Varan, D. (2012). Do GCM's predict the climate ... or macroweather? *Earth Syst. Dynam. Diss.* **3**, 1259–1286, doi 10.5194/esdd-3-1259-2012.

Madec, G., Delecluse, P., Imbard, M., and Lévy, C. (1998). *OPA 8.1 Ocean General Circulation Model Reference Manual.* Laboratoire d'Océanographie DYnamique et de Climatologie.

Manabe, S., and Wetherald, R. T. (1975). The effects of doubling the CO_2 concentration on the climate of a general circulation model. *Journal of the Atmospheric Sciences* **32**, 3–15.

Mandelbrot, B. B., and Wallis, J. R. (1969). Some long run properties of geophysical records. *Water Resources Research* **5**, 228–67.

Manley, G. (1974). Central England temperatures: monthly means 1659–1973. *Quarterly Journal of the Royal Meteorological Society* **100**, 389–495.

Mann, M. E., (1995). Global interdecadal and century-scale climate oscillations during the past five centuries. *Nature* **378**, 268–70.

Mann, M. E., Bradley, R. S., and Hughes, M. K. (1998). Global-scale temperature patterns and climate forcing over the past six centuries. *Nature* **392**, 779–87.

Mann, M. E., Bradley, R. S., and Hughes, M. K. (1999). Northern hemisphere temperatures during the past millennium: inferences, uncertainties, and limitations. *Geophysical Research Letters* **26**, 759–62.

Mann, M. E., Zhang, Z., Hughes, M. K., ***et al.*** (2008). Proxy-based reconstructions of hemispheric and global surface temperature variations over the past two millennia, *Proceedings of the National Academy of Sciences of the USA* **105**, 13252–7. doi:10.1073/pnas.0805721105.

McIntyre, S., and McKitrick, R. (2003). Corrections to the Mann *et al.* (1998) proxy data base and northern hemispheric average temperature series. *Energy & Environment* **14**, 751–71.

McIntyre, S., and McKitrick, R. (2005). Hockey sticks, principal components and spurious signficance. *Geophysical Research Letters* **32**, L03710–14.

Milankovitch, M. (1941). *Canon of Insolation and the Ice Age Problem* [in German]. Belgrade: Serbian Academy of Sciences and Arts, Special Publication 132; Section of Mathematical and Natural Sciences, 33.

Mitchell, J. M. (1976). An overview of climatic variability and its causal mechanisms. *Quaternary Research* **6**, 481–93.

Moberg, A., Sonechkin, D. M., Holmgren, K., Datsenko, N. M., and Karlén, W. (2005). Highly variable Northern Hemisphere temperatures reconstructed from low- and high-resolution proxy data. *Nature* **433**, 613–17.

Mudelsee, M. (2001). The phase relations among atmospheric CO2 contnent, temperature and global ice volume over the past 420 ka. *Quaternary Science Reviews* **20**, 583–9.

Neukom, R., Luterbacher, J., Villalba, R., *et al.* (2010). Multiproxy summer and winter surface air temperature field reconstructions form southern South America covering the past centuries. *Climate Dynamics online.* DOI: 10.1007/s00382-010-0793-3.

North Greenland Ice Core Project (2004). High-resolution record of Northern Hemisphere climate extending into the last interglacial period. *Nature*, **431**, 147–51.

Osborn, T. J., Raper, S. C. B., and Briffa, K. R. (2006). Simulated climate change during the last 1000 years: comparing the ECHO-G general circulation model with the MAGICC simple climate model. *Climate Dynamics* **27**, 185–197.

Pelletier, J. D. (1998). The power spectral density of atmospheric temperature from scales of 10^{-2} to 10^6 yr. *Earth and Planetary Science Letters* **158**, 157–64.

Petit, J. R., Jouzel, J., Raynaud, D., *et al.* (1999). Climate and atmospheric history of the past 420,000 years from the Vostok Ice core, Antarctica. *Nature* **399**, 429–36.

Pielke, R. (1998). Climate prediction as an initial value problem. *Bulletin of the American Meteorological Society* **79**, 2743–6.

Raddatz, T. J., Reick, C. J., Knorr, W., *et al.* (2007). Will the tropical land biosphere dominate the climate-carbon feedback during the twenty-first century? *Climate Dynamics* **29**, 565–74.

Roeckner, E., Arpe, K., Bengtsson, L., *et al.* (1996). *The Atmospheric General Circulation Model ECHAM-4: Model Description and Simulation of Present-day Climate.* Hamburg: Max-Planck Institute for Meteorology.

Rybski, D., Bunde, A., Havlin, S., and von Storch, H. (2006). Long-term persistance in climate and the detection problem. *Geophysical Research Letters* **33**, L06718.

Rybski, D., Bunde, A., and von Storch, H. (2008). Long-term memory in 1000-year simulated temperature records. *Journal of Geophysical Research* **113**, D02106.

Scafetta, N., and Grigolini, P. (2002). Scaling detection in time series: diffusion entropy analysis. *Physical Review E* **66**, 036130.

Scafetta, N., and West, B. J. (2003). Solar flare intermittency and the Earth's temperature anomalies. *Physical Review Letters* **90**, 248701.

Scafetta, N., and West, B. J. (2005). Estimated solar contribution to the global surface warming using the ACRIM TSI satellite composite. *Geophysical Research Letters* **32**, L18713.

Scafetta, N., and West, B. J. (2007). Phenomenological reconstructions of the solar signature in the Northern Hemisphere surface temperature records since 1600. *Journal of Geophysical Research* **112**, D24S03.

Scafetta, N., and West, B. J. (2008). Is climate sensitive to solar variability? *Physics Today* **61**, 50–1.

Schmitt, F., Lovejoy, S., and Schertzer, D. (1995). Multifractal analysis of the Greenland Ice-core Project climate data. *Geophysical Research Letters* **22**, 1689–92.

Shakun, J. D., Clark, P. U., He, F., *et al.* (2012). Global warming preceded by increasing carbon dioxide concentrations during the last deglaciation. *Nature* **484**, 49–54. doi:10.1038/nature10915.

Schmittner, A., Urban, N. M., Shakun, J. D., *et al.* (2011). Climate sensitivity estimated from temperature reconstructions of the Last Glacial Maximum. *Science* **334**, 1385–8.

Shapiro, A. I., Schmutz, W., Rozanov, E., *et al.* (2011). A new approach to long-term reconstruction of the solar irradiance leads to large historical solar forcing. *Astronomy & Astrophysics* **529**, A67.

Solomon, S., Qin, D., Manning, M., *et al.*, eds. (2007). *Climate Change 2007: the Physical Science Basis. Working Group I Contribution to the Fourth Assessment Report of the Intergovernmental Panel on Climate Change.* Cambridge: Cambridge University Press.

Spencer, R. W., and Braswell, W. D. (2008). Potential biases in feedback diagnosis from observational data: a simple model demonstration. *Journal of Climate* **21**, 5624–8.

Steffensen, J. P., Andersen, K. K., Bigler, M., *et al.* (2008). Abrupt climate change happens in few years. *Science* **321**, 680–4.

Steinhilber, F., Beer, J., and Fröhlich, C. (2009). Total solar irradiance during the Holocene. *Geophysical Research Letters* **36**, L19704.

Vieira, L. E. A., Solanki, S. K., Krivova, N. A., and Usoskin, I. (2011). Evolution of the solar irradiance during the Holocene. *Astronomy and Astrophysics* **531**, A6. doi:10.1051/0004-6361/201015843.

Vinther, B. M., Clausen, H. B., Fisher, D. A., *et al.* (2008). Synchronizing ice cores from the Renland and Agassiz ice caps to the Greenland ice core chronology. *Journal of Geophysical Research* **113**, D08115. doi:10.1029/2007JD009143.

von Storch, H., Zorita, E., Jones, J. M., *et al.* (2004). Reconstructing past climate from noisy data. *Science* **306**, 679–82.

Vyushin, D., Zhidkov, I., Havlin, S., Bunde, A., and Brenner, S. (2004). Volcanic forcing improves atmosphere–ocean coupled general circulation model scaling performance. *Geophysical Research Letters* **31**, L10206.

Wahl, E. R., and Ammann, C. M. (2007). Robustness of the Mann, Bradley, Hughes reconstruction of Northern Hemisphere surface temperatures: examination of criticisms based on nature and processing of climate evidence. *Climate Change* **85**, 33–69. doi: 10.1007/510584-006-9105-7.

Wang, X., Auler, A. S., Edwards, R. L., *et al.* (2007). Millennial-scale precipitation changes in southern Brazil over the past 90,000 years. *Geophysical Research Letters* **34**, L23701. doi: 10.1029/2007GL031149.

Wang, Y.-M., Lean, J. L., and Sheeley, N. R. (2005). Modeling the Sun's magnetic field and irradiance since 1713. *Astrophysical Journal* **625**, 522–38.

Watson, A. J., and Lovelock, J. E. (1983). Biological homeostasis of the global environment: the parable of Daisyworld. *Tellus* **35B**, 284–9.

Wetzel, P., Maier-Reimer, E., Botzet, M., *et al.* (2006). Effects of ocean biology on the penetrative radiation on a coupled climate model. *Journal of Climate* **19**, 3973–87.

Witt, A., and Schumann, A. Y. (2005). Holocene climate variability on millennial scales recorded in Greenland ice cores. *Nonlinear Processes in Geophysics* **12**, 345–52.

Wolff, E. W., Basile, I., Petit, J. -R., and Schwander, J. (1999). Comparison of Holocene electrical records from Dome C and Vostok, Antarctica. *Annals of Glaciology* **29**, 89–93.

Wolff, J.-O., Maier-Reimer, E., and Legutke, S. (1997). *The Hamburg Ocean Primitive Equation Model: HOPE.* Hamburg: German Climate Computer Center (DKRZ).

Wunsch, C. (2003). The spectral energy description of climate change including the 100 ky energy. *Climate Dynamics* **20**, 353–63.

Zhang, Y., Wallace, J. M., and Battisti, D. S. (1997). ENSO-like interdecadal variability 1900–93. *Journal of Climate* **10**, 1004–20.

Zhu, X., Fraedrich, K., and Blender, R. (2006). Variability regimes of simulated Atlantic MOC. *Geophysical Research Letters* **33**, L21603.

Index

Printed in the United States
By Bookmasters